T0202304

TOPOLOGICAL QUANTUM

Topological Quantum

Steven H. Simon

Rudolf Peierls Centre for Theoretical Physics
University of Oxford

OXFORD
UNIVERSITY PRESS

Great Clarendon Street, Oxford, OX2 6DP,
United Kingdom

Oxford University Press is a department of the University of Oxford.
It furthers the University's objective of excellence in research, scholarship,
and education by publishing worldwide. Oxford is a registered trade mark of
Oxford University Press in the UK and in certain other countries

Published in the United States of America by Oxford University Press
198 Madison Avenue, New York, NY 10016, United States of America

British Library Cataloguing in Publication Data

Data available

Library of Congress Control Number: 2023905961

ISBN 9780198886723
DOI: 10.1093/oso/9780198886723.001.0001

Printed and bound by
CPI Group (UK) Ltd, Croydon, CR0 4YY

Preface

This book originated as part of a lecture course given at Oxford. The idea for the course was to give a general introduction to topological quantum ideas, including topological quantum field theories, topological quantum memories, topological quantum computing, topological quantum matter, and topological quantum order. After the better part of a decade the course finally became a book.

Topological quantum is vast. Given enough time I could easily have written a book three times as long. In order to finish in a finite amount of time I focused on only certain parts of the field. Emphasis is given to the structure of topological theories, diagrammatic reasoning, and the examples of toric code, loop gases, and string-nets. The book is aimed at a physics audience (in particular I try to avoid the language of category theory), although some mathematicians may also find the perspectives presented here to be useful. While the ideas are very deep, in principle, an advanced undergraduate should be able to follow most of the material. The only prerequisite is a good background in basic quantum mechanics.

Topological quantum lies at an amazing nexus between physics and mathematics and computer science. It has been an absolute pleasure to spend time working in this field and writing about it. The objective of this book is to make the power and beauty of the topic accessible to a wide audience. It is a subject that I truly love, and it is my goal to make those who read this book fall in love with the topic as well. You will have to let me know if I have succeeded.

Let's get started!

Oxford, United Kingdom
June 2023

How to Read This Book

I originally intended this book to be read from beginning to end. That said, it is also possible to jump around a bit depending on your interests. To give the reader an idea of which chapters are more crucial for the development of later chapters, I've included a plot on the next page indicating how much each (source) chapter references any particular other (target) chapter. Where you see a lot of weight in a column, you can assume that the material in the chapter corresponding to that column is important for later chapters. Where you see no red in a column, that chapter can be easily skipped. In the course I teach, I certainly do not assign all of the chapters!

To summarize, the most crucial parts are Chapters 8–10, which set up the structure of anyon theories (i.e. TQFTs). Once you understand these key chapters, you can read most of the other pieces without too much trouble.

Much more detail of the structure of anyon theories is discussed in Chapters 12–17. However, you can probably skip much of the detailed development, and just refer back to Chapter 16 and parts of Chapter 17 if you need to know more about diagrammatic rules later in the book.

In my lecture course I assign Chapters 2–5 as a motivational introduction to anyons. I encourage readers to try these chapters first.

Chapters 18–25 are not crucial to things that come thereafter. However, it is probably worth reading a bit of Chapter 18 just to have a few examples in mind.

Chapter 26 on quantum error correction is almost entirely stand-alone.

When the toric code is introduced in Chapter 27, it is also fairly independent of the prior chapters. Some people might even want to start reading the book there to have an example of a real anyon theory in mind when learning the subject. Although there is some reference to modular S- and T-Matrices (Chapter 17) most of Chapters 27–31 might be read fairly independently from earlier parts of the book.

Chapters 34–36 are more advanced material, although much of it can be read with only a good understanding of the toric code. Chapter 37 gives a very rapid discussion of some of the interesting experiments.

Chapter 39 is an appendix that describes some useful resources, and 40 is an appendix that introduces some basic mathematics that many people may know but I thought should be included for completeness.

In a margin note of my previous book (Simon [2013]), I said that my next book[1] (i.e. this one) would be about two-dimensional electron systems. The topic is covered only briefly in the section on fractional quantum Hall effect in Chapter 37. Unfortunately, I did not really cover this topic, or any experimental aspect of the field in sufficient depth. I promise a future book to remedy this!

[1] I also suggested that I might write a thriller about physicists defeating drug smugglers. For those who are interested, I'm still working on it, but I discovered that writing a novel is pretty hard.

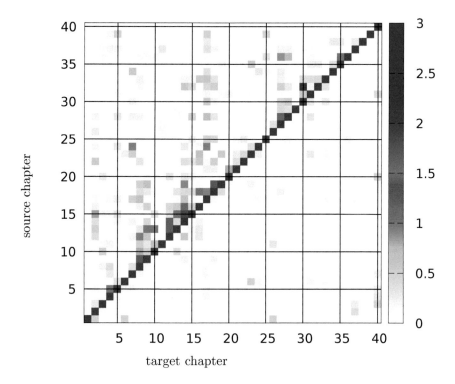

Cross references from source chapter (vertical) to target chapter (horizontal) per page of the source chapter. The blue diagonal represents references from a chapter to itself (the blue diagonal is not the interesting feature of the plot). Consider for example the purple dot just above the diagonal at coordinates (16,18). This indicates that Chapter 18 highly references Chapter 16. Similarly, there is a red dot at coordinate (7,24) showing that Chapter 24 highly references Chapter 7. There is also a lot of orange and red above the diagonal in columns 2, 8, 9, 12, 13, 15, 16, and 17, indicating that these chapters are referenced frequently in later chapters. Chapter 40 is the mathematical appendix.

Acknowledgments

There are many people who helped produce this book in one way or other. First, I credit the many brilliant people who invented this field. I hope I've done justice to your work, and I hope I've cited you properly. I apologize in advance for inevitable oversights or mistakes, which are certainly a reflection of my imperfect understanding and the sheer impossibility of putting all of these wonderful ideas in a single book.

Second, I thank the many people who taught me this subject: my colleagues, collaborators, and competitors. I particularly thank those rare people who wrote readable papers and gave understandable talks. In addition, I thank the many students and postdocs over the years who thought I was teaching them, when usually it was actually the other way around.

Third, there were many people who gave feedback on this book. While I can't possibly mention everyone by name, there are a few whose efforts stand out. Joost Slingerland gave a lot of input in the early stages of this project, as well as being a coauthor who helped to sort out some of the nagging issues in Chapter 14. Fatimah Ahmadi, Julien Vidal, and Aleksander Cianciara all gave very detailed feedback on drafts at various stages of the project. Fiona Burnell gave feedback on some key scientific issues on SPTs and SETs and condensations. Particular thanks go to Kelvin Koor Kai Jie, who did an absolutely amazing job of proofreading and commenting on the beta release version of this book.

Finally, and most importantly, I thank my family: my wife Janina, for more things than I can possibly count, and my daughter Alathea, for being a most excellent little bubakins.

Contents

Introduction: History of Topology, Knots, Peter Tait, and Lord Kelvin

1

Historical Entertainment

The field of quantum topology inhabits a beautiful nexus between mathematics, computer science, and physics. Within the field of physics, it has been fundamental to a number of subfields. On the one hand, topology and topological matter are key concepts of modern condensed matter physics.[1] Similarly, in the field of quantum information and quantum computation, topological ideas are extremely prominent.[2] At the same time much of our modern study of topological matter is rooted in ideas of topological quantum field theories that developed from the high energy physics, quantum gravity,[3] and string theory community starting in the 1980s. These earlier works have even earlier precedents in physics and mathematics. Indeed, the historical roots of topology in physics date all the way back to the 1800s, which is where we will begin our story.

In 1867 Lord Kelvin[4] and his close friend Peter Tait were interested in a phenomenon of fluid flow known as a smoke ring,[5] configurations of fluid flow where lines of vorticity form closed loops as shown in Fig. 1.1. Peter Tait built a machine that could produce smoke rings, and showed it to Kelvin, who had several simultaneous epiphanies. First, he realized that there should be a theorem (now known as Kelvin's circulation theorem) stating that, in a perfectly dissipationless fluid, lines of vorticity are conserved quantities, and the vortex loop configurations should persist for all time. Unfortunately, few dissipationless fluids exist—and the ones we know of now, such as superfluid helium at very low temperatures, were not discovered until the next century.[6] However, at the time, scientists incorrectly believed that the entire universe was filled with a perfect dissipationless fluid, known as luminiferous aether, and Kelvin wondered whether one could have vortex loops in the aether.

At the same time, one of the biggest mysteries in all of science was the discreteness and immutability of the chemical elements. Inspired by Tait's smoke ring demonstration, Kelvin proposed that different atoms corresponded to different knotting configurations of vortex lines in the aether. This theory of "vortex atoms" was appealing in that it gave a

[1] The 2016 Nobel Prize was awarded to Kosterlitz, Thouless, and Haldane for the introduction of topological ideas into condensed matter physics.

[2] We will see this starting in Chapter 26.

[3] See Chapter 6.

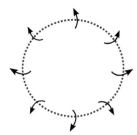

Fig. 1.1 A smoke ring or vortex loop is an invisible ring in space where the fluid flows around the invisible ring as shown by the arrows. The whole thing moves out of the plane of the page at you as the fluid circulates.

[5] Even in 1867, a talented smoker could produce a smoke ring from their mouth.

[6] In fact, Helium was not even discovered yet in 1867!

[4] Actually, in 1867 he was just William Thomson, but he would later be elevated to the peerage and take the name Lord Kelvin after the River Kelvin that flowed by his laboratory.

Fig. 1.2 The simplest few knots made from one strand of string. The top knot, a simple loop, is known as the "unknot," and corresponds to the simple smoke ring in Fig. 1.1. The second knot from the top, known as the trefoil, is not the same as its mirror image (see Exercise 2.1)

reason why atoms are discrete and immutable—on the one hand there are only so many different knots that one can make. (See for example, the list of the simplest few knots you can form from one piece of string shown in Fig. 1.2.) On the other hand, by Kelvin's circulation theorem, the knotting of the vortices in a dissipationless fluid (the aether) should be conserved for all time. Thus, the particular knot could correspond to a particular chemical element, and this element should never change to another one. Hence, the atoms should be discrete and immutable!

For several years the vortex theory of the atom was quite popular, attracting the interest of other great scientists such as Maxwell, Kirchhoff, and J. J. Thomson (no relation). However after further research and failed attempts to extract predictions from this theory, the idea of the vortex atom lost popularity.

Although initially quite skeptical of the idea, Tait eventually came to believe that by building a table of all possible knots (knotted configuration of strands such that there are no loose ends) he would gain some insight into the periodic table of the elements, and in a remarkable series of papers he built a catalogue of all knots with up to seven crossings (the first few entries of the table being shown in Fig. 1.2). From his studies of knots, Tait is viewed as the father of the mathematical theory of knots, which has been quite a rich field of study since that time (and particularly during the last fifty years).

During his attempt to build his "periodic table of knots," Tait posed what has become perhaps *the* fundamental question in mathematical knot theory: how do you know if two pictures of knots are topologically identical or topologically different. In other words, can two knots be smoothly deformed into each other without cutting any of the strands. Although this is still considered to be a difficult mathematical problem, a powerful tool that helps answer this question is the idea of a "knot invariant"—which we will study in Chapter 2. Shortly, it will become clear how this idea is related to physics.

Although Tait invented a huge amount of mathematics of the theory of knots[7] and developed a very extensive table of knots, he got no closer to understanding anything about the periodic table of the atoms. In his later life he became quite frustrated with his lack of progress in this direction and he began to realize that understanding atoms was probably unrelated to understanding knots. Tait died[8] in 1901 not realizing that his work on the theory of knots would be important in physics, albeit for entirely different reasons.

Further Reading: For more history, see Silver [2006].

---

[7]Some of his conjectures were *way* ahead of their time—some being proven only in the 1980s or later! See Stoimenow [2008] for a review of the Tait conjectures proven after 1985.

[8]Peter Tait was also a huge fan of golf and wrote some beautiful papers on the trajectory of golf balls. His son, Freddie Tait, was a champion amateur golfer, being the top amateur finisher in the British Open six times and placing as high as third overall twice. Freddie died very young, at age 30, in the Boer wars in 1900. This tragedy sent Peter into a deep depression from which he never recovered.

Kauffman Bracket Invariant and Relation to Physics

The purpose of this chapter is to introduce you to a few of the key ideas and get you interested in the subject!

2.1 The Idea of a Knot Invariant

Topological equivalence. We say two knots are topologically equivalent if they can be deformed smoothly into each other without cutting.[1] For example, the picture of a knot (or more properly, the picture of the link of two strings) on the left of Fig. 2.1 is topologically equivalent to the picture on the right of Fig. 2.1.

It may appear easy to determine whether two simple knots are topologically equivalent and when they are not. However, for complicated knots, it becomes *extremely difficult* to determine whether two knots are equivalent or inequivalent. It is thus useful to introduce a mathematical tool known as a knot invariant that can help us establish when two knots are topologically inequivalent.

A **knot invariant** is a mapping from a knot (or a picture of a knot) to an output via a set of rules that are cooked up in such a way that two topologically equivalent knots must give the same output (see Fig. 2.2). So if we put two knots into the set of rules and we get two different outputs, we know immediately that the two knots cannot be continuously deformed into each other without cutting.

To demonstrate how knot invariants work, we will use the example of the Kauffman bracket invariant[2,3] (see Kauffman [1987]). The Kauffman

Fig. 2.1 Topological equivalence of two knots. The knot on the left can be deformed continuously into the knot on the right without cutting any strands.

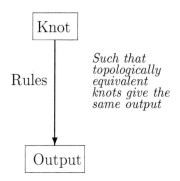

Fig. 2.2 Schematic description of a knot invariant as a set of rules taking an input knot to some mathematical output such that topologically equivalent knots give the same output.

[1]A few pieces of fine print here. (1) I am not precise about knot versus link. Strictly speaking a knot is a single strand, and a link is more generally made of multiple strands. Physicists call them all knots. In either case no dangling ends are allowed. A **knot** can be defined as a particular embedding of a circle (S^1) into a three-dimensional reference manifold such as $\mathbb{R}^3$ (regular three-dimensional space) with no self-intersections. A **link** is an embedding of several circles into the three-dimensional manifold with no intersections. (2) When I say "topologically equivalent" here I mean the concept of **regular isotopy** (see Sections 2.2.1 and 2.6.1). Two knots are isotopic if there is a continuous smooth family of knots between the initial knot and the final knot—although, to be more precise, as we see in Section 2.2.1, we should think of the knots as being thickened to ribbons and we want a smooth family of ribbons.

[2]Be warned: there are multiple things named after Kauffman. The particular normalization of the bracket invariant that we use has been named the *topological bracket* by Kauffman. The more common definition of the bracket is our definition divided by d.

[3]The term "bracket" is due to a common notation where one draws a picture of a knot inside brackets to indicate that one is supposed to evaluate this invariant. We will not draw these brackets.

bracket invariant was essentially invented by Vaughan Jones, who won the Fields medal for his work on knot theoryJones [1985]. Kauffman's important contribution to this story (among his many other contributions in the field of knot theory) was to explain Jones's work in very simple terms.

To define the **Kauffman bracket invariant**, we start with a scalar variable A. For now, leave it just a variable, although later we may give it a value. There are then just two rules to the Kauffman bracket invariant. First, a simple loop of string (with nothing going through it) can be removed from the diagram and replaced with the number[4,5,6]

$$d = -A^2 - A^{-2} \quad . \tag{2.1}$$

The second rule replaces a diagram that has a crossing of strings by a sum of two diagrams where these strings don't cross—where the two possible uncrossings are weighted by A and A^{-1} respectively as shown in Fig. 2.3. This type of replacement rule is known as a *skein* rule.[7,8,9]

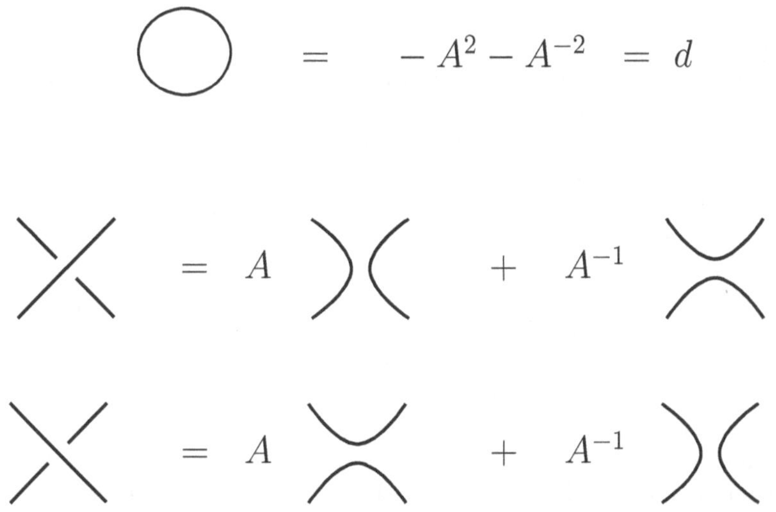

[4]We will eventually see that d stands for "dimension."

[5]There is a hidden assumption that an empty diagram has value 1. This means that the overall value of a diagram with a single loop is d, the overall value of a diagram of two unlinked loops is d^2, and so forth.

[6]Throughout the book we will abuse notation slightly by saying that a knot diagram has a numerical value. Kauffman often puts bracket around the knot to mean we are interested in the corresponding numerical value of the diagram. However, these brackets often become unwieldy so we will not use them.

[7]Most references use the convention given here. However, a non-negligible number of references replace A with A^{-1} in this definition.

[8]It is also common to take the final value of the Kauffman bracket invariant and divide the result by d so that the value of a single loop (an unknot) is just 1 but n loops have value d^{n-1}.

[9]The word "skein" is an infrequently used English word meaning loosely coiled yarn, or sometimes meaning an element that forms part of a complicated whole (probably both of these are implied for our mathematical usage). "Skein" also means geese in flight, but I suspect this is unrelated.

Fig. 2.3 Rules for evaluating the Kauffman bracket invariant. The third line is exactly the same as the middle line except that all the diagrams are rotated by 90 degrees, so it is not an independent rule. However, it is convenient to draw the rule twice to make it easier to compare to other diagrams.

The general scheme is to use the second (and third) rule of Fig. 2.3 to remove all crossings of a diagram. In so doing, one generates a sum of many diagrams with various coefficients. Then once all crossings are removed, one is just left with simple loops, and each loop can just be replaced by a factor of d.

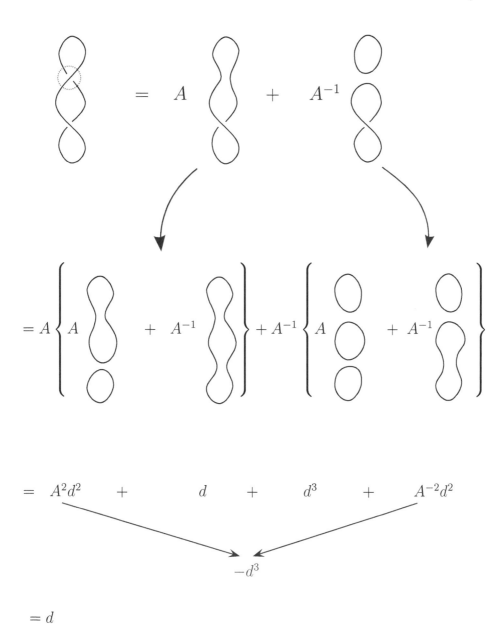

Fig. 2.4 Example of evaluation of the Kauffman bracket invariant for the simple twisted loop in the upper left. The light dotted red circle is meant to draw attention to where we apply the Kauffman crossing rule (the middle line in Fig. 2.3) to get the two diagrams on the right-hand side. After applying the Kauffman rules again (the final line in Fig. 2.3), we have removed all crossings and we are left only with simple loops, which each get the value d. In the penultimate line we have used the definition of d to replace $A^2 + A^{-2} = -d$. The fact that we get d in the end of the calculation is expected since we know that the original knot is just a simple loop (the "unknot") and the Kauffman rules tell us that a loop gets a value d.

[10]To a mathematician the Kauffman invariant is an invariant of regular isotopy—see Section 2.2.1.

[11]The fact that the rules stated in Fig. 2.3 define a knot invariant (i.e. give the same result no matter how many times we fold over the knot) is proven in Exercise 2.3.

[12]The converse is not true. If two knots give the same output, they are not necessarily topologically equivalent. It is an open question whether there are any knots besides the simple unknot (a simple loop) that has Kauffman invariant d. It is also an open challenge to find out whether any combinatoric knot invariants similar to Kauffman can distinguish all topologically inequivalent knots from each other.

To give an example of how these rules work we show evaluation of the Kauffman bracket invariant for the simple knot in the upper left of Fig. 2.4. The output of the calculation is that the Kauffman invariant of this knot comes out to be d. This result is expected since we know that the original knot (in the upper left of the figure) is just a simple loop (the "unknot") and the Kauffman rules tell us that a loop gets a value d. We could have folded over this knot many times[10] and still the outcome of the Kauffman evaluation[11] would be d.

The idea of a knot invariant seems like a great tool for distinguishing knots from each other. If you have two complicated knots and you don't know if they are topologically equivalent, you just plug them into the Kauffman machinery and if they don't give the same output then you know immediately that they cannot be deformed into each other without cutting.[12] However, a bit of thought indicates that things still get rapidly difficult for complicated knots. Figure 2.4 has two crossings, and we ended up with four diagrams. If we had a knot with N crossings we would have gotten 2^N diagrams, which can be huge! While it is very easy to draw a knot with 100 crossings, even the world's largest computer would not be able to evaluate the Kauffman bracket invariant of this knot! So one might then think that this Kauffman bracket invariant is actually not so useful for complicated knots. We return to this issue later in Section 2.4.

2.2 Relation to Physics

There is a fascinating relationship between knot invariants and quantum physics. For certain types of "topological quantum systems" the amplitudes of spacetime processes can be directly calculated via knot invariants such as the Kauffman bracket invariant.

We should first comment that most of what we discuss in this book corresponds to two-dimensional systems plus one dimension of time. There are topological systems in $(3 + 1)$-dimensions (and higher dimensions as well!) but more is known about $(2 + 1)$-dimensions and we focus on that, at least for now.[13]

[13]There is also some discussion of "topological" systems in $(1 + 1)$-dimensions in Chapter 12 for example.

Figure 2.5 shows a particular spacetime process of particle world lines. At the bottom of the figure is shown the shaded two-dimensional system (a disk). At some early time there is a pair creation event—a particle–antiparticle appear from the vacuum, then another pair creation event; then one particle walks around another, and the pairs come back together to try to reannihilate. At the end of the process, it is possible that the particles do reannihilate to the vacuum (as shown in the diagram), but it is also possible that (with some probability amplitude) the particle–antiparticle pairs form bound states that do not annihilate back to the vacuum.

In a topological theory, the quantum amplitude for these processes depends on the topology of the world lines, and not on the detailed geometry (i.e. the probability that the particles reannihilate versus form bound states). In other words, as long as the topology of the world lines looks like two linked rings, it will have the same quantum amplitude as that shown in Fig. 2.5. It should surprise us that systems exist where amplitudes depend only on topology, as we are used to the idea that amplitudes depend on details of things, like details of the Hamiltonian, how fast the particles move, and how closely they come together. But in a topological theory, none of these things matter. What matters is the topology of the spacetime paths.

What should be obvious here is that the quantum amplitude of a process is a knot invariant. It is a mapping from a knot (made by the world lines) to an output (the amplitude) that depends only on the topology of the knot. This connection between quantum systems and knot invariants was made famously by Ed Witten, one of the world's leading string theorists Witten [1989]. He won the Fields medal along with Vaughan Jones for this work.

Such topological theories were first considered as an abstract possibility, mainly coming from researchers in quantum gravity (see Chapter 6). However, now several systems are known in condensed matter that actually behave like this. A brief list of experimental systems where excitations (are thought to) have interesting braiding properties is given in Table 2.1. While not all topological theories are related to the Kauffman bracket invariant, many of them are. There are other knot invariants that occur in physical systems as well—including the so-called HOMFLY invariant Freyd et al. [1985] (see Exercise 2.5). A brief table of some of the physical systems that are believed to be related to nontrivial knot invariants is given in Table 2.2. In Chapter 37 we discuss experimental systems in a bit more detail.

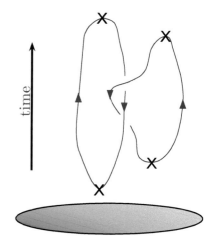

Fig. 2.5 A spacetime process showing world lines of particles for a (2 + 1)-dimensional system (shown as the shaded disk at the bottom). The ✕'s mark the points in spacetime where particles-antiparticle pairs are either pair-created or pair-annihilated. An uparrow represents a particle and downarrow its antiparticle.

- Two-dimensional electron (or possibly boson) systems
  - Fractional quantum Hall effects
  - Fractional Chern insulators
- Quantum spin liquids
- Certain superconductors and superfluids
- Certain quantum error correcting codes

Table 2.1 List of some interesting experimental systems thought to have nontrivial particle braiding. See Chapter 37 for more detail.

(1) $SU(2)_2$ **class**: Knot invariant is (closely related to) the Kauffman bracket invariant using a value $A = i^\alpha e^{\pm i\pi/8}$ with integer α. This is also known as the "Ising" class. Possible physical realizations include

- $\nu = 5/2$ fractional quantum Hall effect (2D electrons at low temperature in high magnetic fields).
- 2D p-wave superconductors.
- 2D films of ^{3}HeA superfluid.[14]
- A host of "engineered" structures that are designed to have these interesting topological properties—particularly those known as "Majorana Materials." Typically these have a combination of spin–orbit coupling, superconductivity, and magnetism of some sort.

(2) $SU(2)_3$ **class**: Knot invariant is (closely related to) the Kauffman bracket invariant using a value $A = i^\alpha e^{\pm i\pi/10}$. The only physical system known in this class is the $\nu = 12/5$ fractional quantum Hall effect.

(3) $SU(2)_4$ **class**: Knot invariant is (closely related to) the Kauffman bracket invariant using a value $A = i^\alpha e^{\pm i\pi/12}$. It is possible that the $\nu = 2 + 2/3$ fractional quantum Hall effect is in this class.

(4) $SU(2)_1$ **class**: Also known as semions. Knot invariant is (closely related to) the Kauffman bracket invariant using a value $A = e^{\pm i\pi/6}$. These are proposed to be realized in rotating boson fractional quantum Hall systems and certain chiral quantum spin liquids.

(5) $SU(3)_2$ **class**: Knot invariant is (closely related to) the HOMFLY knot invariant rather than the Kauffman bracket invariant. It is possible that the unpolarized $\nu = 4/7$ fractional quantum Hall effect is in this class.

[14]Two Nobel Prizes have been given for work on Helium-3 superfluidity.

Table 2.2 The topological properties of a number of interesting experimental systems are closely related to certain knot invariants. Here we show some of the classes of topological models. Within each class there are several possible topological theories, which differ from each other slightly. Each class is labeled by a Chern–Simons theory $SU(N)_k$, which we discuss in Chapter 5, and this Chern–Simons theory is one of the possible theories within the particular class. The slight differences between theories within a class are related to extra quantum mechanical phases that appear due to braiding. More details of the connections of the Kauffman bracket invariants to topological theories are given in Chapter 22. The experimental systems are discussed in more detail in Chapter 37.

2.2.1   Twist and Spin-Statistics

Before moving on, let us do some more careful examination of the Kauff-man bracket invariant. To this end, let us examine a small curl in a piece of string (as shown in Fig. 2.6) and try to evaluate its Kauffman bracket invariant.

$$= \left(A + A^{-1}\left(-A^2 - A^{-2}\right) \right) \Bigg) = -A^{-3} \Bigg)$$

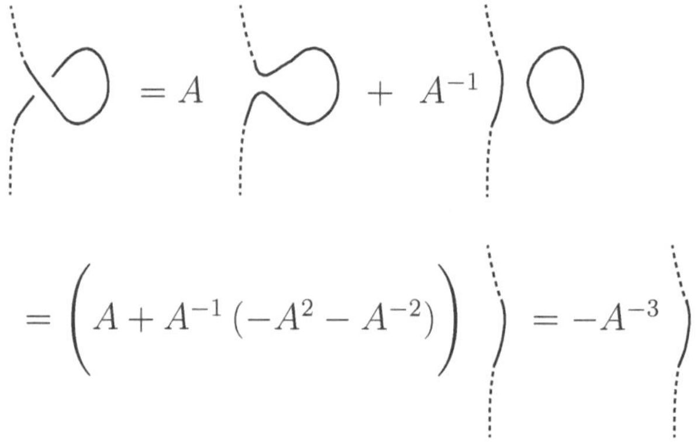

Fig. 2.6 Evaluation of a curl in a string. The dotted lines going off the top and bottom of the diagrams mean that the string will be connected up with itself, but we are not concerned with any part of the knot except for the piece shown. The result of this calculation is that removal of the curl incurs a factor of $-A^{-3}$.

We see from the calculation, that the string with the curl has value of $-A^{-3}$ times the value of string without the curl. But this seems to contradict what we said earlier! We claimed earlier that any two knots that can be deformed into each other without cutting should have the same Kauffman bracket invariant, but they don't!

The issue here is that the curled string on the left and the string on the right are, in fact, *not* topologically equivalent.[15] To see this we should think of the string as not being infinitely thin, but instead having some width, like a garden hose, or a "ribbon."[16] If we imagine straightening a thick string (not an infinitely thin string) we realize that pulling it straight gives a twisted string (see Fig. 2.7)—anyone who has tried to straighten a garden hose will realize this![17]

So the curled string is equivalent to a string with a self twist, and this is then related to a straight string by the factor of $-A^{-3}$. In fact, this is a result we should expect in quantum theory. The string with a self twist represents a particle that stays in place but rotates around an axis. In quantum theory, if a particle has a spin, it should accumulate a phase when it does a 2π rotation, and indeed this factor of $-A^{-3}$ is precisely such a phase in any well-defined quantum theory.

[15]In mathematics we say they are am-bient isotopic but not regular isotopic (see Section 2.6.1).

[16]We should thus think of our knots as not just being a simple embedding of a circle S^1 into a three manifold $\mathbb{R}^3$, but rather an embedding of a ribbon. This is equivalent to specifying an orthogo-nal vector at each point along the knot, which gives the orientation of the rib-bon cross section at each point. When one draws a knot as a line, one must have a convention as to what this means for the orientation of the ribbon. See comment on blackboard framing at the end of this section.

[17]If you have not had this experience with a garden hose, you are not paying enough attention to your garden!

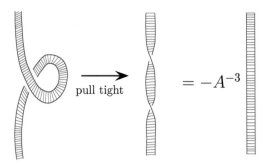

Fig. 2.7 Pulling straight a curl introduces a twist in the string. This twist can be replaced with a factor of $-A^{-3}$.

[18]People often think Lorentz invariance is required for the spin-statistics theorem to hold. Lorentz invariance is sufficient but not necessary. See the comments in Preskill [2004].

[19]In the most interesting case of non-abelian statistics, there may be multiple possible exchange phases for two particles, although this does not affect the equivalence of diagrams stated here. We discuss this more in Chapter 3.

[20]To properly count the self twists, one calculates the "writhe" of the knot (see Section 2.6.2). Give the string an orientation (a direction to walk along the string) and count +1 for each positive crossing and −1 for each negative crossing where a positive crossing is when, traveling in the direction of the string that crosses over, one would have to turn left to switch to the string that crosses under. If we orient the twisted string on the left of Fig. 2.6 as up-going, it then has a negative crossing by this definition.

In fact, Fig. 2.7 is a very slick proof of the famous spin-statistics theorem.[18] In the left picture with the curl, we have two identical particles that change places. When we pull this straight, we have a single particle that rotates around its own axis. In quantum theory, the phases accumulated by these two processes must be identical. As we will see in Chapter 3, in $(2+1)$-dimensions this phase can be arbitrary (not just $+1$, or -1), but the exchange phase (statistical phase) and the twist phase (the spin phase) must be the same.[19]

As a side comment, one can easily construct a knot invariant that treats the curled string on the left of Fig. 2.6 as being the same as the straight piece of string. One just calculates the Kauffman bracket invariant and removes a factor of $-A^{-3}$ for each self twist that occurs.[20] This gives the famed Jones polynomial knot invariant. See Exercise 2.4.

2.2.2   Blackboard Framing

Since it is important to specify when a strand of string has a self twist (as in the middle of Fig. 2.7) it is a useful convention to use so-called *blackboard framing*. With this convention we always imagine that the string really represents a ribbon, and the ribbon always lies in the plane of the blackboard. An example of this is shown in Fig. 2.8. If we intend a strand to have a self twist, we draw it as a curl as in the left of Fig. 2.7 or the left of Fig. 2.6.

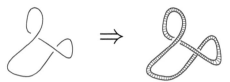

Fig. 2.8 Blackboard framing. The knot drawn on the left represents the ribbon on the right, where the ribbon always lies flat in the plane of the page (i.e. the plane of the blackboard).

2.3 **Bras and Kets**

For many topological theories (the so-called nonabelian theories) the physical systems have an interesting, and very unusual property. Imagine we start in a ground state (or vacuum) of some systems and create two particle-hole pairs, and imagine we tell you everything that you can locally measure about these particles (their positions, their spin, etc.). For most gapped systems (insulators, superconductors, charge density waves), once you know all of the locally measurable quantities, you know the full wavefunction of the system. But this is not true for topological systems.[21] As an example, see Fig. 2.9.

[21] Particles in topological systems seem to "remember" their spacetime history. The reason for this, as we will see in Chapter 4 and thereafter, is that this historical information becomes encoded in the properties of the vacuum; that is, the regions away from the particles where there is nothing but the background, analogous to the vacuum of outer space.

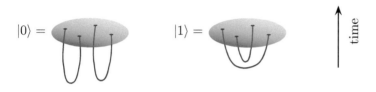

Fig. 2.9 Two linearly independent quantum states that look identical locally but have different spacetime histories. The horizontal plane is a spacetime slice at fixed time, and the diagrams are all oriented so time runs vertically.

Fig. 2.10 Kets are turned into bras by reversing time.

To demonstrate that these two different spacetime histories are linearly independent quantum states, we simply take inner products as shown in Fig. 2.11 by gluing together a ket (Fig. 2.9) with a bra (Fig. 2.10). Since $\langle 0|0 \rangle = \langle 1|1 \rangle = d^2$ but $\langle 0|1 \rangle = d$, we see that $|0\rangle$ and $|1\rangle$ must be linearly independent, at least for $|d| \neq 1$. (We also see that the kets here are not properly normalized, so we should multiply each bra and ket by $1/d$ in order that we have normalized states.)

We can think of the $|0\rangle$ and $|1\rangle$ states as being the result of particle–hole pairs being created from the vacuum, and (up to the issue of having properly normalized states) the inner product produced by graphical gluing a bra to a ket is precisely the inner product of these two resulting states. So for example, the inner product $\langle 0|1 \rangle$ as shown in the bottom of Fig. 2.11 can be reinterpreted as starting from the vacuum, time evolving with the operator that gives $|1\rangle$ then time evolving with the inverse of the operator that produces $|0\rangle$ to return us to the vacuum.

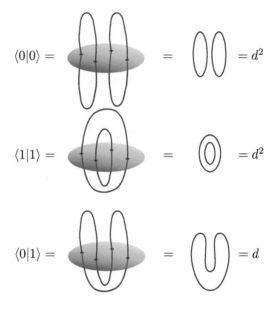

$\langle 0|0\rangle =$ $= d^2$

$\langle 1|1\rangle = $ $= d^2$

$\langle 0|1\rangle = $ $= d$

Fig. 2.11 Showing that the kets $|0\rangle$ and $|1\rangle$ are linearly indepen-dent. For $|d| \neq 1$ the inner products show they must be linearly independent quantities.

Suppose now we insert a braid between the bra and the ket as shown in Fig. 2.12. The braid makes a unitary operation on the two-dimensional vector space spanned by $|0\rangle$ and $|1\rangle$. We can once again evaluate this matrix element by calculating the Kauffman bracket invariant of the resulting knot.

$\langle 0| =$

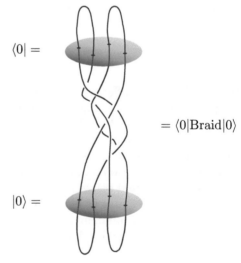

$= \langle 0|\text{Braid}|0\rangle$

$|0\rangle =$

Fig. 2.12 Inserting a braid between the bra and the ket. The braid performs a unitary operation on the two-dimensional vector space spanned by $|0\rangle$ and $|1\rangle$.

2.4 **Quantum Computation with Knots**

Why do we care so much about topological systems and knot invariants? A hint is from the fact that we wrote states above as $|0\rangle$ and $|1\rangle$. This notation suggests the idea of qubits,[22] and indeed this is one very good reason to be interested.

It turns out that many topological quantum systems can *compute* quantities efficiently that classical computers cannot. To prove this, suppose you wanted to calculate the Kauffman invariant of a very complicated knot, say with 100 crossings. As mentioned above, a classical computer would have to evaluate 2^{100} diagrams, which is so enormous, that it could never be done. However, suppose you have a topological system of Kauffman type in your laboratory. You could actually arrange to physically *measure* the Kauffman bracket invariant.[23] The way we do this is to start with a system in the vacuum state, arrange to "pull" particle–hole (particle–antiparticle) pairs out of the vacuum, then drag the particles around in order to form the desired knot, and bring them back together to reannihilate. Some of the particles will reannihilate, and others will refuse to go back to the vacuum (forming bound states instead). The probability that they all reannihilate is (up to a normalization[24]) given by the absolute square of the Kauffman bracket invariant of the knot (since amplitudes are the Kauffman bracket invariant, the square of the Kauffman bracket invariant is the probability). Even estimation of the Kauffman bracket invariant of a large knot is essentially impossible for a classical computer, for almost all values of A. However, this is an easy task if you happen to have a topological quantum system in your lab![25] Thus, the topological quantum system has computational ability beyond that of a classical computer.

It turns out that the ability to calculate the Kauffman bracket invariant is sufficient to be able to do any **quantum computation**.[26] One can use this so-called **topological quantum computer** to run algorithms such as Shor's famous factoring (i.e. code breaking) algorithm.[27] The idea of using topological systems for quantum computation is due to Michael Freedman and Alexei Kitaev.[28]

So it turns out that these topological systems can do quantum computation. Why is this a good way to do quantum computation?[22] First we must ask about why quantum computing is hard in the first place. In the conventional picture of a quantum computer, we imagine a bunch of two-state systems, say spins, which act as our qubits. Now during our computation, if some noise, say a photon, or a phonon, enters the

[22]One of my favorite quotes is "Any idiot with a two-state system thinks he has a quantum computer." The objective here is to show that we are not just any idiot—that quantum computing this way is actually a good idea! We discuss quantum computation more in Chapter 11.

[24]If we pull a single particle–hole pair from the vacuum and immediately bring them back together, the probability that they reannihilate is 1. However, the spacetime diagram of this is a single loop, and the Kauffman bracket invariant is d. The proper normalization is that each pair pulled from the vacuum and then returned to the vacuum introduces a $1/\sqrt{d}$ factor in front of the Kauffman bracket invariant.

[25]The details of this are a bit subtle and are discussed by Aharonov et al. [2009]; Aharonov and Arad [2011]; Kuperberg [2015].

[26]In fact the computational power of being able to evaluate the Kauffman bracket for fixed A is equivalent to the computational power of a quantum computer, with the exception of a few special values of the Kauffman parameter A.

[27]See Nielsen and Chuang [2000], for example, for more detail about quantum computation in general.

[28]Freedman is another Fields medalist, for his work on the Poincaré conjecture in 4D. Alexei Kitaev is one of the most influential scientists alive, a MacArthur winner, Milner Breakthrough Prize winner, etc. Both are very smart people. Freedman is also a champion rock climber.

[23]Perhaps the first statements ever made about a quantum computer were made in 1980 by the Russian mathematician Yuri Manin. He pointed out that doing any calculation about some complicated quantum system with 100 interacting particles is virtually impossible for a classical computer. Say for 100 spins you would have to find the eigenvalues and eigenvectors of a 2^{100}-dimensional matrix. But if you had the physical system in your lab, you could just measure its dynamics and answer certain questions. So in that sense the physical quantum system is able to compute certain quantities, that is, its own equations of motion, that a classical computer cannot. In the following year Feynman started thinking along the same lines and asked the question of whether one quantum system can compute the dynamics of another quantum system—which starts getting close to the ideas of modern quantum computation.

system and interacts with a qubit, it can cause an error or decoherence, which can then ruin your computation. And while it is possible to protect quantum systems from errors (we will see in Chapters 26–27 how you do this), it is very hard.

Now consider what happens when noise hits a topological quantum computer. In this case, the noise may shake around a particle, as shown in Fig. 2.13. However, as long as the noise does not change the topology of the knot, then no error is introduced. Thus the topological quantum computer is inherently protected from errors. (Of course sufficiently strong noise can change the topology of the knot and can still cause errors.)

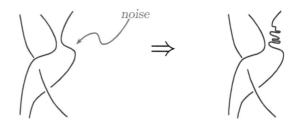

Fig. 2.13 The effect of noise on a topological quantum computation. As long as the noise does not change the topology of the knot, no error is introduced.

2.5   Some Quick Comments about Fractional Quantum Hall Effect

We say a bit more about the fractional quantum Hall effect (FQHE) in Chapter 37. But it is worth saying a few words about FQHE as a topological system now.

FQHE occurs in two-dimensional electronic systems[29] in high magnetic fields at low temperature (typically below 1K). There are many FQHE states that are each labeled by their so-called filling fraction $\nu = p/q$ with p and q small integers. The filling fraction can be changed in experiments by, for example, varying the applied magnetic field (we discuss this later in Chapter 37). The FQHE state emerges at low temperature and is topological.[30]

How do we know that the system is topological? There are not a whole lot of experiments that are easy to do on quantum Hall systems, since they are very low temperature. However, one type of experiment is fairly straightforward—a simple electrical resistance measurement, as shown in Figs. 2.14 and 2.15. In Fig. 2.14 the so-called longitudinal resistance is measured—where the current runs roughly parallel to the voltage. In this case the measured voltage is zero—like a superconductor. This shows that this state of matter has no dissipation, no friction.

The measurement in Fig. 2.15 is more interesting. In this case, the

[29]Electronic systems can be made two-dimensional in several ways. See comments in Chapter 37.

[30]A comment in comparing this paradigm to the common paradigm of high-energy physics: in high-energy there is generally the idea that there is some grand unified theory (GUT) at very high energy scale and it is extremely symmetric, but then when the universe cools to low temperature, symmetry breaks (such as electro-weak symmetry) and we obtain the physics of the world around us. The paradigm is opposite here. The electrons in magnetic field at high temperature have no special symmetry. However, as we cool down to lower temperature, a huge symmetry emerges. The topological theory is symmetric under all diffeomorphisms (smooth distortions) of space and time.

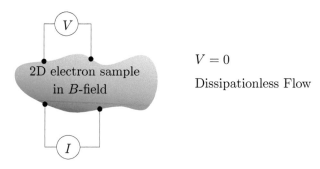

$V = 0$

Dissipationless Flow

Fig. 2.14 Measurement of longitudinal resistance in FQHE experiment.

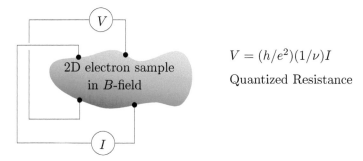

$V = (h/e^2)(1/\nu)I$

Quantized Resistance

Fig. 2.15 Measurement of Hall resistance in FQHE experiment.

Hall voltage is precisely quantized as $V = (h/e^2)(1/\nu)I$ where I is the current, h is Planck's constant, e the electron charge, and $\nu = p/q$ is a ratio of small integers. This quantization of V/I is extremely precise—to within about a part in 10^{10}. This is like measuring the distance from London to Los Angeles to within a millimeter. What is most surprising is that the measured voltage does not depend on details, such as the shape of the sample, whether there is disorder in the sample, where you put the voltage leads, or how you attach them as long as the current and voltage leads are topologically crossed, as they are in the Fig. 2.15, but not in Fig. 2.14. We should emphasize that this is extremely unusual. If you were to measure the resistance of a bar of copper, the voltage would depend entirely on how far apart you put the leads and the shape of the sample. This extremely unusual independence of all details is a strong hint that we have something robust and topological happening here.

Finally we can ask about what the particles are that we want to braid around each other in the FQHE case. These so-called quasiparticles are like the point-vortices of the FQHE superfluid. As we might expect for a dissipationless fluid, the vortices are persistent—they will last forever unless annihilated by antivortices.

So in fact, Kelvin was almost right (see Chapter 1). He was thinking about vortices knotting in the dissipationless aether. Here we are thinking about point vortices in the dissipationless FQHE fluid, but we move the vortices around in time to form spacetime knots!

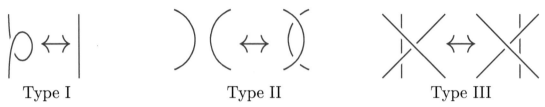

Type I   Type II   Type III

Fig. 2.16 The three Reidemeister moves. Any two knots that can be deformed into each other without cutting (they are "ambient isotopic") can be connected by a series of Reidemeister moves. Strictly speaking the Reidemeister moves include the moves drawn here as well as the front–back mirror-reflections of each of these moves (turn all over-crossings to under-crossings).

2.6   **Appendix: More Knot Theory Basics**

2.6.1   Isotopy and Reidemeister Moves

[31] This is a very old result, by Kurt Reidemeister from 1927. Note that it may take many moves in order to bring a knot into some particular desired form. For example, if there are c crossings in a diagram that is equivalent to the simple unknot (an unknotted loop), the strongest theorem proven so far is that it can be reduced to the simple unknot with $(236c)^{11}$ moves Lackenby [2015].

Two knots (or two pictures of knots) are *ambient isotopic* if one can be deformed into the other without cutting any of the strands. In order for two pictures of knots to be ambient isotopic they must be related to each other by smooth deformations that do not change any of the over- and under-crossings and a series of moves that do change the crossings, known as Reidemeister moves,[31] shown in Fig. 2.16.

In the context of quantum physics, and as elaborated in Section 2.2.1, we are usually concerned with *regular isotopy,* which treats the strands as ribbons. Sometimes to indicate that the knots should be treated as ribbons we say that the knots or links are "framed." When we draw a knot or link with lines (without drawing the whole ribbon) we mean the ribbon to be blackboard framed. Two such ribbons can be deformed into each other (i.e. the knot diagrams are regular isotopic) if their diagrams can be related to each other using only type II and type III moves. A type I move in a diagram inserts a twist in the ribbon (see Fig. 2.7), whereas type II and type III moves do not twist the ribbon.[32]

2.6.2   Writhe and Linking

[32] For regular isotopy of link diagrams one should allow cancelation of opposite ribbon curls, which is sometimes known as a type I′ move (or sometimes type O move) as shown here.

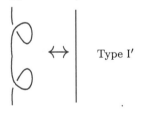

Type I′

Let us put arrows on all strands of our knots and links (so now we have directed lines). For each crossing we define a sign ϵ as shown in Fig. 2.17

$$= -1 \qquad\qquad = +1$$

Fig. 2.17 Defining a sign $\epsilon = \pm 1$ for each crossing of oriented knots and links.

The **writhe** w of an oriented knot (here "knot" means made of a single strand) is the sum of all of the ϵ values of the crossings

$$w(knot) = \sum_{\text{crossings}} \epsilon(\text{crossing}) \quad . \qquad (2.2)$$

Note that type II and III Reidemeister moves preserve the writhe of a knot, whereas type I moves do not. Thus, the writhe is an invariant of regular isotopy but not of ambient isotopy.

For an oriented link made of two strands, the **linking number** lk between the two strands is given by

$$lk(link) \quad = \quad \frac{1}{2} \sum_{\substack{\text{crossings between} \\ \text{two different strands}}} \epsilon(\text{crossing}) \quad . \qquad (2.3)$$

Chapter Summary

- Knot invariants, such as the Kauffman bracket invariant, help distinguish knots from each other.
- The quantum dynamics of certain particles are determined by certain knot invariants.
- Computation of certain knot invariants is computationally "hard" on a classical computer, but not hard using particles whose dynamics are given by knot invariants.
- Computation by braiding these particles is equivalent to any other quantum computer.
- Physical systems that have these particles include fractional quantum Hall effect.

Further Reading:

- The book by Kauffman [2001] is a delightful introduction to knot theory and connections to physics. This was the book that got me interested in the subject back when I was in grad school and changed the course of my life.
- I wrote another easy reading introduction [Simon, 2010] connecting knots to anyons.
- Some nice introductory books on knots include Adams [1994] and Sossinsky [2002]. A beautiful set of course notes on knot theory is given by Roberts [2015].

Exercises

Exercise 2.1 Trefoil Knot and the Kauffman Bracket

Using the Kauffman rules, calculate the Kauffman bracket invariant of the right- and left-handed trefoil[33] knots shown in Fig. 2.18. Conclude these two knots are topologically inequivalent. While this statement appears obvious on sight, it was not proved mathematically until 1914 (by Max Dehn). It is easy using this technique!

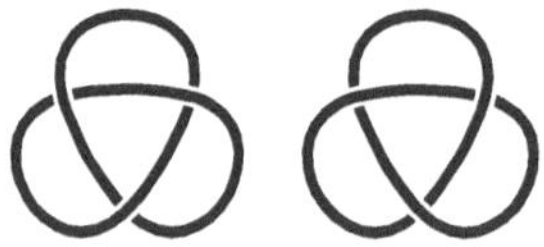

Fig. 2.18 Left- and right-handed trefoil knots (on the left and right, respectively)

---

[33]The word "trefoil" is from the plant trifolium, or clover, which has compound trifoliate leaves.

$$X = e^{i\vartheta}\,)\,($$

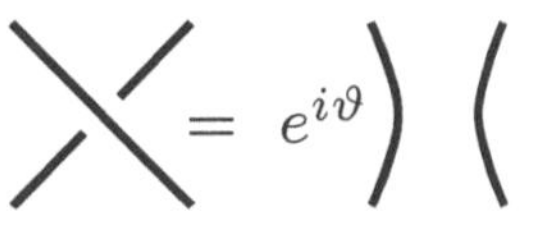

Fig. 2.19   For abelian anyons, exchange gives a phase $e^{i\vartheta}$.

$$\asymp = \pm\,)\,($$

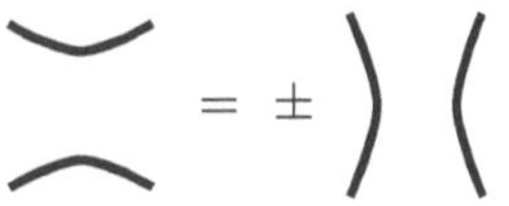

Fig. 2.20   For bosons or fermions the sign in this figure is $+$, for semions the sign is $-$.

Exercise 2.2 Abelian Kauffman Anyons

Particles where the quantum amplitudes of their trajectories are given by the Kauffman bracket invariant with certain special values of the constant A are abelian anyons—meaning an exchange introduces only a simple phase as shown in Fig. 2.19. Here we mean that the vertical direction now means time and the knot or link describes the motion of particles in spacetime.

(a) For $A = \pm e^{i\pi/3}$, show that the anyons are bosons or fermions respectively (i.e. $e^{i\vartheta} = \pm 1$). Further show that for these values of A any diagram calculated with A gives exactly the same result if you use the complex conjugate of A instead.

(b) For $A = \pm e^{i\pi/6}$ show the anyons are semions (i.e. $e^{i\vartheta} = \pm i$). Further show that calculating a diagram using $A = \pm e^{i\pi/6}$ gives exactly the same value as calculating the diagram using $A = \mp e^{-i\pi/6}$.

HINT: For both (a) and (b) show first the identity shown in Fig. 2.20. If you can't figure it out, try evaluating the Kauffman bracket invariant for a few knots with these values of A and see how the result arises.

Case (b) corresponds to the anyons that arise for the $\nu = 1/2$ fractional quantum Hall effect of bosons (We discuss this later in Chapter 37). This particular phase of quantum Hall matter has been produced experimentally [Clark et al., 2020], but only in very small puddles so far and it has not been possible to measure braiding statistics as of yet.

Exercise 2.3 Reidemeister moves and the Kauffman Bracket

Show that the Kauffman bracket invariant is unchanged under application of Reidemeister moves of type II and type III. Thus conclude that the Kauffman invariant is an invariant of regular isotopy.

Exercise 2.4 Jones Polynomial

Let us define the Jones polynomial of an oriented knot as[34]

$$\text{Jones(knot)} = (-A^{-3})^{w(\text{knot})}\, \text{Kauffman(knot)}$$

where w is the writhe (We must first orient the knot, meaning we put arrows on the strands, in order to define a writhe). Show that this quantity is an invariant of ambient isotopy—that is, it is invariant under all three Reidemeister moves.

Exercise 2.5 HOMFLY Polynomial

The HOMFLY[35] polynomial is a generalization of the Jones polynomial which has two variables X and z rather than just one variable. To define the HOMFLY polynomial we must first orient the strings in our knot or link (meaning we put arrows on the lines). The HOMFLY polynomial (Freyd et al. [1985]; Przytycki and Traczyk [1987]) of an oriented link is then defined in terms of two variables X and z by the two rules

$$\bigcirc = \bigcirc = \frac{(X + X^{-1})}{z}$$

$$X \nearrow\!\!\nwarrow + X^{-1}\nearrow\!\!\nwarrow = z \smile$$

[34]The traditional definition of the Jones polynomial writes the result as a function of $t = A^{-4}$. It is also common to divide by an overall factor of d so that the unknot is given the value 1.

[35]HOMFLY is an acronym of the names of the inventors of this polynomial. Sometimes credit is even more distributed and it is called HOMFLYPT.

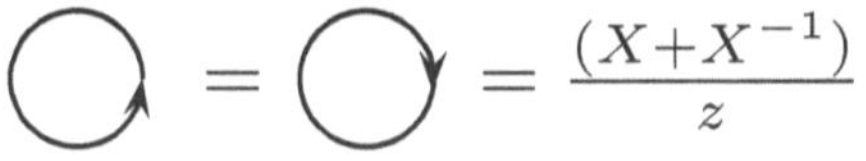

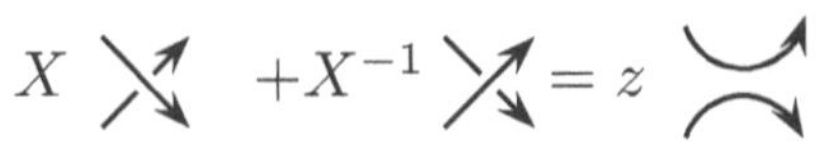

(a) Given the definition of the Jones polynomial in Exercise 2.4, for what value of X and z does the HOMFLY polynomial become the Jones polynomial?

(b) Calculate the HOMFLY polynomial of the right- and left-handed trefoil knots (shown in Fig. 2.18).

Exercise 2.6 Knot Sums

Two knots K_1 and K_2 can be summed together[36] to form a sum, denoted $K_1 \# K_2$, by cutting the knot K_1 to create two endpoints a_1 and b_1 and similarly cutting K_2 to create endpoints a_2 and b_2 then connecting a_1 to a_2 and b_1 to b_2 (Fig. 2.21). Show that

$$\text{Kauffman}(K_1 \# K_2) = d^{-1}\,\text{Kauffman}(K_1)\,\text{Kauffman}(K_2) \quad .$$

[36]Summing a knot with the unknot (a simple loop) leaves the knot unchanged so we think of the unknot as being the identity for this operation. We say a knot K_1 is factored into K_1 and K_2 if $K = K_1 \# K_2$ with neither K_1 nor K_2 being the unknot. There is a notion of a prime knot, which is a knot that cannot be factored. All knots can be uniquely factored into their prime components.

Fig. 2.21 An example of a knot sum.

Part I

Anyons and Topological Quantum Field Theories

Particle Quantum Statistics

Chapter 2 discussed braiding particles around each other, or exchanging their positions. This is often what we call particle statistics (or quantum statistics, or exchange statistics). What we mean by this is "what happens to the many particle wavefunction when particles are exchanged in a certain way."

We are familiar with bosons and fermions.[1,2] If we exchange two bosons the wavefunction is unchanged; if we exchange two fermions the wavefunction accumulates a minus sign. Various arguments have been given as to why these are the only possibilities. The argument usually given in introductory books is as follows:[3]

> If you exchange a pair of particles then exchange them again, you get back where you started. So the square of the exchange operator should be the identity, or one. There are two square roots of one: $+1$ and -1, so these are the only two possibilities for the exchange operator.

In the modern era this argument is considered to be incorrect (or at least not really sufficient). To really understand the possibilities in exchange statistics, it is very useful to think about quantum physics from the Feynman path integral point of view.[4]

3.1 Single Particle Path Integral

Consider a spacetime trajectory of a single non-relativistic particle. We say that we have $\mathbf{x}$ moving in $\mathbb{R}^D$ where D is the dimension of space, so we can write $\mathbf{x}(t)$ where t is time.

Given that we start at position $\mathbf{x}_i$ at the initial time t_i, we can define a so-called propagator, which gives the amplitude of ending up at position $\mathbf{x}_f$ at the final time t_f. This can be written as

$$\langle \mathbf{x}_f | \hat{U}(t_f, t_i) | \mathbf{x}_i \rangle$$

where $\hat{U}$ is the (unitary) time evolution operator.

The propagator can be used to propagate forward in time some arbitrary wavefunction $\psi(x) = \langle \mathbf{x} | \psi \rangle$ from t_i to t_f as follows

$$\langle \mathbf{x}_f | \psi(t_f) \rangle = \int d\mathbf{x}_i \, \langle \mathbf{x}_f | \hat{U}(t_f, t_i) | \mathbf{x}_i \rangle \, \langle \mathbf{x}_i | \psi(t_i) \rangle \quad .$$

If we are trying to figure out the propagator from some microscopic

---

[1] Bose cooked up the current picture of Bose statistics in 1924 in the context of photons and communicated it to Einstein, who helped him get it published. Einstein realized the same ideas could be applied to non-photon particles as well.

[2] Based on ideas by Pauli, Fermi–Dirac statistics were actually invented by Jordan in 1925. Jordan submitted a paper to a journal, where Max Born was the referee. Born stuck the manuscript in his suitcase and forgot about it for over a year. During that time both Fermi and Dirac published their results. Jordan could have won a Nobel Prize (potentially with Born) for his contributions to quantum physics, but he became a serious Nazi and no one really liked him much after that. Born felt terribly guilty about his mistake later in life, stating "I hate Jordan's politics, but I can never undo what I did to him."

[3] The error in this argument is that one has to be much more careful about defining what one means about an "exchange."

[4] If you are familiar with path integrals you can certainly skip down to Section 3.2. If you are not familiar with path integrals, please do not expect this to be a thorough introduction! What is given here is a minimal introduction to give us what we need to know for our purposes and nothing more! See the references at the end of this chapter for a better introduction.

calculation, there are two very fundamental properties it must obey. First, it must be unitary—meaning no amplitude is lost along the way (normalized wavefunctions stay normalized). Secondly, it must obey composition: propagating from t_i to t_m and then from t_m to t_f must be the same as propagating from t_i to t_f. We can express the composition law as

$$\langle \mathbf{x}_f|\hat{U}(t_f,t_i)|\mathbf{x}_i\rangle = \int d\mathbf{x}_m \, \langle \mathbf{x}_f|\hat{U}(t_f,t_m)|\mathbf{x}_m\rangle \, \langle \mathbf{x}_m|\hat{U}(t_m,t_i)|\mathbf{x}_i\rangle \quad .$$

The integration over $\mathbf{x}_m$ allows the particle to be at any position at the intermediate time (and it must be at *some* position). Another way of seeing this statement is to realize that the integral over $\mathbf{x}_m$ is just insertion of a complete set of states at some intermediate time

$$1 = \int d\mathbf{x}_m |\mathbf{x}_m\rangle\langle \mathbf{x}_m| \quad .$$

Feynman's genius[5] was to realize that you can subdivide time into infinitesimally small pieces, and you end up doing lots of integrals over all possible intermediate positions. In order to get the final result, you must sum over all values of all possible intermediate positions, or all possible functions $\mathbf{x}(t)$ as shown in Fig. 3.1. Feynman's final result[6] is that the propagator can be rewritten as

$$\langle \mathbf{x}_f|\hat{U}(t_f,t_i)|\mathbf{x}_i\rangle = \mathcal{N} \sum_{\substack{\text{paths } \mathbf{x}(t) \text{ from} \\ (\mathbf{x_i},t_i) \text{ to } (\mathbf{x_f},t_f)}} e^{iS[\mathbf{x}(t)]/\hbar} \tag{3.1}$$

where $\mathcal{N}$ is some normalization constant. Here, $S[\mathbf{x}(t)]$ is the (classical!) action of the path

$$S = \int_{t_i}^{t_f} dt \, L[\mathbf{x}(t),\dot{\mathbf{x}}(t),t]$$

with L the Lagrangian.

The sum over paths in Eq. 3.1 is often well defined as a limit of dividing the path into discrete time steps and integrating over $\mathbf{x}$ at each time. We often rewrite this sum over paths figuratively as a so-called path integral

$$\langle \mathbf{x}_f|\hat{U}(t_f,t_i)|\mathbf{x}_i\rangle = \mathcal{N} \int_{(\mathbf{x_i},t_i)}^{(\mathbf{x_f},t_f)} \mathcal{D}\mathbf{x}(t) \, e^{iS[\mathbf{x}(t)]/\hbar} \quad . \tag{3.2}$$

Analogous to when we evaluate regular integrals of things that look like $\int dx \, e^{iS[x]/\hbar}$, we can approximate the value of this integral in the small $\hbar$, or classical, limit by saddle point approximation. We do this by looking for a minimum of S with respect to its argument—this is where the exponent oscillates least, and it becomes the term which dominates the result of the integral. Similarly, with the path integral, the piece that dominates in the small $\hbar$ limit is the piece where $S[\mathbf{x}(t)]$ is extremized— the function $\mathbf{x}(t)$ that extremizes the action. This is just the classical principle of least action!

[5]The idea of the path integral was based heavily on earlier work by Dirac. See Dass [2020].

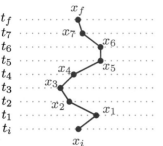

Fig. 3.1 Slicing time into small pieces and integrating over all positions at each intermediate time results in a sum over all possible paths $x(t)$.

[6]The missing piece of this calculation is to show that the propagator for a single very small time step is related to the classical action. See MacKenzie [2000] for example.

3.2 **Two Identical Particles**

We now generalize the idea of a path integral to systems with multiple identical particles, starting with the case of two particles. If the particles are identical there is no meaning to saying that particle one is at position $\mathbf{x}_1$ and particle two is at position $\mathbf{x}_2$. This would be the same as saying that they are the other way around. Instead, we can only say that there are particles at both positions $\mathbf{x}_1$ and $\mathbf{x}_2$. To avoid the appearance of two different states expressed as $|\mathbf{x}_1, \mathbf{x}_2\rangle$ versus $|\mathbf{x}_2, \mathbf{x}_1\rangle$ (which are actually the same physical state![7]), it is then useful to simply agree on some convention for which coordinate we will always write first—for example, maybe we always write the leftmost particle first.[8] For simplicity, we can assume that $\mathbf{x}_1 \neq \mathbf{x}_2$, that is, the particles have hard cores and cannot overlap.[9] For these indistinguishable particles, the Hilbert space is then cut in half compared to the case of two *distinguishable* particles where $|\mathbf{x}_1, \mathbf{x}_2\rangle$ and $|\mathbf{x}_2, \mathbf{x}_1\rangle$ mean physically different things.

We call the space of all states the configuration space $\mathcal{C}$. To construct a path integral, we want to think about all possible paths through this configuration space. The key realization is that the space of all paths through the configuration space $\mathcal{C}$ divides up into topologically inequivalent pieces. That is, certain paths cannot be deformed into other paths by a series of small deformations.

What do these topologically disconnected pieces of our space of paths look like? For example, we might consider the two paths as shown in Fig. 3.2. Here we mean that time runs vertically. It is not possible to continuously deform the path on the left into the path on the right assuming the end points are fixed.

We will call the non-exchange path TYPE $+1$ (left in Fig. 3.2), and the exchange path TYPE -1 (right in Fig. 3.2). The two sets of paths cannot be continuously deformed into each other assuming the end points are fixed. Note that we may be able to further refine our classification of paths—for example, we may distinguish over- and under-crossings, but for now we will only be concerned with exchanges (TYPE -1) and non-exchanges (TYPE $+1$).

Paths can be composed with each other. In other words, we can follow one path first, then follow the second. We can write a multiplication table for such composition of paths (the path types form a *group*, see Section 40.2).

[7]Often books define $|\mathbf{x}_1, \mathbf{x}_2\rangle = -|\mathbf{x}_2, \mathbf{x}_1\rangle$ for fermions. The two kets describe the same state in the Hilbert space only with a different phase prefactor. We should contrast this to the case of distinguishable particles where $|\mathbf{x}_1, \mathbf{x}_2\rangle$ and $|\mathbf{x}_2, \mathbf{x}_1\rangle$ have no overlap for $\mathbf{x}_1 \neq \mathbf{x}_2$

[8]This ordering scheme works in one dimension. In two dimensions we would perhaps say the particle with the smaller x coordinate is written first, but in case of two particles with the same value of x, the particle with smaller y coordinate is written first.

[9]It is sometimes even more convenient to declare $|\mathbf{x}_1 - \mathbf{x}_2| > \epsilon$.

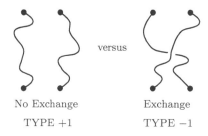

No Exchange   Exchange

TYPE $+1$ TYPE -1

Fig. 3.2 Two possible sets of paths (paths in configuration space) from the same two starting positions to the same two ending positions (we are implying that time runs vertically). We call the non-exchange path TYPE $+1$, and the exchange path TYPE -1.

$$
\begin{array}{llll}
\text{TYPE } +1 & \text{followed by} & \text{TYPE } +1 & = & \text{TYPE } +1 \\
\text{TYPE } +1 & \text{followed by} & \text{TYPE } -1 & = & \text{TYPE } -1 \\
\text{TYPE } -1 & \text{followed by} & \text{TYPE } +1 & = & \text{TYPE } -1 \\
\text{TYPE } -1 & \text{followed by} & \text{TYPE } -1 & = & \text{TYPE } +1
\end{array} \tag{3.3}
$$

So, for example, an exchange path (which switches the two particles) followed by another exchange path (which switches again) results in a net path that does not switch the two particles.

[10]If $|\mathbf{x}_{1i}\mathbf{x}_{2i}\rangle \neq |\mathbf{x}_{1f}\mathbf{x}_{2f}\rangle$, that is, if the initial and final endpoints of the path are not the same, then we need a more general definition of what we call TYPE $+1$ versus TYPE -1. One simple possible definition is to count the number of times the spacetime paths cross given a particular fixed viewing angle. For example, we can use the ordering rule of note 8 and as time evolves, we can count the number of times the ordering changes (which corresponds to a crossing of the world lines as in Fig. 3.2). An even number of such crossings would correspond to a TYPE $+1$ path, and an odd number of crossings would correspond to a TYPE -1 path. Other consistent definitions are also possible as long as the multiplication rule of Eq. 3.3 is maintained.

Now let us try to construct a path integral, or sum over all possible paths. It is useful to think about breaking up the sum over paths into separate sums over the two different classes of paths.[10]

$$\langle \mathbf{x}_{1f}\mathbf{x}_{2f}|\hat{U}(t_f,t_i)|\mathbf{x}_{1i}\mathbf{x}_{2i}\rangle = \mathcal{N}\sum_{\substack{\text{paths} \\ i\to f}} e^{iS[\text{path}]/\hbar} = \tag{3.4}$$

$$\mathcal{N}\left(\sum_{\substack{\text{TYPE }+1\text{ paths} \\ i\to f}} e^{iS[\text{path}]/\hbar} + \sum_{\substack{\text{TYPE }-1\text{ paths} \\ i\to f}} e^{iS[\text{path}]/\hbar}\right).$$

This second line is simply a rewriting of the first having broken the sum into the two different classes of paths.

It turns out, however, that it is completely consistent to try something different. Let us instead write[10]

$$\langle \mathbf{x}_{1f}\mathbf{x}_{2f}|\hat{U}(t_f,t_i)|\mathbf{x}_{1i}\mathbf{x}_{2i}\rangle = \tag{3.5}$$

$$\mathcal{N}\left(\sum_{\substack{\text{TYPE }+1\text{ paths} \\ i\to f}} e^{iS[\text{path}]/\hbar} - \sum_{\substack{\text{TYPE }-1\text{ paths} \\ i\to f}} e^{iS[\text{path}]/\hbar}\right).$$

Notice the change of sign for the TYPE -1 paths.

The reason this change is allowed is because it obeys the composition law. To see this, let us check to see if the composition law is still obeyed. Again, we break the time propagation at some intermediate time[11]

[11]The sum over intermediate states necessarily requires us to include the case discussed in note 10.

$$\langle \mathbf{x}_{1f}\mathbf{x}_{2f}|\hat{U}(t_f,t_i)|\mathbf{x}_{1i}\mathbf{x}_{2i}\rangle =$$

$$\int d\mathbf{x}_{1m}d\mathbf{x}_{2m}\,\langle \mathbf{x}_{1f}\mathbf{x}_{2f}|\hat{U}(t_f,t_m)|\mathbf{x}_{1m}\mathbf{x}_{2m}\rangle\,\langle \mathbf{x}_{1m}\mathbf{x}_{2m}|\hat{U}(t_m,t_i)|\mathbf{x}_{1i}\mathbf{x}_{2i}\rangle$$

$$\sim \int d\mathbf{x}_{1m}d\mathbf{x}_{2m}\left(\sum_{\substack{\text{TYPE }+1 \\ m\to f}} - \sum_{\substack{\text{TYPE }-1 \\ m\to f}}\right)\left(\sum_{\substack{\text{TYPE }+1 \\ i\to m}} - \sum_{\substack{\text{TYPE }-1 \\ i\to m}}\right) e^{iS[\text{path}]/\hbar}$$

where in the last line we have substituted in Eq. 3.5 for each of the two propagators on the right, and we have used a bit of shorthand in writing the result.

Now, when we compose together subpaths from $i \to m$ with those from $m \to f$ to get the overall path, the subpath types multiply according to our above multiplication table Eq. 3.3. For the full path, there are two ways to obtain a TYPE $+1$ path: (1) both subpaths are TYPE $+1$ or (2) both subpaths are TYPE -1. In either case, note that the net prefactor of the overall TYPE $+1$ path is $+1$. (In the case where both subpaths are of TYPE -1, the two prefactors of -1 cancel each other). Similarly, we can consider full paths with overall TYPE -1. In this case, exactly one of the two subpaths must be of TYPE -1, in which case, the overall sign ends up being -1. Thus, for the full path, we obtain exactly the intended form written in Eq. 3.5. That is, under composition of paths, we preserve the rule that TYPE $+1$ paths get a $+1$ sign and TYPE -1 paths get a -1 sign. Thus, this is consistent for quantum mechanics, and indeed, this is exactly what happens in the case of fermions.

The laws of quantum mechanics do not tell us whether we should use a path integral of the type in Eq. 3.4 or that of Eq. 3.5. Both are allowed, and it is a matter of experiment[12] to determine whether a particular type of particle obeys the former (i.e. is a boson) or the latter (i.e. is a fermion).

[12]There may, however, be derivable consistency conditions. For example, if we know something about the spin of a particle, this can constrain us to one or the other possibility.

3.3 Many Identical Particles: Preliminaries

Generalizing this idea, to figure out what is consistent in quantum mechanics, we must do two things:

(a) Characterize the space of paths through configuration space
(b) Insist on consistency under composition.

Let us first discuss our configuration space. If we had N *distinguishable* particles in D dimensions we would have a configuration space $(\mathbb{R}^D)^N$ representing the coordinates $\{\mathbf{x}_1, \mathbf{x}_2, \mathbf{x}_3, \ldots, \mathbf{x}_N\}$. For simplicity we usually assume all of these coordinates are different (i.e. we might imagine that the particles are hard spheres of some very small diameter ϵ). Thus, we write the configuration space as $(\mathbb{R}^D)^N \setminus \Delta$ where Δ represents the so-called *coincidences* where two particles are at the same position and $\setminus \Delta$ means we should remove these coincidences.[13]

In the case of *identical* particles we want to disregard the order in which we write the coordinates. In other words, we have an equivalence relationship $\sim$ between the $N!$ possible orderings of the coordinates

[13]Mathematicians write $(\mathbb{R}^D)^N \setminus \Delta$ to represent removing Δ from the set $(\mathbb{R}^D)^N$. See Section 40.1.6.

$$\{\mathbf{x}_1, \mathbf{x}_2, \mathbf{x}_3, \ldots, \mathbf{x}_N\} \sim \{\mathbf{x}_2, \mathbf{x}_3, \mathbf{x}_7, \ldots, \mathbf{x}_9\} \sim \{\mathbf{x}_3, \mathbf{x}_N, \mathbf{x}_2, \ldots, \mathbf{x}_1\} \sim \ldots \quad .$$

Thus, for indistinguishable particles the configuration space is

$$\mathcal{C} = \left[(\mathbb{R}^D)^N \setminus \Delta \right] / \sim$$

where "$/ \sim$" means that we are "modding out" by the equivalence relationship $\sim$. This just means the order in which we list the coordinates $\{\mathbf{x}_1, \mathbf{x}_2, \mathbf{x}_3, \ldots, \mathbf{x}_N\}$ does not matter (or as described in Section 3.2, we choose some convention for the order, like always writing the left-most first). In the case of two identical particles above, this reduced the Hilbert space by a factor of two. With N identical particles this reduces the Hilbert space by a factor of $(N!)$. This is the same indistinguishability factor as in the Gibbs paradox of statistical mechanics.

We now consider all possible paths through this configuration space $\mathcal{C}$.

[14]The curves are directed because we do not allow them to double back in time as shown in Fig. 3.3, which would represent particle–hole creation or annihilation, which we do not yet consider.

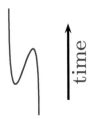

Fig. 3.3 A double back in time is not allowed in our considerations here (and not allowed in the braid group).

[15]In fact what we really want is the *fundamental groupoid,* which allows for the fact that the initial and final positions of particles may not be the same. However, for illustration, the fundamental group will be sufficient.

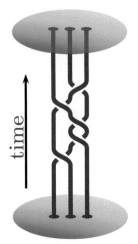

Fig. 3.4 A path through configuration space for three particles in two dimensions (i.e. world lines in (2 + 1)-dimensions) is a braid with three strands.

[16]The identity element 1 of the braid group is everything that is topologically equivalent to the non-braid, that is, particles that do not change their position in space at all. It is easy to see that $\sigma_i \sigma_i^{-1} = 1$.

In other words, we want to consider how these N different points move in time. We can think of this as a set of coordinates moving through time $\{\mathbf{x}_1(t), \ldots \mathbf{x}_N(t)\}$ but we must be careful that the particles are indistinguishable, so the order in which we write the coordinates doesn't matter. We can think of this as N directed curves moving in $(ND+1)$-dimensional space.[14] Since we want to add up all of these possible paths in a path integral it is useful to try to understand the structure of this space of paths better.

Again, the key realization is that the space of all paths through the configuration space $\mathcal{C}$ divides up into topologically inequivalent pieces. That is, certain paths cannot be deformed into other paths by a series of small deformations assuming the endpoints are fixed. The group of paths through $\mathcal{C}$ is familiar to mathematicians and is known as the first homotopy group $\Pi_1(\mathcal{C})$ or fundamental group[15] (see Section 40.3). The reason this is a group is that it comes with a natural operation, or multiplication of elements—which is the composition of paths: follow one path, then follow another path.

3.3.1 Paths in (2 + 1)-Dimensions, the Braid Group

A path through the configuration space of particles in two dimensions is known as a braid. An example of a braid is shown in Fig. 3.4.

A few notes about braids:

(1) Fixing the endpoints, the braids can be deformed continuously, and as long as we do not cut one string through another, it still represents the same topological class, or the same element of the braid group.

(2) We cannot allow the strings to double-back in time as in Fig. 3.3. This would be pair creation or annihilation, which we will consider later, but not now.

The set of braids have mathematical group structure (see Section 40.2): multiplication of two braids is defined by stacking the two braids on top of each other—first do one then do another. All braids can be decomposed into elementary pieces which involve either clockwise or counterclockwise exchange of one strand with its neighbor. These elementary pieces involving single exchanges are known as generators.

The braid group on N strands is usually notated as B_N. The generators of the braid group on four strands are shown in Fig. 3.5. Any braid can be written as a product of the braid generators and their inverses.[16] The "multiplication" of the generators is achieved simply by stacking the generators on top of each other. An expression representing a braid, such as $\sigma_1 \sigma_2 \sigma_3^{-1} \sigma_1$ is known as a "braid word." Typically we read the braid word from right to left (do the operation listed right-most first), although sometimes people use the opposite convention! The important thing is to fix a convention and stick with it!

Note that many different braid words can represent the same braid. An example of this is shown for B_4 in Fig. 3.6. Although a braid can

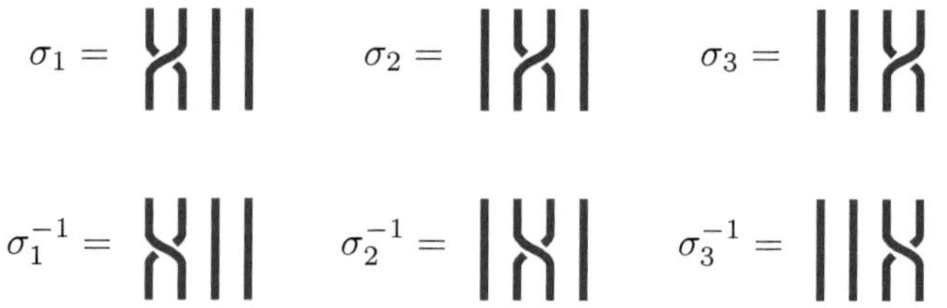

$$\sigma_1 = \text{▯} \quad\quad \sigma_2 = \text{▯} \quad\quad \sigma_3 = \text{▯}$$

$$\sigma_1^{-1} = \text{▯} \quad\quad \sigma_2^{-1} = \text{▯} \quad\quad \sigma_3^{-1} = \text{▯}$$

Fig. 3.5 The three generating elements $\sigma_1, \sigma_2, \sigma_3$ of the braid group on four strands, B_4, and their inverses $\sigma_1^{-1}, \sigma_2^{-1}, \sigma_3^{-1}$. Any braid on four strands (any element of B_4) can be written as a product of the braid generators and their inverses by simply stacking these generators together (see Fig. 3.6 for examples).

$$\sigma_1^{-1}\sigma_2^{-1}\sigma_1 = \text{▯} = \text{▯} = \sigma_2\sigma_1^{-1}\sigma_2^{-1}$$

third   second   first

Fig. 3.6 Two braid words in B_4 that represent the same braid. The figure on the left can be continuously deformed to the one on the right, keeping endpoints fixed. The braidwords are read from right to left indicating stacking the generators from bottom to top.[17]

be written in many different ways,[18] it is possible to define invariants of the braid which do not change under deformation of the braid—as long as the braid is topologically unchanged. One very useful braid invariant is given by the so-called winding number

$$
\begin{aligned}
W &= \text{Winding Number} \\
&= (\text{\# of over-crossings}) - (\text{\# of under-crossings})
\end{aligned}
$$

where an over-crossing is a σ and an under-crossing is a σ^{-1}. As can be checked in Fig. 3.6, the winding number is independent of the particular way we represent the braid. As long as we do not cut one strand through another or move the endpoints (or double-back strands) the winding number, a braid invariant, remains the same.

3.3.2 Paths in (3 + 1)-Dimensions, the Permutation Group

We now turn to consider physics in $(3 + 1)$-dimensions. A key fact is that it is not possible to knot a one-dimensional world line that lives in a four-dimensional space. If this is not obvious consider the following lower-dimensional analogue,[20] shown in Fig. 3.7. In one dimension, two points cannot cross through each other without hitting each other. But

[20]It would be very convenient to be able to draw a diagram in four dimensions!

[17]The observant reader will see the similarity here to Reidemeister moves of type III discussed in Section 2.6.1. Similarly $\sigma_i\sigma_i^{-1} = 1$ is a type II move.

[18]All braid word equivalences can be derived from the identity
$$\sigma_n\sigma_{n+1}\sigma_n = \sigma_{n+1}\sigma_n\sigma_{n+1} \ .$$
For example, try deriving Fig. 3.6 from this. See also Exercise 3.1.

In one dimension:
two objects cannot cross.

In two dimensions:
two objects can go around each other.

Fig. 3.7 **Top:** In one dimension, two points cannot cross through each other without hitting each other. **Bottom:** However, if we allow the points to move in two dimensions they can get around each other without touching. This is supposed to show you that one-dimensional world lines cannot form knots in four-dimensional space.

if we allow the points to move in two dimensions they can move around each other without touching each other. Analogously we can consider strings forming knots or braids in three-dimensional space. When we try to push these strings through each other, they bump into each other and get entangled. However, if we allow the strings to move into the fourth dimension, we can move one string a bit off into the fourth dimension so that it can move past the other string, and we discover that the strings can get by each other without ever touching each other! Hence, there are no knots of one-dimensional objects embedded in four dimensions.

Note that in the permutation group an exchange squared does give the identity. However, in the braid group this is not so—the braid σ_i^2 is not the identity since it creates a nontrivial braid![21]

Given that in $(3 + 1)$-dimensions world lines cannot form knots, the only thing that is important in determining the topological classes of paths is where the strings start and where they end. In other words, we can draw things that look a bit like braid diagrams but now there is no meaning to an over- or under-crossing. If the world line lives in $(3 + 1)$-dimensions, everything can be unentangled without cutting any of the world lines until the diagram looks like Fig. 3.8: indicating only where lines start and end. This is precisely describing the permutation group S_N (see Section 40.2.1).[22]

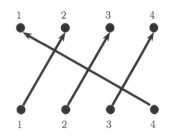

Fig. 3.8 Paths in $(3 + 1)$-dimensions are elements of the permutation group S_N (see Section 40.2.1). Shown here is an element of S_4.

[21] One way to think about the relationship between the permutation group and the braid group is to say that the permutation group S_N is a "truncation" of the braid group B_N, meaning that it obeys the same group properties, except that in S_N, the element σ_i^2 has been identified with the identity. Further, σ_i^2 being the identity is the same as saying that $\sigma_i = \sigma_i^{-1}$, which simply means the over-crossing and under-crossing have been made equivalent in S_N.

[22] S_N is also called the "symmetric group."

3.4   Building a Path Integral: Abelian Case

We now return to the issue of building a path integral. We follow the intuition we gained in the two-particle case, but now we include the information we have discovered about the group of paths through configuration space.

Let us first remind the reader of our approach. Many physicists view the path integral formulation, the sum over histories, to be the most fundamental principle of quantum physics—for each history we sum over e^{iS} for some classical action S. This alone defines most of the properties of many-particle systems. However, quantum mechanics also allows one to add factors out front of this sum as long as (a) these factors are the same for paths that can be continuously deformed into each other and (b) these factors obey composition. We have already seen this in the case of the TYPE +1 and TYPE −1 paths in the two-particle example in Section 3.2. Here we consider a many-particle system, and we would like to determine what possibilities exist for such factors.

Using the notation $\{\mathbf{x}\}$ to denote all of the N particle coordinates, we construct the path integral as

$$\langle\{\mathbf{x}\}_f|\hat{U}(t_f, t_i)|\{\mathbf{x}\}_i\rangle = \mathcal{N} \sum_{g \in G} \rho(g) \sum_{\substack{\text{paths} \in g \\ i \to f}} e^{iS[\text{path}]/\hbar} \qquad (3.6)$$

where $\mathcal{N}$ is a normalization constant. Here, G is the group of paths (the fundamental group of configuration space—or the set of classes of topologically different paths). This group is the permutation group S_N for $(3+1)$-dimensions and is the braid group B_N for $(2+1)$-dimensions. We have split the sum over paths from the into the different classes—the outer sum being a sum over the classes g and the inner sum being the sum over all paths of type g, that is, a set of paths that can be continuously deformed into each other. We have also introduced a factor of $\rho(g)$ out front where ρ is a *scalar unitary representation* of the group G (see Section 40.2.4 on group theory). This means that $\rho(g)$ is a mapping from the elements of G to a complex phase. Since complex phases are commutative, we say this is an *abelian* representation (even though the group G is generally not commutative).

In the case where the initial set of position $|\{\mathbf{x}\}_i\rangle$ and the final set of position $|\{\mathbf{x}\}_f\rangle$ are not the same (similar to the case mentioned in note 10) Eq. 3.6 can still be used, although strictly speaking these are not precisely what we would call braids or permutations (for which initial and final positions are supposed to match). Nonetheless, we can associate an element g of the braid or permutation group to each spacetime path by viewing the motion from some fixed angle and smoothly deforming the paths such that start and endpoints are at some reference positions without introducing any new crossings to the paths.[23]

To show that Eq. 3.6 is allowed by the laws of quantum mechanics, we need only check that it obeys the composition law—we should be able to construct all paths from i to f in terms of all paths from i to m and all paths from m to f.

$$\langle\{\mathbf{x}\}_f|\hat{U}(t_f, t_i)|\{\mathbf{x}\}_i\rangle = \tag{3.7}$$

$$= \int d\{\mathbf{x}\}_m \ \langle\{\mathbf{x}\}_f|\hat{U}(t_f, t_m)|\{\mathbf{x}\}_m\rangle \ \langle\{\mathbf{x}\}_m|\hat{U}(t_m, t_i)|\{\mathbf{x}\}_i\rangle$$

$$\sim \int d\{\mathbf{x}\}_m \left(\sum_{g_1 \in G} \rho(g_1) \sum_{\substack{\text{paths} \in g_1 \\ m \to f}} \right) \left(\sum_{g_2 \in G} \rho(g_2) \sum_{\substack{\text{paths} \in g_2 \\ i \to m}} \right) e^{iS[\text{path}]/\hbar}$$

So we have constructed all possible paths from i to f and split them into class g_2 in the region i to m and then class g_1 in the region m to f. When we compose these paths we will get a path of type $g_1 g_2$. The prefactors of the paths $\rho(g_1)$ and $\rho(g_2)$ then multiply and we get $\rho(g_1)\rho(g_2) = \rho(g_1 g_2)$ since ρ is a representation (the preservation of multiplication is the definition of being a representation! See Section 40.2.4). So the prefactor of a given path from i to f is correctly given by $\rho(g)$ where g is the topological class of the path. In other words, the form shown in Eq. 3.6 is properly preserved under composition, which is what is required in quantum mechanics!

It is important to note that mathematics cannot tell us what representation $\rho(g)$ we should use in our path integral—it can only tell us which

[23] A crossing can be defined as a reordering of coordinates as in note 8.

representations are allowed. Which representation actually occurs is a piece of physics information about the particular system being studied.

3.4.1 (3 + 1)-Dimensions

In $(3 + 1)$-dimensions, the group G of paths through configuration space is the permutation group S_N. It turns out that there are *only two possible*[24] scalar unitary representations of S_N:

[24]See Exercise 3.2. This is a fairly short proof!

- **Trivial rep:** In this case $\rho(g) = 1$ for all g. This corresponds to **bosons**. The path integral is just a simple sum over all possible paths with no factors inserted.

- **Alternating (or sign) rep:** In this case $\rho(g) = +1$ or -1 depending on whether g represents an even or odd number of exchanges. In this case, the sum over all paths gets a positive sign for an even number of exchanges and a negative sign for an odd number. This is obviously **fermions** and is the generalization of the two particle example we considered in Section 3.2 where the exchange was assigned a -1.

3.4.2 (2 + 1)-Dimensions

In $(2 + 1)$-dimensions, the group G of paths through configuration space is the braid group B_N. We can describe the possible one-dimensional representations by a single parameter ϑ. We write the representation

$$\rho(g) = e^{i\vartheta W(g)}$$

[25]There is no reason why this should not have been discovered in the 1930s, but no one bothered to think about it. It is a lucky coincidence that an experimental system of anyons was discovered so soon after the theoretical proposal (fractional quantum Hall effect, discovered by Tsui, Stormer, and Gossard [1982], see Section 37.1), since the original theoretical work was entirely abstract, and they were not thinking about any particular experiment.

[26]The use of the braid group for describing statistics in two dimensions dates back to Goldin et al. [1983]. The use of the permutation group for understanding statistics in three dimensions appears to go back further to Laidlaw and DeWitt [1971].

[27]Among other things, Wilczek coined the term *anyon*. (He also won a Nobel Prize for asymptotic freedom.)

[28]Please also see my *next* book, which will give much more detail on fractional quantum Hall physics!

where W is the winding number of the braid g. In other words, a clockwise exchange accumulates a phase of $e^{i\vartheta}$ whereas a counterclockwise exchange accumulates a phase of $e^{-i\vartheta}$.

- For $\vartheta = 0$ there is no phase, and we simply recover **bosons**.

- For $\vartheta = \pi$ we accumulate a phase of -1 for each exchange no matter the direction of the exchange (since $e^{i\pi} = e^{-i\pi}$). This is **fermions**.

- **Any** other value of ϑ is also allowed. This is known as **Anyons**, or **fractional statistics**. They are also known as **abelian anyons** in contrast with the nonabelian case, which we discuss in a moment.

The fact that this fractional statistics is consistent in quantum mechanics was first point out by Leinaas and Myrheim [1977],[25,26] and then by Goldin et al. [1981] and Wilczek [1982].[27] Soon thereafter, Halperin [1984] and then Arovas, Schrieffer, and Wilczek [1984] showed theoretically that anyons really occur in fractional quantum Hall systems. We discuss these physical systems briefly in Section 37.1.[28] Fractional statistics has recently been observed clearly in several experiments (see Chapter 37)

3.5 **Nonabelian Case**

In Eq. 3.6 we used a scalar (or one-dimensional) representation of the group G. Can we do something more interesting and exotic by using a higher-dimensional representation of the group G of paths in configuration space? Generally in quantum mechanics, higher-dimensional representations correspond to degeneracies, and indeed this is what is necessary.

Suppose we have a system with N particles at a set of positions $\{\mathbf{x}\}$. Even once we fix the positions (as well as the values of any local quantum numbers, like any "color" or "flavor" or "spin" degree of freedom associated with the particle), suppose there still remains an M-fold degeneracy of the state of the system.[29] We might describe the M states as $|n; \{\mathbf{x}\}\rangle$ for $n = 1 \ldots M$. An arbitrary wavefunction of the system can then be expressed as[30]

$$|\psi_{\{\mathbf{x}\}}\rangle = \sum_{n=1}^{M} A_n |n; \{\mathbf{x}\}\rangle \tag{3.8}$$

with the A_n being some complex coefficients. Given the N positions $\{\mathbf{x}\}$, a general wavefunction should be thought of as a vector in M-dimensional complex space. Now that we have a vector, we can use an M-dimensional representation of the braid group in our path integral! We thus write a more general version of Eq. 3.6 as[31]

$$\langle n; \{\mathbf{x}\}_f | \hat{U}(t_f, t_i) | n'; \{\mathbf{x}\}_i \rangle = \mathcal{N} \sum_{g \,\in\, G} [\rho(g)]_{n,n'} \sum_{\substack{\text{paths} \,\in\, g \\ i \to f}} e^{iS[\text{path}]/\hbar}$$

$$\tag{3.9}$$

where now $\rho(g)$ in Eq. 3.9 is an M by M unitary matrix, which is a representation of G, that is, as matrices $\rho(g_1)\rho(g_2) = \rho(g_1 g_2)$. These matrices need to be unitary to assure that probability is conserved. The propagator in Eq. 3.6 should now be thought of as a propagator between the initial ket $|n'; \{\mathbf{x}\}_i\rangle$ and the final bra $\langle n; \{\mathbf{x}\}_f|$.

Let us now consider an experiment where we control the paths of particles, by say, holding them in traps that we move along some specified (many-particle) trajectory $\{\mathbf{x}(t)\}$. Using the usual saddle point approach of only keeping paths that are close to the classical paths (the extrema of the action), we only get contributions to the sum over paths from a single element g of the group G corresponding to the chosen trajectory.[32]

For example, let us consider moving particles along the trajectories shown in Fig. 3.9. Here an initial wavefunction is represented as shown in Eq. 3.8 as a vector $A_n^{(i)}$ multiplying basis states $|n; \{\mathbf{x}\}\rangle$ as in Eq. 3.8. We braid the particles around each other in some braid g and bring them back to the same positions. After braiding, the wavefunction should still

[29]At the moment it may be hard to imagine what this degeneracy is from, but we will see explicit examples starting in Chapter 27.

[30]If we want $|\psi\rangle$ normalized then there is a normalization condition on the A_n coefficients. For example, if the set of $|n; \{\mathbf{x}\}\rangle$ are orthonormal then we need $\sum_n |A_n|^2 = 1$ in order that $|\psi\rangle$ is normalized.

[31]There is a subtlety in this expression. We have chosen to work with a convenient basis for the M-dimensional space $|n, \{\mathbf{x}\}\rangle$ with $n = 1, \ldots, M$ such that the e^{iS} term is diagonal in the matrix indices n, n'. This can always be done for any well-behaved theory. In principle, the basis choice for one set of positions $\{\mathbf{x}\}$ could be chosen independently of the basis choice for a different set of positions, although this would make things messier. This caution is related to notes 10 and 15. See also the related discussion of so-called monodromy and holonomy in Nayak et al. [2008].

[32]In principle all paths are still included in the path integral, but only the ones we specify contribute significantly.

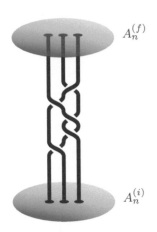

Fig. 3.9 An initial state is described by a vector $A_n^{(i)}$ multiplying the basis states $|n; \{\mathbf{x}\}_i\rangle$ as in Eq. 3.8. The particles are braided around each other in a braid g and brought back to the same set of positions. Since the particles are identical, each particle need not come back to its own initial position. The final state is again described in terms of the same basis vectors but now with coefficients $A_n^{(f)}$, which are obtained from the initial vector by application of the unitary matrix $\rho(g)$ as shown in Eq. 3.10. Here, $\rho(g)$ is a representation of the braid group.

[33]The idea of nonabelian anyons was explored first in the 1980s and early 1990s by several authors in different contexts: Bais [1980] in the context of gauge theories; Goldin et al. [1985], Fröhlich and Gabbiani [1990], and Fredenhagen et al. [1989] in very abstract sense; Witten [1989]; Chen et al. [1989] in the language of topological quantum field theories; and Moore and Read [1991] in the context of quantum Hall effect.

[34]Having $\mathsf{d} > 1$ necessarily implies that the matrices ρ in Eq. 3.10 are non-abelian. See Rowell and Wang [2016].

[35]Because of the possible sign, we distinguish the two quantities by using a different typeface.

[36]And initialization and measurement.

[37]A more concise derivation of the key portion of this result was given using modern category theory techniques by Müger [2007]. While this shorter proof is only 40 pages long, in order to understand the 40 pages you need to read a 400-page book on category theory first!

be composed of the same basis states $|n; \{\mathbf{x}\}\rangle$ since the particles are at the same positions and thus can be written in the form of Eq. 3.8 with a vector $A_n^{(f)}$. Equation 3.9 implies that the final vector is obtained from the initial vector simply by multiplying by the unitary operator, which is the representation of our braid group element g

$$A_n^{(f)} = \sum_{n'} [\rho(g)]_{n,n'} A_{n'}^{(i)} \quad . \tag{3.10}$$

A particle that obeys this type of braiding statistics is known as a **non-abelian anyon**, or **nonabelion**.[33] The word "nonabelian" means non-commutative, and the term is used since generically matrices (in this case the ρ matrices) don't commute.

In general, the Hilbert space dimension M will be exponentially large in the number of particles N. We define a quantity d, known as the **quantum dimension** such that[34]

$$M \sim \mathsf{d}^N \tag{3.11}$$

where the $\sim$ means that it scales this way in the limit of large N. We will see a lot more of this quantity d later. It is not coincidence that we used the symbol d previously in the context of Kauffman anyons! (See Eq. 2.1.) We will see in Section 17.1 that (up to a possible sign) this quantum dimension d is actually the value d of the unknot.[35]

Some Quick Comments on Quantum Computing:

Quantum computing is nothing more than the controlled application of unitary operations to a Hilbert space.[36] Unitary operations is exactly what we can do by braiding nonabelions around each other! That is, we are multiplying a vector by a unitary matrix. Thus we see how braiding of particles, as discussed in Chapter 2, can implement quantum computation. Chapter 11 provides some more explicit descriptions of how one does quantum computation by braiding anyons.

3.5.1 Parastatistics in (3 + 1)-Dimensions

Is it possible to have exotic nonabelian statistics in (3 + 1)-dimensions? Indeed, there do exist higher-dimensional representations of the permutation group, so one can think about particles that obey more complicated statistics even in (3 + 1)-dimensions—which is often known as *parastatistics*. However, it turns out that, subject to some "additional constraints," it is essentially not possible to get anything fundamentally new—all we get is bosons and fermions and possibly some internal additional degrees of freedom. The proof of this statement is due to Doplicher et al. [1971, 1974] and took some 200 pages when it was first proven.[37]

However, we should realize that in making statements like this, the fine print is important. As I mentioned in the previous paragraph we want to add some "additional constraints" and these are what really limit us to just bosons and fermions. What are these additional constraints?

(1) We want to be able to pair create and annihilate. This means we are not just considering the braid group, but rather a more complicated structure that allows not just braiding particles around each other, but also creating and annihilating and even merging particles by bringing them together. This structure is given by category theory, some parts of which we will encounter (in simplified language) starting in Chapter 8.

(2) We also want some degree of locality. If we do an experiment on Earth, while off on Jupiter someone creates a particle–antiparticle pair, we would not want the particles on Jupiter to effect the result of our experiment on Earth at all.

These two restrictions are crucial to reducing the $(3+1)$-dimensional case to only bosons and fermions. We will not go through the full details of how this happens. However, once we see the full structure of anyons in $(2+1)$-dimensions, it ends up being fairly clear why $3+1$ dimensions will be so restrictive. We return to this issue in Section 20.4.

We should note that despite this important result, $(3+1)$-dimensions is certainly not boring—but in order to get "interesting" examples, we have to relax some of our constraints. For example, if we relax the condition that "particles" are point-like, but consider string-like objects instead, then we can have exotic statistics that describe what happens when one loop of string moves through another (or when a point-like particle moves through a loop of string). We would then need to consider the topology of the world sheets describing loops moving through time.

Chapter Summary

- The path integral formulation of quantum mechanics requires us to add up all possible paths in spacetime.
- We can add all of these paths in any way that preserves the composition law and the different possibilities allow for different types of particle statistics.
- The topologically different paths of N particles in spacetime form a group structure (the fundamental group of the configuration space) which is the permutation group S_N in $(3+1)$-dimensions, but is the braid group B_N in $(2+1)$-dimensions.
- Particle braiding statistics must be a representation of this group.
- In $(3+1)$-dimensions we can only have bosons and fermions, but in $(2+1)$-dimensions we can have nontrivial braiding statistics, which may be abelian ("fractional") or nonabelian.
- Quantum computation can be performed by braiding with certain nonabelian representations.

Further Reading:

- For more discussion of particle statistics, a nice, albeit somewhat dated, book is Wilczek [1990].
- A good review discussing many aspects of exotic statistics is Nayak et al. [2008].
- For a basic primer on path integrals see MacKenzie [2000]. The classic reference on the subject is Feynman and Hibbs [1965].

[38]This early paper even noted that two dimensions is different from three dimensions, although it stopped just short of figuring out that fractional statistics exists. Cécile DeWitt (or DeWitt-Morette) was also the founder of the famous Les Houches school of physics that boasts over twenty Nobel laureates among its alumni.

The path integral approach to quantum exchange statistics was pioneered by Laidlaw and DeWitt [1971][38] in the context of bosons and fermions. It was later used by Wu [1984] to describe fractional abelian statistics, and it was pointed out by Goldin et al. [1985] that an extension could be used for nonabelian statistics as well.

Exercises

Exercise 3.1 About the Braid Group

(a) Convince yourself geometrically that the defining relations of the braid group on M particles B_M are:

$$\sigma_i\, \sigma_{i+1}\, \sigma_i = \sigma_{i+1}\, \sigma_i\, \sigma_{i+1} \qquad\qquad 1 \le i \le M-2 \quad (3.12)$$

$$\sigma_i\, \sigma_j = \sigma_j\, \sigma_i \quad \text{for} \quad |i-j| > 1, \qquad 1 \le i,j \le M-1 \quad (3.13)$$

(b) Instead of thinking about particles on a plane, let us think about particles on the surface of a sphere. In this case, the braid group of M strands on the sphere is written as $B_M(S^2)$. To think about braids on a sphere, it is useful to think of time as being the radial direction of the sphere, so that braids are drawn as in Fig. 3.10.

The braid generators on the sphere still obey Eqs. 3.12 and 3.13, but they also obey one additional identity

$$\sigma_1\sigma_2 \ldots \sigma_{M-2}\sigma_{M-1}\sigma_{M-1}\sigma_{M-2} \ldots \sigma_2\sigma_1 = I, \qquad (3.14)$$

where I is the identity (or trivial) braid. What does this additional identity mean geometrically? In fact, for understanding the properties of anyons on a sphere, Eq. 3.14 is not quite enough. We will try to figure out in Exercise 3.3 what is missing.

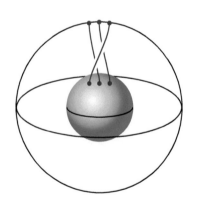

Fig. 3.10 An element of the braid group $B_3(S^2)$. The braid shown here (reading right to left meaning bottom to top in the braid) is $\sigma_2\sigma_1$.

Exercise 3.2 About the Permutation Group

Show that Eqs. 3.12 and 3.13 also hold for the generators of the permutation group S_M on M particles, where σ_i exchanges particle i and $i+1$. In the permutation group we have the additional condition that $\sigma_i^2 = 1$. Prove the statement used in Section 3.4.1 that there are only two one-dimensional (i.e. scalar unitary) representations of the permutation group. Hint: The proof is just a few lines. Use $\rho(\sigma_i)\rho(\sigma_j) = \rho(\sigma_i\sigma_j)$ where ρ is a representation.

Exercise 3.3 Ising Anyons and Majorana Fermions

The most commonly discussed type of nonabelian anyon is the Ising anyon (we discuss this in more depth later). Ising anyons occurs in the Moore–Read quantum Hall state ($\nu = 5/2$), as well as in any chiral p-wave superconductor and in recently experimentally relevant "Majorana" systems (see Chapter 37).

The nonabelian statistics of these anyons may be described in terms of Majorana fermions by attaching a Majorana operator to each anyon. The Hamiltonian for these Majoranas is zero—they are completely noninteracting.

In case you haven't seen them before, Majorana fermions γ_j satisfy the anticommutation relation

$$\{\gamma_i, \gamma_j\} \equiv \gamma_i\gamma_j + \gamma_j\gamma_i = 2\delta_{ij} \tag{3.15}$$

as well as being self-conjugate $\gamma_i^\dagger = \gamma_i$.

(a) Show that the ground-state degeneracy of a system with $2N$ Majoranas is 2^N if the Hamiltonian is zero. Thus conclude that each *pair* of Ising anyons is a two-state system. Hint: Construct a regular (Dirac) fermion operator from two Majorana fermion operators. For example,

$$c^\dagger = \frac{1}{2}(\gamma_1 + i\gamma_2)$$

will then satisfy the usual fermion anticommutation $\{c, c^\dagger\} = cc^\dagger + c^\dagger c = 1$. (If you haven't run into fermion creation operators yet, you might want to read up on this first!) There is more discussion of this transformation in later Exercises 9.7 and 10.2.

(b) When anyon i is exchanged clockwise with anyon j, the unitary transformation that occurs on the ground state is

$$U_{ij} = \frac{e^{i\alpha}}{\sqrt{2}}[1 + \gamma_i\gamma_j] \qquad i < j \tag{3.16}$$

for some real value of α. Show that these unitary operators form a representation of the braid group. (Refer back to Exercise 3.1, "About the Braid Group"). In other words, we must show that replacing σ_i with $U_{i,i+1}$ in Eqs. 3.12 and 3.13 yields equalities. This representation is 2^N-dimensional since the ground-state degeneracy is 2^N.

(c) Consider the operator

$$\gamma^{\text{FIVE}} = (i)^N \gamma_1\gamma_2 \ldots \gamma_{2N} \tag{3.17}$$

(the notation FIVE is in analogy with the γ^5 of the Dirac gamma matrices). Show that the eigenvalues of γ^{FIVE} are ± 1. Further show that this eigenvalue remains unchanged under any braid operation. Conclude that we actually have two 2^{N-1}-dimensional representations of the braid group. We assume that any particular system of Ising anyons is in one of these two representations.

(d) Thus, four Ising anyons on a sphere comprise a single two-state system, or a qubit. Show that by only braiding these four Ising anyons one cannot obtain all possible unitary operation on this qubit. Indeed, braiding Ising anyons is not sufficient to build a quantum computer. Part (d) is not required to solve parts (e) and (f).

(e) [A bit harder] Now consider $2N$ Ising anyons on a sphere (see Exercise 3.1 for information about the braid group on a sphere). Show that in order for either one of the 2^{N-1}-dimensional representations of the braid group to satisfy the sphere relation, Eq. 3.14, one must choose the right abelian phase

α in Eq. 3.16. Determine this phase.

(f) [A bit harder] The value you just determined is not quite right. It should look a bit unnatural as the abelian phase associated with a braid depends on the number of anyons in the system. Go back to Eq. 3.14 and insert an additional abelian phase on the right-hand side, which will make the final result of part (e) independent of the number of anyons in the system. In fact, there should be such an additional factor—to figure out where it comes from, go back and look again at the geometric "proof" of Eq. 3.14. Note that the proof involves a self twist of one of the anyon world lines. The additional phase you added is associated with one particle twisting around itself. The relation between self-rotation of a single particle and exchange of two particles is a generalized spin-statistics theorem.

Exercise 3.4 Small Numbers of Anyons on a Sphere

On the plane, the braid group of two particles is an infinite group (the group of integers describing the number of twists!). However, this is not true on a sphere

(a) First review Exercise 3.1 about braiding on a sphere. Now consider the case of two particles on a sphere. Determine the full structure of the braid group. Show it is a well-known finite discrete group. What group is it?

(b) [Harder] Now consider three particles on a sphere. Determine the full structure of the braid group. Show that it is a finite discrete group. [Even harder] What group is it? It is "well known" only to people who know a lot of group theory. But you can Google to find information about it on the web with some work. It may be useful to list all the subgroups of the group and the multiplication table of the group elements.

(c) Suppose we have two (or three) anyons on a sphere. Suppose the ground state is two fold degenerate (or more generally N-fold degenerate for some finite N). Since the braid group is discrete, conclude that no type of anyon statistics can allow us to do arbitrary $SU(2)$ (or $SU(N)$) rotations on this degenerate ground state by braiding.

Aharonov–Bohm Effect and Charge–Flux Composites

4

Important/Easy

This chapter introduces a simple model of how anyons with fractional statistics can arise. After reviewing the Aharonov–Bohm effect, we describe these exotic particles as charge–flux composites and explore some of their properties. This is a warm-up for our introduction to Chern–Simons theory in Chapter 5.

4.1 Review of Aharonov–Bohm Effect

Let us consider the double-slit interference experiment shown in Fig. 4.1. We all know the result of the double-slit experiment but let us rewrite the calculation in the language of a path integral. We can write

$$\sum_{\text{paths}} e^{iS/\hbar} = \sum_{\text{paths, slit 1}} e^{iS/\hbar} + \sum_{\text{paths, slit 2}} e^{iS/\hbar}$$

$$\sim e^{ikL_1} + e^{ikL_2}$$

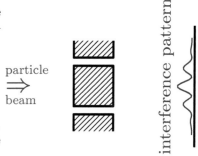

Fig. 4.1 The Young double-slit experiment (not to scale).

where L_1 and L_2 are the path lengths through the two respective slits to whichever point is being measured on the output screen, and k is the wavevector of the incoming wave. In other words, we get the usual double-slit calculation pioneered by Thomas Young in the early 1800s.

Now let us change the experiment to that shown in Fig. 4.2. Here we assume the particle being sent into the interferometer is a charged particle, such as an electron. In this case a magnetic field is added inside the middle box between the two paths. No magnetic field is allowed to leak out of the box, so the particle never experiences the magnetic field. Further, the magnetic field is kept constant so the particle does not feel a Faraday effect either. The surprising result is that the presence of the magnetic field nonetheless changes the interference pattern obtained on the observation screen! This effect, named the Aharonov–Bohm effect, was predicted by Ehrenberg and Siday [1949], then re-predicted independently by Aharonov and Bohm [1959].[1]

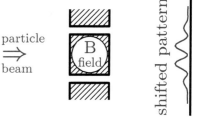

Fig. 4.2 Adding a magnetic field inside the middle box in the Young double-slit experiment. Here the circular region includes a constant magnetic field. No magnetic field leaks out of the box. Nonetheless, if the particle being sent into the interferometer is charged, the interference pattern is changed compared to Fig. 4.1.

---

[1] Possibly the reason it is named after the later authors is that they realized the importance of the effect, whereas the earlier authors pointed it out, but did not emphasize as much how strange it is! The first experimental observation of the effect was by Chambers [1960], although many more careful experiments have been done since.

So why does this strange effect occur? There are several ways to understand it, but for our purpose it will be best to stay with the idea of path integrals and consider the Lagrangian description of particle motion.

We must recall how a charged particle couples to an electromagnetic field in the Lagrangian description of mechanics. We write the magnetic field and electric field in terms of a vector potential

$$\mathbf{B} = \nabla \times \mathbf{A}$$
$$\mathbf{E} = -\nabla A_0 - d\mathbf{A}/dt$$

where A_0 is the electrostatic potential. We can then write the particle Lagrangian as

$$L = \frac{m}{2}\dot{\mathbf{x}}^2 + q(\mathbf{A}(\mathbf{x}) \cdot \dot{\mathbf{x}} - A_0) \quad , \tag{4.1}$$

where q is the particle charge. It is an easy exercise to check that the Euler–Lagrange equations of motion that result from this Lagrangian correctly gives motion under the Lorentz force as we should expect for a charged particle in an electromagnetic field.[2]

We are interested in a situation where we add a static magnetic field to the system. Thus, we need only include $q\mathbf{A}(\mathbf{x}) \cdot \dot{\mathbf{x}}$ in the Lagrangian. The action then becomes

$$S = S_0 + q\int dt\, \dot{\mathbf{x}} \cdot \mathbf{A} = S_0 + q\int \mathbf{dl} \cdot \mathbf{A} \quad , \tag{4.2}$$

where S_0 is the action in the absence of the magnetic field and the integral on the far right is a line integral along the path taken by the particle.

Returning now to the double-slit experiment. The amplitude of the process in the presence of the vector potential can be now rewritten as

$$\sum_{\text{paths, slit 1}} e^{iS_0/\hbar + iq/\hbar \int \mathbf{dl} \cdot \mathbf{A}} + \sum_{\text{paths, slit 2}} e^{iS_0/\hbar + iq/\hbar \int \mathbf{dl} \cdot \mathbf{A}} \quad ,$$

where S_0 is again the action of the path in the absence of the vector potential.

The physically important quantity is the difference in accumulated phases between the two paths. This difference is given by

$$\exp\left[\frac{iq}{\hbar}\int_{\text{slit 1}} \mathbf{dl} \cdot \mathbf{A} - \frac{iq}{\hbar}\int_{\text{slit 2}} \mathbf{dl} \cdot \mathbf{A}\right] = \exp\left[\frac{iq}{\hbar}\oint \mathbf{dl} \cdot \mathbf{A}\right] \quad , \tag{4.3}$$

where the integral on the right is around a loop that goes forward through slit 1, reaches the point on the screen, and then goes backward through slit 2.

Using Stokes's theorem, we have

$$\frac{iq}{\hbar}\oint \mathbf{dl} \cdot \mathbf{A} = \frac{iq}{\hbar}\int_{\text{enclosed}} \mathbf{dS} \cdot (\nabla \times \mathbf{A}) = \frac{iq}{\hbar}\Phi_{\text{enclosed}} \quad ,$$

---

[2]Here are the steps: Start with the Euler–Lagrange equations

$$\frac{d}{dt}\frac{\partial L}{\partial \dot{x}_k} = \frac{\partial L}{\partial x_k}.$$

This gives us

$$\frac{d}{dt}(m\dot{x}_k + qA_k) = q\left(\frac{\partial A_j}{\partial x_k}\dot{x}_j - \frac{\partial A_0}{\partial x_k}\right)$$

with implicit summation over j. Equivalently,

$$m\ddot{x}_k = q\left(\frac{\partial A_j}{\partial x_k}\dot{x}_j - \frac{dA_k}{dt} - \frac{\partial A_0}{\partial t}\right).$$

Now using

$$\frac{dA_k}{dt} = \frac{\partial A_k}{\partial t} + \frac{\partial A_k}{\partial x_j}\dot{x}_j$$

along with the vector identity

$$\dot{\mathbf{x}} \times (\nabla \times \mathbf{A}) = \nabla(\dot{\mathbf{x}} \cdot \mathbf{A}) - (\dot{\mathbf{x}} \cdot \nabla)\mathbf{A}$$

this can be massaged into the Lorentz force

$$m\ddot{\mathbf{x}} = q(\mathbf{E} + \dot{\mathbf{x}} \times \mathbf{B}) \quad .$$

where Φ_{enclosed} is the flux enclosed in the loop. Thus, there is a measurable relative phase shift between the two paths given by $\frac{iq}{\hbar}\Phi_{\text{enclosed}}$. This results in a shift of the interference pattern measured on the observation screen, even though the particle never experienced any magnetic field at its position! Note that although the original Lagrangian Eq. 4.1 did not look particularly gauge invariant, the end result (once we integrate around the full path) is indeed gauge independent.

A few notes about this effect:

(1) If Φ is an integer multiple of the elementary flux quantum

$$\Phi_0 = 2\pi\hbar/q,$$

then the phase shift is an integer multiple of 2π and is hence equivalent to no phase shift.

(2) We would get the same phase shift if we were to move flux around a charge.[3]

(3) More generally for particles moving in spacetime one wants to calculate the relativistically invariant quantity

$$\frac{iq}{\hbar}\oint dl^\mu A_\mu \quad . \tag{4.4}$$

4.2 Anyons as Charge–Flux Composites

We will now consider a simple model of abelian anyons as charge–flux composites. Imagine we have a two-dimensional system with charges q in them, where each charge is bound to an infinitely thin flux tube through the plane, with each tube having flux Φ as shown in Fig. 4.3. We will notate this charge–flux composite object as a (q, Φ) particle. If we drag one such particle around another, we then accumulate a phase due to the Aharonov–Bohm effect. The phase from the charge of particle 1 going around the flux of particle 2 is $e^{iq\Phi/\hbar}$, whereas the phase for dragging the flux of 1 around the charge of 2 is also $e^{iq\Phi/\hbar}$; thus, the total phase for dragging 1 around 2 is given by

(Phase of charge–flux composite 1 encircling 2) $= e^{2iq\Phi/\hbar}$.

Thus we have (as shown in Fig. 4.4)

(Phase for exchange of two charge–flux composites) $= e^{iq\Phi/\hbar}$

and we correspondingly call these particles ϑ-anyons, with $\vartheta = q\Phi/\hbar$. Obviously $\vartheta = 0$ gives bosons and $\vartheta = \pi$ gives fermions, but other values of ϑ are also allowed, giving us abelian anyons as discussed in Chapter 3.

Note that the same type of calculation would show us that taking a composite particle with charge q_1 and flux Φ_1 all the way around a composite particle with charge q_2 and flux Φ_2 would accumulate a phase of $e^{i\varphi}$ with $\varphi = (q_1\Phi_2 + q_2\Phi_1)/\hbar$.

[3]This fact is due to Aharonov and Casher [1984]. To derive it we must write a Lagrangian for a charge at position $\mathbf{x}$ and a flux at position $\mathbf{X}$. The charge–flux coupling term analogous to the coupling term in Eq. 4.1 can be shown to take the form

$$q\mathbf{A}(\mathbf{x} - \mathbf{X})\cdot(\dot{\mathbf{x}} - \dot{\mathbf{X}}) \quad ,$$

where $\mathbf{A}$ is the vector potential associated with the flux (this is the only form possible that will respect Galilean and translational invariance). Using this more general form we can derive the stated result.

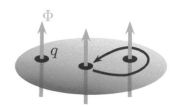

Fig. 4.3 Abelian anyons represented as charges bound to flux tubes through the plane. The charge of each particle is q, the flux of each tube is Φ. Dragging one particle around another incurs a phase both because charge is moving around a flux, but also because flux is moving around a charge.

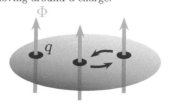

Fig. 4.4 An exchange. This is described as σ_2 in terms of the braid group. Note that the processes shown in Fig. 4.3 is described as σ_2^2. Thus doing this exchange twice is the same as dragging one particle all the way around the other.

[4]Almost any prescription for attaching flux to charge will give the same result. For example we could break the flux into four pieces and attach one piece on each of the four sides (north, south, east, west) of the charge. However, if we try to put the flux and charge at exactly the same position, we get infinities that we don't know how to handle!

Fig. 4.5 Tying flux to charge. We put the flux and the charge at slightly different positions. As a result, when we rotate the particle around its own axis a phase is accumulated as the charge and flux go around each other.

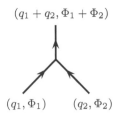

Fig. 4.6 Fusing two anyons to get an anyon of a different type. The anyon that results from this fusion has the sum of the two incoming fluxes and the sum of the two incoming charges.

[5]The vacuum or identity particle can be denoted e, or I or 0 or 1 depending on the context. This nomenclatural problem stems from a similar problem in group theory; see Section 40.2.

Spin of an anyon

Let us see if we can determine the spin of these anyons. Spin refers to properties of the rotation operator, so we need to physically rotate the anyon on its axis. To do this we must think about how the flux is tied to the charge—we must have some microscopic description of exactly where the flux is and where the charge is. It is easiest to put the charge and flux at very slightly different positions as shown in Fig. 4.5.[4] In this case, when we rotate the anyon around its axis we move the charge and flux around each other and we obtain a new phase of

$$e^{iq\Phi/\hbar} = e^{i\vartheta} \quad .$$

This fits very nicely with the spin-statistics theorem—the phase obtained by exchanging two identical particles should be the same as the phase obtained by rotating one around its own axis (see the discussion of Fig. 2.7).

4.2.1 Fusion of Anyons

We can consider pushing two anyons together to try to form a new particle. We expect that the fluxes will add and the charges will add. This makes some sense as the total charge and total flux in a region should be conserved (this is an important principle that we will encounter frequently!). We sometimes will draw a "fusion diagram" as in Fig. 4.6 to show that two anyons have come together to form a composite particle.

A simple example of this is pushing together two particles both having the same charge and flux (q, Φ). In this case we will obtain a single particle with charge and flux $(2q, 2\Phi)$. Note that the phase of exchanging two such double particles is now $\vartheta = 4q\Phi/\hbar$ (since the factor of 2 in charge multiplies the factor of 2 in flux!).

4.2.2 Anti-Anyons and the Vacuum Particle

We now introduce the concept of an anti-anyon. This is a charge–flux composite which instead of having charge and flux (q, Φ) has charge and flux $(-q, -\Phi)$. Fusing an anyon with its anti-anyon results in pair annihilation—the two particles come together to form the vacuum (which we sometimes[5] refer to as the identity I), which has zero total charge and zero total flux, as shown in Fig. 4.7. It may seem a bit odd to call the absence of any charge or any flux a "particle." However, this is often convenient since it allows us to think of pair annihilation (as in the left of Fig. 4.7) in the language of fusion.

In the middle of Fig. 4.7 we show that it is sometimes convenient *not* to indicate the vacuum particle. In this case, we have written the anti-anyon moving forward in time as an anyon moving backward in time. We can continue with our notational simplification and just draw this as a single line with a single arrow as in the right of Fig. 4.7.

If the phase of dragging an anyon clockwise around an anyon is 2ϑ,

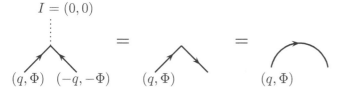

Fig. 4.7 Left: Fusing an anyon and an anti-anyon to get the vacuum, I, drawn as a dotted line. Middle: The anti-anyon moving forward in time is drawn as a downpointing arrow—which looks like an anyon moving backward in time. Right: The process is drawn as a single continuous line.

then the phase of dragging an anti-anyon clockwise around an anti-anyon is also 2ϑ. (The two minus signs on the two anyons cancel—negative flux multiplies negative charge!). However, the phase of dragging an anyon clockwise around an anti-anyon is -2ϑ.

4.3 Anyon Vacuum on a Torus and Quantum Memory

A rather remarkable feature of topological models is that the ground state somehow "knows" what kind of anyons exist in the model (i.e. those that *could* be created), even when they are not actually present. To see this, consider the ground state of an anyon model on a torus (the surface of a doughnut.[6]

We can draw the torus as a square with opposite edges identified as shown in Fig. 4.8. The two nontrivial cycles[7] around the torus are marked as C_1 and C_2.

Let us now construct operators that do the following complicated operations:

T_1 is the operator that creates a particle–antiparticle pair, moves the two in opposite directions around the C_1 cycle of the torus until they meet on the opposite side of the torus and reannihilate.

T_2 is the operator that creates a particle–antiparticle pair, moves the two in opposite directions around the C_2 cycle of the torus until they meet on the opposite side of the torus and reannihilate.

We can also define the inverse of these operators which is just the time-reversed processes. Both of these operators are unitary because they can be implemented (in principle) with some time-dependent Hamiltonian.[8] However, the two operators do not commute. To see this let us consider the operator $T_2^{-1}T_1^{-1}T_2T_1$ where we read time from right to left. This can be interpreted as two particles being created, braiding around each other, and then reannihilating. This procedure is shown in Fig. 4.9.

So what we have now is two operators T_1 and T_2, which do not commute with each other. As derived diagrammatically in Fig. 4.9, we have[9]

[6]See note 1 in Chapter 40.

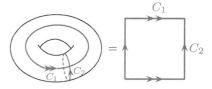

Fig. 4.8 Drawing a torus as a rectangle with opposite edges identified. The two non-contractible cycles around the torus can be considered to be the edges of the square, labeled C_1 and C_2 here.

[7]By "cycle" here we mean a closed curve. A cycle is nontrivial if it cannot be contracted to a point (sometimes it is called a "non-contractible cycle"). Physicists are sometimes sloppy and use the term "cycle" to mean "nontrivial cycle." I will try not to be lazy!

[8]For example, we could insert charges $+Q$ and $-Q$ near to each other which are strong enough to pull a particle–antiparticle pair out of the vacuum, the $-Q$ trapping the $+(q,\Phi)$ and the $+Q$ trapping the $(-q,-\Phi)$. Then we can drag the $\pm Q$ charges around a nontrivial cycle of the torus, dragging the anyons with them.

[9]At least this relation should be true acting on the ground-state space. If some particles are already present, then we have to consider the braiding of the particles we create with those already present, which will be more complicated.

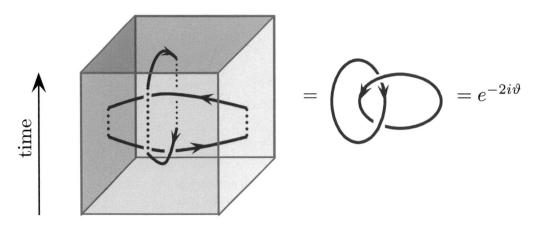

Fig. 4.9 The torus is drawn as a horizontal rectangle with opposite ends identified. Time runs vertically. First create a particle–antiparticle pair at the center of the rectangle and move them in opposite directions, right and left, until they meet at the edges of the rectangle to reannihilate. Note that a particle moving to the right or an antiparticle moving to the left are both drawn as a rightpointed arrow. Next create a particle–antiparticle pair in the center of the rectangle and move them to the front and back walls (which are the same point) to reannihilate. Then the two processes are reversed to give $T_2^{-1}T_1^{-1}T_2T_1$. This procedure can be reduced to one particle wrapping around another which gives a phase of $e^{-2i\vartheta}$. Note that to make the figure on the left look like the linked rings, we should not quite annihilate the particles at the end of the first and second step (turning the dotted lines into solid lines). This is allowed since bringing a particle–antiparticle pair close together looks like they have fused together to the vacuum if we view it from far away—this is true for abelian anyons as it is a special case of the identity in Fig. 12.20.

$$T_2T_1 = e^{-2i\vartheta}T_1T_2 \quad .$$

But both T_1 and T_2 commute with the Hamiltonian (since they start and end with states of exactly the same energy).[10] Whenever you have two operators that don't commute with each other but do commute with the Hamiltonian, it means you have degenerate eigenstates. Let us see how this happens.

Since T_1 is unitary, its eigenvalues must have unit modulus (i.e. they are just a complex phase). Considering the space of possible ground states, let us write an eigenstate of T_1 in this space as

$$T_1|\alpha\rangle = e^{i\alpha}|\alpha\rangle \quad .$$

Note that we are labeling the ket $|\alpha\rangle$ by its eigenvalue under the application of T_1. Now we will generate a new eigenstate with a different eigenvalue of T_1. Consider the state $T_2|\alpha\rangle$. This must also be in the ground-state space since T_2 commutes with the Hamiltonian. But now

$$T_1(T_2|\alpha\rangle) = e^{2i\vartheta}T_2T_1|\alpha\rangle = e^{2i\vartheta}e^{i\alpha}(T_2|\alpha\rangle) \quad .$$

This new ground state $T_2|\alpha\rangle$ has eigenvalue $e^{i\alpha+2i\vartheta}$ under application of T_1. We thus call this new ground state $|\alpha+2\vartheta\rangle = T_2|\alpha\rangle$. We have now generated a new ground state and we can continue the procedure to generate more!

[10]Strictly speaking this means they commute with the Hamiltonian within the ground-state space, or equivalently the commutators $[T_1, H]$ and $[T_2, H]$ both annihilate the ground-state space.

Let us suppose we have a system where the anyons have statistical phase angle

$$\vartheta = \pi p/m$$

where p and m are relatively prime integers (i.e. p/m is an irreducible fraction). Starting with the ground state $|\alpha\rangle$ we can generate a series of ground states by successive applications of T_2,

$$|\alpha\rangle, \quad |\alpha + 2\pi p/m\rangle, \quad |\alpha + 4\pi p/m\rangle, \ \dots, \quad |\alpha + 2\pi p(m-1)/m\rangle.$$

When we try to generate yet another state, we get the phase $\alpha + 2\pi$, which is equivalent to α since it is describing a complex phase, so we are back to the original state. So we now have m independent ground states.[11] Note in particular that the ground-state degeneracy *of the system with no anyons in it* is related to the statistical angle ϑ of the anyons if they were to be created.

4.3.1   Quantum Memory and Higher Genus

The degenerate ground state on the torus can be thought of as a quantum memory. If there are m different orthogonal ground states, the most general ground state wavefunction we can have is some linear superposition

$$|\Psi\rangle = \sum_{n=0}^{m-1} A_n |\alpha + 2\pi np/m\rangle \quad,$$

where the coefficients A_n form an arbitrary (but normalized) complex vector. We can initialize the system in some particular superposition (i.e. some vector A_n) and we can expect that the system remains in this superposition. The only way that this superposition can change is if a T_1 or T_2 operation is performed, or some combination thereof (i.e. if a pair of anyons appears from the vacuum moves around a nontrivial cycle of the torus and then reannihilates). Such a process can be extremely unlikely when the energy gap for creating excitations is large.[12] Hence the quantum superposition is "topologically protected."

In fact, one does not even need to have a system on a torus in order to have a degenerate ground state. It is often sufficient to have an annulus geometry (a disk with a big hole in the middle as shown in Fig. 4.10). In this case, T_1 could correspond to moving an anyon around the loop of the annulus and T_2 could correspond to moving an anyon from the inside to the outside edge.[13]

One can consider more complicated geometries, such as a torus with multiple handles, or a disk with multiple holes cut in the middle. For a theory of abelian anyons (fractional statistics) the ground-state degeneracy for a surface with **genus** g (meaning g handles, or g holes) is m^g (see Exercise 4.1). Thus by using a high genus surface one can obtain a very large Hilbert space in which to store quantum information.

[11]There could be even more degenerate ground states which would be non-generic. What we have proven is that the number of degenerate ground states is m times some integer w, where one generally expects the integer w to be 1, but in principle the integer could be higher due to what is usually called "accidental" degeneracy.

[12]Strictly speaking, for a system of any finite size at any finite temperature, there is a finite time for this process to occur, although it might be very long.

Fig. 4.10 An annulus.

[13]In this case it is often not precisely true that the putative ground states are exactly degenerate, since the multiple states differ in how many anyons are on the inside versus outside edge. Nonetheless, these states can be extremely close to degenerate. A classic example of this is discussed by Gefen and Thouless [1993].

4.3.2   Number of Species of Anyons

Having established multiple vacuum states on a torus, let us now return to study the anyons that we could create in such a system. Again let us consider anyons of statistical angle $\vartheta = \pi p/m$ with p and m relatively prime. We can describe such anyons[14] with a charge–flux composite $(q, \Phi) = (\pi p/m, 1)$. Fusion of n of these elementary anyons will have charge and flux given by[15]

[15]It is only a slight abuse of notation to write the ket $|\text{"}n\text{"}\rangle$ to mean a cluster of n elementary anyons.

$$\text{Fusion of } n \text{ elementary anyons} \;\; = \;\; |\text{"}n\text{"}\rangle = (nq, n\Phi)$$
$$= \;\; (n\pi p/m, n) \;\;.$$

Something special happens when we have a cluster of m of these elementary anyons:

$$|\text{"}m\text{"}\rangle = (\pi p, m) \;\;.$$

[16]As mentioned at the beginning of Section 4.2, the total phase is given by $q_1\Phi_2 + q_2\Phi_1 = (n\pi p/m)m + (\pi p)n$.

If we braid an arbitrary cluster $|\text{"}n\text{"}\rangle = (n\pi p/m, n)$ around one of these $|\text{"}m\text{"}\rangle = (\pi p, m)$ clusters, we obtain a net phase[16] of $2n\pi p$, which is equivalent to no phase at all! Thus we conclude that the cluster of m elementary anyons is equivalent to the vacuum in the sense that all particles get trivial phase if they braid all the way around $|\text{"}m\text{"}\rangle$. That is, if no particle can detect $|\text{"}m\text{"}\rangle$ by braiding around it, this cluster is behaving equivalently to the vacuum, and must therefore be topologically equivalent to the vacuum. Sometimes one says that $|\text{"}m\text{"}\rangle$ is a *transparent* particle.

We might be tempted to conclude that there are exactly m different anyon species in the system. Indeed, this conclusion is often true. However, there is an exception. If both p and m are odd, one obtains a nontrivial sign for exchanging (half braiding, as in Fig. 4.4) a $|\text{"}m\text{"}\rangle = (\pi p, m)$ particle with another $|\text{"}m\text{"}\rangle = (\pi p, m)$ particle. To see this note that a single exchange gives a phase πpm since it is half of the $2\pi pm$ phase for wrapping one particle all the way around the other (as in Fig. 4.3). This means the $|\text{"}m\text{"}\rangle$ particle is a fermion. In fact, this case of p and m both odd is a bit of an anomalous case and is a bit more difficult to handle.[17]

[17]Whenever we have a particle that braids trivially with all other particles (i.e. is transparent), the theory is more complicated. Later on we will call this kind of theory "non-modular." See Section 17.3.1. When the only transparent particle is a single fermion we call it "super-modular" (see Section 17.7).

Neglecting this more complicated case with transparent particles, we are correct to conclude that we have exactly m different species of anyons—and also m different ground states on the torus as calculated above. This connection will occur in any well-behaved topological theory—the number of ground states on the torus will match the number of different species of particles.

Chapter Summary

- The Charge–Flux composite model describes abelian anyons—with the braiding phase coming from the Aharonov–Bohm effect.
- We introduced the ideas of fusion, antiparticles, and spin.
- The vacuum for a system of anyons is nontrivial and can be a quantum memory.

Further Reading:

- A good reference for the charge–flux composite model is John Preskill's lecture notes (Preskill [2004]).
- The idea of the charge–flux composite for exploring fractional statistics was introduced in Wilczek [1982].

Exercises

Exercise 4.1 Abelian Anyon Vacuum on a Two-Handle Torus

Using similar technique as in Section 4.3, show that the ground-state vacuum degeneracy on a two-handled torus is m^2 for a system of abelian anyons with statistical angle $\vartheta = \pi p/m$ for integers p and m relatively prime. Hint: Consider what the independent nontrivial cycles i are on a two-handled torus and determine the commutation relations for operators T_i that create anyon–antianyon pairs, take one of these particles around cycle i and then reannihilate.

Chern–Simons Theory Basics

5.1 Abelian Chern–Simons Theory

It is useful to see how charge–flux binding occurs in a microscopic field theory description of a physical system. The type of field theory we will study, "Chern–Simons" field theory.[1,2] is the main paradigm for topological quantum field theories.

This section considers the simplest type of Chern–Simons theory, which is the abelian type (i.e. it generates abelian anyons, or simple fractional statistics particles). We start by imagining a gauge field a_α, known as the Chern–Simons vector potential, analogous to the vector potential A_α we know from regular electromagnetism. Here we should realize that a_α is not the real electromagnetic vector potential because it lives only in our two-dimensional plane. We should think of it instead as some emergent effective quantity for whatever two-dimensional system we are working with.

Let us write the Lagrangian of our system

$$L = L_0 + \int d^2x \, \mathcal{L} \quad .$$

Here we have written L_0 to be the Lagrangian of our particles without considering the coupling to the (Chern–Simons) vector potential. This might be nothing more than the Lagrangian for free particles—although we could put other things into this part, too, such as interparticle interactions, if we like.

The second term is the integral of a Lagrangian density—and this will be the term that is relevant for the flux-binding and the exchange statistics of the particles. The form of the Lagrangian density is[3]

$$\begin{aligned} \mathcal{L} &= \mathcal{L}_{CS} &&+ \mathcal{L}_{\text{coupling}} \\ &= \frac{k}{4\pi}\epsilon^{\alpha\beta\gamma}a_\alpha\partial_\beta a_\gamma &&- \; j^\alpha a_\alpha \end{aligned} \tag{5.1}$$

where j^α is the particle charge current, k is some coupling constant known as the "level,"[4] and ϵ is the antisymmetric tensor.[5] The indices α, β, γ take values $0, 1, 2$ where 0 indicates the time direction and $1, 2$ are the space directions (and j^0 is the particle charge density). This Lagrangian can easily be generalized to consider multiple gauge fields a_α^J with $J = 1, \ldots, N$, which we explore in detail in Section 5.3.2.

The first term in Eq. 5.1 is the Lagrangian density of the Chern–Simons vector potential itself (it is sometimes known as the "Chern–

[1] Shiing-Shen Chern, often known as the "father or modern differential geometry," was one of the most important mathematicians of the twentieth century. According to a film biography made about him called *Taking the Long View*, in China in the 1980s, he was such a celebrity that "every school child knew his name, and TV cameras documented his every move whenever he ventured forth from the institute."

[2] Jim Simons was a prominent mathematician who wrote the key first paper on what became known as Chern–Simons theory in 1974. Simons was the head of the math department at Stony Brook University at the time. In 1982, he decided to change careers and start a hedge fund. His fund, Renaissance Technologies, became one of the most successful hedge funds in the world. Simons' wealth is now estimated at over 20 billion dollars (as of 2018). More recently he has become a prominent philanthropist, and has donated huge amounts of money to physics and mathematics—now being one of the major sources of funds for the best scientists in the world.

[3] Here (and many other places) we use Einstein summation convention so that repeated indices are summed.

[4] At this point we could have k be any real number, but it will turn out that the theory is particularly well behaved for k an integer.

[5] The antisymmetric tensor is given by $\epsilon^{012} = \epsilon^{120} = \epsilon^{201} = 1$ and $\epsilon^{210} = \epsilon^{102} = \epsilon^{021} = -1$.

Simons term"). The second term in Eq. 5.1 couples the Chern–Simons vector potential to the particles in the system. Its form, $j^\alpha a_\alpha$, may look unfamiliar but it is actually just the expected coupling of the charged particles to a vector potential analogous to what we used when we discussed the Aharonov–Bohm effect in Section 4.1. To see this, let us carefully define the particle charge current j^α. If we have N particles then the particle charge current is

$$j^0(\mathbf{x}) = \sum_{n=1}^{N} q_n \delta(\mathbf{x} - \mathbf{x}_n)$$

$$\mathbf{j}(\mathbf{x}) = \sum_{n=1}^{N} q_n \dot{\mathbf{x}}_n \, \delta(\mathbf{x} - \mathbf{x}_n) \quad .$$

The j^0 component, the particle charge density,[6] is just a delta function peak at the position of each particle with value given by the particle charge q. The 1 and 2 component, $\mathbf{j}$ is a delta function at the position of each particle with prefactor given by the velocity of the particle times its charge. Now when $-j^\alpha a_\alpha$ is integrated over all of space we get[7]

$$\sum_{n=1}^{N} q_n \left[\mathbf{a}(\mathbf{x}_n) \cdot \dot{\mathbf{x}}_n - a_0(\mathbf{x}_n) \right] \tag{5.2}$$

exactly as in Eq. 4.1. So this is nothing more than the regular coupling of a system of charged particles to a vector potential.

As is usual for a gauge theory, the coupling of the particles to the gauge field is gauge invariant once one integrates the particle motion over some closed path (one measures only the flux enclosed, as with the Aharonov–Bohm effect). The Chern–Simons term (the first term in Eq. 5.1) is also gauge invariant, at least on a closed spacetime manifold[8] if we can integrate by parts. To see this, make an arbitrary gauge transformation

$$a_\mu \to a_\mu + \partial_\mu \chi \tag{5.3}$$

for any function χ. Then integrating the Chern–Simons term (by parts if necessary) all terms can be brought to the form $\epsilon^{\alpha\beta\gamma}\chi\partial_\alpha\partial_\beta a_\gamma$ which vanishes by antisymmetry. Note that this gauge invariance holds for any closed manifold, although for a manifold with boundaries, we have to be careful when we integrate by parts as we can get a physically important boundary term. As we discuss in Section 17.3.3 this boundary term indicates the presence of (chiral) excitation modes propagating in one direction[9] around the edge. We will not worry about this further here since we assume a closed manifold.

To determine what the Chern–Simons term does, we need to look at the Euler–Lagrange equations of motion. We have

$$\frac{\partial \mathcal{L}}{\partial a_\alpha} = \partial_\beta \left(\frac{\partial \mathcal{L}}{\partial(\partial_\beta a_\alpha)} \right) \tag{5.4}$$

[6] Again not the real electromagnetic charge, but rather the charge that couples to the Chern–Simons vector potential a_α. Later in this chapter we set $q = 1$ along with $\hbar = 1$ for simplicity of notation.

[7] The change of sign in the first term is due to the Lorentzian signature of the metric, which is not implied in the dot-product in Eq. 5.2 but does occur in $j^\alpha a_\alpha = j^0 a_0 - j^1 a_1 - j^2 a_2$.

[8] We use the term *closed manifold* frequently. This means a manifold without boundary containing all its limit points. More detailed discussions and examples are given in Section 40.1.

[9] See Wen [1992] or Hansson et al. [2017] for discussion of edge dynamics.

which generates the equations of motion[10]

$$j^\alpha = \frac{k}{2\pi} \epsilon^{\alpha\beta\gamma} \partial_\beta a_\gamma \quad . \tag{5.5}$$

This equation of motion demonstrates flux binding. To see this, let us look at the 0th component of this equation. We have[11]

$$j^0 = \sum_{n=1}^{N} q_n \delta(\mathbf{x} - \mathbf{x}_n) = \frac{k}{2\pi} (\partial_1 a_2 - \partial_2 a_1) = \frac{k}{2\pi} b \tag{5.6}$$

where we have defined a "Chern–Simons" magnetic field b to be the two-dimensional version of the curl of the (spatial components of the) Chern–Simons vector potential. In other words, this equation attaches a delta function (infinitely thin) flux tube with flux $2\pi q_n/k$ at the position of each charge q_n. So we have achieved charge–flux binding!

For simplicity, let us now assume all particles are identical with the same charge $q_n = q = 1$. We might expect that the phase obtained by exchanging two such identical charges would be given by the charge times the flux or $\vartheta = 2\pi q^2/k = 2\pi/k$ analogous to Section 4.2. Actually, this is not right! The correct answer is that the statistical phase is[12]

$$\vartheta = \pi/k \quad .$$

To see why this is the right answer, we can multiply our equation of motion Eq. 5.5 by a_α and then plug it back into[13] the Lagrangian 5.1. We then end up with

$$\mathcal{L} = -\frac{1}{2} j^\alpha a_\alpha \quad . \tag{5.7}$$

In other words, the Lagrangian of the Chern–Simons vector potential itself cancels exactly half of the Lagrangian density, and hence will cancel half of the accumulated phase when we exchange two particles with each other! Note that the case $k = 1$ corresponds to a fermion theory[14] with exchange phase of π and no ground state degeneracy on the torus (as we would predict from Section 4.3.2).

Let us abbreviate the paths of all of our N particles as[15]

$$\{\mathbf{r}_n(t)\} = \{\mathbf{r}_1(t), \mathbf{r}_2(t), \dots, \mathbf{r}_N(t)\} \quad .$$

We can now calculate a propagator for the particles by writing

$$\sum_{\text{paths } \{\mathbf{r}_n(t)\}} \quad \sum_{\text{all } a_\mu(\mathbf{x},t)} e^{i(S_0 + S_{CS} + S_{\text{coupling}})/\hbar} \quad . \tag{5.8}$$

Here the first sum is the usual sum over particle paths that we have discussed before. The second sum is the sum over all possible configurations of the field a_μ at all points of spacetime $(\mathbf{x}, t)$. Note that this means we should sum over all configurations in space and time so it is effectively a path integral for a field. (This is potentially everything you ever need to know about field theory!). Often the sum over field

---

[10]It may look like the right result would have $k/(4\pi)$ on the right-hand side, given that it is $k/(4\pi)$ in Eq. 5.1. However, note that when we differentiate with respect to a_α on the left-hand side of Eq. 5.4, we also generate an identical factor of $k/(4\pi)$ and these two add up.

[11]Often one chooses a_μ not to be a real valued field, but rather to be a field valued on a circle, or equivalently valued in the group $U(1)$, the unitary 1 by 1 matrices (i.e. a complex phase of unit magnitude). In this case one obtains the same flux binding modulo a flux of 2π, which gives no net phase when a charge is transported around it. For this reason abelian Chern–Simons theory with coupling constant k is often known as $U(1)_k$. Using the group $U(1)$ also makes better contact with Section 5.2 below where the gauge field will be valued in a general Lie group.

[12]Note here that the choice of unit charge q means that integer k will give us an exchange phase of π times a rational number, which we needed for a well-behaved theory in Section 4.3.

[13]One might worry about whether we are actually allowed to plug the equations of motion back into the Lagrangian when we do a full path integral, as in Eq. 5.8, where we are supposed to integrate over all field configurations, not just those that satisfy equations of motion. While generally in field theory one should not plug equations of motion back into the Lagrangian, it is actually allowed in this case. One can check that integrating over all field configurations actually gives the same result.

[14]In Section 17.7 we will call this case of a single fermion the "sVec" theory.

[15]Apologies for switching the notation for particle positions from $\mathbf{x}$ to $\mathbf{r}$.

configurations is written as a functional integral

$$\sum_{\text{all } a_\mu(\mathbf{x},t)} \to \int \mathcal{D}a_\mu(x) \quad .$$

Formally when we write a functional integral we mean[16] that we should divide space and time into little boxes and within each box integrate over all possible values of a_μ. Fortunately, we will not need to do this procedure explicitly.

At least formally we can thus rewrite Eq. 5.8 as

$$\sum_{\text{paths }\{\mathbf{r_n}(t)\}} e^{iS_0/\hbar} \int \mathcal{D}a_\mu(x)\, e^{iS_{CS}/\hbar}\, e^{i(q/\hbar)\int dl^\alpha a_\alpha} \tag{5.9}$$

where S_0 is the action of the particles following the path but not interacting with the gauge field, S_{CS} is the action of the Chern–Simons gauge field alone (from the first term in Eq. 5.1). The final line integral in the exponential in Eq. 5.9 represents the coupling of the gauge field to the paths $\{\mathbf{r_n}(t)\}$ of the particles[17] analogous to the Aharonov–Bohm phase in Eq. 4.4. This line integral in the exponential is an example of a Wilson-line operator, which we see again in Section 5.2.

As discussed near Eq. 5.7, the equations of motions couple flux to charge and after having enforced this binding, $\mathcal{L}_{CS} + \mathcal{L}_{\text{coupling}}$ can be replaced with $\frac{1}{2}\mathcal{L}_{\text{coupling}}$. Thus, up to a normalization constant, we can rewrite Eq. 5.9 as

$$\sum_{\text{paths }\{\mathbf{r_n}(t)\}} e^{iS_0/\hbar - i(1/2)(q/\hbar)\int dl^\alpha a_\alpha}$$

where now the Chern–Simons field is fixed so that each particle with charge $q = 1$ binds flux $2\pi/k$. The line integral in the exponent, analogous to Eq. 4.4, measures the Aharonov–Bohm phase accumulated when charge wraps around flux (which is bound to the charges). Thus, it measures the winding number of the braids formed by the paths of all of the particles. Thus, we have equivalently

$$\sum_{\text{paths }\{\mathbf{r_n}(t)\}} e^{iS_0/\hbar + i\vartheta W(\text{paths})}$$

where W is the winding number of the braid formed by the spacetime paths of all the particles, and ϑ is the anyon statistical angle. Thus, "integrating out"[18] the Chern–Simons gauge field implements fractional statistics by inserting a phase $e^{\pm i\vartheta}$ for each exchange!

Vacuum Abelian Chern–Simons Theory

Something we have pointed out in Section 4.3 is that the vacuum of an anyon theory knows about the statistics of the particles, even when the particles are not present.[19] That is, the ground-state degeneracy on a torus matches the number of particle species. Thus, in the absence of particles, we will be interested in

$$Z(\mathcal{M}) = \int_{\mathcal{M}} \mathcal{D}a_\mu(x)\, e^{iS_{CS}/\hbar}$$

where $\mathcal{M}$ is the spacetime manifold we are considering.[20]

If we consider a three-dimensional manifold of the form $\mathcal{M} = \Sigma \times S^1$ for a two-dimensional manifold Σ and S^1 represents time (compactified)[21] this integral gives exactly the ground-state degeneracy of the system. As we might expect, this quantity will be a topological invariant of the spacetime manifold. That is, smooth deformations of $\mathcal{M}$ do not change its value (see chapter appendix, particularly Section 5.3.4). This quantity $Z(\mathcal{M})$, often known as the partition function of the theory for the manifold $\mathcal{M}$, will be of crucial importance as we learn more about topological theories in general in Chapter 7.

5.2 Nonabelian Chern–Simons Theory: The Paradigm of TQFT

Among $(2+1)$-dimensional topological quantum systems, pretty much everything of interest is somehow related to Chern–Simons theory—however, we don't generally have the luxury of working with abelian theory as we have been doing so far.

We can generalize abelian Chern–Simons theory by promoting the gauge field a_α to be not just a vector of numbers, but rather a vector of matrices.[22] More precisely, to construct a nonabelian Chern–Simons theory, we consider a vector potential that takes values in a Lie algebra.[23] For example, if we choose to work with the Lie algebra of $SU(2)$ in the fundamental representation we can write a general element of this algebra as a sum of the three generators (proportional to $\sigma_x, \sigma_y, \sigma_z$) so that our Lie algebra valued gauge field is then[24]

$$a_\mu(x) = a_\mu^a(x) \left(\frac{\sigma_a}{2i} \right) \tag{5.10}$$

where σ_a are the Pauli matrices. Now that a_μ is matrix valued it becomes non-commutative and we have to be very careful about the order in which we write factors of a_μ.

The fundamental object that we need to think about is the Wilson loop operator[25]

$$W_L = \mathrm{Tr}\left[P \exp\left(\oint_L dl^\mu a_\mu \right) \right] \tag{5.11}$$

where here the integral follows some closed path L. This object, being

[20]Some spacetime manifolds, such as any two-dimensional manifold Σ cross time (such that $\mathcal{M} = \Sigma \times \mathbb{R}$), seem very natural. However, as we will see in much detail in Chapter 7, we will want to be much more general about the types of manifolds we consider. We should even allow three-dimensional manifolds where the two-dimensional topology of a fixed time slice changes as time evolves! See also the discussion in Chapter 6 and Fig. 6.1.

[21]Compactification of time from $\mathbb{R}$ to S^1 is something that might be familiar from statistical physics where this procedure is used for representing finite temperatures.

[22]If you have studied Yang–Mills theory, you already know about nonabelian vector potentials.

[23]See the introduction to Lie groups and Lie algebras in Section 40.2.3. In brief, a Lie group is a group that is also a continuous manifold. A Lie *algebra* is the algebra of infinitesimal changes in this group. A prime example is the Lie group $SU(2)$ with algebra generated by $i\sigma_j$ with σ_j being the Pauli operators. We write group elements as exponentials of the algebra $g = e^{i\boldsymbol{\sigma}\cdot\mathbf{n}}$.

[25]These are named for Ken Wilson, who won a Nobel Prize for his work on the renormalization group and critical phenomena. There is a legend that Wilson had very few publications when he came up for tenure as a professor at Cornell. Only due to the strong recommendation of his senior colleague Hans Bethe (already a Nobel Laureate at the time) did he manage to keep his job. Bethe knew what Wilson had been working on, and vouched that it would be extremely important. His ground-breaking work on renormalization group was published the next year. Everything worked out for him in the end, but the strategy of not publishing is *not* recommended for young academics trying to get tenure.

[24]For general Lie algebras, we want to write $a_\mu = a_\mu^a T_a$ where T_a are the anti-Hermitian generators of the Lie algebra with $T_a = -T_a^\dagger$. This means that $[T_a, T_b] = f^{abc}T_c$ with f the so-called structure constants of the Lie group, and $\mathrm{Tr}[T_aT_b] \equiv -\frac{1}{2}\delta_{ab}$. In the case of $SU(2)$ in the fundamental representation we have $T_a = -i\sigma_a/2$ with $f^{abc} = \epsilon^{abc}$. Be warned that other normalization conventions do exist, and changing conventions will insert seemingly random factors of 2 or i or worse.

the exponential of an integral of a vector potential, is essentially the nonabelian analogue[26] of the Aharonov–Bohm phase of Eq. 4.3). In Eq. 5.11, the P symbol indicates path ordering—analogous to the usual time ordering of quantum mechanics. The complication here is that $a_\mu(x)$ is a matrix, so when we try to do the integral and exponentiate, we have a problem that $a_\mu(x)$ and $a_\mu(x')$ do not commute. The proper interpretation of the path ordered integral is then to divide the path into tiny pieces of length dl. We then have

$$P \exp\left(\int_L dl^\mu a_\mu\right) = \tag{5.12}$$
$$\dots [1 + a_\mu(x_3)dl^\mu(x_3)] \, [1 + a_\mu(x_2)dl^\mu(x_2)] \, [1 + a_\mu(x_1)dl^\mu(x_1)] \dots$$

where $x_1, x_2, x_3, \dots$ are the small steps along the path L (it does not matter if they are equally spaced or not). If it is a closed path as in Eq. 5.11 the trace (which is invariant under cyclic permutation) will give the same answer regardless of where on the closed path we start.

The proper gauge transformation in the case of a nonabelian gauge field is given by

$$a_\mu \to U^{-1} a_\mu U + U^{-1} \partial_\mu U \tag{5.13}$$

where $U(x)$ is a matrix (which is a function of position and time) that acts on the matrix part of a_μ. Note that this is just the nonabelian analogue[27] of the gauge transformation in Eq. 5.3. To see that this gauge transformation leaves the Wilson loop operators invariant (and hence is the right way to define a gauge transformation!) see Section 5.3.3.

With a_μ a matrix valued quantity, we can write a more general form for the nonabelian Chern–Simons action as

$$S_{CS} = \frac{k}{4\pi} \int_\mathcal{M} d^3x \; \epsilon^{\alpha\beta\gamma} \; \mathrm{Tr}\left[a_\alpha \partial_\beta a_\gamma + \frac{2}{3} a_\alpha a_\beta a_\gamma\right] \tag{5.14}$$

Note that the second term in the brackets would be zero if the a_α were commutative. (In the abelian case above, we have no such term! See Eq. 5.1). We have not derived Eq. 5.14, but we will explain in a moment why it is the only expression we could have written down for the nonabelian generalization of the Chern–Simons action.

The Chern–Simons action is metric independent, which we show explicitly in Section 5.3.4. This means that space and time can be deformed continuously, and the value of the action does not change. While this may not be obvious from looking at the form of the action, a large hint is that the action is written without any reference[28] to the usual spacetime metric $g_{\mu\nu}$.

Since Chern–Simons theory is also a gauge theory, we would like the action to be gauge invariant. It turns out that the action is *almost* gauge invariant, as we will discuss momentarily. At any rate it is close enough to being gauge invariant to be of use for us!

It turns out that the Chern–Simons action is actually unique in being both metric independent and also (at least almost) gauge invariant. In (2

+ 1)-dimensions, no other action can be written down that involves only one gauge field[29] and has these two properties: topological invariance and gauge invariance. This is what makes Chern–Simons theory such a crucial paradigm for topological theories in $(2 + 1)$-dimensions.

Let us now return to this issue of how the Chern–Simons action is only *almost* gauge invariant. First of all, if the manifold has a boundary, we will run into non-gauge-invariant terms as mentioned below Eq. 5.3. For now, let us just assume that our manifold has no boundaries.

More crucially there is another issue with gauge invariance. Under gauge transformation (at least on a closed manifold) as in Eq. 5.13, the Chern–Simons action transforms to (see Exercise 5.2)

$$S_{CS} \to S_{CS} + 2\pi\nu k \qquad (5.15)$$

where

$$\nu = \frac{1}{24\pi^2} \int_{\mathcal{M}} d^3x \ \epsilon^{\alpha\beta\gamma} \operatorname{Tr}\left[(U^{-1}\partial_\alpha U)(U^{-1}\partial_\beta U)(U^{-1}\partial_\gamma U)\right] \qquad (5.16)$$

Surprisingly the complicated expression in Eq. 5.16 (sometimes known as the Pontryagin index or Pontryagin number[30]) is always an integer (see Section 5.3.5 for more detail). The integer ν gives the winding number of the map $U(x)$ from the manifold into the gauge group.[31]

It may now look problematic that our Chern–Simons action is not a true gauge invariant (Eq. 5.15), but we note that the only thing entering our functional integral is $e^{iS_{CS}}$, not the Chern–Simons action itself. Thus, as long as we choose k, the "level," as an integer (and since the winding number ν is also an integer), then we have a well-defined functional integral of the form

$$Z(\mathcal{M}) = \int_{\mathcal{M}} \mathcal{D}a_\mu(x) \ e^{iS_{CS}} \qquad (5.17)$$

where the result $Z(\mathcal{M})$ turns out to be a manifold invariant (see Section 5.3.4), meaning that smooth deformations of space and time do not change its value.

The insertion of the Wilson loop operator into the path integral gives a knot invariant of the link L that the Wilson loop follows. The fact that the result should be a topological invariant should not be surprising given that the Chern–Simons action itself is metric independent and therefore independent under deformations of space and time.[32] Often

---

[29]In Section 5.3.2 we discuss the abelian case with multiple gauge fields.

[31]In the case of the gauge group being $SU(2)$, as mentioned in Section 40.2.3, the gauge group is isomorphic to the manifold S^3. So if the manifold happens to be S^3 then we are looking at mappings from $x \in S^3$ (space) to $U(x) \in S^3$ (group). A mathematician would say that $\Pi_3(S^3) = \mathbb{Z}$, meaning one can wrap S^3 around S^3 any integer number of times. The case of zero winding number is anything that can be continuously deformed to $U = 1$ everywhere. However, we also can consider the identity mapping where S^3 (space) maps into S^3 (group) in the obvious way (every point goes to itself) which gives an $n = 1$ mapping (a 1-to-1 mapping). One can also construct 2-to-1 mappings which have winding $n = 2$ etc. (see Exercise 5.4 for a hint for how to do this!).

---

[30]Lev Pontryagin was a Soviet mathematician who was completely blind after the age of 14. He was also a controversial figure, being accused repeatedly of antisemitism—a charge he repeatedly denied. His name is translated from Russian in at least three different ways: "Pontryagin," "Pontriagin," or "Pontrjagin."

[32]The observant reader will note that we have not specified the "framing" of the knot, that is, if we are to think of the world line as being a ribbon not a line, we have not specified how the ribbon twists around itself (see Section 2.6.1). In field theory language this enters the calculation by how a point-splitting regularization is implemented.

we will think about our link as being embedded in a simple manifold like the three sphere, S^3 (which we define carefully in Section 40.1.1).

So for example, to find the link invariant corresponding to the two linked strings in Fig. 5.1, we have[33]

$$\text{Knot Invariant} = \frac{Z(S^3, L_1, L_2)}{Z(S^3)} = \frac{\int_{S^3} \mathcal{D}a_\mu(x) \, W_{L_1} W_{L_2} \, e^{iS_{CS}}}{\int_{S^3} \mathcal{D}a_\mu(x) \, e^{iS_{CS}}} \quad (5.18)$$

[33]This connection between knot invariants and Chern–Simons theory was famously made by Witten [1989]. This is the work that earned a Fields medal.

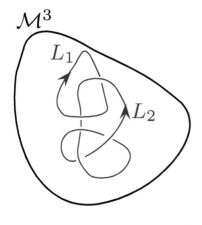

Fig. 5.1 A cartoon of a 3-manifold with a link made of two strands embedded in it.

with W_L being the Wilson loop operators as in Eq. 5.11. Indeed, if we choose to work with the gauge group $SU(2)$ at level k (working with the spin 1/2 representation of the group, that is, with Pauli matrices) we obtain the Kauffman bracket invariant of the knot with $A = \exp(i\pi/(2(k+2))$.

If we keep the same gauge group, but work with a different representation (for example, spin 1, rather than spin 1/2 in Eq. 5.10), we will obtain different "particle types" of the theory.

One can also choose to work with different gauge groups. Using $SU(N)$ and choosing a level k, one obtains the two parameter HOMFLY knot polynomial (the two parameters here being N and k). Similarly, using $SO(N)$ at level k gives a two parameter Kauffman polynomial (not to be confused with the Kauffman bracket). Typically a Chern–Simons theory with gauge group G at level k is notated as G_k (for example, using $SU(2)$ at level 2 we write the theory as $SU(2)_2$). Changing the sign of k corresponds to taking the "mirror image" of the theory (the partition function is complex conjugated).

5.3   Appendix: Odds and Ends about Chern–Simons Theory

5.3.1   Chern–Simons Canonical Quantization for the Abelian Case

One can consider the Chern–Simons theory as a quantum-mechanical theory with wavefunctions and operators (i.e. not in path integral language). To do this, we need to find the commutation relations. Working in the gauge $a_0 = 0$, in the Chern–Simons Lagrangian terms like $\partial_0 a_y$ multiply a_x, and vice versa.[34] This means that $a_y(x)$ is the momentum conjugate to $a_x(x)$, and vice versa. We thus have the commutation relations

[34]Note that for nonabelian Chern–Simons theories working in the $a_0 = 0$ gauge makes the a^3 term of the action vanish!

$$[a_x(\vec{x}), a_y(\vec{x}')] = \frac{2\pi i}{k} \delta(\vec{x} - \vec{x}')$$

with k the level. The arguments $\vec{x}$ here live in two dimensions. Consider now the Wilson loop operators around two different nontrivial cycles of a torus

$$W_j = \exp\left(iq \oint_{L_j} d\vec{l} \cdot \vec{a}\right)$$

where here j indicates we have a loop around either cycle 1 (L_1) or cycle 2 (L_2) of our torus. The two paths must intersect at one point and therefore, due to the above commutation relations, do not commute with each other. We can use a special case of the Baker–Campbell–Hausdorff formula

$$e^A e^B = e^B e^A e^{[A,B]} \quad ,$$

which holds when $[A, [A, B]] = [B, [A, B]] = 0$. This then gives us

$$W_1 W_2 = e^{2\pi i q^2 / k} W_2 W_1 = e^{i\vartheta} W_2 W_1 \quad ,$$

where ϑ is the statistical angle of the theory. Thus the Wilson loop operators act just like operators T_1 and T_2 in Section 4.3 that created particle–antiparticle pairs and moved them around a nontrivial cycle then reannihilated. So even without discussing particles, the ground-state wavefunction of the Chern–Simons theory is degenerate!

5.3.2   Multiple Gauge Fields and the K-Matrix

Abelian Chern–Simons theory can easily be generalized to consider multiple gauge fields.[35]

Here we consider N different gauge fields $a_\mu^J(x)$ where $J = 1 \ldots N$. We write a Lagrangian generalizing Eq. 5.1

$$\mathcal{L} = \frac{K_{IJ}}{4\pi} \epsilon^{\alpha\beta\gamma} a_\alpha^I \partial_\beta a_\gamma^J - j_J^\alpha a_\alpha^J \qquad (5.19)$$

where K_{IJ} is an N-dimensional symmetric invertible matrix of integers, and we have coupled to N different currents j_J, and repeated indices I, J are summed (as are repeated indices α, β, γ as usual). If K is a diagonal matrix, then this simply represents N decoupled theories of the form in Eq. 5.1.

From this Lagrangian we can derive a charge–flux model analogous to what we did in Section 5.1. We obtain equations of motion

$$j_I^\alpha = \frac{K_{IJ}}{2\pi} \epsilon^{\alpha\beta\gamma} \partial_\beta a_\gamma^J \qquad (5.20)$$

which is just the matrix form of Eq. 5.6.

We define N different effective magnetic fields $b^J = (\partial_1 a_2^J - \partial_2 a_1^J)$ corresponding to the N different gauge fields. Looking at the $\alpha = 0$ component of Eq. 5.20 we then have flux binding

$$j_I^0 = \frac{K_{IJ}}{2\pi} b^J \qquad (5.21)$$

or

$$b^J = 2\pi [K^{-1}]^{JI} j_I^0 \quad , \qquad (5.22)$$

which tells us the different fluxes b^J bound to each of the different charges j_I^0. Note that because of the form of the coupling in Eq. 5.19 (the last term), a particle of type J only "sees" the flux of type J.

[35] In fact *any* abelian anyon theory can be expressed in this formalism. See Cano et al. [2014]; Wang and Wang [2020].

Plugging Eq. 5.20 back into the Lagrangian we obtain the effective coupling

$$\mathcal{L} = -\frac{1}{2} j_I^\alpha a_\alpha^I \quad .$$ (5.23)

The combination of Eq. 5.22 and Eq. 5.23 allows us to deduce the exchange statistics of the different types of particles associated with the different currents. The phase for exchanging two identical particles of type J is given by

$$\vartheta_J = \pi [K^{-1}]^{JJ} \quad .$$ (5.24)

We can also derive the *relative* statistics between particle types L and J, that is, the phase accumulated[36] by taking a particle of type L all the way around a particle of type J

$$\vartheta_{LJ} = 2\pi [K^{-1}]^{LJ} \quad .$$ (5.25)

We can construct more types of quasiparticles by fusing together the elementary quasiparticles we have already discussed. The most general quasiparticle we can construct is described by a set of N integers n_J with $J = 1 \dots N$ giving the number of units of charge of type J attached to the quasiparticle, which we often write as vector $\boldsymbol{\ell}$. Fusing together an anyon of type $\boldsymbol{\ell}$ with an anyon of type $\boldsymbol{\ell}'$ gives an anyon of type $\boldsymbol{\ell} + \boldsymbol{\ell}'$ since the charges of each type simply add.

The phase of exchanging two identical particles of type $\boldsymbol{\ell}$ is given by

$$\vartheta_{\boldsymbol{\ell}} = \pi (\boldsymbol{\ell}^T K^{-1} \boldsymbol{\ell})$$ (5.26)

where the quantity inside the parenthesis on the right is just a scalar. Similarly, the phase for wrapping a particle of type $\boldsymbol{\ell}$ around a particle of type $\boldsymbol{\ell}'$ is given by

$$\vartheta_{\boldsymbol{\ell},\boldsymbol{\ell}'} = 2\pi (\boldsymbol{\ell}^T K^{-1} \boldsymbol{\ell}') \quad .$$ (5.27)

As discussed in Section 4.3.2, if you fuse together enough particles of any type of these anyons together you will get back to the vacuum, so not all $\boldsymbol{\ell}$ vectors describe different quasiparticle types. We often define two particle types $\boldsymbol{\ell}$ and $\boldsymbol{\ell}'$ to be the same if they accumulate the same phase when braided around any particle type $\boldsymbol{\ell}''$. With this definition the number of different particle types is given by $|\det K|$, which is also the ground state degeneracy on the torus, analogous to the discussion of Section 4.3.2 (see Exercise 5.6 for proof of these statements).

It is often useful to make a basis change on the K matrix by choosing any N-dimensional matrix W with integer entries and $\det W = 1$ and transforming

$$K \quad \rightarrow \quad WKW^T$$ (5.28)

and one obtains the same anyon theory in another guise (the $\boldsymbol{\ell}$ vectors are correspondingly transformed as $\boldsymbol{\ell} \rightarrow W\boldsymbol{\ell}$).

As with the discussion near Eq. 5.3, the Lagrangian Eq. 5.19 is gauge

[36]In Chapters 10 and 13 we introduce the R-matrix notation for these braiding phases. The phase for exchanging identical particles of type J we call R^{JJ} and the phase for taking particle L all the way around particle J is called $R^{JL}R^{LJ}$.

invariant only on a closed manifold, and the lack of gauge invariance on the boundary indicates that there are N distinct propagating edge modes. There will be n modes propagating in one direction and $N - n$ modes in the opposite direction where n is the number of positive eigenvalues of the matrix K.

The approach described here, the so-called K-matrix formalism,[37] is used heavily in fractional quantum Hall physics, which we explore in a bit more depth in Chapter 37. See also references at the end of the chapter.

[37]This is not related to so-called K-theory. It is just a naming coincidence.

5.3.3   Gauge Transformations with Nonabelian Gauge Fields

Let us define a Wilson-line operator, similar to the Wilson loop but not forming a closed loop, that is, going along a curve C from spacetime point y to spacetime point x, and we do not take the trace here:

$$\tilde{W}_C(x, y) = P \exp\left(\int_C dl^\mu a_\mu\right) \, .$$

Under a gauge transformation function $U(x)$ we intend that the Wilson line operator transforms as

$$\tilde{W}_C(x, y) \to U(x)^{-1} \, \tilde{W}_C(x, y) \, U(y) \quad . \tag{5.29}$$

Clearly this obeys composition of paths, and will correctly give a gauge-invariant result for a closed Wilson loop. Now let us see how gauge field a_μ must transform such that Eq. 5.29 holds. We consider

$$\tilde{W}_C(x, x + dx) = 1 + a_\mu dx^\mu \tag{5.30}$$

and its transformation should be

$$
\begin{aligned}
\tilde{W}_C(x, x + dx) \quad &\to \quad U(x)^{-1}\tilde{W}_C(x, x + dx)U(x + dx) \\
&= \quad U(x)^{-1}[1 + a_\mu dx^\mu]U(x + dx) \\
&= \quad U(x)^{-1}[1 + a_\mu dx^\mu][U(x) + dx^\mu \partial_\mu U(x)] \\
&= \quad 1 + [U^{-1}a_\mu U + U^{-1}\partial_\mu U]dx^\mu \quad . \tag{5.31}
\end{aligned}
$$

By comparing Eq. 5.30 and Eq. 5.31 we see that the gauge transformation rule Eq. 5.13 correctly gives a gauge-invariant Wilson loop operator.

5.3.4   Chern–Simons Action Is Metric Independent

You will often see books state that Eq. 5.14 must be metric independent because you don't see the metric $g_{\mu\nu}$ written anywhere. But that rather misses the point!

A differential geometer would see that one can write the Chern–Simons action in differential form notation

$$S_{CS} = \frac{k}{4\pi} \int \mathrm{Tr}(a \wedge da + \frac{2}{3} a \wedge a \wedge a) \quad ,$$

which then makes it "obvious" that this is metric independent being the integral of a 3-form.

In more detail however, we must first declare how the gauge field transforms under changes of metric. It is a "1-form," meaning it is meant to be integrated along a line to give a reparameterization invariant result, such as in the Wilson loops. In other words, we are allowed to bend and stretch the spacetime manifold, but the flux through a loop should stay constant. Under reparameterization of coordinates we have

$$\int da = \int dx^\mu a_\mu(x) = \int dx'^\mu \frac{\partial x^\nu}{\partial x'^\mu} a_\nu(x') \quad .$$

This means that under reparameterization $x'(x)$ we have

$$a_\mu(x) = \frac{\partial x^\nu}{\partial x'^\mu} a_\nu(x')$$

such that the line integral remains invariant under a reparameterization of the space.

Now, if we make this change on a in the Chern–Simons action we obtain

$$\epsilon^{\alpha\beta\gamma} \ \mathrm{Tr}\left[a_\alpha \partial_\beta a_\gamma + \frac{2}{3} a_\alpha a_\beta a_\gamma\right] \rightarrow$$

$$\epsilon^{\alpha'\beta'\gamma'} \frac{\partial x^\alpha}{\partial x'^{\alpha'}} \frac{\partial x^\beta}{\partial x'^{\beta'}} \frac{\partial x^\gamma}{\partial x'^{\gamma'}} \ \mathrm{Tr}\left[a_\alpha \partial_\beta a_\gamma + \frac{2}{3} a_\alpha a_\beta a_\gamma\right] \quad .$$

But notice that the prefactor, including the ϵ, is precisely the Jacobian determinant and can be rewritten as

$$\epsilon^{\alpha'\beta'\gamma'} \ \det[\partial x/\partial x'] \quad .$$

Thus, the three-dimensional Chern–Simons action integral can be changed to the dx' variables and the form of the integral is completely unchanged and thus depends only on the topological properties of the manifold.

In fact, this feature of the Chern–Simons Lagrangian is fairly unique. Given that we have a single gauge field $a_\mu(x)$ this is the *only* (3-form) gauge-invariant Lagrangian density we can write down that will give a topological invariant!

5.3.5   Winding Number: The Pontryagin Index

We would like to show that the integral in Eq. 5.16 is indeed always an integer independent of the gauge group G, the manifold $\mathcal{M}$ and the function $U(x)$. While doing this rigorously is difficult, it is not too hard to see roughly how it must be done. First, we note that, like the

Chern–Simons action, it is the integral of a 3-form so it does not care about the metric on the manifold (this is not surprising being that this winding number arose from the Chern–Simons action). One can then reparameterize the manifold in terms of coordinates within the group, and convert the integral over space into an integral over the group. The only thing that is left unclear is then in the mapping $U(x) : \mathcal{M} \to G$ how many times the group is covered in this mapping. We then have immediately that the given definition of the winding number must be an integer times some constant. See the references at the end of the chapter for more details. Some further exploration of the Pontryagin index is done in Exercises 5.3, 5.4, and 5.5.

5.3.6 Framing of the Manifold—or Doubling the Theory

There is a bit of a glitch in Chern–Simons theory. We want the Chern–Simons functional $Z(\mathcal{M})$ to be a function of the topology of $\mathcal{M}$ only. This is *almost* true—it is true up to a phase. In order to get the phase, you need to specify one more piece of information, which can be provided in several ways (often called a 2-framing).[38] This additional piece of information is most easily described by saying that you need to specify a bit of information about the topology of the 4-manifold $\mathcal{N}$ that $\mathcal{M}$ bounds (i.e. $\mathcal{M} = \partial \mathcal{N}$). It is a fact that all orientable closed 3-manifolds are the boundary of some 4-manifold—in fact, of many possible 4-manifolds. The phase of $Z(\mathcal{M})$ is sensitive only to the "signature" of the 4-manifold $\mathcal{N}$. (Consult a book on 4-manifold topology such as Gompf and Stipsicz [1999] if you are interested!)

The fact that the Chern–Simons theory should depend on some information about the 4-manifold that $\mathcal{M}$ bounds may sound a bit strange. It is in fact a sign that the Chern–Simons theory is "anomalous." That is, it is not really well defined in three dimensions. If you try to make sense of the functional integral $\int \mathcal{D}a_\mu$, you discover that there is no well-defined limit by which you can break up spacetime into little boxes and integrate over a_μ in each of these boxes.[39] However, if you extend the theory into four dimensions, then the theory becomes well behaved. This is not unusual. We are familiar with lots of cases of this sort. Perhaps the most famous example is the fermion doubling problem. You cannot write down a time reversal invariant theory for a single chirality fermion in D dimensions without somehow getting the other chirality. However, you can think of a system extended into $(D + 1)$-dimensions where one chirality ends up on one of the D-dimensional boundaries and the other chirality ends up on the other D-dimensional boundary.[40]

So to make Chern–Simons theory well-defined, you must either extend into four dimensions, or you can "cancel" the anomaly in three dimensions by, for example, considering two, opposite chirality Chern–Simons theories coupled together (a "doubled" Chern–Simons theory). The corresponding manifold invariant of a doubled theory gets $Z(\mathcal{M})$ from the right-handed theory and its complex conjugate from the left-handed the-

[38] A detailed discussion of 2-framing is given by Atiyah [1990] and by Kirby and Melvin [1999]. This is fairly mathematical stuff!

[39] There are a number of recent attempts to put *abelian* Chern–Simons theory on a lattice. See for example DeMarco and Wen [2021], Jacobson and Sulejmanpasic [2023], or Sun et al. [2015] and references therein. The former has a short discussion of the framing problem.

[40] This is precisely what happens on the surface of materials known as "topological insulators" (or TIs) in three dimensions. The bulk of the system is a gapped insulator, but the surface of the system has a single Dirac fermion (or an odd number of Dirac fermions) and this is impossible to have in a purely two-dimensional system. This will be discussed in my *next* book. See also the brief discussion in Section 35.1.1.

ory, thus giving an end result of $|Z(\mathcal{M})|^2$, which obviously won't care about the phase anyway!

5.3.7 Chern–Simons Theory as the Boundary of a Four-Dimensional Theory

With the considerations of Section 5.3.6, it is interesting to express Chern–Simons theory as the boundary theory of a four-dimensional topological theory. To do this let us define the field strength tensor

$$F_{\mu\nu} = \partial_\mu A_\nu - \partial_\nu A_\mu + [A_\mu, A_\nu] \quad .$$

This definition matches the expression for the electromagnetic field strength in the case where the fields are abelian such that the commutator vanishes. However, more generally for nonabelian gauge theories (including Yang–Mills theory) the additional commutator term must be added.

In four dimensions we can define the dual field strength

$$^*F^{\mu\nu} = \frac{1}{2}\epsilon^{\mu\nu\lambda\rho}F_{\lambda\rho}$$

where ϵ is the antisymmetric tensor. We now consider the following topological action on a 4D manifold $\mathcal{N}$

$$S = \frac{\theta}{16\pi^2} \int_{\mathcal{N}} d^4x \, \mathrm{Tr}\left[F_{\mu\nu}\,^*F^{\mu\nu}\right] \quad .$$

This four-dimensional action is well defined and non-anomalous, meaning it can be regularized and/or treated properly on a lattice.

With a bit of algebra the action can be rewritten as

$$S = \frac{\theta}{8\pi^2} \int_{\mathcal{N}} d^4x \, \partial_\mu G^\mu$$

where

$$G^\mu = 2\epsilon^{\mu\nu\lambda\rho}\,\mathrm{Tr}\left[A_\nu\partial_\lambda A_\rho + \frac{2}{3}A_\nu A_\lambda A_\rho\right] \quad . \tag{5.32}$$

Since the action can be written as the integral of a total derivative, it should give zero when integrated over a closed manifold $\mathcal{N}$. However, when the manifold has a boundary one obtains

$$S = \frac{\theta}{8\pi^2} \int_{\partial N} d^3x \, G^\mu v_\mu$$

where v_μ is the unit vector normal to the boundary. Examining the form of Eq. 5.32 we realize that the action is precisely the Chern–Simons action on the three-dimensional boundary manifold ∂N.

Chapter Summary

- The charge–flux model can be realized in an abelian Chern–Simons theory. Multiple abelian Chern–Simons fields can be handled with the K-matrix formalism.
- We introduced some ideas of nonabelian Chern–Simons theory which is a paradigm for TQFTs. These include:
  - Manifold invariants and vacuum partition functions
  - Wilson loops and relations to knot invariants.

We run into various Chern–Simons theories many times in later chapters, particularly $SU(2)_k$. See, in particular, material in Chapters 18, 21, and 22. See also the resources in Chapters 39 and 17.

Further Reading:

- A good reference for abelian Chern–Simons theory is Wilczek [1990].
- For the more general (abelian) K-matrix formalism, the original work was Read [1990]; Wen and Zee [1991, 1992]. A recent discussion is given in Hansson et al. [2017].

Some good references on nonabelian Chern–Simons theory are:

- Witten [1989]—the paper that really launched many of these ideas. This is the paper that won a Fields medal!
- Nayak et al. [2008]. This has a short discussion of Chern–Simons theory meant to be easily digested.
- Kauffman [2001]. The section on Chern–Simons theory is heuristic, but very useful.
- Treiman et al. [1985]. See particularly the chapters by R. Jackiw.
- Dunne [1998].
- A more detailed discussion of the topology of the Pontryagin index is given in Vandoren and van Nieuwenhuizen [2008], Rajaraman [1982], Coleman [1985]. Also see the above-mentioned book by Treiman et al. [1985].

Exercises

Exercise 5.1 Polyakov Representation of the Linking Number

Consider a link made of two strands, L_1 and L_2. Consider the double line integral

$$\Phi(L_1, L_2) = \frac{\epsilon_{ijk}}{4\pi} \oint_{L_1} dx^i \oint_{L_2} dy^j \frac{x^k - y^k}{|\mathbf{x} - \mathbf{y}|^3} \quad .$$

(a) Show that Φ is equal to the phase accumulated by letting a unit of flux run along one strand, and moving a unit charged particle along the path of the other strand.

(b) Show that the resulting phase is the topological invariant known as the linking number—the number of times one strand wraps around the other: see Section 2.6.2.

This integral representation of linking was known to Gauss.

Exercise 5.2 Gauge Transforming the Chern–Simons Action

Make the gauge transformation Eq. 5.13 on the Chern–Simons action 5.14 and show that it results in the change 5.15. Note that there will be an additional term that shows up, which is a total derivative and will therefore vanish when integrated over the whole manifold $\mathcal{M}$.

Exercise 5.3 Topological Invariance of the Pontryagin Index

Let $U(x)$ be a unitary matrix defined as a function of position x on a manifold $\mathcal{M}$. Show that the Pontryagin index $\nu[U(x)]$ defined by Eq. 5.16 remains unchanged under small deformations of the function $U(x)$, and is thus a topological invariant.

Hints: (a) Starting with some function $U_0(x)$, consider $U = (dU)U_0$ where dU is near the identity matrix $dU(x) = 1 + i\delta H(x)$ for $H(x)$ Hermitian and δ small. Then expand and find the term linear in δ. We want to show that this term linear in δ is zero. (b) Rewrite this term in the form

$$\int_{\mathcal{M}} d^3x \ \epsilon^{\alpha\beta\gamma}(\partial_\alpha A)(\partial_\beta B)(\partial_\gamma C)$$

for A, B, C some functions of U_0 and H. Then integrate by parts to show that this must be zero due to the antisymmetry of $\epsilon^{\alpha\beta\gamma}$.

Exercise 5.4 Adding Pontryagin Indices

Given two mappings $U_1(x)$ and $U_2(x)$ (as defined in Exercise 5.3 and Eq. 5.16) with corresponding Pontryagin indices $\nu[U_1(x)]$ and $\nu[U_2(x)]$. Argue that the Pontryagin index of $U(x) = U_1(x)U_2(x)$ is the sum of the of the Pontryagin indices of the factors

$$\nu[U(x)] = \nu[U_1(x)] + \nu[U_2(x)] \quad .$$

Hint: Don't try to use product rule of derivatives or anything like that (this way lies madness!). Instead, use the fact that the Pontryagin index is a topological invariant and deform the mappings $U_1(x)$ and $U_2(x)$ so that they are constant (or even unity) in large regions.

Exercise 5.5 Calculation of a Pontryagin Index

In this exercise we aim to calculate a Pontryagin index for the example of a simple case to show that it is indeed a non-zero integer. Consider the group $G = SU(2)$ and the manifold $\mathcal{M} = S^3$, which we can write as the set of points $x_1^2 + x_2^2 + x_3^2 + x_4^2 = 1$. Topologically $SU(2)$ and S^3 are identical since we can write any $SU(2)$ matrix as

$$U = \mathbf{1}\, x_4 + i\boldsymbol{\sigma} \cdot \mathbf{x} \quad , \tag{5.33}$$

where $\mathbf{x}$ is the 3-vector (x_1, x_2, x_3) and x on the left is the 4-vector, and $\boldsymbol{\sigma}$ is a vector of the three Pauli matrices. We now consider the identity mapping from the manifold to the group

$$U(x) : S^3 \to SU(2)$$

given by Eq. 5.33 and we wish to evaluate the Pontryagin index in Eq. 5.16.

When evaluating the integral Eq. 5.16 we are free to choose any orthonormal coordinate system we like. In fact, the coordinate systems need not be defined globally, and we can evaluate the integral in patches. Since the sphere (and the mapping) is symmetric, we just need to evaluate the integrand at one convenient point with a convenient coordinate system and multiply by the volume of S^3.

(a) Calculate the volume of S^3 with unit radius.

Now let us choose a convenient point: $x_4 = 1$ (so $\mathbf{x} = \mathbf{0}$) with the three-dimensional coordinate system given by the three usual axes of $\mathbf{x}$.

(b) Evaluate $U^{-1}\partial_{x_i} U$ at this point.

(d) Evaluate the argument of the integral Eq. 5.16 at this point, then multiply by the volume of S^3 to obtain the final result for ν. If you did not make any mistakes, you should get $\nu = 1$.

Exercise 5.6 Properties of the K-Matrix

In Section 5.3.2 we introduced the K-matrix formalism for multiple abelian Chern–Simons gauge fields.

(a) Derive the results Eq. 5.24 and Eq. 5.25 giving the results of exchanging and braiding the elementary charges of such a theory.

Given a theory defined by an N-dimensional K-matrix, the most general quasiparticle is described by a set of N integers ℓ_J with $J = 1 \ldots N$ defining the number of units of charge of type J attached to the quasiparticle (often written as $\boldsymbol{\ell}$).

(b) [Easy] Using Eqs. 5.24 and 5.25 derive Eqs. 5.26 and 5.27.

(c) [Easy] Suppose we have an abelian anyon theory with an anyon of type a with statistical phase ϑ_a for exchanging two anyons of type a with each other. Let $b = a \times a$. Show that for the exchange phase for two particles of type b is given by

$$\vartheta_b = \vartheta_{a \times a} = 4\vartheta_a$$

as we found in Section 4.2.1. Show that more generally if we fuse together n anyons of type a we obtain a particle b with exchange phase

$$\vartheta_b = \vartheta_{\underbrace{a \times a \times a \ldots \times a}_{n \text{ times}}} = n^2 \vartheta_a \quad .$$

(d) If we consider a quasiparticle with $\ell_J = K_{JI}$ for some fixed I, show that this braids trivially around any other quasiparticle. Show that this quasiparticle is a boson if K_{II} is even and is a fermion if K_{II} is odd.

(e) Show further that, given a quasiparticle defined by an arbitrary set of integers ℓ_J, it will have exactly the same braiding properties (i.e. phases accumulated when braiding around any given quasiparticle) as a quasiparticle defined by integers $\ell'_J = \ell_J + K_{JI}$ for any fixed I.

(f) [Harder] Let us define two quasiparticles to be topologically equivalent if they have the same braiding properties (i.e. accumulate the same phase when braided around any other given quasiparticle). Show that the number of topologically inequivalent quasiparticle types is given by $|\det K|$. This is also the ground state degeneracy on a torus. Hint: K is defining a lattice in N-dimensions.

(g) Given an N-dimensional matrix of integers W with $\det W = \pm 1$, show that the matrix $K' = WKW^T$ defines the same anyon theory as the matrix K (with ^{T} indicating transpose).

Short Digression on Quantum Gravity[1]

6

Optional/Motivational

6.1 Why This Is Hard

Little is known about quantum gravity with any certainty. What we do know for sure is the value of some of the fundamental constants that must come into play: the gravitational constant G, the speed of light c, and of course Planck's constant $\hbar$. From these we can put together an energy scale, known as the Planck Scale

$$E_{\text{Planck}} = \sqrt{\frac{\hbar c^5}{G}} \approx 10^{28} \, \text{eV}.$$

The temperature of the world around us is about 0.03 eV. Chemistry, visible light, and biology occur on the scale of 1 eV. The Large Hadron Collider probes physics on the scale of roughly 10^{13} eV. This means trying to guess anything about the Planck scale is trying to guess physics on an energy scale 15 orders of magnitude beyond what any accelerator[2] experiment has ever probed. We must surely accept the possibility that any physical principle we hold dear from all of our experiments on low energy scales could no longer hold true at the Planck scale! The only thing that is *really* required is that the effective low-energy theory matches what we can see at the low energies in the world around us.

6.2 Which Approach?

There are several approaches to quantum gravity. While I will not make any statement about which approaches are promising, and which approaches are crazy and overpublicized,[3] I am comfortable stating that many of these investigations have led to the discovery of incredibly interesting and important things. While in some cases (maybe in most cases) the discoveries may be more about math than about physics, they are nonetheless worthwhile investigations that about which I am enthusiastic.

6.3 Some General Principles?

We have to choose general principles that we believe will always hold, despite the fact that we are considering scales of energy and length

[1] This chapter aims to give context about why people first started studying topological theories. It can be skipped on a first reading (but do come back later to enjoy it!).

[2] Cosmic ray observations have been made at several orders of magnitude higher still—but very little can be deduced from these extremely rare and uncontrolled events. A famous event known as the "Oh my God particle" was apparently 10^{20} eV, still eight orders of magnitude away from the Planck scale.

[3] For some basic information on the wars between some of the different approaches to quantum gravity, see the books *The Trouble With Physics* by Lee Smolin or *Not Even Wrong* by Peter Woit. Or see responses to these, such as the article by J. Polchinski in *American Scientist*. Also enlightening are the online letters between Smolin and Lenny Susskind.

15 orders of magnitude away from anything we have ever observed or measured. Much of the community feels that the most fundamental thing to hold onto is the Feynman picture of quantum mechanics—that all spacetime histories must be allowed. We might write a quantum partition function of the form

$$Z = \sum_{\text{All universes}} e^{iS/\hbar} \tag{6.1}$$

where the sum is now over everything that could happen in all possible histories of the universe—it is the ultimate sum over histories! Obviously such a thing is hard to even contemplate. Several key simplifications will make contemplation easier:

(1) Let us ignore matter. Let us (at least to begin with) try to model only universes that are completely devoid of substance and only contain vacuum.

Thus the universe contains only the spacetime metric. Doing this, the Einstein–Hilbert action[4] for gravity takes the form

$$S_{\text{Einstein}} \sim \int_{\mathcal{M}} d^D x \ R\sqrt{-\det(g)} \tag{6.2}$$

where the integration is over the entire D-dimensional spacetime manifold $\mathcal{M}$, where here g is the spacetime metric tensor and R is the Ricci scalar.[5] One might imagine that we could construct a theory of quantum gravity by plugging the Einstein–Hilbert action into the path integral form of Eq. 6.1. We obtain

$$Z = \int \mathcal{D}g(x) \ e^{iS_{\text{Einstein}}[g(x)]/\hbar} \quad, \tag{6.3}$$

thus summing (or integrating) over all possible spacetime metrics. Even without matter in the universe, the model is very nontrivial because the spacetime metric can fluctuate—these fluctuations are just gravity waves.[6] Even in this limit no one has fully made sense of this type of path integral without many additional assumptions.

(2) Let us simplify even more by considering a $(2 + 1)$-dimensional universe.

We are used to the idea that many things simplify when we go to a lower dimension. Indeed, that is what happens here. In $(2 + 1)$-dimensions, there is an enormous simplification: there are no gravity waves! Why not? In short, there are just not enough degrees of freedom in a $(2 + 1)$-dimensional metric to allow for gravity waves. (For more information about this fact see Section 6.4.) As a result, the only classical solution of the Einstein equations in the vacuum is that $R = 0$ and that is all! That is, the universe is flat and there are no fluctuations.[7] (One can also have a cosmological constant Λ, in which case $R = 2\Lambda g$ is the solution.)

[4]Written down first by Hilbert in 1915.

[5]If you are rusty with your general relativity, recall that the metric tensor g defines the relativistically invariant line element via $ds^2 = g_{\mu\nu}dx^\mu dx^\nu$, and the Ricci scalar R, which is a complicated function of g, is a measure of the curvature of a manifold that compares the volume of a small ball to the volume it would have in flat Euclidean space. In particular for a D-dimensional manifold $\mathcal{M}$ we would consider a D-dimensional ball B^D of radius ϵ and we have

$$\frac{V(B^D) \subset \mathcal{M}}{V(B^D) \subset \mathbb{R}^D} = 1 - \frac{\epsilon^2 R}{6(D+2)} + \dots.$$

[6]Observation of gravity waves by the LIGO experiment won the 2017 Nobel Prize. Long before this we had very strong indirect observation of gravity waves from observation of the Hulse–Taylor binary pulsar, which earned a Nobel Prize in 1993.

[7]In fewer than four dimensions $R = 0$ implies no curvature at all.

One might think that this means that gravity in $(2 + 1)$-dimensions is completely trivial, but it is not. The spacetime manifold, although everywhere curvature free, still has the possibility of having a *nontrivial topology*. Thus, what we are interested in is actually the different topologies that our spacetime manifold might have!

We thus rewrite Eq. 6.1 as

$$
\begin{aligned}
Z &= \sum_{\text{manifolds}\,\mathcal{M}} \int_{\mathcal{M}} \mathcal{D}g(x)\ e^{iS[g(x)]/\hbar} \\
&= \sum_{\text{manifolds}\,\mathcal{M}} Z(\mathcal{M})
\end{aligned}
$$

where $S[g(x)]$ is the Einstein–Hilbert action for a flat universe with metric g, the sum is over all different topologies of manifolds the universe might have, and the integration $\mathcal{D}g$ is an integration over all metrics subject to the condition that the manifold's topology is fixed to be $\mathcal{M}$.

Why would we be interested in such a quantity? In short, suppose we know what the topology is of our (d-dimensional) universe at a fixed time t. We want to know the amplitude that the topology changes as t develops. That is, is the spacetime manifold of our universe of the form $\mathcal{M} = \Sigma \times$ time or does the spacetime manifold split analogous to that shown in Fig. 6.1?

Here is the surprise: the function $Z(\mathcal{M})$ is precisely the Chern–Simons partition function discussed in Section 5.2 for an appropriately chosen gauge group![8] More information about this connection is given in Section 6.5.

Fig. 6.1 An example of a manifold where the topology of a space-like slice (slice at fixed time) changes as time progresses. At the bottom the space-like slice is a single circle, whereas at the top a space-like slice is two circles.

[8] This was first noted by Achúcarro and Townsend [1986] and then was developed further by Witten [1988a] and many others.

6.3.1 Further Comments on Connections to Quantum Gravity

In the "this is not string theory" school of thought for quantum gravity, evaluation of Eq. 6.3 is the main goal. Crucially one needs some variables to describe the metric of the universe. Several different approaches to this seem to converge on some similar structures. One interesting approach, known as loop quantum gravity, uses Wilson loop operators as the elementary variables of the theory (once one has reformulated gravity to look like a gauge theory). Another approach discretizes spacetime and sums over the different possible discretizations.[9] With certain assumptions these approaches appear to be very closely related! In Section 23.3 we return to the issue of discretizing spacetime and how this can result in topological gravity.

[9] Indeed, at length scales as small as the Planck length $l_{Planck} = \sqrt{\hbar G/c^3} = \hbar c/E_{\text{Planck}} \approx 1.6 \times 10^{-35}\,\text{m}$, there is no reason to believe spacetime resembles our macroscopic idea of a smooth manifold. The ratio of the radius of the sun to the radius of an atom is roughly the same as the ratio of the radius of an atom to the Planck length!

6.4   Appendix: No Gravity Waves in (2 + 1)-Dimensions

Why are there no gravity waves in $(2 + 1)$-dimensions? The short argument for this is as follows (taken from Carlip [2005]):

> In n dimensions, the phase space of general relativity is parametrized by a spatial metric at constant time, which has $n(n-1)/2$ components, and its conjugate momentum, which adds another $n(n-1)/2$ components. But n of the Einstein field equations are constraints rather than dynamical equations, and n more degrees of freedom can be eliminated by coordinate choices. We are thus left with $n(n-1) - 2n = n(n-3)$ physical degrees of freedom per spacetime point. In four dimensions, this gives the usual four phase space degrees of freedom, two gravitational wave polarizations and their conjugate momenta. If $n = 3$, there are no local degrees of freedom.

Let us put a bit more detail on this argument. If we write the flat metric as $\eta_{\mu\nu} = \text{diag}[-1, 1, 1, \ldots]$ in any dimension, and we consider small deviations from a flat universe $g = \eta + h$, we can construct the trace-reversed

$$\bar{h}_{\mu\nu} = h_{\mu\nu} - \frac{1}{2}\eta_{\mu\nu}\eta^{\rho\sigma}h_{\rho\sigma} \ .$$

In any dimension, gravitational waves in vacuum take the form $\bar{h}^{\mu\nu}{}_{,\nu} = 0$, and $\Box\bar{h}_{\mu\nu} = 0$, where the comma notation indicates derivatives, and indices are raised and lowered with η.

In any dimension we will have the gravitational wave of the form

$$\bar{h}_{\mu\nu} = \epsilon_{\mu\nu}e^{ik^\rho x_\rho}$$

where the polarization $\epsilon_{\mu\nu}$ is orthogonal to the light-like propagation wavevector, $k^\mu k_\mu = 0$, meaning

$$\epsilon_{\mu\nu}k^\nu = 0 \quad . \tag{6.4}$$

However, one must also worry about gauge freedoms. We can redefine our coordinates and change the form of the metric without changing any of the spatial curvatures. In particular, making a coordinate transformation $x \to x - \xi$, we have

$$\bar{h}_{\mu\nu} \to \bar{h}_{\mu\nu} - \xi_{\nu,\mu} - \xi_{\mu,\nu} + \eta_{\mu,\nu}\xi^\alpha{}_{,\alpha} \quad .$$

Now here is the key: in $(2 + 1)$-dimensions for *any* matrix ϵ you choose, you can always find a

$$\xi_\mu = A_\mu e^{ik^\rho x_\rho}$$

such that

$$\bar{h}_{\mu\nu} = \epsilon_{\mu\nu}e^{ik^\rho x_\rho} = \xi_{\nu,\mu} + \xi_{\mu,\nu} - \eta_{\mu,\nu}\xi^\alpha_{,\alpha} \quad .$$

This means that the wave is pure gauge, and the system remains perfectly flat! That is, if you calculate the curvature with this form of $\bar{h}$, you will find zero curvature.

To be more precise, we find

$$\epsilon_{\mu,\nu} = A_\mu k_\nu - A_\nu k_\mu + \eta_{\mu\nu}A^\sigma k_\sigma$$

and any ϵ that satisfies Eq. 6.4 can be represented with some vector A. It is easy to check this by counting degrees of freedom. ϵ has six degrees of freedom in $(2 + 1)$-dimensions, but Eq. 6.4 is three constraints, and A has three parameters, so we should always be able to solve the equation for A given ϵ.

6.5   Appendix: Relation of Chern–Simons Theory to (2 + 1)-Dimensional GR

The connection between $(2 + 1)$-dimensional general relativity and Chern–Simons theory is developed in detail by Achúcarro and Townsend [1986]; Witten [1988a] and is reviewed well by Carlip [2005].

In the mapping from gravity to Chern–Simons theory, the Chern–Simons gauge field represents the metric, and there are gauge equivalences between different configurations of the metric, which translate into gauge equivalences of the Chern–Simons field.

We consider an Einstein–Hilbert action as in Eq. 6.2, although more generally we allow for a cosmological constant $\Lambda \neq 0$, in which case the term R in the action is replaced by $(R - 2\Lambda)$. We may also consider gravity with a Minkowski (Lorentzian) metric (similar to our universe), which we call $(2 + 1)$-dimensional or a model of gravity for a space with Euclidean metric,[10] which we would call three dimensional.

The simplest and most tractable case to consider is the case of the Euclidean metric, with positive Λ. In this case the Einstein–Hilbert action can be precisely mapped to a Chern–Simons action of the form of Eq. 5.14 where the gauge group is $SO(4) = SU(2) \times SU(2)$. In particular, the resulting theory is the "double" of Chern–Simons $SU(2)_k$, meaning it is the product of $SU(2)_k$ and its mirror image theory $SU(2)_{-k}$ (we will study doubled theories in Chapters 23–24 and 31–33. See in particular Section 23.3 where we discuss quantum gravity again). In this mapping the "level" is related to the cosmological constant as $k = 1/(4G\sqrt{\Lambda})$ with G being Newton's constant.

In Euclidean space, one can also consider the case of $\Lambda < 0$ in which case the gauge group[11] is $SO(3,1)$ or $\Lambda = 0$ where the gauge group is $ISO(3)$, the isometries of three-dimensional space (also known as $E(3)$, the three-dimensional Euclidean group). These two cases are more complicated because the gauge group is non-compact and much less is known about how to handle Chern–Simons theory for non-compact groups.[12]

[10] Admittedly a Euclidean metric is less appropriate for modeling the universe we live in, but then again we are already working in $(2 + 1)$-dimensions!

[11] The special pseudo-orthogonal group (or indefinite special orthogonal group) $SO(p,q)$ is the set of $(p + q) \times (p + q)$ matrices M such that $M^T = \eta M^{-1}\eta$ where η is the diagonal matrix of p entries of 1 and q entries of -1.

[12] In fact, in Bar-Natan and Witten [1991] the authors note that "naive imitation of the usual formulas for compact gauge groups... leads to formulas that are wrong or unilluminating."

Returning to the case of Minkowski (Lorentzian) metrics, we can again consider Λ positive, negative, or zero. For $\Lambda = 0$, the gauge group is the $(2 + 1)$-dimensional Poincaré group (otherwise known as $ISO(2,1)$, the isometry group of $(2 + 1)$-dimensional Minkowski spacetime). For $\Lambda > 0$ the gauge group becomes $SO(3,1)$, whereas for $\Lambda < 0$ the gauge group is $SO(2,2) = SO(2,1) \times SO(2,1)$. Note the similarities of these groups to their Euclidean counterparts. All of these cases with Minkowski metrics also suffer from being non-compact.

Of the three possibilities with Minkowski metrics, the one that is most likely to result in a well-defined theory of quantum gravity appears to be the case of $\Lambda < 0$ (see the discussion in Witten [2007]). In this case again one ends up with a doubled Chern–Simons theory—the product of $SO(2,1)_k$ with $SO(2,1)_{-k}$ where the level is $k = 1/(4G\sqrt{-\Lambda})$ with G being Newton's constant.

More details of the relationship between $(2 + 1)$-dimensional general relativity and Chern–Simons theory are provided in the further reading.

Chapter Summary

- $(2 + 1)$-dimensional quantum gravity is a TQFT. Much of the interest in TQFTs comes from this connection.

Further Reading:

- For a huge amount of information on $(2 + 1)$-dimensional quantum gravity, see Carlip [2005].
- The relationship of $(2 + 1)$-dimensional gravity to Chern–Simons theory was first developed by Ana Achúcarro and Paul Townsend [1986].
- The relationship was further developed by Witten [1988a].
- Years later, the question was revisited by Witten [2007], where doubt is raised as to whether Chern–Simons theory is sufficient to fully describe gravity in $(2 + 1)$-dimensions.
- A (potentially biased) history of various approaches to quantum gravity is given by Rovelli [2000].
- Reviews of loop quantum gravity are given by Rovelli [2008] and Nicolai et al. [2005].
- Discussions of discretization approaches to quantum gravity are given by Regge and Williams [2000] and Lorente [2006].
- The article by Nicolai and Peeters [2007] covers the connections between the loop and discretization approach fairly clearly.

Note that none of these references is particularly easy to digest!

Defining Topological Quantum Field Theory[1]

We already have a rough picture of a topological quantum field theory (TQFT) as a quantum theory that depends on topological properties as opposed to depending on geometric properties. For example, it matters that particle 1 traveled around particle 2, but it doesn't matter how far apart they are.

We can formalize these ideas by saying that the theory should be independent of small deformations of the spacetime metric. We might say that

$$\frac{\delta}{\delta g_{\mu\nu}}\langle \text{any correlator}\rangle = 0 \quad .$$

This is a completely valid way to define a TQFT, but is often not very useful.

Another way to define a $(2 + 1)$-dimensional TQFT is that it is a set of rules that takes an input of a labeled link embedded in a closed 3-manifold[2] and gives an output of a complex number in a way that is invariant under smooth deformations. This definition is quite analogous to our definition of a knot invariant, with two key differences. First, we allow for the lines to be labeled with a "particle type" (and our rules for evaluating the end result will depend on the particular particle type labels). Secondly, the link can be embedded in some arbitrarily complicated 3-manifold.[3] This type of mapping (see Fig. 7.1) is precisely the sort of thing that one gets as an output of Chern–Simons theory, which we called $Z(\mathcal{M}, \text{links})$ as we discussed in Section 5.2. The advantage of thinking in this language is that, strictly speaking, the functional inte-

[1]Many students find this chapter frighteningly abstract. While this chapter sets the stage for a number of ideas that come later, it can also be skipped to a large extent if it seems too difficult. While it may seem a bit cruel to include such a chapter early in the book, I've included it here because it gives the best definition of what a TQFT actually is—which, in one form or another, is what we are studying for the remainder of the book.

[2]Particularly condensed matter physicists might start to wonder why we need to discuss arbitrary, and potentially bizarre sounding, three-dimensional manifolds—what could they possibly have to do with real physical systems? However (besides just being a beautiful mathematical digression), pursuing this direction allows us to understand some of the strong constraints on topological models and their mathematical structure, and this turns out to be important for the analysis of even fairly simple physical systems.

[3]We may also allow world lines of anyons to fuse into other species, as discussed in Section 4.2.

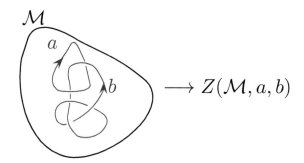

Fig. 7.1 A $(2 + 1)$-dimensional TQFT takes an input of a labeled link in a closed manifold and produces an output of a complex number in a manner which is topologically invariant.

grals of Chern–Simons theory are often not well defined mathematically. Instead, here we bypass the Chern–Simons field theory altogether and define a TQFT simply as a mapping from a manifold with a link to an output.

A more formal definition of TQFTs is given by a set of axioms by Atiyah [1988],[4] which are in some sense much more informative.

7.1   Paraphrasing of Atiyah's Axioms

In this section I give a rough interpretation of Atiyah's axioms of TQFT, suitable for physicists. To begin with, we consider spacetime manifolds with no particles in them. As we have found above, TQFTs are nontrivial even in the absence of any particles. Later on in Section 7.2 we discuss adding particles and moving them around in spacetime too.

We consider a $(D + 1)$-dimensional spacetime manifold,[5,6] which we call $\mathcal{M}$, and a D-dimensional manifold Σ contained in $\mathcal{M}$. We can often think of Σ as being a slice at a fixed time. Almost always we will be thinking of $D = 2$, although the axioms are quite general and can be applied to any D.

Axiom 1: Associated to each D-dimensional manifold Σ (slice of $\mathcal{M}$) is a Hilbert space $V(\Sigma)$ that depends only on the topology[7] of Σ.

We call the space V, which stands for vector space, although sometimes people call it H for Hilbert space.

As an example of what we mean, we have seen that if Σ is a torus, there is a nontrivial Hilbert space coming from the ground-state degeneracy.[8] This degenerate space is the space $V(\Sigma)$. The space $V(\Sigma)$ will depend on the particular anyon theory we are considering. For example, in the case of abelian anyons in Section 4.3, we found a degeneracy of m for a system on a torus with statistical angle $\theta = \pi p/m$.

Note that when we add particles to the system (we will do this in Section 7.2), if the particles are nonabelian, then there will also be a Hilbert space associated with the additional degeneracy that comes with such nonabelian particles.

Axiom 2: The Hilbert space associated with the disjoint union of two D-dimensional manifolds Σ_1 and Σ_2 is the tensor product of the Hilbert spaces associated with each manifold.[9] That is,

$$V(\Sigma_1 \cup \Sigma_2) = V(\Sigma_1) \otimes V(\Sigma_2) \quad .$$

In particular this means that the vector space associated with the null or empty manifold $\varnothing$ must be just the complex numbers. Let us state this mathematically.

Axiom 2 Implies:

$$V(\varnothing) = \mathbb{C} \quad .$$

[4]Sir Michael Atiyah, a Fields medalist who went to primary school in Sudan, was one of the foremost mathematicians of the twentieth century. He specialized in geometry and topology—particularly at the interface between mathematics and physics. You can find videos of him talking about life, physics, and mathematics at webofstories.com.

[5]While it is possible to define certain TQFTs on non-orientable manifolds it is much easier to assume that all manifolds will be orientable—excluding things like Möbius strips and Klein bottles. See Section 40.1.

[6]All manifolds we consider are smooth and compact, possibly with boundaries.

[7]The phrases "depends only on the topology" is something that physicists would say, but mathematicians would not. To a mathematician, topology describes things like whether sets contain their limit points, whether points are infinitely dense, and so forth. Perhaps it would be better to just say that $V(\Sigma)$ does not change under continuous deformation of Σ. This is something mathematicians and physicists would both agree on, and this is what we actually mean here!

[8]The Hilbert space here is the Hilbert space of the manifold in the absence of any particles being present.

[9]This may sound a bit abstract, but it is exactly how the Hilbert spaces of any two systems must combine together. For example, in the case of two spins, the Hilbert space of the union of the two spins is the tensor product of the two Hilbert spaces of the individual spins.

The reason this must be true is because $\varnothing \cup \Sigma = \Sigma$ and $\mathbb{C} \otimes V(\Sigma) = V(\Sigma)$ so the result follows.[10]

Axiom 3: If $\mathcal{M}$ is a $(D+1)$-dimensional manifold with D-dimensional boundary[11] $\Sigma = \partial\mathcal{M}$, then we associate a *particular* element of the vector space $V(\Sigma)$ with this manifold. We write[12,13]

$$Z(\mathcal{M}) \in V(\partial\mathcal{M})$$

where the association (i.e. which particular state in the vector space is chosen) again depends only on the topology of $\mathcal{M}$.

Here we might think of $\partial\mathcal{M}$ as being the space-like slice of the system at a fixed time, and $V(\partial\mathcal{M})$ as being the possible Hilbert space of ground states. The rest of $\mathcal{M}$ (the interior, not the boundary) is the spacetime history of the system, and $Z(\mathcal{M})$ is the particular wavefunction that is picked out by this given spacetime history (see Fig. 7.2). Drawing pictures of these manifolds is a bit of a problem as we point out in the figure and detail in footnote 14.

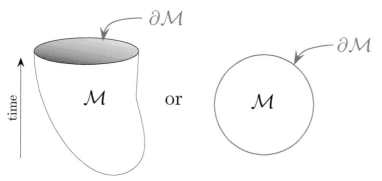

Fig. 7.2 Two depictions of a spacetime manifold $\mathcal{M}$ with boundary $\partial\mathcal{M}$. As per the **caution** in footnote 14, both depictions are problematic. The left depiction is problematic because the only boundary of the manifold is supposed to be the red top surface $\partial\mathcal{M}$ (the black outline of $\mathcal{M}$ is not a boundary, but is just to suggest some manifold attached to the boundary). The right depiction is more accurate, but it depicts a two-dimensional $\mathcal{M}$ and one-dimensional $\partial\mathcal{M}$, which is fairly trivial.

The point of this axiom 3 is to state that the particular wavefunction of a system $Z(\mathcal{M})$, which is chosen from the available vector space, depends on the spacetime history of the system. We have seen this principle before several times. For example, we know that if a particle–antiparticle

[10]If this sounds confusing, remember the space $\mathbb{C}$ of complex scalars is the same as the space of length 1 complex vectors, and tensoring a length n vector with a length m vector gives a size n by m matrix, so tensoring a vector of length n with a length 1 vector gives back a vector of length n.

[11]We use the ∂ to denote boundary. See Section 40.1.4. Note that if $\Sigma = \partial\mathcal{M}$ then $\partial\Sigma = \varnothing$, the empty manifold.

[12]This object Z is a bit tricky. $Z(\mathcal{M})$ can be a scalar as it is in Eq. 7.1 if $\mathcal{M}$ has no boundary, or it can be an element of a vector space as it is here if $\mathcal{M}$ has a boundary. It can even play the role of an operator as in Eq. 7.3 depending on the orientations assigned to the multiple boundaries of $\mathcal{M}$. In math we say that Z is actually a *functor* and what sort of thing it gives as an output depends on the details of the $\mathcal{M}$ it is given as an input. See also Section 7.5.

[13]In analogy with Chern–Simons theory, $Z(\mathcal{M})$ is sometimes known as the *partition function*.

[14]**Caution:** We are mainly interested in three-dimensional manifolds, and it is not really possible to draw most three-dimensional manifolds (particularly when such manifolds cannot even be embedded in $\mathbb{R}^3$). You should therefore view many of the diagrams drawn in this chapter as being suggestive cartoons, but not completely accurate. A manifold without boundary (as in the right-hand side of Fig. 7.4) is sketched with black lines. These black lines are not meant to be a boundary of the manifold! Some "holes" may be sketched to indicate that there may be non-contractible cycles. In this Section (and next section) all boundaries will be marked in red (as in Fig. 7.2). In the figures the red regions themselves have boundaries—this a shortcoming of our depiction; boundaries do not have boundaries.

pair is taken around a nontrivial cycle, this changes which wavefunction we are looking at—this process would be part of the spacetime history.

Axiom 3 Implies: For $\mathcal{M}$ closed, we have $\partial\mathcal{M} = \varnothing$, the empty manifold, so

$$Z(\mathcal{M}) \in \mathbb{C} \qquad (7.1)$$

that is, the TQFT must assign a manifold a topological invariant which is a complex number. This is exactly what we found from Chern–Simons theory.

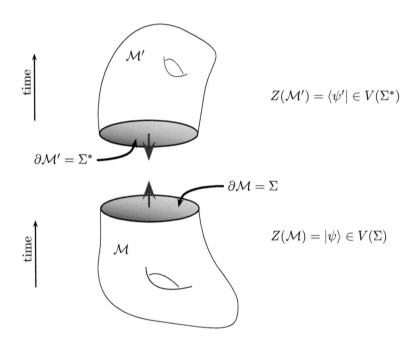

$$Z(\mathcal{M}') = \langle\psi'| \in V(\Sigma^*)$$

$$\partial\mathcal{M}' = \Sigma^*$$

$$\partial\mathcal{M} = \Sigma$$

$$Z(\mathcal{M}) = |\psi\rangle \in V(\Sigma)$$

Fig. 7.3 In this picture $\mathcal{M}$ and $\mathcal{M}'$ are meant to fit together since they have a common boundary but with opposite orientation, $\Sigma = \partial\mathcal{M} = \partial\mathcal{M}'^*$. Here, $\langle\psi'| = Z(\mathcal{M}') \in V(\Sigma^*)$ lives in the dual space of $|\psi\rangle = Z(\mathcal{M}) \in V(\Sigma)$. Note that the normals are oppositely directed.

[15]The words "gluing" and "sewing" are used almost interchangeably.

[16]The notation $\mathcal{M} \cup_\Sigma \mathcal{M}'$ means the union of $\mathcal{M}$ and $\mathcal{M}'$ glued together along the common boundary Σ.

[17]Note that $\mathcal{M}' \cup_\Sigma \mathcal{M}$ and $\mathcal{M} \cup_\Sigma \mathcal{M}'$ are the same manifold but with the opposite orientations. The expressions may look like they should give the same result, but they are not symmetric since we have defined $\Sigma = \partial\mathcal{M}$ rather than $\Sigma = \partial\mathcal{M}'$. From this relationship we can conclude a further result that $Z(\mathcal{M}^*) = Z(\mathcal{M})^\dagger$.

Axiom 4: Reversing orientation

$$V(\Sigma^*) = V^*(\Sigma)$$

where by Σ^* we mean the same surface with reversed orientation, whereas by V^* we mean the dual space—that is, we turn kets into bras. It is a useful convention to keep in mind that the orientation of the normal of $\partial\mathcal{M}$ should be pointing out of $\mathcal{M}$. See Fig. 7.3.

Gluing:[15] If we have two manifolds $\mathcal{M}$ and $\mathcal{M}'$ that have a common boundary $\partial\mathcal{M} = (\partial\mathcal{M}')^*$ we can glue these two manifolds together by taking inner products of the corresponding states as shown in Fig. 7.4. Here, we have $\Sigma = \partial\mathcal{M} = (\partial\mathcal{M}')^*$ so we can glue together the two manifolds along their common boundary to give[16],[17]

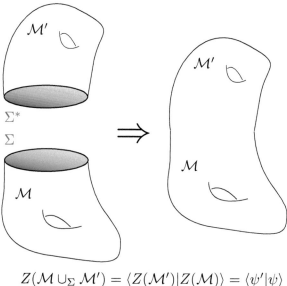

$$Z(\mathcal{M} \cup_\Sigma \mathcal{M}') = \langle Z(\mathcal{M}')|Z(\mathcal{M})\rangle = \langle \psi'|\psi\rangle$$

Fig. 7.4 Gluing two manifolds together by taking the inner product of the wavefunctions on their common, but oppositely oriented, boundaries.

$$Z(\mathcal{M} \cup_\Sigma \mathcal{M}') = \langle Z(\mathcal{M}')|Z(\mathcal{M})\rangle \quad . \tag{7.2}$$

Cobordism: Two manifolds Σ_1 and Σ_2 are called "cobordant" if, after reversing the orientation of one of the manifolds, their disjoint union is the boundary of a manifold $\mathcal{M}$.

$$\partial \mathcal{M} = \Sigma_1^* \cup \Sigma_2 \quad .$$

We say that $\mathcal{M}$ is a cobordism between Σ_1 and Σ_2. See Fig. 7.5 for an example. The mathematical structure of cobordisms provides a very powerful abstract method of defining topological quantum field theories, which we describe very briefly in Section 7.5.

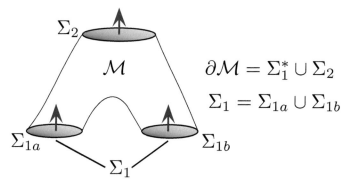

$$\partial \mathcal{M} = \Sigma_1^* \cup \Sigma_2$$
$$\Sigma_1 = \Sigma_{1a} \cup \Sigma_{1b}$$

Fig. 7.5 $\mathcal{M}$ is a cobordism between Σ_1 and Σ_2. That is, $\partial \mathcal{M} = \Sigma_1^* \cup \Sigma_2$. The orientation arrow of Σ_1^* points *out* of $\mathcal{M}$ since it is part of ∂M, hence, the orientation arrow for Σ_1 points *into* $\mathcal{M}$.

When we have a cobordism $\mathcal{M}$ between Σ_1 and Σ_2 we have $Z(\mathcal{M}) \in V(\Sigma_1^*) \otimes V(\Sigma_2)$, so that we can write

$$Z(\mathcal{M}) = \sum_{\alpha\beta} U^{\alpha\beta} |\psi_{\Sigma_2,\alpha}\rangle \otimes \langle\psi_{\Sigma_1,\beta}| \quad , \tag{7.3}$$

where $|\psi_{\Sigma_2,\alpha}\rangle$ is the basis of states for $V(\Sigma_2)$ and $\langle\psi_{\Sigma_1,\beta}|$ is the basis of states for $V(\Sigma_1^*)$. We can thus think of the cobordism $\mathcal{M}$ as being an evolution.[18] For example, the evolution direction in Fig. 7.5 is vertical.

One can view manifolds with multiple boundaries as cobordisms in many different ways. For example, consider the manifold shown in Fig. 7.6 where $\partial\mathcal{M} = \Sigma_1 \cup \Sigma_2 \cup \Sigma_3$. We can choose any subset (including improper subsets) of the disjoint boundaries as the "incoming" boundary set and the remaining boundaries as the "outgoing" boundary set. Three examples are as follows (writing "cob" for a cobordism):

(1) $\mathcal{M}$ is cob between $(\Sigma_1 \cup \Sigma_2)^*$ (incoming) and Σ_3 (outgoing)

$$Z(\mathcal{M}) = \sum_{\alpha\beta\gamma} U^{\alpha\beta\gamma} |\psi_{\Sigma_3,\gamma}\rangle \otimes \left[\langle\psi_{\Sigma_1^*,\beta}| \otimes \langle\psi_{\Sigma_2^*,\alpha}|\right] \quad . \tag{7.4}$$

(2) $\mathcal{M}$ is cob between Σ_1^* (incoming) and $\Sigma_2 \cup \Sigma_3$ (outgoing)

$$Z(\mathcal{M}) = \sum_{\alpha\beta\gamma} U^{\alpha\beta\gamma} \left[|\psi_{\Sigma_2,\alpha}\rangle \otimes |\psi_{\Sigma_3,\gamma}\rangle\right] \otimes \langle\psi_{\Sigma_1^*,\beta}| \quad . \tag{7.5}$$

(3) $\mathcal{M}$ is cob between $\varnothing$ (incoming) and $(\Sigma_1 \cup \Sigma_2 \cup \Sigma_3)$ (outgoing)

$$Z(\mathcal{M}) = \sum_{\alpha\beta\gamma} U^{\alpha\beta\gamma} \left[|\psi_{\Sigma_1,\beta}\rangle \otimes |\psi_{\Sigma_2,\alpha}\rangle \otimes |\psi_{\Sigma_3,\gamma}\rangle\right] \tag{7.6}$$

with $\varnothing$ being the empty manifold.

So, which is the right expression for $Z(\mathcal{M})$? In fact they are all correct! There is an isomorphism between the basis of kets $|\psi_\Sigma, \alpha\rangle$ and the basis of bras $\langle\psi_{\Sigma^*}, \alpha|$. When we glue manifolds with boundary together (say $\mathcal{M}$ to $\mathcal{M}'$ as in Fig. 7.4), it is important that we take inner products by gluing bras to kets. But it does not matter if in $Z(\mathcal{M})$ the wavefunction is written as a bra or a ket as long as in $Z(\mathcal{M}')$ we have the corresponding ket or bra to glue it to.

Identity Cobordism: If we have $\mathcal{M} = \Sigma \times I$ where I is the one-dimensional interval (we could call it the 1-disk, D^1 also) then the boundaries are Σ and Σ^* (see Fig. 7.7), and the cobordism implements a map between $V(\Sigma)$ and $V(\Sigma)$. Since the interval can be topologically contracted to a point (or "infinitesimal thickness"), we can take this map to be the identity:

$$Z(\Sigma \times I) = \sum_{\alpha} |\psi_{\Sigma,\alpha}\rangle \otimes \langle\psi_{\Sigma,\alpha}| = \text{identity}, \tag{7.7}$$

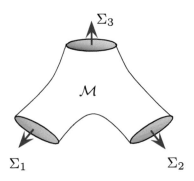

Fig. 7.6 This manifold may be viewed as a cobordism in several ways by choosing which boundary surfaces are "incoming" and which are "outgoing."

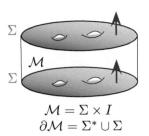

$$\mathcal{M} = \Sigma \times I$$
$$\partial\mathcal{M} = \Sigma^* \cup \Sigma$$

Fig. 7.7 A cobordism that can be topologically contracted to infinitesimal thickness acts as the identity on the Hilbert space $V(\Sigma)$.

where the sum is over the entire basis of states of $V(\Sigma)$.

We can now consider taking the top of the interval I and gluing it to the bottom to construct a closed manifold $\mathcal{M} = \Sigma \times S^1$, where S^1 means the circle (or 1-sphere), as shown in Fig. 7.8. We then have

$$Z(\Sigma \times S^1) = \text{Tr}\left[Z(\Sigma \times I)\right] = \text{Dim}[V(\Sigma)] \quad , \qquad (7.8)$$

where Tr means trace. Thus, we obtain the dimension of the Hilbert space $V(\Sigma)$, or in other words, the ground-state degeneracy of the 2-manifold Σ.

As we have discussed in Section 4.3, for the torus T^2 we have

$$\text{Dim } V(T^2) = \text{number of particle species}, \qquad (7.9)$$

which we argued (at least for modular abelian anyon models) based on non-commutativity of taking anyons around the nontrivial cycles of the torus, and we will justify for nonabelian anyons as well in Section 7.2.1. Similarly, for a 2-sphere S^2, we have

$$\text{Dim } V(S^2) = 1 \qquad (7.10)$$

since there are no non-contractible loops, and this will also hold for both abelian and nonabelian theories. See Section 4.3.1 for discussion of the ground-state degeneracy of abelian theories on higher genus surfaces.

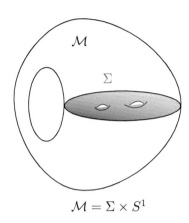

Fig. 7.8 Gluing the top of $\Sigma \times I$ to the bottom we obtain $\mathcal{M} = \Sigma \times S^1$. An important fact is that $Z(\Sigma \times S^1)$ is just the ground-state degeneracy of the 2-manifold Σ

7.2 **Adding Particles**

We now consider extending the ideas of TQFT to spacetime manifolds with particle world lines in them.[19]

Let us imagine that there are different anyon types that we can label as a, b, c, and so forth. The corresponding anti-anyons are labeled with overbar $\bar{a}$, $\bar{b}$, and so forth as in Section 4.2.2. We now imagine a 2-manifold with some marked and labeled points as shown in Fig. 7.9. We call the combination of the 2-manifold with the marked points Σ for brevity. As with the case without particles (axiom 1, in Section 7.1), Σ is associated with a Hilbert space $V(\Sigma)$. The dimension of this Hilbert space depends on the number and type of particles in the manifold (we expect for nonabelian particles, the dimension will grow exponentially with the number of particles). We can span the space $V(\Sigma)$ with some basis states $|\psi_\alpha\rangle$ which will get rotated into each other if we move the marked points around within the manifold (i.e. if we braid the particles around each other).

Similarly a 3-manifold $\mathcal{M}$ is now supplemented with labeled links indicating the world lines of the particles. The world lines should be directed unless the particles are their own antiparticles. The world lines are allowed to end on the boundary of the manifold $\partial\mathcal{M}$. See lower left of Fig. 7.10. Analogously we may sometimes call the combination of the manifold with its world lines $\mathcal{M}$, although sometimes we will write this

[19]For dimension $D > (2+1)$ TQFTs we could have world sheets of moving strings and other higher-dimensional objects as well.

Fig. 7.9 A 2-manifold with particles in it, which are marked and labeled points. We now call the combination (the manifold and the marked points) Σ for brevity.

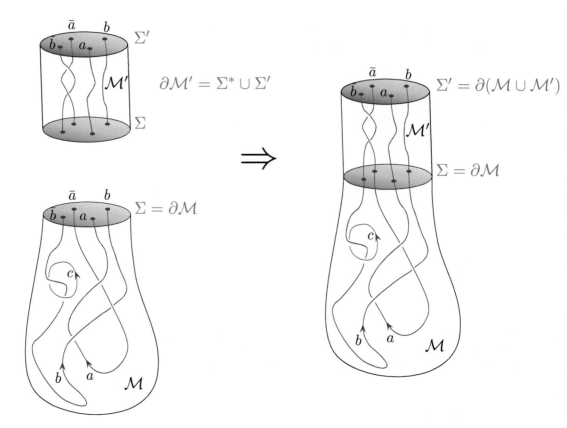

Fig. 7.10 Lower left: A 3-manifold $\mathcal{M}$ with particles in it, which are marked and labeled lines (the lines should be directed unless the particle is its own antiparticle). These world lines may end on the boundary $\Sigma = \partial\mathcal{M}$. The wavefunction on the boundary $\partial\mathcal{M}$ is determined by the spacetime history given by $\mathcal{M}$. Another manifold $\mathcal{M}'$ evolves the positions of the particles in time. Note that by $\mathcal{M}'$ we mean not just the manifold, but the manifold along with the world lines in it. In this particular picture $\Sigma = \Sigma'$ being the same surface with the same types of particles at the same positions.

as $\mathcal{M}; L$ where L indicates the "link" (or knot) of the world lines.

As in the discussion of axiom 3, the spacetime history specifies exactly which wavefunction

$$|\psi\rangle = Z(\mathcal{M}) \in V(\partial\mathcal{M})$$

is realized on the boundary $\Sigma = \partial\mathcal{M}$. If a basis of $V(\partial\mathcal{M})$ is given by wavefunctions $|\psi_a\rangle$ then we can generally write the particular wavefunction $|\psi\rangle$ in this basis

$$|\psi\rangle = \sum_\alpha c_\alpha |\psi_\alpha\rangle \quad .$$

We can now think about how we would braid particles around each other. To do this we glue another manifold $\mathcal{M}'$ to $\partial\mathcal{M}$ to continue the time evolution, as shown in the right of Fig. 7.10. The final wavefunction is written as

$$|\psi'\rangle = Z(\mathcal{M} \cup \mathcal{M}') \in V(\Sigma') \quad .$$

If we put the positions of the particles in Σ' at the same positions as the particles in Σ, then the Hilbert space $V(\Sigma')$ is the same as $V(\Sigma)$, and we can write $|\psi'\rangle$ in the same basis as $|\psi\rangle$

$$|\psi'\rangle = \sum_\alpha c'_\alpha |\psi_\alpha\rangle \quad .$$

We can then think of $Z(\mathcal{M}')$ as giving us a unitary transformation on this Hilbert space—which is exactly what we think of as nonabelian statistics. We can write explicitly the unitary transformation

$$Z(\mathcal{M}') = \sum_{\alpha\beta} U^{\alpha\beta} |\psi_{\Sigma',\alpha}\rangle \otimes \langle\psi_{\Sigma,\beta}| \quad .$$

Thus the gluing operation shown in Fig. 7.10 can be stated as

$$|\psi'\rangle = |Z(\mathcal{M} \cup \mathcal{M}')\rangle = Z(\mathcal{M}') \, |Z(\mathcal{M})\rangle = Z(\mathcal{M}')|\psi\rangle$$

or equivalently

$$c'_\alpha = \sum_\beta U^{\alpha\beta} c_\beta \quad .$$

Note that if the particles stay fixed in their positions (or move in topologically trivial ways) then $\mathcal{M}'$ can be contracted to infinitesimal thickness and we can think of the unitary transformation as being the identity. As with the identity cobordism discussed in Section 7.1, we can take such an identity transformation, glue the top to the bottom, and obtain

$$Z(\Sigma \times S^1) = \mathrm{Dim}[V(\Sigma)] \quad , \tag{7.11}$$

that is, the partition function Z is just the dimension of the Hilbert space of the wavefunction. This holds true even when Σ has marked points, or particles, in it.

7.2.1 Particles or No Particles

In the same way that the ground state of a topological system "knows" about the types of anyons that can exist in the system, it is also the case that the TQFT in the absence of particles actually carries the same information as in the presence of particles.[20] To see this, consider a manifold $\mathcal{M}$ with labeled and directed world lines L_i in them, as shown in Fig. 7.11. Now consider removing the world lines along with a hollow tubular neighborhood surrounding the paths that the world lines follow as shown in the figure. We now have a manifold with a solid torus removed for each world line loop. (Think of a worm having eaten a path out of the manifold.) In this configuration, the boundary $\partial\mathcal{M}$ of the manifold $\mathcal{M}$ now contains the surface of these empty tubes—that is, the surface of a torus T^2 for each world line loop. Note that the empty tube is topologically a solid torus $D^2 \times S^1$ even if the world line forms some knot.[21] The statement that it forms a nontrivial knot is a statement about the embedding of the S^1 loop in the manifold.

Note that the Hilbert space of the torus surface T^2 is in one-to-one

[20]Up to here our discussion has been applicable to TQFTs in any dimension. From here on we specialize to the most interesting case of $D = 2$, that is $(2 + 1)$-dimensions.

[21]D^2 is the usual notation for a two-dimensional disk and S^1 again is the circle.

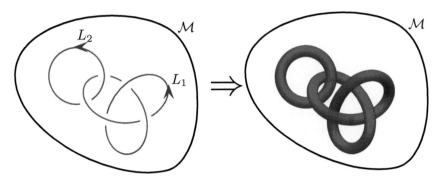

Fig. 7.11 Removing the world lines on the left along with a thickened tube. Imagine a worm burrowing along the path of the world lines and leaving a hollow hole (colored red).

correspondence with the particle types that can be put around the handle of the torus. Indeed, each possible state $|\psi_a\rangle$ of the torus surface corresponds to a picture like that of Fig. 7.12, where a particle of type a goes around the handle. We can think of this solid torus manifold $D^2 \times S^1$ as being a spacetime history where $t = -\infty$ is the central core of the solid torus (the circle that traces the central line of the jelly filling of the donut) and the torus surface is the present time. Somewhere between $t = -\infty$ and the time on the surface of the torus, a particle of type a has been dragged around the handle. Obviously, gluing such a solid torus containing a particle world line (Fig. 7.12) back into the empty solid-torus-shaped tube (right of Fig. 7.11) recovers the original picture of labeled world lines following these paths (left of Fig. 7.11).

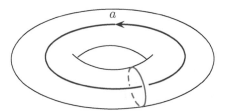

Fig. 7.12 The possible wavefunctions $|\psi_a\rangle$ that we can have on the surface of the torus can be realized by having a world line of a particle of type a going around the handle of the torus. We can call these $Z(\text{solid torus with } a \text{ running around handle}) = |\psi_a\rangle$.

The partition function Z of the manifold with the tori excised from it (the right of Fig. 7.11) contains all of the information necessary to determine the partition function for the left of Fig. 7.11 for *any* particle types that we choose to follow the given world lines. For the manifold on the right there are two surfaces (the two surfaces bounding the holes left where we excised the two tori), so we have

$$Z(\mathcal{M}) = \sum_{i,j} Z(\mathcal{M}; i, j) \, \langle \psi_{L1,i}| \otimes \langle \psi_{L2,j}| \quad,$$

where $Z(\mathcal{M}; i, j)$ is the partition function for the torus with two particle types i, j following the two world line loops L_1 and L_2, and the two wavefunctions are the corresponding boundary condition. Thus, if we want to extract $Z(\mathcal{M}; a, b)$, where the particle lines are labeled with a, b we simply glue in the wavefunction $|\psi_{L1,a}\rangle \otimes |\psi_{L2,b}\rangle$ representing the boundary condition on the two surfaces.

7.3   Building Simple 3-Manifolds

7.3.1 S^3 and the Modular S-matrix

We will now consider building up 3-manifolds from pieces by gluing objects together using the gluing axiom from Section 7.1. The simplest 3-manifold to assemble is the three sphere S^3. Remember that S^3 can be thought of as $\mathbb{R}^3$ compactified with a single point at infinity (the same way that S^2 is a plane, closed up at infinity—think of stereographic projection. See the discussion in Section 40.1). Recall also that a solid torus should be thought of as a disk crossed with a circle $D^2 \times S^1$. I claim that we can assemble S^3 from two solid tori[22]

$$S^3 = (S^1 \times D^2) \cup_{T^2} (D^2 \times S^1) \quad.$$

The notation here is that the two pieces $S^1 \times D^2$ and $D^2 \times S^1$ are joined together on their common boundary which is T^2 (the torus surface).

There is a very elegant proof of this decomposition. Consider the 4-ball B^4. Topologically we have[23]

$$B^4 = D^2 \times D^2 \quad.$$

Now applying the boundary operator ∂ and using the fact that the boundary operator obeys the Leibniz rule (i.e. it distributes like a derivative), we have

$$\begin{aligned} S^3 = \partial B^4 \;\; &= \;\; \partial(D^2 \times D^2) = (\partial D^2 \times D^2) \cup (D^2 \times \partial D^2) \\ &= \;\; (S^1 \times D^2) \cup_{T^2} (D^2 \times S^1) \end{aligned}$$

where we have used the fact that the boundary of a disk is a circle, $\partial D^2 = S^1$. The two solid tori here are glued together along their common $T^2 = S^1 \times S^1$ boundary. To see this note that

$$\partial(S^1 \times D^2) = S^1 \times S^1 = \partial(D^2 \times S^1) \quad.$$

These two solid tori differ from each other in that they have the opposite S^1 (from the $S^1 \times S^1$) filled in to form a D^2.

[22]If you are rusty on these elementary topology manipulations, see the review in Section 40.1

[23]Topologically it is easiest to think about the n-dimensional ball, B^n, as being the interval $I = B^1$ raised to the nth power. The disk (or 2-ball), is topologically a filled-in square $D^2 = B^2 = I \times I$. The usual 3-ball is topologically a cube $B^3 = I \times I \times I$. The 4-ball is topologically a 4-cube $B^4 = I \times I \times I \times I = D^2 \times D^2$.

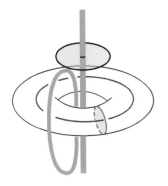

Fig. 7.13 Assembling two solid tori to make S^3. The obviously drawn torus $D^2 \times S^1$ can be thought of as the red disk D^2 crossed with the blue circle S^1. The remainder of space outside of this torus, including the point at infinity is the other solid torus $S^1 \times D^2$. For this "outside" solid torus, the S^1 can be thought of as the vertical green line. This line becomes S^1 by connecting up with itself at the point at infinity. The upper shaded disk is an example of a contractible D^2 that is contained entirely within the outside solid torus. Note that the entire outside solid torus is $S^1 \times D^2$, the vertical green line crossed with disks topologically equivalent to this one. The green loop off to the side (also contained within the outside torus), like the vertical green S^1 loop, is not contractible within the outside solid torus, but can be deformed continuously to the vertical green loop.

[24] Be warned that there is some disagreement in the literature as to which way the arrows on a and b are oriented in the definition of S_{ab}. Some references will have S_{ab} equal to what we have defined as $S_{a\bar{b}}$ for example. Our convention matches that of Kitaev [2006], but disagrees with Bonderson [2007] for example.

[25] Some comments on the S-matrix: (1) Since a linking b is topologically the same as b linking a we should have $S_{ab} = S_{ba}$; (2) While this is not obvious at the moment, it is also true that $S_{\bar{a}b} = [S_{ab}]^*$ where $\bar{a}$ is the antiparticle of a. See Exercise 17.3.

[26] Here we are assuming the theory is *modular* meaning there are no transparent particles. This assumption will be discussed in more depth in Section 17.3.

The two tori are glued together meridian-to-longitude and longitude-to-meridian (i.e. the contractible direction of one torus is glued to the non-contractible direction of the other, and vice versa). A sketch of how the two solid tori are assembled together to make S^3 is given in Fig. 7.13.

Let us think about the partition function Z of these two solid tori, which are glued together on their boundaries to make up S^3. We write the partition function as the overlap between wavefunctions on the outside and inside tori:

$$Z(S^3) = \langle Z(S^1 \times D^2)|Z(D^2 \times S^1)\rangle = \langle \psi_{outside}|\psi_{inside}\rangle$$

where the ψ's are the wavefunctions on the surface of the torus.

We can further consider including world lines around the non-contractible loops of the solid torus, as in Fig. 7.12. There is a different state on the surface of the torus for each particle type we have running around the handle. We then assemble S^3 with these new solid tori and get an S^3 with two particle world lines linked together as shown in Fig. 7.14. Gluing the two tori together we get[24]

$$Z(S^3; a \text{ loop linking } b \text{ loop}) = \langle Z(S^1 \times D^2; b)|Z(D^2 \times S^1; a)\rangle \equiv S_{ab} \;. \tag{7.12}$$

This quantity S_{ab} is known as the **modular S-matrix**, and it is a very important quantity in topological theories as we shall see in Chapter 17.[25]

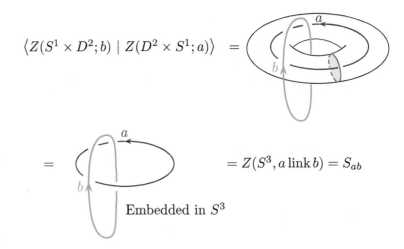

Fig. 7.14 Here we assemble a partition function for S^3 with world lines of a linking b embedded in the S^3. To do this we glue together two solid tori each with a world line running around the handle. The green line marked b runs around the handle of the "outside" torus. The end result is known as the modular S-matrix, and it gives a basis transformation between the two bases, which both span the Hilbert space of the torus surface where the two solid tori are glued together.

Note that the S-matrix is unitary,[26] since it is simply a basis transformation between the two sets of wavefunctions, which both span the vector space $V(T^2)$ of the torus surface T^2 where the two solid tori are glued together. Note also that the element S_{00}, corresponding to the element of the S-matrix where the vacuum particle (no particle at all!) is put around both handles. (Here we are using 0 to mean the vacuum.) This tells us that

$$Z(S^3) = S_{00} \leq 1 \tag{7.13}$$

and in fact, should be strictly less than one unless there are no nontrivial particle types and S is a one-dimensional (scalar) matrix.

Another way of viewing the S-matrix is as a simple link between two strands, as shown in Fig. 7.14. As with the Kauffman bracket invariant, we can construct a set of diagrammatic rules to give a value to knots. In Chapters 8–16 we construct diagrammatic rules to help us "evaluate" knots like this. These rules will be somewhat similar to the rules for the Kauffman bracket invariant, only now we need to keep track of labels on world lines as well.

7.3.2 $S^2 \times S^1$

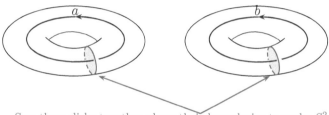

Sew these disks together along their boundaries to make S^2.

Fig. 7.15 Assembling two solid tori to make $S^2 \times S^1$. Here the two contractible disks D^2 are sewed together along their boundaries to make S^2.

There is another way we can put two solid tori together to make a closed manifold.[27] Instead of attaching longitude-to-meridian and meridian-to-longitude, we instead attach meridian-to-meridian and longitude-to-longitude. (This is perhaps a simpler way to put together two solid tori!) See Fig. 7.15. Here we claim that[28]

$$S^2 \times S^1 = (D^2 \times S^1) \cup_{T^2} (D^2 \times S^1) \quad .$$

The gluing together is again done along the common boundary $T^2 = S^1 \times S^1$. The S^1 factors in both solid tori are the same, and both of the D^2 have the same S^1 boundary. Thus we are sewing together two disks D^2 along their S^1 boundaries to make a 2-sphere S^2 (imagine cutting a sphere in half along its equator and getting two disks which are the north and south hemispheres).

As in the previous case, we can put world lines through the handles of the solid tori if we want. If we do so we have[29]

$$\langle Z(D^2 \times S^1; b) \, | \, Z(D^2 \times S^1; a) \rangle = \delta_{ab} \quad .$$

[27] In fact there are an infinite number of ways two tori can be glued together to form a closed manifold. These are discussed in detail in Section 7.4.

[28] One should be warned that $S^2 \times S^1$ cannot be embedded in the usual three-dimensional space $\mathbb{R}^3$, so visualizing it is very hard!

[29] It is worth considering how the world lines, in the case where $a = b$, are positioned within the $S^2 \times S^1$. Each S^2 is assembled from two D^2 disks, one from each $D^2 \times S^1$, sewn together. Each world line around the handle of the torus cuts a D^2 exactly once. But the D^2 disks are sewed together with opposite orientation so the two world lines cut through the sphere in opposite directions. This fits with the principle that a non-zero amplitude of two particles on the surface of a sphere can only occur if the two particles are a particle–antiparticle pair. This is discussed in Section 8.4. Note that each world line can intersect the S^2 exactly once because this S^2 is not the boundary of any 3-manifold.

The reason it is a delta function is that both the bra and ket are really the same wavefunctions (we have not switched longitude to meridian). So except for the conjugation we should expect that we are getting the same basis of states for both tori.

In particular, in the case where we put no particle (the vacuum) around both handles, we have (i.e. $a = b = I = 0$)

$$\langle Z(D^2 \times S^1) | Z(D^2 \times S^1) \rangle = \delta_{ab} = 1 \quad .$$

So we have the result

$$Z(S^2 \times S^1) = 1 \quad . \tag{7.14}$$

Note that this agrees with two of our prior statements. On the one hand Eq. 7.11 says that Z for any two-dimensional manifold crossed with S^1 should be the dimension of the Hilbert space for that manifold; on the other hand, Eq. 7.10 states that the dimension of the Hilbert space on a sphere is 1.

7.3.3 Connected Sums

An important notion in topology is the idea of a connected sum of two manifolds $\mathcal{M}$ and $\mathcal{M}'$, denoted by $\mathcal{M}\#\mathcal{M}'$, which is equal to $\mathcal{M}'\#\mathcal{M}$. To form the connected sum of two manifolds we delete a small ball from each manifold and glue together the two manifolds at the revealed spherical surfaces.[30] Note that for D-dimensional manifolds a D-dimensional sphere acts as the identity for the connected sum

$$\mathcal{M}^D \# S^D = \mathcal{M}^D \quad . \tag{7.15}$$

The reason for Eq. 7.15 is that S^D can be assembled from two balls B^D attached along their S^{D-1} surfaces.[31] So the connected sum in Eq. 7.15 is just deleting B^D from the manifold $\mathcal{M}^D$ and replacing it with the B^D that results from deleting B^D from S^D.

Here we will specialize to the case where the manifolds are three-dimensional[32] and we will derive a simple formula for $Z(\mathcal{M}\#\mathcal{M}')$.

First, let us define $\mathcal{M}\backslash B$ to be the manifold $\mathcal{M}$ with a small ball B removed. Similarly let $\mathcal{M}'\backslash B'$ be the manifold $\mathcal{M}'$ with a small ball B' removed. Gluing together $\mathcal{M}\backslash B$ with $\mathcal{M}'\backslash B'$ results in

$$\mathcal{M}\#\mathcal{M}' = (\mathcal{M}\backslash B) \cup_{\partial(\mathcal{M}\backslash B)} (\mathcal{M}'\backslash B') \quad .$$

Invoking Eq. 7.2 thus gives us

$$Z(\mathcal{M}\#\mathcal{M}') = \langle Z(\mathcal{M}\backslash B) \, | \, Z(\mathcal{M}'\backslash B') \rangle \quad . \tag{7.16}$$

Now we can reconstruct Z for each manifold by gluing the pieces back together

$$Z(\mathcal{M}) = \langle \, Z(\mathcal{M}\backslash B) \, | \, Z(B) \rangle \tag{7.17}$$

$$Z(\mathcal{M}') = \langle \, Z(B') \, | \, Z(\mathcal{M}'\backslash B') \rangle \quad . \tag{7.18}$$

[30]For example, the connected sum of two two-dimensional torus surfaces is a two-handled torus surface:

$$T^2 \# T^2 = \text{two handle torus}.$$

In pictures,

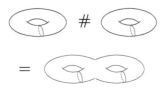

[31]For example, the usual two-dimensional sphere S^2 can be assembled from two B^2 discs (also known as D^2 disks) sewed together along their S^1 circle boundaries.

[32]In three dimensions there is a notion of a prime decomposition of a manifold. A three-dimensional manifold $\mathcal{M}$ is prime if $\mathcal{M} = \mathcal{N}\#\mathcal{P}$ implies either $\mathcal{N} = S^3$ or $\mathcal{P} = S^3$. Any orientable compact 3-manifold has a decomposition into primes

$$\mathcal{M} = P_1 \# P_2 \# \dots P_N$$

which is unique up to reordering the terms and connected sums with S^3 (see Milnor [1962]). In $D = 4$ such decompositions are not unique. In $D = 2$ decompositions of orientable manifolds are unique and the only prime manifold is the torus.

Now it is crucial that the boundaries of these two manifolds $\mathcal{M}\backslash B$ and $\mathcal{M}'\backslash B'$ are spheres S^2 and the Hilbert space on this sphere is one-dimensional, as pointed out in Eq. 7.10. For states $|a\rangle, |b\rangle, |c\rangle, |d\rangle$ in a one-dimensional Hilbert space we have the relationship[33]

$$\langle a|b\rangle\langle c|d\rangle = \langle a|d\rangle\langle c|b\rangle$$

so we have

$$\langle Z(\mathcal{M}\backslash B) \,|\, Z(\mathcal{M}'\backslash B')\rangle\langle Z(B')|Z(B)\rangle \qquad (7.19)$$
$$= \langle\, Z(\mathcal{M}\backslash B) \,\,|\, Z(B)\,\rangle\langle\, Z(B') \,|\, Z(\mathcal{M}'\backslash B')\rangle \quad .$$

Finally, we realize that when we glue together two three-dimensional balls B and B' along their S^2 surfaces we obtain the manifold S^3 (see comment below Eq. 7.15), so that

$$\langle Z(B')|Z(B)\rangle = Z(S^3) \quad . \qquad (7.20)$$

Using Eqs. 7.16, 7.17, 7.18, and 7.20 in Eq. 7.19 we obtain our final result

$$Z(\mathcal{M}\#\mathcal{M}') = \frac{Z(\mathcal{M})Z(\mathcal{M}')}{Z(S^3)} \quad . \qquad (7.21)$$

We can check that this equation is compatible with Eq. 7.15 by setting $\mathcal{M}' = S^3$ for example.

[33]Witten [1989] refers to this relationship as a "wonderful fact."

7.4   Appendix: Gluing Solid Tori Together

While this discussion is a bit outside the main train of thought (being the development of TQFTs), it is interesting to think about the different ways two solid tori may be glued together to obtain a closed manifold.

A solid torus is written as $D^2 \times S^1$. We define the meridian m to be the S^1 boundary of any D^2. That is, the meridian is a loop on the surface around the contractible direction of the solid torus. We define the longitude l as being any loop around the surface of the solid torus that intersects a meridian at one point. This definition unfortunately has some (necessary) ambiguity. A line that loops around the meridian n times as it goes around the non-contractible direction of the torus, is just as good a definition of a longitude (an example of this is Fig. 7.16, which is $n = 1$). We call this line $l + nm$ where n is the number of times it goes around the meridian and l was the original definition of the longitude that did not loop around the meridian. Redefining the longitude this way is known as a "Dehn twist."

Let us choose a meridian m_1 on the surface of one solid torus and choose to sew it to the line $-qm_2 + pl_2$ of the second solid torus (that is, the line that goes p times around the longitude and $-q$ times around meridian, we make $-q$ negative so that the two tori surfaces are opposite oriented for attaching them together. Once the two lines are glued together this uniquely defines how the rest of the two torus surfaces

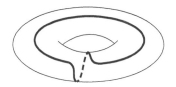

Fig. 7.16 A line that wraps both the longitude and meridian of the torus.

are glued together. The resulting object is known as the "Lens space" $L(q, p)$. In Section 7.3.1 we showed that $L(0, 1) = S^3$ and in Section 7.3.2 we showed that $L(1, 0) = S^2 \times S^1$. Note that due to the ambiguity of definition of the longitude of the torus $-qm_2 + pl_2$, under redefinition of the longitude, goes to $(-q - np)m_2 + pl_2$. Thus $L(q + np, p) = L(q, p)$, and in particular, $L(1, 1) = S^3$ also.

7.5 Appendix: Cobordisms and Category Theory

While I have promised not to resort to the use of category theory, I briefly break this promise in order to describe how many mathematicians view topological quantum field theories. The reason I think this is worth mentioning here is so the interested reader has some minimal exposure to a few of the key notions.

First, we should mention that (roughly) a category is a mathematical structure that has *objects* and *morphisms*. A morphism is a map (or arrow) between objects. Two categories that we need areL (1) the category known as **Vect**, whose objects are vector spaces and whose morphisms are linear maps between vector spaces; and (2) the category known as **Cob**$(D + 1)$, whose objects are closed D-dimensional manifolds (what we called Σ) and whose morphisms are ($D + 1$-dimensional) cobordisms between these manifolds.

A *functor* is a mapping between two categories that preserves certain structures.[34] In particular it should map objects to objects and morphisms to morphisms.

Mathematicians often define a topological quantum field theory to be a functor

$$Z : \mathbf{Cob}(D + 1) \to \mathbf{Vect} \quad .$$

Each manifold (object in **Cob**) is mapped to a vector space (object in **Vect**). Disjoint unions of manifolds become tensor products of vector spaces. Cobordisms between manifolds are mapped to linear maps between vector spaces.

[34]The term *functor* is also used in computer science in functional programming in essentially the same sense.

Chapter Summary

- The Atiyah axioms formalize the idea of a topological quantum field theory.
- A TQFT calculates the "partition function" of a manifold $Z(\mathcal{M})$. If $\mathcal{M}$ has no boundary, then Z is a number. If $\mathcal{M}$ has a boundary then Z is a wavefunction that lives on the boundary. Inner products of wavefunctions glue together manifolds.
- We discussed gluing together manifolds.
- Particle world lines can be added into manifolds.

Further Reading:

- For discussion on the Atiyah axioms, see Atiyah [1988, 1997]

- A discussion of gluing together manifolds (as in Sections 7.3–7.4) is given by Rolfson [1976] or Saveliev [2012]. The book Farb and Margalit [2012] may also be useful. We discuss this type of sewing further in Chapter 24.

- For category theory, a classic reference is Mac Lane [1971], which was written long before the idea of topological quantum field theory was around. A more modern and not too difficult book on the topic is Leinster [2014]. A beautiful Master's thesis by Bartlett [2005] discusses TQFTs from the category perspective.

Exercises

Exercise 7.1 Knot Connected Sums and Manifold Connected Sums
    Consider a manifold $\mathcal{M}$ containing a knotted world line K_a where the quantum number on the world line is a. Similarly, consider a manifold $\mathcal{M}'$ containing a knotted world line K'_a where the quantum number on the world line is a. Generalize the argument of Section 7.3.3 to show the following:
    (a) If we take the connected sum of these two manifolds where the knots are unchanged we have

$$Z(\mathcal{M}\#\mathcal{M}'; K_a \cup K'_a) = \frac{Z(\mathcal{M}; K_a)Z(\mathcal{M}'; K'_a)}{Z(S^3)} \quad .$$

(This part is pretty trivial. The knots just go along for the ride!)
    (b) If we connect the two knots together in a knot sum $K_a\#K'_a$, when we connect the manifolds show that we obtain

$$
\begin{aligned}
Z(\mathcal{M}\#\mathcal{M}'; K_a\#K'_a) &= \frac{Z(\mathcal{M}; K_a)Z(\mathcal{M}'; K'_a)}{Z(S^3 \text{ with an unknotted loop of } a \text{ in it})} \\
&= [S_{0a}]^{-1} Z(\mathcal{M}; K_a)Z(\mathcal{M}'; K'_a) \quad .
\end{aligned}
$$

The definition of a knot sum is given in Exercise 2.6. In short, one breaks open the knot K_a and breaks open K'_a and connects them together to make a single knot.

Part II

Anyon Basics

Anyon basics

Fusion and Structure of Hilbert Space

As discussed in Section 7.1, each two-dimensional surface (a slice of a three-dimensional spacetime manifold) has an associated Hilbert space. In the case where there are particles on this surface, the dimension of the Hilbert space will reflect the nature of the particles. We now seek to understand the structure of this Hilbert space and how it depends on the particles. At the same time we will build up a diagrammatic algebra with the goal of constructing a mapping from world lines of particles to complex numbers (a definition of a TQFT as given in Fig. 7.1). We briefly introduced graphical notation in Section 4.2.1 and we continue that development here. For those who prefer more mathematical detail, in Section 8.6 (as well as in Chapter 12) we introduce tensor description of diagrams and the associated Hilbert spaces.

8.1 Basics of Particles and Fusion—The Abelian Case

Particle types

There should be a finite set of labels, which we call particle types. For now, let us call them a, b, c, etc.

Fusion

World lines can merge which we call fusion, or do the reverse, which we call splitting. If an a particle merges with b to give c, we write $a \times b = b \times a = c$. Fusion and splitting are shown diagrammatically in Fig. 8.1. Sometimes colloquially we call both diagrams "fusion."

It should be noted that we can think of two particles as fusing together even if they are not close together. We only need to draw a circle around both particles and think about the "total" particle type inside the circle. For example, we sometimes draw pictures as shown in Fig. 8.2.

In our charge–flux model of abelian anyons (see Section 4.2), if the statistical angle is $\theta = \pi p/m$ (p and m relatively prime and not both odd) then we have m species $a = (aq, a\Phi)$ for $a = 0 \ldots m - 1$, where $q\Phi = \pi p/m$. The fusion rules are simply addition modulo m. That is, $a \times b = (a + b) \bmod m$.

Fig. 8.1 Fusion (left) and splitting (right) diagrams can be thought of as part of a spacetime history of the particles. If we are describing two separated particles a and b whose overall quantum number is c (sometimes we say "overall fusion channel" is c), we would describe the ket for this state using the right-hand picture—which we can think of as a spacetime description of how the current situation (a on the left b on the right) came about (with time going up). In contrast the left figure represents a single particle c, which came about by fusing a and b. Details of the formal meaning of these diagrams in terms of bras and kets are given in Section 8.6 and Chapter 12.

Fig. 8.2 Another notation to describe the fusion of two particle types to make a third $a \times b = c$. The two particles need not be close to each other. This figure is equivalent to the right of Fig. 8.1.

[1] It is annoying that we have so many different ways to express the identity, but in different contexts different notations seem natural. For example, if our set of particles is fusing by addition (as we discussed in the charge–flux model) the identity should be 0. But if our group fuses by multiplication, the identity is more naturally 1. See note 5 in Chapter 40.

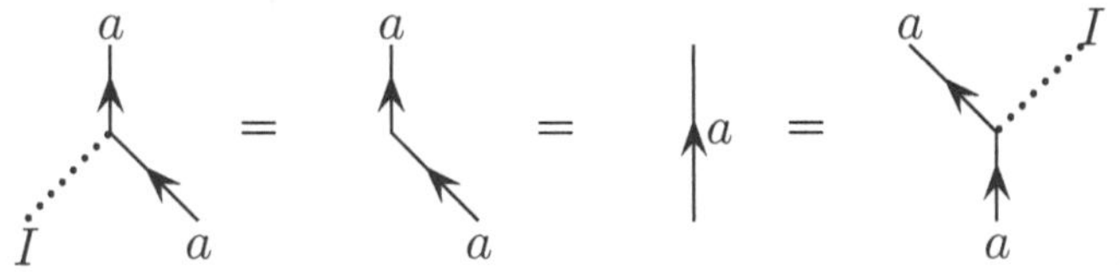

Identity (or Vacuum)

Exactly one of the particles should be called the identity or vacuum. We write this[1] as 1 or 0 or I or e. The identity fuses trivially

$$a \times I = a$$

for any particle type a. In the charge–flux model (Section 4.2) we should think of the identity as being no charge and no flux. Fusion with the identity is depicted schematically in Fig. 8.3. Often we do not draw the identity particle at all, being that it is equivalent to the absence of any (nontrivial) particle.

Fig. 8.3 Fusion of a particle with the identity $a \times I = a$. The dotted line indicates the identity. In some of these pictures the a particle appears to move slightly to the left. However, this is not important for topological properties since the path can be deformed continuously to a particle that does not move.

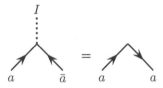

Fig. 8.4 A particle going forward should be equivalent to an antiparticle going backward.

Antiparticles

Each particle a should have a unique antiparticle, which we denote as $\bar{a}$. The antiparticle is defined by $a \times \bar{a} = I$. (There should only be one particle which fuses with a given a to give the identity!) A particle going forward in time should be equivalent to an antiparticle going backward in time as shown in Fig. 8.4. Fusion to the identity can be thought of as a particle turning around in spacetime as shown in Fig. 8.5.

A particle may be its own antiparticle, in which case we do not need to draw arrows on its world lines. An example of this in our charge–flux model from Section 4.2 would be the "2" particle (fusion of two elementary anyons, see Section 4.3) in the case of $\theta = \pi/4$. Also, the identity particle I is always its own antiparticle.

Fig. 8.5 Fusion of an anyon with its anti-anyon to form the identity can be thought of as a particle turning around in spacetime. On the right, we have reversed one arrow and changed $\bar{a}$ to a, and we have not drawn the identity line.

8.2 Multiple Fusion Channels—The Nonabelian Case

For the nonabelian theories as we have discussed previously (for example in Section 3.5), the dimension of the Hilbert space must increase with the number of particles present. How does this occur? In nonabelian models we have multiple possible orthogonal fusion channels

$$a \times b = c + d + \dots \tag{8.1}$$

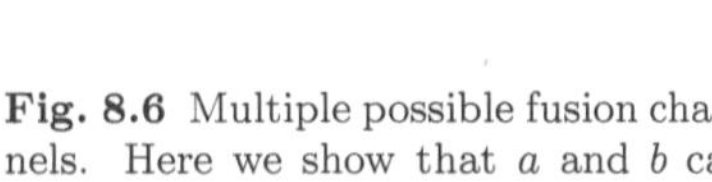

Fig. 8.6 Multiple possible fusion channels. Here we show that a and b can fuse together to give either c or d or other possible results.

meaning that a and b can come together to form either c or d or ..., as shown in Fig. 8.6. A theory is nonabelian if at least one of the fusion equations of the theory has multiple possible fusion channels (i.e. has more than one particle listed on the right-hand side of Eq. 8.1).

If there are s possible fusion channels for $a \times b$, then the two particles a and b have an s-dimensional Hilbert space (part of what we called $V(\Sigma)$).

What is this Hilbert space associated with multiple fusion channels? A slightly imperfect analogy is that of angular momentum addition. We know the rule for adding spin $1/2$,

$$\frac{1}{2} \otimes \frac{1}{2} = 0 \oplus 1$$

which tells us that two spin $1/2$'s can *fuse* to form a singlet or a triplet. As with the case of spins, we can think about the two particles being in a wavefunction such that they fuse in one particular fusion channel or the other—even if the two particles are not close together. The singlet or $J = 0$ state of angular momentum is the identity here: it has no spin at all. The analogy with spins is not exact though—unlike the case of spins, the individual particles have no internal degrees of freedom (analogous to the two states in the spin $1/2$ case), nor do any results of fusion have an m_z degree of freedom (like a triplet would).

Locality

The principle of locality is a predominant theme of anyon physics (if not of physics altogether).

The quantum number (or "charge") of a particle is locally conserved in space. Consider, for example, Fig. 8.7. On the left, a particle a propagates along and suddenly something complicated happens locally. If only a single particle comes out of this region it must also be a particle of type a. (If two particles come out of this region, we could have a split into two other species as in the right of Fig. 8.1). We sometimes call this the **no-transmutation** principle. It allows us to conclude that the complicated picture on the left of Fig. 8.7 must be equal to some constant times the simple propagation of an a particle as shown on the right.

If two particles (maybe far away from each other) fuse together to some overall particle type (in a case where multiple fusion channels are available) it is not possible to determine this fusion channel by measuring only one of the initial particles. In order to determine the fusion channel of the two particles, you have to do an experiment that involves both of the initial particles. For example, one can perform an interference measurement that surrounds both of these particles. We would say that the fusion channel is "local to the pair."

Similarly, if we have some particles, b and c, and they fuse to d (see Fig. 8.8), no amount of braiding b around c will change this overall fusion channel d. The fusion channel is *local* to the pair. If these two then fuse with a to give an overall fusion channel f, no amount of braiding a, b, and c will change the overall fusion channel f. However, if a braids

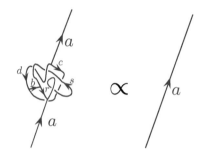

Fig. 8.7 If a particle a goes into a spacetime region, then a net particle charge a must come out. This is also sometimes called the "no-transmutation" principle. From far away, one can ignore any local processes (up to an overall constant).

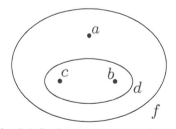

Fig. 8.8 In this picture b and c fuse to d. Then this d fuses with a to give an overall fusion channel of f. No amount of braiding b around c will change the fact that the two of them fuse to d. However, if we braid a with b and c, this can change the fusion of b with c subject to the constraint that the fusion of all three particles will give f.

with b and c, then the fusion of b and c might change, subject to the constraint that the overall channel of all three particles remains f.

Locality gives another important way in which anyons differs from the fusion of spins. With spins, if you can measure two spins individually you can (at least sometimes) determine the fusion channel of the spins. For anyons you must be able to measure a loop that *surrounds* both anyons in order to determine their collective fusion channel—measuring each anyon individually does not tell you the fusion of the two!

Antiparticles in the Case of Multiple Fusion Channels

When we have multiple fusion channels (i.e. for nonabelian theories) we define antiparticles via the principle that a particle *can* fuse with its antiparticle to give the identity, although other fusion channels may be possible:

$$a \times \bar{a} = I + \text{ other fusion channels.}$$

As in the abelian case we use the overbar notation to indicate an antiparticle. It should be the case that for each particle a there is a unique particle that can fuse with it to give the identity, and we call this particle $\bar{a}$. As in the abelian case, a particle may be its own antiparticle if $a \times a = I +$ other fusion channels, in which case we do not put an arrow on the line corresponding to the particle.

8.2.1   Example: Fibonacci Anyons

Perhaps the simplest nonabelian example is the anyon system known as Fibonacci[2] anyons. Something very close to this is thought to occur in the so-called $\nu = 12/5$ quantum Hall state, which we comment on in Chapter 37. Fibonacci anyons are closely related to the $SU(2)_3$ Chern–Simons theory.[3]

In this example the particle set includes only two particles, the identity I and a nontrivial particle which is often called τ.

$$\text{Particle types} = \{I, \tau\}.$$

The fusion rules are

$$
\begin{aligned}
I \times I &= I \\
I \times \tau &= \tau \\
\tau \times \tau &= I + \tau \ .
\end{aligned}
\tag{8.2}
$$

The first two of these rules hardly need to be written down (they are implied by the required properties of the identity). It is the final rule that is nontrivial. This final rule also implies that τ is its own antiparticle $\tau = \bar{\tau}$ which means we do not need to put arrows on world lines.

With two Fibonacci anyons the Hilbert space is two-dimensional, since the two particles can fuse to I or τ, as shown in Fig. 8.9.

With three Fibonacci anyons the Hilbert space is three dimensional,

[2]Fibonacci, also known as Leonardo of Pisa, was born around 1175 AD. Perhaps his most important contribution to mathematics is that he brought Arabic numerals (or Hindu–Arabic numerals) to the western world. The Fibonacci sequence $1, 1, 2, 3, 5, 8, 13, \ldots$ is named after him, although it was known in India hundreds of years earlier!

[3]Fibonacci anyons can be described exactly by the G_2 level 1 Chern–Simons theory. This involves a messy Lie algebra called G_2. The $SU(2)_3$ Chern–Simons theory contains some additional particles besides the Fibonacci particles, but ignoring these, it is the same as Fibonacci.

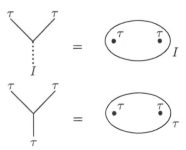

Fig. 8.9 Two different notations for the two different fusion channels of two Fibonacci anyons.

as shown in Fig. 8.10. The key thing to notice is that if the first two particles fuse to τ, then this combination acts as being a single particle of overall charge τ—it can fuse with the third τ in two ways.

There is a single state in the Hilbert space of three anyons with overall fusion channel I. This state is labeled as[4] $|N\rangle$. As shown in Fig. 8.8, due to locality, no amount of braiding among the three particles will change this overall fusion channel (although braiding may introduce an overall phase).

There are two states in the Hilbert space of three anyons with overall fusion channel τ. These are labeled $|1\rangle$ and $|0\rangle$ in Fig. 8.10. Again, as mentioned in Fig. 8.8, due to locality, no amount of braiding amongst the three particles will change the *overall* fusion channel of all three particles from τ. Furthermore, since in these two basis states the first two particles furthest left are in an eigenstate (either I in state $|0\rangle$ or τ in state $|1\rangle$) no amount of braiding of the first (leftmost) two particles will change that eigenstate from $|0\rangle$ to $|1\rangle$ or from $|1\rangle$ to $|0\rangle$. However, as we will see in Section 10.1, if we braid the second particle with the third, we can then change the quantum number of the first two particles and rotate between $|0\rangle$ and $|1\rangle$.

[4]Here $|N\rangle$ stands for "non-computational," since it is not used in many quantum computing protocols that use Fibonacci anyons.

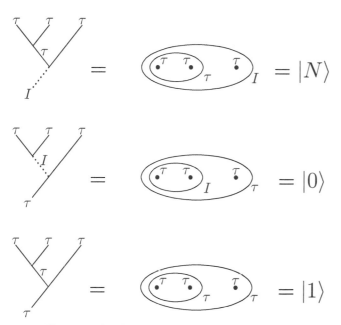

Fig. 8.10 Notations for the three different orthogonal fusion channels of three Fibonacci anyons. The notation $|N\rangle, |1\rangle$, and $|0\rangle$ are common notations for those interested in topological quantum computing with Fibonacci anyons!

Let us now try to figure out the Hilbert space dimension for four Fibonacci anyons. First, we fix a branching structure for our fusion (or more properly splitting) tree. For example, we might consider the tree structure shown in Fig. 8.11. Labeling all of the top branches with τ we need only figure out what all the possible labelings of the remaining

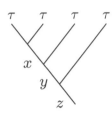

Fig. 8.11   Fusing four Fibonacci anyons in a tree structure.

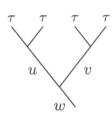

Fig. 8.12   Fusing four Fibonacci anyons in a different tree structure.

[5] Again, the $\sim$ means "scales as."

[6] The name Ising is used here due to the relationship with the Ising conformal field theory, which describes the Ising model in two dimensions at its critical point.

[7] The fusion rules of Ising and $SU(2)_2$ are the same, but there are some spin factors which differ, as well as a Frobenius–Schur indicator—see Sections 14.2 and 18.4.

[8] Another common notation is to use ϵ instead of ψ in the Ising theory. In $SU(2)_2$ the particles I, σ, ψ may be called $0, 1/2, 1$ or $0, 1, 2$.

segments of the tree are (the labels x, y, z in the figure). Using the Fibonacci fusion rules (Eq. 8.2) we know that (a) x can be I or τ; (b) if x is I then y must be τ, and if x is τ then y can be I or τ; (c) if y is I then z must be τ, but if y is τ then z can be I or τ. This leads to five possibilities for the triple (x, y, z) given by $(I, \tau, I), (I, \tau, \tau), (\tau, I, \tau), (\tau, \tau, I), (\tau, \tau, \tau)$.

One might wonder what would happen had we chosen a different tree structure. For example, with four Fibonacci anyons we might have instead chosen the tree shown in Fig. 8.12. Counting the possible labelings of the segments marked $u, v,$ and w and we find that there are again five possibilities consistent with the allowed fusion rules: $(u, v, w) = (I, I, I), (I, \tau, \tau), (\tau, I, \tau), (\tau, \tau, I), (\tau, \tau, \tau)$. So whether we use the fusion tree of Fig. 8.11 or Fig. 8.12 we obtain the same Hilbert space dimension: the order we fuse the anyons together does not change the Hilbert space dimension. Indeed, as we discuss in Section 8.3.1 and particularly Chapter 9, the different trees are different descriptions of the same Hilbert space.

We have now seen for our Fibonacci system, with two particles the Hilbert space is two dimensional. With three particles the Hilbert space is three dimensional. With four particles the Hilbert space is five dimensional, and then with five particles the space is eight dimensional (try showing this!), and so forth. This pattern continues following the Fibonacci sequence, hence, the name "Fibonacci anyons."

Since the Nth element of the Fibonacci sequence for large N is approximately[5]

$$\text{Dim of Hilbert space of } N \text{ Fib anyons} = \text{Fib}_N \sim \left(\frac{1 + \sqrt{5}}{2} \right)^N \qquad (8.3)$$

We say that the *quantum dimension* of this particle is $d_\tau = (1 + \sqrt{5})/2$, the golden mean (see Eq. 3.11).

8.2.2   Example: Ising Anyons

The Ising[6] anyon system is extremely closely related to $SU(2)_2$ Chern–Simons theory,[7] and this general class of anyons is believed to be realized in the $\nu = 5/2$ quantum Hall state, topological superconductors, and other so-called Majorana systems (see Chapter 37).

The Ising theory has three particle types:[8]

$$\text{Particle types} = \{I, \sigma, \psi\}.$$

The nontrivial fusion rules are

$$
\begin{aligned}
\psi \times \psi &= I \\
\psi \times \sigma &= \sigma \\
\sigma \times \sigma &= I + \psi
\end{aligned}
\qquad (8.4)
$$

where we have not written the outcome of any fusion with the identity, since the outcome is obvious. Again, each particle is its own antiparticle $\psi = \bar\psi$ and $\sigma = \bar\sigma$, so we need not put arrows on any world lines.

Fusion of anything with the ψ particle always gives a unique result on the right-hand side. We thus call ψ an abelian particle (despite the fact that the full theory is nonabelian), or we say that ψ is a *simple current* (precise definition in next section). Fusion of many ψ particles is then easy, since each pair fuses to the identity in only one way.

Fusion of many σ particles, however, is nontrivial. The first two σ's can either fuse to I or ψ, but then when the third is included the overall fusion channel must be σ (since fusing σ with either ψ or I gives σ). Then adding a fourth σ to this cluster whose overall fusion channel is σ again gives two possible outcomes. Such a fusion tree is shown in Fig. 8.13.

By counting possible labelings of the tree, we find that the total number of different fusion channels for N particles of type σ is $2^{\lfloor N/2 \rfloor}$ (where $\lfloor N/2 \rfloor$ means we should round down if $N/2$ is not an integer). To see this result in another way, corresponding to the fusion tree structure in Fig. 8.14, we can group σ particles together in pairs where each pair gives either ψ or I, so two σ particles comprises a two-state system, or a qubit. Then the I's and ψ's fuse together in a unique way. Since the Hilbert space dimension is $(\sqrt 2)^N$ the quantum dimension of the σ particle is $\mathsf{d}_\sigma = \sqrt 2$ (see Eq. 3.11).

8.3 **Fusion and the N matrices**

We are well on our way to fully defining an anyon theory. A theory must have a finite set of particles, including a unique identity I, with each particle having a unique antiparticle.

The general fusion rules can be written as

$$a \times b = \sum_c N_{ab}^c \, c \ , \tag{8.5}$$

where the sum over c runs over all particle types in the theory. Here, N_{ab}^c are non-negative integers known as the fusion multiplicities. N_{ab}^c is zero if a and b cannot fuse to c. N_{ab}^c is one if we have $a \times b = c + \dots$, and c only occurs once on the right-hand side. If c occurs more than once on the right-hand side, then N_{ab}^c counts the number of times it occurs.

What does it mean that a particle type can occur more than once in the list of fusion outcomes? It simply means that the fusion result can occur in multiple orthogonal ways[9] in which case a diagram with a vertex showing a and b fusing to c should also contain an index $\mu \in \{1, \dots, N_{ab}^c\}$ at the vertex indicating which of the possible c fusion channels occurs, as shown in Fig. 8.15. A more formal way to describe N_{ab}^c is to say that it is the Hilbert space dimension associated with anyons a and b fusing to total quantum number c.

For most simple anyon theories N_{ab}^c is either 0 or 1 (and in particular N_{ab}^c is never greater than 1 for any abelian anyon theories). We will not

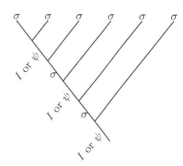

Fig. 8.13 The fusion tree for many σ particles in the Ising anyon theory. The total Hilbert space dimension is eight corresponding to three choices of I or ψ.

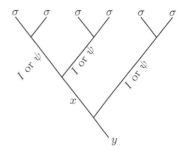

Fig. 8.14 A different fusion tree for many σ particles in the Ising anyon theory. Here the labels x and y are completely fixed by the choices we make of I or ψ in the branches above. Again, the Hilbert space dimension is eight.

[9]While this does not occur for angular momentum addition of $SU(2)$ (and also will not occur in Chern–Simons theory $SU(2)_k$ correspondingly) it is well known among high-energy theorists who consider the fusion of representations of $SU(3)$. In the standard $SU(3)$ notation,

$$8 \otimes 8 = 1 \oplus 8 \oplus 8 \oplus 10 \oplus \overline{10} \oplus 27$$

and the 8 occurs twice on the right.

Fig. 8.15 Multiple fusion channels. In nonabelian theories fusion of a and b to c can occur in multiple orthogonal ways when $N_{ab}^c > 1$. To specify which way they fuse, we add an additional index $\mu \in 1 \dots N_{ab}^c$ at the vertex as shown.

often consider the more complicated case with $N_{ab}^c > 1$ for simplicity, but they are discussed in the chapter appendices for completeness (see Section 9.5.3). It is good to keep in mind that such more complicated cases do exist.

One can use the fusion multiplicity matrices to determine all the possible fusion channels associated with larger groups of anyons. For example, using Eq. 8.5 multiple times we can write

$$
a \times b \times c = \left(\sum_d N_{ab}^d d \right) \times c = \sum_d N_{ab}^d \, (d \times c)
$$

$$
= \sum_{d,e} N_{ab}^d N_{dc}^e \, e \quad .
\tag{8.6}
$$

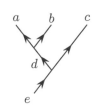

Fig. 8.16 Fusing three particles together.

This corresponds to the fusion tree diagram in Fig. 8.16. The two different factors of N correspond to the two different vertices in the diagram.

One can similarly use the fusion multiplicity matrices for keeping track of the dimension of Hilbert spaces. Here, we simply count all the possible fusion trees

$$
\text{Dim of Hilbert space of } a \times b =
\tag{8.7}
$$

$$
\text{Dim of Hilbert space of } \left(\sum_c N_{ab}^c \, c \right) = \sum_c N_{ab}^c
$$

where we have used Eq. 8.5 in going from the first line to the second.

Similarly, we can find the Hilbert space dimension of larger groups of particles:

$$
\text{Dim of Hilbert space of } a \times b \times c =
\tag{8.8}
$$

$$
\text{Dim of Hilbert space of } \left[\sum_{d,e} N_{ab}^d N_{dc}^e \, e \right] = \sum_{d,e} N_{ab}^d N_{dc}^e
$$

where we used Eq. 8.6 to go from the first line to the second.

Elementary properties of the fusion multiplicity matrices

- Commutativity of fusion $a \times b = b \times a$

$$
N_{ab}^c = N_{ba}^c
\tag{8.9}
$$

- Time reversal

$$
N_{ab}^c = N_{\bar{a}\bar{b}}^{\bar{c}}
\tag{8.10}
$$

- Trivial fusion with the identity

$$
N_{aI}^c = \delta_{ac}
\tag{8.11}
$$

- Existence and uniqueness of inverse

$$
N_{ab}^I = \delta_{a\bar{b}}
\tag{8.12}
$$

A set of particle types with a unique identity and a set of N^c_{ab}s satisfying the axioms listed here is known as a *commutative fusion ring*.[10] All TQFTs are described by commutative fusion rings. However, commutative fusion rings exist that are not consistent with any TQFT.

Some More Useful Definitions

We say a particle a is a *simple current* if for each particle b, the fusion of a and b results in a single particle type c_b as the output: $a \times b = c_b$. Equivalently we can say a is a simple current if for each b there exists one particle type c_b such that $N^{c_b}_{ab} = 1$ and $N^c_{ab} = 0$ for any $c \neq c_b$. Yet another equivalent statement is a is a simple current iff $\sum_c N^c_{ab} = 1$ for each particle b.

There are a number of variants of the notation for N^c_{ab}. Sometimes one will see N^{ab}_c to describe a splitting diagram and N^c_{ab} to describe a fusion diagram (see Fig. 8.1). However, $N^c_{ab} = N^{ab}_c$ so we will almost always use the notation N^c_{ab} which is more common.

It is also possible to define fusion multiplicities for more particles. For example, looking at Eq. 8.6 and Fig. 8.16 we can define a notation

$$N^e_{abc} = \sum_d N^d_{ab} N^e_{dc} \tag{8.13}$$

to indicate the Hilbert space dimension of fusing a, b, and c to total quantum number e. This can be further generalized to consider the fusion of even more particles. For example,

$$N^e_{abcd} = \sum_{fg} N^f_{ab} N^g_{fc} N^e_{gd} \quad .$$

We will not often use this more general notation.

However, a very commonly used notation is

$$N_{abc} \equiv N^I_{abc} = \sum_d N^d_{ab} N^I_{dc} = N^{\bar c}_{ab} \tag{8.14}$$

which is the number of different ways that a, b, and c can fuse to the identity. Here we used Eq. 8.13 in the middle step, and we used the fusion with an inverse to obtain the last expression on the right. An example of this equivalence is shown graphically in Fig. 8.17. The advantage of N_{abc} is that it is fully symmetric in all of its indices: $N_{abc} = N_{bca} = N_{cab} = N_{bac} = N_{acb} = N_{cba}$. For example, using this notation Eq. 8.11 and Eq. 8.12 are actually the same equation (replacing $\bar b$ with c). Further, using Eq. 8.14 along with the symmetry of N_{abc} we can derive identities such as

$$N^c_{ab} = N^{\bar b}_{a\bar c} = N^b_{\bar ac}. \tag{8.15}$$

where in the last step we used Eq. 8.10.

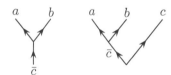

Fig. 8.17 Equivalence of $N^{\bar c}_{ab}$ and N_{abc}. Both vertices have the same fusion multiplicity. Note that the left half of the right picture is exactly equivalent to the left picture: $\bar c$ is entering the vertex from below. On the right, this $\bar c$ "turns over" to become a c going up on the far right. An interpretation of the right picture is that a, b, and c are fusing to the vacuum at the bottom, hence, $N_{abc} = N^I_{abc}$. Note that the vertex at the bottom on the right side (c fusing with $\bar c$ to give the vacuum) could be written as a smooth single line turning over as shown below (analogously to Fig. 4.7).

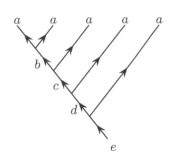

Fig. 8.18 Fusing five type a anyons together.

Fusing Many Anyons

The general strategy for fusing together multiple anyons was introduced in Eqs. 8.6–8.9. Here we use the same strategy, but we consider fusing many anyons of the same type together.

For example, if we are to fuse, say, five anyons of type a together, we can do so via a tree as shown in Fig. 8.18. To find the dimension of the Hilbert space (generalizing Eq. 8.9) we have

$$
\begin{aligned}
\text{Dim of Hilbert space of five } a \text{ anyons} \;&=\; \sum_e N^e_{aaaaa} \\
&=\; \sum_{b,c,d,e} N^b_{aa} N^c_{ba} N^d_{ca} N^e_{da} \\
&=\; \sum_{b,c,d,e} N^b_{aa} N^c_{ab} N^d_{ac} N^e_{ad}
\end{aligned}
$$

where each factor of N corresponds to one of the vertices in the figure. It is convenient to think of the tensor N^c_{ab} as a matrix N_a with indices b and c, that is, we write $[N_a]^c_b$, such that we have

$$
\text{Dim of Hilbert space of five } a \text{ anyons} = \sum_e [(N_a)^4]^e_a \quad .
$$

Similarly, were we to have a larger number p of anyons of type a we would need to calculate $[N_a]^{p-1}$. We recall (see Eq. 3.11) that the quantum dimension d_a of the anyon a is defined via the fact that the Hilbert space dimension should scale as $(\mathsf{d}_a)^p$ where p is the number of a particles fused together. We thus have that[11]

$$
\mathsf{d}_a = \text{largest eigenvalue of } [N_a] \tag{8.17}
$$

which is always real and positive.[12] Note that, given the symmetries of N we have $\mathsf{d}_a = \mathsf{d}_{\bar a}$.

Example of Fibonacci Anyons

The fusion matrix for the τ particle in the Fibonacci theory is

$$
N_\tau = \begin{matrix} & I \;\; \tau \\ \begin{pmatrix} 0 & 1 \\ 1 & 1 \end{pmatrix} & \begin{matrix} I \\ \tau \end{matrix} \end{matrix}
$$

where, as indicated here, the first row and first column represent the identity and the second row and second column represent τ. The first row of this matrix says that fusing τ with the identity gives back τ and the second row says that fusing τ with τ gives I and τ. It is an easy exercise to check that the largest eigenvalue of this matrix is indeed $\mathsf{d}_\tau = (1 + \sqrt{5})/2$, in agreement with Eq. 8.3.

---

[11]To prove this we need to show that elements of the matrix $[N_a]^{p-1}$ are approximately proportional to $(\lambda_1)^p$ for large p where λ_1 is the largest eigenvalue of N_a. Note that the words "scale as" in the previous sentence tell us that we can ignore factors of order unity, and in particular we don't have to worry about the difference between p and $p-1$. To show our result, we write N_a in diagonalized form (see Eq. 8.20)

$$
N_a = U \Lambda U^{-1}
$$

where Λ is a diagonal matrix of the eigenvalues and U is a unitary matrix. Then we have

$$
[N_a]^p = U \Lambda^p U^{-1}. \tag{8.16}
$$

For large p, the largest eigenvalue λ_1 to the pth power is much larger than any other eigenvalue to the pth power, so we can approximate all the other eigenvalues as zero. Then Λ^p becomes a diagonal matrix including $(\lambda_1)^p$ on one entry and all other entries are zero. Plugging this back into Eq. 8.16 tells us that all elements of $[N_a]^p$ can be taken to be proportional to λ_1^p in the large p limit.

[12]In general the eigenvalues of N_a can be complex. However, in Section 17.8 we give a proof that there is always a positive real eigenvalue whose magnitude is greater or equal to that of any other eigenvalue and we can take this as the more precise definition.

Example of Ising Anyons

The fusion matrix for the σ particle in the Ising theory is[13]

$$N_\sigma = \begin{pmatrix} & I & \sigma & \psi \\ 0 & 1 & 0 \\ 1 & 0 & 1 \\ 0 & 1 & 0 \end{pmatrix} \begin{matrix} I \\ \sigma \\ \psi \end{matrix}$$

where the first row and column represent the identity, the second row and column represent σ and the third row and column represent ψ. So, for example, the second row here indicates that $\sigma \times \sigma = I + \psi$. Again, it is an easy exercise to check that the largest eigenvalue of this matrix is $d_\sigma = \sqrt{2}$, as described in Section 8.2.2.

8.3.1 Associativity

It should be noted that the fusion multiplicity matrices N are very special matrices since the outcome of a fusion should not depend on the order of fusion. That is, $(a \times b) \times c = a \times (b \times c)$.

For example, let us try to calculate how many ways $a \times b \times c$ can give an outcome of e (i.e. calculate N^e_{abc}). We can either try fusing $a \times b$ first as on the left of Fig. 8.19 (and also done previously in Eq. 8.6) or we can try fusing b and c first as on the right of Fig. 8.19. Whichever we choose, we are describing the same Hilbert space and we should find the same overall dimension either way. In other words, we should have the same total number of fusion channels. Thus, corresponding to these two possibilities we have the equality[14]

$$\sum_d N^d_{ab} N^e_{cd} = \sum_f N^f_{cb} N^e_{af} \quad . \tag{8.18}$$

Again, thinking of $N^c_{ab} = [N_a]^c_b$ as a matrix labeled N_a with indices b and c, Eq. 8.18 tells us that

$$[N_a, N_c] = 0 \quad . \tag{8.19}$$

Therefore, all of the N matrices commute with each other. In addition the Ns are *normal* matrices, meaning that each N_a commutes with its own transpose (Since $[N_a, N_{\bar{a}}] = 0$ and $N_a = N^T_{\bar{a}}$ by Eq. 8.15). A set of normal matrices that all commute can be simultaneously diagonalized, thus

$$[U^\dagger N_a U]_{xy} = \delta_{xy} \lambda^{(a)}_x \tag{8.20}$$

and all N_as get diagonalized with the same unitary matrix U. Surprisingly (as we see in Section 17.3.1) for well-behaved ("modular"[15] anyon theories) the matrix U is precisely the modular S-matrix we discussed in Eq. 7.12!

[13] For both the Ising and Fibonacci theories, the N_a matrices are symmetric. This is because each particle is its own antiparticle in these simple theories. More generally we have $N^c_{ab} = N^{\bar{b}}_{a\bar{c}}$. So that when $c = \bar{c}$ and $b = \bar{b}$ then $[N_a]^c_b = [N_a]^b_c$. However, if we do not have $c = \bar{c}$ and $b = \bar{b}$ then N_a need not be symmetric.

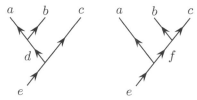

Fig. 8.19 Fusing $(a \times b) \times c$ should be equivalent to $a \times (b \times c)$. On the left a and b fuse to d first then this composite fuses with c to give e. On the right b and c fuse to f first, then this composite fuses with a to give e. Both diagrams represent the same physical Hilbert space. Fixing a, b, c, e the figure on the left spans the Hilbert space with different values of d whereas the figure on the right spans the same space with different values of f.

[14] The left is exactly what we found in Eq. 8.6. Note that we have used $N^e_{cd} = N^e_{dc}$ to rewrite the result very slightly.

[15] For non-modular theories, or even fusion rings which don't even correspond to anyon theories of any type, we can still diagonalize N in the form of Eq. 8.20, and the resulting unitary matrix U is sometimes known as the *mock* S-matrix.

8.4   Application of Fusion: Dimension of Hilbert Space on 2-Manifolds

[16] We are again assuming manifolds are always orientable, so this excludes objects like the Klein bottle or the Möbius strip. Only a subset of TQFTs are well defined in the non-orientable case.

The structure of fusion rules can be used to calculate the ground-state degeneracy of wavefunctions on two dimensional manifolds.[16] Here we again examine the Hilbert space $V(\Sigma)$ where Σ is our 2-manifold, which may or may not have particles in it.

Let us start by considering the sphere S^2, and assume that there are no anyons on the surface of the sphere. As mentioned previously in Eq. 7.11, there is a unique ground state in this situation because there are no non-contractible loops (see Sections 7.1 and 4.3.1). The dimension of the Hilbert space is just 1,

$$\text{Dim } V(S^2) = 1 \quad .$$

This will be the starting point for our understanding. All other configurations (change of topology, adding particles, etc.) will be related back to this reference configuration.

[17] By nontrivial we mean this particle is not the vacuum particle.

Now let us consider the possibility of having a single (nontrivial)[17] anyon on the sphere. In fact, such a thing is not possible because you can only create particles in a way that conserves the overall quantum number. If we start with no particles on the sphere, the total anyon charge must be conserved—that is, everything on the sphere must fuse together to a total quantum number of the identity. Thus, we have

[18] For higher genus surfaces with non-abelian theories it is possible to have a single anyon alone on the surface. An example of this is when $a \times \bar{a} = I + c$. In this case a pair a and $\bar{a}$ may be created, one particle can move all the way around a nontrivial cycle to fuse with its partner, but it may leave behind a single anyon c since some quantum numbers can be changed by the action of moving the anyon around the cycle. If we try this on the sphere (without the handle) we would always find that the pair reannihilates to the vacuum. See further discussion near Eq. 8.22.

$$\text{Dim } V(S^2 \text{ with one (nontrivial) anyon}) = 0 \quad . \tag{8.21}$$

In other words, since only particle–antiparticle pairs can be created from the vacuum, there is no spacetime history that could prepare the state with just a single (non-vacuum) particle on the sphere.[18]

We can however consider the possibility of two anyons on a sphere. We can create an a particle with an $\bar{a}$ particle, and since these two particles must fuse back to the identity in a unique way we have[19]

[19] It is implied that we are counting states here with the particles a and $\bar{a}$ at some given fixed position (all positions being topologically equivalent). If we were to count different positions as different states in the Hilbert space we would have to include this nontopological degeneracy in our counting as well.

$$\text{Dim } V(S^2 \text{ with one } a \text{ and one } \bar{a}) = 1 \quad .$$

The two particles must be antiparticles of each other, otherwise no state is allowed and the dimension of the Hilbert space is zero. This is a general principle: the fusion of all the particles on the sphere must be the vacuum, since these particles must have (at some point in history) been pulled from the vacuum.

Now we could also imagine puncturing the sphere to make a hole where the particles were. In the spirit of what we did in Section 7.2.1 we could refill the hole with any particle type.[20] However, if we refill one hole with a particular particle type a, then the other hole can only get filled in with the antiparticle type $\bar{a}$. Nonetheless, we can conclude that

[20] Since there is a time direction S^1_{time} as well, removing a disk with a particle in it from a spatial manifold Σ is precisely the same as removing a tubular neighborhood with a particle world line in it from the spacetime manifold.

$$\text{Dim } V(S^2 \text{ with two unlabeled punctures}) = \text{Number of particle types.}$$

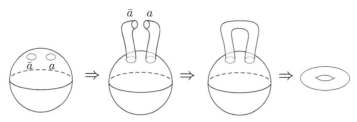

Fig. 8.20 Using surgery to turn the twice-punctured sphere into a torus. This is the gluing axiom in action. Note that we are implicitly assuming the system is trivial in the "time" direction, which we assume to form a circle S^1_{time}.

Now consider the procedure shown in Fig. 8.20. We start with the twice punctured sphere. The two punctures can be labeled with any particle–antiparticle pair labels. We can then deform the sphere to sew the two punctures together in a procedure that is sometimes called *surgery* (we discuss surgery in more detail in Chapter 24). The result of this surgery is the torus surface T^2 and we conclude that

$$\text{Dim } V(T^2) = \text{Number of particle types}$$

as we have already discussed. The general rule of surgery is that two punctures can be sewed together when they have opposing particle types (i.e. a particle and its antiparticle). This is exactly the gluing property of the TQFT. Although we are gluing together pieces along a one dimensional boundary (the edge of the punctures), we should realize that there is also a time direction, which we have implicitly assumed is compactified into S^1_{time}. Thus, we are actually gluing together the 2-surface $(S^1_{puncture} \times S^1_{time})$ with another 2-surface $(S^1_{puncture} \times S^1_{time})$, and the inner product between the two wavefunctions on these 2-surfaces ensures that the quantum number on these two punctures are conjugate to each other.[21]

We can continue on to consider a sphere with three particles. Similarly, we should expect that the three particles should fuse to the identity as shown in Fig. 8.21.   We can then think of the sphere with three particles as being a sphere with three labeled punctures, which is known as a "pair of pants," for reasons that are obvious in Fig. 8.22. It turns out that any orientable two dimensional manifold (except S^2 or T^2, which we have already considered) can be constructed by sewing together the punctures of pants—this is known as a "pants decomposition." For example, in Fig. 8.23 we sew together two pairs of pants to obtain a two-handled torus.

[21]In Eq. 7.9 we had a torus surface which we crossed with an interval of time and we closed up the interval to form a circle, thus giving $\text{Tr}[Z(T^2 \times I_{time})] = Z(T^2 \times S^1_{time}) = \text{Dim } V(T^2)$. In contrast, in Fig. 8.20 we have a cylinder $S^1 \times I$ (topologically the same as a sphere with two holes) crossed with S^1_{time} and we close the cylinder to get $\text{Tr}[Z((S^1 \times I) \times S^1_{time})] = Z(T^2 \times S^1_{time})$.

Fig. 8.21 Three particles that fuse to the identity. There are $N_{abc} = N^{\bar{c}}_{ab}$ different fusion channels.

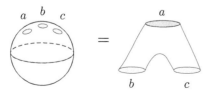

Fig. 8.22 A three-times punctured sphere is known as a "pair of pants."

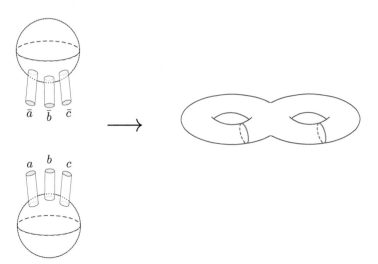

Fig. 8.23 Sewing together two pairs of pants to form a two-handled torus.

To find the ground-state degeneracy of the two-handled torus,

$$\mathrm{Dim}\, V(\text{Two-handled torus}) = Z(\text{Two-handled torus} \times S^1),$$

we assemble the manifold using two pairs of pants as shown in Fig. 8.23 and then we simply need to figure out the number of possible fusion channels where we could satisfy $a \times b \times c \rightarrow I$ (for the bottom pair of pants) and $\bar{a} \times \bar{b} \times \bar{c} \rightarrow I$ (for the top pair of pants). This number of possible fusion channels is given in terms of the fusion multiplicities N_{abc} as shown in Fig. 8.21. Essentially we are just looking at the number of ways we can assign labels to the punctures when we glue the objects together. Thus, we have

$$\mathrm{Dim}\, V(\text{Two-handled torus}) = \sum_{abc} N_{abc} N_{\bar{a}\bar{b}\bar{c}}\quad.$$

Another interesting use of the pants diagram is to determine the degeneracy of a torus T^2 with a single anyon on it labeled a. Unlike the sphere, where one cannot have a single anyon on the surface (see Eq. 8.21) one can have a single anyon on a torus (see note 18 of this chapter). To see how this is possible, take a pants diagram with the holes labeled $b, \bar{b}$, and a. Connect up the b to the $\bar{b}$ to give a torus with a single puncture remaining labeled a. Thus, we conclude that

$$\mathrm{Dim}\, V(T^2 \text{ with one } a) = \sum_{b} N_{b\bar{b}a} \equiv L_a \qquad (8.22)$$

where we have defined this quantity to be called L_a.

One final example is to determine the ground-state degeneracy of a three-handled torus. There are many ways we might cut a three-handled torus into pieces, but a convenient decomposition is the one shown in

Fig. 8.24. Here there are three tori, each with a puncture in it (marked as a red collar), and a single pants in the middle connecting the three. Each torus with a puncture has a Hilbert space dimension L_a where a is the quantum number assigned to the puncture. Thus, the total dimension of the Hilbert space is conveniently written as

$$\text{Dim } V(\text{Three-handled torus}) = \sum_{abc} L_a L_b L_c N_{\bar{a}\bar{b}\bar{c}} \quad . \tag{8.23}$$

Example: Fibonacci Anyons

With the Fibonacci fusion rules, there are five ways we can fuse three particles and get the identity:

$$N_{III} = 1$$
$$N_{\tau\tau I} = N_{\tau I\tau} = N_{I\tau\tau} = 1$$
$$N_{\tau\tau\tau} = 1$$

and all other $N_{abc} = 0$. Thus there are five possible labelings of the punctures in a pants diagram that allow overall fusion to the identity. If we match these together on both top and bottom of the diagram on the left of Fig. 8.23, we conclude that in the Fibonacci theory we have

$$Z(\text{Two-handled torus} \times S^1) = \text{Dim } V(\text{Two-handled torus}) = 5.$$

Similarly, we can consider the degeneracy of states for a torus with a single τ particle on its surface

$$\text{Dim } V(T^2 \text{ with one } \tau \text{ particle on it}) = 1$$

coming from the allowed fusion $N_{\tau\tau\tau} = 1$. Thus, we have $L_I = 2$ and $L_\tau = 1$. It is then easy to plug into Eq. 8.23 to obtain

$$\text{Dim } V(\text{Three-handled torus}) = 15 \quad .$$

8.5 Product Theories

A very common construction is to consider the product of two anyon theories. Given two anyon theories (let us call them T and t) with particle types

$$\begin{aligned} a, b, c, \ldots \ &\in \ t && \text{with } i \text{ the identity} \\ A, B, C, \ldots \ &\in \ T && \text{with } I \text{ the identity} \end{aligned}$$

we consider the product theory $T \times t$. The Hilbert space of the product theory is just the product of the Hilbert spaces of the constituent theories. So any arbitrary particle α in the product theory is composed of one particle from each of the constituent theories

$$\alpha \in T \times t \qquad \Longrightarrow \qquad \alpha = (Y, x) \text{ with } Y \in T \text{ and } x \in t \quad .$$

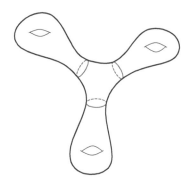

Fig. 8.24 Decomposing a three-handled torus into three copies of a torus with puncture (the puncture is the red collar), and a single pants in the middle. I have resisted the urge to draw a three-handled object as being covered with moss.

Roughly we can think of this as putting both theories in the same place at the same time—particles from T can exist (combined with i from t) and particles from t can exist (combined with I from T) and any combination of particles from both theories can also exist.[22]

For example, in the theory (Ising × Fibonacci), there are six particle types, which we can label as

$$(i, I) \quad (i, \tau) \quad (\sigma, I) \quad (\sigma, \tau) \quad (\psi, I) \quad (\psi, \tau)$$

where we wrote the identity of the Ising theory as i.

The fusion multiplicity matrices N for the product theories are just the product of the N matrices for the constituent theories[23]

$$N^{(C,c)}_{(A,a),(B,b)} = N^C_{A,B} \, N^c_{a,b} \quad .$$

Sometimes we say that a product theory corresponds to *stacking* the two constituent theories—imagining putting two planes, one with each TQFT, right on top of each other.

8.6   Appendix: Tensor Description of Fusion and Splitting Spaces

Let us now try to give a bit more precise mathematical meaning to the idea of fusion as well as to some of the diagrams we have been drawing.[24] For each fusion N^c_{ab} we define a space V^c_{ab} known as a fusion space and a space V^{ab}_c known as a splitting space. Both of these spaces have dimension N^c_{ab}

$$\dim V^c_{ab} = \dim V^{ab}_c = N^c_{ab} = N^{ab}_c \quad .$$

Each of these spaces can be given an orthonormal basis, which we label with an index μ. We can write states in this space as kets, which we draw as diagrams. For example,

$$= \quad |a, b; c, \mu\rangle \in V^{ab}_c \tag{8.24}$$

describes the splitting space with c splitting to a and b in fusion channel $\mu \in \{1, \dots, N^c_{ab}\}$.

The Hermitian conjugate, the corresponding bras, are drawn as fusion diagrams

$$= \quad \langle a, b; c, \mu| \in V^c_{ab} \quad , \tag{8.25}$$

which describes the fusion space. In the most commonly considered case, $N^c_{ab} = 1$, in which case there is a unique state and we do not need to specify μ since it has only one possible value. Cases where $N^c_{ab} = 0$

are non-allowed fusions meaning that the spaces V_{ab}^c and V_c^{ab} are zero-dimensional. A diagram with a non-allowed vertex is given the value zero.

In Eq. 8.24 we described states in the space associated with a single anyon splitting into two, and in Eq. 8.25 we described states in the space associated with two anyons fusing into one. It is possible to also describe the splitting or fusion space for a single anyon splitting or fusing into more pieces. For example, when splitting/fusing into three pieces the relevant spaces are often denoted[25] as V_e^{abc} or the Hermitian conjugate V_{abc}^e. A basis of states in this space can be written diagrammatically with the above described splitting vertices as

$$= |(a,b); d, \mu\rangle \otimes |d, c; e, \nu\rangle \in V_d^{ab} \otimes V_e^{dc} \subseteq V_e^{abc} \cdot \quad (8.26)$$

The insertion of the parenthesis (a, b) in Eq. 8.26 (and similarly the parenthesis (b, c) in Eq. 8.29) are crucial to indicate which splitting is closest to the leaves of the tree (furthest from the root).[26]

The full splitting space V_e^{abc} can thus be described as

$$V_e^{abc} \cong \bigoplus_d V_d^{ab} \otimes V_e^{dc} \quad (8.27)$$

with dimension

$$\dim V_e^{abc} = \sum_d N_d^{ab} N_e^{dc} = N_e^{abc} \quad . \quad (8.28)$$

On the other hand, we could just as well have described a state in this space as

$$= |a, f; e, \lambda\rangle \otimes |(b,c); f, \eta\rangle \in V_e^{af} \otimes V_f^{bc} \subseteq V_e^{abc} \cdot \quad (8.29)$$

In this language the full splitting space V_e^{abc} can be described as

$$V_e^{abc} \cong \bigoplus_f V_e^{af} \otimes V_f^{bc} \quad (8.30)$$

with dimension

$$\dim V_e^{abc} = \sum_d N_e^{af} N_f^{bc} = N_e^{abc} \quad . \quad (8.31)$$

[25] We do not mean e to necessarily be the identity here. See note 1 of this chapter. We use this notation to match that of Chapter 9.

[26] Without the parenthesis one can have ambiguous notation, such as in the Fibonacci theory, where $|\tau\tau; \tau\rangle \otimes |\tau\tau; \tau\rangle$ could mean either the state in Eq. 8.26 or Eq. 8.29. The notation is telling us something important: that the kets in Eq. 8.26 and Eq. 8.29 are living in different, albeit isomorphic, spaces.

We thus have

$$V_e^{abc} \cong \bigoplus_d V_d^{ab} \otimes V_e^{dc} \cong \bigoplus_f V_e^{af} \otimes V_f^{bc} \qquad (8.32)$$

where "$\cong$" means "isomorphic to." In other words, these are two isomorphic descriptions of the same space. Equating the two different expressions (Eq. 8.28 and 8.31) for the dimension of this space recovers the equality Eq. 8.18. The isomorphism between these two descriptions of the same space will be explored in detail in Chapter 9.

One can describe more complicated splitting and fusion spaces in an analogous way. For example, the space V_e^{aaaaa} can be described as

$$V_e^{aaaaa} \cong \bigoplus_{b,c,d} V_b^{aa} \otimes V_c^{ba} \otimes V_d^{ca} \otimes V_e^{da}$$

where each term in the direct sum (i.e. each term with fixed b, c, d) is drawn diagrammatically as in Fig. 8.18.

Chapter Summary

- The ideas of particle types, fusion, and the structure of Hilbert space are some of the most important concepts for understanding topological quantum systems.
- Associativity constrains fusion rules.
- The fusion rules are sufficient information for calculating the Hilbert space dimension on arbitrary two-dimensional surfaces.

Further Reading:

- Bonderson [2007] as well as the appendix of Kitaev [2006] give good formal expositions of anyon theory. The introduction of Preskill [2004] is also nice but comes at the topic from a slightly different angle.
- This type of structure of Hilbert space was long understood in the mathematics and conformal field theory literature. Much of this literature is very hard reading. One early reference that is perhaps a good starting point is Moore and Seiberg [1989b].

Exercises

Exercise 8.1 Quantum Dimension
Let N_{ab}^c be the fusion multiplicity matrices of a TQFT

$$a \times b = \sum_c N_{ab}^c \, c$$

meaning that N_{ab}^c is the number of distinct ways that a and b can fuse to c. (In many, or even most, theories of interest all Ns are either 0 or 1).

The quantum dimension d_a of a particle a is defined as the largest eigenvalue of the matrix $[N_a]_b^c$ where this is now thought of as a two-dimensional matrix with a fixed and b, c the indices.

Show that

$$d_a d_b = \sum_c N_{ab}^c d_c \quad .$$

We prove this formula algebraically in Chapter 17. However, there is a simple and much more physical way to get to the result. Imagine fusing together M anyons of type a and M anyons of type b where M gets very large and determine the dimension of space that results. Then imagine fusing together $a \times b$ and do this M times and then fuse together all the results.

Exercise 8.2 Fusion and Ground-State Degeneracy

To determine the ground-state degeneracy of a 2-manifold in a $(2 + 1)$-dimensional TQFT one can cut the manifold into pieces and glue back together. One can think of the open "edges" or connecting tube-ends as each having a label given by one of the particle types (i.e. one of the anyons) of the theory. Really we are labeling each edge with a basis element of a possible Hilbert space. The labels on two tubes that have been connected together must match (label a on one tube fits into label $\bar{a}$ on another tube.) To calculate the ground-state degeneracy we must keep track of all possible ways that these assembled tubes could have been labeled. For example, when we assemble a torus as in Fig. 8.20, we must match the quantum number on one open end to the (opposite) quantum number on the opposite open end. The ground-state degeneracy is then just the number of different possible labels, or equivalently the number of different particle types.

For more complicated 2D manifolds, we can decompose the manifold into so-called pants diagrams that look like Fig. 8.22. When we sew together pants diagrams, we should include a factor of the fusions multiplicity N_{ab}^c for each pants, which has its three tube edges labeled with a, b and $\bar{c}$.

(a) Show that the general formula for the ground-state degeneracy of an g-handled torus in terms of the N matrices can be written as follows

$$\mathrm{Dim} V(g\text{-handled torus}) = \mathrm{Tr}[M^{g-1}] \quad \text{where} \quad M_{cd} = \sum_{a,b} N_{ab}^{\bar{c}} N_{\bar{a}\bar{b}}^d \quad (8.33)$$

where the sum over a and b are over all particle types (including the identity). This form was first written down in Verlinde [1988].

(b) For the Fibonacci anyon model, find the ground-state degeneracy of a four-handled torus.

(c) Show that in the limit of large number of handles g the ground-state degeneracy scales as $\sim \mathcal{D}^{2g}$ where $\mathcal{D}^2 = \sum_a d_a^2$.

(d) Generalize Eq. 8.33 to the case of a g-handled torus where there are also m particles on the surface of the manifold with quantum numbers $a_1, \ldots, a_m$.

Exercise 8.3 Consistency of Fusion Rules

Show by using commutativity and associativity of fusion along with identity 8.10, that no anyon theory can have a particle a different from the vacuum I such that $a \times a = a$, meaning a fuses with a to form only a and nothing else.

Change of Basis and F-Matrices[1]

9
Important

Let us consider the case of three anyons $a, b,$ and c that fuse together to form an anyon e. As mentioned several times previously (see Fig. 8.19 or Eqs. 8.26 and 8.29), one can describe the state of these three particles in two different ways. We can describe the space by describing how a fuses with b (the value of d on the left of Fig. 9.1), or by how b fuses with c (the value of f on the right of Fig. 9.1). Either of these two descriptions should be able to describe any state of the three anyons $a, b,$ and c fusing to e. However, in the two different cases these states are described in different bases. We define the change of basis using a set of unitary matrices[2,3,4,5] called F, as shown in Fig. 9.1.

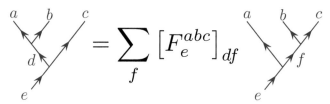

Fig. 9.1 The F-matrix makes a change of basis between the two different ways of describing the space spanned by the fusion of three anyon charges a, b, and c when they all fuse to a total quantum number of e. For fixed $a, b, c,$ and e, the matrix F is unitary in its subscripts d, f. Here F is defined to be zero if the fusion diagram is not allowed, which happens when any of the fusion multiplicities N_{ab}^d, N_{dc}^e, N_{bc}^f, N_{af}^e are zero.

Several brief comments are in order. First, as noted in the caption of Fig. 9.1, the F-matrix element $[F_e^{abc}]_{df}$ is considered to be zero if any of the vertices on either side of the diagram are not allowed vertices of the fusion algebra. Secondly F-moves involving the identity particle (i.e. with a, b, or c being the identity in the figure) are chosen to have a value of unity.[6] In particular this means[7]

$$[F_e^{Ibc}]_{be} = [F_e^{aIc}]_{ac} = [F_e^{abI}]_{eb} = 1 \quad . \tag{9.1}$$

---

[1]This chapter is crucial for the understanding of nonabelian topological quantum systems. If there is one chapter to really study closely, this one is it! Don't worry too much about the section on gauge transformation or the appendices.

[2]For simplicity we are assuming no fusion multiplicities N_{ab}^c greater than 1. In cases where $N_{ab}^c > 1$ (as in Fig. 8.15), each vertex gets an additional index, which ranges from 1 to its multiplicity so that the F-matrix gets additional indices as well. This case is discussed in Section 9.5.3.

[3]The conventions for writing F-matrices used in this chapter match that of Refs. Kitaev [2006] and Bonderson [2007].

[5]Sometimes one calls these F-symbols rather than F-matrices because they have more indices than a matrix usually has.

[6]This involves a gauge choice; see Section 9.4.

[7]We could have also written, for example, $[F_e^{Ibc}]_{bf} = \delta_{ef}$ to emphasize the matrix nature of F.

[4]In the notation of Section 8.6, the F-matrix describes the isomorphism $V_d^{ab} \otimes V_e^{dc} \cong V_e^{af} \otimes V_f^{bc}$. We can write the basis change more algebraically as

$$|(a,b); d\rangle \otimes |d, c; e\rangle = \sum_f [F_e^{abc}]_{df} \, |a, f; e\rangle \otimes |(b, c); f\rangle$$

which represents Fig. 9.1 (where we have again suppressed indices μ at the vertices for simplicity; see Fig. 9.9).

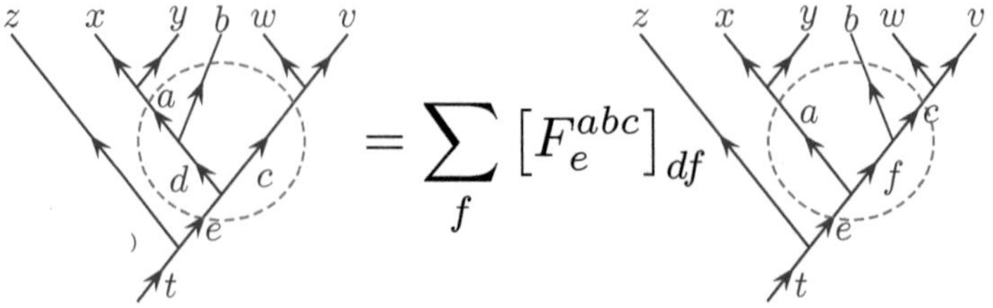

$$\text{(left diagram)} = \sum_f \left[F_e^{abc}\right]_{df} \text{(right diagram)}$$

Fig. 9.2 The F-matrix can be applied inside of more complicated diagrams. Outside of the red circle both diagrams are the same. Inside the circle there is exactly the same transformation as that shown in Fig. 9.1.

Finally, being a change of basis, the F-matrix (for fixed a, b, c, e) is unitary viewed as a matrix with indices d and f.

This idea of change of basis is familiar from angular momentum addition where the F-matrix is known as a $6j$ symbol (note it has six indices). One can combine three objects with L^2 angular momenta values a, b, and c in order to get L^2 angular momentum e, and quite similarly you can describe this space in terms of a combined with b to get d (as in the left of Fig. 9.1) or in terms of b combined with c to get f (as in the right of Fig. 9.1). In fact, even when studying TQFTs, sometimes people refer to F-matrices as $6j$ symbols.

It is important to emphasize that an F-matrix can act on a portion of a diagram, as shown in Fig. 9.2. This allows us to convert any tree structure in a fusion diagram to any other tree structure.

9.1 Example: Fibonacci Anyons

Again we turn to the example of Fibonacci anyons for clarification. We imagine fusing together three τ particles. As shown in Fig. 8.10, there is a single state $|N\rangle$ in which the three fuse to the identity I. It should not matter if we choose to fuse the leftmost two anyons first, or the rightmost two. In either case there is only one possible state in the Hilbert space (i.e. the intermediate edge has only a single possible label, τ). We can thus draw the simple identity shown in Fig. 9.3. Mathematically we would write that $[F_I^{\tau\tau\tau}]_{\tau\tau} = 1$. (And as noted in Eq. 9.1, if any of the three upper indices are the identity, we also obtain unity: $[F_e^{I\tau\tau}]_{\tau e} = [F_e^{\tau I\tau}]_{\tau\tau} = [F_e^{\tau\tau I}]_{e\tau} = 1$).

The more interesting situation is the case where the three Fibonacci anyons fuse to τ. In this case, there is a two-dimensional space of states, and this two-dimensional space can be described in two ways. We can fuse the left two particles first to get either I (yielding overall state $|0\rangle$) or to get τ (yielding overall state $|1\rangle$). See the top of Fig. 9.4. On the other hand, we could fuse the right two particles first to get either I (yielding overall state $|0'\rangle$) or to get τ (yielding overall state $|1'\rangle$). See the bottom of Fig. 9.4.

The space of states spanned by the three anyons is the same in either

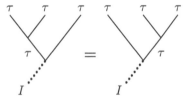

Fig. 9.3 There is only one state in the Hilbert space of three Fibonacci anyons fusing to the identity (we previously called this $|N\rangle$). Thus it does not matter if you fuse the left two first or the right two first—you are describing the same state.

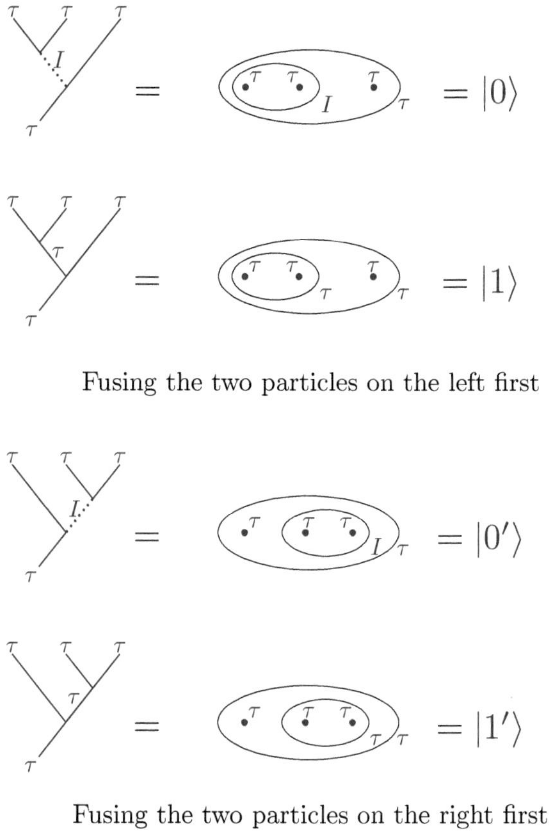

Fusing the two particles on the left first

Fusing the two particles on the right first

Fig. 9.4 Two ways to describe the same two-dimensional space in the case of Fibonacci anyons. The basis $\{|0\rangle, |1\rangle\}$ fuses the left two particles first, whereas the basis $\{|0'\rangle, |1'\rangle\}$ fuses the right two particles first.

description. Thus, there must be a unitary basis transformation given by

$$
\begin{pmatrix} |0\rangle \\ |1\rangle \end{pmatrix} = \begin{pmatrix} F_{00'} & F_{01'} \\ F_{10'} & F_{11'} \end{pmatrix} \begin{pmatrix} |0'\rangle \\ |1'\rangle \end{pmatrix} \quad . \tag{9.2}
$$

Here F is a two by two matrix, and in the notation of the F-matrix defined in Fig. 9.1, this two by two matrix is $[F_\tau^{\tau\tau\tau}]_{ab}$ and the indices a, b should take the values I and τ instead of 0 and 1, but we have used abbreviated notation here for more clarity.

For the Fibonacci theory the F-matrix is given explicitly by[8]

$$
F_\tau^{\tau\tau\tau} = F = \begin{pmatrix} \phi^{-1} & \phi^{-1/2} \\ \phi^{-1/2} & -\phi^{-1} \end{pmatrix} \quad , \tag{9.3}
$$

where $\phi = (\sqrt{5} + 1)/2$ is the golden mean. As one should expect for a change of basis, this matrix is unitary.[9] In Section 9.3 we discuss how this matrix is derived (see also Section 18.2).

[8] We can redefine kets with different gauge choices (see Section 9.4) and this will insert some phases into the off-diagonal of this matrix, but the simplest gauge choice gives the matrix as shown.

[9] If we do not insist on unitarity, one can obtain another consistent theory (a solution of the pentagon, see Section 9.3) by replacing $\phi \to -1/\phi$. This is known as Galois conjugation, and it is interesting mathematically as frequently such transformations can be made (see Gannon [2003]). However, unitarity is required for any quantum mechanical theory, so we will not dwell on this further here.

$= |0\rangle$

$= |1\rangle$

$= |0'\rangle$

$= |1'\rangle$

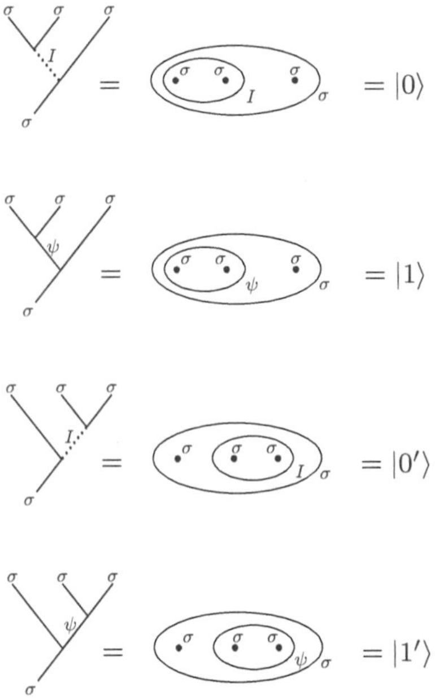

Fig. 9.5 Fusing three σ particles in the Ising theory. In $|0\rangle$ and $|1\rangle$ we fuse the left two particles first, whereas in $|0'\rangle$ and $|1'\rangle$ we fuse the right two particles first.

[10]It is interesting that Eq. 9.5 is a gauge-independent statement, whereas Eq. 9.6 and Eq. 9.4 involve a gauge choice. See Section 9.4 and Exercise 9.1.

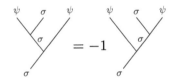

$= -1$

Fig. 9.6 Diagrammatic representation of Eq. 9.5. Both diagrams describe the same state in the Hilbert space, but they differ by a -1 phase.

9.2 Example: Ising Anyons

The situation with Ising anyons is quite similar, so we will be rather brief. Let us fuse three σ particles to an overall fusion channel of σ. There is no other choice, three σ particles can *only* fuse to σ (i.e. there is no $|N\rangle$ state; see Section 8.2.2). The Hilbert space is two dimensional and we can use either of two bases as shown in Fig. 9.5—either fusing the left two particles first or fusing the right two particles first. Analogous to the Fibonacci case we can write an F-matrix that relates the two basis descriptions as in Eq. 9.2. However, here the F-matrix is instead given by[10]

$$F_\sigma^{\sigma\sigma} = F = \frac{1}{\sqrt{2}} \begin{pmatrix} 1 & 1 \\ 1 & -1 \end{pmatrix} \quad , \tag{9.4}$$

which is sometimes known as a *Hadamard* matrix. Deriving this form of the F-matrix is described roughly in Section 9.3, and is done in detail in Sections 18.4 and 22.4 (see also Exercise 9.7).

In the Ising theory we can also look at situations where we have both σ and ψ particles. In this case we have[10]

$$[F_\sigma^{\psi\sigma\psi}]_{\sigma\sigma} = -1 \tag{9.5}$$
$$[F_\psi^{\sigma\psi\sigma}]_{\sigma\sigma} = -1 \quad . \tag{9.6}$$

Equation 9.5 is shown diagrammatically in Fig. 9.6. The other elements of F in the Ising theory that we have not mentioned so far (i.e. those not described by Eqs. 9.4–9.6) are either 1 if all the fusion vertices are allowed, or are zero if any of the fusion vertices are not allowed (see Fig. 9.1 caption).

The presence of the minus signs in Eq. 9.5 (for example) may seem a bit puzzling being that the diagrams on the left and right of Fig. 9.6 are describing the same state in the Hilbert space. However, we explain in the next section that this sign is required in order to have a consistent F-matrix (see Exercise 9.3 for a more detailed calculation).

9.3 Pentagon

It is possible to describe the same Hilbert space in many ways. For example, with three anyons, as in Fig. 8.19, one can describe the state in terms of the fusion channel of the two anyons on the left, or in terms of the two on the right. That is, we can describe $(a \times b) \times c$ or $a \times (b \times c)$, and as in Fig. 9.1, these two descriptions can be related via an F-matrix.

When there are four anyons, there are even more ways to group particles to describe the states of the Hilbert space, and these can also be related to each other via F-matrices (analogous to that shown in Fig. 9.2). The fact that we can change the connectivity of these tree diagrams then allows one to make multiple changes in the trees as shown in Fig. 9.7. Indeed, in this figure one sees that one can go from the far left to the far right of the diagram via two completely different paths (the

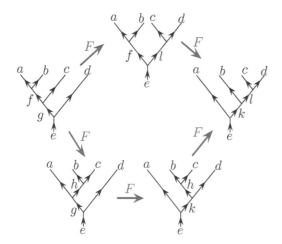

Fig. 9.7 Pentagon diagram. Each step in the diagram is a new description of the same basis of states with the basis transformation being made via the F-matrix.

top and the bottom path) and the end result on the far right should be the same either way. This diagram, known as the pentagon diagram,[11] puts a very strong constraint on the F-matrices, which written out algebraically would be

$$[F_e^{fcd}]_{gl}[F_e^{abl}]_{fk} = \sum_h [F_g^{abc}]_{fh}[F_e^{ahd}]_{gk}[F_k^{bcd}]_{hl} \quad , \qquad (9.7)$$

where the left-hand side represents the top route of the figure and the right-hand side represents the bottom route.[12]

For very simple theories, such as the Fibonacci anyon theory, the fusion rules and the pentagon diagram are sufficient to completely define the F-matrices (up to some gauge convention choices as in Section 9.4). See Exercise 9.4. Further, for *any* given set of fusion rules there are a finite set of possible solutions of the pentagon equation[13]—a property that goes by the name "Ocneanu rigidity."[14] It is also possible that for a given set of particle types and fusion rules (a set of N matrices) there are no solutions of the pentagon equations at all.[15]

One might think that one could write down more complicated trees and more complicated paths through the trees analogous to Fig. 9.7 and somehow derive additional constraints on the F-matrices. A theorem by Mac Lane [1971], known as the "coherence theorem", guarantees that no more complicated trees generate new identities beyond the pentagon diagram.

9.4 **Gauge Transformations**

We have the freedom to make gauge transformations on our diagrams and these will be reflected in the F-matrix. While this is a bit of a

[11]An analogous relation holds for $6j$-symbols of angular momentum addition, known often as the Elliot–Biedenharn identity.

[12]It is very worth working through this to make sure you understand how this equation matches up with the figure! Note that in the equation the F-matrices are written in an order such that those furthest right in Fig. 9.7 are furthest right in the equations.

[13]Here we mean a finite set of *gauge inequivalent* solutions. That is, a gauge transformation of a given solution does not count as a new solution.

[14]Ocneanu did not manage to ever publish this important result. See for example Etingof et al. [2005].

[15]The simplest case of fusion rules without pentagon solutions are theories with only one nontrivial particle type with fusion rules $\tau \times \tau = 1 + n\tau$ with $n > 1$, as discussed by Ostrik [2002]. See also Exercise 17.5 for proof of a closely related result.

technical point, we make frequent use of gauge transformation in some later chapters (particularly Chapter 14) so it is worth discussing it briefly here.

A gauge choice is a choice of a phase associated with the vertices in a diagram. If we change this gauge choice, diagrams are then multiplied by phases.

In particular a gauge transformation multiplies the vertices in a diagram by a phase as shown in Fig. 9.8. The tilde over the vertex on the right notates that we have made a gauge transformation to a tilde gauge[16]

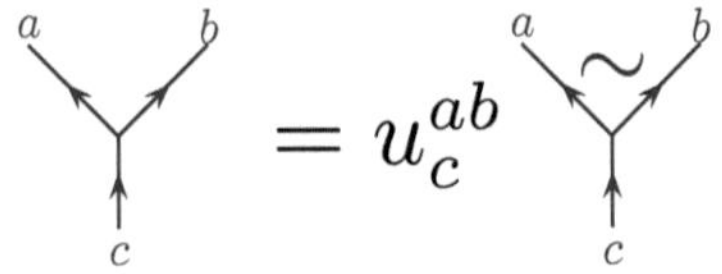

Fig. 9.8 We have the freedom to make a gauge transformation of a vertex by multiplying by a phase u_c^{ab}. The tilde on the right notates that the vertex is in the tilde gauge.

Under such a gauge transformation, the F-matrix must correspondingly transform as

$$\widetilde{[F_e^{abc}]}_{df} = \frac{u_e^{af} u_f^{bc}}{u_d^{ab} u_e^{dc}} [F_e^{abc}]_{df} \quad . \tag{9.8}$$

As described in Section 14.2, some gauge choices are more natural than others, but we should always keep in mind that we have this freedom.

Note that if one of the upper legs is the identity ($a = I$ or $b = I$ in Fig. 9.8) we typically do not allow a gauge transformation of this type of vertex, since the presence of a vertex with the vacuum is the same as the absence of a vertex with the vacuum (i.e. we can add or remove lines labeled by I for free).[17]

9.5 Appendix: F-Matrix Odds and Ends

9.5.1 Product Theories

Given two anyon theories T and t, we can construct the product theory $T \times t$ as in Section 8.5. If the theory T has consistent F-matrices $[F_E^{ABC}]_{DF}$ and the theory t has consistent F-matrices $[F_e^{abc}]_{df}$ ("consistent" here means satisfying the pentagon relation), then the product theory has consistent F-matrices

$$[F_{(E,e)}^{(A,a)(B,b)(C,c)}]_{(D,d),(F,f)} = [F_E^{ABC}]_{DF} \, [F_e^{abc}]_{df}$$

The point here is that in a product theory, the two constituent theories don't "see" each other at all.

---

[16]This is much more easily expressed using the notation of Section 8.6 where we can just write

$$|a, b; c\rangle = u_c^{ab} \widetilde{|a, b; c\rangle} \quad .$$

[17]There can be cases where we *do* want to specify that a vacuum line has branched off at one particular point and we *do* allow choosing a nontrivial gauge for such a vertex (see Lin and Levin [2014] for further discussion of this possibility).

9.5.2 Unitarity of F

The F-matrix relation we defined as

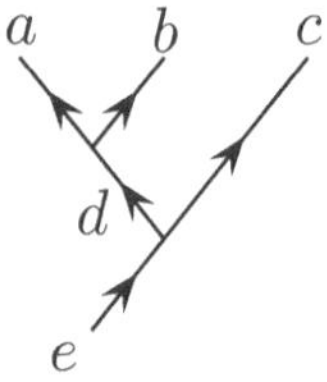

$$\text{(diagram)} = \sum_f \left[F_e^{abc} \right]_{df} \text{(diagram)} .$$

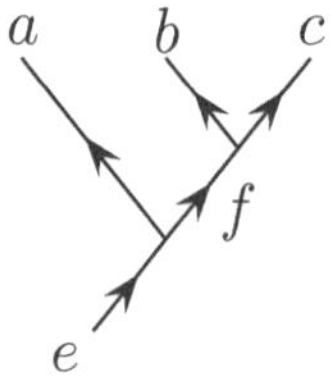

The fact that F is unitary in its indices d and f means we can also write[18]

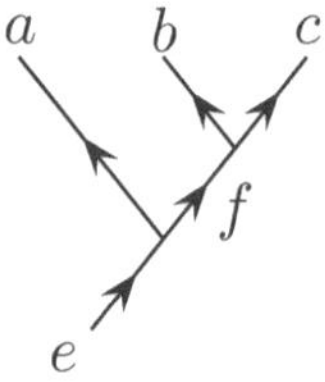

$$\text{(diagram)} = \sum_d \left[F_e^{abc} \right]_{df}^{*} \text{(diagram)} .$$

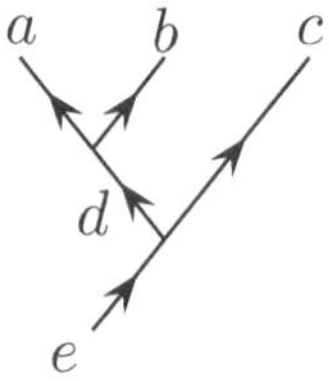

[18] We always use the star to indicate complex conjugation. Note that some mathematicians annoyingly use a star to indicate hermitian conjugation, meaning complex conjugation *and* transposition!

9.5.3 F-Matrix with Higher Fusion Multiplicities

In cases where there are fusion multiplicities N_{ab}^c greater than 1, each vertex gets an additional index as shown in Fig. 8.15. The F-matrix must also describe what happens to these indices under basis transformations. We thus have a more general basis-change equation given in Fig. 9.9.

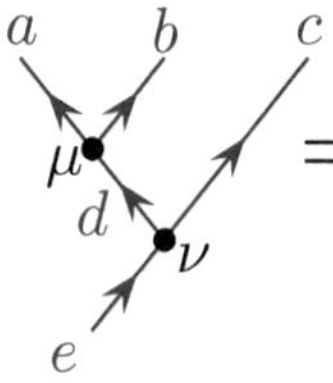

$$\text{(diagram)} = \sum_{f,\alpha,\beta} \left[F_e^{abc} \right]_{(d\mu\nu)(f\alpha\beta)} \text{(diagram)}$$

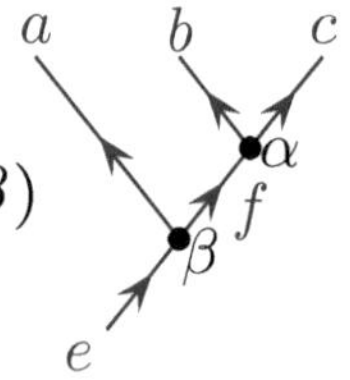

Fig. 9.9 The F-matrix equation with fusion multiplicities greater than one. Here the vertex indices are $\mu \in 1 \dots N_{ab}^d$ and $\nu \in 1 \dots N_{dc}^e$ and $\alpha \in 1 \dots N_{bc}^f$ and $\beta \in 1 \dots N_{af}^e$. The subscripts $(d\mu\nu)$ and $(f\alpha\beta)$ are "super-indices" of the matrix F_e^{abc}. That is, d, μ, and ν are joined together to make a single index.

In the language of Section 8.6 this F-transformation is written as

$$|(a,b);d,\mu\rangle \otimes |d,c;e,\nu\rangle = \sum_{f,\alpha,\beta} \left[F_e^{abc} \right]_{(d\mu\nu)(f\alpha\beta)} |a,f;e,\beta\rangle \otimes |(b,c);f,\alpha\rangle .$$

Gauge Transformations with Higher Fusion Multiplicities

With higher fusion multiplicities $N_{ab}^c > 1$, our diagrams have indices at the vertices. Gauge transformations are generally a unitary matrix within this index space and take the form shown in Fig. 9.10.[19]

[19] Again, this is much more easily expressed using the notation of Section 8.6 where we can just write

$$|a,b;c,\mu\rangle = \sum_{\mu'} [u_c^{ab}]_{\mu\mu'} |a,\widetilde{b;c,\mu'}\rangle .$$

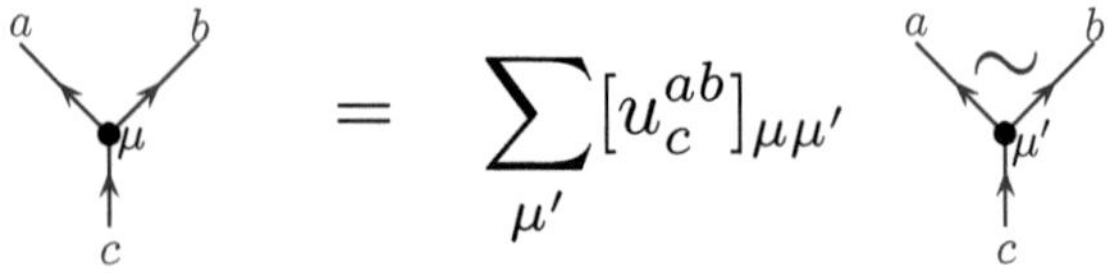

Fig. 9.10 We have the freedom to make a gauge transformation of a vertex by multiplying by a unitary matrix $[u_c^{ab}]_{\mu\mu'}$. The tilde on the right notates that the vertex is in the tilde gauge.

Under such gauge transformations, the F-matrix must correspondingly transform as

$$\widetilde{[F_e^{abc}]}_{(d\mu'\nu')(f\alpha'\beta')} = \tag{9.9}$$
$$\sum_{\alpha,\beta,\mu,\nu} ([u_d^{ab}]^{-1})_{\mu'\mu}([u_e^{dc}]^{-1})_{\nu'\nu}[F_e^{abc}]_{(d\mu\nu)(f\alpha\beta)}[u_e^{af}]_{\beta\beta'}[u_f^{bc}]_{\alpha\alpha'} \quad.$$

Chapter Summary

- The F-matrix implements a change of basis.
- Understanding F-matrices is fairly crucial for understanding topological models.
- The pentagon equations are consistency conditions.

Further Reading:

- Good descriptions of the F-matrix are given by Nayak et al. [2008]; Kitaev [2006]; Bonderson [2007].
- This material has been around for a long time in the mathematics and conformal field theory literature, but is not presented particularly simply in these references.

Exercises

Exercise 9.1 F Gauge Choice
(a) Explain why in the Fibonacci theory, $[F_\tau^{\tau\tau\tau}]_{\tau\tau}$ is gauge independent but $[F_\tau^{\tau\tau\tau}]_{I\tau}$ is gauge dependent.
(b) Explain why in the Ising theory $[F_\sigma^{\psi\sigma\psi}]_{\sigma\sigma}$ is gauge independent, but $[F_\psi^{\sigma\psi\sigma}]_{\sigma\sigma}$ is gauge dependent.

Exercise 9.2 F with the Vacuum Field I
Explain why $[F_e^{aIc}]_{ac} = [F_d^{abI}]_{db} = [F_e^{Ibc}]_{be} = 1$.

Exercise 9.3 Ising Pentagon
Consider a system of Ising anyons. Given the fusion rules, F_w^{xyz} will be a 2 by 2 matrix in the case of $x = y = z = w = \sigma$ (given by Eq. 9.4) and is a simply a scalar otherwise. One might hope that these scalars can all be taken to be unity. Unfortunately this is not the case. By examining the pentagon equation, Eq. 9.7 in the case of $a = b = c = \sigma$ and $d = f = \psi$ show that taking the scalar to always be unity is not consistent. Show further that choosing $[F_\sigma^{\psi\sigma\psi}]_{\sigma\sigma} = -1$ (and leaving the other scalars to be unity) allows a consistent solution of the pentagon equations for $a = b = c = \sigma$ and $d = f = \psi$.

Exercise 9.4 Fibonacci Pentagon
In the Fibonacci anyon model, there are two particle types which are usually called I and τ. The only nontrivial fusion rule is $\tau \times \tau = I + \tau$. With these fusion rules, the F-matrix is completely fixed up to a gauge freedom (corresponding to adding a phase to some of the kets). If we choose all elements of the F-matrix to be real, then the F-matrix is completely determined by the pentagon equations up to one sign (gauge) choice. Using the pentagon equation, determine the F-matrix. (To get you started, note that in Fig. 9.7 the variables a, b, c, d, e, f, g, h can only take values I and τ. You only need to consider the cases where a, b, c, d are all τ).
If you are stuck as to how to start, part of the calculation is given in Nayak et al. [2008].

Exercise 9.5 Pentagon and Fusion Multiplicities
Consider the case of Section 9.5.3 where there are fusion multiplicities $N_{ab}^c > 1$. Write the generalization of the pentagon equation Eq. 9.7.

Exercise 9.6 Gauge Change
(a.i) Confirm that the F-matrix transforms under gauge change as indicated in Eq. 9.8. (a.ii) Show that a solution of the pentagon equation remains a solution under any gauge transformation.
[Harder] Now consider the case of Section 9.5.3 where there are fusion multiplicities $N_{ab}^c > 1$.
(b.i) Analogous to (a.i), confirm Eq. 9.9. (b.ii) Analogous to (a.ii) show that a solution of the pentagon equations remains a solution under any gauge transformation. (You will need to solve Exercise 9.5 first!)

Exercise 9.7 Ising F-Matrix
[Hard] As discussed in the earlier problem, "Ising Anyons and Majorana Fermions" (Exercise 3.3), one can express Ising anyons in terms of Majorana fermions, which are operators γ_i with anticommutations $\{\gamma_i, \gamma_j\} = 2\delta_{ij}$. As discussed there we can choose any two Majoranas and construct a fermion operator
$$c_{12}^\dagger = \frac{1}{2}(\gamma_1 + i\gamma_2)$$
and this fermion orbital can be either filled or empty. We might write this as $|0_{12}\rangle = c_{12}|1_{12}\rangle$ and $|1_{12}\rangle = c_{12}^\dagger|0_{12}\rangle$, the subscript 12 here meaning that we have made the orbital out of Majoranas number 1 and 2. Note, however, that we have to be careful that $|0_{12}\rangle = e^{i\phi}|1_{21}\rangle$ where ϕ is a gauge choice that is arbitrary (think about this if it is not obvious already).

Let us consider a system of four Majoranas, $\gamma_1, \gamma_2, \gamma_3, \gamma_4$. Consider the basis of states

$$
\begin{aligned}
|a\rangle &= |0_{12}0_{34}\rangle \\
|b\rangle &= |0_{12}1_{34}\rangle \\
|c\rangle &= |1_{12}0_{34}\rangle \\
|d\rangle &= |1_{12}1_{34}\rangle
\end{aligned}
$$

and rewrite these states in terms of the basis of states

$$
\begin{aligned}
|a'\rangle &= |0_{41}0_{23}\rangle \\
|b'\rangle &= |0_{41}1_{23}\rangle \\
|c'\rangle &= |1_{41}0_{23}\rangle \\
|d'\rangle &= |1_{41}1_{23}\rangle
\end{aligned}
$$

Hence determine the F-matrix for Ising anyons. Be cautious about fermionic anticommutations: $c_x^\dagger c_y^\dagger = -c_y^\dagger c_x^\dagger$ so if we define $|1_x 1_y\rangle = c_x^\dagger c_y^\dagger |0_x 0_y\rangle$ with the convention that $|0_x 0_y\rangle = |0_y 0_x\rangle$ then we will have $|1_x 1_y\rangle = -|1_y 1_x\rangle$. Note also that you have to make a gauge choice of some phases (analogous to the mentioned gauge choice above). You can choose F to be always real.

Exchanging Identical Particles

10

Important

We would now like to determine what happens when two particles are exchanged with each other. As one might expect for anyons, phases are accumulated from such exchanges. However, one must be cautious because the phases accumulated will generally depend on the fusion channels of the particles being exchanged.

10.1 Introducing the R-Matrix

Let us begin with a simple case where two identical particles of type a are braided around each other. Let us specify that the two a particles fuse together to an overall channel c. Let us call this quantum-mechanical state $|\text{state}\rangle$ as shown in two different notations in Fig. 10.1.

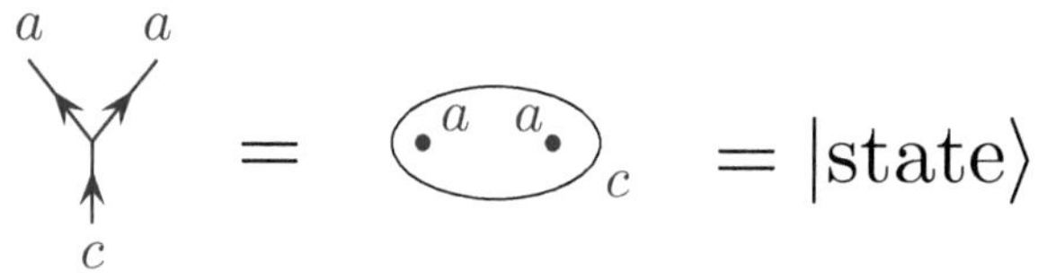

Fig. 10.1 Two a particles fusing to a c particle. More precisely, the diagram on the left indicates two a particles *splitting* from c. Either way, the fusion channel of the two a's is c (see Fig. 8.1).

We then (half)-braid the two particles around each other (counterclockwise observing from above).[1] The final fusion channel of the two a particles is still c (by the locality principle of Section 8.2). However, a phase will be accumulated in the process which we call R_c^{aa} as shown in Fig. 10.2. The inverse phase would be accumulated for an exchange in the opposite direction. If the product $a \times a$ does not include the particle type c (i.e. if $N_{aa}^c = 0$), then R_c^{aa} is defined to be zero.

[1] In the language of the braid group we would call this exchange σ. See Section 3.3.1.

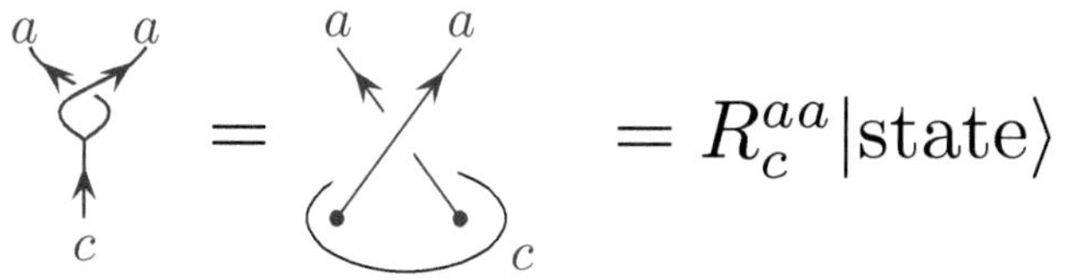

Fig. 10.2 The phase accumulated by exchanging two a particles that fuse to c is called R_c^{aa}.

These so-called R-matrices[2] along with the corresponding F-matrices will allow us to compute the result of braiding any number of a particles around each other in arbitrary ways.

[2] Analogous to F, sometimes the R-matrix is known as an R-symbol.

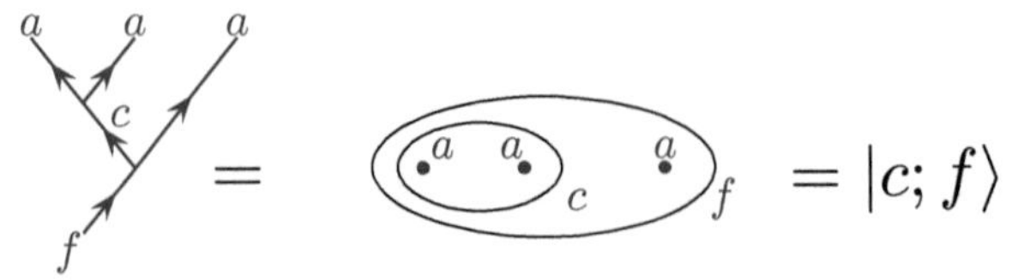

Fig. 10.3 A basis of states for three type a anyons fusing to an overall quantum number f.

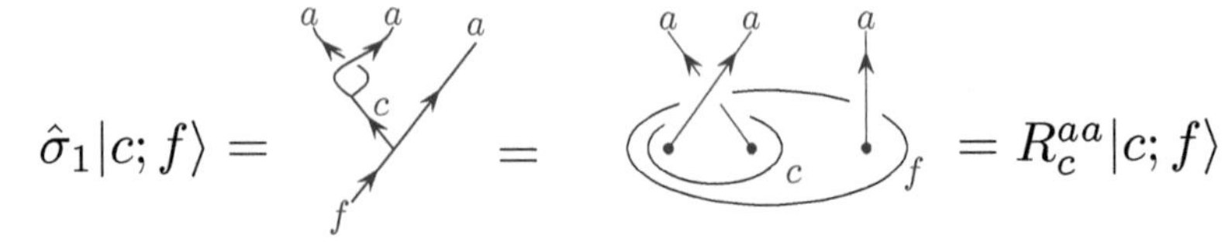

Fig. 10.4 Exchanging the two left particles incurs a phase R_c^{aa}.

Let us consider the case of three anyons of type a. We can write a basis for the possible states of three anyons as shown in Fig. 10.3.

We now consider exchanging the left two particles as shown in Fig. 10.4. (We call the operator that performs this exchange $\hat{\sigma}_1$ in analogy with the braid group discussed in Section 3.3.1.) Since we know the fusion channel of these two particles (c) we know that the phase accumulated in this exchange is just R_c^{aa}. This seems fairly simple as it is precisely the type of exchange we defined in Fig. 10.2.

As with all operators in quantum mechanics that can be implemented as a time evolution, the exchange operator is linear, meaning that it acts on superpositions by acting on each term individually:

$$\hat{\sigma}_1 \sum_c \alpha_c |c; f\rangle = \sum_c \alpha_c R_c^{aa} |c; f\rangle \quad .$$

Let us now instead consider exchanging the right two particles, an operation we call $\hat{\sigma}_2$. Since the right two particles are not in a definite fusion channel we cannot directly apply the R-matrix. However, we can use the F-matrix to rewrite our state as a superposition of states where the right two particles are in a definite fusion channel as shown in Fig. 10.5.

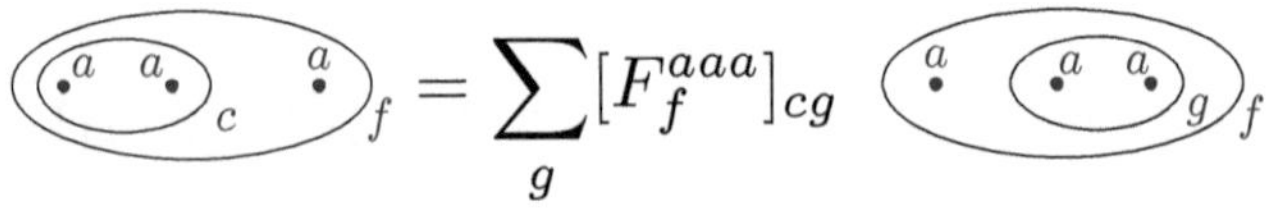

Fig. 10.5 Using an F-move to work in the basis with a known fusion channel of the right two particles.

Once we have transformed to this new basis, we can exchange the right two particles and apply the R-matrix directly to the right two particles as shown in Fig. 10.6. Having established the effect of the exchange, we can (if desired) convert back into the original basis, which describes the fusion of the left two particles using F^{-1}.

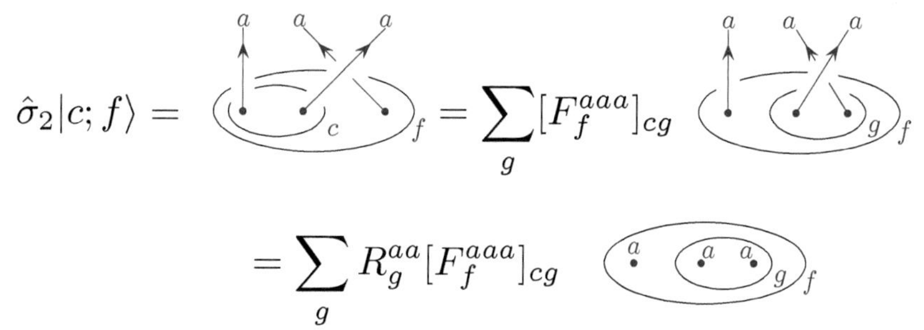

$$\hat{\sigma}_2|c;f\rangle = \left(\text{diagram}\right)_f = \sum_g [F_f^{aaa}]_{cg} \left(\text{diagram}\right)_f$$

$$= \sum_g R_g^{aa}[F_f^{aaa}]_{cg} \left(\text{diagram}\right)_f$$

Fig. 10.6 In order to describe the exchange of the right two particles, we first change to a basis where the fusion channel of those two particles is explicit. We can then apply the R-matrix directly.

The result of this procedure in terms of the original basis is given by

$$\hat{\sigma}_2|c;f\rangle = \sum_{g,z} [F_f^{aaa}]_{cg}\ R_g^{aa}\ [(F_f^{aaa})^{-1}]_{gz}\ |z;f\rangle \ . \tag{10.1}$$

The general principle is that, to evaluate any exchange of identical particles, we use F-matrices to convert to a basis where the fusion channel of the two particles to be braided is known. Once we are working in this basis, we can then apply the R-matrix directly. At the end we can transform back to the original basis if we so desire. This scheme works for *any* set of identical particles given appropriate F- and R-matrices.

10.1.1   Locality

An important principle we will often use is that the result of braiding a group of particles with a given total quantum number c is the same as if that entire group were replaced with just a single particle with the same quantum number c. For example, in Fig. 10.7 when we braid a cluster a,b with overall quantum number c around a cluster x,y,z with overall quantum number w, the phase accumulated should be the same as if we simply braided c around w.

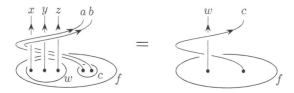

Fig. 10.7 Braiding a cluster of particles with overall quantum number c around a cluster of particles with overall quantum number w should have the same effect as braiding a single c particle, around a single w particle. The result should just be a phase dependent on the quantum numbers c, w, and f. In Chapter 13 we refer to this phase as $R_f^{wc}R_f^{cw}$. Note that a and b do not braid around each other (nor do x,y,z braid around each other). If they did additional phases would be accumulated.

10.2 Some Examples

Since the idea of using the R-matrix is quite important, it is worth working through a few examples explicitly.

10.2.1 Fibonacci Anyons

Recall the properties of Fibonacci anyons (see Section 8.2.1); there is only one nontrivial particle type which we call τ and the only nontrivial fusion rule is $\tau \times \tau = I + \tau$. As we saw in Section 8.2.1, the fusion rule implies that there are two possible states of two Fibonacci anyons: the state where they fuse together to form I and the state where they fuse together to τ. (See Fig. 10.8 and Fig. 8.9 where we previously introduced these two states.) We call these states $|I\rangle$ and $|\tau\rangle$, respectively.

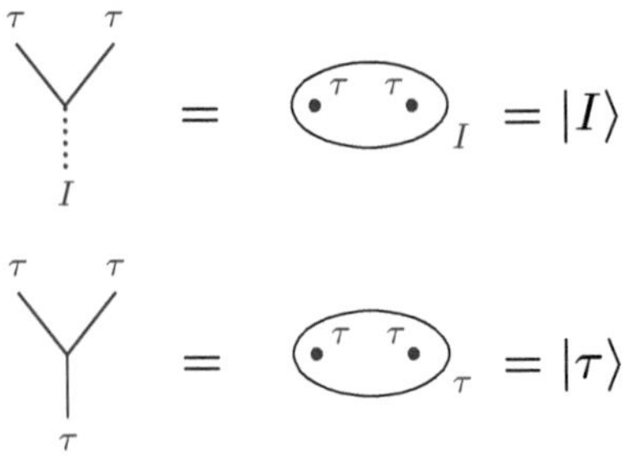

Fig. 10.8 The two possible states of two Fibonacci anyons. Note that we do not draw arrows on the particle lines in the left diagrams since τ is self-dual.

[3]Which set of R-matrices is considered right-handed and which one left-handed is simply a convention. The point here is that there are two solutions related to each other by mirror symmetry or complex conjugation. What we have defined here as right-handed corresponds to the Chern–Simons theory $(G_2)_1$.

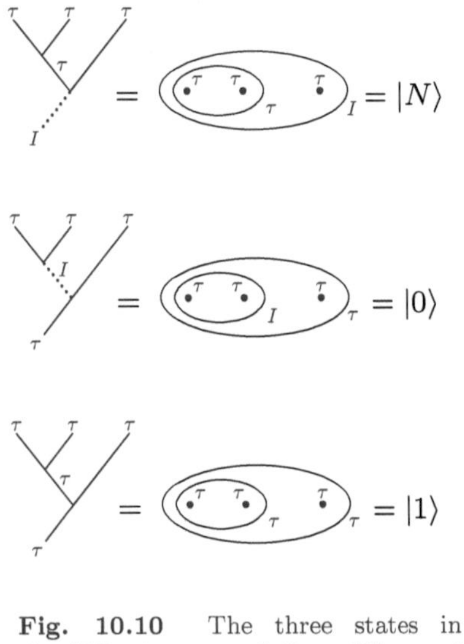

Now consider the operator $\hat{\sigma}$ that exchanges the two Fibonacci anyons counterclockwise as viewed from above, as shown in Fig. 10.9. This operator yields the phase $R_I^{\tau\tau}$ if the fusion channel of the two particles is I or $R_\tau^{\tau\tau}$ if the fusion channel of the two particles is τ.

$$\hat{\sigma}|I\rangle = \quad = \quad = R_I^{\tau\tau}|I\rangle$$

$$\hat{\sigma}|\tau\rangle = \quad = \quad = R_\tau^{\tau\tau}|\tau\rangle$$

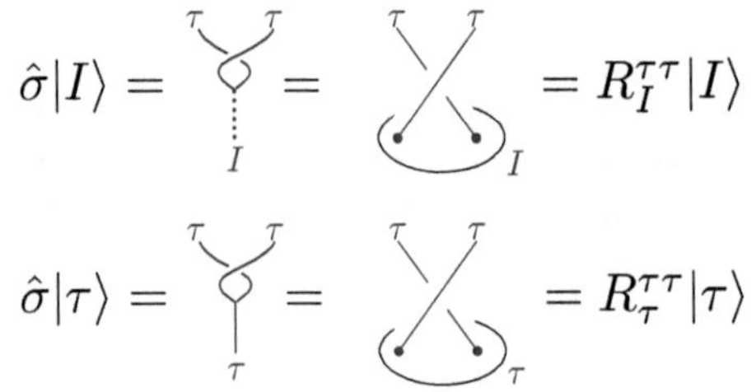

Fig. 10.9 Exchanging two anyons gives a phase dependent on their fusion channel.

In Section 13.3 (see also Exercise 10.6) we explain how we actually compute the phases $R_\tau^{\tau\tau}$ and $R_I^{\tau\tau}$. It turns out there are two possible sets of Rs that describe two different Fibonacci anyon theories—we call these "right-" and "left-handed" Fibonacci anyons. Here we give the values of R for the right-handed Fibonacci anyons[3]

$$
\begin{aligned}
R_\tau^{\tau\tau} &= e^{+3\pi i/5} \\
R_I^{\tau\tau} &= e^{-4\pi i/5} \quad .
\end{aligned}
\tag{10.2}
$$

For the left-handed type of Fibonacci anyons, the values of R have phases that are complex conjugate of those in Eq. 10.2.

As with all operators in quantum mechanics we can act on superpositions by acting on each term individually:

$$\hat{\sigma}\left(\alpha|I\rangle + \beta|\tau\rangle\right) = \alpha R_I^{\tau\tau}|I\rangle + \beta R_\tau^{\tau\tau}|\tau\rangle \quad .$$

Fig. 10.10 The three states in the Hilbert space of three Fibonacci anyons.

If we think of our two states $|I\rangle$ and $|\tau\rangle$ as the two states of a qubit, the $\hat{\sigma}$ operator is what is known as a controlled phase gate in quantum information processing—the phase accumulated depends on the state of the qubit.

Next let us consider the possible states of three Fibonacci anyons. As described in Section 8.2.1, the space spanned by such states is three-dimensional, and we can choose as a basis the three states shown in Fig. 10.10 (we already introduced these states in Fig. 8.10). Now consider an operator $\hat{\sigma}_1$ that braids the two leftmost particles around each other as shown in Fig. 10.11. Here, the phase accumulated depends on the fusion channel of the leftmost two particles, entirely analogous to Fig. 10.9.

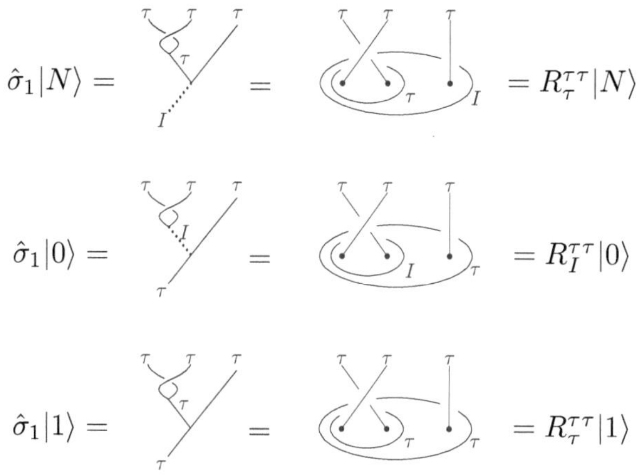

$$\hat{\sigma}_1|N\rangle = \quad = \quad = R_\tau^{\tau\tau}|N\rangle$$

$$\hat{\sigma}_1|0\rangle = \quad = \quad = R_I^{\tau\tau}|0\rangle$$

$$\hat{\sigma}_1|1\rangle = \quad = \quad = R_\tau^{\tau\tau}|1\rangle$$

Fig. 10.11 Exchanging the left two particles.

More interesting is the question of what happens if we exchange the right two particles, as shown in Fig. 10.12. As discussed in Section 10.1, the trick here is to use the F-matrix to change the basis such that we know the fusion channel of the right two particles, and then once we know the fusion channel we can use the R-matrix. If we want, we can then use the F-matrix to transform back to the original basis.

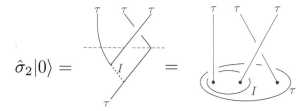

$$\hat{\sigma}_2|0\rangle =$$

Fig. 10.12 Exchanging the two particles on the right for the $|0\rangle$ state where these two particles on the right are not in a definite fusion channel. Note that in the tree diagram on the left, the state below the dashed red line is exactly $|0\rangle$.

To see how this works, recall that we can use the F-matrix to write the $|0\rangle$ state in the basis of the $|0'\rangle$ and $|1'\rangle$ as in Eq. 9.2, the relevant parts of which we reproduce here:

$$|0\rangle = F_{00'}|0'\rangle + F_{01'}|1'\rangle \tag{10.3}$$

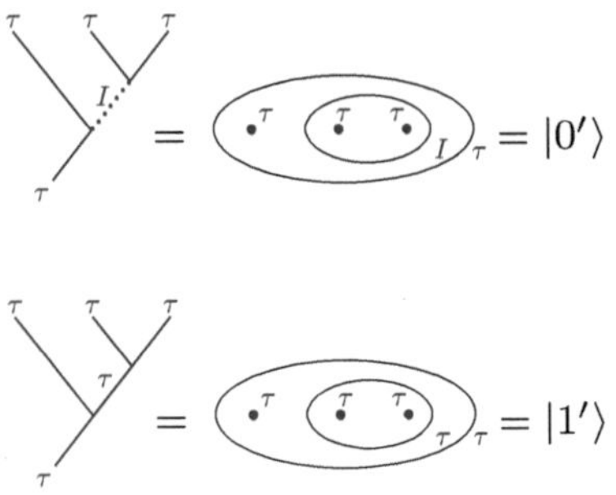

Fig. 10.13 In the primed basis the two particles on the right are in a definite fusion channel.

[4]For this particular case (using Eq. 9.3 for the F-matrix) the matrix F and F^{-1} happen to be the same matrix. However, we write out the inverse explicitly for clarity!

[5]To fully harmonize the notation with that of Eq. 10.1 we should replace $|0\rangle \to |I;\tau\rangle$ and $|1\rangle \to |\tau;\tau\rangle$. The indices 0 and 1 are replaced by I and τ and (as mentioned above) the F_{ab} matrix is really $[F_\tau^{\tau\tau\tau}]_{ab}$.

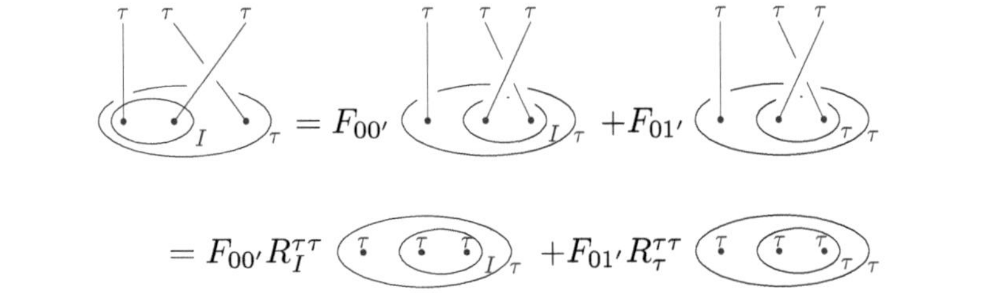

where $|0'\rangle$ and $|1'\rangle$ are shown in Fig. 10.13. Note that here F_{ab} is shorthand for $[F_\tau^{\tau\tau\tau}]_{ab}$.

On the right-hand side of Fig. 10.13 (i.e. in the primed basis) we know the fusion channels of the rightmost two particles, so we can braid them around each other and use the R-matrix to compute the corresponding phases as shown in Fig. 10.14.

Between Eq. 10.4 and 10.5 we have used the inverse F transform to put the result back in terms of the original $|0\rangle$ and $|1\rangle$ basis.[4] The final result, Eq. 10.6, is precisely the same as Eq. 10.1 just written out in detail for the Fibonacci theory.[5]

$$
\begin{aligned}
&= F_{00'} R_I^{\tau\tau} |0'\rangle + F_{01'} R_\tau^{\tau\tau} |1'\rangle &(10.4)\\
&= F_{00'} R_I^{\tau\tau} \left([F^{-1}]_{0'0}|0\rangle + [F^{-1}]_{0'1}|1\rangle\right) &(10.5)\\
&\quad + F_{01'} R_\tau^{\tau\tau} \left([F^{-1}]_{1'0}|0\rangle + [F^{-1}]_{1'1}|1\rangle\right)\\
&= \left(F_{00'} R_I^{\tau\tau}[F^{-1}]_{0'0} + F_{01'} R_\tau^{\tau\tau}[F^{-1}]_{1'0}\right)|0\rangle\\
&\quad + \left(F_{00'} R_I^{\tau\tau}[F^{-1}]_{0'1} + F_{01'} R_\tau^{\tau\tau}[F^{-1}]_{1'1}\right)|1\rangle &(10.6)
\end{aligned}
$$

Fig. 10.14 To exchange the right two particles we first use an F-move so that we know the fusion channel of these two particles, then we can apply R and then F^{-1} to transform back into the original basis.

We can summarize the results of the two possible braiding operations on the three-dimensional Hilbert space. Assuming right-handed Fibonacci anyons and using a basis $|N\rangle, |0\rangle, |1\rangle$ (also notated as $|\tau;I\rangle$, $|I;\tau\rangle$, $|\tau;\tau\rangle$) we have

$$
\hat\sigma_1 = \begin{pmatrix} e^{3\pi i/5} & & \\ & e^{-4\pi i/5} & \\ & & e^{3\pi i/5} \end{pmatrix} \tag{10.7}
$$

$$
\hat\sigma_2 = \begin{pmatrix} e^{3\pi i/5} & & \\ & \phi^{-1}e^{4\pi i/5} & \phi^{-1/2}e^{-3\pi i/5} \\ & \phi^{-1/2}e^{-3\pi i/5} & -\phi^{-1} \end{pmatrix} \tag{10.8}
$$

where $\phi = (1+\sqrt5)/2$ is the golden mean.

10.2.2 Ising Anyons

For Ising anyons the situation is perhaps even simpler since three σ particles (Fig. 10.15) have only two fusion channels (see Section 8.2.2).

Fig. 10.15 A basis for a qubit made from three Ising anyons (see Fig. 9.5).

The appropriate F-matrices are given by Eq. 9.4 and the R-matrices for a *right-handed* Ising theory are given by

$$R_I^{\sigma\sigma} = e^{-i\pi/8} \tag{10.9}$$
$$R_\psi^{\sigma\sigma} = e^{i3\pi/8} = i R_I^{\sigma\sigma} \tag{10.10}$$

with the R-matrices for the left-handed theory being the complex conjugates of these expressions. From the R-matrix, we immediately obtain the form of the exchange operator $\hat{\sigma}_1$, that counterclockwise exchanges the leftmost two Ising anyons

$$\hat{\sigma}_1 = e^{-i\pi/8} \begin{pmatrix} 1 & 0 \\ 0 & i \end{pmatrix} . \tag{10.11}$$

Then using Eq. 10.1 we can evaluate the exchange operator $\hat{\sigma}_2$, which counterclockwise exchanges the rightmost two anyons of the three, giving

$$\hat{\sigma}_2 = \frac{e^{i\pi/8}}{\sqrt{2}} \begin{pmatrix} 1 & -i \\ -i & 1 \end{pmatrix} . \tag{10.12}$$

Chapter Summary

- The R-matrix describes the braiding of anyons around each other.
- Using R and F together one can calculate the effect of moving anyons around to form any spacetime braid.
- Braiding around a cluster of particles is the same as braiding around a single particle with quantum number that is the total quantum number of the cluster.

Further Reading:

- As before, the main references for this material are Nayak et al. [2008]; Kitaev [2006] and Bonderson [2007]. Also, Preskill [2004] is a nice introduction.
- Again, all of this material has been known in other communities for a while. In the context of conformal field theory, see Moore and Seiberg [1989b] for example.

Exercises

Exercise 10.1 Calculating Exchanges
(a) Using known values of F- and R-matrices, use Eq. 10.1 to (i) confirm Eq. 10.12 for the Ising theory, and (ii) confirm Eq. 10.8 for the Fibonacci theory.

(b) Confirm the braiding relation $\hat{\sigma}_1\hat{\sigma}_2\hat{\sigma}_1 = \hat{\sigma}_2\hat{\sigma}_1\hat{\sigma}_2$ in both cases. What does this identity mean geometrically? See Exercise 3.1.

Exercise 10.2 Ising Anyons Redux

In Exercises 3.3 and 9.7 we introduced a representation for the exchange matrices for Ising anyons which, for three anyons, would be of the form

$$\hat{\sigma}_1 = \frac{e^{i\alpha}}{\sqrt{2}}(1 + \gamma_1\gamma_2) \tag{10.13}$$

$$\hat{\sigma}_2 = \frac{e^{i\alpha}}{\sqrt{2}}(1 + \gamma_2\gamma_3) \tag{10.14}$$

where the γs are Majorana operators defined by

$$\{\gamma_i, \gamma_j\} \equiv \gamma_i\gamma_j + \gamma_j\gamma_i = 2\delta_{ij}$$

with $\gamma_i = \gamma_i^\dagger$.

Show that the exchange matrices in Eqs. 10.11–10.12 are equivalent to this representation. How does one represent the $|0\rangle$ and $|1\rangle$ states of the Hilbert space in this language? The answer may not be unique.

Exercise 10.3 Exchanging More Particles

(a) Consider a system of four identical Ising anyons. Use the F- and R-matrices to calculate the braid matrices $\hat{\sigma}_1, \hat{\sigma}_2$, and $\hat{\sigma}_3$. (You should be able to check your answer using the Majorana representation of Exercise 3.3.)

(b) [Harder] Consider a system of four identical Fibonacci anyons. Use the F- and R-matrices to calculate the braid matrices $\hat{\sigma}_1, \hat{\sigma}_2$, and $\hat{\sigma}_3$.

Exercise 10.4 Determinant and Trace of Braid Matrices

Consider a system of N-identical anyons with a total Hilbert space dimension D. The braid matrix $\hat{\sigma}_1, \hat{\sigma}_2, \ldots, \hat{\sigma}_{N-1}$ are all D-dimensional. Show that each of these matrices has the same determinant, and each of these matrices has the same trace. Hint: This is easy if you think about it right!

Exercise 10.5 Checking the Locality Constraint

[Easy] Consider Fig. 10.16. The braid on the left can be written as $\hat{b}_3 = \hat{\sigma}_2\hat{\sigma}_1^2\hat{\sigma}_2$.

(a) For the Fibonacci theory with $a = \tau$ check that the matrix $\hat{b}_3$ gives just a phase, which is dependent on the fusion channel c. That is, show the matrix $\hat{b}_3$ is a diagonal matrix of complex phases. Show further that these phases are the same as the phases that would be accumulated for taking a single τ particle around the particle c.

(b) Consider the same braid for the Ising theory with $a = \sigma$. Show again that the result is a c-dependent phase.

[Hard] Consider the braid shown on the left of Fig. 10.17. The braid can be written as $\hat{b}_4 = \hat{\sigma}_3\hat{\sigma}_2\hat{\sigma}_1^2\hat{\sigma}_2\hat{\sigma}_3$.

(c) Consider Ising anyons where $a = \sigma$. Use the F- and R-matrices to calculate $\hat{\sigma}_3$ (see Exercise 10.3.a). Since the fusion of three σ anyons always gives $c = \sigma$, show that $\hat{b}_4$ is just a phase times the identity matrix. Show further that this phase is the same as the phase accumulated by taking a single σ all the way around another σ.

(d) Consider Fibonacci anyons with $a = \tau$, Use the F- and R-matrices to calculate $\hat{\sigma}_3$ (see Exercise 10.3.b). Check that $\hat{b}_4$ is a diagonal matrix of phases. Show that the phases match the two possible phases accumulated by wrapping a single τ all the way around a single particle c, which can be I or τ.

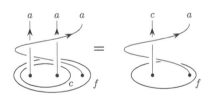

Fig. 10.16 The locality constraint (see similar Fig. 10.7).

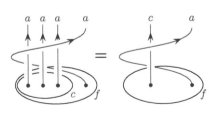

Fig. 10.17 The locality constraint (see similar Fig. 10.7).

Exercise 10.6 Enforcing the Locality Constraint

The locality constraint shown in Fig. 10.16 turns out to be extremely powerful. In this exercise we use this constraint to (almost) derive the possible values for the R-matrix for Fibonacci anyons given the known F-matrix.

Consider an anyon theory with Fibonacci fusion rules and Fibonacci F-matrix as in Eq. 9.3.

(a) [Easy] Confirm the locality constraint shown in Fig. 10.16 (see also Fig. 10.7) given the values of R given in Eq. 10.2. Make sure to confirm the equality for all three cases $f = I, c = \tau$, and $f = \tau, c = I$, and $f = \tau, c = \tau$.

Note that on the left of Fig. 10.16 is the braiding operation $\hat{O} = \hat{\sigma}_2 \hat{\sigma}_1 \hat{\sigma}_1 \hat{\sigma}_2$, whereas the operation on the right is σ^2.

(b) Show that the locality constraint of Fig. 10.16 would also be satisfied by

$$R_I^{\tau\tau} \to -R_I^{\tau\tau} \qquad R_\tau^{\tau\tau} \to -R_\tau^{\tau\tau} \quad . \qquad (10.15)$$

(c) In addition to right- and left-handed Fibonacci anyons and the two additional spurious solutions provided by Eq. 10.15, there are four additional possible sets of R-matrices that are consistent with the F-matrices of the Fibonacci theory given the locality constraint of Fig. 10.16. These additional solutions are all fairly trivial. Can you guess any of them?

If we cannot guess the additional possible R-matrices, we can derive them explicitly (and show that no others exist). Let us suppose that we do not know the values of the R-matrix elements $R_I^{\tau\tau}$ and $R_\tau^{\tau\tau}$.

(d) For the case of $f = I$ and $c = \tau$, show that Fig. 10.16 implies

$$[R_\tau^{\tau\tau}]^4 = [R_I^{\tau\tau}]^2 \quad . \qquad (10.16)$$

(e) [Harder] For the case of $f = \tau$ we have a two-dimensional Hilbert space spanned by $c = I$ and $c = \tau$. Any linear operator on this Hilbert space should be a 2×2 matrix. Thus, the locality constraint Eq. 10.16 is actually an equality of 2×2 matrices. Derive this equality.

(f) Use this result in combination with Eq. 10.16 to find all possible R-matrices that satisfy the locality constraint. You should find a total of eight solutions. Six of these are spurious, as we discuss in Section 13.3.

The calculation you have just done is equivalent to enforcing the so-called hexagon condition, which we discuss in Section 13.3.

Computing with Anyons

Having discussed the basics of anyon theories, we are now in a position to discuss how one might perform quantum computations with braids.

In Chapter 2 we briefly introduced some ideas of topological quantum computation. In Chapter 8 we discussed how we might define a qubit in several simple anyon theories. This chapter briefly discusses how anyons can be used to fulfill the requirements for quantum computation.[1]

11.1 Quantum Computing

To have a quantum computer, we must first have a Hilbert space, and we usually think of this Hilbert space as being built from small pieces, such as qubits or qutrits.[2] This Hilbert space will be the quantum memory upon which the computer acts.

Our model of a quantum computer has four key steps for quantum computation:[3]

(0) Find a Hilbert space to work with.

(1) Initialize system to some known state within the Hilbert space.

(2) Perform a controlled unitary operation on the state.

(3) Measure some degrees of freedom in the Hilbert space.

If the controlled unitary (step 2) is implemented as a series of unitary operations, each of which acts on only small parts of the Hilbert space (such as acting on just a few qubits at a time), we call this scheme for quantum computation the *quantum circuit model*.[4]

We discuss each of the above steps (0)–(3) for our anyon systems in Section 11.2. First, however, we introduce the idea of what it means for a quantum computer to be "universal" in the quantum circuit model.

[1] For more of the basics of quantum computation, a classic reference is Nielsen and Chuang [2000]. We also provide some basic information in Section 26.1.

[2] Qubits are two-state systems (such as spin-$\frac{1}{2}$ particles), qutrits are three-state systems, etc. The general case is known as a qudit. See the introduction to quantum information in Chapter 26.1.

[3] There are variants on this theme. For example, it might be sufficient to initialize to a state that is only partially known, or it might be sufficient to have a somewhat noisy measurement. Most interesting is the issue of whether one can tolerate some amount of imperfection in the system (noise in the system, uncontrolled operations on the Hilbert space, etc.). We discuss this issue further in Chapter 26.

[4] There are other models of quantum computation. We mention in particular the measurement schemes (see Raussendorf and Briegel [2001]; Gross et al. [2007]) where no unitary is explicitly performed, but rather the computation is implemented as a series of measurements on an initial highly entangled state. In the context of topological quantum computation, an important variant is a computation that is implemented by a combination of unitary operations and projective measurements. The earliest proposal for quantum computing with anyons was of this type (see Kitaev [2003]). See also note 17.

11.1.1 Universal Quantum Computing in the Quantum Circuit Model

[5]Recall that a matrix U is unitary if and only if $U^\dagger U = UU^\dagger = 1$.

[6]Quantum-mechanical time evolution is always unitary. This is simply the statement that a normalized ket remains normalized. It is worth noting that we are excluding the possibility of making measurements (which are often considered to be nonunitary)[7] on the system before the end of the computation. This would be outside of the quantum circuit model.

[7]All of quantum mechanics can be viewed as unitary time evolution. Measurements may look like they are nonunitary, but one can always include the measuring apparatus within the system being considered and then the full system (including the measuring apparatus) obeys unitary evolution. The idea of including measurements within your system in order to maintain unitarity is sometimes known as "the church of the larger Hilbert space." See note 3 from Chapter 26.

[8]Recall that the last step of a quantum computation, after applying a unitary U (via some sequence of gates as in Eq. 11.1) to our Hilbert space, we obtain an output "answer" by measuring whether some particular qubits are in the $|0\rangle$ state or the $|1\rangle$ state. The probabilities of these outcomes is completely independent of the overall phase of the U. That is, if we changed $U \to e^{i\phi}U$ we would have the same probabilities of outcomes.

[9]I believe this distance measure was introduced by Fowler [2011]. Other definitions of distance can also be used. Relationships between this distance measure and more conventional operator norms are given by Field and Simula [2018] and Amy [2013].

Let us suppose our Hilbert space consists of N qubits (each qubit being a two-state system). The Hilbert space dimension is then $D = 2^N$. The space of possible unitary[5] operations[6] on these qubits is just the group of D-dimensional unitary matrices—a group known as $U(D)$.

Let us now suppose our quantum computer can implement any one of p different elementary operations (usually called "gates") in a single time step (each gate will act only on a small number of qubits). Each gate corresponds to a particular unitary operation $U_i \in U(D)$ with $i \in \{1,\ldots,p\}$ that is applied to the Hilbert space. One can assemble a sequence of such gates to construct a particular unitary operation that is just a product of these gates (the time order runs from right to left)

$$U = U_{i_t} \ldots U_{i_2} U_{i_1} \tag{11.1}$$

where each $i_x \in \{1,\ldots,p\}$, and the number of gates, t, can be thought of as the "run time" of the computation.

Suppose there is some particular computation we would like to perform, and this computation corresponds to a unitary U which we hope to construct via a series of gates as in Eq. 11.1. Note, however, that in quantum computation we are never worried about the overall phase of our result:[8] if we want to construct some particular unitary U, it is just as good to construct $e^{i\phi}U$ for any value of ϕ.

Unfortunately, even with this freedom of phase, most unitary operations (except for a set of measure zero) are actually impossible to construct exactly from a finite set of elementary gates as in Eq. 11.1. Fortunately, for computational purposes it is good enough to be able to *approximate* the desired unitary operation to some (potentially high) accuracy. A set of gates that enables such an accurate approximation is called *universal*. We will be more precise about the definition of this word in a moment.

Since we will be discussing approximations of desired operations, it is useful to define a distance between two unitary matrices as a measure of the accuracy of our approximation. Given two D-dimensional unitary matrices U and V, we define a phase invariant distance measure between them as[9]

$$\mathbf{dist}(U;V) = \sqrt{1 - \frac{|\mathrm{Tr}[U^\dagger V]|}{D}} \ . \tag{11.2}$$

Note that multiplying either matrix by an overall phase leaves **dist** unchanged, and if U and V are the same up to a phase, then **dist** is zero. We say that V is a good approximation of U up to a phase if $\mathbf{dist}(U;V)$ is small.

Having defined this distance measure, we can be more precise about what we mean by a universal gate set. A gate set $\{U_i \in U(D)\}_{i=1,\ldots,p}$ is universal if for any desired target operation $U_{\mathrm{target}} \in U(D)$ and any desired error tolerance ϵ, we can find a sequence of gates $U_{i_t} \ldots U_{i_2} U_{i_1}$

(which will generally depend on ϵ) such that the phase invariant distance is less than the error tolerance

$$\mathbf{dist}(U_{\text{target}} \; ; \; U_{i_t} \ldots U_{i_2} U_{i_1}) < \epsilon \quad . \tag{11.3}$$

In other words, our gate set can approximate any target unitary as precisely as we want.

We might wonder how long a run time (how many gates) will we typically need to have. A beautiful theorem by Kitaev and Solovay[10] assures us that the run time is not too long.[11] In particular,

$$t \sim \mathcal{O}\left(\log(1/\epsilon)\right) \quad . \tag{11.4}$$

We are thus guaranteed that if we have a universal gate set, then the run time of the computer gets at most logarithmically longer as we try to increase the quality of our approximation of the target operation U_{target}.

The essence of this theorem is as follows. If we consider a sequence of t gates (i.e. a run time of t), if there are p different elementary gates, we can construct roughly p^t different sequences of gates.[12] Thus, as t gets larger, there are exponentially more possible unitaries we can construct and these roughly cover the space $U(D)$ evenly. With the number of points we can construct in this space growing exponentially with t, the distance ϵ of an arbitrary target unitary to the nearest unitary we can construct must drop exponentially with t, hence justifying Eq. 11.4.

It is a nontrivial calculation to determine which set of elementary gates is sufficient to have a quantum computer that is universal. However, an important result is that if one can perform arbitrary rotations on a single qubit and in addition if one can perform *any* entangling two-qubit operation between any of these two qubits (or even just between neighboring bits), then one has a universal quantum computer.[13]

[10]The Kitaev–Solovay theorem, often viewed as one of the most fundamental results of quantum computation, is discussed nicely in Dawson and Nielsen [2006],Harrow [2001], and Harrow et al. [2002].

[11]The usual proof of the Kitaev–Solovay theorem assumes that the gate set must contain inverses. In other words, if U_n is one of the elementary gates, then U_n^{-1} should also be one of the elementary gates.

[12]We will not get exactly p^t different unitaries, since different sequences might generate the same unitary operation.

[13]This important theorem is sometimes known as the Brylinski theorem after its discoverers, Brylinski and Brylinski [2002]. The authors are married. A simpler version of the proof is given by Bremner et al. [2002].

11.2   Topological Quantum Computing

11.2.1   Hilbert Space

With a topological quantum computer, a qubit (or qutrit, etc.) can be formed from multiple anyons that can be put into multiple fusion channels (see Chapter 8). For example, with Fibonacci anyons a qubit might be formed from three Fibonacci anyons fusing to τ, as shown in Fig. 9.4. With the Ising theory, one might use a cluster of three Ising anyons fusing to σ as a qubit, as shown in Fig. 9.5. There are, of course, many more options of how one encodes a qubit in any given theory.[14] For example, in the Ising theory it may be more convenient to work with clusters of four Ising anyons fusing to I, as shown in Fig. 11.1.

Fig. 11.1   A qubit made from four Ising anyons in an overall fusion channel of I. The two states of the qubit are $x = \psi$ and $x = I$. Note that due to the fusion rules of the Ising theory, if the overall state of the four qubits is I, then if the left two anyons are in state x, the right two must also be in state x. Using a qubit made of four anyons has advantages for other topological theories such as $SU(2)_k$ with $k > 4$. See, for example, Hormozi et al. [2009].

[14]Most choices of qubit encodings are relatively "inefficient" as quantum memories. For example, if we use three Fibonacci anyons to encode each qubit, we have a computational Hilbert space of dimension $2^{N/3}$ with N the number of anyons. However, the full Hilbert space dimension is much larger being approximately dimension $(\mathsf{d}_\tau)^N$ with $\mathsf{d}_\tau = (1 + \sqrt{5})/2 > 2^{1/3}$. The reason we sacrifice such a large amount of the Hilbert space is to simplify the construction of useful unitary gates.

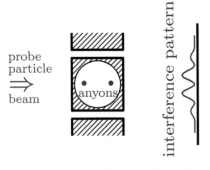

Fig. 11.2 Using Aharonov–Bohm-like interference to measure the fusion channel of two anyons (inside the circle) that are far apart.

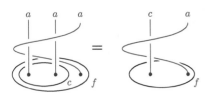

Fig. 11.3 The interference experiment in Fig. 11.2 is equivalent to measuring the phase of wrapping the probe anyon (right) around the two test particles. The general expression for the resulting phase would be $\hat{\sigma}_2\hat{\sigma}_1^2\hat{\sigma}_2$, which is dependent on the fusion channel c. (In Chapter 13. we refer to this phase as $R_f^{ca}R_f^{ac}$.)

11.2.2 Measurement (in Brief) and Initialization

A topological qubit could be measured in several ways, depending on the particular physical system in question. The general principle of locality that we introduced in Section 8.2 (see in particular Fig. 8.8) gives a good idea how such measurements can (or can't) be done.

Let us suppose, for example, we have two anyons of type a and we would like to measure their fusion channel. Given the principle of locality, to measure this fusion channel we must perform an operation which is local to both particles, that is, a measurement that surrounds both.

One way to measure the fusion channel of two anyons is to bring them together to the same point, or at least bring them physically close on a microscopic scale. When two anyons are microscopically close to each other, in essence their wavefunctions mix with each other and in this case the measurement of *almost any* nontrivial operator near that location will suffice to distinguish between the different possible fusion channels. For example, one could measure the energy of the two anyons, or the force between them, which would generally distinguish the fusion channels. Note, however, when the anyons are moved macroscopically far apart *no* local operators can distinguish the fusion channels (we discuss precisely why this is the case in Section 29.2.2).

Another way to measure the fusion channel of two anyons would be to leave the two anyons far apart from each other but implement a measurement that surrounds them both—such as an Aharonov–Bohm-type interference measurement, as shown in Fig. 11.2. Here, a test anyon particle wave is split into two partial waves that travel on opposite sides of the anyons to be measured and then reinterfere with each other. This is entirely analogous to the regular Aharonov–Bohm effect (see Section 4.1 and Fig. 4.2), where the partial waves travel on opposite sides of a flux and then reinterfere. In the usual Aharonov–Bohm effect, the net phase we measure is the phase of wrapping a single test particle all the way around the central region (see Eq. 4.3). Analogously, here we measure the phase of wrapping the probe anyon all the way around the anyons in the central region to measure their fusion channel, as shown in Fig. 11.3. Experiments of this sort have been attempted in quantum Hall systems (see Willett et al. [2019] and Chapter 37).

Once we know how to measure the state of the anyons in our Hilbert space (and assuming we know how to manipulate our qubits) it is then fairly trivial to initialize the Hilbert space. We simply measure the state of a qubit: if it is in the state we want, we have finished. If it is in some other state, we apply the appropriate unitary operation to put it into the desired initial state. We discuss unitary operations next.

11.2.3 Universal Braiding

The most interesting part of topological quantum computation is the idea that we can apply unitary operations to our Hilbert space by braiding anyons around each other. The elementary gates of the system (or elementary unitary operations) are the (counterclockwise) exchanges of two identical anyons, which, in braid group notation,[15] we call $\hat{\sigma}_n$, as well as the inverse (clockwise) exchanges $\hat{\sigma}_n^{-1}$, where $n \in \{1, 2, \ldots, (N-1)\}$ for a system of N identical anyons. Each of these braid operators corresponds to a unitary matrix operating on the Hilbert space.

It turns out that for many types of nonabelian anyon theories, the gate set made up of elementary braiding exchanges is universal in the sense defined in Section 11.1.1.[16] For example, braiding is universal for Fibonacci anyons. Similarly, $SU(2)_k$ Chern–Simons theory is universal for $k = 3$ and $k \geq 5$. In fact, among nonabelian anyon theories, theories where braiding is *not* universal are somewhat of an exception. Ising anyons and the closely related $SU(2)_2$ Chern–Simons anyons are two of these nonuniversal exceptions.[17]

It turns out that any system of N identical anyons that is capable of universal quantum computation by braiding is also capable of universal quantum computation by *weaving*.[18] Here, what we mean by "weave" is that we fix the positions of $N-1$ of the anyons and only move the one remaining anyon around all the other stationary anyons. An example of a weave is shown in Fig. 11.4. The weaves are a very restricted subset of the possible braids, but still the weaves form a universal set of gates for these anyon systems. This result will be important in Section 11.4.1 (see also Exercise 11.5).

In fact, if one is able to measure fusion channels easily,[19] it is also possible to implement universal quantum computation just by making many measurements of fusion channels without physically braiding any particles around any others.[20]

It is crucial to emphasize a key advantage of using braids to implement unitary operations: the outcome is quantized. For a conventional qubit, rotating the qubit by a certain angle may mean that you apply a microwave field to a spin of a certain amplitude for a certain amount of time. However, if you mistakenly apply this field for 1% too long, then you have made a 1% error in your result. In contrast, for a topological qubit, one either does, or does not, make the right braid—you cannot miss by a small amount. You might naively worry that you bring the anyon back to a slightly wrong location after making the braid—however, such an imperfection cannot matter since the actual position where the anyon sits is not important. The only thing that matters in the end is the knot in spacetime of the world line of the particle.

[15]Here we have put hats on $\hat{\sigma}_n$ to indicate that we mean these to be gates—unitary operators acting on the Hilbert space caused by braiding anyons. In contrast σ_n is the braid generator itself. In other words, $\hat{\sigma}_n$ is the representation matrix of the braid group element σ_n.

[16]This result was shown by Freedman et al. [2002a, b] These papers are not particularly easy to read for physicists.

[18]This is proven by Simon et al. [2006]. Publication of this work reduced my Erdős number to its current value of 3. See Batagelj and Mrvar [2000].

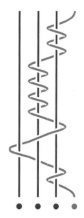

Fig. 11.4 A *weave* is a braid where only one particle moves and all the other particles remain stationary. All the particles in this figure are supposed to be of the same type. The single particle that moves is colored red just for clarity.

[19]Note that measurement schemes of the Aharonov–Bohm type, as in Fig. 11.2, involve braiding a test particle around other particles.

[20]See Bonderson et al. [2008].

[17]$SU(2)_4$ is an interesting case where braiding alone is not universal. However, if we are allowed to go beyond the quantum circuit model and combine braiding with many intermediate projective measurements (i.e. not just making one measurement at the end of the computation), then $SU(2)_4$ anyons can implement universal quantum computation (see Cui and Wang [2015]; Levaillant et al. [2015]). In fact, the first proposal of a topological quantum computer, by Kitaev [2003], described a computation scheme that involved both braiding and projective measurements. A simple discussion of this scheme is given by Preskill [2004] with early extensions of the scheme given by Mochon [2003, 2004].

11.2.4   Computing with Non-Universal Anyons?

For abelian anyons, increasing the number of anyons does not increase the size of the Hilbert space, so one cannot encode qubits in the state of multiple anyons. However, one can still encode qubits in the ground-state degeneracy of a manifold of high genus (meaning a g-handled torus, or more simply a disk with some number of punctures in it, given the right boundary conditions on the puncture. See the discussion of Chapter 27 for example). One can, in principle, use such a system as a quantum memory. It is even possible to construct gates to implement unitary operations on this Hilbert space to build a quantum computer. (Examples of this are given by Lloyd [2000],Fowler et al. [2012], and Pachos [2006].) However, one will not be able to achieve universal quantum computation by braiding. As such, one needs to introduce non-topological gates, which are then not protected from errors.

In the case of nonabelian anyons, the Hilbert space grows exponentially with the number of quasiparticles (Eq. 3.11) making a very convenient quantum memory (one does not have to use a nontrivial manifold). However, as mentioned just above, there are a few cases, such as $SU(2)_2$ or Ising, where braiding alone is not sufficient to perform universal quantum computation, and again one needs to introduce certain non-topological gates. However, in these particular cases, there are still significant advantages to using braiding. In particular, in the case of $SU(2)_2$ or Ising, one can assume that the gates implemented by braiding can be done with very high accuracy since they are topologically protected. This then allows us a very high tolerance on how inaccurately we can perform the remaining non-topological gates! That is, because some gates (those done by braiding) are very accurate, the non-topological gates are allowed to be very inaccurate, and we can still complete the quantum computation. We can think of this as a *partially topological* quantum computer. This is discussed in some depth by Bravyi [2006] and Baraban et al. [2010].

11.3   Fibonacci Example

As an example, we consider the case of Fibonacci anyons, which is potentially the simplest anyon system that is universal for quantum computation.

11.3.1   A Single Fibonacci Qubit

Let us consider a single qubit made of three Fibonacci anyons. We have discussed this several times before in Sections 8.2.1, 9.1, and 10.2.1. To remind the reader, there are three possible states of three Fibonacci anyons, which we label $|N\rangle, |0\rangle, |1\rangle$ (see Fig. 8.10)—which represent a qubit (the states $|0\rangle$ and $|1\rangle$) and one additional "non-computational" state $|N\rangle$ that we will not use for storing quantum information.

We now think about braiding our three anyons. For the braid group

Fig. 11.5 This is the braid written in Eq. 11.6, which gives an approximation of the X-gate on a single qubit made from Fibonacci anyons. As usual, time runs from bottom to top. The distance to the target is **dist** ≈ 0.17.

on three strands, B_3 (see Section 3.3.1), there are two generators, σ_1, exchanging the first two strands counterclockwise, and σ_2, exchanging the second two strands counterclockwise. Any braid of three particles can be constructed as some product of $\sigma_1, \sigma_2, \sigma_1^{-1}$, and σ_2^{-1} in some order as shown, for example, in Fig. 11.5.

The action of these braid operations on the three-dimensional Hilbert space is shown in Eqs. 10.7 and 10.8, which we calculated in Section 10.2.1. By multiplying these matrices together, we can figure out how any complicated braid acts on our Hilbert space. For now we are only interested in how the matrices act on the space of the qubit states $|0\rangle$ and $|1\rangle$ and we will worry about the $|N\rangle$ state in Section 11.4.1.

Example of X-Gate

We are now interested in the following simple quantum computation problem. Given a particular target unitary operation which we might want to perform on our qubit, how should we move the anyons? That is, what braid should we do to implement the target operation?

For example, suppose we want to design a braid that implements an X-gate[21] (just a Pauli σ_x)

$$U_{\text{target}} = X = \begin{pmatrix} 0 & 1 \\ 1 & 0 \end{pmatrix} \tag{11.5}$$

With a very short braid (shown in Fig. 11.5), we can make a fairly poor approximation to this gate (this braid is the best we can do with only five braid operations) given by[22]

$$U_{\text{approx}} = \hat{\sigma}_2^{-1} \hat{\sigma}_1^3 \hat{\sigma}_2^{-1} \approx e^{-3\pi i/5} \begin{pmatrix} 0.073 - 0.225i & 0.972 \\ 0.972 & -0.073 - 0.225i \end{pmatrix}. \tag{11.6}$$

For the approximation given in Eq. 11.6 the phase invariant distance from the target is[23]

$$\mathbf{dist}(U_{\text{target}} \,;\, U_{\text{approx}}) \approx 0.17,$$

which is not a great approximation. However, with a longer braid having nine braid operations, shown in Fig. 11.6, one can make a better approximation with $\mathbf{dist} \approx 0.08$. If we consider braids that are longer and longer, we can get successively better approximations to the desired target as would be expected from the Kitaev–Solovay theorem discussed in Section 11.1.1.

As mentioned in Section 11.2.3, it is possible to find braids that are *weaves*, meaning that only a single anyon moves. For example, we show a weave in Fig. 11.7 that implements an X-gate to precision $\mathbf{dist} \approx 0.18$. Note that due to the restricted weave form of this braid, a slightly larger number of elementary exchanges are required to reach roughly the same precision as the braid shown in Fig. 11.5. In principle, by using longer weaves one can get as close to the target as one likes.

[21] Note that a Z-gate can be implemented exactly as $\hat{\sigma}_1^5$. It is unusual and non-generic that a target can be constructed exactly.

[22] Here $\hat{\sigma}_1$ and $\hat{\sigma}_2$ are the lower 2×2 block of the 3×3 matrices in Eqs. 10.7 and 10.8. We have dropped the first rows and columns because we are not concerned with the $|N\rangle$ state currently.

[23] Recall that in comparing Eq. 11.6 to Eq. 11.5 we are not concerned with the overall phase, so we ignore the prefactor of $e^{-3\pi i/5}$ in Eq. 11.6.

Fig. 11.6 A longer braid gives a more accurate approximation to the desired target X-gate for Fibonacci anyons. This braid, $\sigma_1^{-2}\sigma_2^3\sigma_1^2\sigma_2^{-2}$, having nine braid operations, has distance $\mathbf{dist} \approx 0.08$ to the target.

Fig. 11.7 A weave that approximately implements an X-gate for Fibonacci anyons. All three anyons are meant to be identical. The anyon colored red is mobile whereas the other two are kept stationary. The distance to the target is **dist** ≈ 0.18. Because we have restricted the form of this braid to be a weave, the braid is longer (has more elementary exchanges) than the one in Fig. 11.5 for roughly the same accuracy.

[24]Quantum compiling is analogous to compiling for a conventional computer, which is the task of starting with a high-level programming language and determining which machine-level instructions to implement at the computer chip level. See Harrow [2001] for a discussion of quantum compiling in general.

[25]The field of topological quantum compiling was started by Bonesteel et al. [2005]. A very nice recent review of the topic as well as discussion of a number of other approaches toward topological quantum compiling is given by Field and Simula [2018].

[26]For example, we might not want to search any braids where σ_i and σ_i^{-1} occur in a row since they would cancel.

[27]Here we mean the computational effort for the classical computer that we use to design our quantum algorithm!

11.3.2   Topological Quantum Compiling: Single Qubit

Even if there exists a braid that performs a unitary operation that approximates some target operation within some small error distance ϵ, it is a nontrivial task to figure out what that braid is. In other words, how do you know what braid you should implement on your computer in order to perform the desired operation?

The general task of determining which elementary gates should be performed, and in what order, to implement some desired target unitary is known as *quantum compiling*.[24] For a topological quantum computer, the task of designing a braid is therefore known as *topological quantum compiling*. Here we discuss several approaches to topological quantum compiling in order of their complexity, and their effectiveness.[25] We continue to focus only on compiling braids for a single Fibonacci qubit. Multi-qubit braids are discussed in Section 11.4.

Brute Force Search

If we are willing to accept a fairly poor approximation of our target unitary (a fairly large **dist** between our approximation and the target) we will be able to use a fairly short sequence of elementary gates (i.e. a short braid). In this case we can consider some maximum gate sequence length t (maximum run time) and search all possible gate sequences of length less than t, choosing the one that best approximates our target. We should expect to achieve a distance to the target that drops exponentially with t, as discussed near Eq. 11.4.

If we are considering a single qubit made of three Fibonacci anyons, our elementary gates are the braid generators $\hat{\sigma}_1, \hat{\sigma}_2, \hat{\sigma}_1^{-1}, \hat{\sigma}_2^{-1}$. This means that if we want to search through all braids of length t we have to search roughly 4^t braids. While there are some tricks that allow us to reduce this number somewhat,[26] the computational effort[27] will always grow exponentially with the length t. If one wants to make a highly accurate approximation of a target unitary with a very small error distance, run times t can become large enough that brute-force searching becomes unfeasible.

Kitaev–Solovay Algorithm

Kitaev and Solovay[28] provide us an explicit algorithm to construct very accurate approximations of any desired unitary given a universal set of elementary gates with reasonable (not exponentially growing!) computational effort[27]. The essence of this algorithm is as follows. Let us suppose that by brute-force search we can approximate any unitary operation to within a distance **dist** $\sim \epsilon_0$ with a sequence of elementary gates (elementary braids in the topological case) of length t_0. Let us say

[28]See again Dawson and Nielsen [2006] and Harrow [2001] for discussions of the Kitaev–Solovay algorithm. These works are rigorous about why distances scale as ϵ to certain powers, which we simply state here without proof.

that the classical computational time to achieve this is T_0. Now given a target unitary $U_{\text{target}}^{(0)}$ that we would like to approximate, we start with this brute-force search, and construct our approximation $U_{\text{approx}}^{(0)}$, which is accurate to within **dist** $\sim \epsilon_0$. This is our 0th level approximation of the target. We would next like to repair this approximation with another series of gates to make it more accurate. We thus define

$$U_{\text{target}}^{(1)} \equiv [U_{\text{approx}}^{(0)}]^{-1} U_{\text{target}}^{(0)} \quad .$$

If we could find a series of gates that would exactly give us $U_{\text{target}}^{(1)}$ we could exactly construct the original objective $U_{\text{target}}^{(0)}$ as

$$U_{\text{target}}^{(0)} = U_{\text{approx}}^{(0)} U_{\text{target}}^{(1)} \quad .$$

However, it is not obvious that we have any better way to approximate $U_{\text{target}}^{(1)}$ than we had to approximate $U_{\text{target}}^{(0)}$, so why does this help? The key here is that $U_{\text{target}}^{(1)}$ is necessarily close (**dist** $\sim \epsilon_0$) to the identity. We then decompose

$$U_{\text{target}}^{(1)} = VWV^{-1}W^{-1}$$

with W and V being unitary operations close to the identity (**dist** $\sim \sqrt{\epsilon_0}$). We then have an amazing result, that if we are able to approximate V and W to an accuracy ϵ_0 (which we can do here by brute-force search) we will get $U_{\text{target}}^{(1)}$ accurate to **dist** $\sim \epsilon_0^{3/2}$. Thus we obtain

$$U_{\text{target}}^{(0)} = U_{\text{approx}}^{(0)} VWV^{-1}W^{-1} \tag{11.7}$$

accurate to order $\epsilon_0^{3/2}$. The total sequence of gates is now of length $5t_0$ since each of the factors on the right-hand side of Eq. 11.7 is of length t_0. The classical computational effort to achieve this is roughly $3T_0$ since we must search for $U_{\text{approx}}^{(0)}$ and V and W.

This scheme can then be iterated to make our approximation even better. The only change is that in the next level of approximation, instead of using brute-force search to make approximations good to **dist** $\sim \epsilon_0$, we use the entire above described algorithm to make all of our approximations good to **dist** $\sim \epsilon_0^{3/2}$. When $U_{\text{approx}}^{(0)}$ and V and W are calculated to order $\epsilon_0^{3/2}$ our new approximation for $U_{\text{target}}^{(0)}$ will be accurate to **dist** $\sim (\epsilon_0^{3/2})^{3/2}$.

The entire scheme can be iterated recursively to any level of accuracy. At the nth level of this approximation, we have a gate sequence of length $5^n t_0$ and an accuracy **dist** $\sim \epsilon_0^{(3/2)^n}$ and the computational effort[27] to achieve this scales as $3^n T_0$.

Thus if we want to achieve some overall accuracy ϵ of our operation, the gate sequence will be of length

$$t = \mathcal{O}\left([\log(1/\epsilon)]^{\log(5)/\log(3/2)} \right) = \mathcal{O}\left([\log(1/\epsilon)]^{3.969\ldots} \right) \tag{11.8}$$

and this requires classical computation time

$$T = \mathcal{O}\left([\log(1/\epsilon)]^{\log(3)/\log(3/2)}\right) = \mathcal{O}\left([\log(1/\epsilon)]^{2.710\ldots}\right). \qquad (11.9)$$

While this algorithm produces gate sequences that are longer than one obtains with brute-force searching (which produces gate sequence lengths as in Eq. 11.4) for the same desired accuracy ϵ, it has the advantage that it is computationally feasible[27] for much smaller values of ϵ and can therefore produce more accurate results.

Galois Theory Optimal Compiling:

A rather remarkable scheme for quantum compiling was developed in Kliuchnikov et al. [2014] based on ideas from Galois theory.[29,30] While we cannot review Galois theory here, nor even do justice to the details of the algorithm, we can nonetheless discuss some of the mathematical structure that makes this approach possible.

It turns out that any unitary that can be constructed by braiding three Fibonacci anyons can be written (up to a phase) in the form

$$U(u,v,k) = \begin{pmatrix} u & v^*\omega^k\phi^{-1/2} \\ v\phi^{-1/2} & -u^*\omega^k \end{pmatrix} \qquad (11.10)$$

where $\phi = (1+\sqrt{5})/2$ is the golden mean, k is an integer, $\omega = e^{2\pi i/10}$, and

$$|u^2| + \phi^{-1}|v|^2 = 1 \qquad (11.11)$$

where u and v come from the so-called ring of cyclotomic integers $\mathbb{Z}[\omega]$, which means that

$$u = \sum_{i=0}^{3} a_i\omega^i \qquad v = \sum_{i=0}^{3} b_i\omega^i \qquad (11.12)$$

with coefficients a_i and b_i all being integers. The fact that the unitaries that can be generated by braiding take a very restricted mathematical form is, in fact, a generic property of all anyon theories,[31] although the particular form taken depends on the particular anyon theory.

Further, given values of u, v, and k a relatively simple algorithm is provided that finds a braid[32] that results exactly in this unitary, where the length of the braid is no longer than

$$t \sim \log\left(|\sum_{i=0}^{3} a_i\omega^i|^2 + |\sum_{i=0}^{3} a_i\omega^{3i}|^2\right).$$

This procedure is known as *exact synthesis* as it constructs exactly the desired $U(u,v,k)$ as a series of elementary braid operations.

The remainder of the algorithm is to find values of u, v, k (with u and v of the form in Eq. 11.12) so that Eq. 11.10 approximates any given target unitary. This task can exploit established methods from algebraic

---

[29] Évariste Galois was undoubtedly one of the most interesting and brilliant mathematicians of all time. Being politically active in an era shortly after the French revolution, he spent a decent fraction of his short adult life in prison. His mathematical works (some written while in prison) opened up vast new fields of research. He died at age 20 in a duel.

[30] This approach is not restricted to anyons, but can be applied to compiling many types of one-qubit gate sets. See Kliuchnikov et al. [2015] and Kliuchnikov and Yard [2015].

[31] This is due to the fact that the F- and R-matrices of an anyon theory live in a particular so-called Galois extension of the rationals—meaning that only certain irrational factors can show up in any mathematical expression. This fact can be used to prove various statements about what type of operations can or cannot be done exactly by braiding. See for example Freedman and Wang [2007].

[32] Kliuchnikov et al. [2014] also provide a similar algorithm for generating *weaves*. See Section 11.2.3.

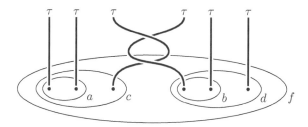

Fig. 11.8 The braid shown here between two Fibonacci qubits entangles the two qubits but also results in leakage error. When we use three Fibonacci anyons as a qubit, we set the overall fusion channel of the three to be τ, so $c = d = \tau$ in this figure. The quantum information is stored in the quantum numbers a and b. The shown braid results in some amplitude ending up in the non-computational space where either c or d is I rather than τ.

number theory. The interested reader is referred to Kliuchnikov et al. [2014].

   The end result of this approach is an algorithm that, although it does not find the absolute optimal braid,[33] is nonetheless *asymptotically optimal* in the sense that it produces braids of length

$$t = \mathcal{O}(\log(1/\epsilon))$$

as in Eq. 11.4. Further, the computational time[27] to achieve this scales only as $T = \mathcal{O}(\log(1/\epsilon)^2)$. Using this approach, it is easy to generate braids with error distances of order 10^{-100} or even better, and these braids are longer than the absolute optimal braid by only a factor of order unity.

11.4   Two-Qubit Gates

Having studied single-qubit operations, we now turn to a brief discussion of two-qubit gates.[34] As mentioned in Section 11.1.1, the Brylinski theorem tells us that to have a universal quantum computer, we need only have single-qubit rotations along with any entangling two-qubit gate. To construct such an entangling two-qubit gate we need to have a braid that physically entangles the world lines of the anyons comprising the two qubits, such as the example shown in Fig. 11.8.

   However, there is a crucial complication with braiding anyons between qubits. If we perform a braid such as that shown in Fig. 11.8, the fusion channel of the anyons comprising each of the qubits (quantum numbers c and d in the figure) are not preserved (see the discussion of locality in Section 8.6) and this means that amplitude can *leak* into the non-computational space.

   To be more explicit for the Fibonacci case, recall that we encode our qubits ($|0\rangle$ or $|1\rangle$) by using three Fibonacci anyons in overall fusion channel τ (see Fig. 8.10). The fusion channel of the three anyons to I is termed non-computational ($|N\rangle$) and is not used for computation. If some of the amplitude of the wavefunction ends up in this non-computational space, it is called *leakage error*, and only very small

[33]The "optimal braid" is the one that would be found by brute-force search if one had the exponentially enormous computational power necessary to find it.

[34]Here we are constructing a two-qubit unitary operation, which we call a two-qubit gate, from our elementary gates—the elementary braid operations.

quantities of such leakage errors can be tolerated for any realistic computation. Braids like the one shown in Fig. 11.8 always produce non-zero leakage error. The problem of leakage error in two-qubit gates is not special to Fibonacci anyons, but is in fact a generic property of all anyon theories that have universal braiding.[35]

[35] Ainsworth and Slingerland [2011] show that it is not possible to design completely leakage-free gates for any universal anyon theory, and leakage can only be made approximately zero.

While we cannot completely eliminate leakage, we can in principle design entangling gates with arbitrarily small (albeit non-zero) leakage. Such braids with low leakage error do exist, but finding them is highly nontrivial. Inconveniently, the Hilbert space of six Fibonacci anyons, as in Fig. 11.8, is thirteen-dimensional.[36] Searching such a large space for particular unitaries with low leakage is numerically unfeasible. We thus need a more clever way to design braids with low leakage.

[36] This space is subdivided into an eight-dimensional subspace with $f = \tau$ and a five-dimensional subspace with $f = I$. No braiding of these six anyons will change the f quantum number. Note, however, that gates intended to work in our computational Hilbert space must have low leakage independent of the value of f.

In designing any computation, it is almost always advantageous to simplify the desired task into smaller tasks that can be addressed one at a time. This "divide and conquer" approach will allow us to tackle the job of designing two-qubit gates. In the next section we give an example of how entangling gates with negligible leakage can be designed.

11.4.1   Controlled Gates

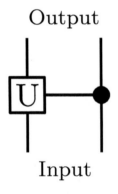

Fig. 11.9 Typical notation for a controlled unitary gate $C(U)$. The second (right) qubit controls the first (left).

[37] As always in quantum mechanics the operation acts linearly on superpositions.

[38] The constructions discussed here were introduced in Bonesteel et al. [2005].

In quantum computation it is often very convenient to use entangling gates that are so-called controlled U-gates, or $C(U)$ where U is a two-dimensional unitary matrix. A controlled U-gate (see Fig. 11.9) acts on two qubits such that one qubit (the "target" qubit) is acted on with a two-dimensional unitary operator U if the other qubit (the "control" qubit) is in the $|1\rangle$ state, whereas the control qubit remains unchanged:

$$
C(U): \quad
\begin{cases}
|0\rangle \otimes |0\rangle & \rightarrow & |0\rangle \otimes |0\rangle \\
|1\rangle \otimes |0\rangle & \rightarrow & |1\rangle \otimes |0\rangle \\
|0\rangle \otimes |1\rangle & \rightarrow & (U|0\rangle) \otimes |1\rangle \\
|1\rangle \otimes |1\rangle & \rightarrow & (U|1\rangle) \otimes |1\rangle \ .
\end{cases}
\tag{11.13}
$$

Thus, the first qubit here is being controlled by the second qubit.[37] A common example of a controlled gate is $U = X$ (see Eq. 11.5), which we call a controlled-X, or more often a controlled-NOT (or CNOT) gate.

The key to our construction of controlled gates[38] is the locality principle of Section 10.1.1. If we are given a cluster of two anyons that are in the τ fusion channel (for example, set $c = \tau$ in Fig. 10.7) and we braid it around some other anyons, this will have the same effect as if we just braided a single τ around the other anyons. However, if the cluster of two anyons is in the trivial (or I) fusion channel, then braiding this cluster around other anyon never does anything, as braiding the vacuum particle is always trivial. Thus we can see that the effect of the braid is "controlled" by the fusion channel of the two anyons.

Controlled $\hat{\sigma}_2^2$ gate

Consider the construction shown in Fig. 11.10. On the far left of this figure, we have shown a *weave*, meaning only a single anyon, the one drawn in red, moves and the other two anyons remain stationary (see the discussion in Fig. 11.4). This weave has been designed to have approximately the same effect as if the two blue anyons are wrapped around each other (exchanged twice counterclockwise), that is, $\hat{\sigma}_1^2$, as shown in the figure. For the particular weave shown, the distance to the target $\hat{\sigma}_1^2$ is **dist** $\approx .12$. We could make the approximation of $\hat{\sigma}_1^2$ more accurate by using a longer weave designed by any of the compiling methods discussed in Section 11.3.2. Note that the equivalence between the weave on the far left and σ_1^2 is true as a 3×3 matrix[39] acting on the full three-dimensional Hilbert space spanned by the three fusion channels of three Fibonacci anyons (i.e. on the space spanned by $|0\rangle, |1\rangle, |N\rangle$, not just $|0\rangle, |1\rangle$).

[39]In fact it will be crucial that the equivalence holds as a 3×3 matrix since the braiding on the right of Fig. 11.10 takes anyons out of their original groups of three so we cannot easily separate the computational and non-computational spaces.

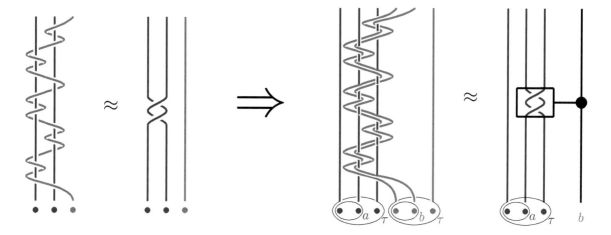

Fig. 11.10 Construction of a controlled gate using Fibonacci anyons. The weave on the far left is designed to have almost the same effect on the three-dimensional Hilbert space as the braiding (two counterclockwise exchanges) of the two blue particles as shown. Using a longer weave one can more closely approximate the braiding of the two blue particles. On the right, we have a system of six anyons representing two qubits. The value of the right (red) qubit is b which takes the value I or τ corresponding to $|0\rangle$ or $|1\rangle$. On the far right the red qubit is drawn a single black line that can have two possible orthogonal quantum states analogous to Fig. 11.9.

Now consider the braiding of six anyons on the right of Fig. 11.10 representing two qubits—the right (red) anyons form the control qubit and the left (blue) anyons form the target qubits. We group the two red anyons in fusion channel b and we move them around as a single unit to form the same weave as shown on the far left (here using the two red anyons and the right two anyons of the blue qubit). If these two red anyons are in the vacuum fusion channel $b = I$, then this braiding has no effect on the Hilbert space (braiding of the vacuum particle is always trivial). On the other hand, if the two anyons are in fusion channel $b = \tau$ then this braid is equivalent to moving a single τ particle through exactly the same weave as on the far left, thus having the same effect as exchanging the two rightmost blue anyons twice counterclockwise. We have thus constructed a controlled operation, $C(\hat{\sigma}_2^2)$, which is notated on the far right of the figure in a manner analogous to Fig. 11.9: the operation implemented on the blue qubit is (approximately) a full braiding of the right two blue strands, if and only if the right qubit (b) is in the τ or $|1\rangle$ state[40].

A crucial feature of this construction is that, to the extent that the weave we use accurately approximates $\hat{\sigma}_1^2$ on the far left of Fig. 11.10, the resulting construction leads to no leakage error. The right (red) qubit (b) is completely unchanged (hence, not creating leakage into the non-computational space of the right three anyons), and the effect on the left (blue) qubit (a) is designed to be equivalent to just braiding two of the blue anyons—which also does not create leakage.

Controlled U-gate

With a bit more work, we can in fact make any controlled gate $C(U)$ for an arbitrary two-dimensional unitary U, as in Fig. 11.9 (up to an overall phase as discussed in Section 11.1.1).

First let us discuss the so-called *injection weave* described in Fig. 11.11. An ideal injection weave is meant to leave the Hilbert space unchanged (it only applies an identity matrix). However, it has the nontrivial effect of rearranging the three strands comprising a qubit. As shown in the figure, the injection weave moves the red strand from the far right at the bottom to the far left at the top. As discussed in Section 11.3.2 we can more precisely approximate the ideal injection by using a longer weave.

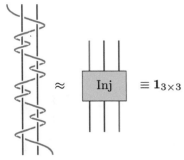

Fig. 11.11 An approximate injection weave is shown on the left. A perfect injection has no effect on the Hilbert space (it applies the identity matrix to the Hilbert space) but moves the red strand from the right to the left of the three anyons. The distance to the target for this particular weave is **dist** $= 0.09$. With a longer weave one can more accurately approximate a perfect injection.

We now construct the braid shown on the left of Fig. 11.12. As in Fig. 11.10, we group together the two red anyons in fusion channel b and we move them around as a group. These two anyons are first put through an injection weave with the rightmost two blue anyons. This moves the group of two red anyons into the middle position of the left qubit. A weave to implement an arbitrary unitary U is then implemented on the three strands furthest left, treating the two red strands grouped together as a single strand. Finally, this injection weave is inverted to bring the two red particles back to their original positions. The braid constructed in this way will implement a controlled U gate $C(U)$, as shown using the notation of Fig. 11.9 on the right of Fig. 11.12: the left (blue) qubit (a) has the unitary U applied to it, if and only if the right (red) qubit (b) is in the $|1\rangle$ or τ state. If we choose U to be an X gate, such that the necessary weave in the middle step is a weave like that shown in Fig. 11.7, we obtain a $C(X)$ or controlled NOT gate (CNOT).

To understand this procedure we realize that the only two anyons moved in this procedure are the two red anyons in state b, and these two are moved together as a group. As in Fig. 11.10, if these two anyons are in fusion channel $b = I$ (or $|0\rangle$), then the Hilbert space is left unchanged. However, if $b = \tau$ (or $|1\rangle$), then there will be an effect on the blue qubit— hence, we have a controlled unitary operation. Let us now consider the case when $b = \tau$, so that we can think of the two red strands as being

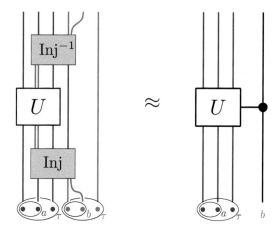

Fig. 11.12 Construction of a controlled U gate $C(U)$ with Fibonacci anyons. The two red anyons in state b are moved as a group and all other anyons are kept stationary. If $b = I$ or $|0\rangle$, then the weave has no effect on the Hilbert space. If $b = \tau$ or $|1\rangle$, then this weave implements a U rotation on the left (a) qubit.

a single τ strand. The injection weaves are designed to have no effect on the Hilbert space—their only effect is to move the red double strand inside of the blue qubit. The unitary operation U is thus the only nontrivial effect on the Hilbert space.

Chapter Summary

- Certain anyon systems can implement universal topological quantum computation—implementing unitary gates arbitrarily accurately by braiding anyons around each other.
- Determining which sequence of braids should be done to implement a given computation is known as "topological quantum compiling."

Further Reading:

- For general information about quantum computation, there are a bazillion references. However, the real "standard reference" of the field is Nielsen and Chuang [2000].
- The original ideas that braiding anyons can do quantum computation is due to Kitaev [2003], but is also credited to Freedman [1998].
- Freedman et al. [2002a] and Freedman et al. [2002b] did the initial work showing that $SU(2)_k$ is universal for $k = 3$ and $k \geq 5$ (these are not easy reading). Many of the same ideas can be generalized for other anyon systems.
- Nayak et al. [2008] is the standard reference for topological quantum computation. A more recent reference that gives a lot of attention to compiling and Fibonacci anyons is Field and Simula [2018].

Exercises

Exercise 11.1 Ising Nonuniversality

The braiding matrices for Ising anyons are given by Eqs. 10.11 and 10.12. Demonstrate that any multiplication of these matrices and their inverses will only produce a finite number of possible results. Thus conclude that Ising anyons are not universal for quantum computation. Hint: Write the braiding matrices as $e^{i\alpha}U_j$ where U_j is unitary with unit determinant, that is, is an element of $SU(2)$. Then note that any $SU(2)$ matrix can be thought of as a rotation $\exp(i\,\hat{n} \cdot \boldsymbol{\sigma}\,\theta/2)$ where here θ is an angle of rotation, $\hat{n}$ is the axis of rotation, and $\boldsymbol{\sigma}$ is the vector of Pauli spin matrices.

Exercise 11.2 Brute-Force Search

Given the braid matrices for Fibonacci anyons in Eqs. 10.7 and 10.8, ignoring the non-computational state $|N\rangle$, and ignoring the overall phase as usual, we would like to find an approximation to the Hadamard gate

$$H = \frac{1}{\sqrt{2}} \begin{pmatrix} 1 & 1 \\ 1 & -1 \end{pmatrix} .$$

(a) Use brute-force searching to find a braid of length three with a phase-invariant distance to target **dist** ≈ 0.084.

(b) Write computer code to find longer braids which are more accurate. How long a braid does one need in order to improve over the length-three braid from part (a)?

Exercise 11.3 Scaling of Kitaev–Solovay Algorithm

Given the discussion just above Eq. 11.8, prove Eqs. 11.8 and 11.9.

Exercise 11.4 About the Injection Weave

One might wonder why we choose to work with an injection weave in Fig. 11.11 that moves the red strand from the far right at the bottom all the way to the far left on the top. Show that for three Fibonacci anyons, there does not exist any injection weave that moves the (red) strand from the far right on the bottom (as in Fig. 11.11) to the *middle* on the top, even up to an overall phase. That is, show that no weave exists starting on the bottom far right ending in the middle on the top whose effect on the three-dimensional Hilbert space is $e^{i\phi}\mathbf{1}_{3\times3}$ for any phase ϕ.

Exercise 11.5 Universal Weaving and the Injection Weave

Consider injection weaves as described in Fig. 11.11. Let us assume that we can construct an injection weave of arbitrary precision. Given such an (approximately) perfect injection weave show that for any number of anyons $N > 3$, a weave can be constructed that performs the same unitary operation on the Hilbert space as any given braid. A more general mathematical proof of the universality of weaving is also given in Simon et al. [2006].

Part III

Anyon Diagrammatics (in Detail)

Part IV

Agent Programming (in Detail)

Planar Diagrams[1]

One of our objectives is to come up with some diagrammatic rules (somewhat analogous to those of the Kauffman bracket invariant) that will allow us to evaluate any diagram of world lines (i.e. a labeled link, possibly now including diagrams where particles come together and fuse, or split apart) and get an output that is a complex number as desired in Fig. 7.1. In Chapters 8–10 we have been putting together some of the necessary pieces for these diagrammatic rules. Here we begin to formalize our diagrammatic algebra a bit more precisely.[2] While we try to physically motivate all of our steps, in essence the rules of this chapter can be taken to be axioms of the diagrammatic algebra.

In this chapter we focus only on planar diagrams—that is, we do not allow lines to cross over and under each other forming braids. We can roughly think of such planar diagrams as being particles moving in $(1 + 1)$-dimensions. Since there are no over- and under-crossings, the only nontrivial possibility is that particles come together to fuse, or they split apart. An example of a planar fusion diagram is shown in Fig. 12.1. It is convenient to draw diagrams so that no lines are drawn exactly horizontally. The reader should be cautioned that there are several different normalizations of diagrams—two in particular that we will discuss. These two normalization conventions are useful in different contexts. We start with a more "physics"-oriented normalization in this chapter but we switch to a more "topologically" oriented normalization in Chapter 14 and in later chapters.

We start by briefly reviewing some of the notions introduced in Chapters 8–9. We assume a set of particle types $a, b, c, \ldots$, which we draw as labeled lines with arrows in our diagrammatic algebra. This set of particles includes a unique identity or vacuum particle I, which may be drawn as a dotted line, or may not be drawn at all since it corresponds to the absence of any particles. Each particle type has a unique antiparticle denoted with an overbar ($\bar{a}$ for the antiparticle of a). As we discussed in Section 8.1, if we reverse the arrow on a line we turn a particle into its antiparticle. If a particle is its own antiparticle we do not draw an arrow on its line.

Fusion rules are given by the matrices N_{ab}^c having the properties discussed in Section 8.3. We also assume a consistent[3] set of F-matrices as discussed in Chapter 9.

[1] This chapter through Chapter 15 develop the diagrammatic algebra in some detail. Those who would like a brief and easier (albeit not as general or axiomatic) introduction to diagrammatic algebra should go straight to Chapter 16.

[2] Formally, some of the mathematical structure of planar diagrams was introduced in Section 8.6. The mathematical structure we define by the rules in this chapter is known as a *unitary fusion category* to mathematicians. If various additional properties are satisfied other names may be used (spherical category, braided category, modular category, etc.).

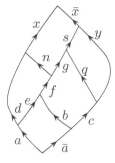

Fig. 12.1 A planar fusion diagram starting and ending with the vacuum.

[3] The word "consistent" here means that the F-matrices satisfy the pentagon equation, Eq. 9.7.

12.1 Diagrams as Operators

If, like Fig. 12.1, a diagram starts at the bottom from the vacuum and ends at the top with the vacuum, we interpret that diagram to represent a complex number, or an amplitude. However, we will also consider diagrams that have "loose ends" (lines sticking off the top or bottom of the page), meaning that they may not begin or end with the vacuum.[4] We can view these diagrams with loose ends as being a sub-diagram of a larger diagram that begins and ends in the vacuum. However, it is also useful to give such diagrams quantum-mechanical meaning in their own right.

Our convention is that, when we draw a diagram with world lines that end pointing upward, we should view these particles as kets (independent of the direction of any arrow drawn on the world line). If world lines end pointing downward, we mean them to be bras. Many diagrams will have world lines that point both up and down, in which case we mean that the diagram has some particles that live in the vector space of kets and some in the dual (bra) space. Such diagrams can be interpreted as operators that take as input the lines coming in from the bottom and give as output the lines going out the top.[5] The lines coming in from the bottom are thus in the bra part of the operator and the lines pointing out the top are the ket part of the operator.[6] If we consider, for example, diagrams with M_{in} incoming lines from the bottom and M_{out} lines going out the top, we can write a general operator[7] as

$$\text{Operator} = \sum_{n,m,q} C_{n,m,q} \, |n, M_{out}; q\rangle\langle m, M_{in}; q| \quad . \tag{12.1}$$

An example of such an operator is shown diagrammatically in Fig. 12.2 with two incoming and three outgoing lines. In Eq. 12.1 the states $|n, M_{out}; q\rangle$ are a complete orthonormal set of states of M_{out} particles where all the particles together fuse to the quantum number q and n runs from 1 to the dimension of the complete set. Similarly, the states $|m, M_{in}; q\rangle$ are a complete orthonormal set of states of M_{in} particles where all the particles together fuse to the quantum number q. The value of the coefficients $C_{n,m,q}$ depend on the details of the diagram being considered. The fact that the operator is necessarily diagonal in the variable q means that the total quantum number of all the incoming particles must be the same as the total quantum number of all the outgoing particles (i.e. they fuse to the same overall charge). This conservation of overall quantum number is a reflection[8] of the locality principle of Section 8.2.

Generally in a diagram, lines will be labeled with particle types and (for non-self-dual particles) arrows. We have not labeled the incoming and outgoing lines in Fig. 12.2, as we assume that these labels and arrows occur inside the hidden box. However, it is sometimes useful to reinstate these labels as in Fig. 12.3. As we discuss in more detail in Section 12.2.1, a label restricts the quantum number of the corresponding line.

[4] Many of the diagrams we have drawn (such as Fig. 8.1 or Fig. 9.1) have not started at the bottom with the vacuum or ended at the top with the vacuum.

[5] It is only for convenience that we think of this as mapping an input from the bottom to an output on the top. We could equally well think of such a diagram as an operator that takes lines as input from the top and gives lines as output to the bottom. Or we could think of such a diagram as taking lines as input from both the top and the bottom and giving a complete diagram with no loose ends as an output. Analogously in terms of Eq. 12.1 we could think of this operator applied to a ket on the right as a mapping from a ket input to a ket output, or applying this operator to a bra on the left it maps a bra input to a bra output, or sandwiching this operator between a bra and a ket it is a map from a bra and a ket both input to a complex number output.

[6] Analogous to some of the ideas of Chapter 7, the bras and kets are meant to be contracted together with bras and kets from other diagrams, pasting together such operators to assemble a picture with no loose ends like Fig. 12.1, which starts and ends in the vacuum.

[7] The only constraint on this operator is that it conserves the total quantum number (or "charge"). One could imagine operators that do not conserve total quantum number. Such operators would be nonphysical and are also outside of what we can express with diagrams.

[8] We do not need an extra axiom for total quantum number conservation, as this will arise as a result of the other rules we introduce in this chapter.

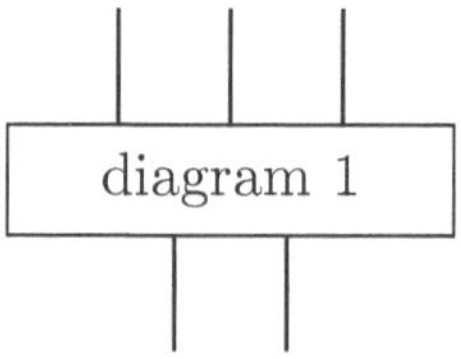

Fig. 12.2 Diagram 1, representing an arbitrary diagram (or linear combination, that is, weighted sum, of diagrams), is understood as part of a larger diagram, and is interpreted as an operator. Incoming lines from the bottom correspond to bras and outgoing lines toward the top correspond to kets. The states $|n, 3; q\rangle$ are a complete set of states for three particles where all the particles together fuse to the quantum number q. Similarly, the states $|m, 2; q\rangle$ are a complete set of states for two particles where all the particles together fuse to the quantum number q. The superscript on $C^{(1)}_{n,m,q}$ indicates that these constants correspond to the particular "diagram 1" in the box. The total quantum number q of all the particles is conserved by the operator due to the locality principle from Section 8.2.

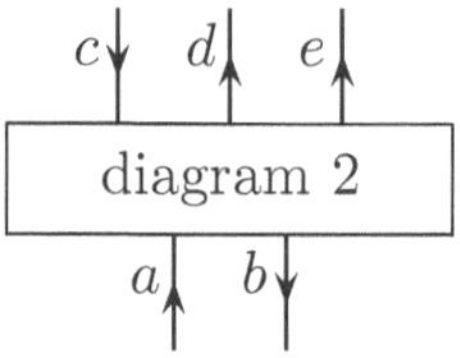

Fig. 12.3 In a figure with labeled incoming and outgoing lines, the quantum numbers on these lines are fixed, as compared to Fig. 12.2, where the diagram may have a superposition of quantum numbers on the external lines.

We now introduce an important diagrammatic principle:

> **Hermitian Conjugation:** Reflecting a diagram around a horizontal axis and then reversing the direction of all arrows implements Hermitian conjugation.[9]

For example, reflecting Fig. 12.3 and then reversing the arrows on all lines results in the Hermitian conjugate diagram

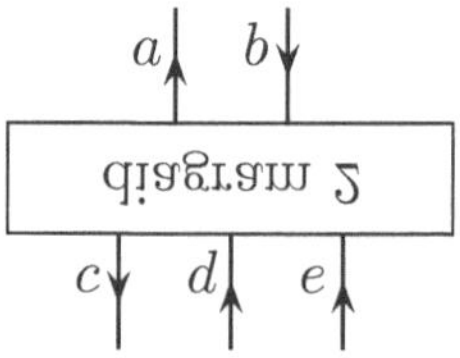

Fig. 12.4 Flipping the diagram in Fig. 12.3 results in the Hermitian conjugate. The coefficients $C^{(2)}_{n,m,q}$ in Fig. 12.3 are complex conjugated to obtain $[C^{(2)}_{n,m,q}]^*$ here.

It is crucial that when we turn a bra into a ket (reflecting the diagram and then reversing the arrows), down-pointing arrows remain down-pointing and up-pointing arrows remain up-pointing. Note, for example, that the arrow on a is pointing up both in Fig. 12.3 and Fig. 12.4.

[9]We have already used this principle as far back as Chapter 2. For example, in Figs. 2.9 and 2.10 we see that we flip over the diagram to turn a ket $|0\rangle$ into a bra $\langle 0|$. In those figures we did not put arrows on lines. However, it is clear that the rule of reflecting then reversing the arrows must be the correct rule if we are to be able to bring the bra and ket together to form an inner product $\langle 0|0\rangle$ as in Fig. 2.11 where we connect up lines with arrows going the same direction.

Diagrams that start from the vacuum at the bottom are an important special case. When there are no incoming lines at the bottom of a diagram the expression becomes

$$|\text{ket}\rangle = \sum_{n,q} C_{n,\varnothing,q} |n, M_{\text{out}}; q\rangle \quad , \tag{12.2}$$

which we can also interpret as an "operator" that accepts the vacuum as an input at the bottom and gives a ket as an output at the top. The symbol $\varnothing$ here means that the m index used in Fig. 12.2 and Eq. 12.1 is just the empty set (nothing summed over), or equivalently that the diagram starts from the vacuum. An example of such a diagram is shown in Fig. 12.5.

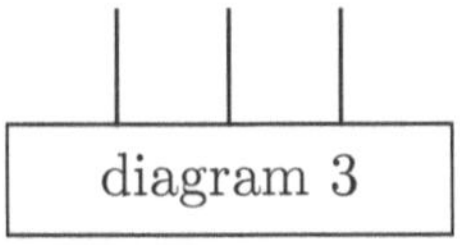

$$= \sum_{n,q} C_{n,\varnothing,q}^{(3)} |n, 3; q\rangle$$

Fig. 12.5 A diagram with no incoming lines at the bottom is interpreted as a ket.

Similarly, we can consider diagrams that end in the vacuum at the top. When there are no outgoing lines at the top of a diagram we have

$$\langle\text{bra}| = \sum_{m,q} C_{\varnothing,m,q} \langle m, M_{\text{in}}; q| \quad , \tag{12.3}$$

which is an operator that accepts a ket as an input and gives a complex number as an output. An example of such a diagram is shown in Fig. 12.6.

$$= \sum_{m,q} C_{\varnothing,m,q}^{(4)} \langle m, 3; q|$$

Fig. 12.6 A diagram with no outgoing lines at the top is interpreted as a bra.

If diagram 3 happens to be the reflection of diagram 4 around a horizontal axis with all arrows reversed, then these two diagrams are Hermitian conjugates of each other and $C_{\varnothing,n,q}^{(4)} = [C_{n,\varnothing,q}^{(3)}]^*$.

12.1.1   Stacking Operators

Stacking operators on top of each other contracts bras with kets in the natural way.[10] For example, if we define the operator, diagram 5, as in Fig. 12.7, we can then stack diagram 5 (Fig. 12.7) on top of diagram 1 (Fig. 12.2) to obtain Fig. 12.8. The resultant operator, diagram 6, on the right is given by

[10]The observant reader will see similarities between this stacking procedure and the stacking of manifolds with boundary discussed in Chapter 7. These similarities are not a coincidence!

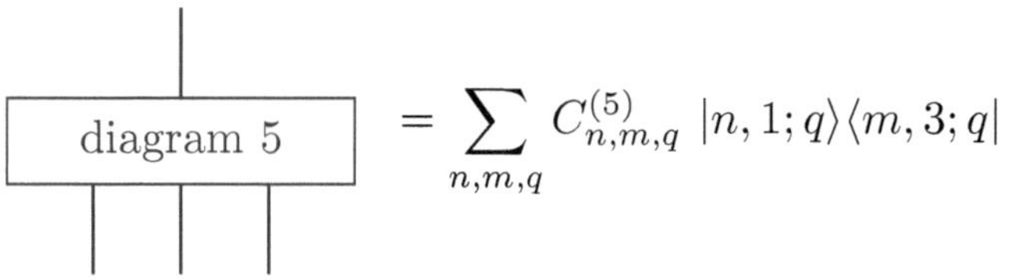

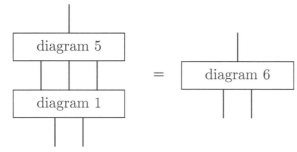

$$\text{diagram 5} \quad = \sum_{n,m,q} C^{(5)}_{n,m,q} \, |n, 1; q\rangle\langle m, 3; q|$$

Fig. 12.7 Another example of an operator.

$$\text{diagram 5} \atop \text{diagram 1} \quad = \quad \text{diagram 6}$$

Fig. 12.8 Stacking operators on top of each other to create new operators.

$$
\begin{aligned}
\text{Operator} \; &= \; \left(\sum_{n,m',p} C^{(5)}_{n,m',p} \, |n, 1; p\rangle\langle m', 3; p| \right) \left(\sum_{n',m,q} C^{(1)}_{n',m,q} \, |n', 3; q\rangle\langle m, 2; q| \right) \\
&= \; \sum_{n,m,q} \left(\sum_{n'} C^{(5)}_{n,n',q} C^{(1)}_{n',m,q} \right) |n, 1; q\rangle\langle m, 2; q|
\end{aligned}
$$

where we have used the orthonormality of the states $|n', 3; q\rangle$, and their corresponding bras $\langle m', 3; p|$, to generate Kronecker deltas $\delta_{p,q}\delta_{m',n'}$. Thus, diagram 6 can be written in the usual form of Eq. 12.1 with coefficients

$$C^{(6)}_{n,m,q} = \sum_{n'} C^{(5)}_{n,n',q} C^{(1)}_{n',m,q} \quad .$$

A particularly important case is that of stacking a bra diagram on top of a ket diagram, which generates a scalar. For example, stacking the bra diagram 4 on top of the ket diagram 3 generates the usual scalar inner product as shown in Fig. 12.9. This fits with our claim at the beginning of this chapter that a diagram that starts and ends in the vacuum should correspond to a complex amplitude.

12.2   Basis of States

In our definition of an operator (see Eq. 12.1) we invoked the existence of a complete orthonormal basis of states $|n, M; q\rangle$ for M particles having total quantum number q. We now specify some details of this basis.

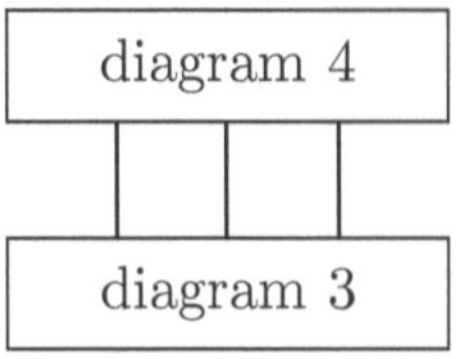

$$\boxed{\text{diagram 4}} \;\; = \left(\sum_{m,p} C^{(4)}_{\varnothing,m,p} \langle m,3;p| \right) \left(\sum_{n,q} C^{(3)}_{n,\varnothing,q} |n,3;q\rangle \right)$$

$$\boxed{\text{diagram 3}} \qquad\quad = \sum_{m,q} C^{(4)}_{\varnothing,m,q} C^{(3)}_{m,\varnothing,q}$$

Fig. 12.9 Stacking a bra operator on top of a ket operator generates a scalar. We have used orthonormality of the kets $|n,3;q\rangle$, and their corresponding bras $\langle m,3;p|$ on the right-hand side.

12.2.1   One Particle

We begin by considering a single particle at a time. For a single particle, a complete orthonormal basis is given by the different particle types[11] $|a\rangle$ (including the vacuum $|I\rangle$). We denote a projector onto a particular particle type as a simple labeled straight line, as shown in Fig. 12.10. The vacuum can be drawn as a dotted line or may not be drawn at all.

$$a \!\!\uparrow = |a\rangle\langle a|$$

Fig. 12.10 A labeled straight line is just a projector onto the particle type.

Since the different particle types are assumed orthonormal $\langle a|b\rangle = \delta_{ab}$, applying two such projectors in a row diagrammatically gives the identity shown in Fig. 12.11.

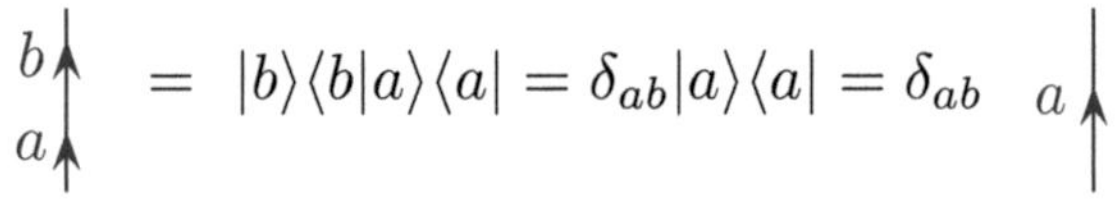

$$\begin{array}{c} b \!\!\uparrow \\ a \!\!\uparrow \end{array} = |b\rangle\langle b|a\rangle\langle a| = \delta_{ab}|a\rangle\langle a| = \delta_{ab}\ \ a \!\!\uparrow$$

Fig. 12.11 Orthogonality of projection operators.

This identity exemplifies the more general rule shown in Fig. 12.12, which also agrees with the fact that the operators in Eq. 12.1 are diagonal in the overall quantum number q. Again this is simply a reflection of the locality, or no-transmutation, principle[12] of Section 8.2 (see in particular Fig. 8.7).

$$\begin{array}{c} b \!\!\uparrow \\ \boxed{\text{anything}} \\ a \!\!\uparrow \end{array} = 0 \quad \text{unless } a = b$$

Fig. 12.12 The locality, or no-transmutation, principle as in Fig. 8.7.

Since we assume the set of particle types is complete, the identity operator is given by the sum over all particle types, including the vacuum particle, as in Fig. 12.13. We represent the identity operator on the right in Fig. 12.13 as a straight unlabeled line.[13] This is convenient since it

[11] In keeping with the notation of Fig. 12.2 the state $|a\rangle$ should be notated $|a,1;a\rangle$ to indicate a single line, but here we use just $|a\rangle$ for simplicity.

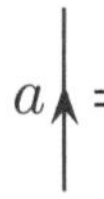

[12] As mentioned in note 8 earlier in this chapter, this principle is not an axiom of our diagrammatics, but rather can be derived from the other rules we introduce in this section. See Exercise 12.2.

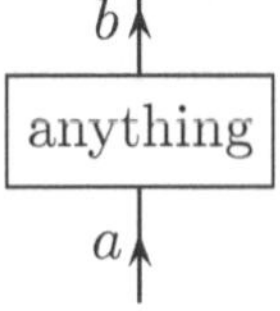

[13] **Caution:** the identity operator is not the vacuum (or identity) particle, but rather an operator that takes any input particle at the bottom and gives out the same particle as output at the top.

allows us to extend labeled lines by appending unlabeled lines as shown in Fig. 12.14.

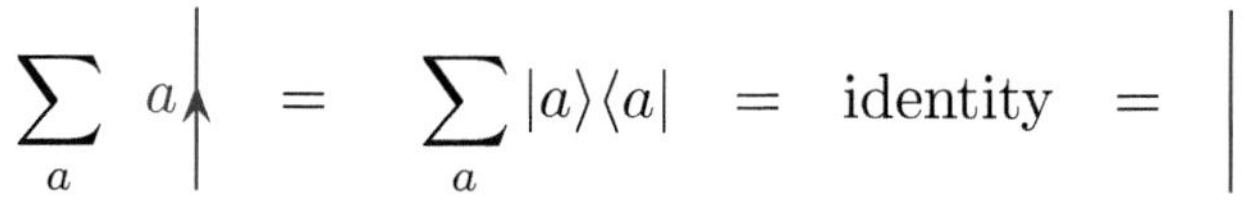

$$\sum_a \; a \Big\uparrow \;\; = \;\; \sum_a |a\rangle\langle a| \;\; = \;\; \text{identity} \;\; = \;\; \Big|$$

Fig. 12.13 The completeness relation for single lines.

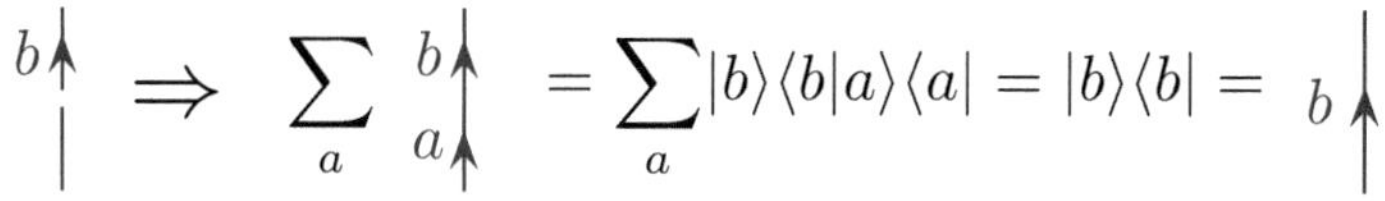

$$b\Big\uparrow \;\; \Rightarrow \;\; \sum_a \; b \Big\uparrow_a \;\; = \sum_a |b\rangle\langle b|a\rangle\langle a| = |b\rangle\langle b| = \; b\Big\uparrow$$

Fig. 12.14 Appending the identity (an unlabeled line) to a labeled line just extends the labeled line. Equivalently, attaching a labeled line (b) to an unlabeled line puts the label (b) onto the unlabeled line.

12.2.2 Two Particles

Let us now move on to the case of two particles. As discussed in Chapter 8, to fully describe the state of two particles, we need to give the quantum number (particle type) of each particle *and* the fusion channel between the two particles. We thus draw the state of two anyons with a vertex diagram[14] as shown in Fig. 12.15.

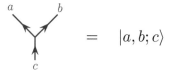

$$= \quad |a, b; c\rangle$$

Fig. 12.15 Particles a and b have fusion channel c.

The notation of the ket[15] $|a, b; c\rangle$ means that the total quantum number of particles a and b is c (or a and b fuse[16] to c). If $N_{ab}^c = 0$, that is, if the diagram is a disallowed fusion, then the value of the diagram is zero. The set of allowed states $|a, b; c\rangle$ for all possible a, b, c is assumed to form a complete orthonormal set of states for two anyons. Note in particular that for $a \neq b$ the ket $|a, b; c\rangle$ is orthogonal to $|b, a; c\rangle$—that is, in our planar diagram algebra, it matters which particle is to the left and which is to the right.[17]

The Hermitian conjugate of the vertex ket in Fig. 12.15 is the corresponding bra shown in Fig. 12.16.

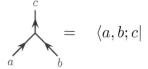

$$= \quad \langle a, b; c|$$

Fig. 12.16 This is the bra which is the Hermitian conjugate of the ket in Fig. 12.15.

The fact that the diagram for the bra looks like the ket upside-down is in accordance with our general principle of Hermitian conjugation[18] introduced in Section 12.1 (see the discussion near Fig. 12.4).

[14]In cases where the fusion multiplicity $N_{ab}^c > 1$, we must also add an index $\mu \in \{1, \ldots, N_{ab}^c\}$ at the vertex as in Eq. 8.24, and we would write the ket as $|a, b; c, \mu\rangle$. We suppress this additional index here for simplicity. It is reinstated in Section 12.5.

[15]In the notation of Fig. 12.2 the state $|a, b; c\rangle$ might be notated $|(a, b), 2; c\rangle$ to indicate there are two outgoing lines. If we wanted to emphasize that there is one incoming line and two outgoing lines we might write $|a, b; c\rangle\langle c|$ instead. Here we use abbreviated notation.

[16]More properly for Fig. 12.15 we should say that a and b split from c, whereas in Fig. 12.16 we should say that a and b fuse to c. Most of the time people are careless in distinguishing fusing and splitting.

[17]In quantum mechanics the state of a particle at position x is orthogonal to the state of the particle at position $y \neq x$, that is, $\langle x|y\rangle = \delta(x - y)$. In TQFTs we throw out the position information and keep only the topological information. For planar diagrams this means we consider only whether particle a is to the left or right of particle b.

[18]In Section 12.1 we treated the statement that flipping the diagram gives Hermitian conjugation as an axiom. However, one could instead treat Fig. 12.16 as the axiom and build up the general principle from only this statement.

[19] Again, if $N^c_{ab} > 1$ there are additional indices μ at the vertices and the kets are orthonormal in these indices as well. See note 14, and Section 12.5.

[20] This inner product between bra and ket does not give a scalar but rather a scalar times a c particle line. This is because the ket in Fig. 12.15 is actually an operator that takes an incoming single line as input and gives two lines as output. (And conversely with the bra in Fig. 12.16.) See also the comment on notation in note 15.

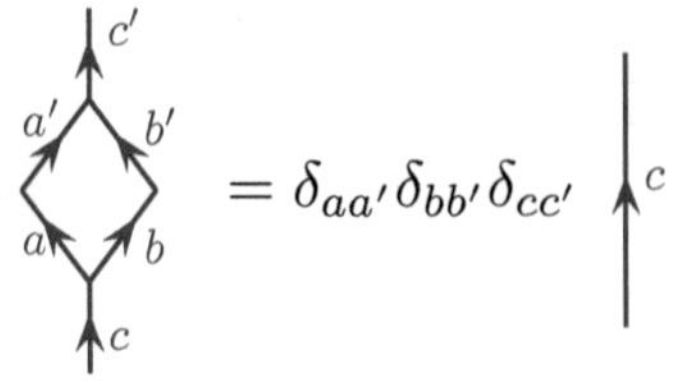

To take inner products between a bra (like Fig. 12.16) and a ket (like Fig. 12.15) we simply stack the bra on top of the ket, in accordance with Section 12.1.1, to produce the diagram[19,20] shown in Fig. 12.17.

Fig. 12.17 The inner product between the bra in Fig. 12.15 and the ket in Fig. 12.16. This gives Kronecker deltas on the right given the physics normalization we are using in this chapter. This normalization will be changed in Chapter 14 (see Fig. 14.7). If $N^c_{ab} = 0$ the value of the diagram on the left is zero. For $N^c_{ab} > 1$ see Section 12.5.

The fact that we obtain delta functions on the right is equivalent to the statement that the kets $|a, b; c\rangle$ form an orthonormal set. The normalization of Fig. 12.17 (i.e. that one gets Kronecker deltas on the right and no numerical constants) is our *physics normalization*. This normalization will be changed in Chapter 14.

Note that the first two delta functions $\delta_{aa'}$ and $\delta_{bb'}$ in Fig. 12.17 can be interpreted as a result of Fig. 12.11 (the lines are angled instead of vertical, but this does not change their meaning). As a result, the diagram in Fig. 12.17 is often written in the simplified form shown in Fig. 12.18.

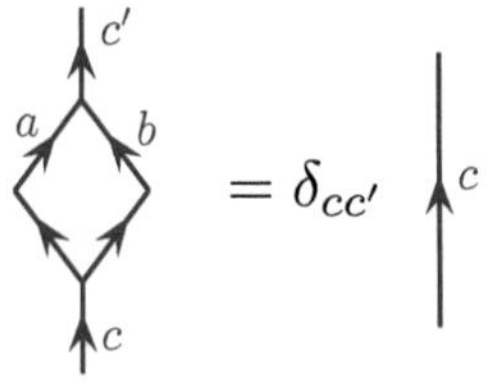

Fig. 12.18 A simplified version of the inner product in Fig. 12.17. This gives a Kronecker delta on the right in the physics normalization we are using in this chapter. The normalization will be changed in Chapter 14. If $N^c_{ab} = 0$ then the value of the diagram on the left is zero. For $N^c_{ab} > 1$ see Section 12.5.

The fact that c must equal c' in Figs. 12.17 and 12.18 is consistent with the no-transmutation principle Fig. 12.12.

The principle of orthonormality of vertices implies the useful result that a loop, as shown in Fig. 12.19, is given the value of unity (this is just Fig. 12.18 where we have set $c = c' = I$ and not drawn the identity line). At the risk of being repetitive we again note that we change this

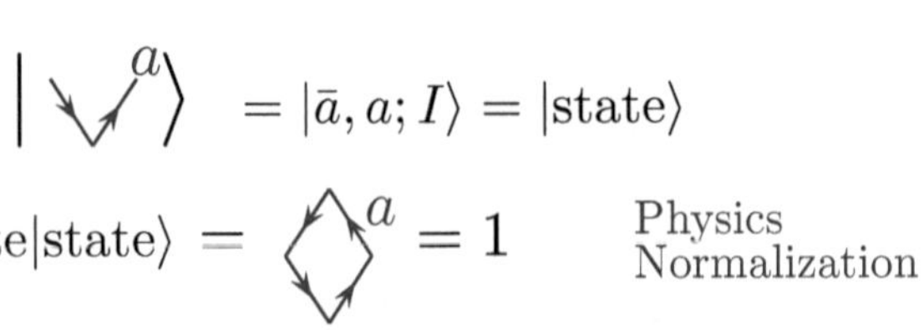

Fig. 12.19 The orthonormality shown in Fig. 12.18, setting $c = I$, implies a particle loop gets a value of 1 if we are using physics normalization.

normalization in Chapter 14 and in later chapters, although it is correct for this section.

Since the vertex diagrams $|a, b; c\rangle$ from Fig. 12.15 form a *complete* set of states for the two particles, we can construct an identity operator for two strands as shown in Fig. 12.20.

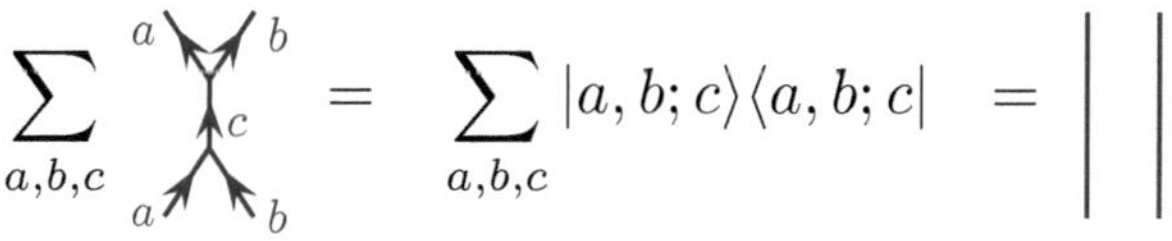

$$\sum_{a,b,c} \quad = \quad \sum_{a,b,c} |a, b; c\rangle\langle a, b; c| \quad = \quad$$

Fig. 12.20 Insertion of a complete set of states. This figure uses physics normalized diagrams. The normalization will be changed in Chapter 14.

By attaching strings with labels x to the left strand and y to the right as in Fig. 12.14 we label the unlabeled strands to obtain Fig. 12.21.

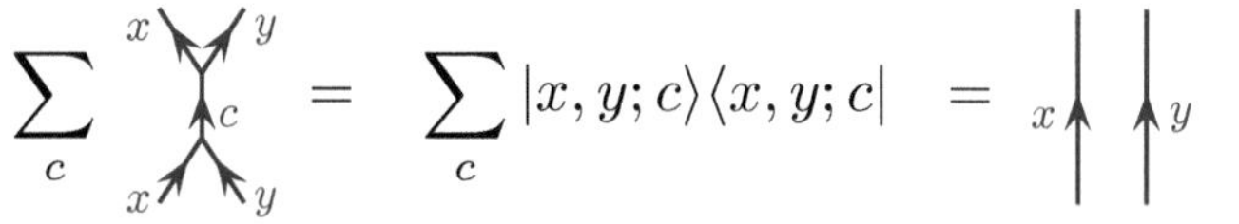

$$\sum_{c} \quad = \quad \sum_{c} |x, y; c\rangle\langle x, y; c| \quad = \quad$$

Fig. 12.21 Insertion of a complete set of states, with fixed quantum numbers x and y on both ends. This figure uses physics normalized diagrams. The normalization will be changed in Chapter 14.

An arbitrary operator with two incoming and two outgoing lines can be written as in Fig. 12.22, where the coefficients $C^{x,y}_{a,b;c}$ depending on the operator we are defining.

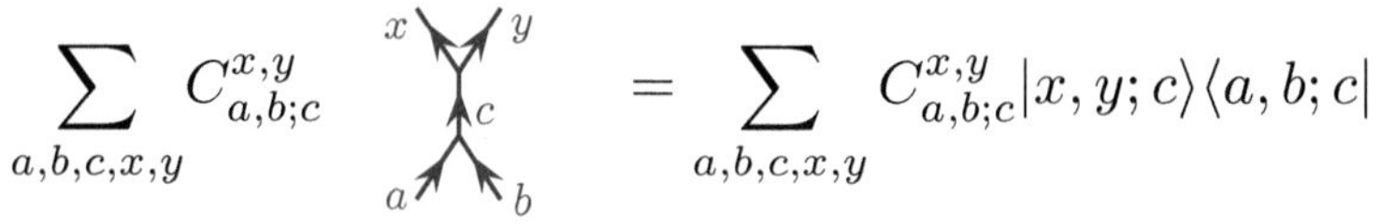

$$\sum_{a,b,c,x,y} C^{x,y}_{a,b;c} \quad = \quad \sum_{a,b,c,x,y} C^{x,y}_{a,b;c} |x, y; c\rangle\langle a, b; c|$$

Fig. 12.22 An arbitrary operator with two incoming and two outcoming lines. The coefficients C are arbitrary.

12.2.3 Three Particles

We can continue on and consider states of three particles. All the same principles apply here. As discussed in Chapter 8, we can write a complete orthonormal set of states for three particles as a fusion tree[21],[22] as in Fig. 12.23. Here we have introduced a new notation

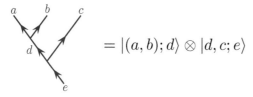

$$= |(a, b); d\rangle \otimes |d, c; e\rangle$$

Fig. 12.23 An orthonormal set of states for three particles can be described as a fusion tree. The ket notation on the right includes parenthesis around (a, b) to indicate that they are further from the root of the tree. This notation matches that of Section 8.6.

[21] In cases where there are fusion multiplicities $N^d_{ab} > 1$ or $N^e_{dc} > 1$ then we must place an additional index μ or λ at the corresponding vertex. See for example, Section 9.5.3.

[22] As mentioned in note 15, although we write this as a ket, it is really an operator, and to emphasize this we might write something like $\left[|(a, b); d\rangle \otimes |d, c; e\rangle\right]\langle e|$ instead.

[23]This is already implied by looking at the individual vertices and considering the rules of a single vertex, as in Fig. 12.15.

If either $N_{ab}^d = 0$ or $N_{dc}^e = 0$ then the corresponding fusion is disallowed and the value of the diagram is zero.[23] The corresponding bras are obtained using the Hermitian conjugation rule of flipping the diagram and reversing arrows, as shown in Fig. 12.24.

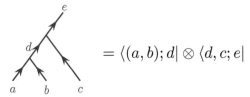

$$= \langle (a,b); d| \otimes \langle d, c; e|$$

Fig. 12.24 The bras corresponding to the kets in Fig. 12.23.

The inner product of such states is given by stacking the bra on top of the ket as in Fig. 12.25.

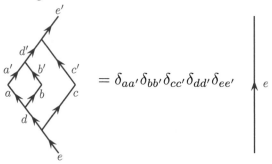

$$= \delta_{aa'} \delta_{bb'} \delta_{cc'} \delta_{dd'} \delta_{ee'}$$

Fig. 12.25 The orthonormality of tree states. This diagram uses physics normalization.

To derive the orthonormality of tree diagrams shown in Fig. 12.25 we start by focusing our attention on the left branches—the edges labeled d, a, b, a', b', d'. This sub-diagram looks exactly like Fig. 12.17 so we immediately obtain $\delta_{aa'} \delta_{bb'} \delta_{dd'}$ and we can replace the small diamond on the left branch with a single d-line. The remaining figure then looks exactly like Fig. 12.17 again and gives us the delta functions $\delta_{cc'} \delta_{ee'}$. Thus, the orthonormality of these tree states is not a separate assumption but can be derived from the orthonormality of two-particle states that we discussed in Section 12.2.2.

The completeness of this set of states similarly can be expressed with diagrams as shown in Fig. 12.26. Once again we can derive this completeness relation from what we know about the two-particle case. We can start in the very center of Fig. 12.26, considering the lines labeled with d, c, and e, and apply the completeness relation Fig. 12.21 resulting in the middle diagram of Fig. 12.26. The sum over the c index gives a single unlabeled line on the right as in Fig. 12.13. The remaining diagram on the left of the middle diagram (with lines labeled a, b, and d) is of the form of Fig. 12.20, which summed over gives two unlabeled lines. Thus, the completeness relation for three particles is not an independent assumption but follows from the completeness of the one- and two-particle cases.

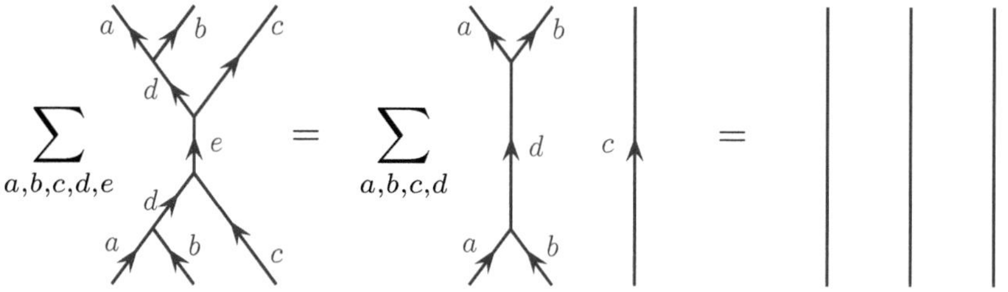

Fig. 12.26 The completeness of tree states for three particles. This diagram uses physics normalization.

One can use these basis states to build arbitrary operators with three particle states. Just as an example, in Fig. 12.27 we show the most general form of an operator that takes two particles as an input and gives three particles as an output.

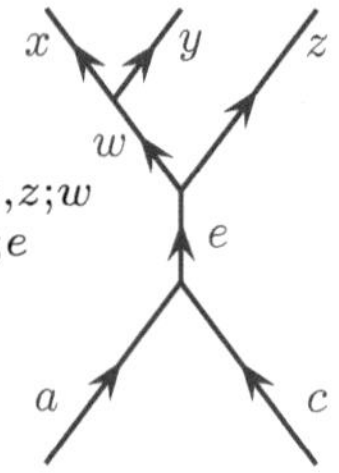

$$\sum_{a,b,e,x,y,z,w} C^{x,y,z;w}_{a,b;e} \quad = \quad \sum_{a,c,e,x,y,z,w} C^{x,y,z;w}_{a,b;e} \left[|(x,y); w\rangle \otimes |w, z; e\rangle \right] \langle a, b; e|$$

Fig. 12.27 An arbitrary operator with two incoming lines and three outgoing lines. The coefficients C depend on the operator we are defining. The kets in the bracket are of the same form as the tree in Fig. 12.23 for example.

12.2.4 *F*-Matrices Again

In defining our three-particle states in Fig. 12.23 we have fused the two particles a and b on the left first to form d and then fused d with c to form e. Our notation to indicate this is[24] $|(a, b); d\rangle \otimes |d, c; e\rangle$. However, we could have chosen to fuse the particles in a different order to form a different tree as shown in Fig. 12.28. Here, b and c fuse together to form f and then a and f fuse together to form e. We notate this state as $|a, f; e\rangle \otimes |(b, c), f\rangle$.

[24] Or, more simply, we just draw the tree!

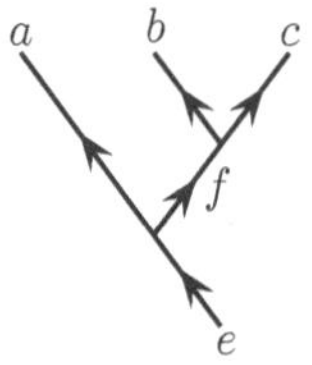

$$= |a, f; e\rangle \otimes |(b, c); f\rangle$$

Fig. 12.28 Another orthonormal set of states for three particles. Compare to Fig. 12.23.

The set of states defined by the fusion trees in Fig. 12.28 also forms a perfectly good (but different) complete orthonormal basis of states for three particles. For example, we have the orthogonality relation shown in Fig. 12.29 (compare Fig. 12.25).

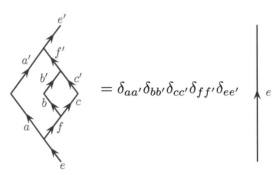

Fig. 12.29 The orthonormality of tree states in a different basis. This diagram uses physics normalization.

As described in detail in Chapter 9, if we draw trees with different branching structure, we are describing the same Hilbert space, but in a different basis—the basis change being implemented by a unitary F-matrix transformation as shown in Fig. 12.30 (see also discussion of Fig. 9.1).

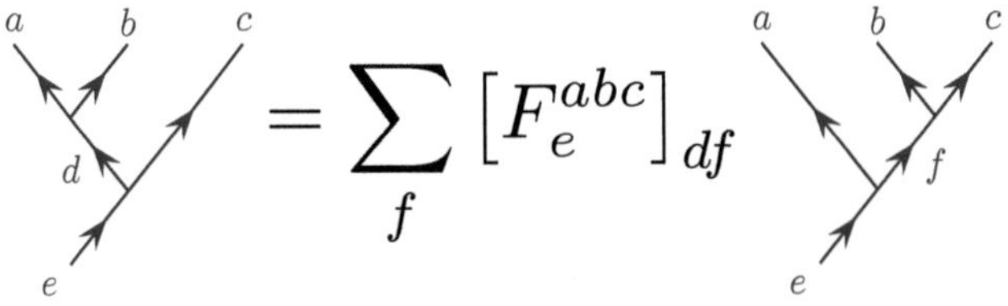

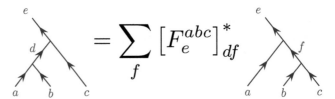

Fig. 12.30 The F-matrix. See Chapter 9.

Similarly, we have the relation between the Hermitian conjugate states as shown in Fig. 12.31.

Fig. 12.31 F-matrix Hermitian conjugated.

Note that because the F-matrix is unitary in its two outside indices (d and f in Fig. 12.30) we have

$$[F_e^{abc}]_{df}^* = ([F_e^{abc}]^\dagger)_{fd} = ([F_e^{abc}]^{-1})_{fd} \quad .$$

12.2.5 More Particles

The principles we have developed for one-, two-, and three-particle states are easily extended to greater numbers of particles. Each branching structure, or shape, of a fusion tree diagram defines a different basis for a complete orthonormal set of states. For example, with four particles, we might choose the tree shape shown on the left of Fig. 12.32, or we might choose the tree shape shown on the right of Fig. 12.32. Either one of these makes a perfectly good orthonormal basis for four particles— and these two bases are related to each other by F-matrices as discussed in Chapter 9.

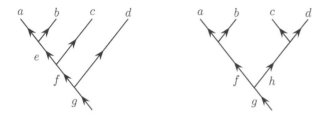

Fig. 12.32 Two (of five) possible bases for describing states of four particles. These bases are related to each other by F-moves (see Fig. 9.7).

The left-hand tree structure in Fig. 12.32, with all of the particles on top branching from a single line going from top left to bottom right, is sometimes known as the "standard basis."

One can use F-moves to evaluate more complicated diagrams. An example of this is shown in Fig. 12.33.

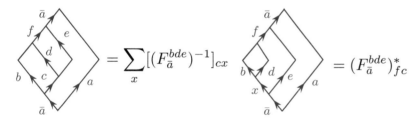

Fig. 12.33 The diagram on the left is evaluated by applying an F-move on the lower left part of the diagram. The resulting diagram is evaluated to a function δ_{xf} due to the orthonormality of tree diagrams. Finally, we use the unitarity of F in the last step. Since this diagram starts and ends with the vacuum it evaluates to a scalar. This diagram is evaluated with physics normalization.

12.3   Causal Isotopy

Keeping with the idea of diagrams that are planar (no over- and under-crossings), we now consider how we may deform these diagrams. When we discussed the Kauffman bracket invariant we were allowed to freely deform any diagram as long as we did not cut any strands. This property is known as *isotopy invariance*.[25] Analogously, if a planar diagram retains the same value for any deformation that does not involve cutting strands or crossing them over each other, we say the theory has *planar isotopy invariance*. Examples of this are shown in Fig. 12.34.

[25]In the Kauffman case we have *regular isotopy* invariance, meaning that we can deform knots freely in three dimensions as long as we treat the strands as ribbons. See Sections 2.2.1 and 2.6.1. Any theory that has regular isotopy invariance has planar isotopy invariance when restricted to the plane.

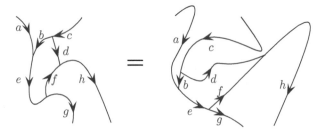

Fig. 12.34 For a theory with planar isotopy invariance, these two diagrams should evaluate to the same result. Planar isotopy invariance allows us to distort the diagram in any way as long as we do not cut any strands or cross lines through each other.

We need to ask how much topological invariance we should really expect from our physical theories. In the mathematical world of TQFTs and knot invariants, it is fine to assume that all directions are equivalent, and we can freely distort a line travelling in the x direction (horizontally) on the page to a line travelling in the t direction (vertically). However, in real physical systems, generically the time direction might be different from the space directions. In this section we discuss topological theories that allow deformation in space, but without allowing one to freely exchange the time and space directions. In particular, some amount of causality might be demanded.

In Chapter 16 we consider a subset of theories that have a much higher level of topological invariance, known as *regular isotopy* invariance,[26] which allows us to freely distort diagrams in either the space or time direction and further allows us to interchange the two.

In this chapter through Chapter 15 we do not assume regular isotopy invariance (or planar isotopy in the case of planar diagrams) but rather assume only what we call *causal isotopy*.[27] Here, we allow deformation of spacetime diagrams as long as we do not change the time-direction orientation of any lines or vertices. In other words, the path of a particle that is moving forward in time should not be distorted such that it is moving backward in time (and vice versa, a particle moving backward should not be distorted so that it is moving forward)—but other than this constraint, any smooth deformation is allowed. Two examples of deformations that are allowed under causal isotopy are shown in Fig. 12.35.

Certain deformation of diagrams are not allowed by causal isotopy. Two examples of such disallowed deformations are given in Fig. 12.36. On the left of the figure we see a particle that turns around in time. This need not be the same as the particle moving straight in time as it involves a particle creation event and a particle annihilation event. On the right of Fig. 12.36 a vertex is altered so instead of an a particle going out of the vertex, an $\bar{a}$ particle goes in. In this case we must have an a with $\bar{a}$ annihilation event in the far-right diagram that does not exist in the simpler diagram where a and c directly fuse to b. Thus, these two diagrams do not necessarily evaluate to the same result. (Although

[26]The term "regular" implies that strands are treated as ribbons, but other than this caveat, all deformations without scissors are allowed. See Section 2.6.1.

[27]This nomenclature was introduced in Simon and Slingerland [2022].

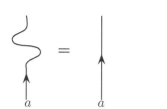 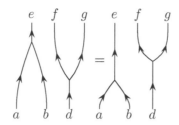

Fig. 12.35 Two examples of deformations that are allowed under causal isotopy. Deformations of the path are allowed as long as they do not require a particle to reverse directions in the time-like direction. In the left example, this deformation is allowed because in both cases the particle continues to move forward in the time direction. In the right example, the temporal order of the vertices does not matter.

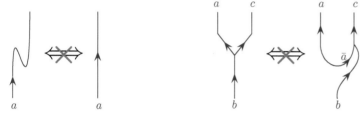

Fig. 12.36 Two examples of transformations that are not necessarily allowed under causal isotopy. In some special theories these transformations will be allowed, but generically they are not allowed. The two diagrams in the left figure are discussed in Chapter 14. The two diagrams in right figure are discussed in Sections 12.4.1 and 14.4.

in some cases, such as in Chapter 16, one may have a simple theory for which the transformations shown in Fig. 12.36 are allowed, such theories are not generic.)

12.4 Summary of Planar Diagram Rules in Physics Normalization

With the principles we have discussed we should be able to evaluate any planar diagram. That is, take a spacetime process that starts and ends in the vacuum and turn it into an amplitude (i.e. a complex number). The same principles can be used to simplify operators such as Eq. 12.2.

Here is a summary of the important rules we have learned for diagram evaluation:

(1) One is free to continuously deform a diagram consistent with causal isotopy as described in Section 12.3. That is, particles must not change their direction in time due to the deformation.

(2) One is free to add or remove lines from a diagram if they are labeled with the identity or vacuum (I). See the example in Fig. 12.37.

(3) Reversing the arrow on a line turns a particle into its antiparticle (see Fig. 8.4).

(4) Regions must maintain their quantum number locally, as in Fig. 12.12.

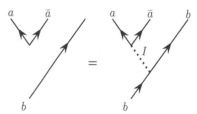

Fig. 12.37 One can always add or remove the identity (or vacuum) line to any diagram.

In particular this means that a line must maintain its quantum number unless it fuses with another line, or splits.

(5) Splitting and fusion vertices are allowed for fusion multiplicities $N_{ab}^c > 0$ (see Section 8.3). This includes particle creation and annihilation as special cases where a particle–antiparticle pair fuses to the vacuum or splits from the vacuum. (An example is shown in Fig. 12.37.)

(6) Hermitian conjugation is given by reflection of a diagram across a horizontal line along with flipping the direction of arrows (see Fig. 12.4 or Fig. 12.24).

(7) One can use F-moves to simplify the structure of fusion trees. For example, in Fig. 12.33.

(8) One can use completeness relations, such as Figs. 12.20 or 12.21.

(9) Once one reduces a diagram into tree structures that have the same branching in the upper and lower half (as on the right of Fig. 12.33) we can use the orthonormality of trees to complete the evaluation. In cases where the diagram starts and ends in the vacuum this reduces the diagram to a complex number (see, for example, Fig. 12.33). More generally operator diagrams can be reduced to simple forms analogous to Fig. 12.27.

With these principles (and given an F-matrix as input information which will depend on the particular physical system we are considering), it is possible to fully evaluate any planar diagram, starting and ending in the vacuum, into a complex number. While there may be many strategies to use these rules to reduce a complicated diagram to a single complex number, the final result is independent of the order in which we apply the rules.[28]

[28] This is guaranteed by the pentagon equation and the Mac Lane coherence theorem.

The mathematical structure we have defined thus far (our Hilbert space and F-matrices) is known as a "unitary fusion category." There is more structure to be uncovered in later chapters that follows from what we have defined so far, and there are many special cases to be discussed. In addition, note that here we have only described planar diagrams, so we have not yet described $(2+1)$-dimensional theories—in order to describe these, we will have to include braiding rules for our diagrams, which we do in Chapter 13.

12.4.1 A Simple Example

As a simple example, let us try to evaluate the diagram shown on the far left of Fig. 12.39. We first work on a small part of the diagram as shown in Fig. 12.38 (note that this is the same as the far right of Fig. 12.36).

The result in Fig. 12.38 can also be reflected across the horizontal axis as in Fig. 12.4 to give the Hermitian conjugate diagram. Using both Fig. 12.38 and its reflection, we obtain the result given in Fig. 12.39.

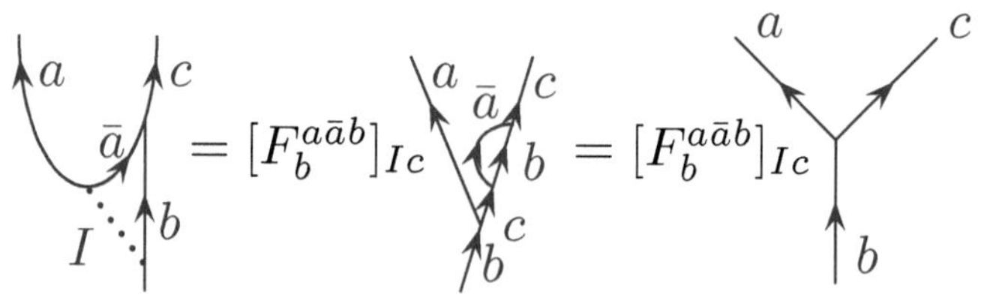

Fig. 12.38 To evaluate the diagram on the left, the vacuum line is inserted and an F-move is made. The bubble is then removed with Fig. 12.17. These diagrams use physics normalization. We will re-examine this diagram using a different normalization in Section 14.4.

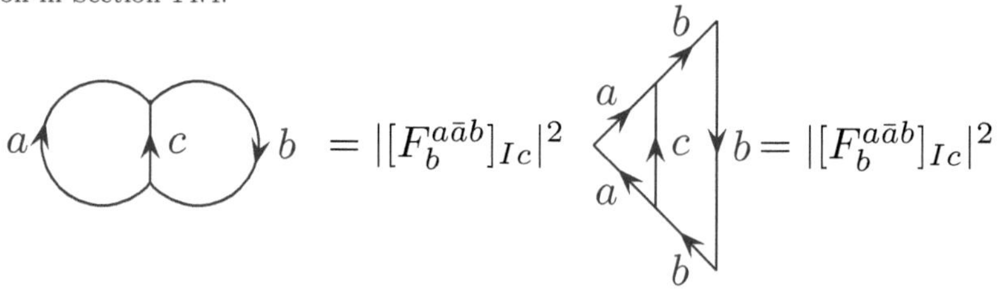

Fig. 12.39 The first step invokes Fig. 12.38 and its Hermitian conjugate. The figure on the right is a tree that evaluates to the identity as long as the fusion vertices are allowed and assuming physics normalizations.

12.5 Appendix: Higher Fusion Multiplicities

When we have a theory with higher fusion multiplicities (i.e. $N_{ab}^c > 1$ for at least one fusion channel), then the vertices must be given indices, and tree states are orthogonal in these indices as well. For example, we would need to modify Figs. 12.18 and 12.21 to the form shown in Figs. 12.40 and 12.41. See also the discussion of the F-matrix with higher fusion multiplicities in Section 9.5.3.

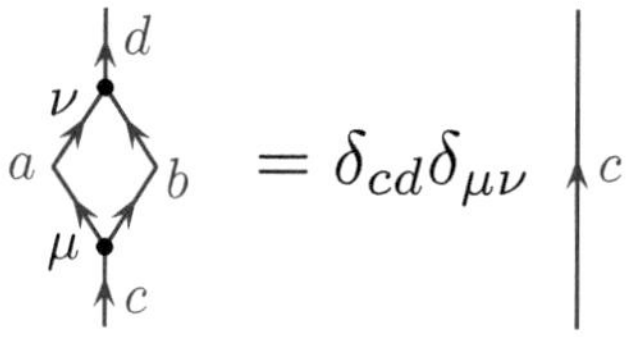

Fig. 12.40 The bubble diagram when there are fusion multiplicities. This diagram is a result of the orthonormality of tree diagrams. The variables at the vertices must match in order for the result to be non-zero. This diagram is drawn in the physics normalization. We will change the normalization in Chapter 16.

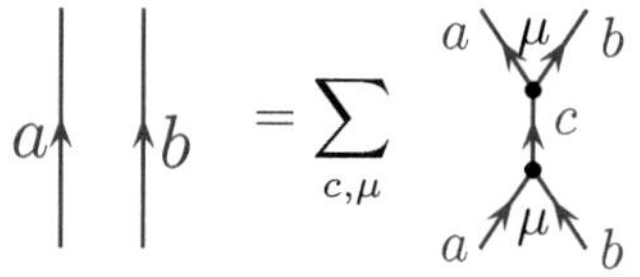

Fig. 12.41 Insertion of a complete set of states with fixed quantum numbers a and b on both ends. When there are fusion multiplicities $\mu \in \{1, \ldots, N_{ab}^c\}$, the index μ must be summed over as well. This diagram is drawn in the physics normalization. We will change the normalization in Chapter 16.

Chapter Summary

- Complete diagrams (no lines having ends) evaluate to numbers. Partial diagrams (having lines ending at top and/or bottom) are interpreted as operators.
- Hermitian conjugation reflects diagrams around a horizontal axis and reverses arrows.
- States of a system can be expressed in multiple bases. Conversion between bases is implemented by F-matrices.
- Certain deformations of a diagram ("causal isotopy") are always allowed without changing the meaning of the diagram.

Further Reading:

- And again, the main references for this material are Nayak et al. [2008],Kitaev [2006], and Bonderson [2007]. Also Preskill [2004] is a nice introduction. The introduction and appendices of Eliëns et al. [2014] are also very nice.
- And again, all of this material was well known in other communities a long time ago but is not presented in nice forms anywhere.

Exercises

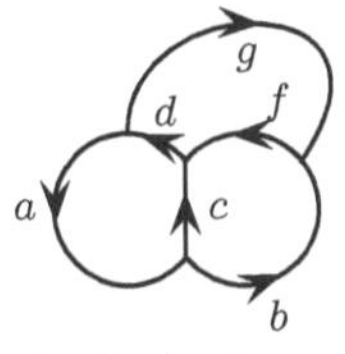

Exercise 12.1 Evaluating Diagrams with F-matrices
Evaluate the diagram in Fig. 12.42, writing the result in terms of Fs.

Exercise 12.2 Locality Principle
Show that the locality principle (Fig. 12.12) is derivable from our other rules for evaluating diagrams and is therefore not an independent assumption.

Exercise 12.3 H-variant of F-move
A variant of the F-move is given by Fig. 12.43. Show that (in physics normalization!)

$$[F^{ab}_{cd}]_{ef} = \left([F^{ceb}_{f}]_{ad}\right)^{*} \quad .$$

Hint: Add vertices at the top and bottom of diagrams to make the connection between this diagram and the usual F-move diagram.

Fig. 12.42 A diagram.

$$\mathord{\text{\raisebox{0pt}{$a\ \ b$}}} \quad = \sum_f [F^{ab}_{cd}]_{ef} \quad \mathord{\text{\raisebox{0pt}{$a\ \ b$}}}$$

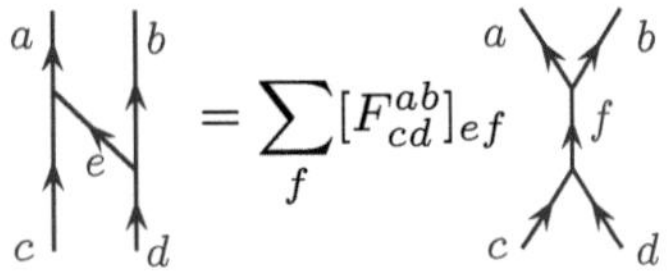

Fig. 12.43 The H-variant of the F-move

Braiding Diagrams[1]

In Chapters 8, 9, and 12 we focused on planar diagrams. These diagrams can be thought of as describing the physics of objects that live in $(1 + 1)$-dimensions. More to the point, the nontrivial physics we discovered is really just a reflection of the nontrivial structure of the Hilbert spaces we are working with.

Here we extend our diagrammatic rules to the $(2 + 1)$-dimensional world. In particular we want to describe what happens when we braid world lines. In Chapter 10 we started to discuss braiding of identical particles and we continue that discussion here.

13.1 Three-Dimensional Diagrams

We begin by generalizing the concept of a diagram that we developed in Chapters 9–12. The diagrams we want to consider now allow over- and under-crossings of lines as in Fig. 13.1. We will end up with a set of rules that are conceptually similar to the knot invariants we discussed way back in Chapter 2—starting with a picture of a generalized knot (like Fig. 13.1), we reduce it to an output number. The generalization here is that the lines have labels $(a, b, c, \ldots)$ and lines can fuse with each other in addition to crossing over and under each other.

We should be somewhat cautious here: when we considered the Kauff-man bracket invariant we had regular isotopy invariance—meaning that, treating strands as ribbons, any deformation of the diagram was allowed as long as we did not cut any strands. In contrast here (while we should still treat strands as ribbons) not all deformations are allowed. In general we will only have the same type of causal isotopy as described in Section 12.3 (that is, we cannot freely deform a particle line going forward in time to one that goes backward in time). Of course there do exist anyon theories with a higher level of isotopy invariance (regular isotopy), which we discuss in Chapter 16, but we should realize that these are not generic.

Our rules for evaluating diagrams with over- and under-crossings will be a consistent extension of the set of rules for evaluating planar diagrams.[2] Our next task is to consider how we handle over- and under-crossings. With this information, used in conjunction with the rules we have already developed for planar algebras, we will be able to evaluate any diagram in $(2 + 1)$-dimensions.

[1]This chapter continues the development of the diagrammatic algebra in some detail. Those who would like a brief and easier (albeit not as general) introduction to diagrammatic algebra should go straight to Chapter 16.

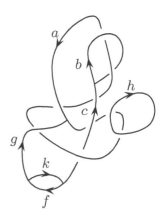

Fig. 13.1 A diagram with over- and under-crossings representing a process in $(2 + 1)$-dimensions.

[2]In mathematical language, the rules introduced in this chapter give additional structure to a unitary fusion category to make it a unitary braided fusion category, or unitary ribbon fusion category (these notions are equivalent).

13.2 Braiding Non-Identical Particles

We now turn to ask what happens if we exchange two different particle species, say a and b. We might be tempted to do something similar to Figs. 10.1 and 10.2—that is, we define a state with two particles in a given fusion channel, then we exchange the two particles and determine the phase accumulated in this process. However, such a scheme cannot work in the case of non-identical particles. The reason this fails is that when the two particles are not identical the initial and final states are fundamentally different and cannot be related to each other by just a phase—for example, the initial state for Fig. 13.2 has a to the left of b whereas the final state has a to the right of b.

Nonetheless, the R-matrix can still be precisely defined even when we are braiding non-identical particles. Diagrammatically we define the R-matrix as shown in Fig. 13.3. On the right of this figure, the particles b and a come from c, with a going off to the right and b to the left. In the left of the figure, the two particles are moved away from each other, b to the right and a to the left, before they are braided around each other. The key here is that in both cases, the final state of the system has b on the left and a on the right, and the two particles fuse to a quantum number c, so that the two processes can be compared to each other and differ from each other only by a phase, which we define[3] to be R_c^{ab}.

Fig. 13.2 One would ideally like a rule for exchanging any two particles. However, this will not generally be just a phase since the initial and final states are fundamentally different from each other.

[3]The notation we use matches that of Bonderson [2007].

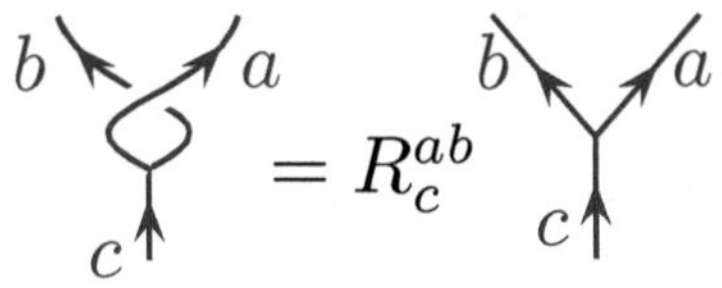

$$\ = R_c^{ab}\ $$

Fig. 13.3 Definition of the R-matrix. (Here b and a may or may not be equal, but for generality we usually assume they are unequal.) It is crucial that the final state of the system on both the left and the right has b on the left and a on the right, and in both cases the two particles fuse to c. However, the left diagram includes an exchange of the two particles. The additional exchange accumulates the phase R_c^{ab}.

In a unitary theory the R-matrix is always just a complex phase (magnitude one complex number). Note that R_c^{ab} is defined to be zero if a and b are not allowed to fuse to c (i.e. if $N_{ab}^c = 0$). Further, note that braiding anything with the identity (vacuum) particle should be trivial,

$$R_a^{Ia} = R_a^{aI} = 1 \quad .$$

A full braid (two exchanges in the same direction) of two particles a and b fusing to c is given by $R_c^{ba} R_c^{ab}$ as shown in Fig. 13.4. Note that in the representation on the far left of the figure, in both the initial and final configurations of particles, the b particle is to the left of the a particle meaning that we can understand this process as simply incurring an Aharonov–Bohm-like phase dependent on the fusion channel.

Similarly, we have the inverse braid shown in Fig. 13.5. Note carefully that the upper legs are labeled in the opposite order comparing Eq. 13.3 to 13.5. The reason for this is that, after a single exchange, the order of

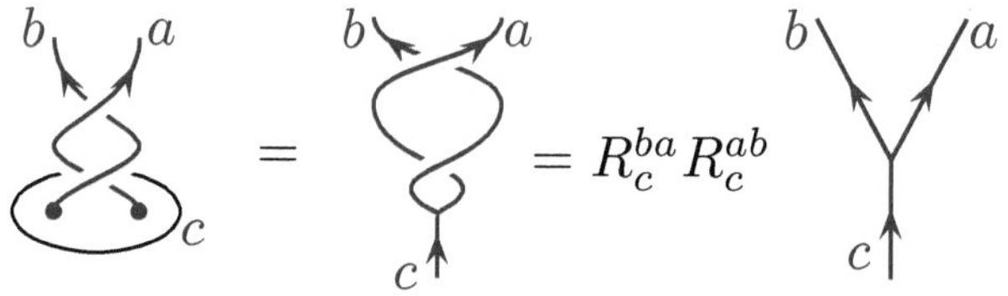

Fig. 13.4 A double exchange is a full braid (one particle wrapping fully around another). If you read two R-matrices in right to left order, they are implemented top to bottom. Note that in the right-hand figure R_c^{ab} acts on the vertex *first* exactly as in Fig. 13.3 before R_c^{ba} acts.

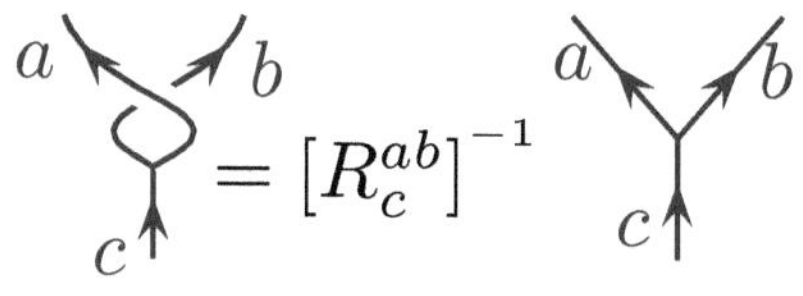

Fig. 13.5 An exchange in the opposite direction from Fig. 13.3 gives $[R_c^{ab}]^{-1}$. Note carefully that compared to Fig. 13.3, the upper legs are labeled in the opposite order.

the two legs at the vertex is switched, so in the application of R^{-1} we act on a state with the legs in the opposite order, as shown in Fig. 13.6.

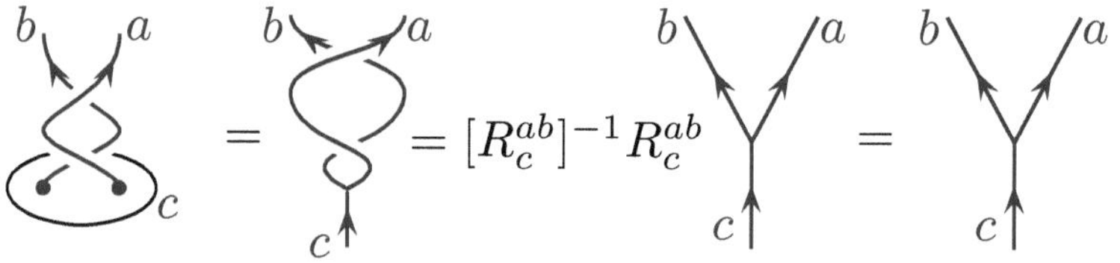

Fig. 13.6 An exchange followed by the inverse exchange. Compare to Fig. 13.4.

If a particle a has trivial full-braiding with all other particles of a theory, that is, if $R_c^{ab} R_c^{ba} = 1$ for all b, c where $N_{ab}^c > 0$, then we call the particle type *transparent*. (The identity, or vacuum particle, is always transparent.)

Taken together with the F-matrices, the R-matrices allows us to calculate the physical result of any braid. The scheme is mostly analogous to the cases we discussed for braiding identical particles in Chapter 10. If we want to exchange two particles we first use the F-matrices to put the system in a basis where those two particles have a known fusion channel. We can then directly apply the R-matrix to describe the exchange.

In particular we can now give a general scheme for evaluating (or "resolving") any crossing of the form shown in Fig. 13.2, which is shown in Fig. 13.7. Using this procedure any diagram with braiding can be reduced to a planar diagram which can then be evaluated using only the F-matrices.

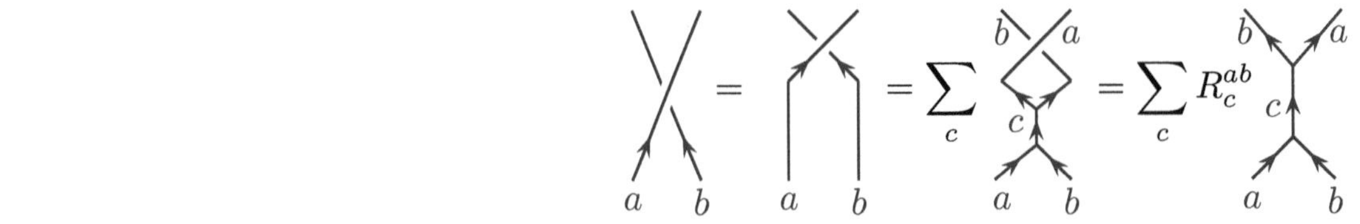

$$\text{X} = \text{Π} = \sum_c \overset{b \quad a}{\underset{c}{\diamondsuit}} = \sum_c R_c^{ab} \overset{b \quad a}{\underset{c}{\text{Y}}}$$

Fig. 13.7 A generic crossing can be reduced to a planar diagram (or "resolved") using the R-matrix. In the second step a complete set of particles c is inserted as in Fig. 12.21. Note this figure uses physics normalization.

$$\text{X} = \sum_c [R_c^{ba}]^{-1} \overset{b \quad a}{\underset{c}{\text{Y}}}$$

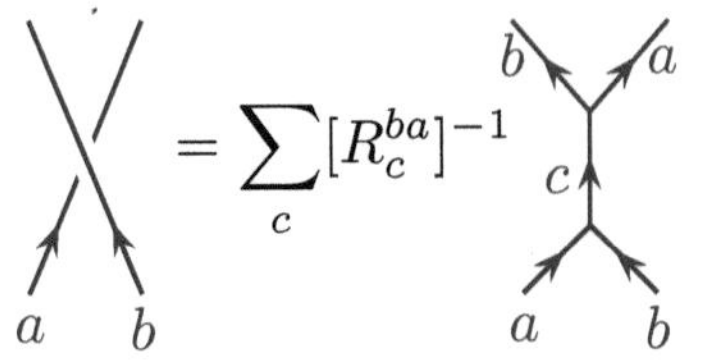

Fig. 13.8 The inverse crossing. This figure uses physics normalization.

13.2.1 Summary of Rules for Evaluating any (2 + 1)-Dimensional Diagram with Physics Normalization

The rules for evaluating any diagram in (2 + 1)-dimensions (working with physics normalization of diagrams) are thus a very simple extension of the rules presented in Section 12.4. We simply add two more rules:

(1) We are allowed to use R-moves as in Fig. 13.3 and 13.5. In particular, this enables the resolving of crossings by using Fig. 13.7 and Fig. 13.8.

(2) Once any diagram is reduced to a planar diagram, we can use the rules of Section 12.4.

As with the case of planar diagrams, there is some degree of deformation of diagrams (causal isotopy; see Section 12.3) that is freely allowed. Here again the rules are similar: any deformation that does not involve cutting lines or changing the time direction of motion is allowed. Without introducing new assumptions, natural moves such as those shown in Figs. 13.9 and 13.10 can be derived (see Exercise 13.6). These are nothing more than the Reidemeister type II and III moves introduced in Section 2.6.1, although here the strands carry labels.

Fig. 13.9 These moves, Reidemeister type II moves, are allowed in any anyon theory. See Section 2.6.1.

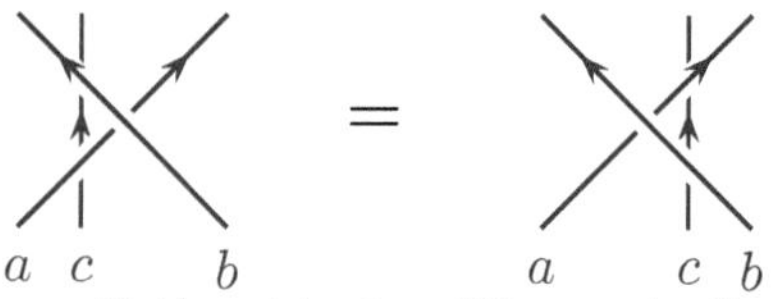

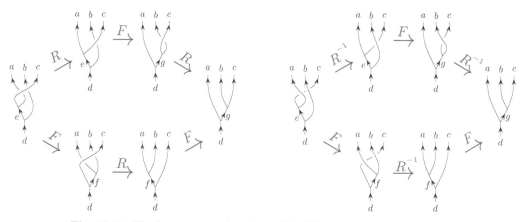

Fig. 13.10 This move, a Reidemeister type III move, is allowed in any anyon theory. See Section 2.6.1.

13.3   **The Hexagon**

Using R-moves and F-moves any $(2+1)$-dimensional diagram (starting and ending in the vacuum) can be reduced to a complex amplitude. One might worry if the rules we have listed for evaluation of diagrams are self-consistent: that is, does it matter in which order we apply the rules? Will we always obtain the same complex amplitude result? Indeed, given an F-matrix, only certain sets of (physically acceptable) R-matrices will have the property that the diagrammatic rules give a unique final result. In fact, it is even possible that for a given set of F-matrices that satisfy the pentagon equations, there may not even exist a set of consistent R-matrices!

When we discussed planar diagrams in Chapter 12, the pentagon equations guarantee self-consistency of F-matrices. Now, given some F-matrices that satisfy the pentagon equations, the consistency equations for R-matrices are known as the hexagon equations and are shown diagrammatically in Fig. 13.11.

Fig. 13.11 The hexagon equations in graphical form.

In equations the hexagon conditions can be expressed as

$$R_e^{ca}[F_d^{acb}]_{eg}R_g^{cb} = \sum_f [F_d^{cab}]_{ef} R_d^{cf} [F_d^{abc}]_{fg} \qquad (13.1)$$

$$[R_e^{ac}]^{-1}[F_d^{acb}]_{eg}[R_g^{bc}]^{-1} = \sum_f [F_d^{cab}]_{ef} [R_d^{fc}]^{-1} [F_d^{abc}]_{fg} \quad .(13.2)$$

The top equation is the left diagram whereas the lower equation is the

[4]This is also sometimes known as a *unitary ribbon tensor category* due to the fact that Eq. 15.5 holds, which is always true for unitary theories with braidings. The unitary braided tensor category is also sometimes known as a *premodular* category.

[5]Solutions that can be obtained from other solutions by gauge transformation are not considered to be different solutions.

[6]The simplest cases where there is a solution to the pentagon equations but not the hexagon equations are certain theories with three particle types discussed by Hagge and Hong [2009].

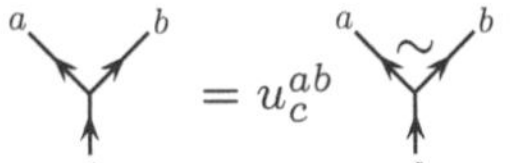

Fig. 13.12 We have the freedom to make a gauge transformation of a vertex by multiplying by a phase u_c^{ab}. The tilde on the right notates that the vertex is in the tilde gauge.

right diagram in Fig. 13.11. The left-hand side of the equation corresponds to the upper path, whereas the right-hand side of the equation corresponds to the lower path.

The structure we have now defined—a consistent set of (unitary) F- and R-matrices satisfying the pentagon and hexagon equations, is known as a *unitary braided fusion category*.[4] All $(2 + 1)$-dimensional anyon theories must be of this form.

Given a set of fusion rules, the pentagon and hexagon equations are very strong constraints on the possible F- and R-matrices consistent with these fusions. We mentioned in Section 9.3 that, given a set of fusion rules, there are only a finite number of solutions of the pentagon equations. Similarly, once we have a solution of the pentagon equations, that is, once the F-matrices are fixed, there are only a finite number of possible solutions of the hexagon equations. Both of these statements are known as *rigidity*. In particular, this means that if you have an anyon theory (a solution of the pentagon and hexagon equations) it is not possible to perturb this solution a small amount and get another solution.[5]

With simple fusion rules, such as Fibonacci fusion rules (as we saw in Exercise 9.4) the fusion rules completely determine the F-matrices of the theory. In this particular case, there are exactly two consistent solutions to the hexagon equations, corresponding to the left- and right-handed types of Fibonacci anyons (see Eq. 10.2 and Exercise 13.1). It can also be the case that given a set of F-matrices satisfying the pentagon equations, there may be *no* valid solutions of the hexagon equation.[6]

13.4 R-Matrix Odds and Ends

13.4.1 Appendix: Gauge Transformations and R

As in Section 9.4, one can make gauge transformations on the vertices of a theory. Given the transformation shown in Fig. 13.12 (see also Fig. 9.8), the R-matrix transforms as

$$\widetilde{R_c^{ab}} = \frac{u_c^{ba}}{u_c^{ab}} R_c^{ab} \quad . \tag{13.3}$$

Note that R_c^{aa} is gauge invariant, as is the full braid $R_c^{ab} R_c^{ba}$ in Fig. 13.4.

13.4.2 Product Theories

Given two anyon theories T and t, we can construct the product theory $T \times t$ as in Section 8.5. If the theory T has consistent R-matrices R_C^{AB} and the theory t has consistent R-matrices R_c^{ab} ("consistent" here means there are F-matrices that satisfy the pentagon relations and the Fs and Rs satisfy the hexagon relations), then the product theory has consistent R-matrices

$$R_{(C,c)}^{(A,a)(B,b)} = R_C^{AB} R_c^{ab} \quad .$$

Again, the point here is that in a product theory, the two constituent theories don't "see" each other at all.

13.4.3 Appendix: Higher Fusion Multiplicities

When we have a theory with higher fusion multiplicities (i.e. $N_{ab}^c > 1$ for at least one fusion channel), then the vertices must be given indices as in Section 12.5. In this case, the R-matrix carries such indices as well (Fig. 13.13), and is a unitary matrix with respect to these indices.

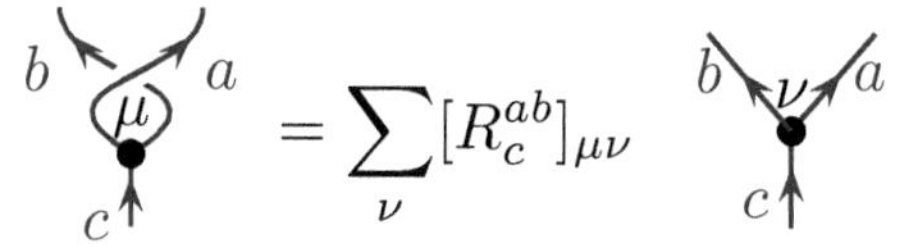

Fig. 13.13 Definition of R-matrix when there are higher fusion multiplicities. Here, the vertices carry indices, and R is a unitary matrix with respect to these indices.

Under a gauge transformation, as in Fig. 9.10, the R-matrix transforms as

$$\widetilde{[R_c^{ab}]}_{\mu'\nu'} = \sum_{\mu,\nu} ([u_c^{ab}]^{-1})_{\mu'\mu} [R_c^{ab}]_{\mu\nu} [u_c^{ba}]_{\nu\nu'} \quad . \tag{13.4}$$

Chapter Summary

- We extend the idea of the R-matrix introduced in Chapter 10 to allow for exchanging non-identical particles.
- The hexagon equations are a set of consistency conditions on R given an F-matrix that satisfies the pentagon.

Further Reading: Same as chapter 12.

Exercises

Exercise 13.1 Fibonacci Hexagon
Once the F-matrices for a TQFT are defined, the consistency of the R-matrix is enforced by the so-called hexagon equations as shown diagrammatically in Fig. 13.11. For the Fibonacci anyon theory, once the F-matrix is fixed as in Eq. 9.3, the R-matrices are defined up to complex conjugation (i.e. there is a right- and left-handed Fibonacci anyon theory—both are consistent). Derive these R-matrices. Confirm Eq. 10.2 as one of the two solutions and show no other solutions exist.

Exercise 13.2 Braiding Abelian Particles
Use the locality principle (see Section 10.1.1) to show that for abelian anyons

$$R_{a^{n+1}}^{a,a^n} R_{a^{n+1}}^{a^n,a} = (R_{a^2}^{a,a})^{2n}$$

where a^n means the result of fusing together n particles of type a. Show more generally that

$$R^{a^m,a^n}_{a^{n+m}} R^{a^n,a^m}_{a^{n+m}} = (R^{a,a}_{a^2})^{2nm} \quad .$$

Note that the locality principle can be proven with the hexagon equations; see Exercise 13.6.

Exercise 13.3 Evaluation of a Diagram

Evaluate the diagram shown in Fig. 13.14 in terms of Rs and Fs. Hint: First reduce the diagram to that shown in Exercise 12.1.

Exercise 13.4 Gauge Transformations of R and Hexagon

(a) Confirm the gauge transformation Eq. 13.3.

(b) Show that a set of F-matrices and R-matrices satisfying the hexagon equations, Eqs. 13.1 and 13.2, remains a solution after a gauge transformation. Remember that both R and F are transformed.

Exercise 13.5 Hexagon with Fusion Multiplicity

Generalize the hexagon equations to the case where $N^c_{ab} > 1$ so that vertices have additional indices as in Section 13.4.3.

Exercise 13.6 Reidemeister Moves

(a) Use the R-matrix, and the completeness relationship, to derive the equivalence shown on the left of Fig. 13.9.

(b) How does the hexagon equation imply the equivalence shown in Fig. 13.15. Hint: This is very subtle, but is almost trivial.

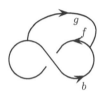

Fig. 13.14 Evaluate this diagram.

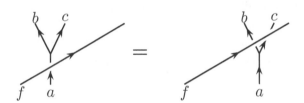

Fig. 13.15 This move is implied by the hexagon equations. (Similar with the straight line f going under the other two, and similar if the left to right slope of f is negative instead of positive.)

(c) Use Fig. 13.15 to show the equality on the right of Fig. 13.9.

(d) Use the result of Fig. 13.15 along with completeness and the R-matrix to demonstrate Fig. 13.10.

This exercise shows that equalities like those shown in Figs. 13.9 and 13.10 are not independent assumptions but can be derived from the planar algebra and the definition of an R-matrix satisfying the hexagon equation.

Seeking Isotopy[1]

When we discussed knot invariants, like the Kauffman bracket invariant, we were allowed to deform a knot in arbitrary ways as long as we didn't cut any strands.[2] This is what we called isotopy invariance. We would very much like the diagrammatic rules of our topological theories to obey such isotopy invariance. However, as we discussed in Section 12.3 we really only expect invariance under a more limited set of moves which we called *causal isotopy*. In this chapter we take steps to achieve the higher level of isotopy invariance necessary for constructing knot and link invariants.

We have already established the invariance of our rules under certain Reidemeister moves (see Figs. 13.9 and 13.10 and Exercise 13.6). However, there is one much more crucial move that we need to have our theory. Whether we are considering a planar theory or a (2 + 1)-dimensional theory, isotopy invariance requires the so-called zig-zag identity shown in Fig. 14.1, which is not a property of theories having only causal isotopy invariance, as shown in Fig. 12.36.

Unfortunately, a set of F-matrices (even if they satisfy the pentagon self-consistency condition Eq. 9.7) does not generically satisfy this zig-zag identity Fig. 14.1. To see this, consider the manipulations shown in Fig. 14.2. With the physics normalization of diagrams we have been using, the zig-zag identity does not hold.

[1]This chapter continues the development of the diagrammatic algebra in some detail. Those who would like a brief and easier (albeit not as general) introduction to diagrammatic algebra should go straight to Chapter 16.

[2]Meaning regular isotopy—that is, we should treat strings as ribbons. See Section 2.2.1.

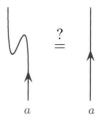

Fig. 14.1 A topological theory with isotopy invariance should have this "zig-zag" identity. However, generically a set of F-matrices will not satisfy this equality (see Fig. 14.2). We can often repair this problem by changing the normalization of kets.

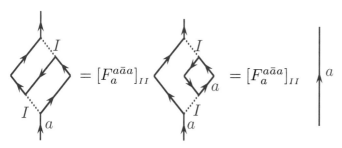

Fig. 14.2 Straightening a zig-zag incurs a factor of F using physics normalization of diagrams. The left of this diagram is the same as the left of Fig. 14.1. In the first step we use an F-move on the lower part of the diagram. We then use orthogonality of the tree to remove the small a bubble. This part of the diagram is just Fig. 12.19. Thus this small a bubble can be removed. We conclude that with the physics normalization we cannot satisfy the zig-zag identity in Fig. 14.1.

14.1 Isotopy Normalization of Diagrams

To fix the zig-zag problem, we take a cue from the Kauffman bracket invariant and change our definition of diagrams just by a small bit. In particular, let a simple loop of particle a, as shown in Fig. 14.3, be given a value d_a, which is sometimes called the "loop weight." This is different from our prior definition where we set the loop weight to one as in Fig. 12.19. The change here only means that we will be working with unnormalized bras and kets. We will call this new normalization "isotopy normalization."

$$\left| \swarrow^{a} \right\rangle = |\text{state}\rangle$$

$$\langle\text{state}|\text{state}\rangle = \diamondsuit^{a} = \bigcirc^{a} = d_a \qquad \text{Isotopy Normalization}$$

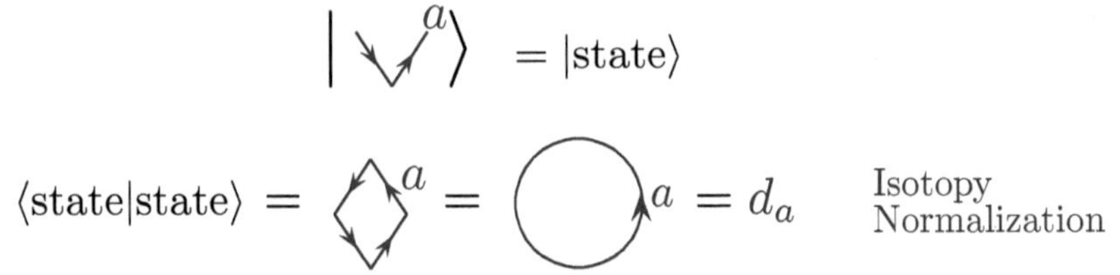

Fig. 14.3 Using a new normalization (which we call "isotopy normalization") of bras and kets. Compare to Fig. 12.19.

For the identity particle $d_I = 1$ since we should be able to add and remove vacuum lines, and vacuum loops, freely.[3] For other particles a, however, we will try to choose d_a so as to achieve isotopy invariance.

[3]Evaluation of an empty diagram also gives unity. We can think of the empty diagram as being equivalent to any number of loops of the vacuum I.

We should not worry about working with unnormalized bras and kets—we are allowed to do this in quantum mechanics. The price for using unnormalized states is that expectation values of operators are now given by

$$\langle\hat{O}\rangle = \frac{\langle\psi|\hat{O}|\psi\rangle}{\langle\psi|\psi\rangle}$$

instead of the usual expression for normalized states which just has the numerator.

We note that we would very much like to have the loop weight

$$d_a = \langle\text{state}|\text{state}\rangle > 0 \tag{14.1}$$

so that we have a positive definite inner product, which is required by quantum mechanics. Often it is said that "negative-normed states break unitarity" since they destroy the concept of the state living in a proper Hilbert space.[4] Note that we will also sometimes discuss d_a negative, despite the problems in doing so!

[4]The definition of a Hilbert space requires a positive definite inner product.

Having allowed an arbitrary loop weight d_a, when we recalculate the value of a zig-zag diagram (Fig. 14.1) we obtain the result shown in Fig. 14.4: straightening a zig-zag incurs a factor of $d_a[F_a^{a\bar{a}a}]_{II}$.

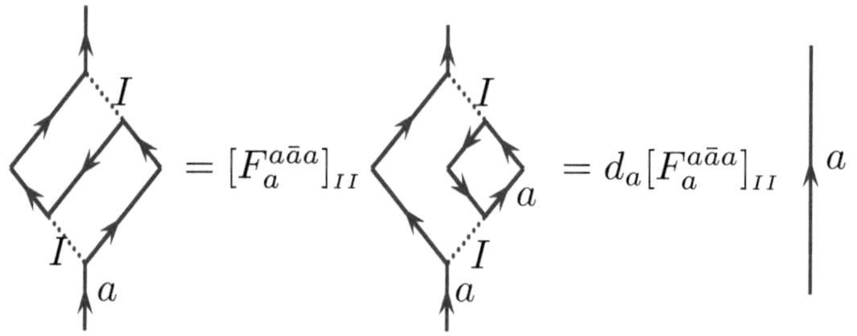

$$\text{(diagram)} = [F_a^{a\bar{a}a}]_{II} \qquad \text{(diagram)} = d_a[F_a^{a\bar{a}a}]_{II} \quad \text{(diagram)}\,a$$

Fig. 14.4 With the new isotopy-invariant normalization of diagrams, straightening a zig-zag incurs a factor of $d_a[F_a^{a\bar{a}a}]_{II}$. We will choose the value of d_a so as to make this factor ±1.

We now choose to work with a loop weight given by

$$d_a = \frac{\epsilon_a}{[F_a^{a\bar{a}a}]_{II}} \tag{14.2}$$

where for ϵ_a, which we call the *zig-zag phase*, we choose one of the two possibilities

$$\epsilon_a = \pm1 \quad . \tag{14.3}$$

This convention (Eq. 14.2) for the loop weight d_a is known as *isotopy normalization*. Note that we can, and will, always arrange that $[F_a^{a\bar{a}a}]_{II}$ is real by using a gauge transformation (see Sections 9.4 and 14.2) so that the loop weight d_a is also real.

We now declare that:

Henceforth, we will use isotopy normalization!

Given this choice of normalization, the product $d_a[F_a^{a\bar{a}a}]_{II}$ in Fig. 14.4 is $\epsilon_a = \pm1$ and the zig-zag identity (Eq. 14.1) generally becomes modified to that shown in Fig. 14.5. The point of choosing this isotopy

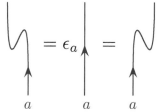

Fig. 14.5 The modified zig-zag identity. Here, ϵ_a, which we call the "zig-zag phase," is always arranged to be $+1$ or -1. The equality on the left is from Fig. 14.4. The equality on the right follows from Hermitian conjugation of the equality on the left (turning the diagrams upside down and reversing the arrows).

normalization is that straightening a zig-zag will now incur at most a phase of ±1. One might think that we should always choose $\epsilon_a = +1$ so that straightening a zig-zag (as Fig. 14.1) would be an allowed transformation without incurring any phase. However, as we show in Section 14.2, choosing $\epsilon_a = +1$ is not always allowed, which is why we allow ±1 more generally. For the moment, let us ignore the possible sign accumulated from a zig-zag (i.e. assume $\epsilon_a = +1$) and return to worry about this sign in Section 14.2.

Thus, we have defined a new normalization of the loop Fig. 14.3 given by the choice of Eq. 14.2. As we see in Section 17.1, the normalization constant d_a will turn out (up to a possible sign) to be the same quantum dimension d_a that we found in Eq. 8.17 from the Hilbert space dimension of fusing anyons together!

Having changed the normalization of our kets, we now choose to change the normalization of fusion and splitting vertices as shown in Fig. 14.6.[5]

[5]We will only need to consider cases where the factor inside the brackets ends up positive (see Eq. 14.5).

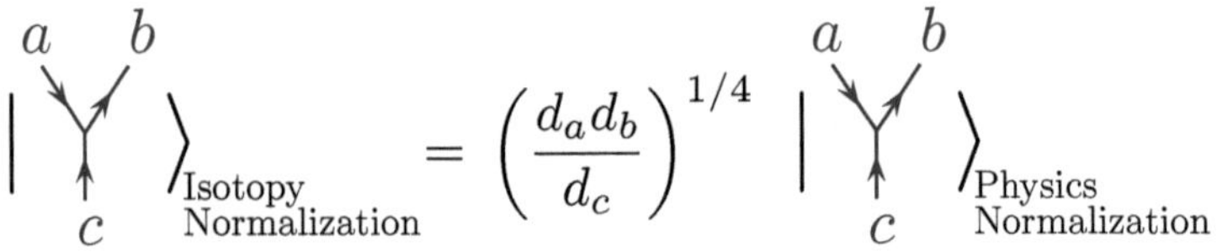

$$\left| \begin{array}{c} a \quad b \\ \mathsf{Y} \\ c \end{array} \right\rangle_{\substack{\text{Isotopy} \\ \text{Normalization}}} = \left(\frac{d_a d_b}{d_c}\right)^{1/4} \left| \begin{array}{c} a \quad b \\ \mathsf{Y} \\ c \end{array} \right\rangle_{\substack{\text{Physics} \\ \text{Normalization}}}$$

Fig. 14.6 New "isotopy" normalization for vertices.[5] Note that this is consistent with Fig. 12.19 by setting $c = I$ with $a = b$ (and note that $d_I = 1$).

We have chosen this normalization for all vertices so that the F-matrix does not need any alteration when we switch from physics normalization to isotopy normalization. One can check that in changing normalizations, both sides of Fig. 9.1 (equivalently Fig. 12.30) are multiplied by the same factor of $(d_a d_b d_c/d_e)^{1/4}$.

With this new normalization, the orthonormality of trees is now different from what we had previously. For example, Fig. 12.33 should now have a factor of $\sqrt{d_a d_b d_e d_d}$ on the right-hand side.

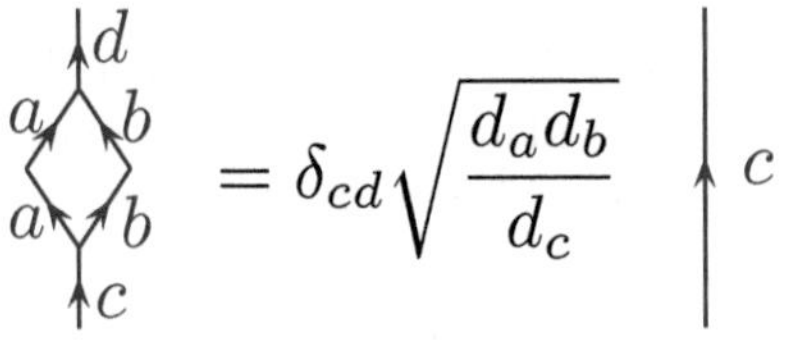

$$\begin{array}{c} d \\ a \bigwedge b \\ a \bigvee b \\ c \end{array} = \delta_{cd} \sqrt{\frac{d_a d_b}{d_c}} \quad \Big\uparrow c$$

Fig. 14.7 Bubble, or orthonormality, diagram with isotopy-invariant normalization of diagrams. See Eq. 14.9 for how to interpret the square root in cases where $d < 0$.

$$a\uparrow \quad \uparrow b = \sum_c \sqrt{\frac{d_c}{d_a d_b}} \begin{array}{c} a \quad b \\ \mathsf{X} c \\ a \quad b \end{array}$$

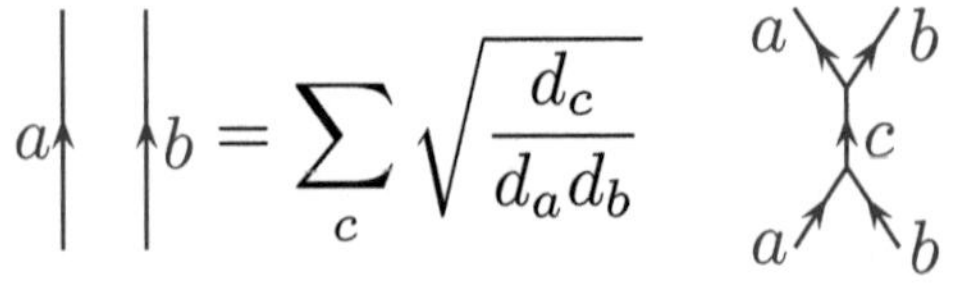

Fig. 14.8 Insertion of a complete set of states with isotopy-invariant normalization of diagrams. See Eq. 14.9 for how to interpret the square root in cases where $d < 0$.

[6]Once again if $N_{ab}^c > 1$ there are additional indices at the vertices and these must match as well. See Section 12.5.

Similarly, our bubble diagram Fig. 12.18 and our completeness diagram Fig. 12.21 need to be modified as shown in Fig. 14.7 and Fig. 14.8.[6]

With this isotopy-invariant normalization the rules for evaluating planar diagrams are exactly the same as those described in Section 12.4

except that loops are now normalized with d_a as in Fig. 14.3 and our completeness and orthonormality relations (Fig. 12.21 and Fig. 12.18) are altered to those shown in Fig. 14.8 and Fig. 14.7. There are of course more general completeness and orthonormality relations for three or more particles analogous to Figs. 12.21 and 12.25. These can be derived simply by converting the diagram in physics normalization to one in isotopy-invariant normalization just by including appropriate factors of loop weights at each vertex, as given by Fig. 14.6.

The same R-matrix rules can be applied to diagrams with over- and under-crossings as in Chapter 13. The use of the R-matrix is unchanged. Be warned, however, that in Figs. 13.7 and 13.8 we have used the completeness relationship Fig. 12.21, which now needs to be modified to Fig. 14.8, so that when we evaluate crossings we now obtain, for example, Fig. 14.9.

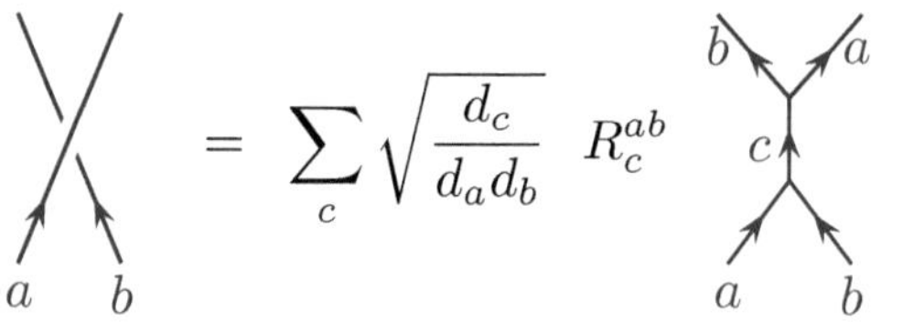

$$ \text{(crossing)} = \sum_c \sqrt{\frac{d_c}{d_a d_b}}\ R_c^{ab}\ \text{(vertex)} $$

Fig. 14.9 Resolving a crossing with isotopy normalization. Compare to Fig. 13.7. See Eq. 14.9 for how to interpret the square root in cases where $d < 0$.

14.2 Gauge Choice and Frobenius–Schur Indicator

Let us now return to the zig-zag in Fig. 14.1 and our choice of the quantity d_a in Eq. 14.2. First, we claimed that we can always arrange to have $[F_a^{a\bar{a}a}]_{II}$ be real. With a gauge choice, we can fix the phase of $[F_a^{a\bar{a}a}]_{II}$ any way we like, at least for cases where $a \neq \bar{a}$. Let us see how this can be done. On the far left of Fig. 14.4 we have a vertex $|a\bar{a}\rangle$ as well as a vertex that we write as $\langle\bar{a}a|$ (compare to Fig. 14.10). Note that, at least when $a \neq \bar{a}$, these two vertices are *not* Hermitian conjugates of each other (recall that when Hermitian conjugating a diagram arrows get reversed as well as reflecting the diagram across a horizontal axis). By making separate gauge transformations on these two states, these kets ($|a\bar{a}\rangle$ and $|\bar{a}a\rangle$) can be redefined by arbitrary phases as discussed in Section 9.4, and these phases then end up in $[F_a^{a\bar{a}a}]_{II}$ (see the transformation in Eq. 9.8). Thus, by a gauge choice we can choose any phase for $[F_a^{a\bar{a}a}]_{II}$, as long as $a \neq \bar{a}$. It is often (albeit not always) convenient to choose $[F_a^{a\bar{a}a}]_{II}$ to be real and positive so that we can have d_a positive and $\epsilon_a = +1$ positive as well.

However, if $a = \bar{a}$, it is not possible to change $[F_a^{aaa}]_{II}$ by gauge transformation. In this case, the kets $|a\bar{a}\rangle$ and $|\bar{a}a\rangle$ are equal and we do not have the freedom to gauge transform them separately. It is easy to show that when $a = \bar{a}$, the factor of $[F_a^{aaa}]_{II}$ must be real (see Section

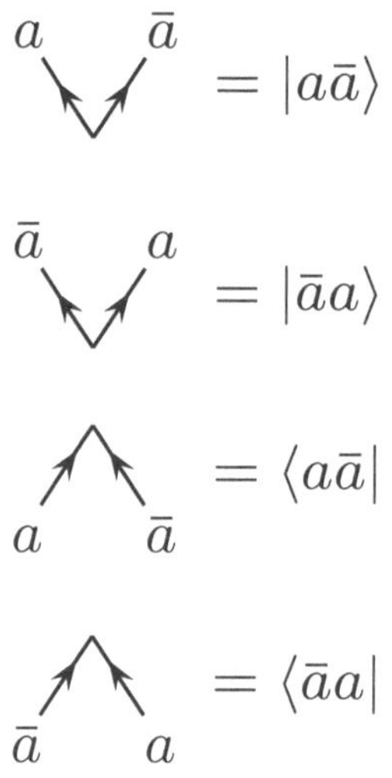

$$ a \quad \bar{a} \qquad = |a\bar{a}\rangle $$

$$ \bar{a} \quad a \qquad = |\bar{a}a\rangle $$

$$ a \quad \bar{a} \qquad = \langle a\bar{a}| $$

$$ \bar{a} \quad a \qquad = \langle\bar{a}a| $$

Fig. 14.10 Cups (top two) and caps (bottom two). The vertex $|a\bar{a}\rangle$ (top figure) and the vertex $|\bar{a}a\rangle$ (second figure) can be assigned different phases as a gauge choice (see Section 9.4). The third figure here is the Hermitian conjugate of the top figure and does not get an independent phase choice. The bottom figure is the Hermitian conjugate of the second figure. In Fig. 14.4 the leftmost figure includes $|a\bar{a}\rangle$ and $\langle\bar{a}a|$, whereas in the middle picture of Fig. 14.4 the phases cancel in the loop formed from $|\bar{a}a\rangle$ and $\langle\bar{a}a|$. Thus, choosing gauges we can choose any phase for $[F_a^{a\bar{a}a}]_{II}$ unless $a = \bar{a}$ (see Section 9.4 for discussion of the effects of gauge transformation on F).

[7]For particles that are not self-dual there are several different definitions of what people call the Frobenius–Schur indicator. Some references just define it to be zero for such particles. Other references define it to be $\epsilon_a \text{sign}[d_a]$ and others define it to be unity! To avoid confusion we will not use the phrase Frobenius–Schur in the context of non-self-dual particles.

[8]Theories with all positive Frobenius–Schur indicators are sometimes called *unimodal* or *unimodular*.

[9]This is the convention chosen by Bonderson [2007].

14.6 for a three line proof). The sign of $[F_a^{aaa}]_{II}$ is then a gauge-invariant quantity, known as the Frobenius–Schur indicator[7]

$$\kappa_a = \text{sign}[F_a^{aaa}]_{II} = \epsilon_a \, \text{sign}[d_a] \quad . \qquad (14.4)$$

If the Frobenius–Schur indicators are positive for all the self-dual particles in a theory (or if there are no self-dual particles in the theory) then we can set $\epsilon_a = +1$ for all particles and we can also have d_a positive for all particles. This means that we can both have a positively normed inner product, and we can freely straighten out zig-zags as in Figs. 14.5 or 14.1 without incurring any minus signs. Theories of this type are fairly simple to work with.[8]

However, when the Frobenius–Schur indicator of a self-dual particle is negative, things are more complicated as it looks like we must give up either isotopy invariance or a positively normed inner product. Unfortunately, many anyon theories, even very simple ones like $SU(2)_1$ (the semion theory) have negative Frobenius–Schur indicators. The Frobenius–Schur sign associated with a zig-zag, though perhaps surprising, is a genuine physical quantity that occurs in more familiar settings as well. In Section 14.7 we show that such a Frobenius–Schur sign actually occurs for simple systems such as spin-half particles! Independent of why this sign occurs, theories that have self-dual particles with negative Frobenius–Schur indicators seem to force us to choose between several imperfect options.

Option A: Negative ϵ_a

Here, for particles with negative Frobenius–Schur indicators, we choose $\epsilon_a = -1$, but $d_a > 0$. With $d_a > 0$ we have a positive definite inner product, and therefore a theory that can properly represent a (necessarily unitary) quantum-mechanical system. On the other hand, with $\epsilon_a = -1$, we incur a minus sign for straightening out any zig-zag as in Fig. 14.5, so our diagrammatic rules are not completely isotopically invariant.[9]

We will typically work with Option A so that all $d_a > 0$ and self-dual particles with negative Frobenius–Schur indicators have $\epsilon_a = -1$ (and all other particles have $\epsilon_a = +1$).

Option B: Negative d_a

Here, for particles with negative Frobenius–Schur indicators, we choose $\epsilon_a = +1$, but $d_a < 0$. Choosing $\epsilon_a > 0$ means we have isotopy-invariant diagram rules. However, negative d_a means a non-positive-definite inner product (see Eq. 14.1) which is inappropriate for quantum-mechanical systems. Nonetheless, one will often see theories with negative d_a in the physics literature attempting to describe quantum-mechanical systems. While this appears problematic, often one can reinterpret this as a valid theory with a certain bookkeeping trick whereby we redefine our inner product and signs are pushed around from one place to another to make it look like we have $d_a < 0$ and $\epsilon_a = +1$, whereas what we actually have is the reverse $d_a > 0$ and $\epsilon_a = -1$, and thus we are describing a valid quantum-mechanical theory.

The interpretation of negative d_a as a bookkeeping trick is elaborated in Section 14.5. It will turn out that doing this bookkeeping gives us a theory that is precisely equivalent to Option A. Nonetheless, it may be convenient at times to work with negative d_a so as to obtain isotopy-invariant rules. A good example of this is discussed in Chapter 32.

The requirement for this bookkeeping trick to be able to remove all zig-zag signs is that we must have

$$\text{sign}[d_a]\,\text{sign}[d_b] = \text{sign}[d_c] \qquad \text{when} \qquad N^c_{ab} > 0 \quad , \qquad (14.5)$$

which is known as a $\mathbb{Z}_2$ grading of our fusion rules. While there are theories that cannot satisfy this condition (in which case we must resort to using the more inconvenient Option A), most of the time[10] this condition is satisfied (see Section 14.5).

Option C?

There is yet a third possibility, although it is physically slightly different from what we have described so far. While the above-described diagram algebra prescriptions are functionally very useful, they actually miss something important about the physics. If we put this piece of physics back in we will find that we can preserve both the positive definite inner product (i.e. describe a unitary theory) *and* maintain isotopy invariance, at the price of some increased diagrammatic complexity. This is what we discuss next.

14.3 Isotopy-Invariant Unitary Rules

Despite the fact that many anyon theories have particles with negative Frobenius–Schur indicators, which appear to break isotopy invariance, it is *always* possible to construct an isotopy-invariant set of rules for evaluating knot and link diagrams while at the same time maintaining unitarity and a proper inner product. Indeed, such rules are natural in the language of category theory (which we will not discuss). To understand what this alternative approach is, it is worth thinking back to what we learned about Chern–Simons theory (which often does have particles with negative Frobenius–Schur indicators).

14.3.1 Isn't Chern–Simons Theory Isotopy Invariant and Unitary?

Yes! All the way back in Chapter 5 we explained that Chern–Simons theory is based on a topologically invariant (diffeomorphism invariant) action, and it is a unitary quantum theory. So why do we now seem to

[10]For example, this is always possible for Chern–Simons theories. Among "modular" theories (i.e. those with braiding and no transparent particles, see Sections 4.3.2 and 17.3.1) the simplest case that I know of that fails to meet this condition has 22 particle types. See Simon and Slingerland [2022].

[11]The discussion of this section is taken from Simon and Slingerland [2022]. See also references at the end of this chapter.

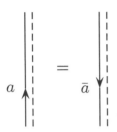

Fig. 14.11 Drawing a particle line as a ribbon rather than a single line. Right-handed framing means the dotted line is to the right of the solid line when walking in the direction of the arrow. Left-handed framing means the dotted line is to the left of the solid line when walking in the direction of the arrow. In the left figure we have a going up, whereas in the right we have $\bar{a}$ going down (these two pictures are equivalent) and in both cases the dotted line is on the right of the solid line. However, since the direction of the arrow switches between the two figures, the left figure is right-handed framed, whereas the right figure is left-handed framed. Thus framing a right-handed is equivalent to framing $\bar{a}$ left-handed.

Fig. 14.12 A right-handed framed a particle (going up) can (only!) annihilate a left-handed framed $\bar{a}$ particle.

[12]In category theory one often attempts to differentiate between a and $\bar{a}$ even if $a = \bar{a}$. One states that taking the dual of a particle and then taking the dual again gives you something *isomorphic to* the original particle, but not necessarily exactly the original particle. If you do it four times, you necessarily get back to the original particle. See for example Bakalov and Kirillov [2001],Turaev and Virelizier [2017], and Bartlett [2016].

have this contradiction that we either have to give up unitarity (i.e. accept a non-positive-definite inner product) or give up isotopy invariance (i.e. accept a sign when we deform certain lines to insert a wiggle as in Fig. 14.5). What went wrong here? Why do negative Frobenius–Schur indicators show up in Chern–Simons theory?[11]

A hint to the answer to this question is hidden in note 32 of Chapter 5. The Chern–Simons invariant of a knot or link is actually an invariant of the *framed* knot or link. That is, one must describe not just the path of the knot, but also how the ribbon twists (as well as specifying a front and back of the ribbon). We tried to take care of this issue by declaring way back in Section 2.2.2 that we would always assume "blackboard framing," meaning we assume the ribbon always lies flat in the plane of the blackboard. However, here this simplification has come back to bite us. What we need here is a slightly more general notation.

Let us draw not just labeled lines with arrows on them, but rather draw a solid line parallel to a dotted line to represent a ribbon as in Fig. 14.11. We have a choice as to whether we put the dotted line to the right or left of the solid line. We say the ribbon is right-handed framed if the dotted line is to the right of the solid line when walking along the line in the direction of the arrow. Framing particle a right-handed is equivalent to framing particle $\bar{a}$ left-handed as shown in Fig. 14.11. A right-handed particle a can (only) annihilate a left-handed particle $\bar{a}$ as shown in Fig. 14.12. Note that we cannot turn right-handed framing into left-handed framing by performing a half-twist (flipping over the ribbon) because the ribbon is meant to have a well-defined front and back (say, two different colors on the two different sides): a right-handed framed ribbon turned over is inequivalent to a left-handed framed ribbon.

For each pair of particle types, a and $\bar{a}$, we can choose a convention such that one of them is framed right-handed, and the other is framed left-handed so that the two can annihilate. However, we run into a problem when we try to establish a convention for particles that are self-dual: since we do not draw arrows on such particle lines, we cannot distinguish right- from left-handed framing. To consistently describe the framing, we need to establish a bit more notation, known as flags.

The scheme here is based on the idea that we should introduce a new degree of freedom whose physical meaning is related to the framing.[12] A version of this scheme is described by Kitaev [2006], which we will follow. (This is essentially equivalent to the discussions of Bakalov and Kirillov [2001] and Turaev and Virelizier [2017].)

While one can apply this scheme to all the particles in the theory, for non-self-dual particles, and self-dual particles with positive Frobenius–Schur indicators, it is easy enough to just choose a gauge so that you can remove zig-zags as in Fig. 14.5 with $\epsilon = +1$. As such, we focus here only on the self-dual particles with negative Frobenius–Schur indicators where the minus signs from straightening zig-zags arise.

We assign each cup (pair creation) and cap (pair annihilation) a big triangular arrow (called a "flag"), which can either point left or right as shown in Fig. 14.13. These are not the same arrows we have been using

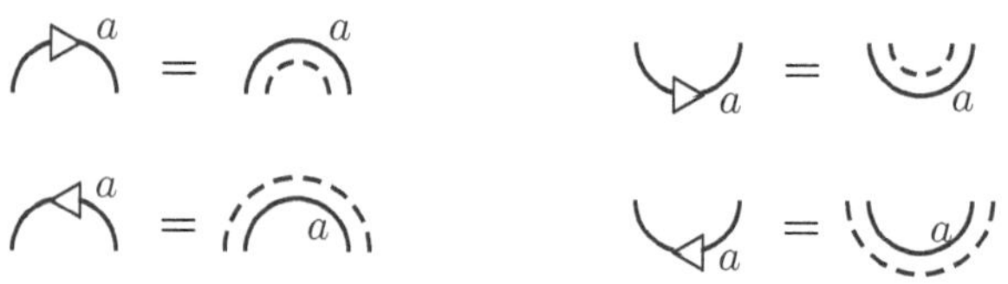

Fig. 14.13 Flags on cups and caps (the large arrows) are used to fix framing conventions. Here we are assuming that a is self-dual so we cannot use the small arrows on lines to fix a framing as in Fig. 14.11.

Fig. 14.14 Hermitian conjugation preserves the direction of the flag.

to distinguish particles from their antiparticles. Indeed, the particles we are focusing on here are self-dual! These flags are supposed to indicate the framing of the world lines.[13]

Hermitian conjugation does not flip the direction of the flag as shown in Fig. 14.14. Thus, the corresponding inner product is defined to be positive definite as shown in Fig. 14.15. Straightening a wiggle in a ribbon corresponds to canceling a cup/cap pair pointing in opposite directions as shown in Fig. 14.16.

[13]In category language the flag tells you if you are using ev versus $\widetilde{\text{ev}}$ for caps, and similarly coev versus $\overline{\text{coev}}$ for cups. See Bakalov and Kirillov [2001] and Turaev and Virelizier [2017].

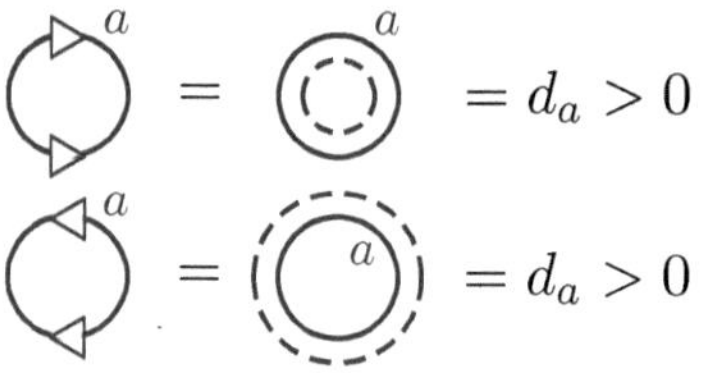

Fig. 14.15 The inner product is positive definite. (Flags must align to join ribbons.)

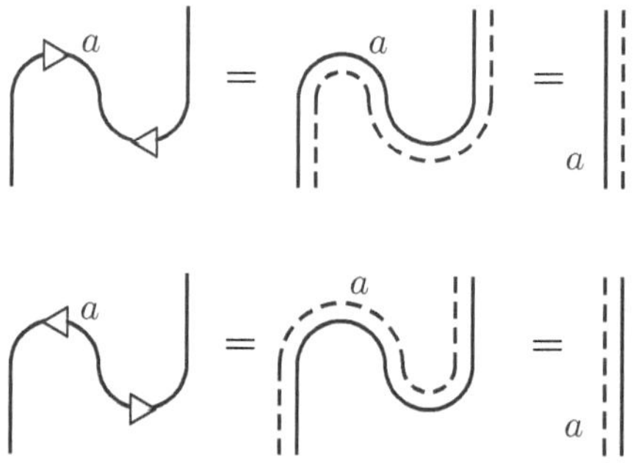

Fig. 14.16 Straightening a wiggle corresponds to canceling a cup/cap pair pointing in opposite directions.

So far, the flags have done little but remind us that we need to frame our world lines in Chern–Simons theory, that is, we must draw a dotted line along with a solid line. However, we now add a rule which is outside of what can be done in Chern–Simons theory, but will allow us to use the same flag notation to describe the options from Section 14.2: let us declare that the direction of a flag can be reversed at the price of a factor of the Frobenius–Schur indicator (which is $\kappa_a = -1$ since these are the only particles we are concerned with here)

$$\text{⌐}_a = \kappa_a \;\; \text{⌐}_a \qquad \kappa_a = -1 \;. \qquad (14.6)$$

Diagrams thus have different values depending on how the flags are assigned to the cups and caps of the diagram (for all self-dual particles with negative Frobenius–Schur indicator). If we make the rule that all flags always point right, we recover the above "Option A" (Section 14.2), that is, removing a zig-zag incurs a minus sign (to remove a zig-zag we flip the direction of a flag using Eq. 14.6, incurring a factor of $\kappa_a = -1$, and then use Fig. 14.16).

However, another possible choice (motivated by Chern–Simons theory) is to say that flags should be put on cups and caps so that they alternate directions as you walk along any line. Because zig-zags with alternating flags can be freely straightened we then have an isotopy-invariant diagrammatic set of rules. This scheme of decorating with alternating flags then provides a way that *any* braided anyon theory can be converted into a knot (or link) invariant with isotopy invariance.[14] Note, however, that it provides a different output from our "gauge-fixed" choice (Option A).

One should be cautious that we defined our F-matrices without ever specifying the direction of any flags (in some F-matrix diagrams with cups and caps, such as in Figs. 14.2 and 14.4, we would be able to decorate these cups and caps with flags). We should realize that in all prior sections, we were implicitly working with the Option A version of diagrams—in other words, our definition of the F-matrix assumes that all flags are pointing to the right. If flags are not all pointing right, we should first flip the flags using Eq. 14.6 before using the F-matrix.

Let us now consider the isotopy-invariant rules for evaluation of a knot or link diagram, as defined in this section (where flags alternate directions). Since we have isotopy invariance we can first use this isotopy invariance to simplify the knot or link as much as possible. Then before applying F-moves, we should flip flags using Eq. 14.6 to get to a gauge-fixed diagram (with all flags pointing right) and continuing our evaluation. Once we have a gauge-fixed diagram we must keep careful track of zig-zags, which may now incur minus signs if they are straightened.

So which convention should one use? The answer is "it depends." For many mathematics applications (certainly for knot theory) one wants to work with an isotopy-invariant set of rules as described in this section. However, for certain diagrammatic calculations, Option A will be more appropriate. We again emphasize that the phases associated with straightening zig-zags can actually be physical, as discussed more in Section 14.7.

[14]Meaning regular isotopy invariance: strands should be treated as ribbons in three dimensions.

14.3.2   What Have We Achieved?

One of our original hopes for defining a TQFT, way back in Chapter 7, was some prescription that would turn a labeled knot or link diagram into a complex amplitude (see Fig. 7.1) where the result would be unchanged by any smooth deformation of three-dimensional spacetime (treating the strands of the knots as ribbons, that is, we are allowed regular isotopy of the diagram, see Section 2.6.1). As we discussed in Chapter 13.3, our anyon theories already have invariance under Reidemeister moves II and III. The only impediment to regular isotopy invariance was the zig-zag identity 14.5 for particles with negative Frobenius–Schur indicators. Using the methods of Section 14.3 we have now achieved this regular isotopy invariance.[15] We should be warned, however, that once we start evaluating diagrams using F-moves and R-moves, we may not have isotopy invariance of diagrams with fusions and splitting, as we will discuss further in Section 14.4—we have only guaranteed isotopy invariance for knot and link diagrams.

Note that in Chapter 7 when we were defining a TQFT we wanted to more generally have a prescription for turning a knot or link *embedded in an arbitrary closed manifold* into a complex number output. This generalization is indeed possible, and we return to this issue in Chapter 24. However, for now we note that our scheme gives unity for an empty diagram (which we can think of as any number of loops of the identity particle with $d_I = 1$) so our diagrammatic evaluation corresponds to

$$\text{diagram} \;=\; \frac{Z(S^3 \text{ with labeled link embedded})}{Z(S^3)} \tag{14.7}$$

$$=\; Z(S^2 \times S^1 \text{ with labeled link embedded})$$

with the caveat that the link does not go around the nontrivial cycle of the S^1 in the latter case. Note that this normalization matches that of Eq. 5.18.

[15] We can also obtain an isotopy-invariant knot invariant by using Option B, meaning using negative d_a, from Section 14.2, but we give up unitarity. See the discussion of Section 14.5.

14.4   Impediments to Isotopy Invariance in Fusion Diagrams

Once we consider diagrams with fusion and splitting vertices, full isotopy invariance can be even harder to obtain. Even neglecting issues with Frobenius–Schur indicators, we are not guaranteed that we can deform lines in any way we like in the plane. For example, the right-hand side of Fig. 12.36 cannot generically be turned into an equality. In Figs. 14.17 and 14.18 we give similar examples (recapitulating the calculation in Fig. 12.38 but now using isotopy normalization) of turning-up transformations that generically incur nontrivial factors.

Thus it seems that our most general theory with fusions and with causal isotopy invariance cannot achieve (even planar) isotopy invari-

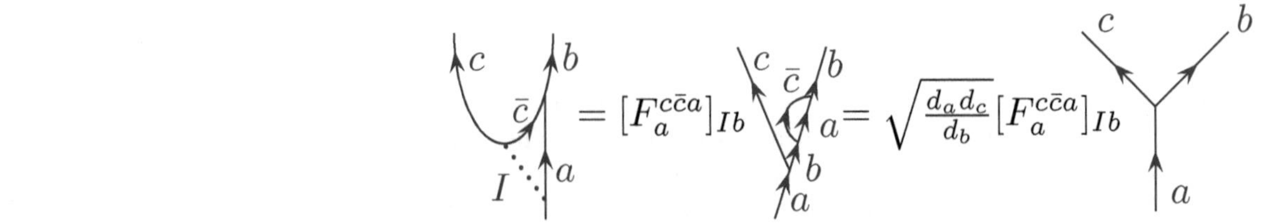

Fig. 14.17 To evaluate the diagram on the left, the vacuum line is inserted and an *F*-move is made. The bubble is then removed with Fig. 14.7. Note that if we were to use the physics normalization, the prefactor of $\sqrt{d_a d_c / d_b}$ would be absent (see Fig. 12.38). Generally we should not expect that the prefactors of $\sqrt{d_a d_c / d_b}$ and F obtained on the right should cancel each other. In Chapter 16 we focus on precisely the theories where this does turn out to be unity as is required for isotopy invariance. More generally, as we discuss in Section 14.8.1, the transformation from left to right in this figure is unitary, meaning the resulting factor on the right $\sqrt{d_a d_c / d_b} [F_a^{c\bar{c}a}]_{Ib}$ is just a magnitude one complex phase.

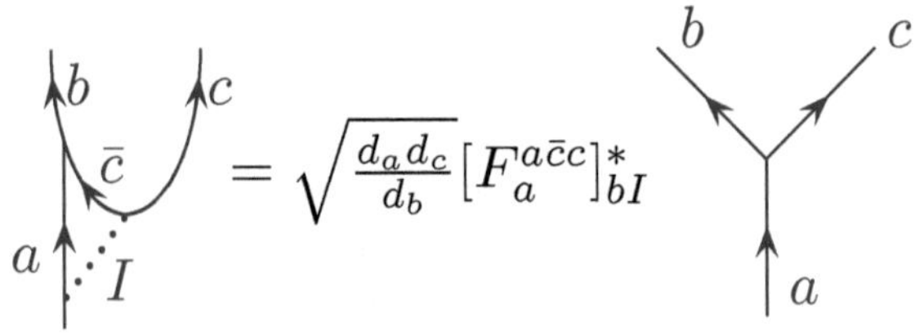

Fig. 14.18 The mirror image of Fig. 14.17. Here we use the fact that F is unitary, so $F^{-1} = [F^*]^T$.

ance. Perhaps this is not surprising. Even if we can deform space-time world lines into each other, we might still expect that there would be some difference between a process on the far left and far right of Fig. 14.17: On the far left, c and $\bar{c}$ are produced from the vacuum, then $\bar{c}$ and a come together to form b, whereas on the far right, a simply turns into c and b. Fortunately, many topological theories are not this complicated: as we will see in Chapter 16, there are many theories where one does have planar isotopy invariance (or even isotopy invariance in three dimensions), and the prefactor incurred in the process shown in Figs. 14.17 and 14.18 turns out to be unity. That is, we can often choose a gauge such that turning-up and turning-down of legs (as in Figs. 14.17 and 14.18) are trivial. If it is not possible to choose such a gauge we say there is an "obstruction." For theories with well-defined braidings and twists, such obstructions can only occur if there are fusion multiplicities $N_{ab}^c > 1$.

To show an example of such an obstruction we consider the *third* Frobenius–Schur indicator,[16] written as $\kappa_a^{(3)}$, which is defined in Fig. 14.19. The third Frobenius–Schur indicator for particle a is non-zero only if $N_{aa}^{\bar{a}} > 0$. We say the third Frobenius–Schur indicator is trivial if W^a is the identity matrix, or equivalently if

$$\kappa_a^{(3)} = N_{aa}^{\bar{a}} \qquad \Longleftrightarrow \qquad \text{"Trivial."} \qquad (14.8)$$

If the third Frobenius–Schur indicator is nontrivial, then there is an obstruction to obtaining an isotopy-invariant diagram algebra. This

[16] The Frobenius–Schur indicator we have been talking about, which we defined in Eq. 14.4, is strictly speaking the *second* Frobenius–Schur indicator, but rarely does anyone speak of the higher cases, so it is often just called *the* Frobenius–Schur indicator without specifying that it is the second.

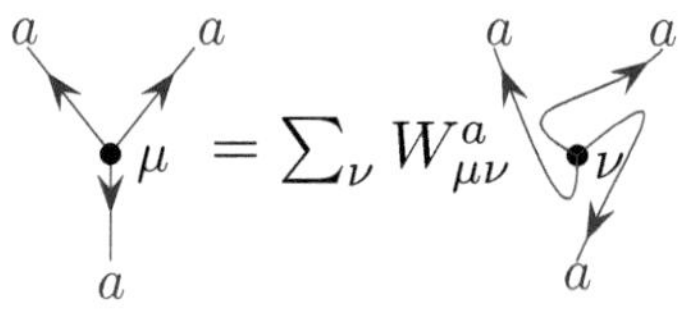

$$\overset{a}{\underset{a}{\mathbf{Y}}}_{\mu} = \sum_{\nu} W_{\mu\nu}^{a} \overset{a}{\underset{a}{\mathbf{V}}}_{\nu} \qquad\qquad \kappa_{a}^{(3)} = \mathrm{Tr}[W^{a}]$$

Fig. 14.19 The definition of the third Frobenius–Schur indicator $\kappa_a^{(3)}$ for particle type a. Here we have defined it in generality, allowing for the possibility of fusion multiplicity (so the vertices have indices μ). If $N_{aa}^{\bar{a}} = 0$ then $\kappa_a^{(3)} = 0$. If $N_{aa}^{\bar{a}} = 1$ then W^a is just a scalar rather than a matrix. As a result of Section 14.8.1, the eigenvalues of the matrix W^a must be third roots of unity. If the matrix W^a is not the identity matrix, then one cannot choose a gauge such that one has an isotopy-invariant diagram algebra.

means the two diagrams in Fig. 14.19 cannot be deformed into each other without incurring a gauge-invariant phase.[17]

We will see simple examples of planar diagram algebras with nontrivial $\kappa^{(3)}$ in Exercise 19.3. Further discussion of obstructions is given in Simon and Slingerland [2022].

[17]The simplest braided theories I know of having a nontrivial third Frobenius-Schur indicator are the Chern–Simons theories $SO(5)_4$ and $(E_8)_5$, which both have 15 different particle types.

14.4.1 Planar Diagrams with Isotopy Invariance

Assuming we do not have such an obstruction (and given that we have satisfied Eq. 14.5 so that we have already removed signs associated with zig-zags), then we can always choose a gauge such that all turning-ups and turning-downs are trivial (Figs. 14.17 and 14.18). In this case we can deform a diagram in any way in the plane (we have not yet worried about over- and under-crossings) without changing its value. When we think about interpreting the diagram in terms of propagation in time, we may choose any direction on the plane to represent "time" and in fact the chosen direction does not have to be the same everywhere in the diagram. Similarly, when we think of diagrams as operators connecting bras and kets (as in Section 12.1) we have freedom to choose any direction as the input and the opposite direction as the output.

Since we can turn up and turn down lines freely it is sometimes useful to define vertices with all lines coming in from one side, as shown in Fig. 14.20.

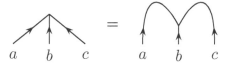

$$\overset{}{\underset{a \quad b \quad c}{\bigwedge}} = \overset{}{\underset{a \quad b \quad c}{\bigcap\bigcap}}$$

Fig. 14.20 When we have planar isotopy we can freely turn up and down the legs of a vertex. We can then define vertices with all three legs pointing in the same direction.

14.4.2   Isotopy in Three-Dimensional Diagrams

Once we start considering over- and under-crossings, we would like to be able to freely deform our diagrams in three dimensions. Even if we've arranged to be able to turn-up and turn-down legs of vertices without incurring phase factors for planar diagrams, this *still* does not guarantee that we have three-dimensional isotopy invariance of fusion and splitting diagrams once we start thinking about diagrams as living in three dimensions.

Let us think of the anyon world lines as ribbons which lay flat in a plane—if there is a vertex, we imagine the ribbon branching into two ribbons. Now imagine painting the front of the ribbons white and the back black. In fact, all of our diagrammatic rules so far maintain that the white side of the ribbons should remain on the front. We have no rules for what value to give the diagram if the ribbons are turned over so that the black side faces forward. In chapter 15 we discuss twisting the ribbons a full 2π (as we have seen previously in Fig. 2.7). However, this does not generally allow us to give value to diagrams with π twists of ribbons.[18] What is guaranteed in our diagrammatic rules is that the value of a ribbon diagram will remain unchanged after any isotopy, as long as the final configuration has all white sides facing forward.

[18]There have been attempts to extend this type of diagrammatic calculus to give values to π twists, but these turn out to be quite complicated. See Bärenz [2017] and Snyder and Tingley [2009].

14.5   Appendix: Bookkeeping Scheme with Negative d

This bookkeeping scheme shows how one can push the minus sign from ϵ_a onto d_a, redefine the inner product and reproduce the results of Option A in Section 14.2 by using Option B.[19]

[19]This scheme is detailed by Simon and Slingerland [2022].

We would like to work with a situation where isotopy invariance is assured, that is, where $\epsilon_a = +1$ for all particle types. For any self-dual particle with $\kappa_a = -1$ we choose $d_a < 0$ to allow $\epsilon_a = +1$. For non-self-dual particles we then choose a gauge such that the condition Eq. 14.5 is satisfied, that is signs of d multiply under fusion. Note that satisfying this condition may involve choosing $d_a < 0$ for certain non-self-dual particles, but we can do this with a gauge choice just as well as we can choose $d_a > 0$. The sign condition Eq. 14.5 assures that the factor $d_a d_b / d_c$ in Fig. 14.6 is positive, so that, for example, the square roots in Figs. 14.7, and 14.8 have positive arguments. This nonetheless leaves an ambiguity as to whether we should take the positive or negative square root. The convention we use below is that

$$\sqrt{\frac{d_a d_b}{d_c}} = \begin{cases} \text{negative} & d_a < 0 \text{ and } d_b < 0 \\ \text{positive} & \text{otherwise} \end{cases} \tag{14.9}$$

It may sound problematic to have some $d_a < 0$ since the negative-normed states in Fig. 14.3 seem like they would violate the principles of unitarity in quantum mechanics. However, with a small reinterpretation of the meaning of our inner product, our diagrammatics can be interpreted as representing a well-behaved unitary theory.

Our reinterpretation of the diagrammatic algebra is quite simple. We evaluate diagrams using the rules given in Section 14.1. That is, we use the rules from Section 12.4 except that loops are now normalized with d_a as in Fig. 14.3 and our orthonormality relationships (Fig. 12.21 and Fig. 12.18) are altered to those shown in Fig. 14.8 and Fig. 14.7 (noting the choice of sign in square roots given by Eq. 14.9). If there are over- and under-crossings, these can be evaluated using the R-matrix as in Fig. 14.9. Crucially, since we have set all $\epsilon_a = +1$, zig-zags like Fig. 14.5 can be freely straightened out.[20]

In the case where there are some $d_a < 0$, we call the result of this evaluation the *nonunitary* evaluation[21], or the $\epsilon > 0$ evaluation, of the diagram as it corresponds to the nonunitary inner product. While we are free to evaluate diagrams this way, they do not correspond to unitary theories. If we want to recover the physics of a unitary theory, we now insert one additional rule into our list:

(0) Before evaluating a diagram, count the number of negative-d caps, and call it n. After fully evaluating the diagram multiply the final result by $(-1)^n$.

Here a negative-d cap occurs when we go forward in time and two particles with $d < 0$ come together to annihilate or form a particle having $d > 0$. (See examples in Figs. 14.21 and 14.22.) Another way of counting the negative-d caps is to imagine erasing all lines in the diagram that have $d > 0$. This leaves only a set of closed loops (due to Eq. 14.5). We then just need to count caps in this set of closed loops of the form shown in the left of Fig. 14.22.

With these new rules, we are now describing a unitary positive-normed quantum theory—we call this evaluation of a diagram, including rule 0, the *unitary* evaluation of the diagram. This will be entirely equivalent to the evaluation of the diagram using the technique of Option A where we switch all d_a to be positive, but the zig-zag phase ϵ_a is given by the Frobenius–Schur indicator for self-dual particles.

To understand the intuition behind rule 0, consider a self-dual particle with negative Frobenius–Schur indicator. For such particles a zig-zag like in Fig. 14.5 is supposed to incur a minus sign—however, in our scheme we have set $\epsilon = +1$ and instead made d negative. Since $\epsilon = +1$ in the diagrammatic algebra, there is no sign associated with straightening a zig-zag. However, the zig-zag in Fig. 14.5 has a negative-d cap, so in the final evaluation of the diagram (applying rule 0) we correctly obtain the required minus sign.

As a simple example, consider the evaluation of a single loop as in Fig. 14.3 where $d_a < 0$. Before evaluating the loop we count that there is a single negative-d cap on the top of the loop (as in the left of Fig. 14.22). We evaluate the diagram, with the rules of Section 16.1.2, to obtain $d_a < 0$ as the nonunitary evaluation. However, applying rule (0) this quantity is then multiplied by -1, giving the final result of $-d_a = |d_a| > 0$. This is the result of the unitary evaluation of the diagram, and it is positive

[20]Recall that we may choose a gauge so that we can turn up and down legs, but only if there is no nontrivial third Frobenius–Schur indicator; see Section 14.4.

[21]Note that even though we don't have a positive-definite inner product, the F-matrices may still be unitary!

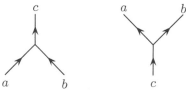

Fig. 14.21 With time going vertically, the left diagram is a negative-d cap if and only if $d_a < 0$ and $d_b < 0$. (The directions of the arrows do not matter, and if the particles are self-dual we do not draw arrows.) The right diagram is never a negative-d cap.

Fig. 14.22 With time going vertically, the left diagram is a negative-d cap if and only if $d_a < 0$. The right diagram is never a negative-d cap. We can think of these diagrams as being the same as the diagrams in Fig. 14.21 with c being the identity. The directions of the arrows do not matter.

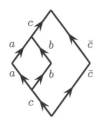

Fig. 14.23   An example of a diagram that should have a positive definite value since it can be written as ⟨state|state⟩.

[22] A simple example of this is the loop gas with $d = -1$. We've run into this previously in Exercise 2.2 or item (4). After implementation of rule 0, this corresponds to the semion model $SU(2)_1$, which we ran into in Table 2.2. We will see this case again many times in later chapters.

as we would hope for a positive definite inner product for a diagram that can be written as ⟨state|state⟩ (see Fig. 14.3).

As a second example, consider the diagram Fig. 14.23, and let us assume that $d_a, d_b < 0$ and $d_c > 0$. The (nonunitary) evaluation of the diagram (without rule 0) gives $-d_c\sqrt{d_a d_b / d_c}$, the square root coming from Fig. 14.7 and the sign from the rule Eq. 14.9 of how to handle square roots with negative d. However, applying rule 0, there is a single negative-d cap (from the vertex with a and b coming in from the bottom, and c going out the top), and hence the unitary evaluation of this diagram is $+d_c\sqrt{d_a d_b / d_c}$. Note that this is positive as it should be for a diagram that can be written as ⟨state|state⟩ analogous to Fig. 14.3.

As a third example, consider the same diagram Fig. 14.23, but consider the case where $d_a, d_c < 0$ and $d_b > 0$. Here, the nonunitary evaluation gives $d_c\sqrt{d_a d_b / d_c}$, but applying rule 0, with a single negative-d cap (the top of the c loop) we obtain a final result of the unitary evaluation given by $-d_c\sqrt{d_a d_b / d_c}$. Note that this is also positive as it should be.

The situation described in this section—having a theory that allows straightening of zig-zags ($\epsilon = +1$), but has negative d_a, is quite common.[22] It is very useful to be able to interpret such theories as being unitary theories with this additional rule 0.

14.6   Appendix: $[F_a^{aaa}]_{II}$ Is Real

Let a be a self-dual particle (i.e. $a = \bar{a}$). Working with the physics normalization we already showed (Fig. 14.2) that

$$a\;\bigvee \;= [F_a^{aaa}]_{II}\;\Big|a\,.$$

Similarly, using an inverse F-move, and the fact that F is unitary (see Section 9.5.2) we derive

$$a\;\bigcap\;= [F_a^{aaa}]^*_{II}\;\Big|a\,.$$

Equivalently the last diagram can be derived as being the Hermitian conjugate of the previous diagram.

Finally, assuming only causal isotopy invariance, the equality

$$[F_a^{aaa}]_{II}\;\overset{a}{\bigcirc}\; = \;^a\!\bigcap\!\!\bigcup\; = \;\bigcap\!\!\bigcup^a\; = [F_a^{aaa}]^*_{II}\;\overset{a}{\bigcirc}$$

then shows that $[F_a^{aaa}]_{II}$ must be real.

14.7 Appendix: Spin-$1/2$ Analogy and Why We Have a Frobenius–Schur Sign

It may seem a bit odd that a zig-zag in a spacetime line (as in Fig. 14.5) can incur a minus sign. While this might appear a bit strange it turns out that there is a familiar analog in angular momentum addition— where the particle types (the labels a, b, c, etc.) correspond to the eigenvalue of the total angular momentum squared J^2.

Consider three spin-1/2 particles that all taken together are in an eigenstate of $J = 1/2$. We can describe the possible states of the system with fusion trees as in Fig. 14.24 (see also Fig. 9.1)—in this case where a, b, c, and e are all labeled with $J = 1/2$. In Fig. 14.24 we can (on the left of the figure) consider either the fusion of the leftmost two particles to some angular momentum $d = 0$ (meaning a singlet) or $d = 1$ (meaning a triplet), or we can (on the right of the figure) consider fusion of the rightmost two particles to either $f = 0$ or $f = 1$. The F-matrix that relates these two descriptions of the same space is given by $[F_{\frac{1}{2}}^{\frac{1}{2}\frac{1}{2}\frac{1}{2}}]_{df}$ which is often known as a $6j$-symbol in the theory of angular momentum addition. The analogy of a negative Frobenius–Schur indicator here is the fact that $[F_{\frac{1}{2}}^{\frac{1}{2}\frac{1}{2}\frac{1}{2}}]_{00}$ is negative.

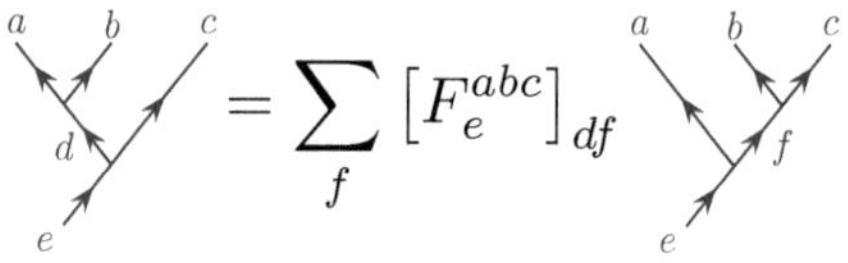

$$\sum_f [F_e^{abc}]_{df}$$

Fig. 14.24 The F-move.

Let us try to see how this happens more explicitly. Given that the total spin is $1/2$ we can focus on the case where the total z-component of angular momentum is $J_z = 1/2$ as well. The state where the leftmost two particles fuse to the identity (or singlet $J = d = 0$) can then be written explicitly as

$$|\psi\rangle = \frac{1}{\sqrt{2}} \left(|\uparrow_1\downarrow_2\rangle - |\downarrow_1\uparrow_2\rangle \right) \otimes |\uparrow_3\rangle \tag{14.10}$$

where the subscripts are the particle labels given in left to right order. This wavefunction is precisely analogous to the lower half (the "ket") of the far left-hand picture in Fig. 14.2.

On the other hand, we could use a basis where we instead fuse the rightmost two particles together first, as in the right-hand side of Fig. 14.24. We can write the state where the right two fuse to $J = f = 0$ analogously as

$$|\psi'\rangle = \frac{1}{\sqrt{2}} |\uparrow_1\rangle \otimes \left(|\uparrow_2\downarrow_3\rangle - |\downarrow_2\uparrow_3\rangle \right) \tag{14.11}$$

which is precisely analogous to (the Hermitian conjugate of) the top half (the "bra") of the left-hand side of Fig. 14.2.

[23]This result of $-1/2$ is precisely the $6j$-symbol

$$\left\{ \begin{array}{ccc} 1/2 & 1/2 & 0 \\ 1/2 & 1/2 & 0 \end{array} \right\} .$$

It is easy to check that the inner product of these two states $|\psi\rangle$ and $|\psi'\rangle$, corresponding to the value of the left diagram of Fig. 14.2 is[23]

$$\langle \psi' | \psi \rangle = -1/2 \quad .$$

By redefining the normalization of these states, we can arrange for this overlap to have unit magnitude. However, the sign cannot be removed. The situation is the same for any two half-odd-integer spins fused to a singlet.

14.8   Appendix: Some Additional Properties of Unitary Fusion Categories

Unitary fusion categories (the theories we have been discussing!) have two useful properties. We do not prove these properties here. More detailed discussion is given by Kitaev [2006]. More detailed discussions are given for example in Jones and Penneys [2017] or Etingof et al. [2005]. These latter references are quite mathematical.

14.8.1   Pivotal Property

[24]We do not say there is a unitary transformation between the two diagrams since the two diagrams operate on different Hilbert spaces—the left diagram having one down leg and two up, whereas the right has one up and two down.

A property that may seem obvious is known as the pivotal property. This states that there should be isomorphisms[24] between a vertex with a downturned line and one with an upturned line, such as that shown in Fig. 14.25. While this seems like a rather small statement (which is a

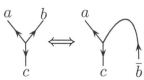

Fig. 14.25 A theory is pivotal if there exist isomorphisms between the states of the Hilbert spaces described by pairs of vertices that differ by downturning and upturning lines.

property of any unitary fusion category) it turns out to be quite powerful. One can deduce from this that the transformations in Figs. 14.17 and 14.18 are unitary—meaning that the constants on the right-hand side have unit magnitude

$$\left| \sqrt{\frac{d_a d_c}{d_b}} [F_a^{c\bar{c}a}]_{Ib} \right| = \left| \sqrt{\frac{d_a d_c}{d_b}} [F_a^{a\bar{c}c}]_{bI} \right| = 1 \quad . \tag{14.12}$$

See Kitaev [2006] and Bonderson [2007]. In the more general case where the fusion multiplicity N_{ab}^c is greater than 1, the vertices have additional indices μ and ν and the transformation is a unitary matrix in these indices. An example of this is given in Fig. 14.26.

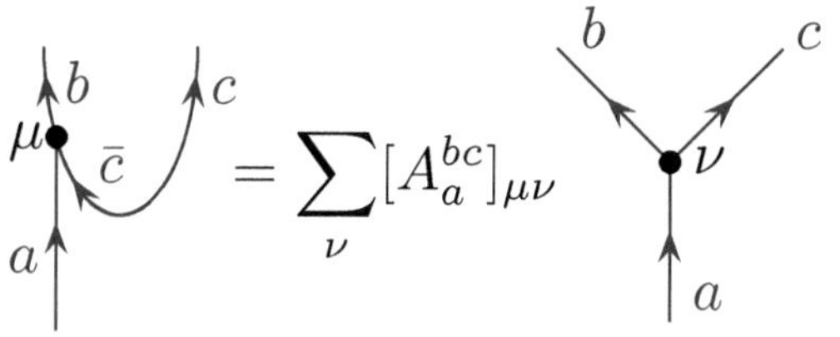

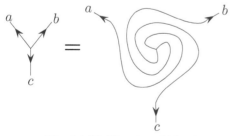

Fig. 14.26 The matrix A_a^{bc} is a unitary matrix in the indices μ and ν. In the simpler case of Fig. 14.18, the prefactor is a unitary 1×1 matrix, meaning it is a magnitude one complex scalar.

From this type of identity one can successively turn up and down legs at vertices to obtain the identity in Fig. 14.27, where the prefactor C in the figure is also a unit magnitude complex scalar (or a unitary matrix in the case where the vertex has an additional index).

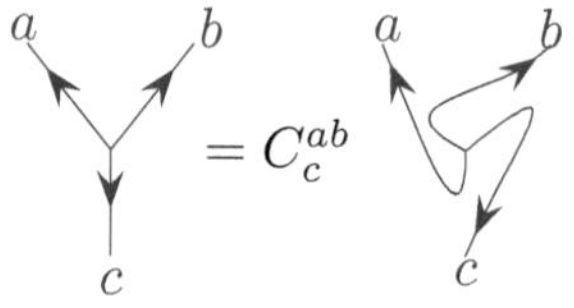

Fig. 14.27 The relationship between these two diagrams is unitary, meaning C is just a phase. For cases where $N_{ab}^c > 1$ the vertices are marked with indices, say μ on the left and ν on the right, and for fixed a, b, and c, the constant C_c^{ab} becomes a unitary matrix $[C_c^{ab}]_{\mu\nu}$ in the indices μ, ν.

Quite a few more identities can also be derived from the pivotal property. Detailed discussions of this property (and its meaning) are given by Kitaev [2006] and Bartlett [2016]. One particularly useful identity is given by applying Fig. 14.27 three times in a row to obtain

$$C_c^{ab} C_a^{bc} C_b^{ca} = 1 \tag{14.13}$$

which diagrammatically is drawn as the "pivotal identity" in Fig. 14.28. In the case where there are additional indices at the vertex, Eq. 14.13 becomes a matrix product that equals the identity matrix. Note in

Fig. 14.28 The pivotal identity.

particular that if $a = b = c$ we obtain the statement from Fig. 14.19 that the eigenvalues of C must be third roots of unity.

The derivation of the pivotal identity is a bit complicated and is given by Kitaev [2006]. However, it can be made a bit more intuitive physically by turning up one of the branches to obtain the alternate form of the pivotal identity, shown in Fig. 14.29. This form can be understood as the statement that the vacuum (or particles fusing to the vacuum) can be rotated freely in spacetime.

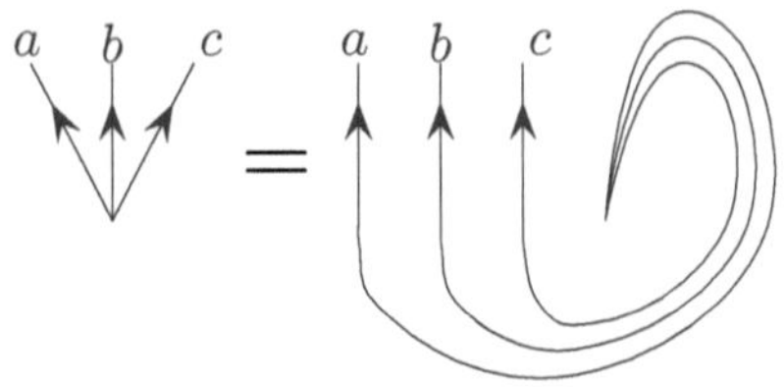

Fig. 14.29 Another version of the pivotal identity. We can derive this from Fig. 14.28 by turning up the *c*-leg.

14.8.2 Spherical Property

Theories that are unitary (describing real quantum-mechanical particles) have an additional property called being "spherical." Consider a diagram X with a line coming out the top and a line coming in the bottom. The so-called left trace is defined by connecting the top line with the bottom line in a loop going to the left, as in the left of Fig. 14.30. The right trace is defined similarly, except that the loop goes to the right of the diagram X, as in the right of Fig. 14.30. If the left trace is always equal to the right trace, we say that the theory is spherical. The name here comes from the idea that we could pull the string around the back of a sphere in order to turn a left trace into a right trace as shown in Fig. 14.31. However, the spherical property is actually stronger than its name (and Fig. 14.31) suggests since it allows us to turn a right trace into a left trace even when there are other objects on the sphere that might prevent us from dragging a string all the way around the back of the sphere. An obvious result of the spherical property is that $d_a = d_{\bar{a}}$.

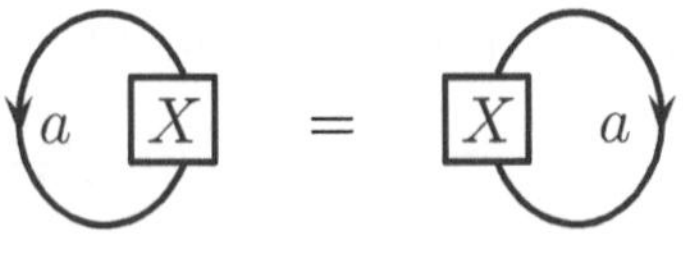

Fig. 14.30 The spherical property sets the left trace equal to the right trace, as shown in the picture.

14.9   Appendix: Higher Fusion Multiplicities

When we have a theory with higher fusion multiplicities (i.e. $N_{ab}^c > 1$ for at least one fusion channel), then the vertices must be given indices. The first two figures of this appendix (Figs. 14.32 and 14.33) are identical to that of Section 12.5 except that here we have changed the normalization from physics normalization to isotopy-invariant normalization.

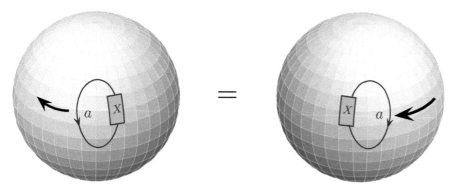

Fig. 14.31 The naming "spherical" comes from the idea that we can pull the string around the back of a sphere (as indicated by the black arrows) to turn a left trace into a right trace.

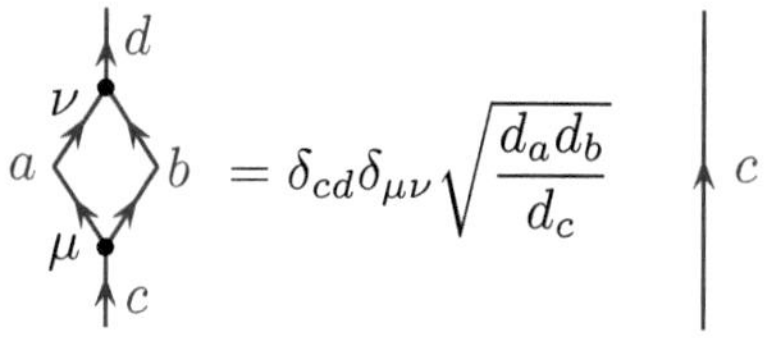

$$a \underset{\mu}{\overset{\nu}{\diamondsuit}} b = \delta_{cd}\delta_{\mu\nu}\sqrt{\frac{d_a d_b}{d_c}} \qquad \uparrow c$$

Fig. 14.32 The bubble (or orthonormality) diagram when there are fusion multiplicities. This diagram is drawn in the isotopy-invariant normalization. Compare to Fig. 12.40.

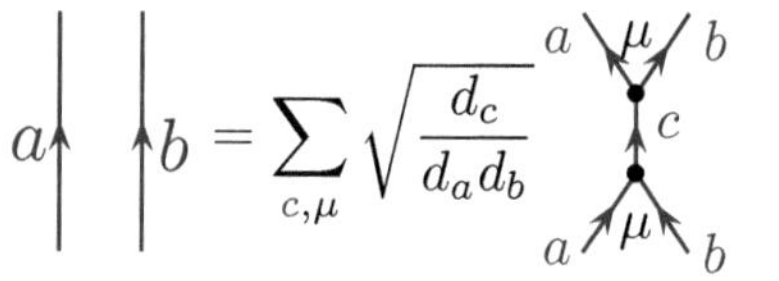

$$a\uparrow \quad \uparrow b = \sum_{c,\mu} \sqrt{\frac{d_c}{d_a d_b}} \quad a\overset{\mu}{\underset{\mu}{\bowtie}}b \quad c$$

Fig. 14.33 Insertion of a complete set of states. When there are fusion multiplicities, these must be summed over $\mu \in N_{ab}^c$ as well. This diagram is drawn in the isotopy-invariant normalization. Compare to Fig. 12.41.

We can combine Fig. 14.33 with Fig. 13.13 to find the generalization of Fig. 14.9.

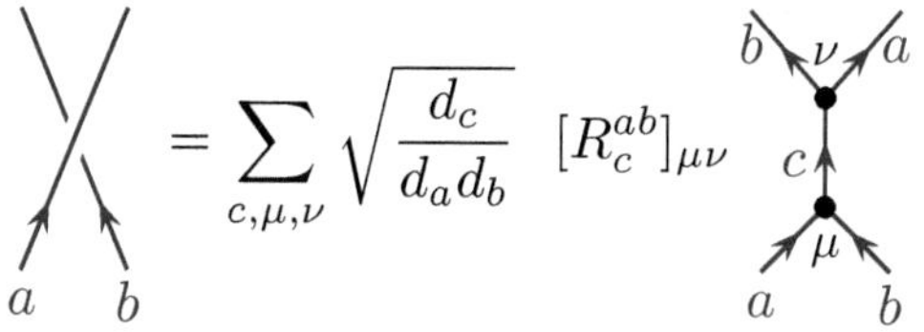

$$\times\!\!\!\!\!\!\!\underset{a \quad b}{} = \sum_{c,\mu,\nu} \sqrt{\frac{d_c}{d_a d_b}} \, [R_c^{ab}]_{\mu\nu} \quad \overset{b \, \nu \, a}{\underset{a \, \mu \, b}{\bowtie}} c$$

Fig. 14.34 Resolving a crossing with isotopy normalization and higher fusion multiplicity. See Eq. 14.9 for how to interpret the square root in cases where $d < 0$.

Chapter Summary

> - Our diagrammatic calculus can be used to construct isotopy invariant rules that allow any smooth deformation of a diagram without changing its values.
> - The obstruction to isotopy invariance often comes from a nontrivial Frobenius–Schur indicator, which can be handled in a number of different ways, some of which have slightly different physical meanings.

- Much of this material is not presented well anywhere in the literature (although, as usual, some of it is implied in Kitaev [2006]). I have tried to clean it up in Simon and Slingerland [2022], which this chapter echoes.

Exercises

Exercise 14.1 Higher Fusion Multiplicities
Derive Eq. 14.34.

Exercise 14.2 Knot Sums
Consider a knot K labeled with a particle quantum number a. Let the diagrammatic evaluation of this knot be called $\langle K \rangle_a$. Consider also a knot K', also labeled with a particle quantum number a and let the diagrammatic evaluation of this knot be called $\langle K' \rangle_a$. Construct the knot sum $K \# K'$ (see Exercise 2.6 for the definition of a knot sum) and show that the diagrammatic evaluation of the resulting knot, when labeled with a, is given by

$$\langle K \# K' \rangle_a = d_a^{-1} \langle K \rangle_a \langle K' \rangle_a \quad .$$

You should use diagrammatic evaluation conventions that are isotopy invariant. Hint: Use the completeness relation.

Compare your result to that of Exercise 2.6 and also that of Exercise 7.1.

Exercise 14.3 H-variant of F-move
Show that in isotopy normalization, the H-shaped F shown in Fig. 14.35 is given by

$$[F^{ab}_{cd}]_{ef} = \left([F^{ceb}_f]_{ad} \right)^* \sqrt{\frac{d_e d_f}{d_a d_d}} \quad .$$

See Exercise 12.3 for a hint. In the case where some d are taken negative, when should we take the sign of the square root negative?

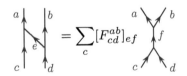

$$\includegraphics \quad = \sum_c [F^{ab}_{cd}]_{ef}$$

Fig. 14.35 The H-variant of the F-move.

Twists

Recall from Chapter 2 that we considered the procedure of pulling tight a curl in a ribbon as shown in Fig. 15.1 (compare Fig. 2.7). Pulling tight results in a twisted ribbon, which (viewing time as going vertically) corresponds to a particle rotating around its own axis, while at a fixed point in space, as time progresses. The twist in the ribbon can be removed at the cost of a complex phase, which we call θ_a, known as the particle's **twist factor**.[1] In other words, the particle rotating around its own axis accumulates a phase θ_a compared to a particle that does not rotate. We should expect this phase for any particle with a spin, since rotating a spin in quantum mechanics accumulates a phase.

In our diagrammatic notation, we do not draw ribbons. Rather, to represent a particle twisting around itself we use blackboard framing as discussed in Section 2.2.2 and we always imagine the ribbon lying flat in the plane. Thus, we formally define the twist factor[2] as given in Fig. 15.2.

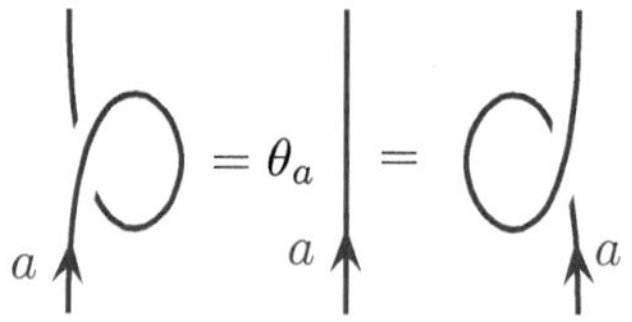

[1] In Chapters 3–5 we used the notation $e^{i\vartheta}$ to describe the exchange of two anyons. This is not the same symbol (ϑ versus θ). For abelian anyons $\theta = e^{i\vartheta}$.

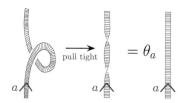

Fig. 15.1 Pulling tight a curl in a ribbon results in a twist. This twist in a ribbon of particle type a can be removed at the cost of a phase factor of θ_a. See also Fig. 2.7.

Fig. 15.2 The definition of the twist factor θ_a drawn using blackboard framed diagrams. The curled strings should be thought of as ribbons lying in the plane (as in Fig. 15.1), which are equivalent to ribbons twisting around their own axis.

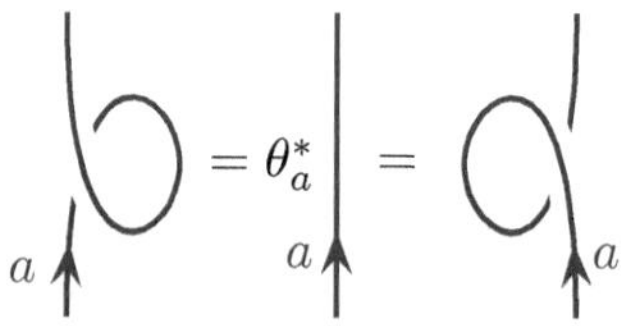

Fig. 15.3 The mirror image diagrams to those of Fig. 15.2.

Invoking the Hermitian conjugation principle (if we reflect a diagram around a horizontal axis, and reverse the arrows so they remain pointing

[2] If a nonunitary theory with $d_a < 0$ is used, but is meant to represent a unitary theory, via the "Option B" of Section 14.2 (detailed in Section 14.5), we must be careful of the fact that removing the curl also removes a cap. This means that

$$\theta_a^{(d<0,\epsilon=+)} = -\theta_a^{(d>0,\epsilon=-)} \qquad . \tag{15.1}$$

Because of this potential confusion, it is often good to avoid discussion of θ_a, using $R_I^{a\bar{a}}$ instead (related to θ_a via Eq. 15.4), which is unambiguous and is the same for either Option A or "Option B" (i.e. it is the same for both the "unitary" and "nonunitary" evaluations of a diagram).

[3]One might think that θ_a has been acting strange for a while—but it is just a phase. Ha ha!

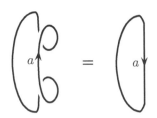

Fig. 15.4 This equality establishes $\theta_a \theta_a^* = 1$, hence, $|\theta_a| = 1$. We can evaluate the diagram on the left by removing the two curls and getting $\theta_a \theta_a^*$ times the diagram on the right. On the other hand, the diagram on the left can also be turned directly into that on the right just by using moves that we know are allowed such as Fig. 13.9. (see also Exercise 15.4).

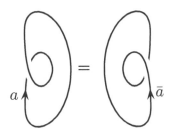

Fig. 15.5 The equality of these diagrams establishes $\theta_a = \theta_{\bar{a}}$. See Exercise 15.4.

[4]Note that θ_a is gauge independent, whereas R_c^{ab} is generally gauge dependent. (Although R_c^{aa} is gauge independent as well.)

[5]If $d_a < 0$ as in the case of using Option B from Section 14.2, this equation correctly gives $\theta_a^{\text{nonunitary}} = \text{sign}(d_a)\theta_a^{\text{unitary}}$ as mentioned in note 2.

in the same direction, we complex conjugate its amplitude; see Fig. 12.4), we similarly have the mirror image diagrams shown in Fig. 15.3.

It is easy to confirm that the twist factor θ_a can only be a phase[3] (a unit magnitude complex number) as expected. The proof of this is given in Fig. 15.4.

One further interesting fact about the twist factor is that

$$\theta_a = \theta_{\bar{a}}$$

which can be seen from the equality of the diagrams shown in Fig. 15.5.

The twist factor is related to the so-called **topological spin**, or **conformal scaling dimension**, usually called h_a, via the relation

$$\theta_a = e^{2\pi i h_a} \quad .$$

This phase accumulated from a 2π rotation is what we typically get in quantum mechanics from the operator $e^{2\pi i \hat{S}}$ with $\hat{S}$ the spin operator and we set $\hbar = 1$. The vacuum, or identity particle, should have zero scaling dimension, $h_I = 0$.

Note that many quantities of interest will depend only on the twist factor θ_a, that is, the fractional part of the topological spin, h_a mod 1. Indeed, we will see that many of the topological properties of a system are independent of the integer part of the topological spin, and depend only on the fractional part.

Recall also the famous spin-statistics theorem (as discussed near Fig. 2.7), which tells us that the twist factor should give us the phase for exchanging two identical particles, and is thus intimately related to the anyonic statistics of particles. Of course two cases are very well known to us: if the spin h_a is an integer, then $e^{2\pi i h_a}$ is the identity, and the particle is a boson. If h_a is a half-odd-integer, then the phase is -1 and the particle is a fermion.

15.1   Relations between θ and R

Braiding and twisting are very closely related to each other. In fact, twist factors θ are related to the R-matrices we introduced in Chapters 10 and 13 in several different ways.

First, let us try to evaluate the curled ribbon in Fig. 15.2 using the R-matrix as in Fig. 15.6.

This manipulation establishes the relation[4,5]

$$\theta_a = \sum_c \frac{d_c}{d_a} R_c^{aa} \tag{15.2}$$

where we are assuming $N_{aa}^c = 0$ or 1 (see Section 15.2 for the case where $N_{aa}^c > 1$). Note in particular that in the case of an abelian theory where N_{aa}^c is non-zero for only a single value of c and $d_a = d_c = 1$ this

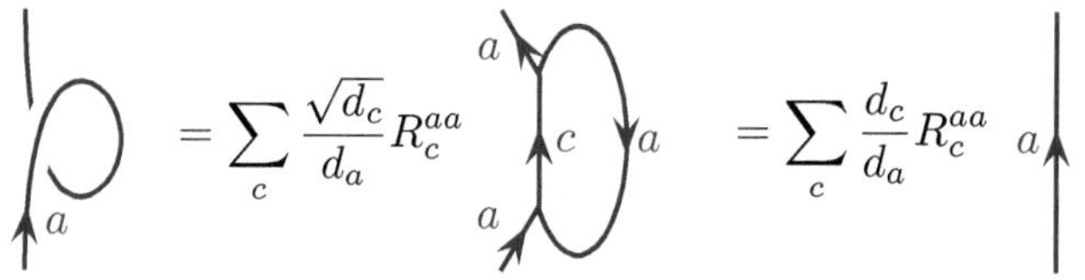

$$\includegraphics{fig15_6}$$

Fig. 15.6   Relation of the twist factor to the R-matrix. In the first step we use Fig. 14.9. In the second step we use Fig. 14.18 along with Eq. 14.12 and finally Eq. 14.7. We have assumed $N_{aa}^c = 0$ or 1 only. See Eq. 15.6 for the more general case.

establishes

$$\theta_a = R_{a \times a}^{aa} \qquad \text{(abelian anyons)}. \qquad (15.3)$$

A second, and different, relationship can be derived via the manipulations shown in Fig. 15.7. One might be tempted to identify the left of

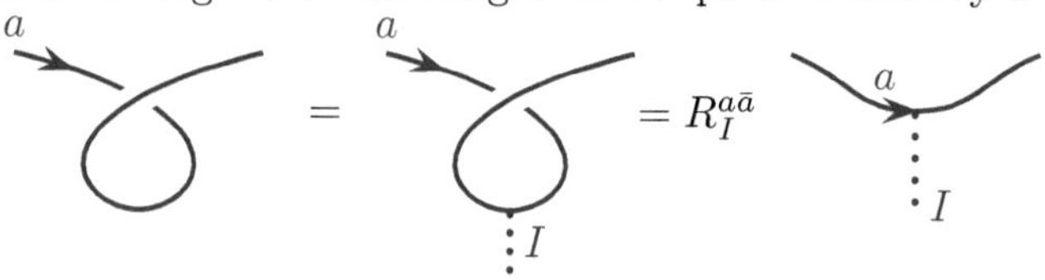

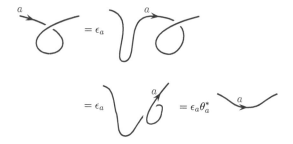

Fig. 15.7   This curl is evaluated with the R-matrix. Note that this diagram is not precisely equal to θ_a^* (see Fig. 15.8).

Fig. 15.7 with the twist factor θ but this is not quite right when we look at it carefully, as shown in Fig. 15.8. From Fig. 15.7 and 15.8 we derive

Fig. 15.8   Removing curl in a string turned sideways accumulates a twist factor θ_a^* along with a zig-zag factor ϵ_a. The first step introduces a zig-zag in the curve and we incur a factor of ϵ_a as in Fig. 14.5. The second step is an allowed smooth deformation (see Exercise 15.4). In the last step we remove the curl and obtain a factor of θ_a^* as in Fig. 15.3. In isotopy-invariant cases we have chosen ϵ_a to be unity.

$$R_I^{a\bar{a}} = \epsilon_a \, \theta_a^* \qquad . \qquad (15.4)$$

where ϵ_a is the zig-zag factor (see Fig. 14.5). As noted in note 2 of this chapter, $R_I^{a\bar{a}}$ is the same whether we are working with Option A or Option B from Section 14.2. However, if we push a minus sign from the zig-zag phase ϵ_a onto the loop weight d_a, as pointed out in Eq. 15.1, then θ_a changes sign.

The final relationship between R and θ, called the ribbon identity, is a gauge-invariant statement

$$R_c^{ba} R_c^{ab} = \frac{\theta_c}{\theta_a \theta_b} \tag{15.5}$$

which can be derived by the geometric manipulations in Fig. 15.9.

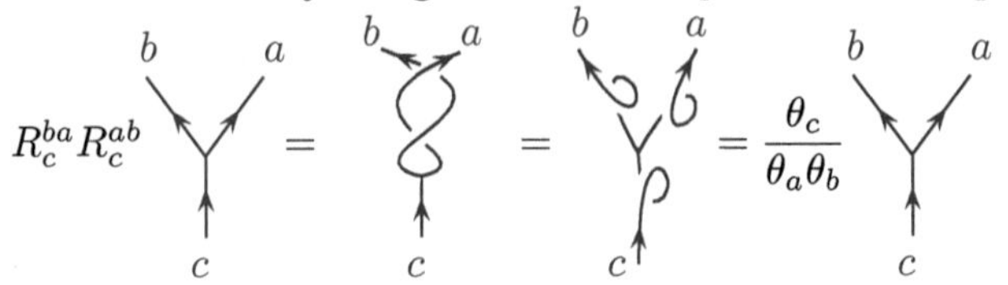

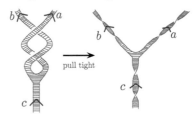

Fig. 15.9 Deriving the ribbon identity. The middle is the non-obvious geometric step. See also Exercise 15.5.

The middle step in Fig. 15.9 is perhaps non-obvious, but is clarified if viewed as a ribbon diagram as in Fig. 15.10. See also Exercise 15.5.

Fig. 15.10 The middle step of Fig. 15.9 viewed as a ribbon diagram.

15.2 Appendix: Higher Fusion Multiplicities

In the case where we have fusion multiplicities $N_{ab}^c > 1$, then we have the following more general equations. Eq. 15.2 is generalized to

$$\theta_a = \sum_{c,\mu} \frac{d_c}{d_a} [R_c^{aa}]_{\mu\mu} \tag{15.6}$$

and the ribbon identity Eq. 15.5 is generalized to

$$\sum_{\nu} [R_c^{ba}]_{\mu\nu} [R_c^{ab}]_{\nu\lambda} = \frac{\theta_c}{\theta_a \theta_b} \delta_{\mu\lambda} \quad . \tag{15.7}$$

Chapter Summary

- The twist factor θ_a gives the value for removing a curl in an a line in a diagram.
- The twist factor is related to the R-matrix in several ways.
- There can be confusion in the value of θ_a if one is using $d_a < 0$. For safety, stick with quoting R instead.

Further Reading: Same as that of Chapter 12.

Exercises

Exercise 15.1 Twists in Abelian Theories
(a) By using the ribbon identity, show that for any abelian anyons, the twist factor of $a \times a$ is related to the twist factor of a via

$$\theta_{a \times a} = (\theta_a)^4$$

in agreement with the exchange phase in Eq. 4.2.1, and Exercise 5.6 (see note 1 of this chapter).
(b) Use Exercise 13.2 and the ribbon identity to show that for abelian anyons

$$\theta_{\underbrace{a \times a \times a \ldots \times a}_{n \text{ times}}} = (\theta_a)^{n^2} \quad .$$

Exercise 15.2 Twists in Product Theories
Consider two anyon theories T_1 and T_2 with twist factors $\theta_a^{(T_1)}$ and $\theta_\alpha^{(T_2)}$ respectively. Show that in the product theory $T_1 \times T_2$ for particle type (a, α) we have a twist factor

$$\theta_{(a,\alpha)}^{(T_1 \times T_2)} = \theta_a^{(T_1)} \theta_\alpha^{(T_2)} \quad .$$

Exercise 15.3 Fibonacci Twists
For Fibonacci anyons, the twist factor is $\theta_\tau = e^{\pm 4\pi i/5}$ (With $\pm$ being for right- or left-handed theories, respectively). Check that these twist factors agree with the R-matrices (Eq. 10.2) and Eq. 15.2 and Eq. 15.4.

Exercise 15.4 Using Geometric Moves I
(a) Using the allowed moves such as Fig. 13.9 and Fig. 13.10 and Fig. 13.15, show the equivalence of the left and right diagrams of Fig. 15.4.
(b) Similarly, show the equivalence of the left and right diagrams of Fig. 15.5.
(c) Similarly, show the equivalence of the middle two diagrams in Fig. 15.8.

Exercise 15.5 Using Geometric Moves II
Demonstrate the middle step of Fig. 15.9 by using allowed geometric moves such as Fig. 13.9 and Fig. 13.10 and Fig. 13.15. You may also need the pivotal identity Fig. 14.28.

Exercise 15.6 Gauge Independence of Ribbon Identity
Show that the ribbon identity Eq. 15.5 is gauge independent.

Exercise 15.7 Higher Fusion Multiplicities
Derive Eq. 15.6 and Eq. 15.7.

Nice Theories with Planar or Three-Dimensional Isotopy

In Chapters 12 through 15 we carefully developed the principles of anyon diagrammatics in quite a bit of generality. In this chapter we aim for a slightly simplified and abbreviated, but still extremely useful, version of the diagrammatic rules developed (roughly axiomatically) in the prior chapters.

Our original intent for a TQFT was to develop rules that would map a labeled knot or link diagram into a complex amplitude output (as in Fig. 7.1) in a way that would be invariant under any smooth deformations (isotopy) of spacetime. Most generally in topological theories, we found that there could be some restrictions on what sort of deformations of spacetime would leave the output unchanged (see Section 14.4). In this section we focus on a simpler class of theories where these impediments are lifted. In particular, the topological theories of this chapter have the property that they give the same output amplitude for *any* smooth deformation of spacetime, treating world lines as ribbons. We say such theories have "ribbon isotopy invariance" (or "regular isotopy" invariance if we are considering knots and links as in Section 2.6.1). An example of such isotopy invariance is shown in Fig. 16.1.

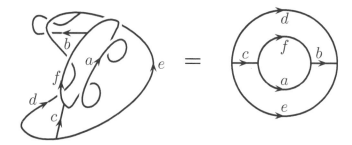

Fig. 16.1 For a theory with ribbon isotopy invariance (isotopy invariance in three dimensions treating world lines as ribbons) these two diagrams must evaluate to the same result since one can be continuously deformed into each other.

16.1 Planar Diagrams

We start by simplifying even further and considering only planar diagrams, so we do not allow over- and under-crossings (which we reintroduce in Section 16.2). Here we specialize only to theories with planar isotopy invariance. As a result, our rules for diagrammatic manipulation will be slightly easier than those in Chapter 12.

As in Chapter 12 there is still a bra and ket interpretation of diagrams. Roughly we can think of cutting a diagram in half and viewing one side as a bra and the other as a ket.[1] We can also roughly think of these diagrams as being world lines of particles moving in $(1 + 1)$- dimensions.

[1]We can think of any direction as being time, although it is sometimes most convenient to think of time as up. The fact that we can choose any direction as "up" is a reflection of the fact that there is an isomorphism between states, which allows us to interpret incoming lines as either bras or kets. See the discussion of Section 14.4.1.

16.1.1 Planar Diagrammatic Rules

When considering theories with planar isotopy invariance, any deformation of lines in a diagram is allowed (as long as we remain in the plane). In particular, we are freely allowed to make the zig-zag deformation shown in Fig. 16.2. In the language of Fig. 14.5 we are assuming all $\epsilon_a = +1$. Further, we allow turning up and down of edges at vertices (see the discussion of Section 14.4), and we can even have vertices where all three incident edges point up or all three point down (see Fig. 14.20). An example of planar isotopy invariance is shown in Fig. 16.3

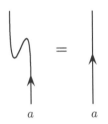

Fig. 16.2 For the isotopy-invariant theories considered in this chapter, this deformation is allowed.

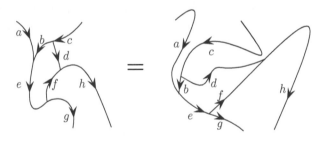

Fig. 16.3 For a theory with planar isotopy invariance, these two diagrams evaluate to the same result. Planar isotopy invariance allows us to distort the diagram in any way as long as we do not cut any strands or cross lines through each other.

As in previous chapters we would like to use F-matrices to help us convert one diagram into another. Although we previously found that bending lines up and down (as in Fig. 14.17) can incur nontrivial factors, in this chapter we instead assume no such nontrivial factors so we may turn up and down legs freely. Our F-matrix can thus be written as in Fig. 16.4. Note that the conventions we use in this chapter are different from that of previous chapters but instead match those introduced by Levin and Wen [2005].

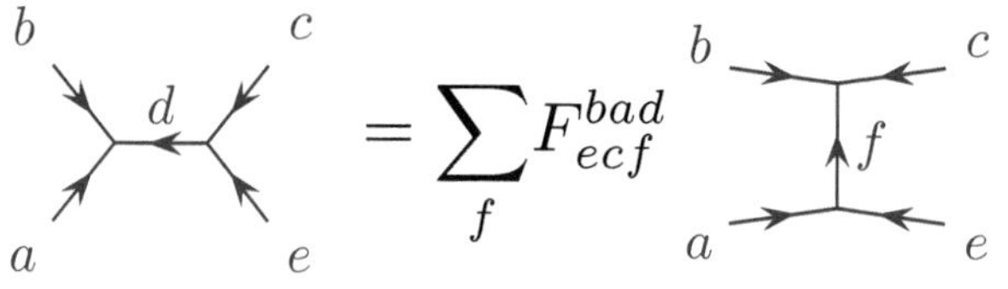

Fig. 16.4 The definition of the F-matrix for fully isotopy-invariant theories. This notation uses the conventions of Levin and Wen [2005]. For a unitary theory the F-matrix with fixed indices a, b, c, d is unitary in the indices d and f. For this F-matrix to be non-zero, the vertices in the pictures must be allowed fusions—that is, $N_{abd} = N_{ce\bar{d}} = N_{bcf} = N_{ae\bar{f}} = 1$. The case with fusion multiplicities $N_{abc} > 1$ is considered in Section 16.4.

In this chapter, the orientation of this diagram (how we direct the legs compared to some direction we call time) does not matter. Further, we can freely rotate the diagrams in Fig. 16.4, and we can bend legs up and down freely as well. For example, the same F-matrix as in Fig. 16.4 applies to Fig. 16.5.

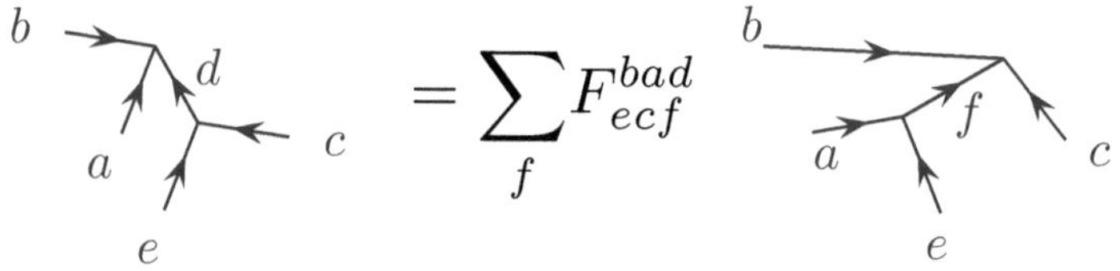

Fig. 16.5 For fully isotopy-invariant theories, the F-moves can be deformed in arbitrary ways. For example the same F-matrix governs the transformation in Fig. 16.4 as in this figure.

We can compare the definition of F-matrix in Fig. 16.4 to our prior definition of the F-matrix shown in Fig. 9.1. Since we now assume that we can bend legs up and down freely, we can bend legs in Fig. 9.1 and reverse arrows to make it look like Fig. 16.4 and we thereby derive the relation between the two definitions

$$F^{bad}_{ecf} = [F^{\bar{a}\bar{b}\bar{c}}_{e}]_{df} \quad . \tag{16.1}$$

Again the idea of the F-matrix is to write a single diagram (on the left of Eq. 16.4) as a sum of diagrams on the right. By successively applying such F-moves to parts of complicated diagrams we can restructure any given diagram in a multitude of ways.

There are several further useful rules for diagram evaluation. First, we need to give a value to the labeled loop as in Fig. 16.6. As in the case of the Kauffman bracket invariant, the value of the loop or "loop weight"[2] will be called d, here indexed with a subscript a for each possible particle type a.

[2] We have not yet shown the relationship between this definition of the quantum dimension d_a and the definition of d_a, which we called "quantum dimension," given in Eq. 3.11. In Section 17.1 we show that these two definitions are in fact the same up to a possible sign!

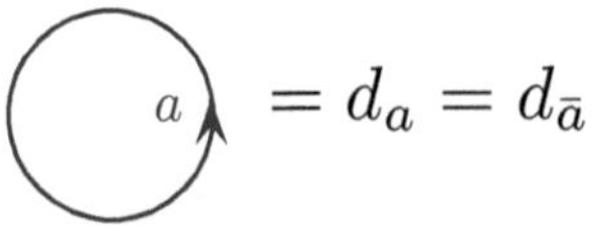
a
$= d_a = d_{\bar{a}}$

Fig. 16.6 The value of a loop labeled a is given by the quantum dimension d_a. Here we have invoked the spherical assumption to give us $d_a = d_{\bar{a}}$.

It is always true that $d_I = 1$, meaning that loops of vacuum can be freely added or removed from a diagram. As emphasized in Section 14.1, giving the loop this normalization implies we are working with non-normalized kets (see Fig. 14.3, and also note 24 of Chapter 2). We will allow d_a to be either positive or negative and we discuss the meaning of negative d_a further in Section 16.1.3.

Secondly we define the contraction of a bubble as shown in Fig. 16.7.

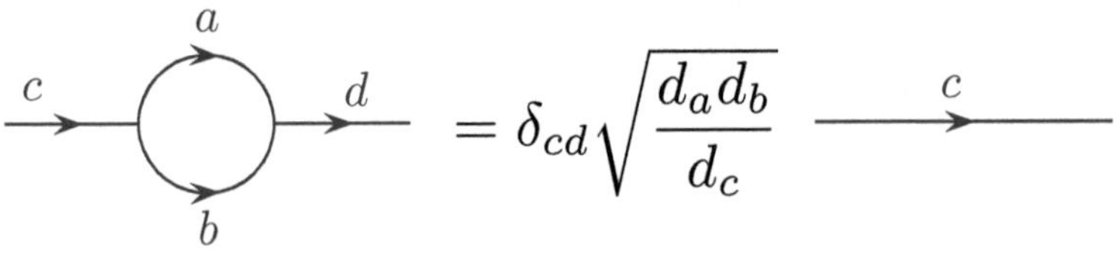
a
c
d
b
$= \delta_{cd} \sqrt{\frac{d_a d_b}{d_c}}$
c

Fig. 16.7 Contraction of a bubble for isotopy-invariant theories (also the "locality" principle and orthonormality condition). In cases where some d are negative we interpret the sign outside the square root as negative if and only if both $d_a < 0$ and $d_b < 0$. We must have $N^c_{ab} = 1$ or the value of the diagram is zero. The case with fusion multiplicities $N^c_{ab} > 1$ is given in Fig. 16.29.

This identity is the same as the orthonormality diagram Fig. 14.7 only written sideways (in this chapter the orientation of the diagram on the page does not matter). Physically we should think of this as being a version of the locality rule of Section 8.2—looked at from far away, one does not see the bubble. In particular this locality rule implies the "no-tadpole" rule,[3] that any diagram of the sort shown in Fig. 16.8 must vanish unless the incoming line is the vacuum.

[3]Famously, *Physical Review* did not allow the use of the name "spermion" for diagrams of the sort shown in Fig. 16.8.

a
c

Fig. 16.8 Picture of a tadpole. (Apparently this picture is supposed to look like a tadpole.) The locality principle Fig. 16.7 implies that any diagram containing a tadpole must vanish unless the incoming line is labeled with the vacuum (i.e., unless there is no incoming line).

We also have again the completeness relation as shown in Fig. 16.9, which is precisely the same as Fig. 14.8, only now we can orient the diagram in any direction.

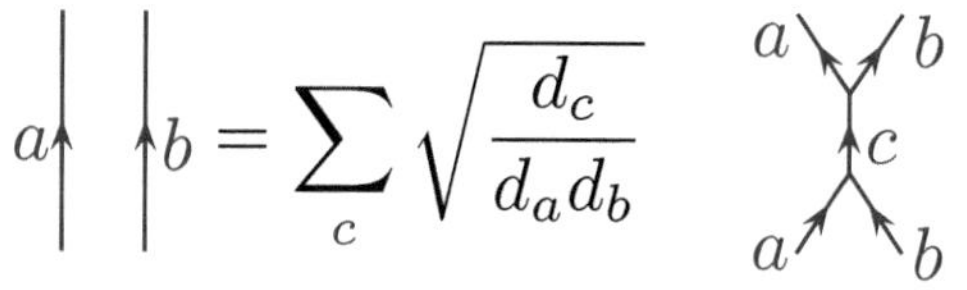

$$a{\uparrow} \quad {\uparrow}b = \sum_c \sqrt{\frac{d_c}{d_a d_b}} \quad \text{(diagram with } a, b, c\text{)}$$

Fig. 16.9 Insertion of a complete set of states. In cases where some d are negative we interpret the sign outside the square root as negative if and only if both $d_a < 0$ and $d_b < 0$. See Fig. 16.30 for the more general case where there are fusion multiplicities $N^c_{ab} > 1$.

16.1.2 Summary of Diagram Rules Planar Isotopy-Invariant Theories

Given the rules established in Section 16.1.1, we can evaluate any planar diagram[4] and turn it into a complex scalar number made up of factors of F and d—very similar to what we did with the Kauffman bracket invariant, only without over- and under-crossings here. Here are a summary of the rules for diagram evaluation in the case of isotopy-invariant planar theories. These rules are analogous to those presented in Section 12.4, only here the rules are simpler.

(1) One is free to continuously deform a diagram in any way as long as we do not cut any strand (for this section we assume no over- or under-crossings).

(2) One is free to add or remove lines from a diagram if they are labeled with the identity or vacuum (I). See the example in Fig. 16.10.

(3) Reversing the arrow on a line turns a particle into its antiparticle (see Fig. 8.4).

(4) A line must maintain its quantum number unless it fuses with another line, or splits.

(5) Vertices are allowed for multiplicities $N^c_{ab} > 0$ (see Section 8.3). This includes particle–antiparticle creation and annihilation processes where $N^I_{a\bar{a}} = 1$ (an example is shown in Fig. 16.10).

(6) One can use F-moves to change the structure of diagrams.

(7) One can use relations Fig. 16.9 and 16.7 to change the structure of diagrams.

(8) Every diagram can be reduced to a set of loops that can each be evaluated to give d_a for each loop of type a.

16.1.3 Negative d_a and Unitarity

We have allowed theories with d_a negative. We will assume, however, that

$$\text{sign}[d_a]\text{sign}[d_b] = \text{sign}[d_c] \quad \text{whenever} \quad N^c_{ab} > 0 \tag{16.2}$$

as we described in Section 14.5 (and as mentioned there, this is not a particularly stringent condition).

[4]Any planar diagram with no loose ends. As described in detail in Section 12.1 a diagram with loose ends should be considered a bra or ket or operator.

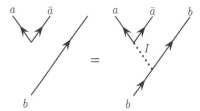

Fig. 16.10 One can always add or remove the identity (or vacuum) line to any diagram.

As discussed extensively in Section 14.2, if we think of a loop as being an inner product ⟨state|state⟩ as in Fig. 14.3, a negative d_a implies a non-positive-definite inner product, which is forbidden in quantum mechanics. This apparent problem is discussed in detail in Sections 14.1–14.2. Here, we very briefly adapt the scheme discussed in Section 14.5 to the current situation.

Here we accept that our diagrammatic algebra has negative d_as. Evaluation of such diagrams with negative d_as we call the *nonunitary* evaluation. However, with a small reinterpretation of the diagrams we can still think of these diagrams as describing a unitary theory.

To implement this reinterpretation of the diagrams, as discussed in Section 14.5, we add two simple rules to our list for evaluation of diagrams.

(0′) We must break the spacetime symmetry and define a time direction (often up on the page).

(0) Before evaluating a diagram, count the number of negative-d caps, and call it n. After fully evaluating the diagram multiply the final result by $(-1)^n$.

Fig. 16.11 With time going vertical, the left diagram is a negative-d cap if and only if $d_a < 0$ and $d_b < 0$. (The directions of the arrows do not matter, and if the particles are self-dual we do not draw arrows). The right diagram is never a negative-d cap.

Recall from Section 14.5, a negative-d cap occurs when we go forward in time and two particles with $d < 0$ come together to annihilate or form a particle having $d > 0$. (see examples in Figs. 16.11 and 16.12).

These modifications guarantee we are describing a unitary theory. For example, if we take as simple loop like Fig. 16.6 with $d_a < 0$, the naive evaluation (before application of rule 0) gives a negative result. However, the diagram has one negative-d cap and so the result is multiplied by $(-1)^1$ thus giving a positive result as we should expect for a diagram that can be interpreted as ⟨state|state⟩ as in Fig. 14.3. More examples of how these evaluations work appear in Section 14.5.

The situation described in this section—having a theory that is isotopy invariant but has negative d_a—is quite common. Fundamentally, as discussed in Sections 14.1–14.2, the need to use negative d comes from negative Frobenius–Schur indicators. There are many topological theories of this type—including very simple theories such as the semion theory $SU(2)_1$ and more generally theories like $SU(2)_k$.

Fig. 16.12 With time going vertical, the left diagram is a negative-d cap if and only if $d_a < 0$. The right diagram is never a negative-d cap. We can think of these diagrams as being the same as the diagrams in Fig. 16.11 with c being the identity. The directions of the arrows do not matter.

16.1.4 Constraints and Examples

There are many constraints on our diagrammatic algebras for planar isotopy-invariant theories. This section describes these constraints and explains where they all come from.

Constraint: The Pentagon

The consistency condition on F-matrices given in Eq. 9.7 can be converted to the notation of this chapter (see Eq. 16.1) to give[5]

[5]In deriving Eq. 16.3 from Eq. 9.7 we have taken $a, b, c, d \to \bar{a}, \bar{b}, \bar{c}, \bar{d}$ for ease of notation.

$$F_{edl}^{c\bar{f}g} F_{e\bar{l}k}^{baf} = \sum_h F_{gch}^{baf} F_{edk}^{\bar{h}ag} F_{kdl}^{cbh} \quad . \tag{16.3}$$

Constraint: Relating F to d

For any theory with planar isotopy the value of d_a should be fixed by the F-matrices:

$$d_a = \frac{1}{F^{\bar{a}aI}_{\bar{a}aI}} \quad . \tag{16.4}$$

This is demonstrated by the manipulations of Fig. 14.4, converted into the notation of the current chapter. Recall that we are assuming $\epsilon_a = +1$ so one can straighten zig-zags freely.

Constraint: Inversion

One can perform an F-move on the right-hand side of Fig. 16.4 to bring it back into the form on the left. We obtain the diagrammatic relation shown in Fig. 16.13,

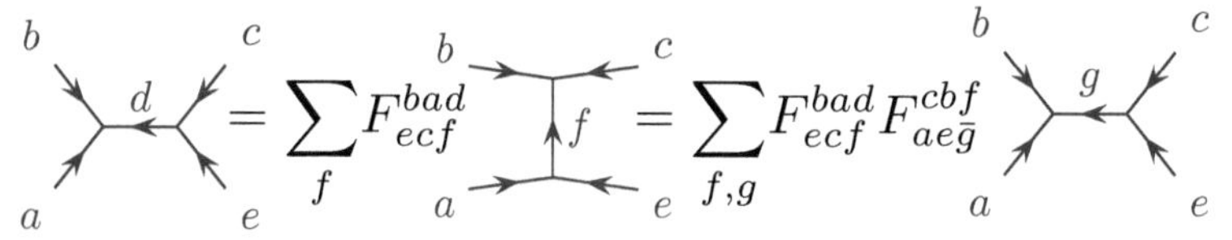

Fig. 16.13 In the second step we apply the same F-matrix equation from Fig. 16.4, but the diagram is rotated by 90 degrees.

which necessarily implies the consistency condition

$$\sum_f F^{bad}_{ecf} F^{cbf}_{ae\bar{g}} = \delta_{dg} \quad . \tag{16.5}$$

Constraint: Rotation

Rotating the diagram in Fig. 16.4 by 180 degrees and comparing it to the original diagram, one derives

$$F^{bad}_{ecf} = F^{ec\bar{d}}_{ba\bar{f}} \quad . \tag{16.6}$$

Constraint: Turning Up and Down

For a theory to be fully isotopy invariant, we must be able to freely make the moves shown in Fig. 14.17. As shown there, this requires $1 = \sqrt{(d_a d_c)}/d_b [F^{c\bar{c}a}_a]_{Ib}$, or in the notation of this chapter

$$F^{c\bar{c}I}_{a\bar{a}b} = \sqrt{\frac{d_b}{d_a d_c}} \tag{16.7}$$

whenever $b \times c = a + \dots$, with the sign of the square root taken negative if and only if d_a and d_c are both negative.

Constraint: Unitarity

As mentioned in Fig. 16.4, the F-matrix, being a change of basis, must be unitary. This means that

$$\sum_f F^{bad}_{ecf}[F^{bad'}_{ecf}]^* = \delta_{dd'} \tag{16.8}$$

$$\sum_d F^{bad}_{ecf}[F^{bad}_{ecf'}]^* = \delta_{ff'} \tag{16.9}$$

or equivalently $[F^{ba}_{ec}]^\dagger = [F^{ba}_{ec}]^{-1}$. Comparing the former to Eq. 16.5 we obtain

$$[F^{bad}_{ecf}]^* = F^{cbf}_{ae\bar{d}} \qquad . \tag{16.10}$$

Constraint: Hermitian Conjugation

Using reflection across the horizontal axis as in Fig. 12.4, we can reflect the F-matrix equation Fig. 16.4 and compare the reflected to the unreflected diagram to obtain

$$F^{bad}_{ecf} = [F^{\bar{a}\bar{b}\bar{d}}_{\bar{c}\bar{e}f}]^* \tag{16.11}$$

$$= [F^{\bar{c}\bar{e}d}_{\bar{a}\bar{b}f}]^* \tag{16.12}$$

$$= F^{\bar{e}\bar{a}f}_{\bar{b}\bar{c}d} \tag{16.13}$$

where in the second line the first line has been used in combination with Eq. 16.6, whereas in the third line the first line has been used in combination with Eq. 16.10.

Constraint: Reflection

An independent condition that is very often imposed is that the F-matrix should be invariant under left–right reflection. Compare the diagram shown in Fig. 16.14 to that of Fig. 16.4.

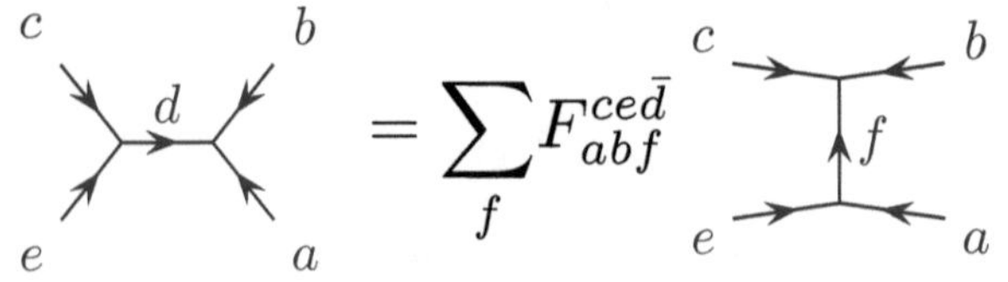

Fig. 16.14 The diagrammatic equation in Fig. 16.4 after being left–right reflected. This is necessary for a isotopically invariant $(2+1)$-dimensional theory, but is an independent assumption for a planar diagram algebra.

If a theory has left–right reflection symmetry, then we must have a further constraint

$$F^{bad}_{ecf} = F^{ce\bar{d}}_{abf} \qquad . \tag{16.14}$$

This additional condition is not required but it is often assumed.[6]

[6]See the discussion by Simon and Slingerland [2022].

Using Eq. 16.14 along with Eq. 16.13 gives us the natural seeming constraint

$$F^{bad}_{ecf} = [F^{\bar{b}\bar{a}\bar{d}}_{\bar{e}\bar{c}\bar{f}}]^* \quad . \tag{16.15}$$

Example: Evaluating a Bubble

As an example of showing how further constraints are derived, let us use F-moves to evaluate the bubble shown in Fig. 16.15.

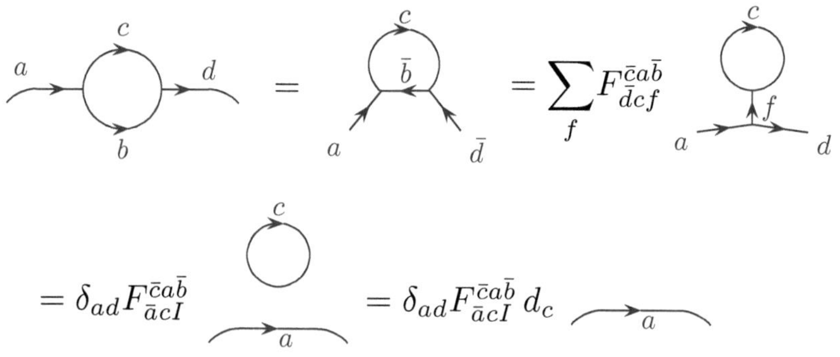

$$= \delta_{ad} F^{\bar{c}a\bar{b}}_{\bar{a}cI} \qquad\qquad = \delta_{ad} F^{\bar{c}a\bar{b}}_{\bar{a}cI} d_c$$

Fig. 16.15 Evaluation of a bubble diagram. In the first step, as usual we can flip the direction of an arrow and turn a particle into its antiparticle. In the second step we apply an F-move (compare to Fig. 16.4). Then by the no-tadpole (locality) rule (Fig. 16.8), we can set f to the vacuum particle I and hence, $a = d$.

However, we also know the value of the diagram in Fig. 16.15 from Fig. 16.7, which gives us $\sqrt{d_c d_b / d_a}$. Thus, we derive $F^{\bar{c}a\bar{b}}_{\bar{a}cI} d_c = \sqrt{\frac{d_c d_b}{d_a}}$, or equivalently (while replacing b with $\bar{b}$ for simplicity and using $d_b = d_{\bar{b}}$) we have

$$F^{\bar{c}ab}_{\bar{a}cI} = \sqrt{\frac{d_b}{d_a d_c}} \tag{16.16}$$

whenever $c \times b = a + \dots$ where the sign of the square root is taken negative if and only if d_a and d_b are both negative. Note that Eq. 16.16 could also be obtained from Eq. 16.7 with Eq. 16.13.

Example: The Theta Diagram

A commonly considered diagram is the Theta diagram $\Theta(a, b, c)$ shown in Fig. 16.16. This diagram is easily evaluated by using Fig. 16.7 along with the value of a single bubble Fig. 16.6.

$$\Theta(a, b, c) \quad = \qquad\qquad = \qquad\qquad = d_c \sqrt{\frac{d_a d_b}{d_c}} = \sqrt{d_a d_b d_c}$$

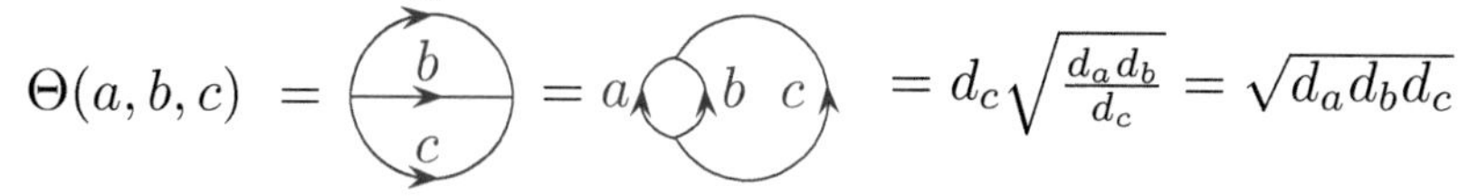

Fig. 16.16 The Theta diagram. This is evaluated by using Fig. 16.7 along with the value of a single bubble Fig. 16.6. The sign on the square root on the far right is taken negative unless all three d_a, d_b, and d_c are positive.

Example: The Tetrahedral Diagram

Let us consider one more evaluation known as the tetrahedral diagram as shown in Fig. 16.17. At this point we are considering this as a planar diagram even though it looks three-dimensional! However, we usually consider diagrams to be well defined if they live on the surface of a sphere, so if we want to think about this as being three-dimensional, we should think of this as living on a spherical surface.

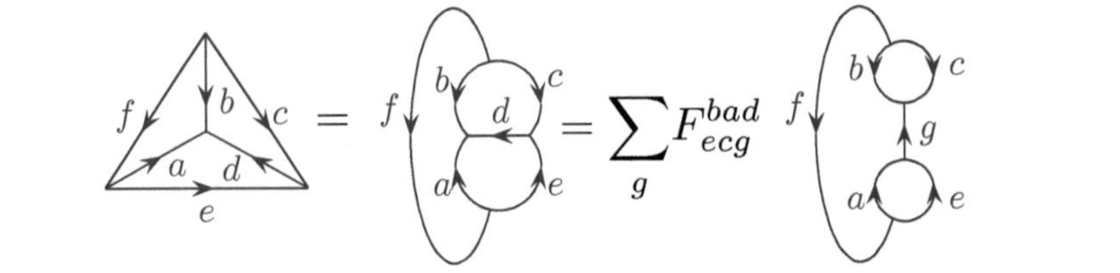

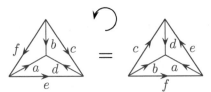

$$= F^{bad}_{ecf}\, d_f \sqrt{\frac{d_b d_c}{d_f}} \sqrt{\frac{d_a d_e}{d_f}} \equiv G^{bad}_{ecf}$$

Fig. 16.17 Evaluation of the tetrahedral diagram. The first step is just smooth deformation. The second step is application of an F-move. Using Fig. 16.7, the index g must be equal to the index f and we obtain some factors of $\sqrt{d}$. Finally, we are left with a single loop of f that gives a factor of d_f to give the final result, which we give the name G. As in Fig. 16.7 the square roots are taken negative if and only if both factors of d in the numerator of the square root are negative.

Fig. 16.18 An obvious rotational symmetry of the tetrahedral diagram.

For theories with full planar isotopy, the tetrahedral diagram has some obvious symmetries. For example, we should have rotational symmetry in the plane as shown in Fig. 16.18, which implies the identity (note the definition of G in Fig. 16.17)

$$G^{bad}_{ecf} = G^{dba}_{f\bar{e}\bar{c}} \quad . \tag{16.17}$$

Another symmetry comes from Eq. 16.6

$$G^{bad}_{ecf} = G^{ec\bar{d}}_{ba\bar{f}} \quad , \tag{16.18}$$

which we draw as shown in Fig. 16.19.

Although the diagram shown in Fig. 16.19 is a planar diagram, from the far left to the far right, it appears as if it is a rotation in three dimensions. Using Figs. 16.18 and 16.19 we can rotate this tetrahedron in any way we like. If one assumes the reflection symmetry, Eq. 16.14, then one can also take the mirror image of the tetrahedron as well to obtain an equivalence between 24 tetrahedral diagrams related by symmetries.[7]

---

[7] For an example of a spherical category that cannot be put in a form with full tetrahedral symmetry, see Hong [2009]. There are, however, even fairly simple modular categories that cannot be put in a form with full tetrahedral symmetry, although to my knowledge, they all have fusion multiplicities $N^c_{ab} > 1$. See Simon and Slingerland [2022].

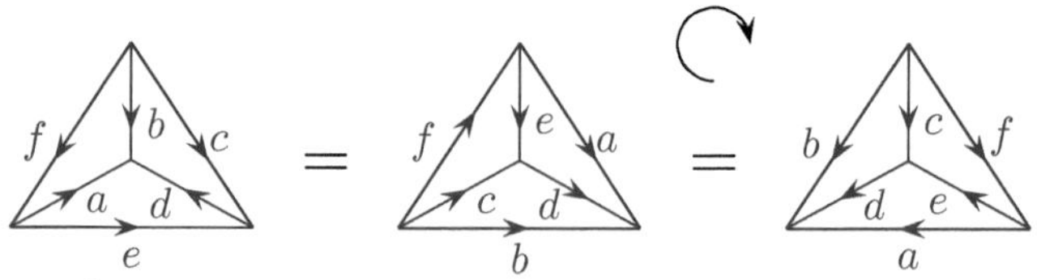

Fig. 16.19 The first step is the identity in Eq. 16.18 and the second step is a rotation as in Fig. 16.18. Although this is actually a planar diagram it appears as a rotation in three dimensions.

16.2 Braiding Diagrams Revisited

So far in this chapter we have considered planar theories only. Extension to fully (regular) isotopy-invariant three-dimensional[8] theories follows the expositions of Chapter 13. Here, we recapitulate the key points.

First, all deformations of the diagram that can be done within the plane of the board are allowed as a result of the planar isotopy.

Second, we can "regular isotope" diagrams, meaning deform them in three dimensions, treating the lines as ribbons. If we blackboard frame the diagrams (i.e. lay the ribbons flat in the plane) and paint the front of the ribbons white and the back of the ribbons black, any deformation of the ribbons is allowed that ends up blackboard framed, as long as the white side of all ribbons are on the front in the end.

Any of these deformations can be done to the diagram without changing its value. Often this sort of manipulation can turn a diagram with braiding into a planar diagram that can then be evaluated using the rules of Section 16.1.2. An example of this is shown in Fig. 16.1.

Most generally, however, we will not be able to eliminate all over- and under-crossing of lines just by using isotopy (i.e. by deforming a diagram). To handle crossings, we can invoke the R-matrix discussed in Chapter 13. The basic moves we need are summarized in Fig. 16.20 (which just repeats the previously defined moves from Figs. 13.3 and 13.5). Note that if we think of these strands as being ribbons, on both sides of the equalities in Fig. 16.20 the same side of the ribbon faces forward (i.e. a twist is not introduced with these R-moves).

[8] Here we mean $(2 + 1)$-dimensional theories, but we sometimes may not specify a particular time direction.

Fig. 16.20 The two basic uncrossing moves.

Note that in this chapter, where we are considering isotopy-invariant theories, the diagram may be oriented in any direction on the page, so for example, one may use the R-moves as shown in Fig. 16.21.

From the definition of the R matrix along with the completeness relation (16.9), we may derive the uncrossing identities shown in Fig. 16.22. For isotopy-invariant theories, these relations can be used even when the diagrams are rotated in any direction on the page. So for example, we have Fig. 16.23.

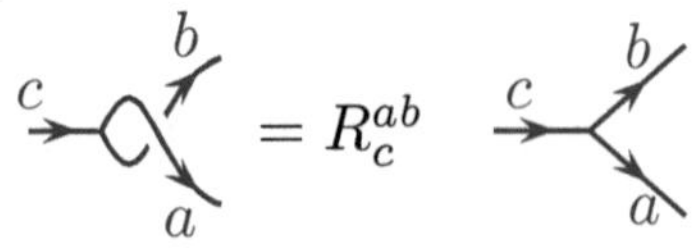

$$\text{(diagram)} = R_c^{ab}$$

Fig. 16.21 For isotopy-invariant theories, the R-moves can be oriented in any direction on the page.

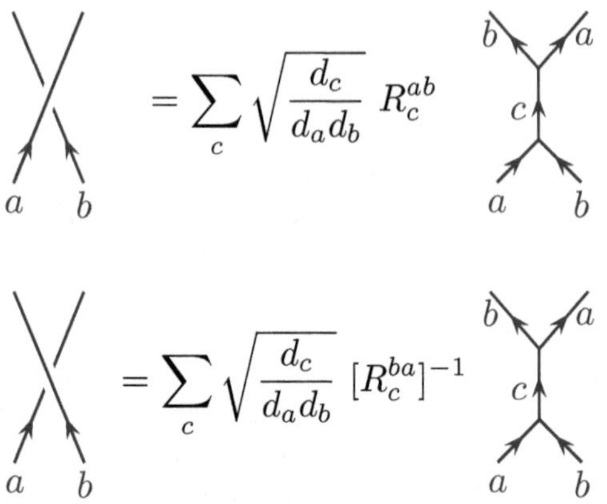

$$\text{(crossing)} = \sum_c \sqrt{\frac{d_c}{d_a d_b}}\, R_c^{ab}$$

$$\text{(crossing)} = \sum_c \sqrt{\frac{d_c}{d_a d_b}}\, [R_c^{ba}]^{-1}$$

Fig. 16.22 Resolving a crossing with isotopy normalization. The square roots are taken negative if any only if d_a and d_b are both negative. The case with fusion multiplicities $N_{ab}^c > 1$ is given in Section 16.4.

[9]Turning a crossing into a planar diagram is known as "resolution of a crossing."

The R-matrix moves in Figs. 16.20–16.22 allow us to take any diagram with over- and under-crossings and turn it into a planar diagram,[9] which can then be evaluated using the rules of Section 16.1.2.

Summarizing the results of this section, we add two rules for evaluation of diagrams in three dimensions to our previously stated rules for planar diagrams of Section 16.1.2:

(1′) One is free to continuously deform diagrams in three dimensions as long as we do not cut any strands, and strands are treated as ribbons. For knots and links, we have regular isotopy—meaning any smooth deformation of the ribbons allowed. For diagrams with fusion and splitting, it is more complicated. All planar isotopies are always allowed. For non-planar deformations we need to keep track of the sides of the ribbon. The ribbons have two sides, white and black. The ribbons should start in blackboard framing with the white side forward, and after a deformation they should also be blackboard framed with the white side forward. (This rule replaces rule (1) from the list in Section 16.1.2.)

(9) Over- and under-crossings can be turned into planar diagrams using the R-matrix as in Figs. 16.20–16.22.

16.2.1   Constraints

There are several further constraints on the braiding diagrammatic algebra.

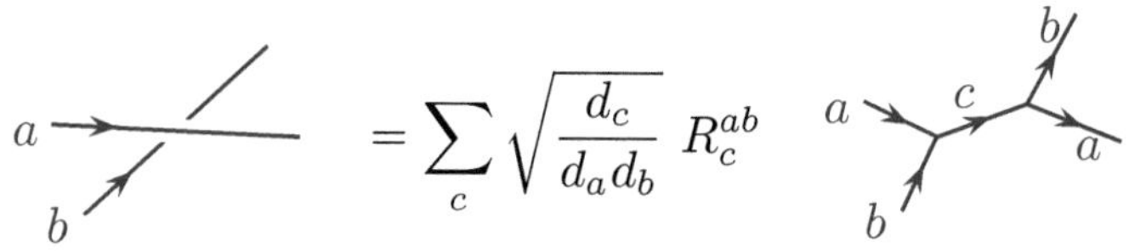

Fig. 16.23 With isotopy-invariant theories, we can rotate diagrams freely. Thus, the uncrossing formula here is identical to that in Fig. 16.22 (top).

Constraint: Rotation and Conjugation

As mentioned in Fig. 16.23, we can rotate crossings freely. By turning a crossing entirely upside-down and then using Hermitian conjugation (as described in Section 12.1), we obtain an identity that holds for isotopy-invariant theories

$$[R_c^{ab}]^* = [R_{\bar c}^{\bar a \bar b}]^{-1} \quad . \tag{16.19}$$

Constraint: Hexagon Equations

As discussed in Section 13.3, there are consistency conditions between F- and R-matrices known as hexagon equations (Eqs. 13.1 and 13.2). In the notation of the current chapter these can be written as

$$R_e^{ca} F_{d\bar bg}^{\bar c \bar a e} R_g^{cb} = \sum_f F_{d\bar bf}^{\bar a \bar c e} R_d^{cf} F_{d\bar cg}^{\bar b \bar a f} \tag{16.20}$$

$$[R_e^{ac}]^{-1} F_{d\bar bg}^{\bar c \bar a e} [R_g^{bc}]^{-1} = \sum_f F_{d\bar bf}^{\bar a \bar c e} [R_d^{fc}]^{-1} F_{d\bar cg}^{\bar b \bar a f} \quad . \tag{16.21}$$

Relation to Twists

As detailed in Chapter 15, each particle has a twist factor $\theta_a = \theta_{\bar a}$ (with $\theta_I = 1$) describing curled strands, which we show again here for completeness in Fig. 16.24.

Fig. 16.24 Definition of twist factors (see Chapter 15 for more details).

In this chapter the direction the twist is drawn on the page is not important (only its chirality). Note that since we can relate twists to R-matrices (Fig. 15.6) it is not necessary to have this twist relation (Fig. 16.24) as its own axiom, or rule, in our list of rules.

As noted in note 2 of Chapter 15 we must be careful when we work with $d_a < 0$ since this then changes the sign of θ_a compared to the Option A approach, where we give up isotopy invariance but have all

$d_a > 0$. As a result we sometimes prefer to work with $R_I^{\bar{a}a} = \epsilon_a \theta_a^*$ which is unambiguous (see Eq. 15.4).

There are some additional relations between R and twists that occur for isotopy-invariant theories, such as

$$R_c^{ab} = R_I^{a\bar{a}}[R_{\bar{b}}^{a\bar{c}}]^{-1} = R_I^{\bar{b}b}[R_{\bar{a}}^{\bar{c}b}]^{-1} \tag{16.22}$$

which is derived by the sequence of moves shown in Fig. 16.25 (see also Exercise 16.2).

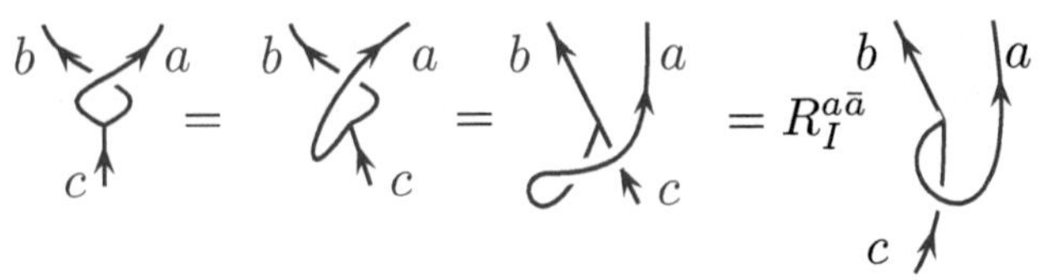

Fig. 16.25 The sequence of moves used to derive Eq. 16.22 (see also Exercise 16.2).

16.3 Gauge Transformations

As in Sections 9.4 and 13.4.1, it is possible to make gauge transformations of vertices, which makes resulting changes on F-matrices and R-matrices (see Sections 9.5.3 and 13.4.1) which we show in Fig. 16.26.

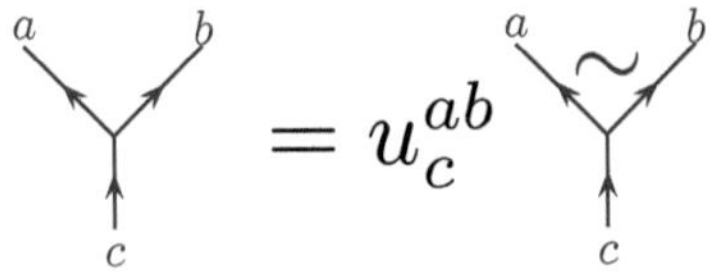

Fig. 16.26 We have the freedom to make a gauge transformation of a vertex by multiplying by a phase u_c^{ab}. The tilde on the right notates that the vertex is in the tilde gauge.

However, since in an isotopy-invariant theory we have Fig. 16.27, we conclude that we must have $u_c^{ab} = u_{\bar{a}}^{b\bar{c}} = u_{\bar{b}}^{\bar{c}a}$.

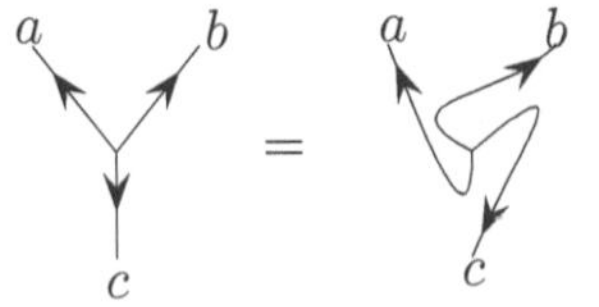

Fig. 16.27 This is an allowed identity in isotopy-invariant theories. (Compare the more general Fig. 14.27.

Given such a gauge transformation, the F- and R- matrices transform as

$$\widetilde{F_{ecf}^{bad}} = \frac{u_f^{\bar{b},\bar{c}} u_e^{\bar{a}f}}{u_d^{\bar{a}\bar{b}} u_e^{d\bar{c}}} F_{ecf}^{bad} . \tag{16.23}$$

$$\widetilde{R_c^{ab}} \;=\; \frac{u_c^{ba}}{u_c^{ab}} R_c^{ab} \quad . \tag{16.24}$$

16.4   Appendix: Higher Fusion Multiplicities

As in Section 9.5.3, when fusion multiplicities are greater than one, the vertices have additional indices that we label with Greek indices $\mu, \nu, \ldots$. For example, if a and b fuse to c with $N_{ab}^c > 1$, then the vertex will have an additional index $\mu \in \{1 \ldots N_{ab}^c\}$. In the conventions of the current chapter we would then have:

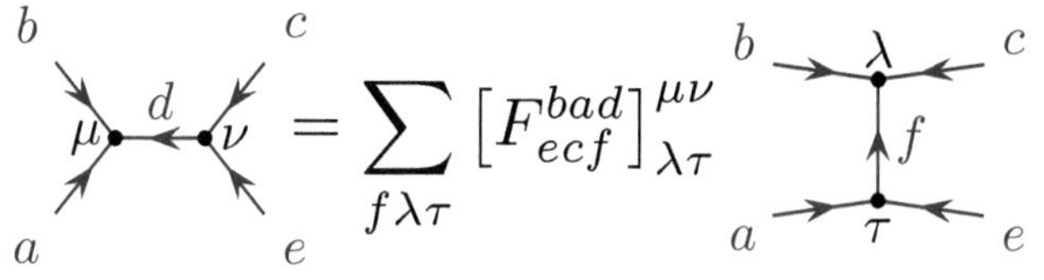

Fig. 16.28   *F*-matrix for isotopy-invariant theories with fusion multiplicity.

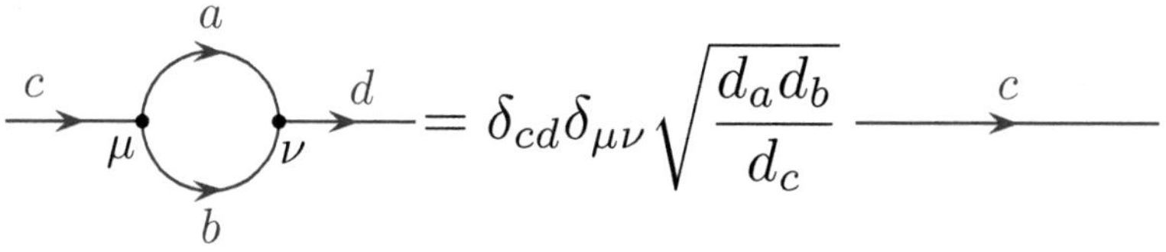

Fig. 16.29   The locality principle (and orthonormality) for the isotopy-invariant diagrammatic algebra with fusion multiplicity.

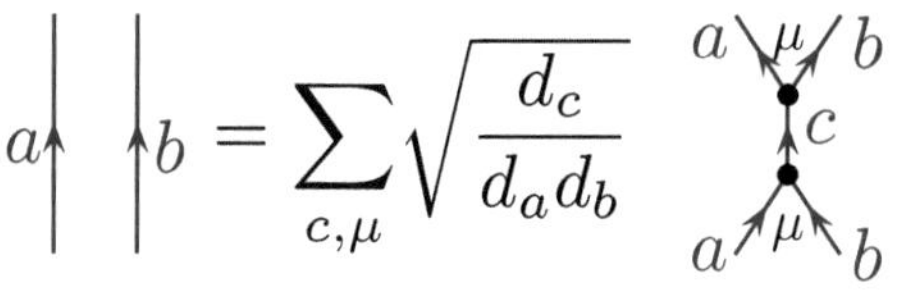

Fig. 16.30   Insertion of a complete set of states with isotopy normalization and fusion multiplicity.

Chapter Summary

- This chapter describes the diagrammatic rules for "nice" theories. First we discuss theories with planar isotopy invariance. Then we discuss theories with three-dimensional isotopy invariance. Almost all braided theories can be put into such a form.
- With some change of notation, this is a summary of the diagrammatic rules from Chapters 9–15.

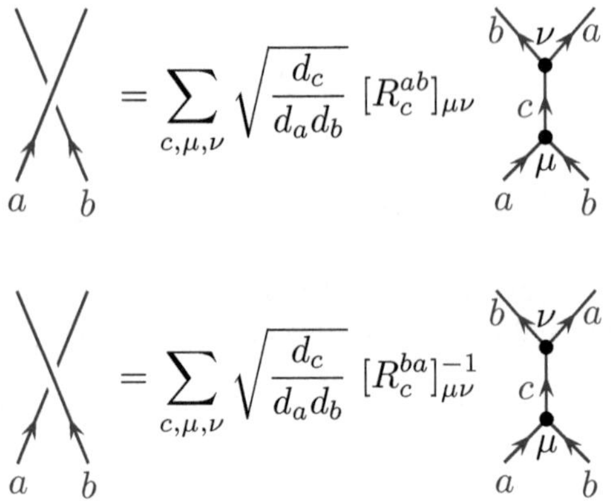

$$\text{(crossing diagram)} = \sum_{c,\mu,\nu} \sqrt{\frac{d_c}{d_a d_b}} \, [R_c^{ab}]_{\mu\nu} \quad \text{(vertex diagram)}$$

$$\text{(crossing diagram)} = \sum_{c,\mu,\nu} \sqrt{\frac{d_c}{d_a d_b}} \, [R_c^{ba}]_{\mu\nu}^{-1} \quad \text{(vertex diagram)}$$

Fig. 16.31 Resolving a crossing with isotopy normalization and fusion multiplicity. The square roots are taken negative if any only if d_a and d_b are both negative.

Further Reading:

- The F-matrix notation of this chapter matches that of Levin and Wen [2005].

Exercises

Exercise 16.1 Triangle Bubble Collapse
A useful lemma is the collapsing of a triangular bubble.

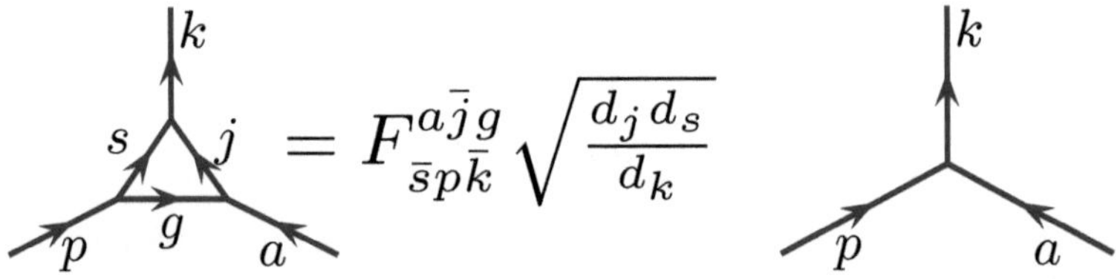

$$\text{(triangle diagram)} = F_{\bar{s}p\bar{k}}^{a\bar{j}g} \sqrt{\frac{d_j d_s}{d_k}} \quad \text{(Y diagram)}$$

Derive this lemma.

Exercise 16.2 Relating Twist to R
(a) Use the manipulations of Fig. 16.25 to derive Eq. 16.22.
(b) Derive the same result starting with the hexagon equations, setting a particular variable to I, and being careful to note certain properties of F that must hold for isotopically invariant theories.

Further Structure[1]

17
Important

In this chapter we explore some further structure that is inherent in topological theories.

[1] The discussions in this chapter, particularly the properties of the S-matrix and the Kirby Ω strand, are referred to many times later in the book.

17.1 Quantum Dimension

Recall that we defined d_a, the quantum dimension, in terms of how fast the Hilbert space grows as we fuse together many a particles (see Eq. 3.11):

$$\text{Dim of } M \text{ anyons of type } a \sim d_a^M \quad .$$

An alternative definition (Eq. 8.17) is[2]

$$d_a = \text{the largest eigenvalue of the matrix } N_a \quad . \qquad (17.1)$$

[2] Recall that N_a is defined as the matrix with components $[N_a]_b^c = N_{ab}^c$ which are the fusion multiplicities.

We have claimed several times that this quantum dimension is (up to a possible sign) equal to the value of a loop d_a in our diagrammatic algebra (the "loop weight"). In this section we finally prove this important result.

To make a connection to our diagrammatic algebra, let us consider fusing two loops labeled a and b as shown in Fig. 17.1.[3]

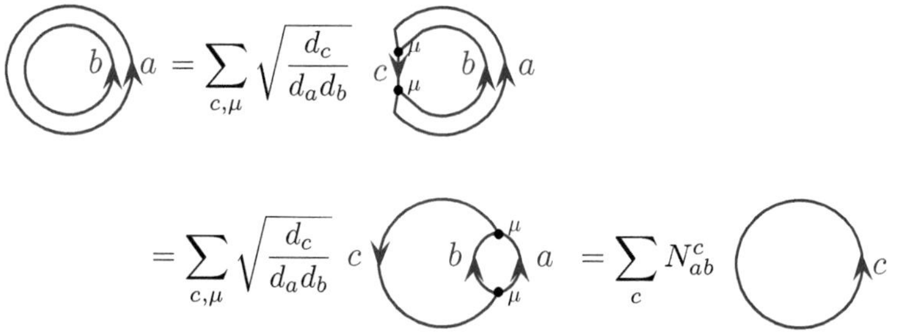

Fig. 17.1 Fusing two loops into a single loop. In the first line we use the completeness relation Fig. 14.33, then we deform to the second line, and finally we remove the bubble using Fig. 14.32.

The result seems rather natural, that a and b can fuse together to form c in all possible ways. The derivation uses the completeness relation in

[3] This result holds very generally. One might worry that for general theories, without isotopy invariance, going from the first line to the second line might be problematic. However, it turns out that one does not need full isotopy invariance; just the pivotal property is enough to get to the second line (see Section 14.8.1 and Exercise 17.2).

the first line (Fig. 14.8), then we deform to get to the second line, and finally, we remove the bubble using Fig. 14.7.

Now the value of a loop in our diagrammatic algebra is d_a. However, if d_a is negative we must implement rule 0 from Section 14.5 and we always obtain a positive result $|d_a|$ for a single loop (appropriate for a diagram that can be interpreted as $\langle \text{state}|\text{state}\rangle$; see Fig. 14.3). Thus, the diagrammatic manipulations of Fig. 17.1 give us[4]

[4] One says that the quantities $|d_a|$ form a representation of the fusion algebra. Compare to Eq. 8.5.

$$|d_a|\,|d_b| = \sum_c N_{ab}^c\,|d_c| \quad . \tag{17.2}$$

Eq. 17.2 has an interesting interpretation if we define a vector $\vec{d}$ to have components $|d_c|$ with the vector index being c. We can then rewrite Eq. 17.2 as an eigenvalue equation

$$|d_a|\,\vec{d} = N_a\,\vec{d} \quad .$$

[5] The Perron–Frobenius theorem is usually applied to matrices of all positive numbers and is slightly weaker when applied to the non-negative case. See the detailed discussion in Section 17.8.

Since N_a is a matrix of non-negative numbers,[5] and $\vec{d}$ is a vector of positive numbers, we identify $\vec{d}$ as the so-called Perron–Frobenius eigenvector of the matrix N_a, and its eigenvalue $|d_a|$ is guaranteed by the Perron–Frobenius theorem (see Section 17.8) to be the largest[6] eigenvalue of N_a which, by Eq. 17.1 gives us

[6] Strictly speaking, N_a may have several eigenvalues of the same absolute magnitude, with the Perron–Frobenius eigenvalue being the only one that is real and positive. This does not change our conclusion.

$$\mathsf{d}_a = |d_a|$$

as claimed. This derivation does not assume that the theory has a well-defined braiding.

17.2 The Unlinking $\tilde{S}$-Matrix

First let us use the locality principle (or no-transmutation) principle (see Fig. 8.7) to show that a closed loop of type a around a world line of type x gives some constant, which we call $\tilde{S}_{ax}$ as shown in Fig. 17.2. (In fact using the R-matrix we can explicitly derive this identity and evaluate $\tilde{S}_{ax}$ in terms of twist factors θ, the fusion multiplicities N_{ax}^c and the quantum dimensions d. See Exercise 17.4. However, we will not need this explicit expression.) Note in particular that $\tilde{S}_{Ix} = 1$ since the identity loop can be removed for free, and $\tilde{S}_{aI} = \mathsf{d}_a$ since a single loop[7] of a gives d_a.

[7] Here, if d is negative we include the sign associated with rule 0 of Section 14.5 in the evaluation of the diagram so that we are discussing a unitary theory and we obtain a positive quantum dimension d.

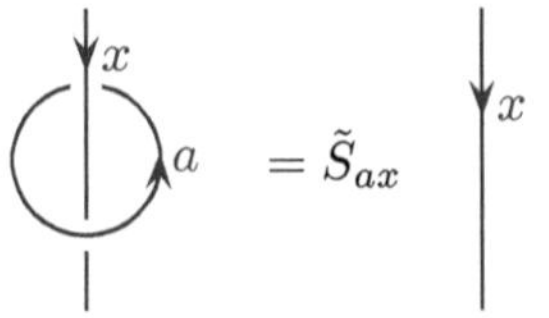

Fig. 17.2 The locality principle tells us that the value of a loop of a around a world line x is some number that we call $\tilde{S}_{ax}$. (Indeed, we can use the R-matrix to calculate $\tilde{S}_{ax}$. See Exercise 17.4.)

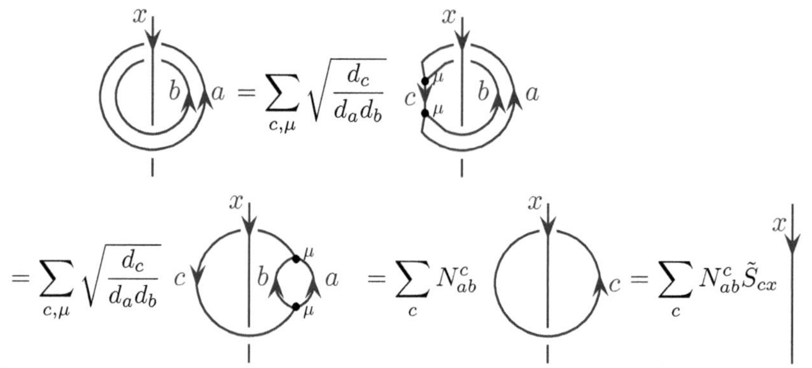

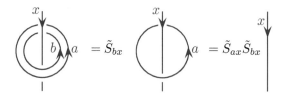

$$= \sum_{c,\mu} \sqrt{\frac{d_c}{d_a d_b}} \; c \; \left(\begin{array}{c} \mu \\ b \\ \mu \end{array} \right) a \; = \sum_c N_{ab}^c \; \left(\begin{array}{c} x \\ \\ \end{array} \right) c = \sum_c N_{ab}^c \tilde{S}_{cx} \; \left| \begin{array}{c} x \\ \\ \end{array} \right.$$

Fig. 17.3 Similar reasoning as in Fig. 17.1 allows us to write this diagrammatic relationship.

$$\left(\begin{array}{c} x \\ b \\ \end{array} \right) a = \tilde{S}_{bx} \qquad \left(\begin{array}{c} x \\ \\ \end{array} \right) a = \tilde{S}_{ax} \tilde{S}_{bx} \qquad \left| \begin{array}{c} x \\ \\ \end{array} \right.$$

Fig. 17.4 Application of $\tilde{S}$ twice (compare to Fig. 17.2).

Now, if we have two loops a and b around x, we can fuse the two loops to all possible loops c as shown in Fig. 17.3. This identity is entirely analogous to that of Fig. 17.1. In essence we are fusing a and b to form all possible strands c and then in the last step we apply $\tilde{S}$. On the other hand, we could also evaluate the left-hand side of Fig. 17.3 by applying the identity of Fig. 17.2 twice in a row, as in Fig. 17.4.

Equating the result of Fig. 17.3 to that of Fig. 17.4 we obtain

$$\tilde{S}_{ax} \tilde{S}_{bx} = \sum_c N_{ab}^c \tilde{S}_{cx} \quad . \tag{17.3}$$

This result holds for any anyon theory with a well-defined braiding (i.e. that satisfies the hexagon relationship). Note that in the special case where x is the identity we just recover Eq. 17.2.

17.3   The (Modular) S-Matrix

Recall from Section 7.3.1 that we defined the S-matrix (Eq. 7.12) in several ways. On the one hand we defined

$$S_{ab} = Z(S^3; a \text{ loop linking } b \text{ loop}) \tag{17.4}$$

whereas on the other hand, we said that S was (under certain conditions that the theory has no transparent particles) a unitary transformation between two different bases for describing the Hilbert space of a torus.

Now recall our normalization of diagrams is given by

$$\text{normalized diagram} = Z(S^3; \text{diagram})/Z(S^3)$$

(see Eq. 14.7), so we can write[8]

[8] As mentioned near Fig. 7.14, there is no agreement in the literature as to which way the arrows should point in the definition of S_{ab}. Our convention matches that of Kitaev [2006], which seems to be more common.

$$S_{ab} = \qquad \times \frac{1}{\mathcal{D}} \qquad (17.5)$$

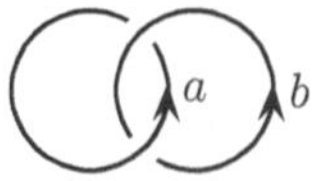

where we have defined

$$\mathcal{D} = \frac{1}{Z(S^3)} \quad . \qquad (17.6)$$

Using isotopy of diagrams and Hermitian conjugation it is easy to establish (see Exercise 17.3) that

$$S_{ab} = S_{ba} = S_{\bar{a}\bar{b}} = S_{\bar{b}\bar{a}} = S_{\bar{a}b}^* = S_{b\bar{a}}^* = S_{a\bar{b}}^* = S_{\bar{b}a}^* \quad . \qquad (17.7)$$

In particular, S_{ab} is real if either a or b is self-dual.

Further, by setting one of the indices to the vacuum I, we are left with a single loop that evaluates to the quantum dimension,[9] hence, giving us

[9] Again, if $d_a < 0$, we use rule 0 from Section 14.5 so that we are working with a unitary theory, and a single loop always evaluates to a positive number $d_a = |d_a|$ independent of the sign of d_a. If we do not apply rule 0, the diagram on the right-hand side of Eq. 17.5 divided by $\mathcal{D} > 0$ will instead give us $S_{ab} \text{sign}(d_a d_b)$.

$$S_{Ia} = S_{aI} = d_a/\mathcal{D} \quad . \qquad (17.8)$$

Let us now evaluate the S-matrix in terms of our unlinking matrix $\tilde{S}$. By bending the top of x in Fig. 17.2 and forming a closed loop with the bottom of x, we construct linked rings as shown in Fig. 17.5.

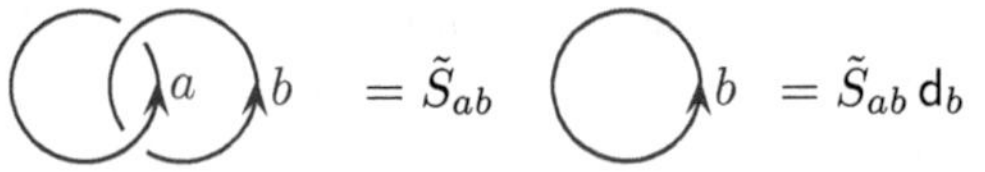

Fig. 17.5 Evaluation of linked rings. In the case where $d_b < 0$, we have applied rule 0 from Section 14.5 so that the single loop gives us a positive d_b quantum dimension.

Comparing to the definition of S in Fig. 17.5 we obtain

$$S_{ab} = \tilde{S}_{ab} d_b/\mathcal{D} = \tilde{S}_{ab} S_{Ib}$$

or equivalently

$$\tilde{S}_{ab} = \frac{S_{ab}}{S_{Ib}} \quad .$$

Plugging this into Eq. 17.3 gives us

$$\frac{S_{ax} S_{bx}}{S_{Ix}} = \sum_c N_{ab}^c S_{cx} \quad . \qquad (17.9)$$

Again, this is generally true for any braided anyon theory, that is, any theory that satisfies the pentagon and hexagon relations.

17.3.1   Unitary S = Modular

When S is unitary, we say the theory is *modular*.[10] It turns out that S is unitary if and only if the vacuum (identity) particle is the only *transparent* particle in the theory.[11] A particle a is said to be transparent if braiding a all the way around any (non-vacuum) particle accumulates no phase. Equivalently, in terms of the S-matrix we can write

$$a \text{ is transparent} \quad \Leftrightarrow \quad S_{ax} = \frac{\mathsf{d}_a \mathsf{d}_x}{\mathcal{D}} \quad \text{for all } x, \qquad (17.10)$$

that is, a full braiding of a with any (non-vacuum) particle is trivial.[12]

It is clear that an S-matrix cannot be unitary if there is any transparent particle a besides the identity (since the row S_{ax} and the row S_{Ix} would be proportional to each other). What is not obvious is that the absence of any transparent particle guarantees S is unitary. This statement can be proven using our axiomatic diagrammatic principles (i.e. not invoking any of the topological discussion of Section 7.3.1). However, the proof is a bit complicated, and we refer the reader to Kitaev [2006] and Etingof et al. [2015] for details.

As mentioned a number of times, in some sense all "well-behaved" anyon theories are modular (we say they are *modular tensor categories*). Unfortunately, there are common theories that are not modular, for example, a simple theory of a single fermion that obtains a minus sign under exchange. A full braiding gives a plus sign exactly like the vacuum, and is hence non-modular (although it may have a well-defined braiding). It is not unusual to have fermions in a theory that braid identically to the vacuum and thus prevent the theory from being modular.

Let us assume for the remainder of this section that we have a modular theory, that is, that S is unitary, as we had stated in Section 7.3.1. First, unitarity implies that

$$1 = \sum_a |S_{aI}|^2 = \frac{1}{\mathcal{D}^2} \sum_a \mathsf{d}_a^2$$

which thus allows us to identify[13]

$$\mathcal{D} = +\sqrt{\sum_a |\mathsf{d}_a|^2} \qquad (17.11)$$

which is usually called the *total quantum dimension*. Note in particular that this implies

$$S_{II} = 1/\mathcal{D} = Z(S^3) \quad . \qquad (17.12)$$

Again assuming a unitary S-matrix we can multiply Eq. 17.9 by $S^{-1} = S^\dagger$ on the right to obtain the often-quoted Verlinde formula[14]

$$N_{ab}^c = \sum_x \frac{S_{ax} S_{bx} [S^{-1}]_{xc}}{S_{Ix}} = \sum_x \frac{S_{ax} S_{bx} S_{cx}^*}{S_{Ix}} \qquad (17.13)$$

[10]We explain the meaning of the word "modular" in Section 17.3.2.

[11]We ran into this concept as far back as Section 4.3.2.

[12]An equivalent statement using the ribbon identity Eq. 15.5 is that a is transparent if and only if $\theta_c/(\theta_a \theta_b) = 1$ for all b and c whenever $N_{ab}^c \neq 0$.

[13]We conventionally choose the positive square root. Choosing the negative square root describes a theory whose central charge is different by 4 from the positive square root case, which we can see from Eq. 17.16.

[14]Verlinde [1988] derived this in the context of conformal field theories. In a different context it was derived earlier by Pasquier [1987].

which tells us that all the information about the fusion algebra is contained entirely within the S-matrix!

Alternatively, one can multiply Eq. 17.9 by $S^{-1} = S^\dagger$ on the left to obtain (with S and N treated as matrices on the left)

$$[S^\dagger N_a S]_{xy} = \delta_{xy} \left(\frac{S_{ax}}{S_{Ix}} \right) \ .$$

This means that the S-matrix (at least for modular theories) is the unitary matrix that simultaneously diagonalizes all of the N_a fusion multiplicity matrices (we mentioned this previously after Eq. 8.20).

A useful quick application of the Verlinde formula, Eq. 17.13, is to write an expression for the conjugation[15] matrix

[15]The word "conjugation" is meant to evoke charge conjugation, which changes positive charges to negative charges.

$$C_{ab} \equiv \delta_{a\bar{b}} \ .$$

This is simply a permutation matrix that permutes each particle with its antiparticle. Obviously $C^2 = \mathbb{1}$ is the identity.

We can find a relationship between C and the S-matrix by writing C as the fusion multiplicity matrix of two particles fusing to the identity

$$C_{ab} = N_{ab}^I = \sum_x S_{ax} S_{bx} = [SS^T]_{ab}$$

where we have used the Verlinde formula to evaluate N_{ab}^I along with S_{Ix} being real. Finally, using the fact that S is symmetric we obtain

$$C = S^2 \ .$$

17.3.2   Modular Group and Torus Diffeomorphisms

Let us define one more matrix, which is the diagonal matrix of the twist factors[16]

[16]We should assume here that θ_a is defined with $d_a > 0$, that is, using Option A from Section 14.5.

$$T_{ab} = \theta_a \delta_{ab} \ . \tag{17.14}$$

It turns out to be more useful to absorb an additional complex phase into this matrix, so let us also define[17]

[17]Most references mean $\tilde{T}$ when they say "T-matrix." However, a few mean T.

$$\tilde{T} = T e^{-2\pi i c/24} \tag{17.15}$$

where c is a real constant, known as the *chiral central charge*. The central charge modulo 8 can be calculated from the twist factors and quantum dimensions via (Fröhlich and Gabbiani [1990]; Rehren [1990])

$$e^{2\pi i c/8} = \frac{1}{D} \sum_a d_a^2 \theta_a \ . \tag{17.16}$$

[18]We do not need C as an independent generator since $C = S^2$.

[19]Be warned there are several closely related groups that are sometimes known as the modular group.

We discuss the central charge further in Section 17.3.3.

The set of operations generated by $\tilde{T}$ and S form[18] a group known as the *modular group*.[19]

$$S^2 = C \qquad C^2 = \mathbb{1} \qquad (S\tilde{T})^3 = C \ . \tag{17.17}$$

For a proof of the last identity we again refer the reader to Kitaev [2006] and Etingof et al. [2015]. These relations are equivalent to the group algebra of $SL(2, \mathbb{Z})$, the group of 2×2 matrices with integer coefficients and unit determinant that has generators

$$\underline{S} = \begin{pmatrix} 0 & -1 \\ 1 & 0 \end{pmatrix} \qquad \underline{T} = \begin{pmatrix} 1 & 1 \\ 0 & 1 \end{pmatrix} \qquad (17.18)$$

It is easy to check that Eq. 17.17 are satisfied by these 2×2 matrices.

The modular group has a beautiful topological interpretation: it is the group of topologically distinct[20] orientation-preserving diffeomorphisms of the torus surface. To see how this works, we consider a torus to be a plane $\mathbb{R}^2$ with the lattice of integers $\mathbb{Z}^2$ modded out so that each lattice point is identified with every other lattice point:

$$T^2 = \mathbb{R}^2 / \mathbb{Z}^2 \quad .$$

Any transformation on the plane that one-to-one maps lattice points to lattice points gives a representative diffeomorphism of the torus. Such transformations are given by the elements of the group $SL(2, \mathbb{Z})$ just by mapping points $\vec{v}$ in the plane to $A\vec{v}$ where A is a member of $SL(2, \mathbb{Z})$.[21] Graphical descriptions of the $\underline{S}$ and $\underline{T}$ transformations are shown in Fig. 17.6.

[20]What we mean here is the "mapping-class group" of the torus surface. That is, two mappings of the torus surface to the torus surface are considered to be the same if one can be smoothly deformed into the other.

[21]It is obvious that such a transformation of the plane corresponds to a diffeomorphism of the torus. What is a bit less obvious is that *all* diffeomorphisms of the torus are topologically equivalent to (i.e. can be smoothly deformed into) a linear map of this sort. See, for example, Rolfson [1976] and Farb and Margalit [2012] for detailed discussions of this point.

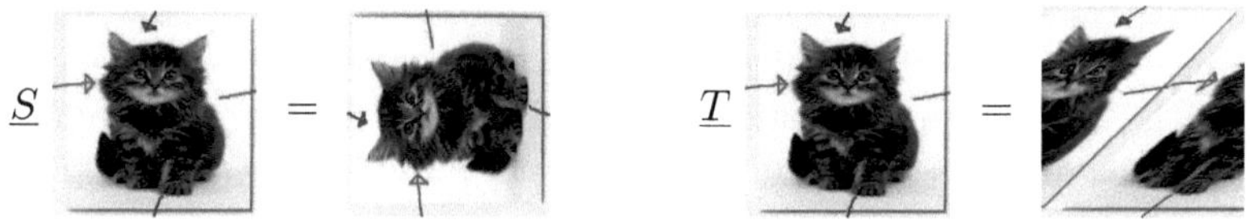

Fig. 17.6 The $\underline{S}$ and $\underline{T}$ transformations on the surface of a torus. Here, one interprets the picture as being put on the surface of a torus with opposite sides of the picture identified. The matrices from Eq. 17.18 are applied to the coordinates in the plane.

This analogy with the diffeomorphisms of the torus is certainly not coincidental! Let us think back to the discussion of Sections 7.3 and 7.4. We considered the solid torus $D^2 \times S^1$ with a particle world line of type a around the handle. We wrote the wavefunction on the surface as

$$|\psi_a\rangle = |Z(D^2 \times S^1; a)\rangle$$

which forms an orthonormal basis $\langle \psi_a | \psi_b \rangle = \delta_{ab}$. As mentioned there, this inner product corresponds to gluing together the two solid tori to create $S^2 \times S^1$. However, we could also have glued the tori together after exchanging meridian and longitude to create S^3, and the inner product then becomes (see Eq. 7.12)

$$\langle Z(S^1 \times D^2; b) | Z(D^2 \times S^1, a) \rangle = S_{ab} \quad .$$

A different way of thinking of this is that we make a diffeomorphism on the surface of the torus before gluing the two halves back together.

Thus, we could have said

$$\langle\psi_b|\hat{S}|\psi_a\rangle = S_{ab}$$

where $\hat{S}$ is the operator that makes the diffeomorphism on the surface (the diffeomorphism precisely exchanges meridian and longitude).

Note further that $\hat{C} \equiv \hat{S}^2$ exchanges the meridian and longitude twice, giving a net effect of rotating the torus by 180 degrees. If we think back to the solid torus with a world line of type a around the handle, the rotation of the torus surface by 180 degrees changes the relative direction of the embedded world line and thus changes a to $\bar{a}$, implementing the conjugation operation C.

Finally, we can define a $\hat{T}$ operation on the torus surface, which implements a Dehn twist as described in Section 7.4. Analogously the matrix elements of the $\hat{T}$ operation will give the matrix $\tilde{T}$. The fact that the $\tilde{T}$ operation on the state $|\psi_a\rangle$ corresponds to a twist factor θ_a is fairly obvious from looking at Fig. 7.16. The presence of the additional complex factor related to the central charge in Eq. 17.15 is a subtle point and stems from a change in the 2-framing discussed in Section 5.3.6.

The (unitary) matrices S and T (or $\tilde{T}$) are known as the *modular data* of a TQFT. If you are given the modular data of a TQFT, usually this is enough to fully define the TQFT—that is. the modular data is usually enough to uniquely determine[22] F-matrices and R-matrices, and so forth, and the central charge is determined[23] modulo 8. However, there are cases known where the modular data does not completely define the TQFT—that is, more than one TQFT shares the same modular data. The smallest known case where this occurs has 49 particle types (see Mignard and Schauenburg [2021]).

We should also be cautioned that the TQFT properties of a system—the braiding, twisting, fusion, and central charge properties of the anyons—does not completely define a physical system. One can have many different theories that share all of these topological properties and yet differ in other respects—for example, by having fundamentally different edge theories. This is discussed in some detail for abelian anyon theories by Cano et al. [2014], and is not fully explored in the nonabelian case.

17.3.3 Central Charge and Relation to Conformal Field Theory

In Eqs. 17.15 and 17.16 we introduced the idea of the central charge of a TQFT. The central charge stems from the relationship between $(2+1)$-dimensional TQFTs and $(1+1)$-dimensional conformal field theories. This relationship has some deep physics buried in it and is particularly useful in the context of fractional quantum Hall effect (briefly discussed in Sections 2.5 and 37.1).[24]

Roughly, $(1+1)$-dimensional conformal field theories describe the physics of gapless one-dimensional systems.[25] For a chiral system (meaning that flow only goes in one direction) the central charge c is related

[22] "Uniquely determined" does not mean that the F-matrices and R-matrices are easily calculated, just that only one set of Fs and Rs exist given S and T.

[23] The modular data only specifies the central charge mod 8 via Eq. 17.16. However, by some definitions, central charge mod 24 is often considered to be a property of the TQFT as well. This is a piece of information beyond the modular data.

[24] My *next* book will give more attention to the details of fractional quantum Hall physics.

[25] "Gapless" means having arbitrarily low-energy excitations in the large system size limit. This is the opposite of "gapped."

to (i.e. can be *defined* by) the heat capacity per unit length c_v of the system via $c_v = \pi k_B^2 Tc/(3\hbar v)$ where v is the speed of light (the velocity of flow at low energy), T is temperature, and k_B is Boltzmann's constant. If one makes a small change in temperature, the change in heat flowing along the edge is then

$$\delta J^q = (\pi k_B^2 Tc/(3\hbar))\delta T \quad . \tag{17.19}$$

TQFTs are gapped systems:[25] there are no low-energy excitations near zero energy in the bulk. Thus, the low-temperature heat capacity must be zero. However, in a finite system (say on a disk geometry) the edge of a TQFT system may have gapless modes that can carry heat, and this heat transport defines the central charge of the TQFT (Cappelli et al. [2002]).[26] For example, one can consider a finite rectangular strip of the TQFT and attach two opposite ends to two thermal reservoirs differing in temperature by δT. The heat flowing between the reservoirs is given by Eq. 17.19, where c is the central charge of the bulk. Note that heat always flows from the hot reservoir to the cold reservoir[27] even though the central charge can have either sign. The sign of the central charge simply determines in which direction (which chirality, clockwise or counterclockwise) around the edges the heat is flowing.

Thus, we have a relationship that $(1 + 1)$-dimensional conformal field theories exist at the edges of TQFTs with $c \neq 0$. However, the relationship is more universal than this. In fact, *every $(1 + 1)$-dimensional (unitary, rational)[28] conformal field theory defines a $(2 + 1)$-dimensional TQFT that it bounds!* An enormous amount is known about conformal field theories, and we will not venture to properly introduce the subject here (see instead Di Francesco et al. [1997] for a good introduction).

The coefficient in parenthesis in Eq. 17.19 is known as the thermal Hall coefficient[29] κ_{xy} in analogy with the electrical Hall coefficient σ_{xy} which we discussed[30] in Section 2.5 (see also Section 37.1).

Given two different TQFTs, say A and B, as described in Section 8.5, we can construct a product theory $A \times B$, which can roughly be thought of as just putting the two theories in the same place at the same time. The central charge of the two theories necessarily add since the total thermal current will just be the sum of the two thermal currents of the two theories.

The fact that the central charge in Eq. 17.16 is only defined modulo 8 has an interesting explanation: There exists a particular conformal field theory, the Chern–Simons theory $(E_8)_1$ (meaning we use the exceptional Lie group E_8 at level 1),[31] which has *no particles* in it except the identity

---

[26] The fact that TQFTs can (and sometimes must) have nontrivial boundaries is something we have hinted at before, for example, near Eq. 5.3. There we pointed out that Chern–Simons theory is not fully gauge invariant on a manifold with boundary. The cure for this illness is to supplement the Chern–Simons theory with a nontrivial edge theory, which restores gauge invariance, but at the same time gives edge dynamics as well. See for example Wen [1992] and Hansson et al. [2017].

[27] One must obey the second law of thermodynamics!

[28] "Unitary" here means that it properly describes quantum mechanics. "Rational" means that there are only a finite number of different particle types.

[29] It is also known as the Righi-Leduc coefficient, after Augusto Righi and Sylvestre Anatole Leduc, who independently discovered the thermal Hall effect in 1888.

[30] Strictly speaking, σ_{xy} is the ratio of current density in one direction to electric field in the perpendicular direction. In Section 2.5 we discussed total voltage drop divided by total current in the perpendicular direction.

---

[31] There is a beautiful description of this CFT using the K-matrix formalism of Section 5.3.2. We set K equal to an 8×8 matrix with (i) integer entries, (ii) unit determinant, (iii) all eigenvalues positive, and (iv) even entries along the diagonal. It turns out that eight is the smallest dimension matrix for which (i), (ii), (iii), and (iv) can all be true and the corresponding matrix is known as the E_8 matrix (the Cartan matrix of the E_8 Lie algebra). Since the matrix is even along diagonals there are no transparent fermions, as discussed in Exercise 5.6.d) and the theory is modular. Since this matrix has all positive eigenvalues, as discussed in Section 5.3.2, there are eight co-propagating edge modes (each having unit central charge). However, the fact that it has unit determinant means that it has a unique ground state and no nontrivial quasiparticle types.

particle I, but has central charge $c = 8$. Given any TQFT, call it A, we can construct the product theory $A \times (E_8)_1$ and it would have exactly the same particle content, and hence, the same result on the right of Eq. 17.16, but the central charge would be increased by 8 compared to the theory A. Similarly the opposite chirality theory $(E_8)_{-1} = \overline{(E_8)_1}$ has central charge $c = -8$. Such a TQFT that has no particle types is known as *invertable* topological order.[32]

[32]The order is "invertable" because $(E_8)_1$ can be completely erased (or inverted) by merging it with an $(E_8)_{-1}$.

17.4   Tables of TQFTs

As mentioned in Sections 9.3 and 13.3, anyon theories (or TQFTs) are extremely constrained by the pentagon and hexagon equations. Indeed, for any given set of fusion rules there are only a finite number of possible solutions up to gauge equivalence (Etingof et al. [2005]; see also the discussions of rigidity in Sections 9.3 and 13.3). Once one includes additional conditions, such as the theory being modular, the number of possible solutions drops even more. This makes it possible to build a table[33] of all possible TQFTs—that is, a complete list of all consistent modular solutions of pentagon and hexagon equations. The procedure for building this table is to hypothesize that there are only n different particle types (we call this the "rank" of the theory), with n a small number. With fixed n one can constrain the possible fusion rules, then find solutions of the pentagon equations, and finally solutions of the hexagon equations. For modular tensor categories this program has been carried out by Rowell et al. [2009] for up to four particle types.[34] Some of the key results from this table are presented in Table 17.1. In Chapter 18 we will explore some of the basic principles used for compiling such a table. The case of abelian theories has been essentially completed by Galindo and Jaramillo [2016] (see also Wang and Wang [2020] and Cano et al. [2014]), and we return to explain this briefly in Section 18.6.

[33]There is a temptation to call this a "periodic table," since, like the periodic table of the elements, it includes all possible cases. However, it is not really "periodic," so "table" is more appropriate.

[34]See also earlier work by Gepner and Kapustin [1995] as well as Bonderson [2007].

Extensions of the idea of building tables of TQFTs have been made in a number of directions. References to these works are given in the further reading.

17.5 Ω Strand (Kirby Color)

A particularly useful object to consider is a weighted sum of particle types known as an Ω strand,[35] or sometimes "Kirby color" strand,[36,37] as shown (purple) in Fig. 17.7.

[35]The Ω strand can be defined for any planar diagram algebra, even without a well-defined braiding.

[36]Sometimes the Ω strand is normalized with an additional factor of $\mathcal{D}$ or $\mathcal{D}^{-1}$ out front. See the definition of $\tilde{\Omega}$ in Fig. 17.11.

[37]One can also define Ω with a prefactor of $d_a/\mathcal{D}$ instead of $\mathsf{d}_a/\mathcal{D}$. This is appropriate if one is not applying rule 0, that is, if one is making a nonunitary evaluation (as discussed in Section 14.5). The killing property, Fig. 17.9, will then hold for the nonunitary evaluation of the diagram. See for example, note 17 in Chapter 24.

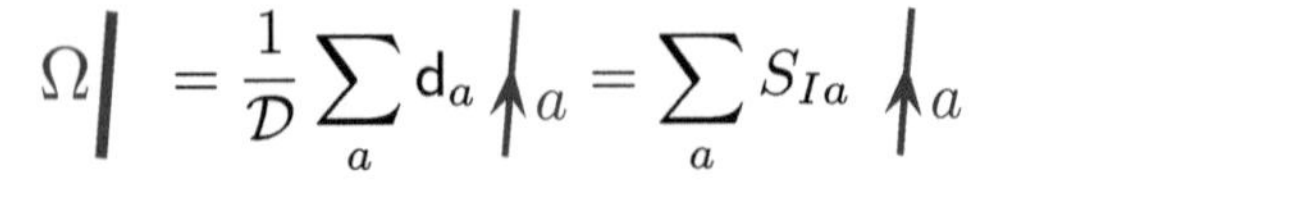

$$\Omega\, \Big| \;=\; \frac{1}{\mathcal{D}} \sum_a \mathsf{d}_a \,\Big\uparrow a \;=\; \sum_a S_{Ia}\, \Big\uparrow a$$

Fig. 17.7 A string of Kirby color (Ω strand) is a weighted superposition of all anyon string types.[36] Note that the Kirby color string does not have an arrow on it since it is an equal sum over all particles and their antiparticles. Here, d is the quantum dimension, S is the modular S-matrix, and I is the identity, or vacuum, particle.

Table 17.1

Particle types	Fusion rules	Examples	$c \bmod 8$	$h \bmod 1$
I		$(E_8)_1$	0	0
I, X	$X^2 = I$	$SU(2)_1 = \mathbb{Z}_2^{(\frac{1}{2})} = $ Semion	1	$0, \frac{1}{4}$
I, X	$X^2 = I + X$	$(G_2)_1 = $Fibonacci	14/5	$0, \frac{2}{5}$
I, X, Y	$X^2 = Y,\ XY = I,\ Y^2 = X$	$SU(3)_1 = \mathbb{Z}_3^{(1)} \simeq \overline{(E_6)_1}$	2	$0, \frac{1}{3}, \frac{1}{3}$
I, X, Y	$X^2 = I + Y,\ XY = X,\ Y^2 = I$	Ising $\simeq \overline{(E_8)_2}$	1/2	$0, \frac{1}{16}, \frac{1}{2}$
		$SU(2)_2$	3/2	$0, \frac{3}{16}, \frac{1}{2}$
		$(C_2)_1$	5/2	$0, \frac{5}{16}, \frac{1}{2}$
		$(B_3)_1$	7/2	$0, \frac{7}{16}, \frac{1}{2}$
I, X, Y	$X^2 = I + X + Y,\ XY = X + Y$ $Y^2 = I + X$	$SU(2)_5/\mathbb{Z}_2$	8/7	$0, \frac{2}{7}, \frac{6}{7}$
I, X, Y, Z	$X^2 = Z^2 = Y,\ XZ = Y^2 = I$ $XY = Z,\ ZY = X$	$(D_9)_1 \simeq \mathbb{Z}_4^{(\frac{1}{2})}$	1	$0, \frac{1}{8}, \frac{1}{2}, \frac{1}{8}$
		$SU(4)_1 = \mathbb{Z}_4^{(\frac{3}{2})}$	3	$0, \frac{3}{8}, \frac{1}{2}, \frac{3}{8}$
I, X, Y, Z	$X^2 = Y^2 = Z^2 = I$ $XY = Z,\ XZ = Y,\ YZ = X$	$D(\mathbb{Z}_2) = $ Toric Code	0	$0, 0, 0, \frac{1}{2}$
		$(D_4)_1$	4	$0, \frac{1}{2}, \frac{1}{2}, \frac{1}{2}$
		$SU(2)_1 \times SU(2)_1$	2	$0, \frac{1}{4}, \frac{1}{4}, \frac{1}{2}$
		$SU(2)_1 \times \overline{SU(2)_1}$	0	$0, \frac{1}{4}, \frac{3}{4}, 0$
I, X, Y, Z	$X^2 = Z^2 = I + X,\ Y^2 = I$ $XZ = Y + Z,\ XY = Z,\ YZ = X$	$(G_2)_1 \times SU(2)_1$	19/5	$0, \frac{2}{5}, \frac{1}{4}, \frac{13}{20}$
		$(G_2)_1 \times \overline{SU(2)_1} = SU(2)_3$	9/5	$0, \frac{2}{5}, \frac{3}{4}, \frac{3}{20}$
I, X, Y, Z	$X^2 = I + X,\ Y^2 = I + Y,\ XY = Z$ $XZ = Y + Z,\ YZ = X + Z,$ $Z^2 = I + X + Y + Z$	$(G_2)_1 \times (G_2)_1$	28/5	$0, \frac{2}{5}, \frac{2}{5}, \frac{4}{5}$
		$(G_2)_1 \times \overline{(G_2)_1}$	0	$0, \frac{2}{5}, \frac{3}{5}, 0$
I, X, Y, Z	$X^2 = I + X + Y,\ XY = X + Y + Z$ $XZ = Y + Z,\ Y^2 = I + X + Y + Z$ $YZ = X + Y,\ Z^2 = I + X$	$SU(2)_7/\mathbb{Z}_2 \simeq \overline{(G_2)_2}$	10/3	$0, \frac{2}{9}, \frac{2}{3}, \frac{1}{3}$

Table of Rank ≤ 4 Modular Categories. From Rowell et al. [2009]. Here G_2, E_8, C_2, B_3, D_4, D_9, and $SU(N)$ are Lie groups. Theories listed, such as $(G_2)_2$, are Chern–Simons theories with the last subscript indicating the level of the theory. Further details of these theories are easily obtained numerically, as discussed in Chapter 39. Details of the $\mathbb{Z}_N^{(n)}$ and $\mathbb{Z}_N^{(n+\frac{1}{2})}$ theories appear in Section 18.5.2. An overline on a theory indicates we should take the mirror image theory (c and h change signs), which is equivalent to changing the sign of the level. Here, c is the central charge and h is the topological spin such that the twist factor θ_a is $\text{sign}(d_a)\theta_a = e^{2\pi i h_a}$. The list of h values are given for the particles in order I, X, Y, Z. The theory $D(\mathbb{Z}_2)$ is the quantum double of $\mathbb{Z}_2$, known as the toric code, which we discuss extensively starting in Chapter 27. $SU(2)_7/\mathbb{Z}_2$ means start with the Chern–Simons theory $SU(2)_7$ and consider only the "even" subset of anyons, which form a complete theory by themselves (and similarly for $\overline{SU(2)_5}/\mathbb{Z}_2$). For every theory with $c \neq 0$ or 4 mod 8, one can obtain a distinct theory by taking the mirror image. The equivalences between theories marked with $\simeq$ is an equivalence of modular data—the central charges may differ by eight times an integer and h values may differ by integers between the two theories. Note that $(E_8)_1$ is not entirely trivial since it has $c = 8$, but it appears trivial (having no particles and $c = 0$) when we examine c mod 8 only. It happens that for this table we do not have any fusion multiplicity $N_{ab}^c > 1$.

This weighted sum will occur in several different contexts, including in Sections 24.3, 30.2.3, 31.4, 32.2, 33.1.3, 33.2.1, and 33.2.1.

The Kirby color strand allows for writing a number of interesting identities diagrammatically. For example, we can write the central charge mod 8 (Eq. 17.16) in terms of a loop of Ω with a curl as shown in Fig. 17.8.

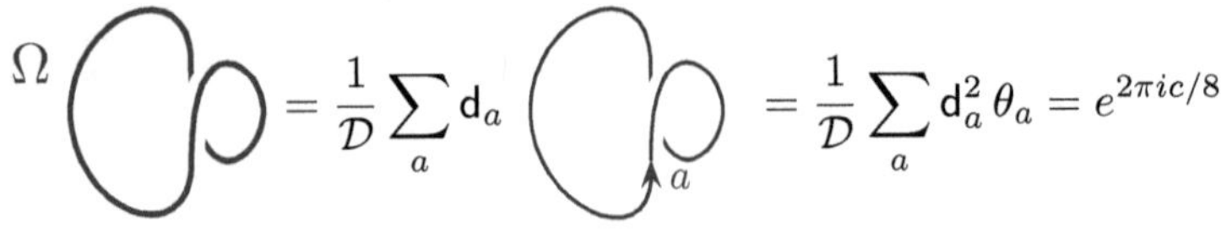

$$\Omega \quad = \frac{1}{\mathcal{D}} \sum_a \mathsf{d}_a \qquad = \frac{1}{\mathcal{D}} \sum_a \mathsf{d}_a^2\, \theta_a = e^{2\pi i c/8}$$

Fig. 17.8 A loop of Ω strand with a curl gives the central charge mod 8. The last step is Eq. 17.16.

The Ω strand has some additional interesting properties that we will find useful in later chapters. For example, in Fig. 17.9 we show the "killing property" of a loop of Ω for modular theories—a loop of Ω allows only the identity to go through it: all other particle types are "killed." A useful corollary of the killing property is given in Fig. 17.10. Given the factors of $\mathcal{D}$ that occur in these expressions, it is sometimes useful to define an Ω strand with a different normalization as in Fig. 17.11.

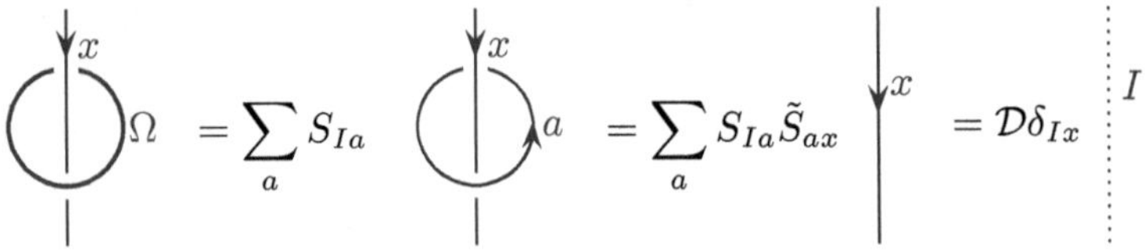

$$\Omega \quad = \sum_a S_{Ia} \qquad a = \sum_a S_{Ia}\tilde{S}_{ax} \qquad = \mathcal{D}\delta_{Ix}$$

Fig. 17.9 The killing property of the Ω strand for modular theories. A loop of Kirby color (Ω) allows no particles through it except the identity. In the final step we use unitarity of S (and the fact that $S_{Ia} = S_{Ia}^*$).

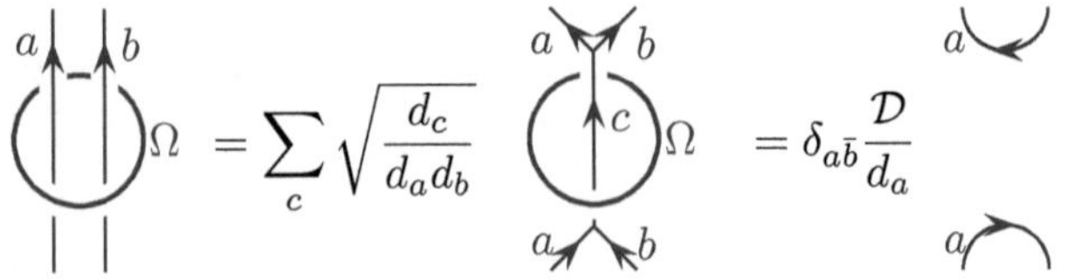

$$\Omega \quad = \sum_c \sqrt{\frac{d_c}{d_a d_b}} \qquad \Omega = \delta_{a\bar{b}}\frac{\mathcal{D}}{d_a}$$

Fig. 17.10 The Ω strand joins two lines due to the killing property, which we use in the second step to force c to be the identity.

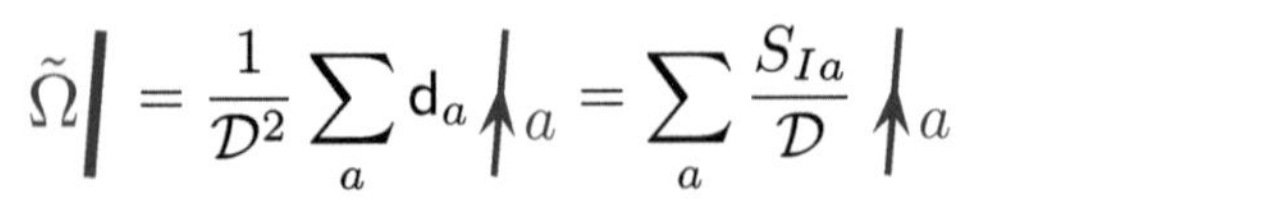

$$\tilde{\Omega} = \frac{1}{\mathcal{D}^2} \sum_a \mathsf{d}_a\, a = \sum_a \frac{S_{Ia}}{\mathcal{D}}\, a$$

Fig. 17.11 A different normalization of the Ω strand. Compare Fig. 17.7. A loop of $\tilde{\Omega}$ is normalized to have value unity.

17.6 **Still Further Structure**

Modular (and many non-modular) theories have a great deal of extra structure that we have not even touched on. The theories are obviously very highly constrained, so it is rather natural to expect that there will be many nontrivial relationships between the quantities we have discussed. A useful relationship assigned as Exercise 17.4 is[38]

$$S_{ab} = \frac{1}{\mathcal{D}} \sum_c N_{a\bar{b}}^c \frac{\theta_c}{\theta_a \theta_b} d_c = \frac{1}{\mathcal{D}} \sum_c \mathrm{Tr}[R_c^{\bar{b}a} R_c^{a\bar{b}}] d_c \qquad (17.20)$$

where the trace refers to cases where there are fusion multiplicities (otherwise the Rs are just scalars). For abelian theories, this formula comes out particularly simple

$$S_{ab} = \frac{1}{\mathcal{D}} \frac{\theta_{a \times \bar{b}}}{\theta_a \theta_b} \qquad \text{(abelian)} \qquad (17.21)$$

with $\mathcal{D}$ being the (positive) square root of the total number of particles in the theory.

Another interesting result is a theorem by Bantay [1997], which gives us the following nontrivial relationship between the Frobenius–Schur indicator κ_k of a particle k and the modular S-matrix:

$$\sum_{i,j} N_{ij}^k S_{0i} S_{0j} \left(\frac{\theta_i}{\theta_j} \right)^2 = \begin{cases} \kappa_k = \epsilon_k \mathrm{sign}(d_k) & k = \bar{k} \\ 0 & k \neq \bar{k} \end{cases} . \qquad (17.22)$$

Replacing the exponent 2 on the left of Eq. 17.22 with 3 yields the third Frobenius–Schur indicator, as defined in Fig. 14.19 (see Ng and Schauenburg [2007]).

A beautiful theorem by Vafa [1988] tells us that for any braided unitary theory (modular or not) all the spin factors θ_a must be an nth root of unity so that $\theta_a^n = 1$ where the integer n is determined only by the fusion multiplicity matrices N_{ab}^c. For example, we have

$$\prod_b \theta_b^{X_{ab}} = 1 \qquad (17.23)$$

with

$$X_{ab} = -2N_{a\bar{a}}^b N_{a\bar{a}}^{\bar{b}} - N_{aa}^b N_{\bar{a}\bar{a}}^{\bar{b}} + 4\delta_{ab} \sum_q N_{aa}^q N_{\bar{a}\bar{a}}^{\bar{q}} .$$

17.7 **Fermions and Super-Modular Theories**

Most of our discussion has been devoted to modular theories—TQFTs with unitary S-matrices. Modular theories are "well-behaved" TQFTs. However, as mentioned in Section 4.3.2, even some very simple anyon theories are not modular. The typical reason why a theory fails to be modular is that it contains a transparent fermion f such that[39]

[38] Don't miss the fact that the subscript on N is $\bar{b}$ not b (!) See note 8 of this chapter.

[39] It is of course possible to imagine theories with more transparent particles. The case where all particles are transparent is known as a symmetric tensor category. See the discussion in Section 20.4.

Table 17.2

Particle types	Fusion rules	Examples	$c \bmod 1/2$	$h \bmod 1$
I, f, X, fX	$f^2 = I,\ X^2 = I + X + fX$	$SU(2)_6/\mathbb{Z}_2$	$1/4$	$0, \frac{1}{2}, \frac{1}{4}, \frac{3}{4}$
I, f, X, Y, fX, fY	$f^2 = I,\ X^2 = I + X + Y$	$SU(2)_{10}/\mathbb{Z}_2$	0	$0, \frac{1}{2}, \frac{1}{6}, \frac{1}{2}, 0, \frac{2}{3}$
	$XY = X + Y + fY$			
	$Y^2 = 1 + X + Y + fX + fY$			

Table of Rank ≤ 6 Non-Split Super-Modular Categories. From Bruillard et al. [2020]. Note that the central charge, which is inherited from the modular extension of the theory, is shown only mod $1/2$ since multiple modular extensions exist, but they all have the same central charge mod $1/2$. Not all fusion rules are given, but the remainder can be easily found by associativity. See also the comments on Table 17.1.

$$f \times f = I \qquad \theta_f = -1 \ . \tag{17.24}$$

The simplest such theory is a TQFT that contains only the identity I and the fermion f. This theory is sometimes known as sVec.[40] The S- and T-matrices for sVec are given by

$$S_{\mathrm{sVec}} = \frac{1}{\sqrt{2}} \begin{pmatrix} 1 & 1 \\ 1 & 1 \end{pmatrix} \qquad T_{\mathrm{sVec}} = \begin{pmatrix} 1 & 0 \\ 0 & -1 \end{pmatrix} \ . \tag{17.25}$$

A theory where there is a single transparent particle in addition to the identity, and this transparent particle is a fermion (satisfies Eq. 17.24), is known as "super-modular." All super-modular theories have the nice property that $S = \underset{\sim}{S} \otimes S_{vec}$ and $T = \underset{\sim}{T} \otimes T_{vec}$ for some matrices $\underset{\sim}{S}$ and $\underset{\sim}{T}$ and some version of the Verlinde formula exists for $\underset{\sim}{S}$ (see Bruillard et al. [2017] and Bruillard et al. [2020]).

Sometimes super-modular categories are simple product theories with sVec:

$$\mathcal{C}_{\mathrm{split-super-modular}} = \mathcal{C} \times \mathrm{sVec}$$

where $\mathcal{C}$ is some modular theory, in which case $\underset{\sim}{S}$ and $\underset{\sim}{T}$ are simply the S- and T-matrices of $\mathcal{C}$. Theories of this form are known as "split." Super-modular theories that do not factorize this way are called "non-split."

Tables have been constructed of all possible super-modular categories (see references at the end of the chapter). Even more so than the modular case, there are a rather limited number of theories with low number of particles in them (i.e. of low "rank"). For example, up to six particles, the only non-split super-modular theories are $SU(2)_{4k+2}/\mathbb{Z}_2$ with $k = 1, 2$ (also called "$PSU(2)_{4k+2}$"). We write some of their properties in Table 17.2. Here (as in Table 17.1) the notation $/\mathbb{Z}_2$ means one should take only the "even" subset of particles from the theory. In fact, sVec is also of this form with $k = 0$. To see this explicitly we consider $SU(2)_2$ which has three particle types (see Table 17.1) with topological spins 0, $\frac{3}{16}$, and $\frac{1}{2}$. The particle with spin $\frac{1}{2}$ is a fermion f with $f \times f = I$. The "even" subalgebra includes only this fermion and the identity (with spin 0). Thus, $SU(2)_2/\mathbb{Z}_2$ is exactly sVec.

---

[40] Sometimes sVec is also denoted as $\mathcal{F}_0$. sVec stands for "super vector space."

As might be suggested by the case of $SU(2)_{4k+2}/\mathbb{Z}_2$, a very convenient way of understanding super-modular theories is to find a modular extension of the super-modular theory—that is, a modular theory that fully contains the super-modular theory as a subalgebra. It has been conjectured (Bruillard et al. [2020]) that given a super-modular category $\mathcal{C}$ there always exists a extension $\mathcal{C}_{ext}$ such that the total quantum dimensions satisfy

$$\mathcal{D}^2_{\mathcal{C}_{ext}} = 2\mathcal{D}^2_{\mathcal{C}} \quad . \tag{17.26}$$

Unfortunately, such a modular extension is not unique. Indeed, it has been conjectured (Bruillard et al. [2017]) that there are exactly 16 different modular extensions satisfying Eq. 17.26 and these different extensions have central charges that differ by multiples of $1/2$ (this is explored further in Exercise 25.7). For example, if we consider the super-modular category sVec, we can extract from Table 17.1 16 different theories that all have $\mathcal{D}^2 = 4$ and all contain sVec as a subcategory. These theories, known as the 16-fold way (a phrase coined by Kitaev [2006]) are all shown in Table 17.3.

Analogous to the case of the $(E_8)_1$ theory for modular categories, there exists an "invertable" topological order for super-modular theories—sometimes this is known as $p_x \pm ip_y$ order in reference to the superconducting state with this type of symmetry (see Wen [2017]). This means that given a super-modular theory one can immediately generate another one with identical properties, except with a central charge increased or decreased by $1/2$ (see Exercise 25.7). Thus, the central charge in Table 17.2 is given only modulo $1/2$.

Super-modular categories are extremely common in physical systems. In fact, *all* fractional quantum Hall systems made of one species of electron[41] are super-modular, where the electron itself is the transparent fermion. Given that the fractional quantum Hall effect is the best physical example we have of anyon theories, super-modular categories are particularly important to consider carefully. We discuss the fractional quantum Hall effect in a bit more depth in Section 37.1.

The fact that the fermion is transparent has some ramifications in quantities such as the ground state degeneracies[42]. The general rule (which we found as far back as Eq. 7.9) is usually stated that the ground-state degeneracy on the torus should be equal to the number of particle species. However, the transparent fermion does not "count" in this bookkeeping. Indeed, we've already seen a simple example of this! In Section 4.3 when we considered the charge–flux model of anyons we found that for elementary anyons with statistical angle $\theta = \pi p/m$ with p and m relatively prime, the ground-state degeneracy on a torus is m. If pm is even, then we have a modular theory, there are m particle species, and the ground-state degeneracy on a torus is m. However, if pm is odd

theory	c mod 8
Toric Code $= D(\mathbb{Z}_2)$	0
Ising $\simeq \overline{(E_8)_2}$	1/2
$(D_9)_1$	1
$SU(2)_2$	3/2
$SU(2)_1 \times SU(2)_1$	2
$(C_2)_1$	5/2
$SU(4)_1$	3
$(B_3)_1$	7/2
$(D_4)_1$	$4 = -4$
$\overline{(B_3)_1}$	-7/2
$\overline{SU(4)_1}$	-3
$\overline{(C_2)_1}$	-5/2
$\overline{SU(2)_1} \times \overline{SU(2)_1}$	-2
$\overline{SU(2)_2}$	-3/2
$\overline{(D_9)_1}$	-1
$\overline{\text{Ising}} \simeq (E_8)_2$	-1/2

Table 17.3 The 16-fold way. All of these modular theories have total quantum dimension $\mathcal{D}^2 = 4$. All of these theories contain sVec, the simplest super-modular theory, as a subalgebra.

[41] One species of electron means spin-polarized electrons without valley degeneracies. This is the case for a large fraction of known fractional quantum Hall states.

---

[42] One must be cautious when a theory has a (Dirac) fermion since this forces one to define a "spin-structure" on the manifold. In two dimensions, this amounts to specifying whether the boundary condition for this fermion are either periodic (P) or antiperiodic (A) around each nontrivial cycle (see the discussion in Ginsparg [1990]). Generally, we simply consider the periodic case, but strictly speaking we should worry about the other sectors as well.

one finds a transparent fermion, and we have a super-modular theory with $2m$ species of particles, yet the ground-state degeneracy remains m. Effectively the particles which are related to each other by fusion of the fermion (a and $a \times f$) should be treated as a single particle for the counting of the ground-state degeneracy!

17.8 Appendix: Perron–Frobenius Theorem

The Perron–Frobenius theorem states that, for a matrix with all positive entries, there is a unique eigenvector with all positive entries (up to multiplication by an overall constant) known as the Perron–Frobenius eigenvector. The corresponding eigenvalue is the largest-magnitude eigenvalue, and is positive. Further, for a unitarily diagonalizable non-zero matrix with all non-negative entries, if there exists an eigenvector with all positive entries (up to multiplication by an overall constant), its eigenvalue is positive and is of magnitude greater than or equal to all other eigenvalues. The application of the Perron–Frobenius theorem to the fusion matrix N_a is a bit tricky since the theorem is stronger when the matrix being considered has strictly positive entries but the N_a matrices are only guaranteed to have non-negative entries. To avoid this problem, construct an arbitrary sum of the N_a matrices $M = \sum_a \alpha_a N_a$ where all the coefficients α_a are positive. Since all the N_a's have common eigenvectors (because they are normal matrices and they commute with each other; see Section 8.3.1), these are also the eigenvectors of the matrix M. Further, all the elements of M are strictly positive,[43] so we may apply the Perron–Frobenius theorem for positive definite matrices to M. We thus obtain a Perron–Frobenius eigenvector of M with strictly positive entries (up to a multiplicative constant). But the eigenvectors of N_a match those of M so we have an eigenvector of N_a with all positive entries which then must be the Perron–Frobenius eigenvector whose eigenvalue is greater than or equal to any eigenvalue of N_a.

[43] M_b^c is positive if any a exists so that $c \in a \times b$ so that $[N_a]_b^c > 0$. However, for any b, c we can construct $a = \bar{b} \times c$ so that $c \in a \times b$.

17.9 Appendix: Algebraic Derivation of the Verlinde Form

In this section we show that we do not need the structure of braiding in order to derive an equation of the Verlinde form, analogous to Eq. 17.13. Let us begin by recalling from Section 8.3.1 that for any topological theory[44],[45] the fusion rules are described by fusion multiplicity matrices N_{ab}^c, which can be viewed as a set of square normal matrices N_a with indices b and c. As discussed in Section 8.3.1 these normal matrices commute with each other and therefore can be simultaneously diagonalized by a matrix that we will call U (see Eq. 8.20):

$$N_a = U \lambda^{(a)} U^\dagger \tag{17.27}$$

[44] Here we can mean any planar algebra (unitary fusion category) or any $(2 + 1)$-dimensional topological theory with a braiding (unitary braided fusion category).

[45] All we actually need is a commutative fusion ring with a unique identity and inverses.

where $\lambda^{(a)}$ is a diagonal matrix for each a. We note again that we will discover below that for a so-called modular braided theory we will find that U is the modular S-matrix. More generally we call U the *mock S-matrix*.

From Eq. 17.27, the columns of U are the simultaneous eigenvectors of the N matrices which we can make explicit as

$$\sum_c [N_a]_b^c U_{cd} = U_{bd} \lambda_d^{(a)} \tag{17.28}$$

and no sum on d implied. Note, at this point, the columns of U may be multiplied by an arbitrary phase (i.e. a phase redefinition of the eigenvectors).

Since there is a particle type called the vacuum I (or identity) which fuses trivially with all other particles, we have $[N_a]_I^c = \delta_a^c$ so we have

$$U_{ad} = \sum_c [N_a]_I^c U_{cd} = U_{Id} \lambda_d^{(a)}$$

so that

$$\lambda_d^{(a)} = \frac{U_{ad}}{U_{Id}} . \tag{17.29}$$

Substituting back into Eq. 17.27 we get the Verlinde formula

$$[N_a]_b^c = \sum_x U_{bx} \frac{U_{ax}}{U_{Ix}} U_{cx}^* . \tag{17.30}$$

This result is extremely general[45].

17.10 Appendix: Algebraic Derivation that Quantum Dimensions Form a Representation of the Fusion Algebra

Recall that the columns of U are the simultaneous eigenvectors of the N_a matrices. Invoking the Perron–Frobenius theorem, there must be a particular index z such that[46] the eigenvector U_{bz} has all positive entries (up to an overall multiplicative constant). This is a common eigenvector of all the N_a matrices, and the corresponding eigenvalues are $\lambda_z^{(a)}$. By the Perron–Frobenius theorem, since the eigenvector has all positive entries, $\lambda_z^{(a)}$ must be the largest eigenvalue of N_a. Recalling (Eq. 8.17) that the quantum dimension can also be defined as the largest eigenvalue of N_a, we have

$$\mathsf{d}_a = \lambda_z^{(a)} .$$

Now let us multiply the Verlinde relation Eq. 17.30 on both sides by

[46]This index z must be I in well-behaved modular anyon theories, but more generally in fusion rings it could be another index (Gannon [2003]).

$d_c = \lambda_z^{(c)} = U_{cz}/U_{Iz}$ (see Eq. 17.29) and sum over c. We have

$$
\begin{aligned}
\sum_c N_{ab}^c d_c &= \sum_{x,c} U_{bx} \frac{U_{ax}}{U_{Ix}} U_{cx}^* \frac{U_{cz}}{U_{Iz}} \\
&= \sum_x U_{bx} \frac{U_{ax}}{U_{Ix} U_{Iz}} \delta_{xz} \\
&= \frac{U_{bz}}{U_{Iz}} \frac{U_{az}}{U_{Iz}} = d_a d_b
\end{aligned}
$$

where we have used the fact that U is unitary. Thus, we conclude

$$
d_a d_b = \sum_c N_{ab}^c d_c \quad .
$$

Chapter Summary

- The loop weights $|d_a|$ are the quantum dimensions d_a.
- The diagrammatic expression for the S-matrix element S_{ab} is just two linked rings, one labeled a and one labeled b.
- The Verlinde formula connects the fusion rules to the S-matrix for modular theories. For the non-modular case, there is a "Verlinde-like" relation using the "mock" S-matrix.
- The S- and T-matrices describe modular transformations of a torus.
- TQFTs have an enormous amount of structure that allows us to make tables of all possible theories starting with those having very few particle types.

Further Reading:

- Kitaev [2006] gives a detailed discussion of the modular relations (and fills in some of the steps we have left out). Moore and Read [1991] established the relationship between bulk and edge theories in the quantum Hall context. This relation is reviewed in the conformal field theory context in Nayak et al. [2008] and Hansson et al. [2017]. Simon in Halperin and Jain [2020] gives a modern view of these relations mostly outside the context of conformal field theory.
- The effort to build tables of anyon theories started with Gepner and Kapustin [1995]. The table from Rowell et al. [2009] was extended to five particle types in Bruillard et al. [2015]; Hong and Rowell [2010], and to six in Creamer [2019] and Green [2019]. The modularity condition was relaxed to build a table of all unitary braided theories for up to four particles in Bruillard [2016] and five

in Bruillard and Ortiz-Marrero [2018]. A table for theories with
fermions was built in Bruillard et al. [2020] (see also Bruillard et al.
[2017]). There have also been tables built by Lan et al. [2017, 2016]
and Wen [2015].

- Various online resources are available that list all possible (small)
theories with certain nice properties. See Chapter 39 for a list
of such resources, and particularly for a discussion of the useful
computer program *Kac*.

Exercises

Exercise 17.1 Fibonacci S-matrix
Recall the Fibonacci theory introduced in Sections 8.2.1 and 10.2.1.
(a) First let us pretend that we have not calculated the R-matrices or θ_τ,
that is, we do not know the braiding phases or the twist factors. We only know
the fusion rules $\tau \times \tau = I + \tau$. Using the quantum dimensions, we can obtain
three out of four elements of the 2×2 S-matrix. Determine the remaining
element of the S-matrix by enforcing unitarity.
(b) Given the twist factor $\theta_\tau = e^{\pm 4\pi i/5}$ (with $\pm$ being for right- or left-
handed theories), calculate the S-matrix explicitly by using Eq. 17.20.

Exercise 17.2 Using the Pivotal Property
Use the pivotal property (Section 14.8.1) to demonstrate the identity shown
in Fig. 17.12. You should not assume full isotopy invariance. Nor should you
assume $\epsilon = +1$ for any of the particles.

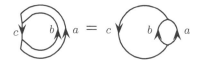

Fig. 17.12 This identity can be shown
without full isotopy invariance by using
the pivotal property.

Exercise 17.3 Symmetries of S
Use isotopy of diagrams and Hermitian conjugation of diagrams to show
the identities in Eq. 17.7.

Exercise 17.4 Evaluation of the S-link
(a) Use the R-matrices and Eq. 15.5 to derive the value of the matrix
$\tilde{S}_{ax}$ (see Fig. 17.2) in terms of fusion multiplicities, twist factors θ_a, and the
quantum dimensions d_a.
(b) From your result show that

$$ \left(\bigcirc a \right) b = \sum_c N_{ab}^c \frac{\theta_c}{\theta_a \theta_b} d_c \quad . $$

Note that this diagram differs from S_{ab} by a factor of $Z(S^3) = 1/\mathcal{D}$.

Exercise 17.5 Theories With One Nontrivial Particle
Consider an anyon theory with only the identity and one nontrivial particle
type s having twist factor θ_s. The only possible fusion rules are $s \times s = I + ms$
for some integer m (the semion model is $m = 0$, the Fibonacci model is $m = 1$).
Calculate d_s and $\mathcal{D}$ from the fusion rules. Use Eq. 17.20 to calculate the S-
matrix in terms of θ_s. Show that this matrix cannot be unitary for any $m > 1$.

This justifies that on Table 17.1 there are only two types of theories with one nontrivial particle.

Exercise 17.6 Product Theories[Easy]
Given two anyon theories A and B with corresponding S-matrices S_A and S_B:
(a) Show that the product theory $A \times B$ has S-matrix $S_A \otimes S_B$.
(b) Show that $A \times B$ is modular if and only if both A and B are modular.
(c) Show that the central charge of the product theory is the sum of the central charges of the constituent theories. That is,

$$c_{A \times B} = (c_A + c_B) \bmod 8 \quad . \tag{17.31}$$

In fact, central charges strictly add in product theories.

Exercise 17.7 Ground-State Degeneracy Redux
(a) Use the Verlinde formula to write the ground-state degeneracy on a g-handled torus (Eq. 8.33 in Exercise 8.2.a) as

$$\mathrm{Dim}V(g\text{-handled torus}) = \sum_x (S_{Ix})^{2-2g} \quad .$$

This result was first written down by Verlinde [1988].
(b) Consider the case of a g-handled torus where there are also m different particles on the surface of the manifold with quantum numbers $a_1, \ldots, a_m$. Show that

$$\mathrm{Dim}V(g\text{-handled torus with particles } a_1, \ldots, a_m \text{ on the surface})$$
$$= \sum_x S_{a_1,x} S_{a_2,x} \ldots S_{a_m,x} (S_{Ix})^{2-2g-m} \quad . \tag{17.32}$$

Hint: Solve Exercise 8.2.d first.
This result was first written down in Moore and Seiberg [1989b] citing also T. Banks. Note that the case of $m = 3$ and $g = 0$ is actually just the regular Verlinde formula!

Exercise 17.8 Unlinking $\tilde{S}$-Matrix
Show that the unlinking matrix $\tilde{S}_{ax}$ has the property that

$$\tilde{S}_{ax} \leq \tilde{S}_{Ix}$$

with I the vacuum.

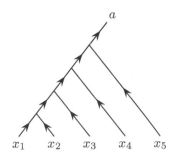

Fig. 17.13 Fusing many anyons with labels $x_1, x_2, \ldots$ to a final result a.

Exercise 17.9 Probability of Fusion Channels
(a) Consider fusion trees with many incoming anyons $x_1, x_2, \ldots, x_N$ as shown in Fig. 17.13 (but assume more incoming anyons than shown in the figure). For a fixed set of incoming anyons at the bottom, count up all possible fusion trees (i.e. all possible labeling of the internal lines in the tree) and show that the number of possible fusion trees with output a at the top is

$$\text{Number of trees} \approx C(x_1, \ldots, x_N) \, \mathsf{d}_a$$

where C is some constant dependent on the incoming $x_1, \ldots, x_N$ but independent of a. You might find it convenient to assume the anyon theory is modular, but in fact you do not need this.

(b) Consider a modular anyon theory on a sphere with a very large number of quasiparticles of all types. Divide these anyons randomly into two large groups. Show that the probability that the two groups have overall fusion channels a and $\bar{a}$ is given by

$$p(a, \bar{a}) = \mathsf{d}_a^2/\mathcal{D}^2 \quad .$$

(c) Now divide the anyons randomly into three large groups. Show that the probability that the groups have overall fusion channels a, b, c is given by

$$p(a, b, c) = N_{ab}^{\bar{c}} \mathsf{d}_a \mathsf{d}_b \mathsf{d}_c / \mathcal{D}^4$$

or equivalently the probability that the groups a, b, and c fuse together to the vacuum in a particular fusion channel $\mu \in 1 \ldots N_{ab}^{\bar{c}}$ is given by

$$p(a, b, c, \mu) = \mathsf{d}_a \mathsf{d}_b \mathsf{d}_c / \mathcal{D}^4 \quad .$$

(d) Finally, try four large groups. Show that the probability that the groups have overall fusion channels a, b, c, d is given by

$$p(a, b, c, d) = N_{abcd} \mathsf{d}_a \mathsf{d}_b \mathsf{d}_c \mathsf{d}_d / \mathcal{D}^6$$

where $N_{abcd} = \sum_e N_{ab}^e N_{ec}^{\bar{d}}$ is the number of ways a, b, c, and d can fuse to the vacuum.

Part IV

Some Examples: Planar Diagrams and Anyon Theories

Some Simple Examples

In this chapter we consider a few simple examples of anyon theories. Our general strategy, shown in Fig. 18.1, will recapitulate the main content of Chapters 8–17,

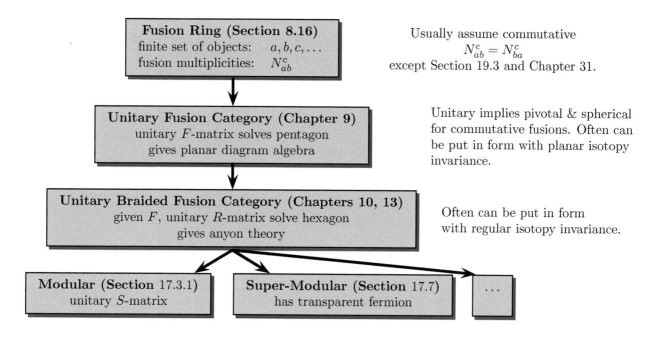

Fusion Ring (Section 8.16)
finite set of objects: $a, b, c, \ldots$
fusion multiplicities: N_{ab}^c

Usually assume commutative
$$N_{ab}^c = N_{ba}^c$$
except Section 19.3 and Chapter 31.

Unitary Fusion Category (Chapter 9)
unitary F-matrix solves pentagon
gives planar diagram algebra

Unitary implies pivotal & spherical for commutative fusions. Often can be put in form with planar isotopy invariance.

Unitary Braided Fusion Category (Chapters 10, 13)
given F, unitary R-matrix solve hexagon
gives anyon theory

Often can be put in form with regular isotopy invariance.

Modular (Section 17.3.1)
unitary S-matrix

Super-Modular (Section 17.7)
has transparent fermion

. . .

Fig. 18.1 The general structure of anyon theories. The side comment that we can "often" put theories in a form with isotopy invariant is discussed at length in Chapter 14. In this chapter we focus on such theories without obstruction to isotopy invariance. The ... at the bottom right represents braided theories that are neither modular nor super-modular.

Let us summarize the content of this diagram. First, we decide on a set of fusion rules. From this we examine the possible planar diagram algebras. To a mathematician these are known as spherical tensor categories (see Section 14.8.2). Once we have found the possible planar algebras we look for possible braidings to build full anyon theories.

In this chapter we focus on theories that enjoy full isotopy invariance ($\epsilon_a = +1$). If there are nontrivial Frobenius–Schur indicators we use negative d_a as discussed in Chapter 16.

18.1 $\mathbb{Z}_2$ **Fusion Rules**

Let us start with the simplest system of particles we can imagine: an identity 0 and a nontrivial particle 1. The simplest fusion rules we can have are[1]

$$1 \times 1 = 0 \qquad (18.1)$$

which tells us that 1 is its own antiparticle $1 = \bar{1}$ so we do not draw arrows on the corresponding line. Eq. 18.1 is known as $\mathbb{Z}_2$ fusion rules and is shown in Fig. 18.2. The corresponding fusion multiplicity matrix is $N^0_{11} = N^0_{00} = N^1_{10} = N^1_{01} = 1$ and $N^1_{11} = N^1_{10} = N^0_{01} = N^1_{00} = 0$. Note that since all particles in this theory are self-dual, we can raise and lower indices on the multiplicity matrix freely, $N^c_{ab} = N_{abc}$. It is convenient to work with the lowered indices because N_{abc} is invariant under any permutation of the indices. Thus, we can write the fusion multiplicity matrices more concisely as $N_{110} = 1$ and $N_{111} = N_{100} = 0$.

With 0 being the identity, the only nontrivial vertices we can have with these fusion rules is where one particle 1 comes in and one particle 1 also goes out, as shown in Fig. 18.3. If one does not draw the identity particle, diagrams must then be just a so-called *loop gas*, as shown in Fig. 18.4. The constraint $N_{100} = 0$ means that loops cannot end, and $N_{111} = 0$ means that loops cannot intersect.

[1]We have switched notations—here the vacuum is 0 not I, and the nontrivial particle is 1. I hope this does not cause confusion!

Fig. 18.2 Fusing two 1 particles to the vacuum, shown in two notations.

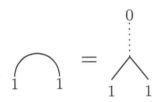

Fig. 18.3 Examples of allowed vertices for the $\mathbb{Z}_2$ fusion rules. A 1 particle (drawn solid) comes into the vertex and the 1 particle must also go out of the same vertex. The 0 particle, the identity, is drawn dotted, but it need not be drawn at all.

Fig. 18.4 A loop gas has $\mathbb{Z}_2$ fusion rules. The loop gas drawn here is planar—there are no over- or under-crossings.

Since the largest eigenvalue of $[N_a]^c_b$ is 1, we have quantum dimensions $d_1 = d_0 = 1$ (see Eq. 8.17).

The $\mathbb{Z}_2$ loop gases were studied in Exercise 2.2 (where we allowed over- and under-crossings in addition to just planar diagrams), and we consider them again in Section 22.1.

With these fusion rules, there are two sets of F-matrices that give consistent isotopy-invariant planar algebras. These two solutions correspond to a loop weight $d_1 = \pm 1$, which are the only two options given that $d_1 = |d_1| = 1$.

18.1.1 $d = +1$ **Loop Gas**

Here, we choose loop weight $d_1 = +1$ for the nontrivial particle, in which case every F which is non-zero is +1 (i.e., every F diagram where all vertices are consistent with the fusion rules. See Fig. 16.4). In other words, $F_{ecf}^{bad} = 1$ for every case where $N_{abd} = N_{ced} = N_{bcf} = N_{aef} = 1$ and F is zero otherwise. We can write out explicitly the non-zero elements

$$\begin{align}
F_{000}^{000} &= 1/d_0 = 1 & (18.2)\\
F_{110}^{110} &= 1/d_1 = 1 & (18.3)\\
F_{001}^{110} &= F_{111}^{000} = 1 & (18.4)\\
F_{101}^{101} &= F_{011}^{011} = 1 & (18.5)\\
F_{010}^{101} &= F_{100}^{011} = 1 \quad . & (18.6)
\end{align}$$

The first two lines are required from Eq. 16.4. Equation 18.4 is from Eq. 16.7. Equation 18.5 and Eq. 18.6 can be derived from Eq. 18.3 and Eq. 18.4 by the tetrahedral symmetry equation Eq. 16.17. Examples of these F-moves are shown in Figs. 18.5 and 18.6.

The $d = 1$ planar loop gas turns out to be a trivial diagrammatic algebra. The value of every allowed diagram is unity, or is zero if there is anything disallowed in the diagram, such as the intersection of loops.

We now turn to consider the possible braidings that we can impose on this planar algebra. The only nontrivial R-matrix element is R_0^{11}. Using the hexagon equation (Eq. 16.20, setting $a = b = c = d = 1$) we obtain

$$[R_0^{11}]^2 F_{110}^{110} = [F_{110}^{110}]^2 R_1^{10} \quad . \qquad (18.7)$$

The R on the right is unity, and the Fs are all unity. Thus,

$$[R_0^{11}]^2 = 1 \quad . \qquad (18.8)$$

This limits us to two possible anyon theories for the $d = +1$ loop gas.

Boson Theory

We choose the $R_0^{11} = +1$ case. This gives us no phases or signs with F-moves or braiding.[2] The corresponding twist factor (via Eq. 15.2) is $\theta_a = +1$, which corresponds to a boson. This theory is also denoted as $\mathbb{Z}_2^{(0)}$ in Section 18.5.3.

Fermion Theory

We choose the $R_0^{11} = -1$ case. This gives us minus sign under exchange of identical particles. The corresponding twist factor (via Eq. 15.2) is $\theta_a = -1$ which corresponds to a fermion. In fact, this is precisely the sVec fermion theory we encountered in Section 17.7, and it is denoted as $\mathbb{Z}_2^{(1)}$ in Section 18.5.3.

For both bosons and fermions the S-matrix describing the braiding is

$$S = \frac{1}{\sqrt{2}} \begin{pmatrix} 1 & 1 \\ 1 & 1 \end{pmatrix} \qquad (18.9)$$

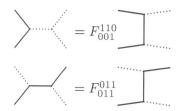

Fig. 18.5 These F-moves for the $\mathbb{Z}_2$ loop gas simply deform the path of the particles. These are known as "isotopy" moves.

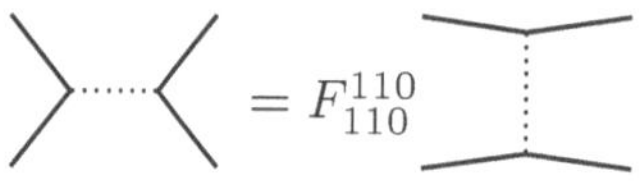

Fig. 18.6 This F-move for the $\mathbb{Z}_2$ loop gas reconnects the paths of particles. This is known as a "surgery" move.

[2] Since the boson braids identically to the vacuum, by some definitions it is just another copy of the vacuum. One should specify carefully if we want to think of this as a separate particle or not.

which is not unitary, so neither of these two cases are modular,[3] although the fermion case is, in fact, super-modular (see Section 17.7) and is what we called sVec previously.

18.1.2 $d = -1$ Loop Gas

Here, we choose the loop weight $d_1 = -1$ for the nontrivial particle, in which case every F that is consistent with the fusion rules is ± 1. The signs of the non-zero elements of F are given as follows:

$$F_{000}^{000} = 1/d_0 = 1 \tag{18.10}$$
$$F_{110}^{110} = 1/d_1 = -1 \tag{18.11}$$
$$F_{001}^{110} = F_{111}^{000} = 1 \tag{18.12}$$
$$F_{101}^{101} = F_{011}^{011} = 1 \tag{18.13}$$
$$F_{010}^{101} = F_{100}^{011} = 1 \ . \tag{18.14}$$

As with the $d = +1$ loop gas, the first two lines are required from Eq. 16.4. Equation 18.4 is from Eq. 16.7. Equation 18.5 and Eq. 18.6 can be derived from Eq. 18.3 and Eq. 18.4 by the tetrahedral symmetry equation Eq. 16.17. Note in particular how the signs work in Fig. 16.17 in the definition of the tetrahedral diagram.

It is worth looking at the two different signs that F can take. (If necessary, refer back to Fig. 16.4 for details of how the F-matrix is defined). Moves such as shown in Fig. 18.5 simply deform the path of the particle and do not incur a sign. However, the moves shown in Fig. 18.6 perform "surgery" on the paths and reconnect loops and do change the sign. Such a surgery always changes the number of loops in the diagram by one. The value of any loop diagram is thus given by

$$\begin{array}{c}\text{Value of } (d = -1)\\ \text{loop diagram}\end{array} = (-1)^{\text{number of loops}} \ . \tag{18.15}$$

As discussed in detail in Section 14.2 (see also Section 16.1.3), while this is a perfectly consistent planar diagrammatic algebra, it has non-positive definite inner products and therefore is not appropriate for describing quantum mechanics. That is, this is the *nonunitary* evaluation of the diagram in the language of Section 14.5.

However, as discussed in Section 14.5 we can make a proper unitary theory out of the $d = -1$ loop gas just by implementing rule 0. So the unitary evaluation of a diagram is

$$\begin{array}{c}\text{Value of } (d = -1)\\ \text{loop diagram}\\ \text{including rule 0}\end{array} = (-1)^{\text{number of loops+number of caps}} \ . \tag{18.16}$$

For example, in Fig. 18.4 there are 10 loops and 14 caps, so the full value of the diagram is $+1$.

The nontrivial particle here has a negative Frobenius–Schur indicator.

Here we have chosen to do our bookkeeping by working with a negative loop weight d, maintaining isotopy invariance of diagrams (with $\epsilon = +1$) and implementing rule 0 to obtain a unitary theory. Equivalently, we could have worked with $\epsilon = -1$ and $d = +1$ and associated a minus sign for a spacetime zig-zag such as Fig. 14.5.

Semions

We now consider possible braidings for the $d = -1$ loop gas. As in the case $d = +1$, we can apply the hexagon equation to obtain Eq. 18.7. In this case, however, $F_{110}^{110} = -1$ so we obtain

$$[R_0^{11}]^2 = -1 \quad.$$

Again, there are two solutions $R_0^{11} = \pm i$ corresponding to right- and left-handed semions. In either case, wrapping a semion all the way around another gives -1, so the S-matrix is given by

$$S = \frac{1}{\sqrt{2}} \begin{pmatrix} 1 & 1 \\ 1 & -1 \end{pmatrix} \quad. \tag{18.17}$$

This is unitary, telling us that the theory is modular. For the right-handed semion theory $R_0^{11} = i$. If we continue to work with negative d, then this right-handed semion has twist factor $\theta = -i$ (see Eq. 15.4). However, if we switch to use $\epsilon = -1$ and $d = +1$ then the twist factor is $\theta = +i$ (see Eq. 15.1). This right-handed semion theory is equivalent to the Chern–Simons theory $SU(2)_1$ or the theory $\mathbb{Z}_2^{(\frac{1}{2})}$ that we introduce in Section 18.5.4. It is also equivalent to the abelian Chern–Simons theory $U(1)_2$ (Eq. 5.1 with coupling constant $k = 2$).[4]

18.2 Fibonacci Fusion Rules: The Branching Loop Gas

We now consider Fibonacci fusion rules as discussed in Sections 8.2.1 and 9.1. Here, the nontrivial fusion rule is[5,6]

$$\tau \times \tau = I + \tau \quad.$$

Again, $\tau = \bar{\tau}$ is self-dual. These fusion rules allow vertices with three τ particles (one coming from each direction, as shown in Fig. 18.7) so the loop gas can have branches, as shown in Fig. 18.8.

[4]Caution: there are multiple conventions as to how you label the subscript in a $U(1)$ theory.

[5]We have switched back to the notation of τ for the nontrivial particle and I for the vacuum. Using 1 and 0 is also common.

[6]In Exercise 17.5 we considered the more general $\tau \times \tau = I + n\tau$ and showed that the only possibility for a modular theory is $n = 1$.

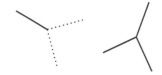

Fig. 18.7 An allowed fusion vertex (right) and a disallowed fusion vertex (left) for the Fibonacci fusion rules. The solid line is τ and the dotted line is the identity. The vertices shown in Fig. 18.3 are also allowed.

Fig. 18.8 A Fibonacci branching loop diagram allows intersections of loops, but no loop ends.

The fusion multiplicity matrix N_{ab}^c is zero if exactly one of the indices is τ and the other two are I. Otherwise, $N_{ab}^c = 1$. We can establish the non-zero components of the F-matrices for these fusion rules:

$$
\begin{align}
F_{III}^{III} &= 1/d_I = 1 \tag{18.18} \\
F_{\tau\tau I}^{\tau\tau I} &= 1/d_\tau \tag{18.19} \\
F_{II\tau}^{\tau\tau I} &= F_{\tau\tau\tau}^{III} = 1 \tag{18.20} \\
F_{\tau\tau\tau}^{\tau\tau I} &= 1/\sqrt{d_\tau} \tag{18.21} \\
F_{\tau\tau I}^{\tau\tau\tau} &= 1/\sqrt{d_\tau} \tag{18.22} \\
F_{\tau I\tau}^{\tau I\tau} &= F_{I\tau\tau}^{I\tau\tau} = 1 \tag{18.23} \\
F_{I\tau I}^{\tau I\tau} &= F_{\tau II}^{I\tau\tau} = 1 \tag{18.24} \\
F_{I\tau\tau}^{\tau\tau\tau} &= F_{\tau I\tau}^{\tau\tau\tau} = F_{\tau\tau\tau}^{I\tau\tau} = F_{\tau\tau\tau}^{\tau I\tau} = 1 \tag{18.25} \\
F_{\tau\tau\tau}^{\tau\tau\tau} &= -1/d_\tau \quad . \tag{18.26}
\end{align}
$$

As with the case of the $\mathbb{Z}_2$ loop gases, the first two lines are required from Eq. 16.4. Equations 18.20 and Eq. 18.21 are from Eq. 16.7. Equation 18.22 comes from Eq. 18.21 and Eq. 16.10. Equations 18.23, 18.24, and 18.25 can be derived from Eqs. 18.19, 18.20, 18.21, and 18.22 by the tetrahedral symmetry equation Eq. 16.17. Finally, Eq. 18.26 comes from the requirement that the 2×2 matrix $[F_{\tau\tau}^{\tau\tau}]$ is a unitary matrix (see Fig. 16.4), which we write out as[7]

$$
F_{\tau\tau}^{\tau\tau} = \begin{pmatrix} 1/d_\tau & 1/\sqrt{d_\tau} \\ 1/\sqrt{d_\tau} & -1/d_\tau \end{pmatrix} \quad . \tag{18.27}
$$

The unitarity requirement on this matrix also gives us

$$
\frac{1}{d_\tau^2} + \frac{1}{d_\tau} = 1 \quad . \tag{18.28}
$$

The solution to this is

$$
d_\tau = \frac{1 + \sqrt{5}}{2}
$$

which matches the expected quantum dimension d_τ given in Eq. 8.3 as it must, given the considerations of Section 17.1. Equation 18.28 also has a solution with $d_\tau < 0$. However, we cannot accept this solution because it would violate Eq. 14.5.[8]

As in the case of the $\mathbb{Z}_2$ loop gases, many of the F-matrix elements correspond to simple deformations of paths (isotopy) as in Fig. 18.5. The nontrivial F-moves (corresponding to the matrix F in Eq. 18.27) are summarized in Fig. 18.9.

18.2.1   Braidings for Fibonacci Anyons

To determine the possible braidings for Fibonacci fusion rules, we must solve the hexagon equations given the F-matrices we just derived. This is assigned as Exercise 13.1. There are two possible solutions: a right-

[7]This F-matrix matches our previous claim in Eq. 9.3. With any proposed F-matrix, one should always check that one has a valid solution of the pentagon equation Eq. 9.7 (or Eq. 16.3). See Exercise 9.4 for the Fibonacci case!

[8]Since $\tau \times \tau = \tau + \dots$, we need to have $\text{sign}[d_\tau] \times \text{sign}[d_\tau] = \text{sign}[d_\tau]$.

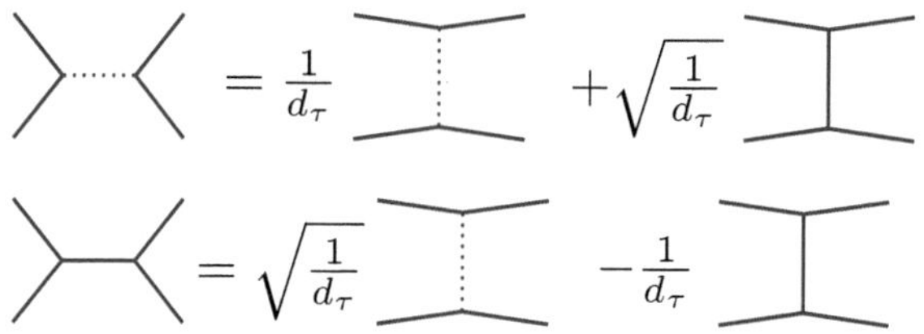

Fig. 18.9 The F-moves for the Fibonacci branching loop gas. Note that the first line is actually the insertion of a complete set of states as in Fig. 16.9.

handed solution given in Eq. 10.2, and a left-handed solutions that is the complex conjugate of the right-handed solution. These are the only solutions of the hexagon equations for the Fibonacci fusion rules. The solution with $(R_I^{\tau\tau})^* = \theta_\tau = e^{4\pi i/5}$ matches the Chern–Simons theory $(G_2)_1$ whereas the case with $(R_I^{\tau\tau})^* = \theta_\tau = e^{-4\pi i/5}$ matches $(F_4)_1 = \overline{(G_2)_1}$.

18.3 The $\mathbb{Z}_3$ Abelian Theories

Here, we consider the possibility of having an abelian theory with two nontrivial particle types. Let us tentatively call these particles I, X, Y. It is easy to show[9] that for an abelian theory we must have $X = \bar{Y}$. It is then convenient to rename the particles $0, +1, -1$ (with 0 the identity) such that the fusion rule is simply addition modulo 3. We can draw such particles as lines with arrows—if you travel with the arrow, the particle is $+1$, and if you travel against the arrow, the particle is -1. The fusion rules are modulo-3 fusions: vertices are allowed but each vertex must either have three arrows incoming or three arrows outgoing, as shown in Fig. 18.10. An example of a loop gas for the $\mathbb{Z}_3$ theory is shown in Fig. 18.11.

[9] The only other possibility would be that $X = \bar{X}$ and $Y = \bar{Y}$. However, if we had that, then what would $X \times Y = Y \times X$ be? It cannot be the identity, since each particle has a unique antiparticle and both X and Y are their own antiparticles. So we must have something like $X \times Y = X$. But then we can premultiply by $\bar{X}$ on both sides to obtain the contradiction $Y = I$.

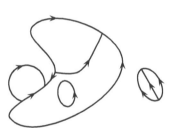

Fig. 18.11 In the $\mathbb{Z}_3$ loop gas lines have arrows and can intersect at vertices. At each vertex the three arrows must either all be pointing in or all be pointing out.

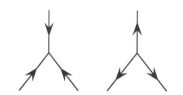

Fig. 18.10 The two allowed vertices in the $\mathbb{Z}_3$ abelian theories.

Since there are no self-dual particles, we do not have to worry about Frobenius–Schur indicators. Insisting on isotopy invariance, we use Eq. 16.16 and tetrahedral invariance and it is very easy to show that all non-zero F-matrix elements must be unity (see exercise 18.1). This is a special case of a more general $\mathbb{Z}_N^{(n)}$ theory we discuss in Section

18.5.3.

We now consider the possible solutions of the hexagon Eqs. 16.20–16.21. Since F is unity we have

$$R^{ca}_{c \times a} R^{cb}_{c \times b} = R^{cf}_{c \times f}$$

where $a \times b = f$. There are exactly three solutions to this

$$R^{ca}_{c \times a} = e^{2\pi i a c n / 3} \tag{18.29}$$

for $n = 0, 1, 2$. From Eq. 15.3 we calculate the twist factors $\theta_a = e^{2\pi i a^2 n / 3}$.

For the $n = 0$ solution, all braiding and twisting is trivial—and this case is not modular (in fact, it is a so-called symmetric tensor category, meaning all particles are transparent).

The $n = 1, 2$ solutions are mirror images of each other and are both modular. Given the twist factors we can work out the corresponding S-matrices (see Eq. 18.49), which are of the form

$$S = \frac{1}{\sqrt{3}} \begin{pmatrix} 1 & 1 & 1 \\ 1 & \omega & \omega^2 \\ 1 & \omega^2 & \omega \end{pmatrix} \tag{18.30}$$

where $\omega = e^{2\pi i n / 3}$ with $n = 1, 2$. In fact, since we are generally interested in modular theories, we might have started by searching for a valid unitary S-matrix. Here, we need a symmetric matrix with all entries of unit magnitude, where there are 1s on the first row and first column, and when multiplied by the prefactor of $\frac{1}{\sqrt{3}}$ for normalization, the matrix is unitary. It is very easy to show that the two matrices given in Eq. 18.30 are the only two such matrices.

The $n = 1$ case is actually the theory $\mathbb{Z}_3^{(1)}$ which we meet in Section 18.5.3, where we also discuss a few further properties of the theory. The $n = 2$ case is the mirror image, or $\mathbb{Z}_3^{(2)}$.

18.4   Ising Fusion Rules

As discussed in Section 8.2.2, the Ising fusion rules (also known as $SU(2)_2$ fusion rules) are given by

$$\begin{aligned} \psi \times \psi &= I \\ \psi \times \sigma &= \sigma \\ \sigma \times \sigma &= I + \psi \end{aligned}$$

with both particle types being self-dual $\psi = \bar{\psi}$ and $\sigma = \bar{\sigma}$. These rules describe a loop gases with two non-vacuum particles ψ (which we draw as blue lines and loops in Fig. 18.12) and σ (which we draw red loops in Fig. 18.12). The rule of this loop gas is that one may have a vertex with two sigmas and one ψ, which appears as a blue line splitting off from a red loop.

Fig. 18.12 A diagram with Ising fusion rules. Here, σ is red and ψ is blue.

Looking at the first fusion rule, $\psi \times \psi = I$, we realize this rule alone is simply a $\mathbb{Z}_2$ fusion rule. Indeed, this tells us immediately that we have

$$1/d_I = 1/d_\psi = 1$$
$$= F^{III}_{III} = F^{\psi\psi I}_{\psi\psi I} = F^{\psi\psi I}_{II\psi} = F^{III}_{\psi\psi\psi} = F^{\psi I\psi}_{\psi I\psi} = F^{I\psi\psi}_{I\psi\psi} = F^{\psi I\psi}_{I\psi I} = F^{I\psi\psi}_{\psi II}$$

as given in Eqs. 18.2–18.6. One might wonder why we do not consider $d_\psi = -1$. This is for the same reason why we could not consider negative d_τ in the Fibonacci case. Here, we must have $\text{sign}[d_\sigma]\text{sign}[d_\sigma] = \text{sign}[d_\psi]$, so we must have d_ψ positive.

Very similarly, we have

$$
\begin{aligned}
F^{\sigma\sigma I}_{\sigma\sigma I} &= 1/d_\sigma \\
F^{\sigma\sigma I}_{II\sigma} &= F^{III}_{\sigma\sigma\sigma} = 1 \\
F^{\sigma I\sigma}_{\sigma I\sigma} &= F^{I\sigma\sigma}_{I\sigma\sigma} = 1 \\
F^{\sigma I\sigma}_{I\sigma I} &= F^{I\sigma\sigma}_{\sigma II} = 1 \quad .
\end{aligned}
$$

The first equation is from Eq. 16.4, and the second from Eq. 16.7. The last two are derived from the first two via the tetrahedral symmetry Eq. 16.17.

Further, using Eqs. 16.7 and 16.16 we obtain

$$F^{\sigma\sigma I}_{\sigma\sigma\psi} = F^{\sigma\sigma\psi}_{\sigma\sigma I} = 1/d_\sigma \tag{18.31}$$
$$F^{\sigma\sigma I}_{\psi\psi\sigma} = F^{\psi\psi I}_{\sigma\sigma\sigma} = F^{\sigma\psi\sigma}_{\psi\sigma I} = F^{\psi\sigma\sigma}_{\sigma\psi I} = 1 \quad . \tag{18.32}$$

Enforcing unitarity on the 2×2 matrix $[F^{\sigma\sigma}_{\sigma\sigma}]$ we get

$$F^{\sigma\sigma\psi}_{\sigma\sigma\psi} = -1/d_\sigma \tag{18.33}$$

giving the 2×2 matrix the form

$$[F^{\sigma\sigma}_{\sigma\sigma}] = \begin{pmatrix} 1/d_\sigma & 1/d_\sigma \\ 1/d_\sigma & -1/d_\sigma \end{pmatrix} \quad . \tag{18.34}$$

The unitarity condition also gives us the condition that

$$d_\sigma = \pm\sqrt{2}$$

which is expected from Section 8.2.2 since the fusion rules give us $d_\sigma = |d_\sigma| = \sqrt{2}$. Both of these roots are viable solutions of the pentagon equations.

As discussed in detail in Section 14.2 (see also Section 16.1.3) if we choose negative d_σ the evaluation of the diagram is *nonunitary* and we must apply rule 0 in order to recover a unitary theory.

The remaining non-zero elements of F are obtained from Eq. 18.31–18.33 by using tetrahedral symmetry Eq. 16.17 to obtain

$$1 = F^{\sigma I \sigma}_{\sigma \psi \sigma} = F^{I \sigma \sigma}_{\psi \sigma \sigma} = F^{\sigma \psi \sigma}_{\sigma I \sigma} = F^{\psi \sigma \sigma}_{I \sigma \sigma} \tag{18.35}$$

$$= F^{\sigma I \sigma}_{\psi \sigma \psi} = F^{I \sigma \sigma}_{\sigma \psi \psi} = F^{\psi I \psi}_{\sigma \sigma \sigma} = F^{I \psi \psi}_{\sigma \sigma \sigma} \tag{18.36}$$

$$= F^{\psi \sigma \sigma}_{\sigma I \psi} = F^{\sigma \sigma \psi}_{I \psi \sigma} = F^{\sigma \sigma \psi}_{\psi I \sigma} = F^{\sigma \psi \sigma}_{I \sigma \psi} \tag{18.37}$$

$$-1 = F^{\sigma \psi \sigma}_{\sigma \psi \sigma} = F^{\psi \sigma \sigma}_{\psi \sigma \sigma} \; . \tag{18.38}$$

The nontrivial F-moves corresponding to the matrix Eq. 18.34 are shown in Fig. 18.13.

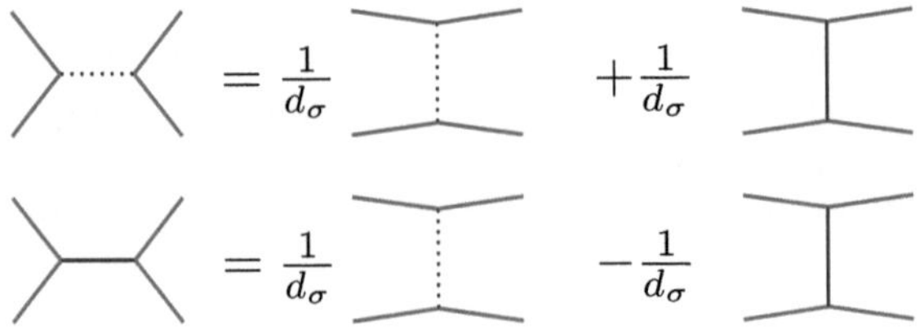

Fig. 18.13 The nontrivial F-moves for the Ising fusion rules. Note that the first line is actually the insertion of a complete set of states, as in Fig. 16.9.

18.4.1   Braidings For Ising Fusion Rules

The most straightforward way to find all the possible braidings for the Ising fusion rules is to explicitly solve the hexagon Eqs. 13.1–13.2. We here outline how we proceed (Exercise 18.2 asks you to work out the details). For each possible set of variables a, b, c, d in the hexagon equation (see Fig. 18.14 for the starting configuration of the hexagon), we derive a different identity. For each of the following cases, the F-matrices are simple scalars (1 and -1 only) so we derive

$$a = \psi, \; b = c = \sigma, \; d = I \; \Rightarrow \; R^{\sigma \psi}_\sigma R^{\sigma \sigma}_\psi = R^{\sigma \sigma}_I \tag{18.39}$$

$$\text{mirror image diagram} \; \Rightarrow \; R^{\psi \sigma}_\sigma R^{\sigma \sigma}_\psi = R^{\sigma \sigma}_I \tag{18.40}$$

$$a = b = \sigma, \; c = \psi, \; d = I \; \Rightarrow \; [R^{\psi \sigma}_\sigma]^2 = R^{\psi \psi}_I \tag{18.41}$$

$$a = b = \sigma, \; c = d = \psi \; \Rightarrow \; [R^{\psi \sigma}_\sigma]^2 = -1 \; . \tag{18.42}$$

Fig. 18.14 Labeling the edges on the first step of the hexagon diagram. See Fig. 13.11.

For the following case one uses the 2×2 F-matrix, meaning we are working with a two-dimensional vector space and the hexagon equations give us two identities

$$a = b = c = d = \sigma, \ e = I \Rightarrow \begin{cases} [R_I^{\sigma\sigma}]^2 = \frac{1}{d_\sigma}(1 + R_\sigma^{\sigma\psi}) \\ R_I^{\sigma\sigma} R_\psi^{\sigma\sigma} = \frac{1}{d_\sigma}(1 - R_\sigma^{\sigma\psi}) \end{cases} . \quad (18.43)$$

These equations are enough to pin down all of the possible solutions for the R-matrix. From Eqs. 18.41 and 18.42 we obtain

$$R_I^{\psi\psi} = -1$$

which also implies $\theta_\psi = -1$ from Eq. 15.4. Note that since ψ is a $\mathbb{Z}_2$ field with $d_\psi = 1$, comparing to our above discussion of the $\mathbb{Z}_2$ fusion rules we already know that ψ has to be either a fermion or a boson (we could re-establish this by looking at the hexagon equation with only ψ and I fields). The hexagon equation including the σ field now establishes ψ to be a fermion!

From Eqs. 18.39, 18.40, and 18.42 we establish

$$R_\sigma^{\sigma\psi} = R_\sigma^{\psi\sigma} = \pm i$$

This sign is an additional free choice we can make (in addition to the choice of $d_\sigma = \pm\sqrt{2}$). To keep these independent choices straight we will use the notation

$$d_\sigma = \overset{1}{\pm} \sqrt{2} \quad (18.44)$$

$$R_\sigma^{\psi\sigma} = \overset{2}{\pm} i . \quad (18.45)$$

We now plug in our choices for d_σ and $R_\sigma^{\psi\sigma}$ into the first line of Eq. 18.43 to solve for $R_I^{\sigma\sigma}$. This gives yet another independent choice of sign for a square root which we label with $\overset{3}{\pm}$. We thus obtain

$$R_I^{\sigma\sigma} = \exp\left[\frac{2\pi i}{16}\left(6 \overset{1}{\mp} 2 \overset{2}{\pm} 1 \overset{3}{\pm} 4\right)\right] \quad (18.46)$$

from which we see there are a total of eight possible choices. For the record, from Eq. 18.39, we also have $R_\psi^{\sigma\sigma} = \overset{2}{\mp} iR_I^{\sigma\sigma}$. We should also check that none of the other hexagon relations are violated for any of these eight solutions. Remarkably, perhaps, all eight solutions solve all the hexagon relations with no violations (see Exercise 18.2).

The eight possible solutions all have the same S-matrix (see Exercise 18.3) and are all modular.

$$S = \frac{1}{2}\begin{pmatrix} 1 & \sqrt{2} & 1 \\ \sqrt{2} & 0 & -\sqrt{2} \\ 1 & -\sqrt{2} & 1 \end{pmatrix} \quad (18.47)$$

We summarize the eight possible anyon theories with Ising fusion in Table 18.1. A table of this sort has previously been worked out by Kitaev [2006] using similar methods.

Table 18.1 Table of anyon theories with Ising fusion rules. Here we have described theories with negative Frobenius–Schur indicator, as having negative d_σ, but no zig-zag phase $\epsilon_a = +1$. Often one sees theories described with $d_\sigma > 0$ but $\epsilon_a = -1$. In that language the twist factor of the σ particle is always $\theta_\sigma = e^{2\pi i c/8}$ where c is the central charge.

$\underset{=\,\mathrm{sign}(d_\sigma)}{\overset{1}{\pm}}$	$\underset{=\,\mathrm{sign}(-iR_\sigma^{\psi\sigma})}{\overset{2}{\pm}}$	$\underset{\text{see Eq. 18.46}}{\overset{3}{\pm}}$	$R_I^{\sigma\sigma}$	name	$c \bmod 8$
$+$	$+$	$+$	$e^{2\pi i(-7/16)}$	$(B_3)_1$	$\frac{7}{2}$
$+$	$+$	$-$	$e^{2\pi i(+1/16)}$	$\overline{\text{Ising}} = (E_8)_2$	$-\frac{1}{2}$
$+$	$-$	$+$	$e^{2\pi i(+7/16)}$	$\overline{(B_3)_1}$	$-\frac{7}{2}$
$+$	$-$	$-$	$e^{2\pi i(-1/16)}$	$\text{Ising} = \overline{(E_8)_2}$	$\frac{1}{2}$
$-$	$+$	$+$	$e^{2\pi i(-3/16)}$	$\overline{(C_2)_1}$	$-\frac{5}{2}$
$-$	$+$	$-$	$e^{2\pi i(+5/16)}$	$SU(2)_2$	$\frac{3}{2}$
$-$	$-$	$+$	$e^{2\pi i(-5/16)}$	$\overline{SU(2)_2}$	$-\frac{3}{2}$
$-$	$-$	$-$	$e^{2\pi i(+3/16)}$	$(C_2)_1$	$\frac{5}{2}$

18.5 More Abelian Theories

[10] A different approach to abelian anyons is given by the K-matrix approach of section 5.3.2. Classification of abelian anyons then reduces to classification of the physically different possible K-matrices. See Cano et al. [2014]; Wang and Wang [2020].

For abelian anyons[10] much can be worked out fairly easily (see also the discussion in Sections 19.1 and 19.4). We simplify the discussion by looking only at theories with valid braidings, which we do by examining the possible S-matrices first.

A few formulae will be particularly useful to us. First, if an abelian particle a has twist factor θ_a, then the result of fusing n-copies of a together gives a particle with twist factor (see Exercise 15.1)

$$\theta_{\underset{n \text{ times}}{a \times a \times a \dots \times a}} = (\theta_a)^{n^2} \quad . \tag{18.48}$$

In particular, this tells us that if n copies of a fuse together to the identity, the twist factor of a must be an (n^2)-root of unity, which now greatly limits the possibilities we need to consider.

Secondly, once we know all the twist factors, the S-matrix can be obtained via Eq. 17.21

$$S_{ab} = \frac{1}{\mathcal{D}} \frac{\theta_{a \times \bar{b}}}{\theta_a \theta_b} \tag{18.49}$$

with $\mathcal{D}$ the square root of the number of particles in the theory. Finally, we note that in abelian theories fusion must be described by an abelian group (see Section 40.2.1), and the quantum dimension $d = 1$ for all particle types.

18.5.1 $\mathbb{Z}_2 \times \mathbb{Z}_2$ Fusion Rules

Here, we will have three self-dual particle types, which we will call X, Y, Z with fusion $X^2 = Y^2 = Z^2 = 1$ and $XY = Z$, $XZ = Y$, and $YZ = X$. This is called $\mathbb{Z}_2 \times \mathbb{Z}_2$ fusion rules since the fusion algebra is the same as the group $\mathbb{Z}_2 \times \mathbb{Z}_2$ (we can think of X and Y as being independent generators, each with $\mathbb{Z}_2$ fusion rules and then we define Z in terms of X and Y).

There are several obvious cases built from theories we already worked out for $\mathbb{Z}_2$ fusion rules. The direct product of any combination of the boson, fermion, left-handed semion, or right-handed semion theories discussed in Section 18.1 will have $\mathbb{Z}_2 \times \mathbb{Z}_2$ fusion rules. The resulting S-matrices in each case would just be the outer product of the S-matrices of the constituent theories (see Exercise 17.6).

For example, we might choose boson $\times$ boson (which would have all equal positive entries in the S-matrix and would be non-modular). More interesting examples are the three different modular cases[11] one can get as simple product theories

$$
\begin{aligned}
SU(2)_1 \times SU(2)_1 &= \text{right-handed semion} \times \text{right-handed semion} \\
SU(2)_1 \times SU(2)_{-1} &= \text{right-handed semion} \times \text{left-handed semion} \\
SU(2)_{-1} \times SU(2)_{-1} &= \text{left-handed semion} \times \text{left-handed semion.}
\end{aligned}
$$

[11]We have not listed $SU(2)_{-1} \times SU(2)_1$ because this is the same as $SU(2)_1 \times SU(2)_{-1}$ just relabeling the particles.

There are also (split)-super-modular possibilities $SU(2)_1 \times$ Fermion and $SU(2)_{-1} \times$ Fermion (see Section 17.7).

We now look for cases that do not factorize. Since each of these particles has $\mathbb{Z}_2$ fusion rules, from Eq. 18.48 (and also as we derived in Section 18.1), its twist factor must be a fourth root of unity: ± 1 or $\pm i$. However, since all the particles are self-dual the S matrix must be real (see Eq. 17.7), so using Eq. 18.49 we find that every entry of the S matrix must be $\pm 1/\mathcal{D} = \pm 1/2$. It is quite easy to look through all the possible sign choices to find that there is one more possible unitary S-matrix outside of the three cases discussed above:

$$
S = \frac{1}{2}\begin{pmatrix} 1 & 1 & 1 & 1 \\ 1 & 1 & -1 & -1 \\ 1 & -1 & 1 & -1 \\ 1 & -1 & -1 & 1 \end{pmatrix} \qquad \begin{array}{c} \text{toric code or} \\ \text{fermionic toric code.} \end{array}
$$

Since the diagonals are positive, it is easy to establish that the twist factors can only be ± 1 (one can either use Eq. 18.49 or just look at each $\mathbb{Z}_2$ subalgebra and follow the arguments of Section 18.1). Again, using Eq. 17.20 we find that up to permutation of the names of the particle types there are exactly two possibilities for the twist factors:

$$
[\theta_I, \theta_X, \theta_Y, \theta_Z] = [+1, +1, +1, -1] \tag{18.50}
$$

or

$$
[\theta_I, \theta_X, \theta_Y, \theta_Z] = [+1, -1, -1, -1] \quad . \tag{18.51}
$$

Equation 18.50 defines the "toric-code" TQFT,[12] which we discuss at length starting in Chapter 27. It is also known as the $D(\mathbf{Z}_2)$ TQFT, which stands for the "quantum double" of the group $\mathbb{Z}_2$.

[12]The standard names for particle types in the toric code are $[I, e, m, f]$ where the f is the fermion.

Equation 18.51 is sometimes colloquially called the "fermionic" toric code (since the three nontrivial particles are all fermions). It is also given by the Chern–Simons theory $(D_4)_1$ meaning we use the Lie group D_4 at level 1.

We say a bit more about these two toric code cases in Section 18.6, and then we say much more about the (non-fermionic) toric code starting in Chapter 27.

18.5.2   General $\mathbb{Z}_N$ Fusion Rules

We now turn to the case of $\mathbb{Z}_N$ fusion rules. We label the particle types as $a \in \{0, 1, 2, 3, \ldots, (N-1)\}$ with fusions $a \times b = (a+b) \bmod N$. Using the fact that a group of N copies of the 1 particle fuse together to make the identity, we have from Eq. 18.48 $(\theta_1)^{N^2} = 1$ or $\theta_1 = e^{2\pi i \alpha / N^2}$ for some integer α. Again using Eq. 18.48 this means that $\theta_a = e^{2\pi i a^2 \alpha / N^2}$ Note now that the particle 1 is the antiparticle of $N-1$, so they must have the same twist factor $\theta_1 = \theta_{N-1}$, which then implies two possible solutions: $\alpha = Nn$, valid for any N; or $\alpha = N(n + \frac{1}{2})$, valid only if N is even, where in both cases n can be chosen as any integer $0 \leq n < N$. We consider these cases one at a time.

18.5.3 $\mathbb{Z}_N^{(n)}$ Anyons

Here, we choose $\alpha = Nn$ and $n \in \{0, 1, \ldots, (N-1)\}$ so that

$$\theta_a = e^{2\pi i n a^2 / N} \quad . \tag{18.52}$$

Using Eq. 18.49 we have the S-matrix

$$S_{a,b} = \frac{1}{\sqrt{N}} \exp\left[-\frac{4\pi i n}{N} ab\right]$$

where ab on the right is actual multiplication, not the group operation, which is addition modulo N. The mirror image (opposite-handedness) of a theory is obtained by taking $n \to N - n$.

It is easy to see that this is modular if and only if n and N are coprime (have no common factors besides 1) and N is also odd. In the case where N is even, $N/2$ is odd, n is odd, and $N/2$ and n are coprime, we have a (split) super-modular theory (see Section 17.7) of the form $Z_N^{(n)} = Z_{N/2}^{(n)} \times$ fermion.

We discuss the F matrices in more depth in Section 19.2. However, it turns out that in an appropriately chosen gauge, the F-matrices are trivial—they are all unity for an allowed diagram (or zero if there is a disallowed fusion in the diagram). To see that this is consistent with the S-matrix above, let us look for solutions of the hexagon Eq. 13.1 to find all possible bradings consistent with $F = 1$.

Plugging $F = 1$ into the hexagon equations (see also Eqs. 19.7–19.8), and using the notation $[a]_N$ for a modulo N, we have

$$
\begin{aligned}
R^{c,a}_{[c+a]_N} R^{c,b}_{[c+b]_N} &= R^{c,[a+b]_N}_{[c+a+b]_N} \\
R^{a,c}_{[c+a]_N} R^{b,c}_{[c+b]_N} &= R^{[a+b]_N,c}_{[c+a+b]_N}
\end{aligned} \qquad (18.53)
$$

There are exactly N solutions[13] of this system, which we label with $n \in \{0, \dots, N-1\}$ and that are given, in a convenient gauge, by

$$
R^{a,b}_{[a+b]_N} = \exp\left[\frac{2\pi i n}{N} ab\right] .
$$

Using Eq. 15.3 we see that this then is compatible with our choice for the twist factors Eq. 18.52 and hence, with our choice for the S.

The theories discussed in this section are sometimes known as $\mathbb{Z}_N^{(n)}$ anyons (see Bonderson [2007][14]). For $n = (N-1)/2$ with N odd, the modular theory is the Chern–Simons theory $SU(N)_1$.

18.5.4 $\mathbb{Z}_N^{(n+\frac{1}{2})}$ with N even

Here, we consider the possibility of $\alpha = N(n + \frac{1}{2})$, with N being even and $n \in \{0, 1, \dots, (N-1)\}$. We then have[15]

$$
\theta_a = e^{2\pi i (n+\frac{1}{2})a^2/N} \qquad (18.54)
$$

which then, via Eq. 18.49, implies the S-matrix

$$
S_{a,b} = \frac{1}{\sqrt{N}} \exp\left[-\frac{4\pi i (n+\frac{1}{2})}{N} ab\right] . \qquad (18.55)
$$

The mirror image theory being obtained by taking $n \to N - 1 - n$.

When $(2n + 1)$ and N are coprime (have no common factors), the theory is modular. These are known as $\mathbb{Z}_N^{(n+\frac{1}{2})}$ anyon theories for obvious reasons. When $n = (N/2 - 1)$ the modular theory matches $SU(N)_1$ with N even, and when $n = 0$ the modular theory matches[16] $U(1)_N$.

Note that the particle $a = N/2$ is self-dual. When the theory is modular, one can use Eq. 17.22 to establish that the Frobenius–Schur indicator of this particle is $\kappa_{N/2} = (-1)^{N/2}$ (see Exercise 18.5).

In Sections 19.2 and 19.4.2 we revisit these theories and work out the F- and R- matrices in some detail.

$\mathbb{Z}_4$ Fusion Rules

It is worth briefly pointing out that for $\mathbb{Z}_4$ fusion rules the possible modular theories are as follows: $\mathbb{Z}^{(\frac{1}{2})}$ corresponds to abelian Chern–Simons theory $U(1)_4$ or equivalently the theory $(D_9)_1$. Its mirror image is $\mathbb{Z}^{(3+\frac{1}{2})}$. The theory $\mathbb{Z}^{(1+\frac{1}{2})}$ is the Chern–Simons theory $SU(4)_1$, and $\mathbb{Z}^{(2+\frac{1}{2})}$ is its mirror image.

[13]From the first equation let $\rho_c(a) = R^{c,a}_{[a+c]_N}$ we see that $\rho_c(a)$ is a group representation of the group $\mathbb{Z}_N$, which can only be of the form $\exp(2\pi i p(c)a/N)$ where p is some function of c. Similarly, with the second equation, this fixes the only possible forms of the result.

[14]Note that the S-matrix quoted in that reference is defined differently $S_{ab} \to S_{a\bar{b}}$.

[15]Note that in cases where there is a negative Frobenius–Schur indicator in this equation we assume a convention of $d_a > 0$. Otherwise, this formula becomes somewhat uglified.

[16]Be cautioned that there is some disagreement in the literature as to how you label the level of a $U(1)$ Chern–Simons theory. Sometimes one writes $U(1)_{N/2}$.

18.6   **All Braided Abelian Theories**

[17] As in the discussion of Section 17.4, here we are describing anyon theories at the level of their modular data. That is, the central charge is specified only mod 8.

A beautiful way to describe all[17] *modular* abelian theories has been given by Galindo and Jaramillo [2016]. See also Wang and Wang [2020] and Cano et al. [2014]. In this approach one determines the "prime" abelian theories and then *any* modular abelian theory can be written as a product of these prime abelian theories, although the decomposition into primes is not generally unique—a given modular abelian theory can be written as a product of primes in more than one way. For example, the toric code and the fermionic toric code are both prime theories. However, (toric code) ×(toric code) = (fermionic toric code) × (fermionic toric code); see Exercise 18.10.

The prime theories are of four types.

$\mathbb{Z}^{(n)}_{p^m}$

For each prime number $p > 2$ and each positive integer m we can write exactly two distinct prime abelian theories of the form $\mathbb{Z}^{(n)}_N$ (see Section 18.5.3) where we take $N = p^m$. In fact, regardless of how we choose n there are exactly two different possible theories that arise (for fixed $N = p^m$). Either the two theories are mirror images of each other or they are each mirror images of themselves. For definiteness, we can choose $n = 2$ as one of the two distinct theories, and then the other value of n we choose needs to give a different theory. This requires finding any integer n such that no integer $x \in \{0, \ldots, p-1\}$ gives $x^2 = 2n \bmod p$ (roughly half of the possible n values will satisfy this).

$\mathbb{Z}^{(n+\frac{1}{2})}_{2^m}$

For each positive integer m we can write a theory of the form $\mathbb{Z}^{(n+\frac{1}{2})}_N$ (see Section 18.5.4) with $N = 2^m$. For $m = 1$ there are two possible prime anyon theories—the right- and left-handed semion theories. Interestingly, among all of the prime abelian anyon theories, these are the only two that have any nontrivial Frobenius–Schur indicators! For $m \geq 2$ there are four prime abelian anyon theories for each value of m. These can be taken as $n = 0$, its complex conjugate $n = (N-1)$, as well as $n = 2$ and its complex conjugate $n = (N-3)$.

$\mathbb{Z}_{2^m}$ **Toric Code**

We have briefly discussed the toric code, which has $\mathbb{Z}_2 \times \mathbb{Z}_2$ fusion rules, in Section 18.5.1, and we discuss it more extensively starting in Chapter 27. A generalization of the toric which we introduce in detail in Sections 27.6 and 28.5 is known as the $\mathbb{Z}_N$ toric code, which has $\mathbb{Z}_N \times \mathbb{Z}_N$ fusion rules. We summarize the results of this model here. Each particle type can be labeled with two integers $n, k \in \{0, \ldots, N-1\}$, which we write as $[n, k]$. The fusion rules explicitly are

$$[n_1, k_1] \times [n_2, k_2] = [(n_1 + n_2)\bmod(N), (k_1 + k_2)\bmod(N)]$$

and twist factors are

$$\theta_{[n,k]} = e^{-2\pi i nk/N}$$

In an appropriately chosen gauge, the F-matrices here are all unity for any allowed fusion, and the R-matrices are

$$R^{[n_1,k_1],[n_2,k_2]}_{[n_1+n_2,k_1+k_2]} = e^{-2\pi i n_1 k_2/N}$$

where the lower indices of R are understood to be modulo N. This then results in an S-matrix

$$S_{[n'k'],[n,k]} = \frac{1}{N} e^{2\pi i (nk'+n'k)/N}$$

which is modular for all possible N. Having briefly defined the $\mathbb{Z}_N$ toric code, we now state the result that for $N = 2^m$ for any positive integer m the $\mathbb{Z}_N$ toric code is a prime abelian theory.

$\mathbb{Z}_{2^m}$ Toric Code Variant

We also discussed the fermionic toric code in Section 18.5.1. We can generalize this theory to a $\mathbb{Z}_N$ version analogous to what we just did for the (non-fermionic) toric code. For lack of a better name, we will call it a toric code variant. The fusion rules are the same as the $\mathbb{Z}_N$ toric code. However, the twist factors are now

$$\theta_{[n,k]} = e^{-2\pi i (n^2+k^2-nk)/N} \quad .$$

Again, in an appropriately chosen gauge, the F-matrices are all unity for any allowed fusion, and the R-matrices are

$$R^{[n_1,k_1],[n_2,k_2]}_{[n_1+n_2,k_1+k_2]} = e^{-2\pi i (n_1 n_2 + k_1 k_2 - n_1 k_2)/N}$$

where again the lower indices of R are understood to be modulo N. This then results in an S-matrix

$$S_{[n'k'],[n,k]} = \frac{1}{N} e^{2\pi i (2nn'+2kk'-nk'-n'k)/N} \quad .$$

This is not modular for N a multiple of 3. The case of $N = 2^m$ for any positive integer m is a prime abelian theory.

18.6.1   Prime Non-Modular Theories

A beautiful result stated by Cano et al. [2014] is that any non-modular abelian anyon theory can be written as a modular abelian anyon theory times some number of trivial fermion (sVec) factors. Thus, we simply need to add one more prime theory—a transparent fermion to our list in order to encompass all abelian anyon theories!

Chapter Summary

> - We construct some simple braided anyon theories. Usually we follow the prescription of Fig. 18.1: Start with fusion rules, find consistent F-matrices, then find consistent R-matrices.
> - We describe all possible braided abelian theories.

Further Reading:

- Kitaev [2006] has a detailed discussion of all the theories with Ising fusion using different techniques.

Exercises

Exercise 18.1 F-matrix for $\mathbb{Z}_3$ Fusion Rules
Assuming isotopy invariance and $\mathbb{Z}_3$ fusion rules as in Section 18.3 show that all non-zero F-matrix elements are unity. Do we need the assumption of isotopy invariance?

Exercise 18.2 Using the Hexagon for Ising Fusion Rules
Use the hexagon relations to derive Eqs. 18.39–18.43. Confirm that the eight solutions we find give no violations of any hexagon relations.

Exercise 18.3 S-matrix for Ising Fusion Rules
Explicitly derive the S-matrix for all eight solutions of the hexagon equation for the Ising fusion rules and confirm that they all give Eq. 18.47. Thus, confirm that all eight solutions are modular. Hint: It might be easier to use Eq. 17.20.

Exercise 18.4 Frobenius–Schur Indicator for Ising Fusion Rules
Use Eq. 17.22 to calculate the Frobenius–Schur indicator for the σ particle in each of the eight possible solutions of the hexagon equation for the Ising fusion rules. Show that the Frobenius–Schur indicator is negative exactly when d_σ is negative.

Exercise 18.5 Frobenius–Schur Indicator for $\mathbb{Z}_N^{(n+\frac{1}{2})}$
Use Eq. 17.22 to show that, if the theory is modular, then the self-dual particle $a = N/2$ of the $\mathbb{Z}_N^{(n+\frac{1}{2})}$ theory has Frobenius–Schur indicator $(-1)^{N/2}$.

Exercise 18.6 Evaluating Diagrams I
Show that evaluation of the diagram in Fig. 18.8 gives $-d_\tau^{9/2}$.

Exercise 18.7 Evaluating Diagrams II
Show that evaluation of the diagram in Fig. 18.12 gives $d_\sigma^3 \kappa_\sigma$.

Exercise 18.8 Deriving an F-matrix

(Hard) Consider a theory containing three particle types, I, S, V where I is the identity. Let the nontrivial fusion rules be given by

$$
\begin{aligned}
S \times S &= I \\
S \times V &= V \\
V \times V &= I + S + V \quad ,
\end{aligned}
$$

Let us assume we have a theory with full isotopy invariance and full tetrahedral symmetry. There is only one set of F-matrices for these fusion rules. Find these F-matrices. Hint: Make use of unitarity.

Exercise 18.9 $\mathbb{Z}_N$ Toric Code for N Odd

(a) Show that for odd N the $\mathbb{Z}_N$ toric code (see Section 18.6) is equivalent to the product of theories $\mathbb{Z}_N^{(1)} \times \mathbb{Z}_N^{(N-1)}$. This decomposition should not be surprising considering that the toric code is only prime for $N = 2^m$.

(b) There is similar decomposition for the toric code variant, with N odd, but not a multiple of 3. To what products of $\mathbb{Z}_N^{(n)}$ theories is this toric code variant equivalent?

Exercise 18.10 Toric Code $\times$ Toric Code

Show that (Toric Code) $\times$ (Toric Code) is the same as (Fermionic Toric Code) $\times$ (Fermionic Toric Code). Hint: Think about what set of particles can be used to generate all particles in both cases.

Exercise 18.11 Ising Modular Relations

Confirm the modular relations, Eq. 17.17, for all eight of the theories with Ising fusion rules listed in Table 18.1.

Anyons From Discrete Group Elements

In this chapter, and in Chapters 20 and 21, we will use the structure of groups to build diagram rules for possible anyon theories.[1] In this chapter we focus on discrete groups, meaning groups with a finite number of elements (see Section 40.2 for a very brief review of group theory).

Our method for defining consistent planar diagram algebras is to choose a discrete group G and associate a particle type with each element $g \in G$ with the identity element I of the group being the vacuum.

Thus, let us consider diagrams where each line is labeled by a group element $g \in G$. Reversal of a line corresponds to inversion of the group element as shown in Fig. 19.1 analogous to reversing an arrow in order to turn a particle into its antiparticle: that is, $\bar{g} = g^{-1}$.

Fusion rules can now be defined to follow the rules for group multiplication. That is, for $g, h \in G$

$$g \times h = gh$$

which we draw as shown in Fig. 19.2.

In cases where the group is abelian we have $g \times h = gh = hg = h \times g$ which is what we required for fusion of particle types in Section 8.1 above. In Section 19.3 we consider the possibility of using nonabelian (non-commutative) groups where $gh \neq hg$, but for now we will assume the group is abelian. We thus have fusion rules given by group multiplication

$$N_{g,h}^a = \delta_{a,gh} = \delta_{a,hg} \quad .$$

Since the result of any fusion is always uniquely defined by group multiplication (one never has a sum on the right-hand side, such as $g \times h = a + b$), the quantum dimension of every particle is $d_g = 1$, meaning the Hilbert space size does not grow with the number of particles.

An example of a planar diagram with this type of group multiplication is shown in Fig. 19.3.

19.1 Group Cohomology

We now have the task of trying to construct consistent F-matrices for our planar diagram algebra. This is an extremely well-studied problem in the field of *group cohomology*.[2]

Consider a general group G. A so-called 3-cocycle of the group is

[1]In Chapter 31 (also Section 23.4) we discuss another construction of an anyon theory from a discrete group, known as the quantum double construction. We defer discussion of that construction for now.

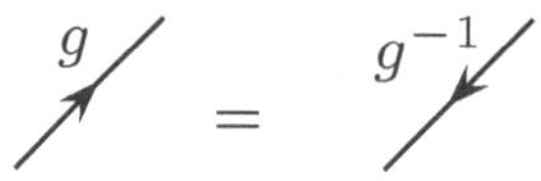

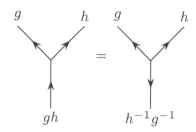

Fig. 19.1 Reversing an arrow inverts the group element.

Fig. 19.2 Fusion is defined by group multiplication. On the right we show the three particles oriented outward, all leaving the vertex. With this orientation when the three particles are multiplied together in clockwise order, they should fuse to the identity $gh(h^{-1}g^{-1}) = h(h^{-1}g^{-1})g = (h^{-1}g^{-1})gh = I$.)

[2]Group cohomology is a very general framework that we will not delve into more than is necessary. However, it is worth knowing that it enters prominently in a number of topological theories.

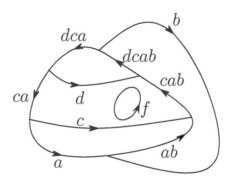

Fig. 19.3 A planar diagram with fusion being defined as group multiplication. For each vertex, if all arrows are pointed out of the vertex, then going around the vertex clockwise, the group elements multiply to the identity, as shown in Fig. 19.2.

given by a function of three variables $\omega(a, b, c)$ where $a, b, c \in G$ that satisfies

$$\omega(a, b, c)\omega(a, bc, d)\omega(b, c, d) = \omega(ab, c, d)\omega(a, b, cd) \quad . \tag{19.1}$$

Generally we will consider cases of ω being a $U(1)$-valued complex phase. In group-cohomology notation we say that

$$\omega \in H^3(G, U(1)) \quad . \tag{19.2}$$

Eq. 19.1 may look obscure, but it is actually just a translation of the pentagon equation! Let us make the identification, in the notation of Chapter 9,

$$[F^{a,b,c}_{(abc)}]_{(ab),(bc)} = \omega(a, b, c)$$

so that we have diagrammatically Fig. 19.4

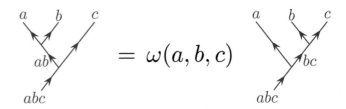

Fig. 19.4 The 3-cocycle is precisely an F-matrix. Compare to Fig. 9.1.

Examining the pentagon equation Eq. 9.7 and Fig. 9.7 we see that this is precisely the same as Eq. 19.1 in a different language. Note that there is no sum over indices here (like the sum over possible elements h in Eq. 9.7) since the fusion of any two group elements always gives a unique group element as an outcome.

As with F-matrices, it is possible to choose different gauges (see Section 9.4). In particular, given a 3-cocycle (i.e. a solution of the pentagon equation) we can multiply each a, b vertex by a phase $u(a, b)$ as shown in Fig. 19.5 to transform the cocycle by

$$w(a, b, c) \rightarrow \frac{u(a, bc)u(b, c)}{u(a, b)u(ab, c)}\, w(a, b, c) \quad . \tag{19.3}$$

By making such a gauge transformation we generate additional solutions of the pentagon equation. We view different solutions that are gauge transformations of each other as being physically equivalent, We will typically work with just one representative 3-cocycle for each equivalence class by choosing a convenient gauge. It is useful to always work with a so-called normalized gauge, where $w(a, b, c) = 1$ whenever $a = I$ or $b = I$ or $c = I$. (That is, fusing with the vacuum gives no phase.) Further, we want to only consider gauge transformations that maintain this normalized gauge, so we must insist on $u(I, g) = u(g, I) = u(I, I) = 1$. Given this restriction to normalized gauge, however, one still has a large additional gauge freedom.

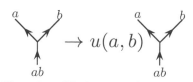

Fig. 19.5   We have the freedom to make a gauge transformation of a vertex by multiplying by a phase $u(a, b)$.

The 3-cocycle (pentagon) equation Eq. 19.1 typically will have more than one gauge-inequivalent solution. Further, if we have two different 3-cocycles w and w', we may multiply these together to generate another solution ww' and we may invert w to generate another solution. Thus, the space of 3-cocycles $H^3(G, U(1))$ in Eq. 19.2 is itself a group, known as the *third cohomology group of G with coefficients in $U(1)$*.

A trivial 3-cocycle $w(a, b, c) = 1$ for all $a, b, c \in G$ is always possible. In this case all diagrams have value 1. However, for any group (beyond the trivial group with only one element), there are always other possible 3-cocycles as well. Such 3-cocycles, and group cohomology in general, have been studied extensively in the mathematics and physics communities and it is possible to simply look up the form of the possible 3-cocycles (see the end of the chapter for good references).

While all 3-cocycles provide a solution to the pentagon equation, they do not always allow for isotopy invariance as discussed in Chapter 16. Indeed, for any 3-cocycle w, we will need to check whether it satisfies all the requirements for isotopy invariance. For example, we want to be able to freely turn up and down legs of a vertex as shown in Fig. 19.6.

Thus, for isotopy invariance (and allowing for d both $+1$ and -1) we need to have

$$s(a, b)w(a, a^{-1}, b) = 1 \tag{19.4}$$
$$s(a, b)w(a, b^{-1}, b) = 1 \tag{19.5}$$

for all a, b in the group with

$$s(a, b) = \begin{cases} -1 & d_a = d_b = -1 \\ +1 & \text{otherwise.} \end{cases} \tag{19.6}$$

While this condition seems quite restrictive, the gauge freedom Eq. 19.3 allows us often to achieve this.

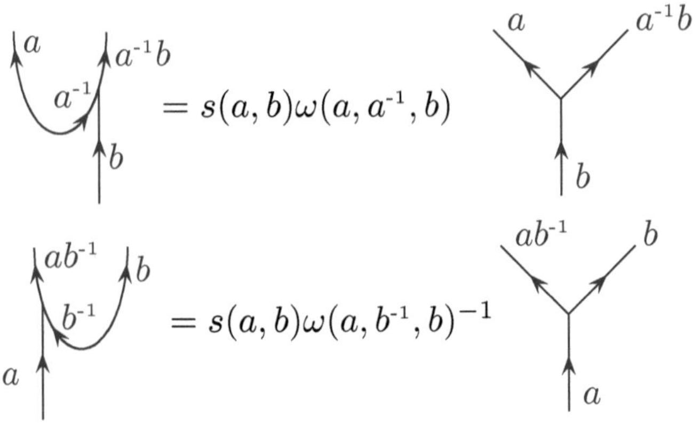

Fig. 19.6 Turning up and down relations (analogous to Fig. 14.17). The prefactor s comes from the proper interpretation of the sign of the $\sqrt{d}$ factors in 14.17. See Section 14.5.

A further item to note is that

$$d_a = \omega(a, a^{-1}, a)$$

and if a is self-dual ($a = a^{-1}$) this is the gauge-invariant Frobenius–Schur indicator. In the case where the Frobenius–Schur indicator is -1, if possible we will choose to work with negative d_a keeping ϵ_a positive as discussed in Section 14.5 so as to maintain isotopy invariance (this is not always possible, but often is).

19.1.1   Braidings for Abelian Group

Once we have a planar algebra we can look for possible solutions to the hexagon equations. Note that the hexagon equations implicitly assume that $a \times b = b \times a$, that is, that the fusion (and hence, the group) is abelian. Assuming an abelian group, we then have (see Exercise 19.2)

$$R_{ca}^{c,a}\, \omega(a,c,b)\, R_{cb}^{c,b} \;=\; \omega(c,a,b)\quad R_{cab}^{c,ab}\quad \omega(a,b,c) \quad (19.7)$$
$$[R_{ac}^{a,c}]^{-1}\omega(a,c,b)[R_{bc}^{b,c}]^{-1} \;=\; \omega(c,a,b)[R_{abc}^{ab,c}]^{-1}\omega(a,b,c) \quad .(19.8)$$

19.2   Simple Examples with $G = \mathbb{Z}_N$

For example, let us take a simple case of the group $G = \mathbb{Z}_N$, the group of integers modulo N with the group operation being addition modulo N. Since this group is abelian, we have[3] $gh = hg$ as required for the fusion of particle types as described in Chapter 8. Anyons based on this type of fusion have already been discussed in Section 18.5.2. Here we re-examine them from a slightly different point of view.

The inequivalent 3-cocycles of the group $\mathbb{Z}_N$ can be written as (see

[3]Confusingly, $gh = g \times h$ here mean addition of g and h modulo N.

references at the end of the chapter)

$$w(a, b, c) = \exp\left(\frac{2\pi i p}{N^2} a(b + c - [b + c]_N)\right) \qquad (19.9)$$

where here $a, b, c \in \{0, \dots, N-1\}$, and the brackets $[b+c]_N$ means $b+c$ modulo N where the result is chosen to lie in the range $\{0, \dots, N-1\}$. Here, the index p is an integer in the range $\{0, \dots, N-1\}$ describing the N different gauge-inequivalent 3-cocycles.

The trivial 3-cocycle is given by $p = 0$, which gives $w = 1$ always. The nontrivial 3-cocycles are more interesting.

$\mathbb{Z}_2$

Let us consider the simple case of $\mathbb{Z}_2$ fusion rules. Here, the group elements are $g = 0, 1$ and the group operation is addition modulo 2. One has the trivial 3-cocycle $p = 0$ in Eq. 19.9, giving $w = 1$, or all F-matrix elements equal to 1, which we identify as being exactly the same as the $d = 1$ loop gas from Section 18.1.1.

The only nontrivial 3-cocycle is the $p = 1$ case. Here, using Eq. 19.9 we determine the 3-cocycle is of the form

$$w(a, b, c) = \begin{cases} -1 & a = b = c = 1 \\ +1 & \text{otherwise.} \end{cases} \qquad (19.10)$$

We recognize this as being exactly the case of the $d = -1$ loop gas from Section 18.1.2. This translates to saying that the F-matrix is -1 if and only if all four incoming legs a, b, c, and abc in Fig. 19.4 are in the 1 state as in Eq. 18.11, and note that abc here means multiplication with the group operation, so is really $(a + b + c) \bmod 2$. It is easy to check that the conditions for isotopy invariance (Eq. 19.4 and Eq. 19.4) are satisfied, and the nontrivial F matrix simply assigns a minus sign to a "surgery" move on a diagram, as in Fig. 18.6.

$\mathbb{Z}_3$ and beyond

Generalizing the $\mathbb{Z}_2$ fusion to $\mathbb{Z}_3$, we now have $g = 0, 1, 2$ with the group operation being addition modulo 3. In this case we have three different 3-cocycles, the trivial 3-cocycle ($p = 0$ in Eq. 19.9) and two nontrivial 3-cocycles ($p = 1$ and $p = 2$ in Eq. 19.9). We discussed the case of the trivial cocycle (all non-zero elements of the F-matrix are unity) for $\mathbb{Z}_3$ in Section 18.3 which resulted in three possible braided theories, two of which are modular.

The nontrivial cocycles also provide a valid solution to the pentagon equation 19.1 (or Eq. 9.7) but they are not in a form where they enjoy isotopy invariance—nor can they be given a consistent braiding. One can use gauge transformations Eq. 19.3 to try to put the cocycles in different forms, but it is not possible to find a gauge where both Eq. 19.4 and Eq. 19.5 are satisfied at the same time. This is precisely the $\kappa^{(3)}$ obstruction discussed in Section 14.4 (see Exercise 19.3). Nonetheless, these cocycles still provide a consistent planar diagram algebra, although not an isotopy invariant one. Thus, the only isotopy-invariant case is the trivial cocycle $p = 0$.

More generally, there are two possible impediments to obtaining an isotopy-invariant diagram algebra. (i) The case of $\mathbb{Z}_N$ for any N a multiple of 3 is similar to the case of $\mathbb{Z}_3$: any p that is not a multiple of 3 cannot be made isotopy invariant. (ii) For $\mathbb{Z}_N$ with N a multiple of 4, only even p can be isotopy invariant (see Exercise 19.3). However, if neither of these two impediments (i or ii) occurs, then we should be able to choose a gauge where the diagram algebra is isotopy invariant. Of all the cocycles, only the $p = 0$ cases (for all N) and $p = N/2$ (for even N) can be given a consistent braiding (see Exercise 19.2). More details of these particular cases are provided in Sections 19.4 and 18.5.2.

19.3   Using Non-Commutative Groups?

[4]Here, we use the word *non-commutative* to mean when $g \times h \neq h \times g$. Often the word "nonabelian" is used to describe this. However, previously (see Section 8.2) we used the word nonabelian to describe fusion rules where there is more than one fusion channel, such as $g \times h = a + b + \ldots$.

In the case where the group is non-commutative, we deviate from what was done when we discussed fusion of particle types in Section 8.1. In the discussion of fusion of particle types, we have always assumed $g \times h = h \times g$ and with a non-commutative[4] group gh may not be the same as hg.

Why did we insist in Chapter 8 that particle fusion should satisfy $g \times h = h \times g$? If we think about particles living in three dimensions, when we bring two particles, g and h, together, looking at the system from one angle it looks like g is to the right of h; however, looking at the two particles from another angle, it looks like h is to the right of g. Thus, there is no way to decide whether the pair fuses to gh or hg.

However, if we are only concerned with a planar diagram algebra (or a diagram algebra on the surface of sphere) then there is no ambiguity! The surface we are considering is assumed to be oriented so we can always unambiguously decide which particle is clockwise of which other particle at a vertex. Thus, we can make the general rule that for a vertex to be an allowed fusion, the three particles *leaving* the vertex must multiply in clockwise order to the identity as shown in the right of Fig. 19.2. Thus, at least for planar diagrams we can generalize our rules for particle fusion to allow non-commutative fusions.

All of the figures we have drawn in this section (Figs. 19.2–19.6) have been drawn so as to be consistent with our rule for non-commutative groups—that is, if all of the arrows are outgoing, when you multiply the group elements clockwise around the vertex you obtain the identity.

As mentioned in Section 19.1, the hexagon equations implicitly assume commutative fusion, so the use of non-commutative groups allows us to construct planar diagram algebras, but we cannot give these proper braidings.

19.4 Appendix: Isotopy-Invariant Planar Algebras and Anyon Theories from $G = \mathbb{Z}_N$ Cohomology

We start with the cocycles for $\mathbb{Z}_N$ given in Eq. 19.9 and we look for isotopy-invariant cases. We have two general types of solution that admit a braiding: the $p = 0$ (trivial cocycle) solution, which exists for any N, and the $p = N/2$ case, which exists for even N only. Let us discuss each of these in a bit more detail.

19.4.1 Trivial Cocycle: $\mathbb{Z}_N^{(n)}$ Anyons

For any $\mathbb{Z}_N$ one can always choose $p = 0$ in Eq. 19.9, which gives the trivial cocycle $\omega(a, b, c) = 1$ for all a, b, c, and we correspondingly have $d_a = 1$ for all a. The solutions of the hexagon equation were already discussed in Section 18.5.3, so we will not discuss them further here.

19.4.2 Nontrivial Cocycle: $\mathbb{Z}_{N=2p}^{(n+\frac{1}{2})}$

Here, we consider $N = 2p$. The cocycle in Eq. 19.9 again has N consistent solutions of the hexagon equations, given by $n = 0, \ldots, (N - 1)$ given by

$$R_{[a+b]_N}^{a,b} = \exp\left[\frac{2\pi i(n + \frac{1}{2})}{N} ab\right]$$

with all $d_a = +1$ and $a, b \in \{0 \ldots (N - 1)\}$, resulting in the twist factors and S-matrix given in Section 18.5.4.

These anyon theories, in the gauge given by Eq. 19.9, are not generally isotopy invariant. We can generally make transformations to put these results in potentially simpler forms.

Case I: p-odd

With $N = 2p$ and p odd, the pth particle is self-dual and has Frobenius–Schur indicator -1, which is a gauge-invariant quantity (see Exercise 18.5). We will thus need to push this sign onto d in order to have an isotopy-invariant theory. As discussed in Section 14.5, we choose a gauge where

$$d_a = (-1)^a$$

(and as usual we are working with all $\epsilon_a = +1$ for an isotopy-invariant theory). Note that the composition rule Eq. 14.5 is satisfied. In this gauge the cocycle can be written as[5]

$$\omega(a, b, c) = \begin{cases} -1 & a, b, c \text{ all odd} \\ +1 & \text{otherwise.} \end{cases} \quad (19.11)$$

In this gauge the hexagon equations again has N solutions, which we

[5]If we try to use the same rule for $N = 2p$, with p even where we still set $d_a = (-1)^a$, this actually gives us the trivial cocycle discussed above in a less convenient gauge. Note that the self-dual particle has $d = +1$ in the p even case indicating that we do not need to push any signs onto d.

index as $n = 0, \ldots, N - 1$,

$$R^{a,b}_{[a+b]_N} = \exp\left[2\pi i \left(\frac{n}{N} + \frac{1}{2}\right) ab\right]$$

with the notation $[a]_N$ meaning a modulo N. To recover the correct S-matrix Eq. 18.55 we must invoke rule 0 from Section 14.5 since now we are working with some negative d_a. Again, the advantage of using this gauge for the description of $\mathbb{Z}_N^{(n+\frac{1}{2})}$ anyons is isotopy invariance (with $N = 2p$ and p odd).

Note that, considering the discussion of Section 18.6, we realize that for p-odd, the theory $\mathbb{Z}_{N=2p}^{(n+\frac{1}{2})}$ is actually of the form of a semion theory multiplied by a $\mathbb{Z}_p^{(n')}$ theory, which explains the simple form of Eq. 19.11.

Case II: p-even

In this case, the Frobenius–Schur indicator of the self-dual particle is $+1$ so we can choose to work in a gauge where all $d_a = +1$. While it is possible to make a gauge transformation that puts the theory into an isotopy-invariant form, the transformation is not particularly transparent. For this reason it is often convenient to stay with the gauge given in Eq. 19.9. However, it is not too hard to transform to an isotopy-invariant gauge if we would like.

As an example of this, let us consider the nontrivial cocycle for the group $\mathbb{Z}_4$. Here, we can make a gauge transformation (see Exercise 19.3) such that

$$\omega(a, b, c) = \begin{cases} -1 & (a, b, c) = (1, 1, 1); (1, 2, 3); (2, 1, 2); \\ & \quad (2, 3, 2); (3, 2, 1); (3, 3, 3) \\ 1 & \text{otherwise} \end{cases} \qquad (19.12)$$

and with $d = 1$ for all particles 0,1,2,3. This gives a fully isotopy-invariant theory. There are correspondingly four solutions of the hexagon equations given by[6] $n = 0, 1, 2, 3$ with

$$R^{a,b}_{[a+b]_4} = \exp\left[\frac{2\pi i(n + \frac{1}{2})}{4} ab\right] (-1)^{r(a,b)}$$

where $r(a, b) = 1$ for $(a, b) = (1, 2), (1, 3), (2, 1), (3, 1)$ only and is zero otherwise. As mentioned in Section 18.5.4, $n = 0$ is the abelian Chern–Simons theory $U(1)_4$ or equivalently the theory $(D_9)_1$. Its mirror image is $n = 3$; $n = 1$ is the Chern–Simons theory $SU(4)_1$, and $n = 2$ is its mirror image.

19.5   Appendix: Cocycles for S_3

To give an example of a non-commutative group, let us look at the case of the group S_3. To remind the reader, this group has six elements which

---

[6]This n is the same as that of Eq. 18.55.

can be written in terms of two generators, x and r, with multiplication rules $x^2 = r^3 = e$ and $xr = r^{-1}x$ with e the identity. The six elements can be written as $e, r, r^{-1}, x, xr, xr^{-1}$. Let us write them as $(A, a) = x^A r^a$ with $A = 0, 1$ and $a = -1, 0, 1$. There are six independent 3-cocycles described by $p = 0, \ldots 5$ in the equation (see references at the end of the chapter)

$$\omega((A, a), (B, b), (C, c)) = \tag{19.13}$$
$$\exp\{i\pi pABC\} \exp\left\{\frac{2\pi ip}{9}(-)^{B+C}a\left\{(-)^C b + c - [(-)^C b + c]_3\right\}\right\}$$

where the bracket $[]_3$ indicates modulo 3 where the result is assumed to be in the range $-1, 0, 1$.

Note that within S_3 there is a $\mathbb{Z}_2$ subgroup consisting of e and x, or $a = 0$ with $A = 0, 1$. The first term on the right-hand side, $\exp(i\pi pABC)$, matches the two possible 3-cocycles from the $\mathbb{Z}_2$ group. For even p it is the trivial cocycle, whereas for odd p we have a ω being -1 only when A, B, C are all in the 1 state, equivalent to Eq. 19.10. The second factor looks similar to the $\mathbb{Z}_3$ cocycles but only when $C = 0$. Setting $C = 0$ for a moment, the same argument as in the $\mathbb{Z}_3$ case shows that we cannot have full isotopy invariance unless $p = 0$ or $p = 3$, in which case the second factor on the right-hand side of Eq. 19.13 is trivial. Thus, this case of $p = 3$ gives an isotopy-invariant cocycle that essentially ignores the a variable of (A, a) and is equivalent to Eq. 19.10 for the A variables with $d_{(A,a)} = (-1)^A$.

Chapter Summary

- A discrete group G can be used to define a planar diagram algebra with the fusion rules given by group multiplication.
- A 3-cocycle gives F-matrices for this planar diagram algebra.
- For abelian (commutative) groups, some 3-cocycles allow a consistent braiding. For non-commutative groups, no consistent braidings are possible.

Further Reading:

- Good references on 3-cocycles and group cohomology are Hu et al. [2013], de Wild Propitius [1995], and Chen et al. [2013]. These are readable by physics audiences. They also include various references to the mathematics literature for those who want to learn group cohomology properly. Chapter 23 discussed how this concept is also used by Dijkgraaf and Witten [1990].
- See Bonderson [2007] for brief discussions of the $\mathbb{Z}_N$ anyon models.

Exercises

Exercise 19.1 Cocycle Equation
(a) Show that the 3-cocycle given by Eq. 19.9 satisfies cocycle condition Eq. 19.1 and thus represents a valid cocycle.

(b) Show that Eq. 19.13 also satisfies Eq. 19.1. You may want to use a computer to help with your bookkeeping!

Exercise 19.2 Cocycle Hexagon
(a) Derive the hexagon equations Eqs. 19.7–19.8 for abelian groups. Explain why one cannot solve a hexagon equation for a non-commutative (nonabelian) group.

(b) Use the hexagon equations, and the 3-cocycle Eq. 19.9 to show that $\mathbb{Z}_N$ can only have solutions of the hexagon equation for $p = 0$ or $p = N/2$ when N is even.

Exercise 19.3 Isotopy Invariance of Cocycles
(a) Show that the cocycle Eq. 19.9 can represent an isotopy-invariant planar diagram algebra only in the following cases:

(i) For N a multiple of 3, only when p is a multiple of 3.

(ii) For N a multiple of 4, only when p is even.

For (i), one can show that there is no isotopy-invariant diagram algebra when N is a multiple of 3 and p is not a multiple of 3 by either showing a nontrivial $\kappa^{(3)}$ or equivalently by showing that Eqs. 19.4 and 19.5 cannot be satisfied at the same time. Can you also show that these two conditions are equivalent?

For (ii) consider the quantum number $N/4$, which fuses with itself to give a self-dual quantum number $(N/2)$. Calculate the Frobenius–Schur indicator of the $N/2$ quantum number. Show that one cannot satisfy Eq. 14.5.

(b) For the case of $n = 4$ and $p = 2$, find the gauge transformation that transforms Eq. 19.9 into Eq. 19.12. You might want to use a computer to help with messy bookkeeping.

Bosons and Fermions from Group Representations: Rep(G)

Another way to construct a consistent planar diagram algebra is to work with representations of discrete groups.[1]

Suppose we have irreducible representations $\mathcal{R}_i$ of a group G. A tensor product of two of these irreducible representations will necessarily decompose into a direct sum of irreducible representations (irreps). That is, we have[2]

$$\mathcal{R}_a \otimes \mathcal{R}_b \simeq \mathcal{R}_c \oplus \mathcal{R}_d \oplus \ldots \tag{20.1}$$

with the sum on the right-hand side being finite. We thus propose to label a particle type for our diagrammatic algebra with an irreducible group representation, and have the fusion relations be given by these tensor-product decompositions. Thus, we interpret the tensor product equation Eq. 20.1 as a particle fusion relation

$$a \times b = c + d + \ldots$$

and accordingly particle a corresponding to representation $\mathcal{R}_a$ has antiparticle $\bar{a}$ corresponding to the dual representation which we write as $\mathcal{R}_{\bar{a}} = \mathcal{R}_a^*$. This fusion category (this set of fusion rules with the associated F-matrices[3]) using the representations of the group G is known as Rep(G).

It is fairly easy using some tricks of group theory to determine the fusion rules for discrete group representations. Recall that a representation $\mathcal{R}$ is a homomorphism[4] from each group element g to a matrix $\rho_{mn}^{\mathcal{R}}(g)$ (see Section 40.2.4). The trace of the representation matrix is known as its character, usually denoted by χ

$$\chi^{\mathcal{R}}(g) = \mathrm{Tr}[\rho^{\mathcal{R}}(g)] \quad .$$

One can either work out the characters of a group explicitly or (much more commonly) just look them up on character tables, which can be found in any group theory book or on the web.

Since $\mathrm{Tr}(ab) = \mathrm{Tr}(ba)$ we have $\chi^{\mathcal{R}}(g) = \chi^{\mathcal{R}}(hgh^{-1})$, meaning that the character depends only on the so-called conjugacy class of the group element g, which is the set of elements that can be written in the form hgh^{-1} for any $h \in G$.

Characters combine in fairly simple ways under both direct product

[1]To remind the reader, each discrete group has a finite number of irreducible representations, and any representation of the group can be decomposed into a direct sum of irreducible representations. See Section 40.2.4.

[2]If we write $M \otimes N = P$ we mean the following. If M_{ab} is a matrix of dimension m and N_{cd} is a matrix of dimension n then P is defined as $P_{(ac),(bd)} = M_{ab}N_{cd}$ and is of dimension nm where (ac) is a compound index taking nm values. If we write $P = N \oplus M$ we mean that P is block-diagonal with blocks N and M and total dimension $n + m$. Finally, note that the relation in Eq. 20.1 is an isomorphism, not an equality. One can choose a basis such that the right-hand side is block-diagonal, although this is not the natural basis for the left.

[3]It may not be clear at the moment that there exist any F-matrices that solve the pentagon equations. We construct them explicitly in Section 20.2.

[4]Meaning a mapping where the group operation is preserved: $\rho^{\mathcal{R}}(g_1)\rho^{\mathcal{R}}(g_2) = \rho^{\mathcal{R}}(g_1 g_2)$.

and direct sum

$$
\begin{align}
\chi^{\mathcal{R}_a \oplus \mathcal{R}_b}(g) &= \chi^{\mathcal{R}_a}(g) + \chi^{\mathcal{R}_b}(g) \tag{20.2}\\
\chi^{\mathcal{R}_a \otimes \mathcal{R}_n}(g) &= \chi^{\mathcal{R}_a}(g)\chi^{\mathcal{R}_b}(g) \quad . \tag{20.3}
\end{align}
$$

Further we have orthonormality relations for irreducible representations:[5]

$$
\frac{1}{|G|} \sum_{g \in G} [\chi^{\mathcal{R}_a}(g)]^* \, \chi^{\mathcal{R}_b}(g) = \delta_{\mathcal{R}_a, \mathcal{R}_b} \tag{20.4}
$$

where the sum is over all elements g of the group G and $|G|$ is *order* of the group, that is, the total number of elements in the group. We can thus deduce the tensor product decomposition[6],[7]

$$
\mathcal{R}_a \otimes \mathcal{R}_b \simeq \bigoplus_{c \in \text{irreps}} N_{ab}^c \, \mathcal{R}_c \tag{20.5}
$$

where

$$
N_{ab}^c = \frac{1}{|G|} \sum_{g \in G} [\chi^{\mathcal{R}_c}(g)]^* \, \chi^{\mathcal{R}_a}(g)\chi^{\mathcal{R}_b}(g) \tag{20.6}
$$

or, in our fusion product language

$$
a \times b = b \times a = \sum_c N_{ab}^c \, c \quad .
$$

Note that in the case where the group is abelian, the representations themselves are also an abelian group (meaning $N_{ab}^c = N_{ba}^c$ and for any a and b where there is only one value of c which makes N_{ab}^c unity, and it is otherwise zero).

It is not hard to show (see Exercise 20.1) that the quantum dimension of a representation $\mathcal{R}_a$ is given by[8]

$$
\mathsf{d}_a = \chi^{\mathcal{R}_a}(e) \tag{20.7}
$$

where e is the identity element of the group.

20.1   **Some Examples**

20.1.1   **Representations of S_3**

As a simple example, let us consider the representations of the group S_3, which can also be thought of as the symmetries of a triangle. To remind the reader[9] this group has six elements which can be written in terms of two generators x (a reflection) and r (a rotation) with multiplication rules $x^2 = r^3 = e$ and $xr = r^{-1}x$ with e the identity. The six elements can be written as $e, r, r^{-1}, x, xr, xr^{-1}$. There are three conjugacy classes, which we will call the identity (e), the rotations $(r, \, r^{-1})$, and the reflections (x, xr, xr^{-1}).

There are also three irreducible representations.[10] The group has a

---

[5]This orthonormality is derived trivially from the grand orthogonality theorem, Eq. 40.5. Since the character $\chi(g)$ is a function of only the conjugacy class of g it is sometimes more convenient to replace the sum over all elements with a sum over conjugacy classes where we then also include a factor of the number of elements in the class. So the left-hand side would read instead

$$
\sum_{\text{classes } C} \frac{|C|}{|G|} [\chi^{\mathcal{R}_a}(g \in C)]^* \, \chi^{\mathcal{R}_b}(g \in C)
$$

with $|C|$ meaning the number of elements in class C.

[6]The $\bigoplus$ symbol here means a direct sum of all the arguments. The prefactor N_{ab}^c here means the $\mathcal{R}_c$ representation occurs N_{ab}^c times in the direct sum.

[7]We have $\mathcal{R}_a \otimes \mathcal{R}_b \simeq \mathcal{R}_b \otimes \mathcal{R}_a$ meaning the two tensor products are isomorphic, but they are not equal. The two matrices have their entries in different places. See the definition in note 2.

[8]This is just the dimension of the representation, which must be an integer.

[9]The group S_3 is also sometimes known as the dihedral group with six elements, often denoted D_3 or sometimes D_6. See Section 40.2 for a few more details of this group.

[10]The number of irreducible reps is always equal to the number of conjugacy classes.

	identity 1 element	rotations 2 elements	reflections 3 elements
trivial rep (I)	1	1	1
sign rep (S)	1	1	-1
2d rep (V)	2	-1	0

Table 20.1 Character table for the group S_3. Notice the orthogonality of rows as defined by Eq. 20.4.

character table as given in Table 20.1. It is then easy to use Eq. 20.6 to determine the fusion rules for the representations, which are given by

$$I \times I = I, \qquad I \times S = S, \qquad I \times V = V \qquad (20.8)$$
$$S \times S = I, \qquad S \times V = V \qquad (20.9)$$
$$V \times V = I + S + V \qquad (20.10)$$

from which we see that I plays the role of the vacuum particle. Just as an example, let us derive Eq. 20.10. From the character table we have $\chi^V = (2, -1, 0)$ and so $\chi^{V \otimes V} = \chi^V \chi^V = (4, 1, 0) = (1, 1, 1) + (1, 1, -1) + (2, -1, 0) = \chi^I + \chi^S + \chi^V$.

20.1.2 Quaternion Group $\mathbb{Q}_8$

The quaternion group[11] can be defined as the eight 2×2 matrices $\pm 1, \pm i\sigma_x, \pm i\sigma_y, \pm i\sigma_z$. The group has five conjugacy classes $1, -1, \{\pm i\sigma_x\}, \{\pm i\sigma_y\}, \{\pm i\sigma_z\}$, and correspondingly five representations. The character table is given in Table 20.2. From the character table it is easy to use Eq. 20.6 to derive the nontrivial fusion rules (again I plays the role of the vacuum particle and we do not write its fusions):

$$\mathcal{R}_i \times \mathcal{R}_i = I \qquad i = x, y, z \qquad (20.11)$$
$$S \times \mathcal{R}_i = S \qquad i = x, y, z \qquad (20.12)$$
$$\mathcal{R}_x \times \mathcal{R}_y = \mathcal{R}_z \quad \text{(and cyclic permutations)} \qquad (20.13)$$
$$S \times S = I + \mathcal{R}_x + \mathcal{R}_y + \mathcal{R}_z \quad . \qquad (20.14)$$

Note that S is the two-dimensional representation given by the defining 2×2 matrices.

[11]The quaternions were famously discovered by Hamilton. He was so excited by this discovery that he carved them into the stone of Brougham (Broom) Bridge in Dublin. There is a plaque there today to commemorate this event.

class	1	-1	$\{\pm i\sigma_x\}$	$\{\pm i\sigma_y\}$	$\{\pm i\sigma_z\}$
elements	1	1	2	2	2
I	1	1	1	1	1
$\mathcal{R}_x$	1	1	1	-1	-1
$\mathcal{R}_y$	1	1	-1	1	-1
$\mathcal{R}_z$	1	1	-1	-1	1
S	2	-2	0	0	0

Table 20.2 Character table for the group $\mathbb{Q}_8$. Notice the orthogonality of rows as defined by Eq. 20.4.

20.2 F-Matrices in Rep(G)

With a bit of work, the F-matrices (often known as $6j$-symbols in this context) can also be derived using group theoretic methods. In general this can be a bit complicated, but the principle is straightforward group theory. Other discussions of similar techniques are given by Buerschaper and Aguado [2009], Butler [1981], and Wang et al. [2020].

We are familiar with group theory applied to angular momentum. In that case the representation is j and the state within the representation is notated by m, so we have kets $|jm\rangle$. Similarly, for a discrete group we have a representation $\mathcal{R}_a$, which we will label as a and the state within the representation we notate by m_a, so we write a ket as $|a\, m_a\rangle$ where the index m_a ranges between 1 and the dimension of the representation d_a. Note that for any one-dimensional representations (including the identity I rep) we do not need to write m since it can take only one value.

With angular momentum algebra, we are familiar with Clebsch–Gordan coefficients that couple $|j_1 m_1\rangle$ and $|j_2 m_2\rangle$ into $|jm\rangle$ and we write the coupling coefficient as $\langle jm|j_1 m_1; j_2 m_2\rangle$. Similarly, for representations $a, b, c, \ldots$ of a group G we have Clebsch–Gordan coefficients, which we now write as a vertex in our diagram algebra as follows:

$$\langle c\, m_c|a\, m_a; b\, m_b\rangle \quad = \quad \text{(vertex diagram with } a, b \text{ incoming and } c \text{ outgoing)}.$$

In Section 20.5.3 we discuss how the Clebsch–Gordan coefficients are calculated for discrete groups in general.

As usual, reflecting the diagram across a horizontal axis and reversing the arrows gives the complex conjugate coefficient (see Section 12.1), which we write as $\langle a\, m_a; b\, m_b|c\, m_c\rangle$. Note that we do not indicate the m values in the diagram. Further, when we assemble a diagram by connecting vertices, all the m values are assumed to be summed over.

As in angular momentum addition, the Clebsch–Gordan coefficients satisfy nice relations such as

$$\sum_{c, m_c} \langle a\, m_a; b\, m_b|c\, m_c\rangle\langle c\, m_c|a\, m_a'; b\, m_b'\rangle \;=\; \delta_{m_a, m_a'}\delta_{m_b, m_b'}$$

$$\sum_{m_a, m_b} \langle c\, m_c|a\, m_a; b\, m_b\rangle\langle a\, m_a; b\, m_b|c'\, m_c'\rangle \;=\; \delta_{m_c m_c'}\,\delta_{c c'}$$

which can be interpreted diagrammatically as the completeness relation of Fig. 12.21 and orthogonality relation Fig. 12.18, respectively. A special case of the latter is

$$a\,\diamondsuit\,\bar{a} \quad = \quad \sum_{m_a, m_{\bar{a}}} \langle I|am_a\bar{a}m_{\bar{a}}\rangle\langle am_a\bar{a}m_{\bar{a}}|I\rangle = 1 \quad ,$$

which tells us that our diagrams here are using "physics normalization."

We may now evaluate F-matrices by evaluating diagrams directly such as the one shown in Fig. 20.1 (compare to the similar diagram 12.33). The value of this diagram (in physics normalization) defines the F-matrix $[F_e^{abc}]_{df}$. We evaluate it using Clebsch–Gordan coefficients to give the following expression

$$[F_e^{abc}]_{df} = \sum_{m_a,,m_b,m_c,m_d,m_e,m_{e'},m_{\bar{e}},m_f} \langle I|em_e;\bar{e}m_{\bar{e}}\rangle\langle em_{e'};\bar{e}m_{\bar{e}}|I\rangle \times$$

$$\langle em_e|dm_d;cm_c\rangle\langle dm_d|am_a;bm_b\rangle\langle bm_b;cm_c|fm_f\rangle\langle am_a;fm_f|em_{e'}\rangle \quad .$$

$$(20.15)$$

Note that the resulting F may not be isotopy invariant, and may have nontrivial Frobenius–Schur indicators (see Section 20.5.2). It may also be convenient to then transform the gauge (see Section 9.4) to simplify F (potentially giving it isotopy invariance). We should be rather familiar with both of these issues by now!

20.3   Some Simple Braidings for Rep(G)

So far we have only discussed consistent fusion of representations, that is, fusion rules that will satisfy the pentagon equation. Given a set of F-matrices, we can then look for braidings, or R-matrices, that satisfy the hexagon equation. There will typically be multiple solutions of the hexagon equations. Some of these braidings can be stated easily for any group G.

20.3.1   "Trivial" Braidings: Bosons

We can assign a "trivial" braiding to Rep(G) by defining a braiding to simply be a reordering of the tensor factors. So if we start with particle a to the left of particle b we have $\mathcal{R}_a \otimes \mathcal{R}_b$ and if we braid b with a (either in front of or behind) we end up with b to the left of a or $\mathcal{R}_b \otimes \mathcal{R}_a$. We can assign[12]

$$R_c^{ab} = \begin{cases} 1 & a \neq b \\ 1 & a = b \text{ and } \mathcal{R}_c \text{ occurs in} \\ & \quad \text{symmetric part of the space } \mathcal{R}_a \otimes \mathcal{R}_a \\ -1 & a = b \text{ and } \mathcal{R}_c \text{ occurs in} \\ & \quad \text{antisymmetric part of the space } \mathcal{R}_a \otimes \mathcal{R}_a \end{cases}$$

$$(20.16)$$

In the case where $a = b$ we need to know a bit about how the c representation is related to product of two a representations. This is not hard to do, and we give details in Section 20.5.1. The intuition for the minus sign in the last case of Eq. 20.16 is clear: if we have an antisymmetric combination of $\mathcal{R}_a$ and $\mathcal{R}_a$ (like a spin-singlet made of two spin $\frac{1}{2}$s), exchanging the two will incur a minus sign.[13]

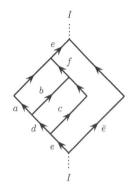

Fig. 20.1 The value of this diagram, in physics normalization, is $[F_e^{abc}]_{df}$.

[12]Note that R_c^{ab} with $a \neq b$ is gauge dependent, so Eqs. 20.16 and 20.17 imply a gauge choice. However, $R_c^{ab} R_c^{ba}$ is gauge independent, R_c^{aa} is gauge independent, and θ_a is gauge independent. Note however that θ_a depends on the sign we choose for the loop weight d_a and sometimes one chooses this sign negative if the Frobenius–Schur indicator κ_a is negative (see Eq. 15.1). Here let us instead choose the loop weight to be always positive.

[13]Since $\theta_a = 1$, using Eq. 15.4, we find that the Frobenius–Schur indicator is $\kappa_a = -1$ exactly when $\mathcal{R}_a$ and $\mathcal{R}_a$ fuse antisymmetrically to the identity representation, again analogous to the spin $\frac{1}{2}$ example of Section 14.7. See Section 20.5.2

This rule (Eq. 20.16) provides a solution of the hexagon equations for $\text{Rep}(G)$ for any group G. If we choose this braiding we are describing particles that are bosons and have trivial spin $\theta_a = 1$, with internal quantum numbers given by the representations of the group G.

20.3.2   Fermions and Bosons

If it so happens that G contains a central element[14] z such that $z^2 = e$ with e the identity of the group, then each representation $\mathcal{R}_a$ can be assigned a degree $p(a)$, which is 0 or 1 depending on whether z acts as the identity in the representation or acts as -1 in the representation.[15]

A consistent braiding is then given by (see Exercise 20.3)

$$[R_c^{a,b}]_{new} = (-1)^{p(a)p(b)}[R_c^{a,b}]_{\text{as in Eq. 20.16}} \quad . \tag{20.17}$$

In other words, we have altered the braiding defined in Eq. 20.16 to account for the possibility of both bosons and fermions. We have declared a particle a to be bosonic if $p(a) = 0$ (if $\rho_{R_a}(z)$ acts as 1) or fermionic if $p(a) = 1$ (if $\rho_{R_a}(z)$ acts as -1). The fermionic particles have $\theta_a = -1$ whereas the bosonic particles have $\theta_a = +1$. Sometimes this braided solution to the hexagon equation is called $\text{Rep}(G, z)$ compared to the fully bosonic solution, which can then be called $\text{Rep}(G, e)$. That is, if you set $z = e$, the identity, then you set all particles to be bosons.

A nontrivial example of $\text{Rep}(G, z)$ is given by the quaternion group $\mathbb{Q}_8$ (see Section 20.1.2) where we choose $z = -1$. It is easy to check that this makes the $\mathcal{R}_i$ representations bosons, but the S representation is a fermion (note the negative sign on the character $\chi^S(-1)$).

Note that for both $\text{Rep}(G)$ and $\text{Rep}(G, z)$ all of the particle types are *transparent*—that is, they braid trivially with all other particles in the theory. This is what is known as a *symmetric* tensor category.

20.3.3   Other Braidings

Once we have our F-matrices, it may or may not be possible to assign a nontrivial braiding to these particles (one that is not just bosons and fermions, but rather has some sort of anyons). It is sometimes tricky to figure out when this can occur. However, there are some cases where it is easy to show that no braidings besides bosons are possible, in particular if the fusion category $\text{Rep}(G)$ does not have any valid subcategories and G is nonabelian then only bosons are possible.[16]

20.4   Parastatistics Revisited

Way back in Section 3.5.1 we asked why we could not have exotic statistics in $(3 + 1)$-dimensions. While there are nontrivial representations of the permutation group that would satisfy the quantum-mechanical composition rule, we stated that additional constraints—such as particle creation and annihilation and locality—limits us to just bosons and fermions. We are now at the point where we can discuss exactly what we mean by this statement.

The structure we have built up for anyons in $(2 + 1)$-dimensions is that of a braided unitary category: a set of particles with fusions, F-matrices satisfying the pentagon equations, and R-matrices satisfying the hexagon equations. If we try to do something similar in $(3 + 1)$-dimensions we will no longer have nontrivial braiding of world lines since, as discussed in Section 3.3.2, in $(3 + 1)$-dimensions no knots can be formed in one-dimensional world lines. Thus, we must impose the restriction shown in Fig. 20.2 that all particles are transparent. In equations this can be stated as

$$R_c^{ab} R_c^{ba} = 1 \qquad (20.18)$$

for all a, b, c such that $N_{ab}^c > 0$. If the condition Eq. 20.18 holds and yet we have a solution of the hexagon equation, we say we have a *symmetric tensor category*.[17] Thus, if we are to construct an anyon theory with point particles in $(3 + 1)$-dimensions, it must be described by a symmetric tensor category.

In fact, we have already given two examples of symmetric tensor categories in Section 20.3: (a) the theory $\text{Rep}(G)$ (or $\text{Rep}(G, e)$) which describes bosons having internal quantum numbers given by the representations of the group G; and (b) the theory $\text{Rep}(G, z)$ where some of the particles are instead declared fermions depending on how their corresponding representation transforms under the action of the element z. The crucial theorem we mentioned in Section 3.5.1, originally due to Doplicher and Roberts (see also Deligne [2002] and Müger [2007]), is that there are no other possibilities: any symmetric tensor category is equivalent to $\text{Rep}(G) = \text{Rep}(G, e)$ if it has no fermions, or $\text{Rep}(G, z)$ if it has fermions. In other words, for point particles in $(3 + 1)$-dimensions, there are only bosons and fermions—nothing else!

20.5 Appendix: Further Group Theory

20.5.1 Symmetry of Fusion Products in $a \otimes a$

When we fuse two representations $a \otimes a$ there is a beautiful formula for determining which part of this fusion product is symmetric between the two pieces and which is antisymmetric. We can construct the character of the symmetric part and antisymmetric parts separately[18]

$$\chi_{\text{sym}}^{a \otimes a}(g) = \left([\chi^a(g)]^2 + \chi^a(g^2) \right) / 2 \qquad (20.19)$$

$$\chi_{\text{asym}}^{a \otimes a}(g) = \left([\chi^a(g)]^2 - \chi^a(g^2) \right) / 2 \quad . \qquad (20.20)$$

As with regular characters, these characters depend only on the conjugacy class of g.

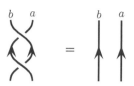

Fig. 20.2 In $(3 + 1)$-dimensions these two pictures are topologically equivalent. Thus, all particles are transparent. This implies Eq. 20.18.

[17] A symmetric tensor category is in some sense the exact opposite of a modular tensor category. For a modular tensor category no particles are transparent except the identity, whereas for a symmetric tensor category all particles are transparent.

[18] The proof of this formula is not too hard. Thinking in terms of d-dimensional representation matrices $\rho(g)$ for some representation (see Section 40.2.4), let the basis states for this representation be e_i with $i = 1, \ldots, d$ and let the eigenvalues of $\rho(g)$ be $\lambda_1 \ldots \lambda_d$. The eigenvalues of $\rho(g^2)$ are then $\lambda_1^2 \ldots \lambda_d^2$. The character is the sum over all eigenvalues. So $\chi(g) = \sum_i \lambda_i$ and $\chi(g^2) = \sum_i \lambda_i^2$. The tensor product $\rho(g) \otimes \rho(g)$ can be subdivided into the so-called symmetric square and the antisymmetric (or alternating) square. The symmetric part has basis states $e_i \otimes e_j + e_j \otimes e_i$ with $1 \le i \le j \le d$ whereas the antisymmetric part has basis states $e_i \otimes e_j - e_j \otimes e_i$ with $1 \le i < j \le d$. The eigenvalues of the symmetric part of $\rho(g) \otimes \rho(g)$ are then just $\lambda_i \lambda_j$ for $1 \le i \le j \le d$ whereas the eigenvalues of the antisymmetric part are just $\lambda_i \lambda_j$ with $1 \le i < j \le d$. Thus, $\chi_{\text{sym}}(g) = \sum_{1 \le i \le j \le d} \lambda_i \lambda_j$ whereas $\chi_{\text{asym}}(g) = \sum_{1 \le i < j \le d} \lambda_i \lambda_j$ which can be massaged into the given result of Eqs. 20.19 and 20.20. Try it!

Example: S_3

As an example, let us consider the group S_3 again. Using the character Table 20.1 along with the group multiplication described in Section 20.1.1, let us attempt to decompose $V \otimes V$ into its symmetric and antisymmetic parts.

From the Table 20.1, the characters of the rep V are $\chi^V = \{2, -1, 0\}$. Thus $[\chi^V]^2 = \{4, 1, 0\}$. To calculate $\chi^V(g^2)$ we have to determine the square of elements g in each conjugacy class. The conjugacy class of the identity includes only the identity, and its square is the identity so $\chi^V(g^2) = 2$. The second conjugacy class is rotations, and the square of a rotation is also a rotation so $\chi^V(g^2) = -1$. The final conjugacy class is a reflection, and the square of a reflection is the identity so $\chi^V(g^2) = 2$. So we have $\chi^V(g^2) = \{2, -1, 2\}$. Thus, using use Eqs. 20.19 and 20.20 we have $\chi^{V \otimes V}_{\text{sym}} = (\{4, 1, 0\} + \{2, -1, 2\})/2 = \{3, 0, 1\} = \chi^V + \chi^I$ whereas $\chi^{V \otimes V}_{\text{asym}} = (\{4, 1, 0\} - \{2, -1, 2\})/2 = \{1, 1, -1\} = \chi^S$. As stated in Eq. 20.10 (and worked out explicitly just thereafter), we have the fusion rules

$$V \times V = I + S + V \quad .$$

Here, we have shown that on the right-hand side of this equation I and V are symmetric and S is antisymmetric.

20.5.2   Frobenius–Schur Indicator

Let us consider $a \otimes a$ and consider whether its fusion to the identity is in the antisymmetric or symmetric channel. We first construct the symmetric and antisymmetric characters as in Eqs. 20.19 and 20.20. Then, to determine whether the identity occurs in the symmetric or antisymmetric channel, we use Eq. 20.4, multiply by the identity character $\chi^I(g) = 1$, and sum over the group. Thus, we have $\frac{1}{|G|} \sum_{g \in G} \chi^{a \otimes a}_{\text{sym}}(g)$ giving 1 if the identity occurs symmetrically in $a \otimes a$ and 0 if it occurs otherwise. And similarly we have $\frac{1}{|G|} \sum_{g \in G} \chi^{a \otimes a}_{\text{asym}}(g)$ giving 1 if the identity occurs antisymmetrically in $a \otimes a$ and 0 otherwise. Subtracting these from each other (and using Eqs. 20.19 and 20.20) we obtain an expression for

$$\kappa_a = \frac{1}{|G|} \sum_{g \in G} \chi^a(g^2) = \begin{cases} 1 & \text{identity occurs in} \\ & \text{symmetric part of } a \otimes a \\ -1 & \text{identity occurs in} \\ & \text{antisymmetric part of } a \otimes a \\ 0 & a \text{ not self-dual} \end{cases}$$

(20.21)

[19]Here we obtain a value of 0 for the case where a is not self-dual. In other places in this book we have avoided defining Frobenius–Schur indicators for the non-self-dual case.

which we recognize as the Frobenius–Schur indicator[19] for the "particle type" a. The discussion of Section 14.7 shows how objects that fuse to the identity in antisymmetric combinations must have negative Frobenius–Schur indicators.

The so-called *third* Frobenius–Schur indicator (see Section 14.4) can be obtained with a similar formula, simply replacing g^2 with g^3.

20.5.3   Clebsch–Gordan Coefficients of Discrete Groups

Details of working out Clebsh-Gordan coefficients are explained by standard books on group theory, such as Butler [1981] and Hamermesh [1989]. Here, we have tried to explain an equivalent version of the procedure as simply as possible.

We start with a d_a-dimensional representation $\rho^a(g)$ where we label the rows and columns with an index $m_a = 1, \ldots, d_a$.

We then write the tensor-product decomposition of two representations (Eq. 20.5) as

$$a \otimes b \simeq \bigoplus_{c\in \text{ irreps}} N^c_{ab}\, c \quad . \tag{20.22}$$

This equation is an isomorphism of representations, not an equality. We can make an equality of the corresponding representation matrices with a unitary transform

$$\rho^a(g) \otimes \rho^b(g) = U \left(\begin{array}{c|c|c|c} \rho^{c_1}(g) & \mathbf{0} & \cdots & \mathbf{0} \\ \hline \mathbf{0} & \rho^{c_2}(g) & \cdots & \mathbf{0} \\ \hline \vdots & \vdots & \ddots & \vdots \\ \hline \mathbf{0} & \mathbf{0} & \cdots & \rho^{c_n}(g) \end{array} \right) U^\dagger \tag{20.23}$$

where each block corresponds to one of the terms on the right-hand side of Eq. 20.22 (if $N^c_{ab} > 1$ then there are N^c_{ab} identical blocks with irrep c). Note that the same unitary matrix block-diagonalizes this equation for all $g \in G$.

On the left of this equation, the rows and columns of the tensor product are indexed by the $d_a d_b$ values of the pair (m_a, m_b) with $m_a = 1, \ldots, d_a$ and $m_b = 1, \ldots, d_b$. The elements of the rows and columns of each block c_i are indexed by $m_{c_i} = 1, \ldots, d_{c_i}$.

For simplicity of notation we will write the block-diagonal matrix as $\rho^{c_1}(g) \oplus \rho^{c_2}(g) \oplus \ldots \oplus \rho^{c_n}(g)$, so that Eq. 20.23 reads

$$\rho^a(g) \otimes \rho^b(g) = U \left(\rho^{c_1}(g) \oplus \rho^{c_2}(g) \oplus \ldots \oplus \rho^{c_n}(g) \right) U^\dagger \quad . \tag{20.24}$$

The entries in the unitary matrix in Eq. 20.24 are precisely the Clebsch–Gordan coefficients of the group—they indicate how elements of reps a and b fuse together to form elements of a given rep c. In particular, for fixed a, b we have

$$U_{(m_a m_b, c_i m_{c_i})} = \langle a m_a; b m_b | c_i m_{c_i} \rangle \quad .$$

It may sound difficult to actually determine the coefficients of the matrix U. However, there is a beautiful trick (I believe first used by Sakata [1974]) that enables quick calculation of this matrix at least in the case where no $N^c_{ab} > 1$. We choose any random matrix A and calculate

$$\tilde{U} = \sum_{g \in G} \left(\rho^a(g) \otimes \rho^b(g) \right) A \left(\rho^{c_1}(g) \oplus \rho^{c_2}(g) \oplus \ldots \oplus \rho^{c_n}(g) \right)^\dagger \quad (20.25)$$

this matrix $\tilde{U}$ is equivalent to U except that the columns of $\tilde{U}$ are un-normalized. For the case where $N_{ab}^c > 1$, slightly more tricky techniques are necessary and are described by van Den Broek and Cornwell [1978] or Butler [1981].

Chapter Summary

- Discrete group representations can be used as "particle types" to define a fusion algebra and F-matrices. Only bosons and fermions may be defined this way.
- In $(3 + 1)$-dimensions, the restriction to trivial braiding of point particles implies that all fusion algebras are (at most) of this discrete group type.

Further Reading:

- There are many good group theory books, but not many that work out $6j$ symbols in detail. See Butler [1981] and Hamermesh [1989].
- I do not know of any references on $\mathrm{Rep}(G)$ that are aimed at physicists.

Exercises

Exercise 20.1 Quantum Dimension of a Representation Prove Eq. 20.7. Hint: Remember that quantum dimension d_a tells you how the Hilbert-space dimension grows as you fuse together the particle a many times. Try fusing together many representations $\mathcal{R}^a \otimes \mathcal{R}^a \otimes \mathcal{R}^a \ldots$ and imagine decomposing the result into irreducible representations using the orthogonality theorem for characters. Note that for characters $\chi(e) \geq \chi(g)$ for e the identity representation.

Exercise 20.2 Frobenius–Schur Indicators in $\mathrm{Rep}(G)$
(a) Use Eq. 20.21 to calculate the Frobenius–Schur indicators of the representations for the groups S_3 and $\mathbb{Q}_8$.
(b) The dihedral group with eight elements, D_8 (sometimes called D_4) is the group of symmetries of a square. Look up the properties of this group. It turns out that it has exactly the same character table as $\mathbb{Q}_8$ (!!). Show that the Frobenius–Schur indicators do not match that of $\mathbb{Q}_8$.

Exercise 20.3 Bosons and Fermions in Rep(G, z)

Let z be a central element of the group G (i.e. $zg = gz$ for all $g \in G$) such that $z^2 = e$, the identity. As in Section 20.3.2, for a representation $\mathcal{R}_a$ set $p(a) = 0$ if z acts as the identity in representation $\mathcal{R}_a$ and set $p(a) = 1$ if z acts as -1 in representation $\mathcal{R}_a$.

(a) Show that if $\mathcal{R}_a \otimes \mathcal{R}_b = \mathcal{R}_c \oplus \ldots$, then $p(a) + p(b) = p(c) \mathrm{mod} 2$. Hint: Consider the characters $\chi^{\mathcal{R}}(e)$ and $\chi^{\mathcal{R}}(z)$.

(b) Given that setting all particles to bosons (i.e. Eq. 20.16) solves the hexagon equation, show that Eq. 20.17 also provides a solution to the hexagon equation.

Exercise 20.4 F-matrix Elements For Representations of S_3 [Hard]

Let us consider the simplest nonabelian group S_3, which we discuss in Sections 19.5, 20.1.1, and 40.2.1.

We remind the reader that this group has six elements which can be written in terms of two generators x and r with multiplication rules $x^2 = r^3 = e$ and $xr = r^{-1}x$ with e the identity. The six elements can be written as e, r, r^2, x, xr, xr^2 that are grouped into conjugacy classes $\{e\}, \{r, r^2\}, \{x, xr, xr^2\}$ (see Table 20.1).

The three representations are as follows. The trivial representation has $\rho^I(g) = 1$ for all g in the group. The sign rep has $\rho^S(g) = 1$ for $g \in \{e, r, r^2\}$ and $\rho^S(g) = -1$ for $g \in \{x, xr, xr^2\}$. (Note that since both these reps are one-dimensional, they are completely defined by the character table). We write the two-dimensional representation in a unitary form as

$$\rho^V(x) = \begin{pmatrix} -1 & 0 \\ 0 & 1 \end{pmatrix} \qquad \rho^V(r) = \frac{-1}{2}\begin{pmatrix} 1 & \sqrt{3} \\ -\sqrt{3} & 1 \end{pmatrix}$$

with $\rho^V(e)$ the identity matrix and all other matrices $\rho^V(g)$ for the other elements g in the group can be generated by using the group multiplication properties. (Do this first, you will need it later!)[20]

Note that we already know the fusion rules for these representations as they are given in Eqs. 20.8–20.10.

In this exercise we will calculate some F-matrix elements by focusing on the most interesting case, where all four incoming lines in Fig. 16.4 are in the two-dimensional V representation. Thus, we are interested in the unitary matrix $[F_V^{VVV}]_{ef}$.

(a) Using Eq. 20.25 calculate the unitary matrix that diagonalizes $V \otimes V$ into direct sums of $I \oplus S \oplus V$. This allows us to extract the needed Clebsch–Gordan coefficients $\langle c|Vm; Vm'\rangle$ with $c = I, S$ and $\langle Vm|Vm'; Vm''\rangle$ and each $m = 1, 2$. We will also need $\langle Vm|Vm'; c\rangle$ and $\langle Vm|c; Vm'\rangle$ with $c = I, S$.

(b) Finally, use Eq. 20.15 to calculate

$$[F_V^{VVV}]_{ef} = \frac{1}{2}\begin{pmatrix} 1 & 1 & \sqrt{2} \\ 1 & 1 & -\sqrt{2} \\ \sqrt{2} & -\sqrt{2} & 0 \end{pmatrix}$$

up to a gauge choice. You may need to gauge transformation to obtain a result agreeing with this.

[20] It may be useful to use a computer to multiply matrices (Mathematica, matlab, octave, and python are all fairly convenient), since there are a lot of matrix manipulations in this problem and a single error will destroy the result.

Quantum Groups (in Brief)

21.1 Continuous (Lie) Group Representations?

In Chapter 20 we considered associating particle types with representations of discrete groups. One can imagine that instead of looking at the representations of discrete groups, we consider instead the representations of continuous (or Lie) groups (see Section 40.2.3). For example, the different representations of the group $SU(2)$ are the different values of the spin quantum number j, and these fuse together via the usual angular momentum addition rules, entirely analogous to the discussion of Section 20.2).

While such a scheme makes a perfectly good planar diagram algebra, one problem is that there are an infinite number of different irreducible representations (for the case of $SU(2)$ the angular momentum j can be infinitely large) and this violates our rule of having a finite number of "particle types" for our diagrammatic algebra. Such algebras can be problematic when used for physical purposes. For example, as we discuss in Section 23.3, using a diagrammatic algebra with an infinite number of irreducible representations for construction of a TQFT results in divergences.

Furthermore, if we were to use the representations j of $SU(2)$ as our particle types, the only braidings possible would be the fairly trivial braidings discussed in Section 20.3. If we are interested in building anyon theories with nontrivial braidings, this is not the right place to look.

One interesting approach that can address both of these problems is to "deform" the representations of the Lie group. In doing so, one can arrange that only a finite number of the representations occur, and further, one can arrange nontrivial braidings. The structure we are looking for that does exactly this are the representations of the "quantum group" built from our Lie group, rather than the representations of the original Lie group. These representations in many cases correspond exactly to the Chern–Simons theories, which use the particular Lie group as a gauge group.

We should be clear that a quantum group is not actually a group, but rather is a more sophisticated mathematical structure—and it is beyond the scope of this book to even explain what a quantum group actually is (see references at the end of this chapter). However, if we focus on just the representation theory of these quantum groups, they seem extremely similar to the regular representation theory of the Lie groups upon which they are based. Indeed, they are just "deformations" of the Lie group representations. I emphasize that this chapter is not meant to give real derivations, but rather to just give a rough idea of how this approach works and some of its results.

21.2 $U_q(sl_2)$: **The Deformation of Representations of** $SU(2)$

For simplicity we focus on the Lie group $SU(2)$, which is probably familiar to most people, with its representation theory being taught as angular momentum algebra in most quantum mechanics courses. Extension of the same principles to other Lie groups follows somewhat similarly (briefly described in Section 21.3).

21.2.1   **Representation Theory of** $SU(2)$

Let us start by reviewing the representation theory for the Lie group $SU(2)$. We write the usual angular momentum operators as $\hat{L}_x, \hat{L}_y, \hat{L}_z$. We find it convenient to rewrite $\hat{L}_x$ and $\hat{L}_y$ as

$$\hat{L}_{\pm} = \hat{L}_x \pm i\hat{L}_y$$

so that we can write the standard commutation relation of angular momentum operators as

$$[\hat{L}_z, \hat{L}_{\pm}] = \pm\hat{L}_{\pm} \tag{21.1}$$
$$[\hat{L}_+, \hat{L}_-] = 2\hat{L}_z \quad . \tag{21.2}$$

We also have the quadratic Casimir operator

$$\widehat{J^2} = \hat{L}_x^2 + \hat{L}_y^2 + \hat{L}_z^2 = \hat{L}_+\hat{L}_- + L_z(L_z - 1) \tag{21.3}$$

which commutes with $\hat{L}_z, \hat{L}_+, \hat{L}_-$ (see Exercise 21.1.a.i).

We choose basis states such that they are eigenstates of $\hat{L}_z$. A "highest-weight" state is one that is annihilated by $\hat{L}_+$ and a "lowest-weight" state is annihilated by $\hat{L}_-$. Let us consider a highest-weight state having eigenvalue j of $\hat{L}_z$ (we do not yet insist that $2j$ is an integer). We write this state as $|j, j\rangle$.

Applying the lowering operator successively to the highest-weight states we obtain a normalized basis of states $|j, m\rangle$ satisfying (see Exercise 21.1.a.ii–iii)

$$\hat{L}_z|j, m\rangle = m|j, m\rangle \tag{21.4}$$
$$\hat{L}_{\pm}|j, m\rangle = \sqrt{(j \mp m)\,(j \pm m + 1)}\,\,|j, m \pm 1\rangle \tag{21.5}$$
$$\widehat{J^2}|j, m\rangle = j(j + 1)|j, m\rangle \quad . \tag{21.6}$$

If we are to obtain a finite-dimensional representation, we must reach a lowest-weight state, which happens when $2j$ is an integer (so that the factor on the right of Eq. 21.5 will vanish both when $\hat{L}_+$ is applied on the highest-weight state and when $\hat{L}_-$ is applied on the lowest weight state). Thus, we have representations with j integer or half-integer, and we have $m = -j, \ldots, j$.

If we are so interested, we can go further to work out details of fusion of representations, and calculate quantities such as Clebsch–Gordan coefficients. We will not do that here, but we trust that most readers have been through this exercise before.[1] Once we have Clebsch–Gordan coefficients, we can put them into Eq. 20.15 to generate $6j$ symbols, or F-matrices for fusing representations, the fusion rules being given by the usual angular momentum addition rules

$$ j_1 \otimes j_2 = |j_1 - j_2| \oplus (|j_1 - j_2| + 1) \oplus \ldots \oplus (j_1 + j_2) \quad . \qquad (21.7) $$

Finally, we can give these representations a braiding using R-matrices as in Eq. 20.16, which in this case amounts to

$$ R_c^{ab} = (-1)^{a+b-c} \qquad (21.8) $$

which is a fairly trivial braiding, but this is what we should expect from the representation of a group.

21.2.2 q-Deformations

The type of deformation we want is known as a q-deformation (or sometimes q-analog). The idea is to take any mathematical object and make it a function of a parameter q in such a way that the limit $q \to 1$ recovers the original (undeformed) quantity. One can define many different deformations, but certain deformations are particularly useful. Consider the following definition[2,3] of a deformation of an integer n,

$$ \begin{aligned} [n]_q &= \frac{q^{n/2} - q^{-n/2}}{q^{1/2} - q^{-1/2}} \\ &= q^{-n/2+1/2} + q^{-n/2+3/2} + \ldots + q^{n/2-3/2} + q^{n/2-1/2} \end{aligned} $$

with $[1]_q = 1$ and $[0]_q = 0$ and generally $[-n]_q = -[n]_q$. The limit $q \to 1$ recovers the integer n. That is, $\lim_{q \to 1}[n]_q = n$.

It is also useful to define q-factorials as

$$ [n]_q! = [n]_q \, [n-1]_q \, [n-2]_q \, \ldots \, [2]_q \, [1]_q $$

and $[0]_q!$ is defined to be 1.

21.2.3 Deformed Representation Theory

Let us now make a particular deformation[4] of the representation theory of $SU(2)$. Let us consider

$$ [\hat{L}_z, \hat{L}_\pm] = \pm \hat{L}_\pm \qquad (21.9) $$

[1] Potentially painfully if your first quantum mechanics course was anything like mine!

[2] In the physics literature some works define q differently (I will call this different definition $\tilde{q}$) such that $\tilde{q} = q^{1/2}$ and

$$ [n]_q = \frac{\tilde{q}^n - \tilde{q}^{-n}}{\tilde{q} - \tilde{q}^{-1}} $$

instead. In the combinatorics literature one often finds a different definition of the q-integer given by

$$ [n]_q = \frac{1 - q^n}{1 - q}. $$

We will never use this definition, but it saves a lot of grief to realize that a lot of papers do use it!

[3] This particular deformation satisfies a number of identities familiar from the usual integers, such as

$$ [n+m]_q[n-m]_q = [n]_q^2 - [m]_q^2 \quad . $$

[4] This deformation is formally known as $U_q(sl_2)$. This is as compared to the representation theory of regular $SU(2)$, which is known as $U(sl_2)$.

$$[\hat{L}_+, \hat{L}_-] \;=\; [2\hat{L}_z]_q \tag{21.10}$$

and the quadratic Casimir operator

$$\widehat{J^2} = \hat{L}_+\hat{L}_- + [L_z]_q[L_z - 1]_q = \hat{L}_-\hat{L}_+ + [L_z]_q[L_z + 1]_q \tag{21.11}$$

which commutes with $\hat{L}_z, \hat{L}_+, \hat{L}_-$ (see Exercise 21.1.b.i). These equations look exactly like Eqs. 21.1–21.3 except that the $\hat{L}_z$ on the right-hand side of Eqs. 21.10 and 21.11 has been q-deformed.

We now follow analogously to the discussion of the undeformed representations. We choose basis states such that they are eigenstates of $\hat{L}_z$. A "highest-weight" state is one that is annihilated by $\hat{L}_+$ and a "lowest-weight" state is annihilated by $\hat{L}_-$. We consider a highest-weight state having eigenvalue j of $\hat{L}_z$, which we write as $|j, j\rangle$.

Applying the lowering operator successively to the highest-weight states we obtain a normalized basis of states $|j, m\rangle$ satisfying (see Exercise 21.1.b.ii–iii)[5]

$$\hat{L}_z|j, m\rangle \;=\; m|j, m\rangle \tag{21.12}$$

$$\hat{L}_\pm|j, m\rangle \;=\; \sqrt{[j \mp m]_q \, [j \pm m + 1]_q} \; |j, m \pm 1\rangle \tag{21.13}$$

$$\widehat{J^2}|j, m\rangle \;=\; [j]_q \, [j + 1]_q \, |j, m\rangle \tag{21.14}$$

Entirely analogous to the situation with undeformed $SU(2)$ we start with the highest-weight state and apply $\hat{L}_-$. If we are to reach a lowest-weight state, $2j$ must be an integer. Thus, we again have representations with j integer or half-integer, and we have $m = -j, \ldots, j$.

We can continue on to formulate q-deformed Clebsch–Gordan coefficients, and then use an analog of Eq. 20.15 to generate $6j$ symbols, or F-matrices.[6] The fusion rules follow the $SU(2)$ fusion rules of Eq. 21.7. For completeness, we give the general formula for the F-matrices here.[7]

For $a \le b + c$, $b \le a + c$, $c \le a + b$ and $a + b + c$ an integer we define

$$\Delta(a, b, c) = \sqrt{\frac{[a + b - c]_q! \, [a - b + c]_q! \, [-a + b + c]_q!}{[a + b + c + 1]_q!}} \quad . \tag{21.15}$$

Then we have

$$[F_d^{abc}]_{ef} \;=\; (-1)^{(a+b+c+d)}\Delta(a, b, e)\Delta(c, d, e)\Delta(b, c, f)\Delta(a, d, f)\sqrt{[2e + 1]_q}\sqrt{[2f + 1]_q}$$

$$\sum_n \frac{(-1)^n [n + 1]_q!}{[a + b + c + d - n]_q! \, [a + c + e + f - n]_q! \, [b + d + e + f - n]_q!}$$

$$\times \frac{1}{[n - a - b - e]_q! \, [n - c - d - e]_q! \, [n - b - c - f]_q! \, [n - a - d - f]_q!} \tag{21.16}$$

where the sum over n is over non-negative integers, such that $\max(a+b+e, c+d+e, b+c+f, a+d+f) \le n \le \min(a+b+c+d, a+c+e+f, b+d+e+f)$. In the limit of $q \to 1$ this reduces to the usual formula for angular

---

[5]The very observant reader will notice that the right-hand side of Eq. 21.13 may not have a real quantity inside the argument of the square root for arbitrary q, which can complicate matters because $\hat{L}_+$ would then not be $\hat{L}_-^\dagger$. However, for q as chosen in Eq. 21.19 the argument is positive at least for $2j \le k + 1$, which is all we will need in the end.

[6]The necessary analog of Eq. 20.15 looks a bit strange because some of the terms on the right-hand side appear to be incorrectly complex conjugated (see Ardonne and Slingerland [2010]). This occurs (roughly) because the inner product that is used in quantum groups is (strictly speaking) an analytic continuation of q from the real axis to the complex plane and this modification is necessary so that the resulting F-matrix remains unitary, at least for the case of q being a primitive root of unity, which is all we will be concerned with in the end.

[7]This result was first worked out by Kirillov and Reshetikhin [1988].

momentum addition (the so-called $6j$-symbol) that we would use for the undeformed $SU(2)$ representations.

One might think that one could also use Eq. 20.16 (or Eq. 21.8) to give a braiding for these representations. However, the deformation of the group representation means that the "trivial" braiding (Eq. 21.8) no longer satisfies the hexagon equation once q is not equal to unity. However, there is a fairly simple fix for this. Instead, we write

$$R_c^{ab} = (-1)^{a+b-c} q^{\frac{1}{2}\{c(c+1)-a(a+1)-b(b+1)\}} \qquad (21.17)$$

and this now satisfies the hexagon!

For completeness, the twist factors of particles here are

$$\theta_j = q^{j(j+1)} \qquad (21.18)$$

and one can check from the F-matrices that the Frobenius–Schur indicators[8] are $\kappa_j = (-1)^{2j}$. Here, our diagram algebra is not isotopy invariant, due to the Frobenius–Schur indicator, but we have already discussed how to deal with this in Section 14.5 for example.

[8]Since quantities must change continuously with q, quantities that can only take quantized discrete values, such as the Frobenius–Schur indicator, must be q-independent.

21.2.4 q at Roots of Unity: $SU(2)_k$

For "generic" values of q, the representation theory of deformed $SU(2)$ is very similar to the representation theory of regular (undeformed) $SU(2)$. Both of these cases have an infinite number of possible irreducible representations (j can be arbitrarily large). However, for certain special values of q, it is more complicated, and indeed, these are the more interesting cases because for these values the number of irreducible representations becomes finite.

We will be concerned here with values of q that are primitive roots of unity. In particular let us set[9]

$$q = e^{2\pi i/(k+2)} \quad . \qquad (21.19)$$

[9]As per margin note 2, if one is working with the other definition of q, we have $\tilde{q} = e^{i\pi/(k+2)}$.

Note that as k gets very large q approaches unity and we should recover the "undeformed" $SU(2)$.

Having chosen q in this way, we now have

$$[k+2]_q = [p(k+2)]_q = 0$$

for any integer p. This has an interesting effect. For any $|j, m\rangle$ we have

$$(\hat{L}_\pm)^{k+2} |j, m\rangle = 0$$

since after raising or lowering enough times we will eventually hit a term of $[p(k+2)]_q$ inside of the factors on the right-hand side of Eq. 21.13. For $j \le k/2$, this doesn't change anything from the "generic" values of q (not roots of unity), since it is already the case that $\hat{L}_\pm^{k+2}$ annihilates any state with this value of j (indeed, this is the same as for regular undeformed $SU(2)$). However, for $j \ge (k+1)/2$ we now have a situation where we have new highest- and lowest-weight states—that is, states that are annihilated by $\hat{L}_+$ or $\hat{L}_-$, which would not be annihilated for generic values of q. This is a sign that the representations at these values of j are not irreducible. As a result, our list of irreducible representations now only includes those with

$$j \leq k/2$$

which is exactly what we were hoping for: a finite list of irreducible representations!

At this special value of q, the fusion rules (Eq. 21.7) now become truncated to

$$j_1 \otimes j_2 = |j_1 - j_2| \oplus (|j_1 - j_2| + 1) \oplus \ldots \oplus \min(j_1 + j_2, k - j_1 - j_2) \tag{21.20}$$

so that representations with $j > k/2$ never appear. Indeed, if we try to plug in $j > k/2$ into Eq. 21.16 the F-matrix will necessarily vanish. Rather surprisingly (and crucially), the resulting q-deformed F-matrices remain unitary so that they are acceptable to us for describing unitary quantum mechanical anyon systems.

With these fusion rules, the quantum dimensions take the simple form

$$\mathsf{d}_j = [2j + 1]_q \quad . \tag{21.21}$$

In fact, inserting the values of q given by Eq. 21.19 into Eqs. 21.16 and 21.17 gives exactly the F- and R-matrices for the Chern–Simons theory $SU(2)_k$ (!) with particle types $j \in 0, \frac{1}{2}, 1, \frac{3}{2}, \ldots, \frac{k}{2}$.

The mirror image theory may be obtained by complex conjugating q everywhere. The F-matrices will be unchanged, but R becomes complex conjugated.

Where Did m Go?

In the structure of the quantum group representations, we have states $|j, m\rangle$ just like we have in regular angular momentum representations. As in regular $SU(2)$ representations, in the F-matrices, everything is written in terms of j values and there is no reference to m but we know that for regular $SU(2)$ the m variable is still a physical variable.

However, in anyon theories we have only j, the particle label—we don't generally have any internal quantum numbers. What happened to the m degree of freedom? In fact, this degree of freedom is gone. We say that the anyon system has a "hidden" quantum group symmetry.

It is in fact crucial that there is no m degree of freedom. Let us look at regular $SU(2)$ spins. Suppose we have two spin $\frac{1}{2}$ particles, $j = 1/2$, forming a singlet $j_{\text{total}} = 0$ and we move the two particles apart from each other. Since these $j = 1/2$ objects have an internal m degree of freedom, we can consider a local operator that changes the local m (flips over the spin $\frac{1}{2}$ for example). This local operation[10] can now turn the

---

[10] Starting with the singlet state

$$\frac{1}{\sqrt{2}}(|\uparrow_1\downarrow_2\rangle - |\downarrow_1\uparrow_2\rangle)$$

we can apply σ_x on the first spin to obtain

$$\frac{1}{\sqrt{2}}(|\downarrow_1\downarrow_2\rangle - |\uparrow_1\uparrow_2\rangle)$$

which is a superposition of two triplet states with $m = -1$ and $m = +1$.

singlet $j_{\text{total}} = 0$ into a triplet $j_{\text{total}} = 1$. However, with anyons we do not expect anything like this to be possible due to the locality principle (see Section 8.6 for example). Suppose we have $SU(2)_k$ anyons (with $k > 1$), and we put two of the $j = 1/2$ anyons into a fusion channel with overall $j_{\text{total}} = 0$ and move the two anyons apart. There should be no operator that acts on only one of the anyons that can change this fusion channel. If such an operator existed it would violate the locality principle!

21.3 Other Lie Groups

One can attempt the same type of quantum group deformation with other Lie groups, analogous to what we outlined here for $SU(2)$. First, one starts with the representation theory of whatever Lie group we are interested in. Lie group representation theory is well known, but it would be a substantial digression to discuss it here (see for example Zee [2016]). In short, while generally a bit more complicated than that of $SU(2)$ one can always define an appropriate set of operators and commutators analogous to Eqs. 21.1–21.2 that then can be used to find all the irreducible representations, Clebsch–Gordan coefficients, and $6j$-symbols or F-matrices. This algebraic approach can then be appropriately q-deformed to give a quantum group, corresponding F-matrices, along with an appropriate braiding R-matrices.

While we do not give detailed descriptions of how it is done, the quantum group approach can be used to describe any Chern–Simons theory—although detailed derivation of quantities such as the F-matrices becomes algebraically very complicated for all but the simplest cases (see references at the end of the chapter).

Nonetheless, we may state some general useful results. First, we choose a Lie algebra $\mathfrak{g}$ (see Table 21.1). The representation of the quantum group with deformation parameter q is known as $U_q(\mathfrak{g})$. Again for values of q given by certain roots of unity we obtain a finite number of irreducible representations. In particular, if we use

$$q = e^{2i\pi/\ell}$$

where $\ell = m(\mathfrak{g})(k + \breve{h}(\mathfrak{g}))$ and the values of $m(\mathfrak{g})$ and $\breve{h}(\mathfrak{g})$ are given in Table 21.1 we obtain precisely the Chern–Simons theory with gauge group Lie algebra $\mathfrak{g}$ at level $k > 0$, which we write as $\mathfrak{g}_k$. Complex conjugating q will give the mirror image theory (or equivalently gives level $-k$). The central charge at level $k > 0$ is given by

$$c(\mathfrak{g}_k) = \frac{k \dim(\mathfrak{g})}{k + \breve{h}(\mathfrak{g})}$$

where the dimension (dim) and dual Coxeter number ($\breve{h}$) for each Lie algebra is given in Table 21.1.

One can also choose q to be a non-primitive root of unity: $q = e^{2i\pi z/\ell}$

Table 21.1 Properties of simple lie algebras. Dim is the dimension of the algebra, $\check{h}$ is the dual Coxeter number, and $\sqrt{m}$ is the ratio of long-to-short root distance.

$\mathfrak{g}$	A_r	B_r	C_r	D_r	E_6	E_7	E_8	F_4	G_2
$\dim(\mathfrak{g})$	$r^2 + 2r$	$2r^2 + r$	$2r^2 + r$	$2r^2 - r$	78	133	248	52	14
$\check{h}(\mathfrak{g})$	$r + 1$	$2r - 1$	$r + 1$	$2r - 2$	12	18	30	9	4
$m(\mathfrak{g})$	1	2	2	1	1	1	1	2	3

$$A_r = su(r+1) \qquad B_r = so(2r+1) \qquad C_r = sp(2r) \qquad D_r = so(2r)$$

with z an integer. If z is relatively prime with ℓ one obtains a theory with the same fusion rules as $z = 1$, albeit not necessarily a modular or unitary theory. There are certain cases where one can use z, which is a multiple of $m(\mathfrak{g})$ (these are known as "nonuniform" cases) that can give theories completely unrelated to the Chern–Simons theory $\mathfrak{g}_k$. Details of this are beyond the current discussion, so we refer the reader to Rowell [2005].

Calculating the F-matrices for general quantum groups (and Chern–Simons theories) can be extremely complicated. Software is available for doing this generally (see references at the end of the chapter). However, the calculation of S- and T-matrices is actually not too complicated analytically (see Rowell [2005] and Di Francesco et al. [1997] for example for the Chern–Simons case). Doing so requires some more detailed discussion of the structure of Lie algebras, so we do not do this here.

Chapter Summary

- Representations of quantum groups provide a way to describe many anyon theories/TQFTs. Chern–Simons theories can generally be described this way.
- The representation theory of a quantum group is a "deformation" of the representation theory of classical Lie algebras.

Further Reading

- There is a long mathematical history developing the technology of quantum groups. Some very complete books on the topic are available, such as Kassel [1995], although this is not particularly easy reading for physicists.
- Much of the structure of quantum groups is used heavily in the study of integrable condensed matter systems. A very nice (and fairly readable) book on this application of quantum groups is given by Goméz et al. [1996].
- A very nice introduction to quantum groups "for pedestrians" is given within Slingerland and Bais [2001].

- An informal (and slightly old) overview of quantum algebras is given by Zachos [1991].

- A nice survey of quantum group constructions for unitary theories is given by Rowell [2005].

- Detailed discussion of how one calculates F-matrices for quantum groups is given by Ardonne and Slingerland [2010], along with detailed calculations for $SU(3)_2$, $SU(3)_3/\mathbb{Z}_3$, $SO(5)_1$, and $(G_2)_1$. Based on this work Mathematica code was produced to calculate F-matrices from quantum groups more generally. This code is available at

   `https://github.com/ardonne/affine-lie-algebra-tensor-category`

   Quite a bit of useful information is also in the README files.

- The program *Kac* allows for automated calculation properties such as the S- and T-matrices (and more) for Chern–Simons theories. See the detailed discussion of its usage in Chapter 39.

Exercises

Exercise 21.1 $SU(2)$ **Representations and Their Deformations**
   First we review the representation theory of the usual $SU(2)$.
   (a.i) Show that the quadratic Casimir operator $\widehat{J^2}$ (in Eq. 21.3) commutes with $\hat{L}_z, \hat{L}_+,$ and $\hat{L}_-$.
   (a.ii) Show that for a highest-weight state $|j,j\rangle$ defined such that $\hat{L}_z|j,j\rangle = j|j,j\rangle$, we have $\widehat{J^2}|j,j\rangle = j(j+1)|j,j\rangle$. Hint: Use the commutator $[\hat{L}_+, \hat{L}_-]$ and the fact that this is a highest-weight state.
   (a.iii) Prove that Eq. 21.5 generates a normalized basis of states $|j,m\rangle$ from a highest-weight state $|j,j\rangle$.
   Now we turn to the deformed $SU(2)$ algebra.
   (b.i) Show the equality shown in Eq. 21.11. Show that the quadratic Casimir operator $\widehat{J^2}$ commutes with $L_z, L_+,$ and L_-. Hint: First prove $[a]_q[b+1]_q - [b]_q[a+1]_q = [b-a]_q$ and use $[-a]_q = -[a]_q$.
   (b.ii) Show that for a highest-weight state $|j,j\rangle$ defined such that $L_z|j,j\rangle = j|j,j\rangle$, we have $\widehat{J^2}|j,j\rangle = [j]_q[j+1]_1|j,j\rangle$ where.
   (b.iii) Prove that Eq. 21.13 generates a normalized basis of states $|j,m\rangle$ from a highest-weight state $|j,j\rangle$.

Exercise 21.2 F- **and** R-**matrices for** $SU(2)_k$
   (a) Use Eqs. 21.16 and 21.17 to calculate the F- and R-matrices of $SU(2)_1$. Compare to results stated in Section 18.1.2.
   (b) Use Eqs. 21.16 and 21.17 to calculate the F- and R-matrices of $SU(2)_2$. Compare to results stated in Section 18.4.

Exercise 21.3 Quantum Dimensions of $SU(2)_k$
   (a) Consider the fusion rules of $SU(2)_k$ given by Eq. 21.20. Write the fusion multiplicity matrix $N_{ab}^c = [N_a]_b^c$ with a fixed to be the smallest-spin particle $j = \frac{1}{2}$. Show the largest eigenvalue of this matrix (and hence, the quantum

dimension of $j = \frac{1}{2}$ is

$$d_{\frac{1}{2}} = 2\cos\left(\frac{\pi}{k+2}\right).$$

(b) Using the fact that

$$d_a d_b = \sum_c N_{ab}^c d_c$$

(as in Eq. 17.2) and knowing the fusion rules of $SU(2)_k$ and knowing the value of $d_{\frac{1}{2}}$, show that

$$d_j = \frac{\sin\left(\frac{(2j+1)\pi}{k+2}\right)}{\sin\left(\frac{\pi}{k+2}\right)} = [2j+1]_q \tag{21.22}$$

for $j \in 0, \frac{1}{2}, 1, \frac{3}{2}, \ldots, \frac{k}{2}$ where we use $q = \exp(2\pi i/(k+2))$.
   (c) Show that the total quantum dimension squared is given by

$$\mathcal{D}^2 = \sum_j d_j^2 = \frac{k+2}{2\sin^2[\pi/(k+2)]}.$$

Exercise 21.4 *S*-matrix of $SU(2)_k$
   The *S*-matrix for $SU(2)_k$ is given by

$$S_{j,j'} = \sqrt{\frac{2}{k+2}}\,\sin\left(\frac{\pi(2j+1)(2j'+1)}{k+2}\right) \tag{21.23}$$

where $j, j' \in 0, \frac{1}{2}, \ldots, \frac{k}{2}$.
   (a) Confirm that this is a unitary matrix.
   (b) For $k = 1, 2, 3$ calculate the *S*-matrix explicitly using Eqs. 21.17 and 21.21 and confirm that the form Eq. 21.23 is correct.

Exercise 21.5 $SU(2)_k/\mathbb{Z}_2$
   For $SU(2)_k$ the "even" subalgebra (the particles with j integer, but not j half-integer), form a well-defined theory by themselves that we call $SU(2)_k/\mathbb{Z}_2$. Given the form of the *S*-matrix for $SU(2)_k$ given in Eq. 21.23 for k odd, show that $SU(2)_k/\mathbb{Z}_2$ is modular and find its *S*-matrix. For k even show that this theory is super-modular.

Exercise 21.6 Central Charges
   One can often guess equivalences between theories by establishing that they have the same central charge. Show that the central charge for $\overline{(E_8)_1}$ matches the central charge for the Ising TQFT (at least mod 8). Show that the central charge for $(G_2)_1$ matches the Fibonacci theory (at least mod 8). Use Eq. 17.16 to determine the central charges for Ising and Fibonacci.

Temperly–Lieb Algebra and Jones–Kauffman Anyons

22

Optional

Let us look back at the Kauffman bracket invariant that we introduced in Chapter 2. In this chapter we want to make use of these rules and determine some of the properties of the corresponding anyons in the language we have been developing since Chapter 8. Our strategy will be to first consider a planar diagram algebra in detail before considering braiding properties in Section 22.5.

We start by considering a planar version of the Kauffman bracket. That is, we only consider diagrams with no over- and under-crossings. Our diagrams are isotopically invariant in the plane and the only additional rule then is that a loop is given a value d, as shown in Fig. 22.1. As compared to the diagrammatic algebra we have constructed over the last few chapters (roughly starting in Chapter 8, and continuing through Chapter 16), one thing that was missing in the discussion of the Kauffman bracket invariant is the idea of multiple particle types and fusion rules. In this chapter we try to construct particle types, fusion rules, and F-matrices given only the rule 22.1 as a starting point. The planar algebra of loops that we construct is known as the Temperly–Lieb algebra. (When we reintroduce braiding to our theory the resulting theory is called Jones–Kauffman, or Temperley-Lieb-Jones–Kauffman.)

Let us start by thinking a bit about what kind of particle types we already have in our theory. Certainly we have the simple string[1] which we will call "1"; and we always have a vacuum particles, which we will call "0." Now we would like to ask whether we can fuse two of these 1 strings together to make another particle.

Several things are immediately obvious. Consider the fact that two 1 particles can fuse to the vacuum, or in other words, a 1-string can go up and then turn down, as shown in Fig. 22.2. This tells us immediately that

$$1 = \bar{1} \quad .$$

The fact that 1 is its own antiparticle is why we do not draw arrows on the 1 string. For simplicity, if a string is not labeled we will assume it is a 1 string. Given that loop of 1 string is assigned the value d, we identify this d_1 (which is often called the quantum dimension, although we have been reserving the words "quantum dimension" for $\mathsf{d}_1 = |d_1|$).

We might also consider the possibility that two of these 1 particles can fuse to something besides the vacuum, in a way similar to that shown in Fig. 22.3. This is a good idea, but it isn't yet quite right. If

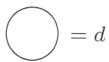

Fig. 22.1 The loop rule for the Kauffman bracket invariant and the Temperly–Lieb algebra.

[1]It is admittedly confusing that 1 is not identity, but this is the usual notation! It is (not coincidentally!) similar to spins where spin 0 is the identity (no spin), and spin 1 is nontrivial.

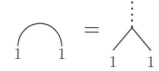

Fig. 22.2 Fusing two 1 particles to the vacuum.

Fig. 22.3 Attempting to fuse two 1 particles to something different from the vacuum.

the two strings fuse to some object besides the vacuum 0, we have to make sure that this new object is appropriately "orthogonal" to 0. This orthogonality must be in the sense of the locality, or no-transmutation rule (see Fig. 8.7): a particle type must not be able to spontaneously turn into another particle type (without fusing with some other particle or splitting). In Fig. 22.3 it looks like the two strings brought together could just fuse together to form the vacuum as in Fig. 22.2, and this would then turn the collection of two strings into the vacuum. To prevent such transmutation, we will work with operators known as Jones–Wenzl projectors.

22.1 Jones–Wenzl Projectors

The general definition of a projector is an operator P such that $P^2 = P$. This means that P has eigenvalues 0 and 1. Let us think of a string diagram as an operator that takes as an input strings coming from the bottom of the page, and gives as an output strings going toward the top of the page (compare Fig. 12.2). Now consider a set of n-strings traveling together in the same direction (in what is often called a *cable*). The Jones–Wenzl projection operator P_n operates on a set of n such strings—it takes n-strings in and gives n-strings out—and it is defined such that attaching a cup or a cap to the bottom or top of the operator gives a zero result (see Fig. 22.4). The n-particle Jones–Wenzl projector P_n acting on a cable of n-strings should be interpreted as the nth particle species.

The purpose of the Jones–Wenzl projector is to fix the problem we discovered with Fig. 22.3. That is, if a cable of two strings forms a nontrivial particle (the particle we will call 2), we should not be able to put a cap on the top of these two strings and transmute the 2 particle to the vacuum. That is, adding a cap should make the entire diagram vanish, and this is the property we are looking for in the 2 string Jones–Wenzl projector.

Let us now try to construct the 2 string Jones–Wenzl projector P_2 out of two incoming 1 particles[2] (two elementary strings). To do this we first construct a different projector $\bar{P}_2$ that forces the two incoming particles to fuse to the vacuum[3] as shown in Fig. 22.5.

Fig. 22.4 A cup (left) and a cap (right).

[2]The Jones–Wenzl projector, if you want to define one, for a single string is the trivial operator. That is, one string comes in and the same string comes out unchanged.

[3]The astute reader will notice that a particle "turning around" as in Fig. 22.2 is not quite the same as projecting to the 0 particle, due to the prefactor $1/d$. We return to this issue in Section 22.3.

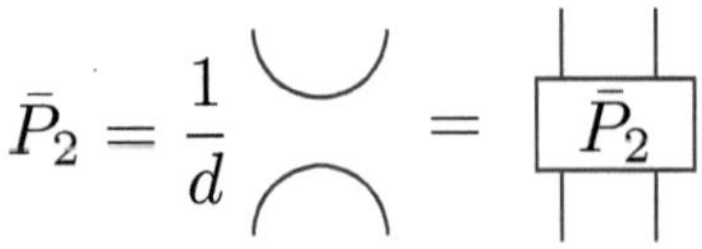

$$\bar{P}_2 = \frac{1}{d}\ \bigcup_{\bigcap}\ = \boxed{\bar{P}_2}$$

Fig. 22.5 The projector of two strings to the vacuum $\bar{P}_2$. This figure should be thought of as an operator that takes as an input two strings coming in from the bottom and gives as an output two strings going out the top. Sometimes the operator is represented as a labeled box as shown on the right.

To establish that this $\bar{P}_2$ operator is a projector we need to check that $[\bar{P}_2]^2 = \bar{P}_2$. To apply the $\bar{P}_2$ operator twice we connect the two strings coming out the top of the first operator to two strings coming in the bottom of the second operator. As shown in Fig. 22.6, using the fact that a loop gets value d, we see that $[\bar{P}_2]^2 = \bar{P}_2$, meaning that $\bar{P}_2$ is indeed a projector.

$$[\bar{P}_2]^2 = \frac{1}{d} \quad \frac{1}{d} \quad = \frac{1}{d^2} \quad = \frac{1}{d} \quad = \bar{P}_2$$

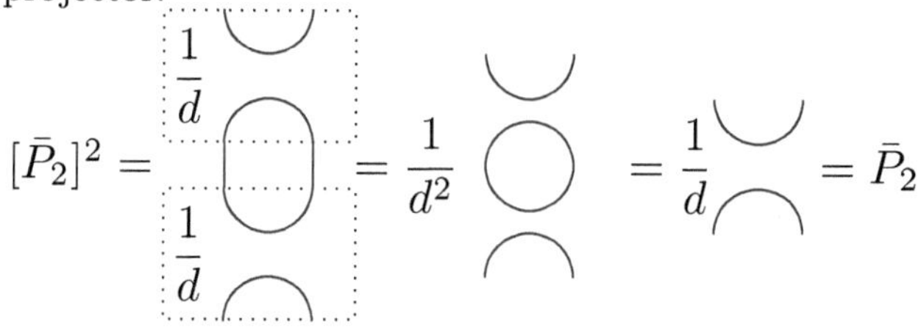

Fig. 22.6 Checking that $[\bar{P}_2]^2 = \bar{P}_2$. In the second step we have used the fact that a loop gets the value d.

The Jones–Wenzl projector P_2 for two strings is the complement of the operator $\bar{P}_2$ we just found, meaning $P_2 = I - \bar{P}_2$ where I is the identity operator, or just two parallel strings. Diagrammatically we have Fig. 22.7. Since the $\bar{P}_2$ operator projects the two strings onto the vacuum, the P_2 operator projects the two strings to a different orthogonal particle type, which we call 2.

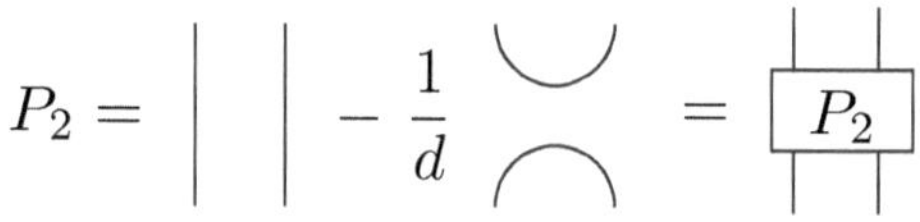

$$P_2 = \quad \Big| \Big| \quad - \frac{1}{d} \quad = \boxed{P_2}$$

Fig. 22.7 The projector of two strings to the nontrivial particle made of two strings $P_2 = I - \bar{P}_2$. Sometimes this projector is drawn as a labeled box, as on the right.

We can algebraically check that P_2 is indeed a projector

$$P_2^2 = (I - \bar{P}_2)(I - \bar{P}_2) = I - 2\bar{P}_2 + \bar{P}_2^2 = I - \bar{P}_2 = P_2$$

and also we can check that P_2 is orthogonal to $\bar{P}_2$, by

$$\bar{P}_2 P_2 = \bar{P}_2(I - \bar{P}_2) = \bar{P}_2 - \bar{P}_2^2 = 0$$

and similarly $P_2\bar{P}_2 = 0$.

Often it is convenient to draw these projection operators as a labeled box, as shown on the right of Figs. 22.5 and 22.7. Sometimes instead of drawing two lines with a projector $\bar{P}_2$ or P_2 inserted, we simply draw a single line with a label, 0 or 2 respectively, as in the right of Fig. 22.10 or the left of Fig. 22.8.

It is useful to calculate the value of the 2 string loop.[4] This is shown in Fig. 22.8.

[4]In many references d_2 is called Δ_2 (and similarly d_n is called Δ_n). We stick with d to fit with the notation in the rest of this book.

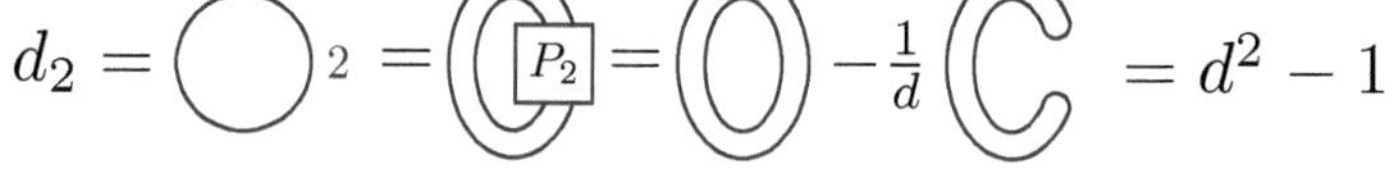

$$d_2 = \bigcirc 2 = \boxed{P_2} = \bigcirc - \frac{1}{d} \, \mathcal{C} = d^2 - 1$$

Fig. 22.8 Evaluating the 2 string loop.

22.1.1 $\mathbb{Z}_2$ Loop Gas

In the case where $d = \pm 1$ it is easy to prove (see Exercise 2.2 and 22.1) that two horizontal strings equals d times two vertical strings as shown in Fig. 22.9. In this case, notice that the projector $P_2 = 0$ since the two terms in the projector in Fig. 22.7 are equal with opposite signs. Correspondingly, note that for $d = \pm 1$ (and only for these values), the value of the 2 string loop is $d_2 = 0$ as shown in Fig. 22.8, meaning that no such 2 particle exists. Thus, the only possible outcome of fusion of two 1 strings is the vacuum as shown in Fig. 22.2. Therefore the entire fusion rules of these theories are

$$1 \times 1 = 0$$

For $d = \pm 1$:

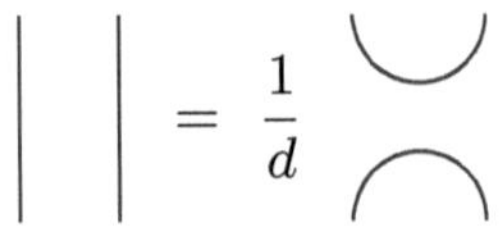

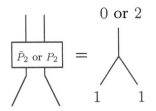

Fig. 22.9 Two cases where the Kauffman bracket invariant rules become very simple. If you have not convinced yourself of these rules, try to do so! (See Exercise 2.2.) Note that $d = 1$ occurs for bosons or fermions and $d = -1$ occurs for semions.

where again 0 is the identity or vacuum. These abelian fusion rules result in abelian braiding statistics.

These two possible cases here obviously correspond to the $d = \pm 1$ loop gases we studied in Sections 18.1.1 and 18.1.2. When braidings are considered we obtain bosons or fermions for $d = 1$ and left- or right-handed semions $(SU(2)_{\pm 1})$ for $d = -1$. This has been discussed in depth in Section 18.1 and we return to discuss it further in Section 22.6.1.

22.1.2 Two Strands in the General Case

For values of d not equal to ± 1, the projector P_2 does not vanish. This means that two 1 strands can fuse to either 0 or 2 as shown in Fig. 22.10. We can write the fusion rule as

$$1 \times 1 = 0 + 2 \quad .$$

We might ask whether it is possible to assemble a third type of particle with two strands. It is obvious this is not possible since $\bar{P}_2 + P_2 = I$, which means these two particle types form a complete set ($\bar{P}_2$ projects the two particles to the vacuum, and P_2 projects to the 2 particle type).

Fig. 22.10 Two possible fusions of two 1 strands, drawn in two different notations. A single line labeled 2 is interpreted as two 1 strands traveling together with a P_2 operator inserted. The label 0 means the two strands fuse to the vacuum, as in Fig. 22.2.

22.1.3 Three Strands in the General Case

We can move on and ask what kind of particles we can make if we are allowed to fuse three strands together. We want to try to construct a three-legged projector. The most general three-legged operator we can construct is of the form in Fig. 22.11.

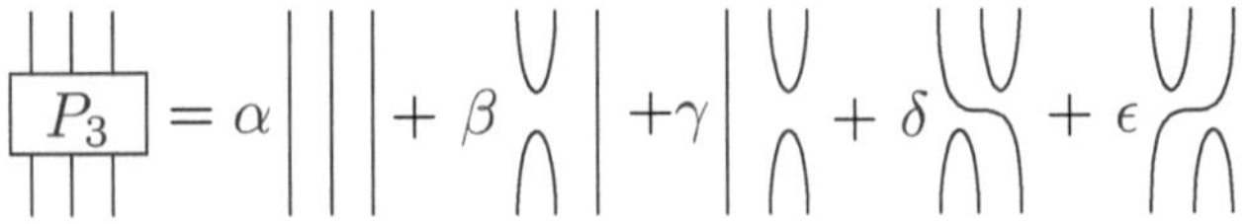

Fig. 22.11 The form of the most general three-legged operator we can construct. Here, $\alpha, \beta, \gamma, \delta, \epsilon$ are arbitrary constants.

We would like to find the 3 string operator that is a projector. So we should enforce $P_3^2 = P_3$. However, there are other things we want to enforce as well. Since 0 is the identity, we want $0 \times 1 = 1$, which means we should not be able to fuse $\bar{P}_2$ (the projector of two strings onto the vacuum) with a single strand to get P_3. Diagrammatically, this means we must insist on relations like Fig. 22.12.

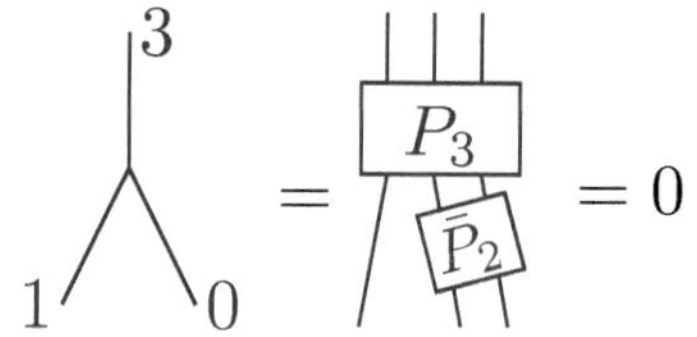

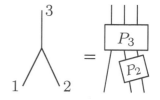

Fig. 22.12 Insisting that 0×1 does not give 3.

This and analogous constraints allow us to insist on the conditions shown in Fig. 22.13.

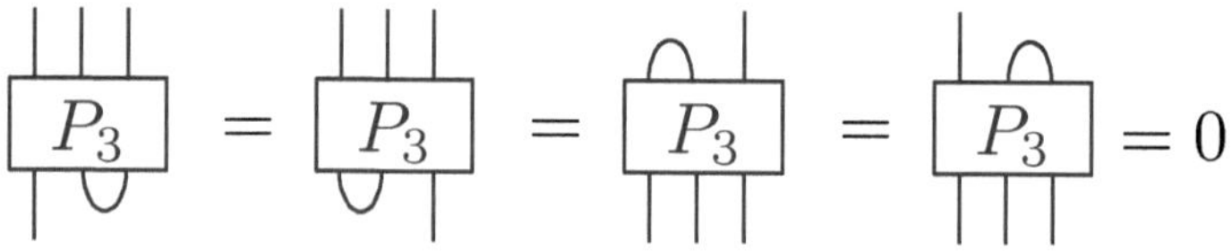

$$\boxed{P_3} = \boxed{P_3} = \boxed{P_3} = \boxed{P_3} = 0$$

Fig. 22.13 Four conditions that come from the fusion condition shown in Fig. 22.12.

However, we should allow fusions of the form $1 \times 2 = 3$, as shown in Fig. 22.14. Enforcing the condition in Fig. 22.13, along with $P_3^2 = P_3$ gives the form of P_3 shown in Fig. 22.11 with the results that (see Exercise 22.1)

$$\alpha = 1$$
$$\beta = \gamma = -\frac{d}{d^2 - 1}$$
$$\delta = \epsilon = \frac{1}{d^2 - 1} .$$

Fig. 22.14 We allow $1 \times 2 = 3$.

We can do a short calculation in the spirit of Fig. 22.8 to obtain the value of a loop of 3 string,[4] giving the result (see Exercise 22.1) shown in Fig. 22.15.

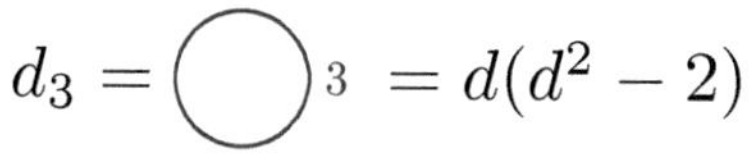

$$d_3 = \bigcirc_3 = d(d^2 - 2)$$

Fig. 22.15 Evaluating the 3 string loop.

22.1.4 Ising Anyons

In the case where $d = \pm\sqrt{2}$ (and only in these cases) the 3 string loop has $d_3 = 0$ meaning that there is no 3 string particle. Equivalently it is possible to show that P_3 vanishes when evaluated in any diagram (see

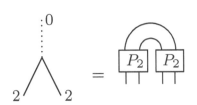

Fig. 22.16 $2 \times 2 = 0$.

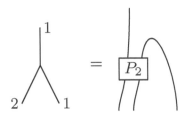

Fig. 22.17 $2 \times 1 = 1$. We recognize this as the fusion $1 \times 1 = 2$ from Fig. 22.10 just turned on its side.

For $d = \pm\sqrt{2}$:

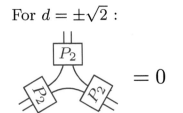

Fig. 22.18 For $d = \pm\sqrt{2}$ we have 2×2 not fusing to 2.

Exercise 22.1). It is similarly possible to show that $P_4 = 0$, and so forth. Thus, for the case of $d = \pm\sqrt{2}$ there are only three particle types: 0, 1, and 2. In addition to the fusions we have already determined, we have $2 \times 2 = 0$, as shown in Fig. 22.16 and $2 \times 1 = 1$, as shown in Fig. 22.17. Note that showing $2 \notin 2 \times 2$ (see Fig. 22.18) requires another explicit calculation, not shown here! See Exercise 22.1.

We thus have the full set of nontrivial fusion rules

$$
\begin{aligned}
1 \times 1 &= 0 + 2 \\
2 \times 2 &= 0 \\
1 \times 2 &= 1
\end{aligned}
\tag{22.1}
$$

which we recognize as Ising fusion rules (see Sections 8.2.2 and 18.4) where $1 = \sigma$ and $2 = \psi$ and 0 is the vacuum I

Recall our discussion of Ising anyons in Sections 8.2.2 and 18.4. There we found that $d_\sigma = \pm\sqrt{2}$ and $d_\psi = 1$. This indeed agrees with the present discussion: We obtain Ising fusion rules for $d_1 = d_\sigma = \pm\sqrt{2}$ and evaluating using Fig. 22.8, we also have $d_\psi = d_2 = 1$. Thus, our string algebra recovers details of the Ising fusion algebra.

22.2   General Values of d

The generalization of the above discussions for $d = \pm 1$ and $d = \pm\sqrt{2}$ is fairly straightforward. One can generally show the following properties (see Kauffman and Lins [1994] and Exercise 22.2). First, the Jones–Wenzl projector for $n + 1$ strands can always be written in terms of the projector for n strands, as shown in Fig. 22.19 (see Exercise 22.2).

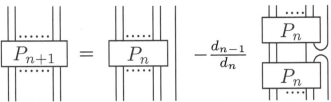

Fig. 22.19 Recursion relation for Jones–Wenzl projectors.

Note in particular that if P_n vanishes, we can conclude that P_m vanishes for all $m > n$ as well. We define d_n, the loop weight of particle type n, by connecting n strings coming from the bottom of projector P_n to those coming from the top, as shown in Fig. 22.20.

Using the recursion shown in Fig. 22.19 and the definition of d_n in Fig. 22.20 we obtain the recursion relation (you can do this in your head!)

$$
d_{n+1} = d\, d_n - d_{n-1}
\tag{22.2}
$$

where we define $d_{-1} \equiv 0$ and $d_0 = 1$ and hence, $d_1 = d$. This recursion has the general solution

$$
d_n = U_n(d/2)
\tag{22.3}
$$

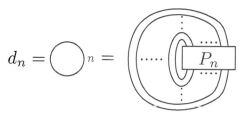

$$d_n = \bigcirc \, n = \text{(figure)}$$

Fig. 22.20 Evaluating the quantum dimension of the n string particle. We connect the n strings coming from the top of the projector P_n to those coming from the bottom. Often this quantity is notated as Δ_n.

where U_n is the nth Chebyshev polynomial of the second kind, which are defined by (see Exercise 22.2)

$$U_n(\cos\theta)\sin\theta = \sin[(n+1)\theta] \quad . \tag{22.4}$$

A theory has a finite number of particle types if $d_n = 0$ for some $n = k+1$ (such that P_n vanishes for all $n \geq k+1$). Solutions to this condition are given by

$$d = d_1 = 2\cos\left(\frac{p\pi}{k+2}\right) \tag{22.5}$$

for values of $p = 1, \ldots, k+1$. For values of d that are not of this form, one can construct an infinite number of orthogonal particle types (n-strand projectors with different values of n), which indicates a badly behaved theory (i.e. the algebra never "closes"). If we plug this value of d back into Eq. 22.3 we can generate the loop weights d_n of the theory for $n = 0, \ldots, k$. For $p = 1$ we generate all positive loop weights.[5]

$$d_n^{(p=1)} = \frac{\sin\left[\frac{(n+1)\pi}{k+2}\right]}{\sin\left[\frac{\pi}{k+2}\right]} = [n+1]_q \quad \text{with} \quad q = e^{\pm 2\pi i/(k+2)}$$

where we have used the quantum deformed integer notation $[]_q$ from Section 21.2.2.

For $p = k+1$ we generate exactly the same loop weights as for $p = 1$ except with alternating signs

$$d_n^{(p=k+1)} = (-1)^n d_n^{(p=1)} \quad .$$

These theories with negative loop weights can be interpreted as unitary if we invoke rule 0 from Sections 14.5 and 16.1.3.[6] For other values of p the theories can be rejected. For cases where p and $(k+2)$ have a common integer factor x we are obviously just describing a theory with $p' = p/x$ and $(k'+2) = (k+2)/x$ so we need not consider these. For other values of p, the theories will need to be rejected for another reason. When $p \neq 1$, $k+1$ and p is coprime with $(k+2)$ the theories have both positive and negative loop weights but we will find that they violate the necessary condition 14.5 needed in order to reinterpret the nonunitary theory as being unitary after applying rule 0. Thus, we will limit our

[5]Comparing this to the quantum dimensions of $SU(2)_k$, given in Eq. 21.21 (see also Exercise 21.3) shows that the quantum dimension here for particle n is precisely the same as that of the particle type $j = n/2$ for the $SU(2)_k$ Chern–Simons theory.

[6]Without rule 0, when we have negative loop weights, the diagram algebras are still self-consistent, but they cannot represent quantum-mechanical systems.

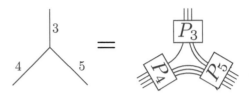

Fig. 22.21 The general vertex in the Temperly–Lieb algebra. Here, the vertex is shown for 4 and 5 fusing to 3.

Fig. 22.22 A general vertex between particle types (a, b, c) with $a, b, c \geq 0$.

attention to $p = 1$ or $k + 1$ corresponding to the values

$$d = d_1 = \pm 2 \cos \left(\frac{\pi}{k+2} \right) \qquad (22.6)$$

Once one constructs the appropriate n strand projectors, the general vertex between three different particle types can be constructed analogous to that shown in Fig. 22.21. Consider a vertex between particle types (a, b, c) with $a, b, c, \geq 0$ as in Fig. 22.22. The number of strings going between the projectors (as in the right of Fig. 22.21) is given by

$$
\begin{aligned}
m &= (a + b - c)/2 = \text{strings between } a \text{ and } b & (22.7) \\
n &= (a + c - b)/2 = \text{strings between } a \text{ and } c & (22.8) \\
p &= (b + c - a)/2 = \text{strings between } b \text{ and } c & . \quad (22.9)
\end{aligned}
$$

These quantities must be non-negative, and we must have all of these quantities integer, which is assured if

$$(a + b + c) \text{ is even.} \qquad (22.10)$$

Note that $a, b,$ or c are allowed to have the value 0, meaning no strings come out that edge. These variables are also allowed to have the value 1 meaning a single string comes out the edge (and no projector is needed, see note 2).

One can further show that a vertex between particle types (a, b, c) can be non-zero only if the projector

$$P_{(a+b+c)/2} \quad \text{is non-zero.} \qquad (22.11)$$

This final condition is nontrivial, and we will not prove it in all generality here (see for example, Kauffman and Lins [1994] for a proof). However, Fig. 22.18 is an example of this condition. When $d = \pm\sqrt{2}$, we have shown that P_3 vanishes and this implies the vertex $(2, 2, 2)$ must also vanish.

The conditions we have just described for a vertex (m, n, p non-negative integers and $P_{(a+b+c)/2}$ non-zero) gives us the fusion relations for the theory which are given by

$$a \times b = |a - b|, \; |a - b| + 2, \; \ldots, \; \min(a + b, 2k - a - b) \qquad (22.12)$$

where k is the largest integer such that P_k is non-zero.[7] Note in particular that the particle with the highest label $a = k$ is always a simple current, meaning that it fuses with any other particle b to give a unique fusion product $k \times b = (k - b)$.

With this definition of a vertex we can evaluate any planar diagram. A particularly useful diagram is the version of the Theta diagram shown in Fig. 22.23. The value of this diagram can be derived generally and is given by

$$\Delta(a, b, c) = (d_{(a+b+c)/2})! \, \frac{(d_{n-1})! \, (d_{m-1})! \, (d_{p-1})!}{(d_{a-1})! \, (d_{b-1})! (d_{c-1})!} \qquad (22.13)$$

where we have defined

$$(d_n)! \equiv d_n d_{n-1} d_{n-2} \ldots d_2 d_1$$

with $d_1 = d$ and $d_0 = d_{-1} = 1$. From Eq. 22.13 we see that $\Delta(a, b, c)$ is symmetric in exchanging any of its arguments. Further, we see that the quantity vanishes when $d_{(a+b+c)/2}$ vanishes, which agrees with the condition Eq. 22.11.

While the most general derivation of Eq. 22.13 is somewhat complicated (see Kauffman and Lins [1994] or Lickorish [1993]) it is easy enough to confirm it is correct for a few examples (see Exercise 22.3).

The value, Eq. 22.13, of the Theta diagram does not match what we would have expected, given the rules in Chapter 16. Comparing to Fig. 16.16 we would have expected the Theta diagram in Fig. 22.23 to have a value $\sqrt{d_a d_b d_c}$. However, it is easy to calculate the value of $\Delta(a, b, c)$ shown in Fig. 22.23 and we discover that it does not generally have this desired value—this means we have somehow normalized our theory incorrectly. In Section 22.3 we fix this normalization so that our theory is a proper (unitary) diagram algebra as we have defined in prior chapters such as Chapter 14.

[7]Equation 22.12 is entirely equivalent to the $SU(2)_k$ Chern–Simons fusion rules given in Eq. 21.20 where we have now made the identification that the quantum number a in the Temperly–Lieb-Jones–Kauffman theory is $2j$ for a spin j particle of the Chern–Simons theory. Note further that in the case where k is infinitely large, these fusion rules match the angular momentum addition rules of regular $SU(2)$. See also note 5.

$$\Delta(a, b, c) =$$

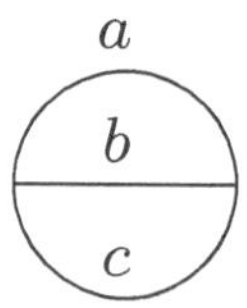

Fig. 22.23 The Theta diagram in the Temperly–Lieb-Jones–Kauffman theory.

22.3 Unitarization

The diagrammatic algebra we have constructed so far in this chapter is a perfectly self-consistent algebra (see Kauffman and Lins [1994] for a large amount of detail of this algebra). However, this algebra does not fit the rules we have established in prior chapters. In Section 22.2 we just found that the value of the Theta diagram does not match the expectation from Chapter 16. If we tried to work out further details of the diagrammatic algebra, we would find other failures as well—for example, we would find the F-matrices to be nonunitary! Fortunately, it is not hard to modify the theory a small amount (as proposed by Kauffman and Lomonaco [2007]) so that it fits within our existing framework from Chapter 16.

Let us define a new vertex, which is a constant multiple of the old

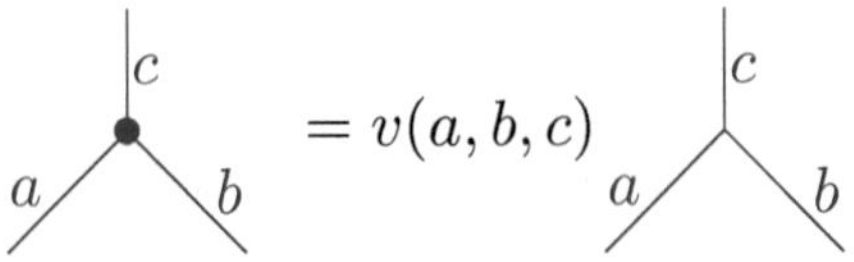

$$\mathbin{\text{\large$\diagup$}}\ = v(a,b,c)\ \mathbin{\text{\large$\diagup$}}$$

Fig. 22.24 A renormalized vertex between particle types (a, b, c) with $a, b, c \geq 0$ marked with a blue dot on the left is defined in terms of the original vertex on the right. We assume here that the vertex on the right, defined analogous to Fig. 22.21, is non-zero.

vertex as shown in Fig. 22.24. We define the rescaling factor as

$$\Theta(a, b, c) = \text{\large$\ominus$}$$

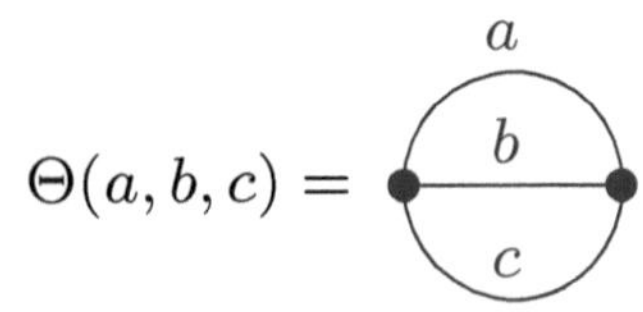

Fig. 22.25 The Theta diagram with renormalized vertices.

$$v(a, b, c) = \sqrt{\frac{\sqrt{d_a d_b d_c}}{\Delta(a, b, c)}}$$

such that the value of the Theta diagram in Fig. 22.25 is now $\Theta(a, b, c) = \sqrt{d_a d_b d_c}$ as we expect from Fig. 16.16. It turns out that this simple modification is sufficient to make the theory fit into the framework developed in Chapter 16. Note that the argument of the square roots are positive in all cases and for either choice of the sign of d in Eq. 22.6.

22.4   *F*-Matrices

We can now determine the F-matrices directly from the graphical algebra. As a simple example, consider the F-matrices $F^{11\alpha}_{11\beta}$ (which we abbreviate as F^{α}_{β}) as shown in Fig. 22.26. Note that for this equation we use renormalized vertices as defined in Eq. 22.24 and notated by dots on the vertices.

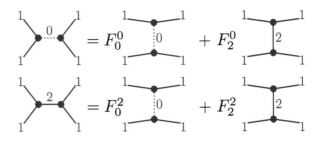

Fig. 22.26 The F-matrix in the Temperly–Lieb-Jones–Kauffman theory is unitary when we use renormalized vertices, indicated by dots. Here we have abbreviated $F^{11\alpha}_{11\beta}$ as F^{α}_{β} for brevity.

This F-matrix equation is that of Fig. 16.4 for four incoming 1 string particles. The F-matrix is nontrivial since there is more than one fusion channel when we fuse the 1 strings together: $1 \times 1 = 0 + 2$, as long as $d \neq \pm 1$ (in which case the 2 string particle vanishes). We can now rewrite the F-matrix equation in terms of string diagrams as in Fig. 22.27. Note that in Fig. 22.27, the prefactors of $d/\sqrt{d_2}$ come from the vertex renormalization factors $v(1, 1, 2)^2$, and the quantities in brackets are P_2

projectors that force the two strings to fuse to the 2 particle.

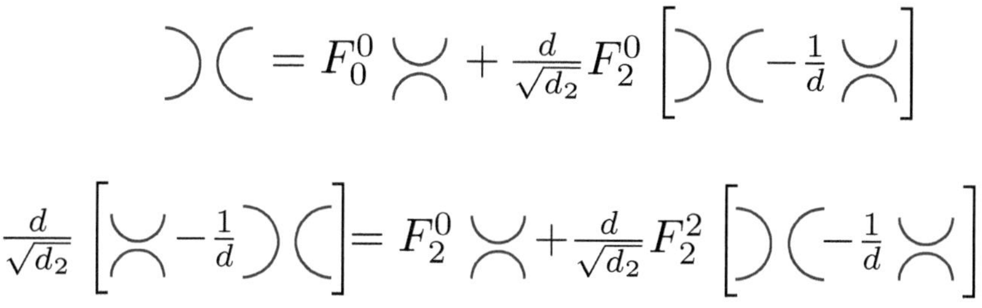

$$
)(= F_0^0 \smile\frown + \frac{d}{\sqrt{d_2}} F_2^0 \left[ )(- \frac{1}{d} \smile\frown \right]
$$

$$
\frac{d}{\sqrt{d_2}} \left[\asymp - \frac{1}{d})(\right] = F_2^0 \smile\frown + \frac{d}{\sqrt{d_2}} F_2^2 \left[ )(- \frac{1}{d} \smile\frown \right]
$$

Fig. 22.27 Explicitly writing out the F-matrix equations of Fig. 22.26. The prefactors terms in brackets are P_2 projectors. The prefactors d_2 is from the vertex renormalization factors $v(1, 1, 2)^2 = d^2/d_2$. The other renormalization factor $v(1, 1, 0) = 1$.

We then match up terms on the right and left of the graphical equations in Fig. 22.27. In the first line we see that the diagram on the left is topologically like the first term in the brackets on the right, so we have $F_2^0 = \sqrt{d_2}/d$. Similarly, the first term on the right is topologically the same as the second term in the brackets, so $F_0^0 = 1/d$. Then in the second line, the second term in brackets on the left is topologically the same as the first term in brackets on the right, so we have $F_2^2 = -1/d$. Then, among the remaining terms, the first term in brackets on the left, the first term on the right, and the second term in brackets on the right, are all topologically the same, so we have $d/\sqrt{d_2} = F_2^0 - (1/\sqrt{d_2})F_2^2$ or $F_2^0 = (1/d)(d^2 - 1)/\sqrt{d_2}$. Finally, using $d_2 = (d^2 - 1)$ (see Fig. 22.8) we obtain the full form of the F-matrix (and returning the 11 superscripts and subscripts we have suppressed)

$$
[F_{11}^{11}] = \begin{pmatrix} \frac{1}{d} & \frac{\sqrt{d^2 - 1}}{d} \\ \frac{\sqrt{d^2 - 1}}{d} & -\frac{1}{d} \end{pmatrix} \tag{22.14}
$$

Note that this matrix is properly unitary for any value of d. For $d = \pm\sqrt{2}$ the matrix matches our expectation for the Ising fusion rules given in Eq. 18.34.

With similar diagrammatic calculations, we can work out the F-matrices for any incoming and outgoing n string particles. Detailed calculations are given in Kauffman and Lins [1994]. However, note that the results given there are nonunitary expressions due to the use of unrenormalized vertices.

22.5   Twisting and Braiding

So far we have not yet used the braiding rules of the Kauffman bracket invariant—we have only used the loop rule and we have only considered planar diagrams. We finally can reintroduce the braiding rules for the Kauffman invariant for evaluating crossings, as in Fig. 2.3, and thus we are now considering a full anyon theory.

Recall now the values of d that give us a well-behaved theory from Eq. 22.6. Using the Kauffman rule (Fig. 2.3)

$$d = -A^2 - A^{-2}$$

this then gives us the possibilities[8]

$$A = i^{\alpha} \exp\left[\frac{\pm i\pi}{2(k+2)}\right] \tag{22.15}$$

with $\alpha = 0, 1, 2, 3$ and $k \geq 1$. Here, the cases of $\alpha = 0, 2$ correspond to negative d and $\alpha = 1, 3$ correspond to positive d.

As shown in Fig. 2.6, comparing to Fig. 15.3 we see that the twist factor of the single strand is[9]

$$R_0^{11} = \theta_1^* = -A^{-3} \quad .$$

It is not hard (Exercise 22.5) to use the Kauffman diagrammatic algebra to show that the twist factor of particle type a is generally given by[9,10]

$$\theta_a = (-1)^a A^{a(a+2)} \quad . \tag{22.16}$$

One can also use the diagrammatic Kauffman rules to evaluate the R-matrices. Just to do a simple example, let us evaluate R_2^{11} as shown in Fig. 22.28.

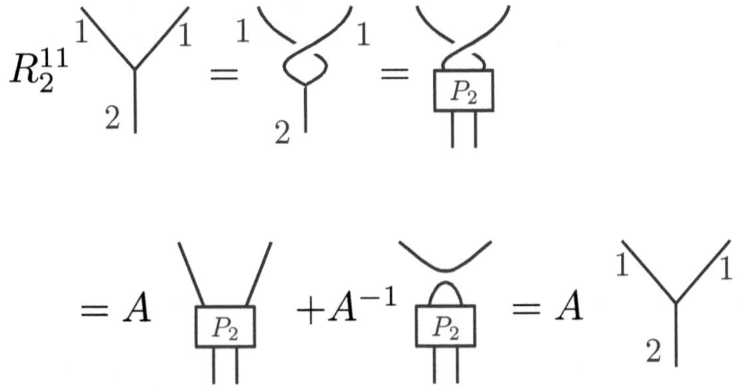

Fig. 22.28 Evaluation of $R_2^{11} = A$ using the Kauffman bracket invariant. In going from the first line to the second we invoke the bracket rules from Fig. 2.3. The last step invokes the fact that P_2 is orthogonal to the turn-around thus killing the term with coefficient A.

More generally one can derive (see Kauffman and Lins [1994]) that[10]

$$R_c^{ab} = (-1)^{\frac{1}{2}(a+b-c)} A^{\frac{1}{2}\{c(c+2)-a(a+2)-b(b+2)\}} \quad . \tag{22.17}$$

A short calculation (see Exercise 22.6) shows that when α and k in Eq. 22.15 are both odd, then the simple current particle (the particle

[8] In much of the literature $k+2$ is often called r.

[9] Here, we are quoting θ for the isotopy invariant diagram algebra. Note that in the case where d is negative, this differs by a sign from θ for the corresponding unitary theory. See note 2 from Chapter 15.

[10] Note the similarity of Eqs. 22.16 and 22.17 to Eqs. 21.18 and 21.17, noting that in this chapter we label particles with integers a whereas in Section 21.2 we used j integer or half-integer, so $a = 2j$. These equations are not identical, although in the case where the Temperly–Lieb-Jones–Kauffmann theory is describing the same $SU(2)_k$ as the quantum groups are describing, R_c^{ab} agrees between the two approaches. For θ_a there is a distinction, but it is just notational. See note 9.

with label $a = k$) is transparent (it braids trivially with all other particles) and thus, the theory is non-modular.

22.6 **Examples**

For each value of k we can generate eight possible values of A in Eq. 22.15. Here, we make connection between these values and theories that we have already discussed.[11]

22.6.1 $\mathbb{Z}_2$ **Loop Gases Again**

Let us consider the case of $k = 1$ in Eq. 22.15. Here, we obtain $d = (-1)^{\alpha+1}$. This corresponds to the $\mathbb{Z}_2$ loop gases that we investigated in Sections 18.1 and 22.1.1. The correspondence between these values of A and each theory is given in Table 22.1. Note that, unusually, for $k = 1$, there are only four different possible theories since taking the mirror image of any theory on the table (complex conjugating A) gives another theory on the table (boson and fermion are both unchanged under complex conjugation, whereas $SU(2)_1$ and $\overline{SU(2)}_1$ transform into each other; see Exercise 2.2).

Table 22.1 Kauffman anyon theories for $k = 1$. When d is negative we invoke rule 0 (Section 14.5) to obtain a unitary theory. Complex conjugation of A gives the mirror image theory. The latter two are not modular, although fermion (also known as sVec) is super-modular.

A	sgn(d)	name
$e^{i\pi/6}$	$-$	$SU(2)_1$
$-e^{i\pi/6}$	$-$	$\overline{SU(2)}_1$
$ie^{i\pi/6}$	$+$	fermion
$-ie^{i\pi/6}$	$+$	boson

22.6.2 **Ising Fusion**

Let us now consider the case of $k = 2$ in Eq. 22.15, which corresponds to $d = (-1)^{\alpha+1}\sqrt{2}$, which, as we identified in Eq. 22.1, yields Ising fusion rules. The possible values of A corresponding to these values of d are given by Eq. 22.15 and are listed in Table 22.2. Along with the mirror image of these theories (complex conjugation of A) these correspond to the eight anyon theories we found in Table 18.1. All of these are modular.

Table 22.2 Kauffman anyon theories for $k = 2$. When d is negative we invoke rule 0 (Section 14.5) to obtain a unitary theory. Complex conjugation of A gives the mirror image theory.

A	sgn(d)	name
$e^{i\pi/8}$	$-$	$SU(2)_2$
$-e^{i\pi/8}$	$-$	$\overline{(C_2)}_1$
$ie^{i\pi/8}$	$+$	$(B_3)_1$
$-ie^{i\pi/8}$	$+$	$\overline{\text{Ising}}$

22.6.3 $SU(2)_3$ **Fusion**

For $k = 3$ we have $d = (-1)^{\alpha+1}\phi$ with $\phi = (1 + \sqrt{5})/2$. For this value of k, and in fact for any odd $k > 1$, the resulting theory always has a factorization into two simpler theories (see Exercise 22.6). Here, the particle labeled $3 = k$ is the simple current that generates one factor of the theory, and the particle labeled 2 generates the other factor of the theory. The two factors are easily identified by calculating R_0^{aa} where $a = 2, 3$. We thus establish Table 22.3.

Table 22.3 Kauffman anyon theories for $k = 3$. When d is negative we invoke rule 0 (Section 14.5) to obtain a unitary theory. Complex conjugation of A gives the mirror image theory. Note that $(G_2)_1$ is the (right-handed) Fibonacci anyon theory.

A	sgn(d)	name	
$e^{i\pi/10}$	$-$	$(G_2)_1 \times \overline{SU(2)_1} = SU(2)_3$	modular
$-e^{i\pi/10}$	$-$	$(G_2)_1 \times SU(2)_1$	modular
$-ie^{i\pi/10}$	$+$	$(G_2)_1 \times$ fermion	super-modular
$ie^{i\pi/10}$	$+$	$(G_2)_1 \times$ boson	

Chapter Summary

- The diagrammatic rules of the Kauffman bracket invariant can be used to build up various anyon theories.
- The planar diagram algebra version of these rules is known as the Temperley-Lieb algebra.

Further Reading:

- The book by Kauffman [2001] is a very good introduction to the Kauffman bracket invariant. The book Kauffman and Lins [1994] fills in many of the details of calculations. Kauffman and Lomonaco [2007] discusses unitarization. Wang [2010] discusses the relationship between the Kauffman-Jones categories and other TQFTs.
- The original work of Temperley and Lieb [1971] was in the context of solving a variety of classical statistical mechanics problems. Around the same era, some similar diagrammatic ideas were pursued by Penrose [1971]

Exercises

Exercise 22.1 Jones–Wenzl Projectors P_0, P_2, and P_3

For two strands one can construct two Jones–Wenzl projectors P_0 and P_2 as shown in Figs. 22.5 and 22.7.

(a) Show that these projectors satisfy $P^2 = P$, so their eigenvalues are 0 and 1. Further show that the two projectors are orthogonal $P_0 P_2 = P_2 P_0 = 0$. (This should be easy it is mostly in the book!)

(b) Show that for $d = \pm 1$ we have $P_2 = 0$ in the evaluation of any diagram. The result means that in these models there is no new particle that can be described as the fusion of two elementary anyons. Why should this be obvious? Hint: Look back at Exercise 2.2.

(c) The 3 strand Jones–Wenzl projector must be of the form shown in Fig. 22.11.

The coefficients $\alpha, \beta, \gamma, \delta, \epsilon$ are defined by the projector condition $P_3^2 = P_3$ and also by the condition that P_3 is orthogonal to P_0, which is shown in Figs. 22.12 and 22.13.

Calculate the coefficients $\alpha, \beta, \gamma, \delta$ in P_3. Calculate the quantum dimension d_3 shown in Fig. 22.15.

(d) Choosing $d = \pm\sqrt{2}$ show that $P_3 = 0$ in the evaluation of any diagram. We can then conclude that in this model there is no new particle that is the fusion of three elementary strands. Hint: Try evaluating some simple diagrams that include a P_3 in them.

(e) For the case of $d = \pm\sqrt{2}$ show that, when evaluated in any diagram, $2 \times 2 \not\ni 2$. In other words, prove Fig. 22.18.

Exercise 22.2 More General Jones–Wenzl Projectors

(a) A Jones–Wenzl projector for n strands is defined both by $P_n^2 = P_n$ as well as by being orthogonal to P_0 analogous to Fig. 22.13. Assuming these properties are satisfied for P_n, show that they are satisfied for P_{n+1} given by Fig. 22.19. Hint: Use the fact that connecting up a single string from P_{n+1} from top to bottom as in Fig. 22.29 must give something proportional to P_n (Why?).

(b) Using Fig. 22.19 derive Eq. 22.2. Show that the solution to this equation is given by Eqs. 22.3 and 22.4. Confirm the condition for d_n to vanish given in Eq. 22.5.

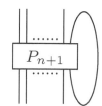

Fig. 22.29 This figure, with n strands going in the bottom and n strands coming out the top, must be proportional to P_n.

Exercise 22.3 Theta Diagram

(a) Show $\Delta(a+1, a, 1) = d_{a+1}$. Hint: Use Fig. 22.29.

(b) More generally show $\Delta(a+k, a, k) = d_{a+k}$. Hint: Generalize Fig. 22.29 to the case where k strands are connected in a loop from the top to the bottom.

Exercise 22.4 F-matrix Diagrammatics

Using the diagrammatic algebra, determine $F_{12\beta}^{21\alpha}$ and $F_{21\beta}^{21\alpha}$ for arbitrary d. Confirm that your results are unitary matrices.

Exercise 22.5 Twists of Kauffman Anyons

Use the Kauffman bracket rules to calculate θ_a for the a-type Kauffman anyon (i.e. the particle type that is a cable of a strands and a P_a projector). Show that

$$R_0^{aa} = (-1)^a A^{-a(a+2)} \quad .$$

This quantity is the same as θ_a^* but see note 9 of this chapter. Hint: Try $a = 2$ then $a = 3$ to figure out the pattern. Use the properties of the projector.

Exercise 22.6 Factoring Kauffman Anyons For Odd k

Consider the case of the Kauffman A parameter given as in Eq. 22.15 where α and k are both odd,

(a) Show that $\theta_{k-n} = \theta_n\theta_k$, and further show that $\theta_k = \pm 1$.

(b) Show that $S_{a,n} = S_{k-a,n}$. Thus, conclude that the S-matrix is of the form of $\underset{\sim}{S} \otimes S_{\text{sVec}}$ where S_{sVec} is given by Eq. 17.25. In cases where $\theta_k = -1$, this is a super-modular theory. Hint: Use Eq. 17.20.

Consider now the case where k is odd, but α is even.

(c) Show that $\theta_{k-n} = \theta_n\theta_k(-1)^n$.

(d) Show that the S-matrix factorizes into $S = \underset{\sim}{S} \otimes S_{\text{semion}}$ where S_{semion} is the S-matrix of the semion theory.

Part V

Applications of TQFT Diagrammatics

State-Sum TQFTs

23

Optional

Having learned about planar diagram algebras we are now in a position to explicitly construct a three-dimensional TQFT.[1] There are several steps in this idea. We start by considering a closed three-dimensional manifold $\mathcal{M}$, which we discretize into tetrahedra (a so-called *simplicial decomposition* of the manifold). Next we construct a model, similar in spirit to statistical mechanics, which sums a certain weight over all quantum numbers on all edges of all tetrahedra. The weights being summed are defined in terms of our planar diagram algebra. The result of this sum is the desired TQFT partition function $Z(\mathcal{M})$ which we discussed extensively earlier in the book, and particularly in Chapter 7.

This discretization of a manifold into tetrahedra is very commonly used in certain approaches to quantum gravity, which we discuss in Section 23.3.

[1] The input for the construction in this chapter is a planar diagram algebra, or spherical category—*we do not have to specify any sort of R-matrix or braiding.* It is a bit surprising that one only needs a planar algebra to make a three-dimensional TQFT! In Section 23.2 we input a category with commutative fusion, $a \times b = b \times a$, whereas in Section 23.4 we input a group and a 3-cocycle.

23.1 Simplicial Decomposition and Pachner Moves

We start by considering a so-called simplicial decomposition of our manifold. Such decompositions can be made of smooth manifolds in any number of dimensions.[2]

23.1.1 Two Dimensions

As a warm-up let us think about two-dimensional manifolds. In two dimensions, the elementary 2-simplex is a triangle, so this decomposition is the familiar idea of triangulation shown in Fig. 23.1.

Since we are only concerned with the topology of the manifold, and not the geometry, the precise position of vertex points we use is irrelevant—

[2] It is interesting (but beyond the scope of this book) that manifolds exist in dimension $d \geq 4$ that cannot be smoothed, and cannot be decomposed into simplices.

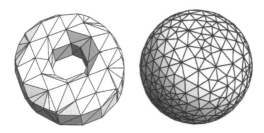

Fig. 23.1 Some triangulations of 2-manifolds.

only the connectivity of the points is important, that is, the topological structure of the triangulation network. Furthermore, a particular manifold, like a sphere, can be triangulated in many different ways. It turns out that any two different triangulations can be related to each other by a series of elementary "moves" known as two-dimensional *Pachner moves*,[3,4] which are shown in Figs. 23.2 and 23.3.

[3]I encourage you to play with these two moves and see how you can restructure triangulations by a series of Pachner moves.

[4]It is interesting to note that two-dimensional Pachner moves can be thought of as viewing a three-dimensional tetrahedron from two opposite directions. We can thus think of two-dimensional Pachner moves as a *cobordism* (see Chapter 7) in three-dimensions between a surface triangulated with the initial triangulation and a topologically equivalent surface triangulated with the final triangulation.

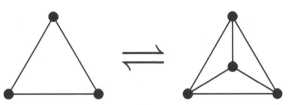

Fig. 23.2 The 1–3 Pachner move in two dimensions corresponds to adding or removing a point vertex from the triangulation. This turns one triangle into three, or vice versa. The added vertex need not be in the center of the original triangle.

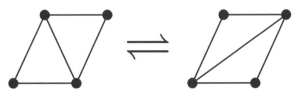

Fig. 23.3 The 2–2 Pachner move in two dimensions corresponds to replacing two adjacent triangles with two complementary triangles. This turns two triangles into two different triangles.

Thus, if we want to construct a manifold invariant (like $Z(\mathcal{M})$ we discussed in Chapter 7) for a manifold represented in terms of a triangulation we only need to find some function of the triangulation that is invariant under these two Pachner moves.

23.1.2   Three Dimensions

[5]For now let us focus on closed manifolds. We briefly discuss manifolds with boundary in Section 23.2.2.

[6]Analogous to the two-dimensional case (see note 4), the three-dimensional Pachner moves can be thought of as viewing a four-dimensional simplex (a so-called pentachoron) from two opposite directions.

The story is quite similar in three dimensions. Since we have focused on $(2 + 1)$-dimensional TQFTs we mostly discuss three-dimensional manifolds. We discretize any closed three-dimensional manifold[5] by breaking it up into tetrahedra (otherwise known as three-dimensional simplices). Any two discretizations are topologically equivalent to each other if they can be related to each other by a series of three-dimensional Pachner moves,[6] which are shown in Figs. 23.4 and 23.5. Again, the key point here is that if we can find some function of the network structure that is invariant under the Pachner moves, we will have constructed a topological invariant of the manifold.

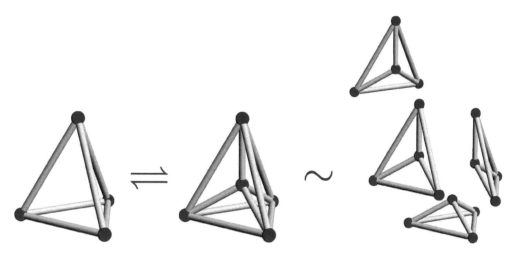

Fig. 23.4 The 1–4 Pachner move in three dimensions corresponds to adding or removing a point vertex to the tetrahedron decomposition. This turns a single tetrahedron into four, or vice versa. On the far right we show the four tetrahedron separated for clarity.

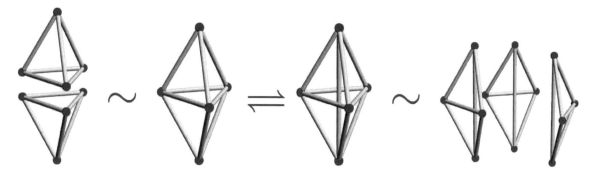

Fig. 23.5 The 2–3 Pachner move in three dimensions corresponds to resplitting a double tetrahedron (left) into three tetrahedra (right). This turns two tetrahedron into three, or vice versa. On the far left we show the two tetrahedra separated for clarity; and on the far right we have the three tetrahedra separated for clarity.

23.2   The Turaev–Viro State Sum

The idea of the Turaev–Viro state sum is to build a three-dimensional manifold invariant from one of the planar diagram algebras we discussed in Chapters 8–22.

First, let us choose any particular planar diagram algebra. We take any decomposition of an orientable three-dimensional manifold into tetrahedra. Let each edge of this decomposition be labeled with one of the quantum numbers (the particle labels) from the diagrammatic algebra.[7] We then consider the following sum[8]

$$Z_{TV}(\mathcal{M}) = \mathcal{D}^{-2N_v} \sum_{\text{all edge labelings}} W(\text{labeling}) \qquad (23.1)$$

[7] As we have been doing all along, when we label an edge with a quantum number we must put an arrow on the edge unless the particle type is self-dual.

[8] If there are fusion multiplicities $N_{ab}^c > 1$ in the diagram algebra these degrees of freedom must also be summed over. For simplicity we ignore this complication.

where N_v is the number of vertices in the decomposition, and

$$\mathcal{D} = \sqrt{\sum_n |d_n|^2}$$

is the total quantum dimension (see Eq. 17.11). In Eq. 23.1, W is a weight assigned to each labeling of all the edges.[9] We consider the following definition of a weight assigned to a given labeling of edges

$$W(\text{labeling}) = \frac{\prod_{\text{tetrahedra}} \tilde{G}(\text{tetrahedron}) \prod_{\text{edges}} d_{\text{edge}}}{\prod_{\text{triangles}} \tilde{\Theta}(\text{triangle})} . \quad (23.2)$$

Thus, each tetrahedron is given a weight $\tilde{G}$, depending on its labeling, each edge labeled a is given a weight d_a and each triangle is given a weight $\tilde{\Theta}^{-1}$, depending on its labeling.

The weights $\tilde{G}$ and $\tilde{\Theta}$ are very closely related to quantities G and Θ we studied[10] in Chapter 16 for example.[11] The functions $\tilde{G}$ and $\tilde{\Theta}$ are given by[12]

$$\tilde{\Theta}\left(\begin{array}{c} c \quad b \\ a \end{array} \right) = \Theta(a,b,c) = \sqrt{d_a d_b d_c} \quad (23.3)$$

and

$$\tilde{G}\left(\begin{array}{c} d \quad e \\ c \quad a \quad f \\ b \end{array} \right) = G^{bad}_{ecf} = F^{bad}_{ecf} d_f \sqrt{\frac{d_b d_c}{d_f}} \sqrt{\frac{d_a d_e}{d_f}} . \quad (23.4)$$

Note that the tetrahedron shown here is different from the one shown in Fig. 16.17, which defines G from a planar diagram (or perhaps more properly a diagram drawn on the surface of a sphere). In fact, the two tetrahedra are *dual* to each other. For example, in Fig. 16.17 the lines $f, e, \bar{c}$ form a loop, whereas $f, \bar{e}, \bar{a}$ meet at a point. On the other hand, in the diagram in Eq. 23.4, $f, e, \bar{c}$ meet at a point where $f, \bar{e}, \bar{a}$ form a loop. In Eq. 23.4 the three edges around any face must fuse together to the vacuum. That is, we have the four conditions

$$N_{bad} > 0 \qquad N_{c\bar{d}e} > 0 \qquad N_{f\bar{e}\bar{a}} > 0 \qquad N_{\bar{c}\bar{f}b} > 0$$

or else $\tilde{G}$ will vanish. Note that, like G, the value of $\tilde{G}$ is unchanged under any rotation of the tetrahedron.

It is important to note that Z_{TV} is *gauge invariant*. While we can make a gauge transformation on the F-matrices (and hence, $\tilde{G}$) as in Eq. 16.23, such gauge factors will cancel out in the overall weight Eq. 23.2 (see Exercise 23.4).

[9] In the language of statistical physics we can think of W as a Boltzmann weight for each edge label configuration, although it need not be positive, or even real.

[10] Many works, including the original work by Turaev and Viro [1992], use the diagrammatic algebra based on Temperly–Lieb which we discussed in Chapter 22. However, these use the nonunitary version of the diagrammatic algebra without the vertex renormalization, which we introduce in Section 22.3. In such an approach $\Theta(a,b,c)$ is replaced by $\Delta(a,b,c)$, for example (see Eq. 22.13). It is easy to show that these vertex-renormalization factors completely cancel and the end value of the Turaev–Viro invariant is independent of whether or not the renormalization factors are included. Indeed, it is not necessary to have a fully unitary algebra for the Turaev–Viro construction to give a well-behaved manifold invariant. See also note 11.

[11] In Chapter 16 we insist on a fully isotopy invariant algebra with commutative fusion, $a \times b = b \times a$, and tetrahedral symmetry, and we continue to assume those simplifications here. However, for constructing a Turaev–Viro invariant it turns out to be sufficient to have a spherical tensor category as we discuss in Chapter 12. Full isotopy invariance is not required. In fact, one can even have non-commutative fusion rules with $a \times b \neq b \times a$, thus subsuming the special case of Dijkgraaf–Witten discussed in section 23.4. The most general construction is discussed by Barrett and Westbury [1996].

[12] See the comments in Chapter 16 about how to choose the signs of the square roots in cases where some d values are chosen negative.

23.2.1   Proof Turaev–Viro Is a Manifold Invariant

The proof that $\mathcal{Z}_{TV}(\mathcal{M})$ is a manifold invariant is not difficult—one only needs to show that it is unchanged under the 1–4 and 2–3 Pachner moves. This is basically an exercise in careful bookkeeping (see Exercise 23.2). Roughly, it is easy to see how it is going to work.

Let us first examine the 2–3 Pachner move shown in Fig. 23.5. On the left we have two tetrahedra (call them 1 and 2) that are joined along a triangle (call it α). On the right we have three tetrahedra (call them 3, 4, and 5) which are joined along three triangles (call them β, γ, and δ) with the three triangles intersecting along a new edge down the middle (shown vertical in the figure), which we label with the quantum number n. To show that the Z_{TV} remains invariant we need to show that

$$\frac{\tilde{G}(1)\tilde{G}(2)}{\tilde{\Theta}(\alpha)} = \sum_n \frac{\tilde{G}(3)\tilde{G}(4)\tilde{G}(5)d_n}{\tilde{\Theta}(\beta)\tilde{\Theta}(\gamma)\tilde{\Theta}(\delta)} \quad .$$

The factors of $\tilde{\Theta}$ are simply factors of $\sqrt{d_a}$ and these cancel some factors of $\sqrt{d_a}$ in the definition of $\tilde{G}$ in Eq. 23.4. After this cancellation what remains is a relationship between two Fs on the left and a sum over three Fs on the right. The relationship that remains is exactly the pentagon equation Eq. 16.3 (or Eq. 9.7)! Thus, any diagrammatic algebra that satisfies the pentagon equation will result in a Turaev–Viro partition function (Eq. 23.1) that is invariant under the 2–3 Pachner move!

The case of the 1–4 Pachner move is only a bit harder, and we sketch the calculation here. The large tetrahedra on the left of Fig. 23.4 (let's call this large tetrahedron 1) needs to be equivalent to the four smaller tetrahedra on the right (let's call these small tetrahedra 3, 4, 5 on top and 6 on the bottom) once we sum over the quantum numbers on the four internal edges on the right. The three tetrahedra 3, 4, and 5 share a common edge, and this is entirely analogous to the three tetrahedra we considered in the case of the 2–3 Pachner move. Summing over the quantum number of this common edge, and using the same pentagon relation replaces the three tetrahedra 3, 4, 5 with two tetrahedra 1 and 2, where 1 is the large tetrahedron and 2 includes exactly the same edges as the remaining small tetrahedron 6. The tetrahedra 2 and 6 have three edges that are not shared with tetrahedron 1—these are the remaining internal edges that need to be summed over. Summing over one of these internal edges, one invokes the consistency condition Eq. 16.5 to create a delta function, which then kills one of the two remaining sums. The last remaining sum just yields a factor of $\mathcal{D}^2 = \sum_n d_n^2$, which accounts for the prefactor in Eq. 23.1 being that we have removed one vertex from the lattice.

23.2.2   Some TQFT Properties

The Turaev–Viro state sum has all the properties we expect of a TQFT. Although we need to discretize our manifold, the resulting "partition function" $Z_{TV}(\mathcal{M})$ for a manifold $\mathcal{M}$ is a complex number that is indeed independent of the discretization and only depends on the topology of the manifold.

As we discuss at length in Section 7.1, we would also like $Z_{TV}(\mathcal{M})$ to represent a *wavefunction* if $\mathcal{M}$ is a manifold with boundary. To remind the reader, the point of this construction is that, when we glue together two manifolds with boundaries to get a closed manifold, this corresponds to taking the inner product between the two corresponding wavefunctions to get a complex number.

To see how this occurs let us consider discretizing a manifold with a boundary. Here, the three-dimensional bulk of the manifold $\mathcal{M}$ should be discretized into tetrahedra, and the two-dimensional boundary surface $\Sigma = \partial M$ should be discretized into triangles. We divide the edge degrees of freedom into bulk and boundary where we say that an edge is a boundary edge if and only if both vertices are on the boundary. All other edges we call bulk edges. We define $Z(\mathcal{M})$ of such a discretized manifold with boundary as a sum like Eq. 23.1 where the sum is only over the edges in the bulk, leaving fixed (un-summed) the quantum numbers for the edges that live entirely on the boundary (i.e. both vertices on the boundary). Thus, for manifolds with boundary we more generally write

$$Z_{TV}(\mathcal{M}; a_1, \ldots, a_N) = \mathcal{D}^{-2N_v - n_v} W'(a_1, \ldots, a_N) \sum_{\text{bulk labelings}} W(\text{bulk labels})$$

where N_v is the number of vertices in the bulk and n_v is the number of vertices on the boundary. The weight function W is exactly the same as the weight function in Eq. 23.2 but only including edges, triangles, and tetrahedra in the bulk (all tetrahedra are considered bulk, and a triangle is considered a boundary triangle only when all three vertices are on the boundary). Here, $a_1, \ldots, a_N$ are the quantum numbers of the edges on the boundary, and these are not included in the sum over bulk labels. An additional weight is included, which is a function of these boundary edge labels

$$W'(a_1, \ldots, a_N) = \sqrt{\frac{\prod_{\text{boundary edges}} d_{\text{edge}}}{\prod_{\text{boundary triangles}} \tilde{\Theta}(\text{triangle})}} \ .$$

The partition function $Z_{TV}(\mathcal{M}; a_1, \ldots, a_N)$ is now a function of the edge variables and is interpreted as a wavefunction[13] $|Z(\mathcal{M})\rangle$ that lives on the boundary $\Sigma = \partial M$.

It is then quite natural to glue the two manifolds together along a common boundary, as in Fig. 7.3, where we have a closed manifold $\mathcal{M} \cup_\Sigma \mathcal{M}'$ where $\mathcal{M}$ and $\mathcal{M}'$ are manifolds with boundary joined along their common boundary $\Sigma = \partial M = [\partial M']^*$. When we glue together $\mathcal{M}$ and $\mathcal{M}'$ we obtain the partition function for the full manifold, as in Eq. 7.2, where we obtain the inner product by summing over the degrees of freedom of the wavefunction—which in this case means summing over the quantum numbers $a_1, \ldots, a_N$ of the edges on the boundaries. In

[13]The wavefunction here takes some complex scalar value as a function of the physical variables, which are the quantum numbers on the edge.

other words, we have

$$Z_{TV}(\mathcal{M} \cup_\Sigma \mathcal{M}') = \langle Z_{TV}(\mathcal{M}')|Z_{TV}(\mathcal{M})\rangle$$

$$= \sum_{a_1,\ldots,a_N} [Z_{TV}(\mathcal{M}'; \bar{a}_1, \ldots, \bar{a}_N)]^* \, Z_{TV}(\mathcal{M}; a_1, \ldots, a_N) \quad (23.5)$$

$$= \sum_{a_1,\ldots,a_N} Z_{TV}(\mathcal{M}'; a_1, \ldots, a_N) \, Z_{TV}(\mathcal{M}; a_1, \ldots, a_N) \quad (23.6)$$

where in the second line the edge variables in the first term are inverted because the surface of $\mathcal{M}'$ has the opposite orientation from the surface of $\mathcal{M}$. Going from the second to third line is an easy exercise (see Exercise 23.1). The final result is easily seen to be the correct expression for the Turaev–Viro invariant for the full manifold $\mathcal{M} \cup \mathcal{M}'$. That is, it now sums over all the quantum numbers in both bulks and on the common boundary.

As in Section 7.2, one can generalize the idea of a TQFT to include particle world lines (labeled links) as a well as the underlying spacetime manifold $\mathcal{M}$. As explained there, we can roughly think of these world lines as internal boundaries, and we just fix the quantum number of edges along these hollow tubes to describe different world line types (see references at the end of the chapter).

The TQFT that arises from an input set of F-matrices (from an input *category*) is known as the *Drinfel'd double*, *Drinfel'd center*, or *quantum double* of the input. We discuss quantum doubles again in Chapters 31–33.

23.2.3 Connection to Chern–Simons Theory

There is a remarkable connection between the Turaev–Viro partition function and Chern–Simons theory. If we build a Turaev–Viro theory from the F-matrices of a Chern–Simons theory, it turns out that the partition function of the Turaev–Viro theory is related to that of the Chern–Simons theory via[14]

$$Z_{TV}(\mathcal{M}) = |Z_{CS}(\mathcal{M})|^2 \quad.$$

This is known as the Turaev–Walker theorem (Turaev [1992, 1994]; Walker [1991]), and we outline its derivation in Section 24.3.2. This formula gives at least one reason why the name *quantum double* is appropriate at least in this case—Turaev–Viro is equivalent to two copies of the Chern–Simons theory (of opposite handedness).

23.3 Connections to Quantum Gravity Revisited

The Turaev–Viro invariant is a descendant of one of the earliest approaches to quantum gravity pioneered by Penrose [1971][15] and Ponzano

---

[14]In Section 24.3.2 we call the Chern–Simons partition function Z_{WRT}.

[15]Roger Penrose (currently at Oxford) is a brilliant mathematician who won the 2020 Nobel Prize for his work on black holes. He has done highly influential work on a wide range of topics in physics and mathematics and has written a number of popular books on science. Despite being a genius, he has presented some extremely "controversial" opinions about human consciousness.

and Regge [1968]. Indeed, much of the continued interest in state-sum invariants is due to the connection to quantum gravity.

An interesting approach to macroscopic general relativity, used for example, in numerical simulation, is to discretize spacetime into simplices—tetrahedra in three dimensions or four-dimensional simplices (sometimes known as pentachora) in four dimensions.[16] The curvature of the spacetime manifold (the metric) is then determined by the lengths assigned to the edges.[17]

If one then turns to quantum gravity, one wants to follow the Feynman prescription and perform a sum over all possible metrics, as discussed in Section 6.3. We can write a quantum partition function as

$$Z = \int \mathcal{D}g \; e^{iS_{Einstein}[g]/\hbar} \quad . \tag{23.7}$$

We can imagine performing such as sum for a discretized system by integrating over all possible lengths of all possible edges. However, not all triangle edge lengths should be allowed—one must obey the crucial constraint of the triangle inequality[18]

$\Rightarrow \quad |l_1 - l_2| \le l_3 \le (l_1 + l_2) \quad . \tag{23.8}$

The key observation is that the triangle inequality is precisely the same as the required inequality for regular angular momentum addition

$$j_1 \otimes j_2 = |j_1 - j_2| \oplus |j_1 - j_2| + 1 \oplus \ldots \oplus |j_1 + j_2| \quad . \tag{23.9}$$

Thus, it is natural to label each edge with a quantum-mechanical spin, and sum over all possible spins. Such an approach is known as a *spin network*. We thus imagine building a Turaev–Viro model (Eq. 23.1) with a planar diagram algebra built from angular-momentum addition rules: quantum numbers are the angular momenta j, the fusion rules are as given in Eq. 23.9, and the F-matrices are given by the regular $6j$-symbols of angular momenta addition.[19] Such a model turns out to be very precisely[20] the quantum gravity partition function Eq. 23.7 (up to the fact that one still needs an additional sum over topologies of the spacetime manifold if one wants a full sum over all possible histories)! As we expect from the discussion in Chapter 6, the resulting description of quantum gravity in (2 + 1)-dimensions is a TQFT.

There is, unfortunately, one clear problem with this approach. Because there are an infinite number of different representations of $SU(2)$—that is, an infinite number of different values for the angular-momentum quantum number j—the partition function sum formally diverges. This divergence becomes regularized if we find a way to consistently cut off the sum over angular momenta at some maximum value k. Using the diagrammatic rules of $SU(2)_k$ (the same diagrammatic rules we built up in Chapters 21–22, see in particular note 7 of Chapter 21) implements this cutoff and yields a divergence-free result.[21]

[16]It is also possible to discretize space and leave time continuous. This leaves some concerns with Lorentz invariance but may have other advantages. Other discretization approaches also exist; see Regge and Williams [2000].

[17]All of general relativity can be reformulated in this discrete language. This is known as *Regge calculus*. See Regge [1961].

[18]These inequalities must hold even with a curved spatial metric.

[19]Building a diagrammatic algebra based on a Lie group ($SU(2)$ in this case) is mentioned in Section 21.1.

[20]Although the idea of spin networks as a toy model for quantum gravity goes back to Penrose [1971], and was pursued further by Ponzano and Regge [1968], it was only much later that Hasslacher and Perry [1981] showed a more precise equivalence of the model to gravity.

[21]As mentioned in Section 23.2.3, the Turaev–Viro model built from the $SU(2)_k$ diagrammatic rules is equivalent to Chern–Simons theory $SU(2)_k \otimes SU(2)_{-k}$. As mentioned in Section 6.5, such a Chern–Simons theory is equivalent to (2 + 1)-dimensional gravity with a cosmological constant $\lambda = (4\pi/k)^2$. Taking the limit of large k then gives the classical limit of simple $SU(2)$ angular-momentum addition corresponding to a universe with no cosmological constant.

23.4 Dijkgraaf–Witten Model

Another state-sum model of some interest is the Dijkgraaf–Witten model (Dijkgraaf and Witten [1990]).[22] As with Turaev–Viro,[23] this model discretizes space into simplices and sums over possible labels of all the edges

In the Dijkgraaf–Witten model we choose a group G and we label the (directed) edges of the simplices with elements from that group. The general idea is very similar to that of Turaev–Viro just using the multiplication properties of the group to give us a set of fusion rules (as in Chapter 19) and we use a 3-cocycle in place of the F-matrix.[24] These fusion rules require that multiplication of the group elements around every triangle must result in the identity as shown in Fig. 23.6. This is the analog of Eq. 23.3 where three quantum numbers around a triangle must fuse to the identity. This condition is known as a "flatness" condition, with the name coming from lattice gauge theory, which we look at in more detail in Section 31.7.

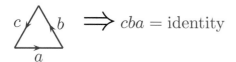

Fig. 23.6 Multiplying group elements around a triangle in Dijkgraaf–Witten theory results in the identity. This is known as the "flatness" condition.

As mentioned in Section 19, when we use group multiplication for fusion rules, the quantum dimensions[25] of all the particles are all $d_a = 1$. This means that in Eq. 23.2 both the d_a factor and the $\tilde{\Theta}$ factor are trivial. We are thus left with only the tetrahedron factor and the Dijkgraaf–Witten partition function looks like a simplified version of the Turaev–Viro case in Eqs. 23.1 and 23.2 given by[26]

$$Z_{DW}(\mathcal{M}) = |G|^{-N_v} \sum_{\text{labelings}} \prod_{\text{tetrahedra}} \tilde{G}(\text{tetrahedron}) \tag{23.10}$$

where N_v is the number of vertices, $|G|$ is the number of elements in the group G, and the sum is only over labelings that satisfy the flatness condition (Fig. 23.6).

The tetrahedral symbol $\tilde{G}$ is a bit more complicated than in the case we discussed for the Turaev–Viro invariant, where we assumed some nice properties such as commutative fusion and tetrahedral symmetry. Here, we do not generally have full-tetrahedral symmetry so it could matter which way we orient the tetrahedron when we evaluate $\tilde{G}$. In order to define the tetrahedral symbol $\tilde{G}$ properly we do the following: First we label each vertex in the system with a unique integer (it will not matter which vertex gets which label!). Given a tetrahedron with

[22]Robbert Dijkgraaf is a very prominent theoretical physicist and string theorist. His surname is likely to be difficult to properly pronounce for those who are not from the Netherlands because the "g" is a guttural sound that only exists in Dutch. However, those from the south of the Netherlands don't use the guttural "g" and instead pronounce it as Dike-Hraff, which is probably about the closest most English speakers will get to the right result. The word "Dijkgraaf" refers to an occupation: A Dijkgraaf is the person in charge of making sure that water stays in the ocean and does not flood the cities and the rest of the Netherlands. Robbert Dijkgraaf is also officially a "Knight of the Order of the Netherlands Lion."

[23]In fact Dijkgraaf–Witten in three-dimensions is really a special case of the most general construction of Turaev–Viro. See Barrett and Westbury [1996].

[24]For the case of an abelian group, Dijkgraaf–Witten is a special case of the Turaev–Viro construction we have already discussed. However, more generally here we allow non-commutative fusion rules where $g \times h \neq h \times g$. The fusion need not be abelian since we only need to have an algebra that is consistent on a plane (or sphere) in order to define its value on a tetrahedron (see the comments in Section 19.3).

[25]In Chapter 19 we considered also the possibility of $d_a = -1$ but this is a choice. We are always entitled to chose $+1$ instead at the cost of possibly losing isotopy invariance.

[26]With apologies for using G and $\tilde{G}$ in the same equation to mean completely different things!

vertices i_1, i_2, i_3, i_4 we sort these vertices in ascending order so that

$$[j_1, j_2, j_3, j_4] = \text{sort}[i_1, i_2, i_3, i_4] \quad \text{such that} \quad j_1 < j_2 < j_3 < j_4$$

we then define

$$\tilde{G}\left(\begin{array}{c} i_1 \\ i_4 \\ i_2 \quad i_3 \end{array} \right) = \omega(g_{j_2,j_1}, g_{j_3,j_2}, g_{j_4,j_3})^{s(j_1,j_2,j_3,j_4)} \quad . \quad (23.11)$$

Here, $g_{k,l}$ is the group element on the edge directed from vertex k to vertex l, and ω is the chosen 3-cocycle. The exponent $s(j_1, j_2, j_3, j_4)$ is either $+1$ or -1, depending on whether the orientation of the tetrahedron defined by the ordered set of vertices $[j_1, j_2, j_3, j_4]$ has the same or opposite orientation as the manifold we are decomposing.[27] This prescription gives a manifold invariant (the Dijkgraaf–Witten invariant) for any choice of 3-cocycle even if the corresponding diagrammatic algebra does not have isotopy invariance. Analogous to the Turaev–Viro invariant, the partition function here is gauge invariant if we transform the cocycles with Eq. 19.5 (see Exercise 23.4).

23.4.1   Other Dimensions

An interesting feature of Dijkgraaf–Witten theory is that essentially the same recipe builds a Dijkgraaf–Witten TQFT in any number of dimensions. One discretizes the D-dimensional manifold into D-dimensional simplices (segments in one dimension, triangles in two dimensions, tetrahedra in three dimensions, pentachora in four dimensions) and labels each edge with a group element $g \in G$ and each vertex is assigned an integer label. The flatness condition is always the same as that shown in Fig. 23.6—multiplying the group elements around a closed loop must give the identity. In D dimensions we build the partition function by multiplying a weight for each D-simplex, where the weight is given now by a so-called D-cocycle,[28] which we call $\omega_D(g_1, g_2, \ldots, g_D)$, and which is now a function of D arguments. Finally, one builds a partition function by summing over all possible labelings

$$Z_{DW}(\mathcal{M}_D) = |G|^{-N_v} \sum_{\text{labelings}} \prod_{D\text{-simplices}} \omega_D(g_{j_2,j_1}, \ldots, g_{j_{D+1},j_D})^{s(j_1,\ldots,j_{D+1})} \,.$$

$$(23.12)$$

As with the three-dimensional case, the arguments of the cocycle $g_{k,l}$ are the group elements along the edges of the simplex from vertex k to vertex l and we always write them ordered such that $j_1 < j_2 < \ldots < j_D$. Finally, the exponent s is always ± 1 depending on whether or not the orientation of simplex described by the ordered set $[j_1, \ldots, j_{D+1}]$ matches that of the underlying manifold.

As a quick example, let us consider the two-dimensional case. The definition of a 2-cocycle ω_2 is any function that satisfies the condition[29,30]

[27] To find the orientation of a tetrahedron, place j_1 closest to you and see if the triangle $[j_2, j_3, j_4]$ is oriented clockwise or counterclockwise.

[28] I won't give the most general definition of cocycle as this takes us too far afield into group cohomology. However, as with the 3-cocycle it is simply a function satisfying a particular cocycle condition. See Eq. 19.1 for the three-dimensional case and Eq. 23.14 for the two-dimensional case.

[29] The 2-cocycle condition is equivalent to the consistency condition for a "projective representation" of the group. For projective representations we have the multiplication rule $\rho(g)\rho(h) = \omega_2(g, h)\rho(gh)$, whereas for regular group representations we have $\omega_2 = 1$. See Section 40.2.4.

[30] As with 3-cocycles one can gauge transform $U(1)$-valued 2-cocycles via

$$\omega_2(g, h) \to \frac{\beta(gh)}{\beta(g)\beta(h)}\omega_2(g, h) \quad (23.13)$$

for any $U(1)$-valued function $\beta(g)$. This does not change the value of the Dijkgraaf–Witten invariant for a 2-manifold without boundary.

$$\omega_2(g,h)\omega_2(gh,k) = \omega_2(h,k)\omega_2(g,hk) \quad . \qquad (23.14)$$

In the partition function, Eq. 23.12, each triangle gets a weight given by the cocycle. It is then easy to see that the cocycle condition is precisely the condition necessary to make the partition function invariant under the 2–2 Pachner move, as shown in Fig. 23.7. It is not a hard exercise to demonstrate invariance under the 3–1 Pachner move as well (see Exercise 23.3).

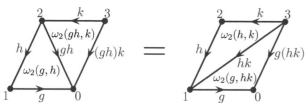

Fig. 23.7 Each triangle satisfies the flatness condition Eq. 23.6, meaning multiplying all three edges in order gives the identity. In the partition function each triangle gets a weight given by the corresponding cocycle ω_2 as written in black text. All of the triangles in the figure are oriented positively $s = +1$. The cocycle condition Eq. 23.14 guarantees that the product of the cocycles on the left equals the product of the cocycles on the right.

23.4.2 Further Comments

One particularly interesting special case of Dijkgraaf–Witten theory is the case of the trivial 3-cocycle where ω is always unity. In this case, the argument of the sum in Eq. 23.10 (or more generally Eq. 23.12) is unity so the partition function just counts the number of flat field configurations (see Fig. 23.6) and then divides by $|G|^{N_v}$. This partition function is exactly that of lattice gauge theory, as we see in Section 31.7 below, and the resulting topological quantum field theory is known as the quantum double of the group G. The more general case, with a nontrivial cocycle, is correspondingly sometimes known as "twisted" gauge theory, where the cocycle is thought of as some sort of twist to the otherwise simple theory. Dijkgraaf–Witten theories are sometimes called *twisted quantum doubles*. We discuss twisted theories again in Section 31.4.1.

A further interesting relationship is that Dijkgraaf–Witten theory can be thought of as the result of symmetry breaking an appropriately chosen Chern–Simons theory (see Dijkgraaf and Witten [1990] and de Wild Propitius [1995]). One might imagine, for example, breaking a compact $U(1)$ Chern–Simons gauge theory into a discrete $\mathbb{Z}_n$ group—like breaking the symmetry of a circle into an n-sided regular polygon. The particular cocycle one gets in the resulting Dijkgraaf–Witten theory depends on the choice of the coefficient (the "level") of the Chern–Simons term.

Rather surprisingly, subject to some simple conditions about point-like particles being bosonic, it turns out that *all* TQFTs in $(3+1)$-dimensions are of the Dijkgraaf–Witten form (Lan et al. [2018]; Johnson-Freyd [2022]).

Dijkgraaf–Witten theory has had extensive recent applications within quantum-condensed matter physics where it turns out that a classification of so-called symmetry-protected topological (SPT) phases is given in terms of Dijkgraaf–Witten theories. We discuss SPT phases in Chapter 35.

Chapter Summary

- State-sum TQFTs build partition functions on triangulated manifolds by summing over quantum numbers on edges, analogous to many statistical mechanics models.
- The Turaev–Viro state sum is built from any input F-matrices satisfying the pentagon equations
- The Dijkgraaf–Witten state sum in d dimensions is built from an input group and a d-cocycle. The trivial cocycle makes Dijkgraaf–Witten into a simple lattice gauge theory.

Further Reading:

- The Turaev–Viro invariant was introduced in Turaev and Viro [1992]. Rather interestingly, Turaev and Viro were apparently unaware of the earlier work by Penrose, Ponzano, Regge, and others when they first discussed these state sums! The work was extended by Barrett and Westbury [1996] to include all spherical fusion categories. A recent rather complete discussion is given by Turaev and Virelizier [2017]. Unfortunately, these references and many other works in the field are written in rather mathematical language that is not particularly transparent for most physicists!

- The original work by Dijkgraaf and Witten [1990] is very mathematical in places, but much of it is fairly readable. Altschuler and Coste [1993] gives a very abbreviated but clear discussion of Dijkgraaf–Witten theory and provides examples of calculating Dijkgraaf–Witten manifold invariants (Chapter 24 may be useful for understanding the latter parts of this paper).

- It is worth commenting that the state-sum approach to quantum gravity has been extended in a multitude of ways, and continues to be an active area of research. Among the key directions are extension to $(3 + 1)$-dimensions (Ooguri [1992] and Crane and Yetter [1993] for example), and extensions to Lorentzian signature (Barrett and Crane [2000]). A nice general discussion of discrete approaches to gravity is given by Regge and Williams [2000].

- One very popular extension of the spin-network modes, known as a *spin-foam*, is to discretize space but allow the discretization to change as a function of time. A nice review of this direction is given by Lorente [2006].

- More recently there have been other extensions of the state-sum approach. Notably, the work of Douglas and Reutter [2018] constructs a (maximally?) general state sum in $(3+1)$-dimensions. Walker [2021] unifies this, and many other state sums, under a single formalism. Both of these recent works are in fairly mathematical language unfortunately.

- An online database of modular data (S- and T-matrices) for Dijkgraaf–Witten theories (i.e. twisted quantum doubles) is given by

  `https://tqft.net/web/research/students/AngusGruen/`

  and is also mirrored on my web page.

Exercises

Exercise 23.1 Some More Facts about Turaev–Viro
Consider a manifold $\mathcal{M}$ with boundary Σ that has been discretized into tetrahedra in the bulk and triangles on the surface. Let the edges on the surface be labeled by $j_1, \ldots, j_N$. Assuming that the theory has reflection symmetry as in Eq. 16.14, show that

$$[Z_{TV}(\mathcal{M}; \bar{a}_1, \ldots, \bar{a}_N)]^* = Z_{TV}(\mathcal{M}; a_1, \ldots, a_N)$$

and as a result show that for a closed manifold $Z(\mathcal{M})$ is real.

Exercise 23.2 Details of Turaev–Viro
Work carefully through the details of the proof that the Turaev–Viro partition function is invariant under Pachner moves.

Exercise 23.3 Two-Dimensional Dijkgraaf–Witten
The invariance of the two-dimensional Dijkgraaf–Witten partition function under the 2–2 Pachner move is established in Section 23.4.1. Show that the partition function is also invariant under the 3–1 Pachner move.

Exercise 23.4 Gauge Invariance
(a) Show that Z_{TV} is unchanged if we make a gauge transformation of the type of Eq. 16.23 (Hint: Each plaquette appears in two F-matrices in the product Eq. 23.2).
(b) Show that Z_{DW} is unchanged if we make a gauge transformation of the type shown in Fig. 19.5.

Exercise 23.5 Lattice Gauge Theory on S^3
The manifold S^3 has two very simple one-tetrahedron triangulations (see Jaco and Rubinstein [2003]). These are shown in Fig. 23.8. First convince yourself that these triangulations describe closed manifolds (i.e. check that each face occurs twice in opposite orientations).

Dijkgraaf–Witten theory with a trivial cocycle ($\omega = 1$) is equivalent to lattice gauge theory. By performing the sum in Eq. 23.10 calculate $Z_{DW}(S^3)$ in the case of the trivial cocycle for both triangulations of S^3 and check that they give the same result.

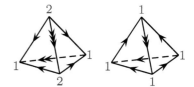

Fig. 23.8 The manifold S^3 has two very simple triangulations each involving only a single tetrahedron. On the left, two vertices (1 and 2) are connected together with three different edges (single, double, and triple arrow). The single arrow edge is drawn more than once though. On the right, a single vertex (1) is connected to itself with two different edges (single and double arrow).

Exercise 23.6 Turaev–Viro on S^3

(a) Consider a Turaev–Viro model built from any F-matrices with the property that there is no particle a, except the identity, for which $a \times a \times \bar{a}$ gives the identity. Using a simple triangulation of S^3, calculate $Z_{TV}(S^3)$. (Hint: This is much easier than it seems if you use the single vertex triangulation from Fig. 23.8). Is this consistent with the statements of Section 23.2.3, at least for the case where the F-matrices correspond to those of a Chern–Simons theory? (More generally, see the statements in Exercise 33.8).

(b) [Harder] Repeat this calculation without the special condition on $a \times a \times \bar{a}$. Hint: You may have to use a completeness relation to get rid of a sum.

Formal Construction of TQFTs from Diagrams: Surgery and More Complicated 3-Manifolds[1]

24

Optional/Hard

Having constructed diagrammatic algebras in $(2 + 1)$-dimensions,[2] we have almost all we need to define a TQFT based on these diagrams. As discussed in Section 14.3.2, our diagrammatic algebra gave us a way to evaluate a "partition function" Z(labeled link in S^3)$/Z(S^3)$, or equivalently, Z(labeled link in $S^2 \times S^1$) with the caveat that no link goes around the nontrivial cycle of S^1. However, a TQFT should be able to evaluate Z(labeled link in $\mathcal{M}$) for *any* arbitrary manifold $\mathcal{M}$. Indeed, in the simplest case we might dispense with the labeled link, and we might want to find a partition function $Z(\mathcal{M})$ of the manifold $\mathcal{M}$ alone.

In this chapter we develop a prescription for handling more complicated manifolds. One important thing this will achieve is to give a formal definition to Chern–Simons theory, which we like to think of as being defined as some sort of functional integral, but (as pointed out in Section 5.3.6) is often not really well defined in that language since such integrals may not actually converge.

The way we handle more complicated manifolds is by sewing pieces of manifolds together with a procedure known as surgery.

24.1 Surgery

In Chapter 7 we saw two examples of assembling manifolds by gluing together pieces. We found that we could assemble together two solid tori $(D^2 \times S^1)$ into either S^3 or $S^2 \times S^1$ depending on how we glue together the $S^1 \times S^1$ surfaces. (In fact, one can consider gluing together the surfaces in yet other ways to get even more interesting results,[3] but we do not need that for the moment). We would like to use this sort of trick to study much more complicated three-dimensional manifolds.

The understanding of three-dimensional manifolds is a very rich and beautiful topic.[4] In order to describe complicated manifolds it is useful to think in terms of so-called surgery. Similar to what we were discussing

[4]Many important results on three-dimensional manifolds have been discovered recently. Perelman's[5] proof of the Poincaré conjecture, along with the methods he used, are apparently extremely revolutionary and powerful. But this is *way* outside the scope of our book!

[5]Grigori Perelman is a brilliant but startlingly puzzling character. He famously declined the million-dollar Millenium Prize offered to him for proving the Poincaré conjecture in three dimensions. He turned down the Fields Medal as well.

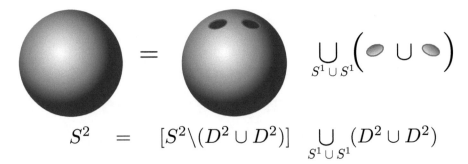

$$S^2 \quad = \quad [S^2 \backslash (D^2 \cup D^2)] \underset{S^1 \cup S^1}{\bigcup} (D^2 \cup D^2)$$

Fig. 24.1 Writing a sphere $\mathcal{M} = S^2$ as the union of two manifolds $\mathcal{M}_1$ and $\mathcal{M}_2$ glued along their boundaries. $\mathcal{M}_2$ is the union of two disks $D^2 \cup D^2$. $\mathcal{M}_1 = S^2 \backslash (D^2 \cup D^2)$ is the remainder.[8] The two manifolds are glued along their common boundary $S^1 \cup S^1$.

[6]Prehensile toes could be useful I suppose!

[7]We will only be concerned with orientable manifolds

in Section 7.3—assembling a manifold by gluing pieces together—the idea of surgery is that we remove a part of a manifold and we glue back in something different. Imagine replacing someone's foot with a hand![6] By using successive surgeries we will be able to construct any three-dimensional manifold.[7]

The general scheme of surgery is to first write a manifold as the union of two manifolds-with-boundary, glued together along their common boundaries. If we have a closed manifold $\mathcal{M}$ that we would like to alter, we first split it into two pieces $\mathcal{M}_1$ and $\mathcal{M}_2$ such that they are glued together along their common boundary $\partial \mathcal{M}_1 = \partial \mathcal{M}_2^*$. So we have

$$\mathcal{M} = \mathcal{M}_1 \cup_{\partial \mathcal{M}_1} \mathcal{M}_2 \quad .$$

We then find another manifold with boundary $\mathcal{M}_2'$ whose boundary matches $\mathcal{M}_2$, that is

$$\partial \mathcal{M}_2 = \partial \mathcal{M}_2' \quad .$$

We can then replace $\mathcal{M}_2$ with $\mathcal{M}_2'$, to construct a new closed manifold $\mathcal{M}'$ as

$$\mathcal{M}' = \mathcal{M}_1 \cup_{\partial \mathcal{M}_1} \mathcal{M}_2' \quad .$$

We say that we have performed surgery on $\mathcal{M}$ to obtain $\mathcal{M}'$. In other words, we have simply thrown out the $\mathcal{M}_2$ part of the manifold and replaced it with $\mathcal{M}_2'$.

24.1.1 Simple Example of Surgery on a 2-Manifold

[8]The backslash notation here mean "remove." That is, $\mathcal{M} \backslash \mathcal{M}'$ means the manifold (with boundary) given after removal of $\mathcal{M}'$ from $\mathcal{M}$.

[9]Strictly speaking, the boundary of the cylinder and the boundary of the two disks are oppositely oriented.

To give an example of surgery consider the sphere $\mathcal{M} = S^2$ as shown in Fig. 24.1. Here, we write the sphere as the union of two disks $\mathcal{M}_2 = D^2 \cup D^2$ and the remainder[8] of the sphere $\mathcal{M}_1 = S^2 \backslash (D^2 \cup D^2)$. These are glued along their common boundary $S^1 \cup S^1$.

Now we ask the question of what other 2-manifolds have the same boundary $S^1 \cup S^1$. There is a very obvious one: the cylinder surface! Let us choose the cylinder surface $\mathcal{M}_2' = S^1 \times I$ where I is the interval (or D^1). This also has boundary[9] $\partial \mathcal{M}_2' = S^1 \cup S^1$, as shown in Fig. 24.2.

$$\partial \quad \boxed{} \quad = \partial(\bullet \cup \bullet) = \quad \circ \cup \circ$$

$$\partial(S^1 \times I) = \partial(D^2 \cup D^2) = S^1 \cup S^1$$

Fig. 24.2 The boundaries of the cylinder surface is the same as the boundary of the two disks. Both boundaries are two circles.[9] This means that we can remove two disks from a manifold and glue in the cylinder.

Thus, we can glue the cylinder surface in place where we removed $\mathcal{M}_2 = D^2 \cup D^2$, as shown in Fig. 24.3. The resulting manifold $\mathcal{M}'$ is the torus T^2.

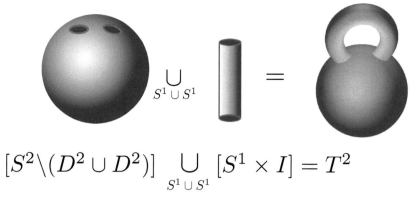

$$[S^2 \backslash (D^2 \cup D^2)] \underset{S^1 \cup S^1}{\cup} [S^1 \times I] = T^2$$

Fig. 24.3 Gluing the cylinder surface $\mathcal{M}_2' = S^1 \times I$ to the manifold $\mathcal{M}_1 = S^2 \backslash (D^2 \cup D^2)$ along their common boundary $S^1 \cup S^1$ gives the torus T^2. Note that the object on the right is topologically a torus.

Thus, we have surgered a sphere and turned it into a torus. Note that there is another way to think of this procedure. If $\mathcal{M} = \partial \mathcal{N}$ then surgery on $\mathcal{M}$ is the same as attaching a handle to $\mathcal{N}$. In the case we just considered we would take $\mathcal{N} = B^3$ the 3-ball (sometimes denoted D^3), and we attach a handle $D^2 \times I$, the solid cylinder. We obtain the new manifold $\mathcal{N}'$, which is the solid torus, whose boundary is T^2, the torus surface. This is written out in the diagram Fig. 24.4

$$\mathcal{N} = B^3 \qquad\qquad\qquad \partial \mathcal{N} = \mathcal{M} = S^2$$

$$\downarrow \text{ Add Handle} \qquad\qquad\qquad \downarrow \text{ Surgery}$$

$$\text{Solid Torus} \qquad\qquad\qquad \partial(\text{Solid Torus}) = T^2$$

Fig. 24.4 Handle attaching on the manifold $\mathcal{N}$ is the same as surgery on a manifold $\mathcal{M} = \partial \mathcal{N}$.

24.1.2 Surgery on 3-Manifolds

[10]This is the part that is guaranteed to make your head explode.

We can also perform surgery on three-dimensional manifolds.[10] Start with a simple closed 3-manifold $\mathcal{M}$, such as S^3 (or, even simpler to think about, consider $\mathcal{M} = \mathbb{R}^3$ and let us not worry about the point at infinity). Now consider a solid torus

$$\mathcal{M}_2 = D^2 \times S^1$$

embedded in this manifold (and $\mathcal{M}_1 = \mathcal{M}\backslash\mathcal{M}_2$ will be the remainder as usual). The surface $\partial\mathcal{M}_2 = S^1 \times S^1 = T^2$ is a torus surface. Now, there is another solid torus with exactly the same surface:

$$\mathcal{M}_2' = S^1 \times D^2 \quad .$$

These two solid tori differ in that they have opposite circles filled in. Both have the same $S^1 \times S^1$ surface, but $\mathcal{M}_2$ has the first S_1 filled in, whereas $\mathcal{M}_2'$ has the second S_1 filled in.

[11]Stop here. Think about what we have done. Collect the pieces of your exploded head.

The idea of surgery is to remove $\mathcal{M}_2$ and replace it with $\mathcal{M}_2'$ to generate a new manifold $\mathcal{M}'$ with no boundary.[11] This is difficult to visualize because if we start with a very simple space like $\mathcal{M} = \mathbb{R}^3$, the new structure $\mathcal{M}'$ is not embeddable within the original manifold $\mathcal{M}$.

This procedure, torus surgery on a 3-manifold, is called Dehn surgery. Another way to describe what we have done is that we have removed a solid torus, switched the meridian and longitude (switched the filled-contractible and the unfilled-uncontractible) and then glued it back in. In fact, one can make more complicated transformations on the torus before gluing it back in (and it is still called Dehn surgery; see Section 7.4) but we will not need this.

It is worth noting that the solid torus we removed could be embedded in a very complicated way within the original manifold—that is, it could follow a complicated, even knotted, path, as in the figure on the right of Fig. 7.11. As long as we have a closed loop S^1 (possibly following a complicated path) and it is thickened to D^2 in the direction transverse to the S^1 path, it is still a solid-torus topologically.

24.2 Representing Manifolds with Knots

[12]In Witten's ground-breaking paper on the Jones polynomial (Witten [1989]), he states the theorem without citation and just says "It is a not too deep result... " Ha! Seriously though, the proof is actually not *too* difficult. Although some of the details are a bit tricky the main idea is fairly understandable. See the references at the end of the chapter.

24.2.1 Lickorish–Wallace Theorem

An important theorem[12] of topology is due to Lickorish [1962] and Wallace [1960].

Theorem: Starting with S^3 one can obtain any closed connected orientable 3-manifold by performing successive torus surgeries, where these tori may be nontrivially embedded in the manifold (i.e. they may follow some knotted path).

One has the following procedure. We start with a link (some knot possibly of several strands), embedded in S^3. Thicken each strand to a solid torus. Excise each of these solid tori, and replace them by tori with longitude and meridian switched.[13] Any possible 3-manifold can be obtained in this way by surgering an appropriately chosen link. We summarize with the mapping

$$\text{Link in } S^3 \;\; \overset{\text{surger}}{\longrightarrow} \;\; \text{Some } \mathcal{M}^3 \;\; . \tag{24.1}$$

We can thus represent any three-dimensional manifold as a link in S^3. If we think of a topological quantum field theory as being a way to assign a complex number to a three-dimensional manifold, that is, $Z(\mathcal{M})$, we realize that what we are now looking for is essentially a knot invariant— each manifold is represented by a knot, and we assign a number to that knot. We elaborate on this in Section 24.3.

24.2.2 Kirby Calculus

It turns out that not all topologically different links, when surgered, give topologically different manifolds. Fortunately, the rules for which different knots give the same manifolds have been worked out by Kirby [1978]. These rules, known as Kirby calculus, are stated as a set of transformation moves on a link that change the link, but leave the resulting manifold implied by Eq. 24.1 unchanged. There are several different sets of moves that can be taken as "elementary" moves that can be combined together to make more complicated transformations. Perhaps the simplest set of two elementary basic moves are known as Kirby moves.[14] We will not rigorously prove that these moves leave the manifold unchanged, but we give rough arguments instead.

Kirby Move 1: Blow Up/Blow Down:[15]

One can add or remove a loop with a single curl, as shown in Fig. 24.6, to a link and the manifold resulting after surgery remains unchanged.

Addition or Removal of

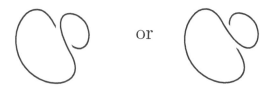

Fig. 24.6 Blow up/ Blow down. Addition or removal of an unlinked loop with a single curl leaves the 3-manifold represented by surgery on the knot unchanged.

Argument: First let us be a bit more precise about the surgery prescription. Given a link, we think of this link as being a ribbon (usually we draw it with blackboard framing; see Section 2.2.2). Thicken each

[13]See also Section 7.4.

[14]If one does not start with the knot embedded in S^3, one may need a third move known as "circumcision." This says that if any string loops only once around another string (without twisting around itself and without looping around anything else), both strings may be removed. That is, in Fig. 24.5, both strings may be removed (independent of how the string going off to the left forms any knot).

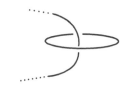

Fig. 24.5 A circumcision. Both strings can be removed. This is a third Kirby move which is implied by the first two if you start with a link embedded in S^3 but is more generally an independent move that is required. See for example Roberts [1997].

[15]The nomenclature is obscure when discussing 3-manifolds, but makes sense when one discusses 4-manifolds. See any of the books on 4-manifold topology listed at the end of the chapter.

Fig. 24.7 A line that wraps both the longitude and meridian of the torus. If we thicken the knot shown in Fig. 24.6 to a torus and draw a line around the longitude of the torus and then try to straighten the torus out to remove the twist, the straight line ends up looking like this.

[16]The nomenclature "handle-slide" comes from an interpretation of this move as sliding handles around on a manifold. Consider the example used in Section 24.1.1 where we attached a handle to a ball and obtained a solid torus. We could also attach two handles and get a two-handled solid torus. Here it doesn't matter where the handles are attached to the sphere – they can be slid around. Indeed, they can even be slid over each other (where one handle attaches to some point on the other handle). It is the sliding of a handle over another handle that gives this move its name.

strand into a solid torus, and draw a line around the surface of this torus that follows one of the edges of the ribbon. Remove this solid torus, but the torus surface that remains still has the line drawn around it. Re-attach a new solid torus where the new meridian (the circle surrounding the contractible direction) follows precisely this line.

Now consider doing this procedure for the curled loop in Fig. 24.6 embedded in S^3. As shown in Fig. 2.7 a string with a small curl as in Fig. 24.6 can be thought of as a ribbon with a twist (but no curl) in it. Let us use this description instead. Thicken the loop to a torus, and then the edge of the ribbon traces out a line as shown in Fig. 24.7 on the torus surface. We remove the solid torus and insert a new torus where the meridian follows the twisted line on the surface of the hole that is left behind. This is exactly the construction of $L(1,1) = S^3$ described in Section 7.4 (it is $(1,1)$ since the blue line goes around each cycle once), thus showing an example of how surgery on the curled loops in Fig. 24.6 does nothing to the manifold.

Kirby Move 2: Handle-Slide:[16]

A string can be broken open and pulled along the full path of another string, and then reconnected, and the resulting manifold remains unchanged. See Fig. 24.8 or 24.9.

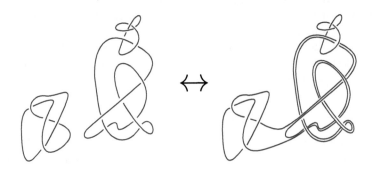

Fig. 24.8 A handle-slide move. (See Fig. 24.9 for another example.) Both left and right sides of this picture represent the same 3-manifold after surgery. Note that we should always view both strings as ribbons, and we need to keep track of how many self twists the ribbon accumulates when it is slid over another string.

Argument: Consider the simple handle-slide shown in Fig. 24.9. Let us think about what happens when we surger the horizontal loop. First, we thicken the horizontal loop into a torus (as shown), and then we exchange the contractible and non-contractible directions. In this procedure, the longitudinal direction (the long direction) of the torus is made into something contractible. This means (after surgery) we can pull the far left vertical line through this torus without touching the three vertical purple lines. Thus, the right and left pictures must describe the same manifold. Although less obvious, this principle remains true even if the torus is embedded in the manifold in a complicated way, as in Fig. 24.8.

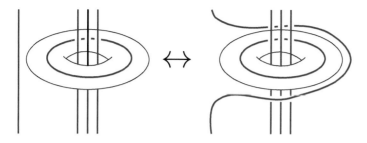

Fig. 24.9 An example of a simple handle-slide move.

Two links in S^3 describe the same 3-manifold if and only if one link can be turned into the other by a sequence of these Kirby moves[14] as well as any smooth deformation of links (i.e. regular isotopy).

Note that if we have two disconnected links L_1 and L_2 in S^3, which surgered give two manifolds, $\mathcal{M}_1$ and $\mathcal{M}_2$, respectively,

$$L_1 \text{ in } S^3 \overset{\text{surger}}{\longrightarrow} \mathcal{M}_1$$
$$L_2 \text{ in } S^3 \overset{\text{surger}}{\longrightarrow} \mathcal{M}_1$$

then if we consider a link $L_1 \cup L_2$ in which is the disconnected union of the two links (i.e. the two links totally separated from each other) it is fairly easy to see that we obtain the so-called connected sum of the two manifolds, which we write as follows:

$$L_1 \cup L_2 \text{ in } S^3 \overset{\text{surger}}{\longrightarrow} \mathcal{M}_1 \# \mathcal{M}_2 \quad . \tag{24.2}$$

See the discussion of connected sums in Section 7.3.3.

24.3 Witten-Reshetikhin-Turaev Invariant

By using the ideas of surgery, we are now in a position to use our diagrammatic algebra to handle complicated manifolds. Recall that one of the definitions of a TQFT is a mapping from a manifold $\mathcal{M}$ to a complex number $Z(\mathcal{M})$ in a way that depends only on the topology of the manifold (for example, Eq. 5.17 or Fig. 7.1 but without the embedded link). By using surgery (Eq. 24.1) we can describe our manifolds as links in S^3. If we can then find a link invariant that is unchanged under Kirby moves, we will effectively have something we can use as a manifold invariant.

We want to have a link invariant that is fully isotopy invariant (since Kirby calculus is isotopy invariant). To this end we use the isotopy invariant diagram algebra we have developed (see the discussion of Section 14.3.2, for example).[17]

The key to this construction is to now consider a link of the Ω (Kirby color) strands discussed in Section 17.5. This link made of Ω represents the link to be surgered, and thus represents our manifold. Let us now

[17]In cases where we have negative Frobenius–Schur indicators, there are several distinct ways to obtain isotopy invariant rules. We can either use the rules of Section 14.3 to insure isotopy invariance. Or we can work with negative d_a values as discussed in Section 14.5 and *not* apply rule 0, that is, performing a nonunitary evaluation of the diagram. Both of these are done in the literature. In the latter case, note that the Kirby color Fig. 17.7 is then defined with $d_a/\mathcal{D}$, rather than $\mathsf{d}_a/\mathcal{D}$. The identities of Fig. 17.9 and 17.8 then do hold for this nonunitary evaluation.

consider a manifold invariant defined as

$$Z_{WRT}(\mathcal{M}) = \qquad\qquad\qquad\qquad\qquad\qquad (24.3)$$
$$\frac{1}{\mathcal{D}} \left[e^{-2\pi i c/8} \right]^{\sigma} \left(\begin{array}{c} \text{Evaluate link made of } \Omega \text{ strands where} \\ \text{surgery on link in } S^3 \text{ gives } \mathcal{M} \end{array} \right)$$

where σ is the so-called signature of the link, defined to be the number of positive eigenvalues minus the number of negative eigenvalues of the matrix of linking numbers lk_{ij} between the (possibly multiple) strands of the link (the diagonal element lk_{ii} is just the self-linking or writhe of strand i in blackboard framing; see Section 2.6.2 for definition of linking number).[18]

It is not so obvious that the definition in Eq. 24.4 should provide a manifold invariant. What we would need to show is that $Z_{WRT}(\mathcal{M})$ gives the same output for *any* link that describes the same $\mathcal{M}$. In other words, we have to show that the expression on the right-hand side of Eq. 24.4 is unchanged when we make Kirby moves on the link.

Let us consider the first Kirby move—the addition of a twisted loop as in Fig. 24.6. Using Fig. 17.8, adding such a twisted loop multiplies the value of the link (the final term in Eq. 24.4) by $e^{\pm 2\pi i c/8}$ ($\pm$ depending on which way the loop is twisted). However, the addition of the twisted loop also changes the signature of the link σ by ± 1, thus precisely canceling this factor. Thus, the expression in Eq. 24.4 is certainly unchanged under the first Kirby move,[19] the Blow Up/Blow Down.

We now turn to the second Kirby move. Here, we show a rather remarkable property of the Ω strand—it is invariant under handle-slides (up to phases that are properly corrected by the prefactor of Eq. 24.4)! The derivation of this result is given in Fig. 24.10. One must be a bit cautious in applying this handle-slide law, as the strand being slid (say the left strand in Fig. 24.10) can develop self twists if it slides over a strand (say the right, Ω strand in Fig. 24.10) that itself has twists. However, the phase prefactor of Eq. 24.4 is designed to precisely account for this. Thus, Eq. 24.4 is unchanged under Kirby moves and therefore gives an invariant of the manifold.

The manifold invariant Eq. 24.4 is known as the Witten-Reshetikhin-Turaev invariant and was invented by Reshetikhin and Turaev [1991]. The reason it also gets named after Witten is that it gives a rigorous re-definition of the Chern–Simons manifold invariants (Eq. 5.17) discussed by Witten [1989]. This is a rather important result since the Chern–Simons functional integral is not well defined as an integral (see the comments in Section 5.3.6)![20]

---

[18]Note that to calculate a linking matrix, we must orient all of the strands (i.e. put arrows on them). It does not matter which way these arrows point.

[19]The killing property of Fig. 17.9 also makes Eq. 24.4 invariant under the third Kirby move; see Fig. 24.5.

[20]We mention in Section 5.3.6 that the Chern–Simons partition function, among other reasons for being ill-defined, actually depends on a so-called 2-framing of the manifold. The Reshetikhin-Turaev invariant corresponds to choosing so-called *canonical* framing. This is discussed in depth by Atiyah [1990] and Kirby and Melvin [1999].

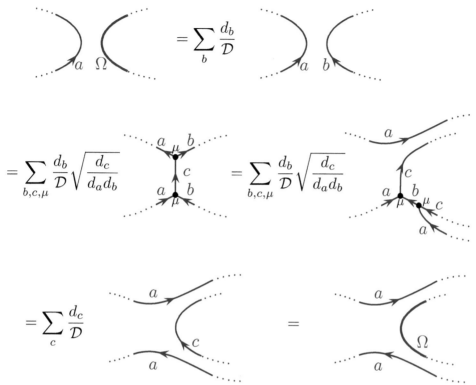

Fig. 24.10 Proof that the Ω strand satisfies the handle-slide. Here we show that any strand a can freely slide over the Ω strand in the sense of Fig. 24.8. The Ω strand on the right is meant to be connected up to itself in some way in a big (potentially knotted or linked) loop, which we don't draw. In going from the first to the second line, and also in going from the second to the third line, we have used the completeness relation Eq. 16.30. The equality in the second line is just sliding the vertex from the top (where a and b split from c) all the way around the b strand on the right until it almost reaches the bottom a, b, c vertex. Note that this derivation does not even require a well-defined braiding. We have written d rather than d because we may have chosen a nonunitary theory with negative d values.

The normalization of the Witten-Reshetikhin-Turaev invariant is such that the empty diagram, which represents the (unsurgered) manifold S^3 properly gives us $1/\mathcal{D}$ as we would expect from Eq. 17.12. This means that the Witten-Reshetikhin-Turaev invariant obeys the multiplication law for connected sums of manifolds

$$Z_{WRT}(\mathcal{M}_1 \# \mathcal{M}_2) = \mathcal{D} \, Z_{WRT}(\mathcal{M}_1) Z_{WRT}(\mathcal{M}_2) \quad . \tag{24.4}$$

This multiplication law is from the fact that surgery on a disjoint union of links gives a connect sum of manifolds (Eq. 24.2) and the evaluation of the disjoint union of links gives the product of the individual evaluation of the two links.[21]

Further, one can extend these manifold invariants to give a topological invariant partition function of a labeled link within a manifold, as in Fig. 7.1 (this was one of our general definitions of what we expected from a TQFT). To make this extension we simply define

[21]Some references redefine Z_{WRT} without the factor of $1/\mathcal{D}$ out front in Eq. 24.4 such that $Z(\mathcal{M}_1 \# \mathcal{M}_2) = Z(\mathcal{M}_1) Z(\mathcal{M}_2)$ instead.

$$Z_{WRT}(\mathcal{M}; \text{labeled link}) = \qquad\qquad (24.5)$$

$$\frac{1}{\mathcal{D}}\left[e^{-2\pi ic/8}\right]^{\sigma} \begin{pmatrix} \text{Evaluate link made of} \\ \left(\Omega \text{ strands where} \right. \\ \text{surgery on link in } S^3 \text{ gives } \mathcal{M} \\ \left. \cup \text{ labeled link}\right) \end{pmatrix}.$$

In other words, we simply include the labeled link into the diagram to be evaluated.

Without ever saying the words "path integral" or "Chern–Simons action" we think of an anyon theory as simply a way to turn a link of labeled world lines into a number (like evaluating a knot invariant, but with rules for labeled links), and surgery on Ω strands allows us to represent complicated manifolds.

24.3.1 Some Examples

It is worth working through a few examples of calculating the Witten-Reshetikhin-Turaev invariant for some simple manifolds.

As mentioned, for $\mathcal{M} = S^3$, we don't need to surger the manifold at all, so we don't need any Ω link at all. The value of the (empty) link is normalized to unity and including the prefactor in Eq. 24.4 (with signature zero) we obtain

$$Z_{WRT}(S^3) = 1/\mathcal{D}$$

which matches our expectations given Eqs. 17.12 and 7.13. Note that given this relationship (see Eq. 17.6) Eq. 24.4 agrees with Eq. 7.21.

For $\mathcal{M} = S^2 \times S^1$ we need to surger a single loop in S^3 to obtain $S^2 \times S^1$ (see Exercise 24.1). Thus, we need to evaluate a single loop of Ω string. It is an easy calculation to evaluate a loop of Ω

$$\bigcirc = \sum_a \frac{d_a}{\mathcal{D}} \bigcirc a \; = \sum_a \frac{d_a^2}{\mathcal{D}} = \mathcal{D} \; . \quad (24.6)$$

Thus, including the prefactor in Eq. 24.4 (the signature of the link is zero) we obtain

$$Z_{WRT}(S^2 \times S^1) = 1$$

which is in agreement with Eq. 7.14.

Finally let us consider the three-torus manifold $\mathcal{M} = S^1 \times S^1 \times S^1 = T^2 \times S^1 = T^3$. First, we note that surgery on the Borromean rings[22] (Fig. 24.11) yields the 3-torus (see Exercise 24.3). To evaluate the link

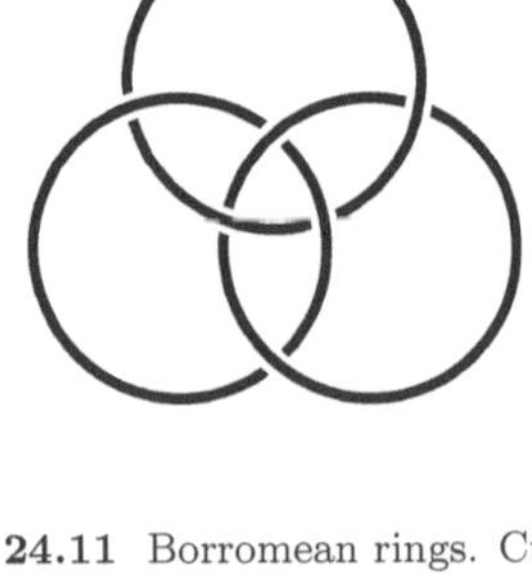

Fig. 24.11 Borromean rings. Cutting any one strand disconnects the other two. Surgery on this link in S^3 creates the 3-torus $S^1 \times S^1 \times S^1$.

[22]The rings are named for the crest of the royal Borromeo family of Italy, who rose to fame in the fourteenth century. However, the knot (in the form of three linking triangles) was popular among Scandinavian runestones five hundred years earlier and was known as the "Walknot" or "Valknut," or "the knot of the slain." However, in other runestones, the "knot of the slain" is a trefoil instead.

we use the corollary of the killing property of the Ω strand, Fig. 17.10, to show (see Exercise 24.4)

$$Z_{WRT}(T^3) = \text{number of particle species}, \qquad (24.7)$$

which matches the prediction from Eq. 7.9 along with Eq. 7.11.

24.3.2   Turaev–Viro Revisited: Chain-Mail and the Turaev–Walker–Roberts Theorem

Using the ideas of surgery, Roberts [1995] produced a beautiful geometric proof of the Turaev–Walker theorem (Turaev [1992, 1994]; Walker [1991]) that relates the Turaev–Viro invariant to the Chern–Simons (Witten-Reshetikhin-Turaev) invariant of a manifold. The result is, given a modular tensor category (a modular anyon theory) we have

$$Z_{TV}(\mathcal{M}) = |Z_{WRT}(\mathcal{M})|^2 \quad . \qquad (24.8)$$

We do not give the full proof here, only the general idea.

First, we require one more minor corollary. Similar to Fig. 17.10 we have the identity shown in Fig. 24.12 (see Exercise 24.6).

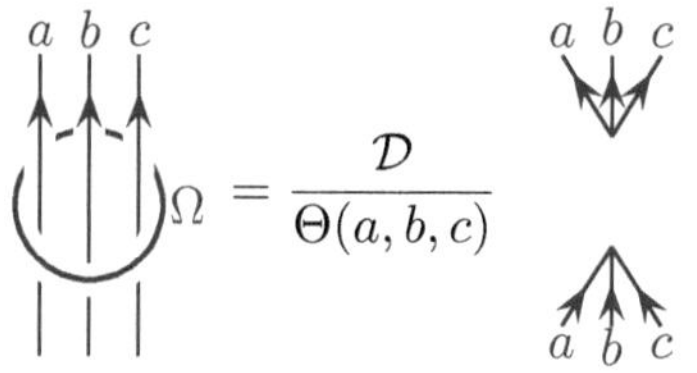

Fig. 24.12 The Ω strand fuses three lines due to the killing property. Here, we have assumed an isotopy-invariant theory as discussed in Chapter 16, so we can draw vertices with all three lines pointing in the same direction so a, b, c fuse together to the identity. Here, $\Theta(a, b, c) = \sqrt{d_a d_b d_c}$ as in Eq. 23.3.

We now want to construct a link of Ω strands that evaluates to the same value as the Turaev–Viro invariant discussed in Chapter 23. Recall that to define the Turaev–Viro invariant, we first make a simplicial decomposition of the manifold, breaking it up into tetrahedra, we label each edge, and we sum a certain weight over all possible labelings as given in Eq. 23.1.

Given our simplicial decomposition we construct a link of Ω strands via the following procedure: put one loop of Ω following the edges of each triangular face (colored gold in Fig. 24.13), and one loop of Ω around the waist of each edge (colored purple in Fig. 24.13) in such a way that the two types of strands link with each other. Such a link is known as *chain-mail*.[23] We then define the so-called *chain-mail invariant* of the manifold $\mathcal{M}$ as

$$CH(\mathcal{M}) = \mathcal{D}^{-N_v - N_{tet}} \begin{pmatrix} \text{Evaluate Chain-Mail Link of } \Omega \text{ strands} \\ \text{for simplicial decomposition of } \mathcal{M} \end{pmatrix} \qquad (24.9)$$

[23]When I have given talks on this subject I have been surprised to discover that many people don't know that chain-mail is a medieval type of armor made of linked metal loops. Of course those who had misspent youth playing *Dungeons and Dragons*, or reading the *Lord of the Rings* are very familiar with the concept and can tell you why Mithril is the best type of chain-mail.

[24]More generally, the chain-mail link can be defined for any *handlebody decomposition* of the manifold where Ω loops are put around 1-handles and 2-handles and N_v is then the 0-cells and N_{tet} is the 3-cells.

where N_v is the number of vertices in the simplicial decomposition and N_{tet} is the number of tetrahedra.[24]

First, it is extremely easy to prove that the chain-mail invariant is independent of the particular simplicial decomposition (and hence, is a manifold invariant as claimed). We only need to show that it is unchanged under the Pachner moves (Figs. 23.4 and 23.5). This can be done entirely geometrically using only the killing property (Fig. 17.10) and the handle-slide property (Fig. 24.10) of the Ω strand (this is Exercise 24.7).

Moreover, it is not hard to show that the chain-mail invariant is actually equal to the Turaev–Viro invariant. To do this we directly evaluate the chain-mail link. We start by using identity 24.12 on each Ω strand attached to each face (those drawn as gold in Fig. 24.13). This generates a factor of $\mathcal{D}/\Theta(a,b,c)$ for each face. The remaining Ω strands (purple in Fig. 24.13) are decomposed into sums of all quantum numbers as per the definition of Ω in Fig. 17.7 each weighted by $d_a/\mathcal{D}$. This leaves one tetrahedron of particle strings per simplex as shown on the right of Fig. 24.13. (Note that the remaining tetrahedron of strings to be evaluated is a tetrahedral diagram *dual* to the original tetrahedron, in agreement with the discussion below Eq. 23.4.)

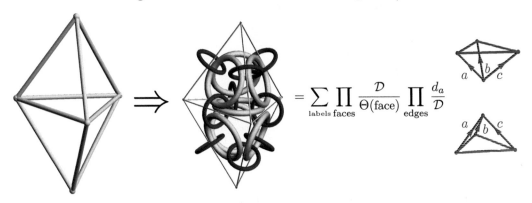

$$= \sum_{\text{labels}} \prod_{\text{faces}} \frac{\mathcal{D}}{\Theta(\text{face})} \prod_{\text{edges}} \frac{d_a}{\mathcal{D}}$$

Fig. 24.13 The chain-mail invariant is equivalent to the Turaev–Viro invariant. We start with a simplicial decomposition on the left. To form the chain-mail link we put one Ω-loop around each triangular face (gold in the figure) and one Ω loop around each edge (purple in the figure) such that the gold and purple are linked. Let the gold loops "kill" the three purple strands that go through them using Fig. 24.12 to leave only tetrahedra (blue on the right) dual to the original tetrahedra.

Putting together the factors we have obtained leaves us with the chain-mail invariant (including the prefactor in the definition) being given by

$$CH(M) = \frac{\mathcal{D}^{N_f}}{\mathcal{D}^{N_v+N_e+N_{tet}}} \sum_{\text{edge labels}} \frac{\prod_{\text{tetrahedra}} \tilde{G}(\text{tetrahedron}) \prod_{\text{edges}} d_{\text{edge}}}{\prod_{\text{triangles}} \tilde{\Theta}(\text{triangle})}$$

with N_v, N_e, N_f, N_{tet} being the number of vertices, edges, faces (triangles), and tetrahedra, respectively. Finally using the well-known topological fact that in three dimensions, the Euler characteristic $N_{tet} -$

$N_f + N_e - N_v$ is zero, the factors of $\mathcal{D}$ are reassembled to give exactly the definition of the Turaev–Viro invariant Eq. 23.1, thus deriving

$$CH(M) = Z_{TV}(M) \quad .$$

Lastly, we briefly discuss the derivation of the Turaev–Walker theorem Eq. 24.8. The key to this derivation is the fact[25] that if one uses a particularly simple decomposition of the manifold, surgery on the chain-mail link generates the connected sum of the original manifold $\mathcal{M}$ and its mirror image $\overline{\mathcal{M}}$

$$\text{Chain-mail link for } \mathcal{M} \quad \overset{\text{surger}}{\to} \quad \mathcal{M} \,\#\, \overline{\mathcal{M}} \quad . \tag{24.10}$$

Evaluating the chain-mail link is therefore essentially equivalent to evaluating $Z_{WRT}(\mathcal{M} \,\#\, \overline{\mathcal{M}})$

Using the equivalence between chain-mail and the Turaev–Viro invariant we thus have (Eq. 24.4)

$$Z_{TV}(\mathcal{M}) \sim Z_{WRT}(\mathcal{M} \,\#\, \overline{\mathcal{M}}) \sim Z_{WRT}(\mathcal{M}) Z_{WRT}(\overline{\mathcal{M}}) \sim |Z_{WRT}(\mathcal{M})|^2 \quad .$$

We have written this equation with $\sim$ rather than an equality because we have dropped factors of $\mathcal{D}$. To get these right we have to know more details about the particular decomposition of the manifold for which Eq. 24.10 holds so that we can keep track of the factors of $\mathcal{D}$ in the definition of the chain-mail invariant (Eq. 24.9). Keeping track of these factors carefully, one obtains the desired Eq. 24.8.

[25]This key fact is not too hard to prove—it requires only about two paragraphs in the original work (Roberts [1995]). However, it requires some knowledge of handlebody theory, so we will not discuss it here.

Chapter Summary

- Any orientable 3-manifold can be described as surgery on a link within some reference manifold (usually S^3). Two links that differ from each other by Kirby moves describe the same manifold.
- The Reshetikhin-Turaev invariant uses diagrammatic rules to construct a link invariant that is unchanged under Kirby moves and therefore can be used as a manifold invariant. This invariant matches what we expect from the Witten–Chern–Simons invariant and can be considered a rigorous construction of Witten's theory.
- The chain-mail construction geometrically relates the Turaev–Viro to the Chern–Simons theory in the case where the theory input into Turaev–Viro is modular.

Further Reading:

- For more detailed discussion of surgery, the Lickorish–Wallace theorem, Kirby calculus, and a nice discussion of manifold invariants; see Prasolov and Sossinsky [1996]; Saveliev [2012] or Lickorish [1997]. Other good references include Gompf and Stipsicz

[1999],Kirby [1989], and Akbulut [2016], although these emphasize four-dimensional topology.

- The chain-mail method introduced by Roberts [1995] can also be extended to describe the four-dimensional Crane-Yetter invariant. An application of this method to the physics of the Levin–Wen model (introduced in Chapter 33) is given by Burnell and Simon [2010, 2011].

Exercises

Exercise 24.1 Surgery on a Loop
Beginning with the three-sphere S^3, consider the "unknot" (a simple unknotted circle S^1 with no twists) embedded in this S^3. Thicken the circle into a solid torus ($S^1 \times D^2$) that has boundary $S^1 \times S^1$. Now perform surgery on this torus by excising the solid torus from the manifold S^3 and replacing it with another solid torus that has the longitude and meridian switched. That is, replace $S^1 \times D^2$ with $D^2 \times S^1$. Note that both of the two solid tori have the same boundary $S^1 \times S^1$ so that the new torus can be smoothly sewed back in where the old one was removed. What is the new manifold you obtain? (This should be easy because it is in the book!).

Exercise 24.2 Surgery on the Hopf Link [Not hard if you think about it right!]
Consider two linked rings, known as the Hopf link (see Fig. 24.14). Consider starting with S^3 and embedding the Hopf link within the S^3 with "blackboard framing" (i.e. don't introduce any additional twists when you embed it). Thicken both strands into solid tori and perform surgery on each of the two links exactly as we did above. Argue that the resulting manifold is S^3.

Fig. 24.14 A Hopf Link

Exercise 24.3 Surgery on the Borromean Rings [Hard]
Consider the link shown in Fig. 24.11 known as the Borromean rings. Consider starting with S^3 and embedding the Borromean rings within the S^3 with "blackboard framing." Thicken all three strands into solid tori and perform surgery on each of the three links exactly as we did in the previous two exercises. Show that one gets the 3-torus as a result. Hint 1: Think about the group of topologically different loops through the manifold starting and ending at the same point, the "fundamental group" or first homotopy group (see Section 40.3). Hint 2: If we say a path around the meridian of one of the three Borromean rings (i.e. threading though the loop) is called a and the path around the meridian of the second ring is called b, then notice that the third ring is topologically equivalent to $aba^{-1}b^{-1}$. Hint 3: In some cases the fundamental group completely defines the manifold! (Don't try to prove this, just accept this as true in this particular case.)

Exercise 24.4 Evaluation of Borromean Ring Ω-Link
Use Fig. 17.10 to evaluate the Ω-link of Borromean rings shown in Fig. 24.11. Use this to establish Eq. 24.7. Note that the signature of the link is zero.

Exercise 24.5 Product of Blow Up and Blow Down

Use the handle-slide and the killing property of Ω to prove that the diagram made of two oppositely twisted Ω loops, as shown in Fig. 24.15, gives the identity.

Exercise 24.6 Killing Three Strands With Ω

Prove the relationship shown in Fig. 24.12.

Fig. 24.15 The product of these two oppositely twisted Ω loops gives the identity.

Exercise 24.7 Pachner Moves and the Chain-Mail Invariant

Using killing moves (Fig. 17.10) and handle-slides (Fig. 24.10), show that the chain-mail invariant Eq. 24.9 is unchanged under Pacher moves (Fig. 23.4 and 23.5). The answer is given by Roberts [1995], but it is a fun exercise. Looking up the answer spoils the fun!

Anyon Condensation

A commonly discussed mechanism that derives one anyon theory from another is known as anyon condensation. The idea is modeled on the notion of conventional Bose-Einstein condensation. Under certain conditions one can imagine anyons forming a superfluid state akin to a Bose-Einstein condensate. One can imagine making a condensate from either continuously reducing the temperature with a fixed Hamiltonian, or by continuously changing the Hamiltonian at fixed (perhaps zero) temperature.[1] If one begins with a consistent anyon theory before the condensation, the system after the condensation will also be a consistent anyon theory, which we call the *condensed* theory.[2] It is believed that *all* continuous phase transitions that can occur between different anyon theories can be described in terms of anyon condensation.[3]

Many of the ideas of anyon condensation were first explored in a somewhat different language of conformal field theory (CFT; see Moore and Seiberg [1989a]; Schellekens [1999]). The realization that these mathematical manipulations correspond to a physical condensation is due to Bais and Slingerland [2009]. This latter work also has the advantage that it also avoids many of the complexities of CFT. Here we give an abbreviated discussion of this work, along with a few explicit examples.

Let us review some aspects of Bose condensation (see Leggett [2006] or Annette [2004] for much more information about the physics of superfluids and condensates). Recall that in a Bose condensate a macroscopic number of the particles reside in one special lowest-energy single-particle eigenstate, which we call the condensate wavefunction. For a uniform system (say with periodic boundary conditions) the wavefunction for this single-particle eigenstate is just a constant

$$\psi(\mathbf{r}) = \frac{1}{\sqrt{V}} \qquad (25.1)$$

with V the volume of the system. It is crucial that bosons accumulate no phase or sign when they are braided around each other or exchanged with each other. If they were to accumulate any phase or sign, this would prevent them from remaining in the eigenstate, Eq. 25.1, which is everywhere real and positive. This gives us:

Principle 1: Bosons must experience no net phase or sign when they move around then return to the same configuration— that is, when bosons exchange or braid with other particles in the condensate.

[1] A phase transition that occurs at zero temperature as some parameter of the Hamiltonian is changed is often known as a "quantum phase transition."

[2] It is sometimes possible that the condensed theory is a trivial theory—a gapped system having only the vacuum particle type, and zero central charge (at least mod 8). We should think of that as just being an uninteresting insulator. Strictly speaking, this is a TQFT, just a very trivial one. It is also possible to have phase transitions from topological theories to non-topological phases of matter such as gapless phases or Fermi liquids.

[3] First-order, or discontinuous, phase transitions can always occur between any two phases of matter.

[4]It is possible that the condensate wavefunction has a spatial structure such as $e^{i\phi(\mathbf{r})}$, which happens when there is, say, a vortex within the condensate. What is crucial is that when bosons move around within the condensate, when they return to the same many-particle configuration, the phase is the same as when they started.

[5]For systems with strictly fixed total number of particles this expectation would be zero and one instead looks at $\langle \hat{\psi}^\dagger(\mathbf{r})\hat{\psi}(\mathbf{r}')\rangle$ in the limit of $\mathbf{r}$ very far from $\mathbf{r}'$. This is known as "off-diagonal long-ranged order" or ODLRO.

[6]In the language of conformal field theory, condensation of a simple current is known as "extension of the chiral algebra."

[7]This condition is also equivalent to either of the following also equivalent statements:
  (a) $J \times \bar{J} = I$.
  (b) $d_J = 1$.

Indeed, accumulating no sign when exchanging with other identical particles is the very definition of a boson.[4]

With interacting bosons, one does not strictly have Bose condensation (not all of the bosons occupy the same single particle eigenstate, since interactions kick the particles out of this eigenstate). Nonetheless, interacting bosons can condense to form superfluids that share many of the properties of Bose condensates. In particular, one still has the idea of a condensate wavefunction (or *order parameter*), and in order to form a condensate, no phase or sign must be accumulated when the particles exchange and braid.

To describe a condensate wavefunction (or order parameter) microscopically, one writes[5]

$$\phi(\mathbf{r}) = \langle \hat{\psi}(\mathbf{r})\rangle$$

where $\hat{\psi}(\mathbf{r})$ is the (second quantized) operator that annihilates a particle at position $\mathbf{r}$. For non-interacting bosons, where many bosons are in the single-particle wavefunction Eq. 25.1, we obtain $|\phi|^2 = N_0/V$ where N_0 is the number of bosons in that single-particle eigenstate.

The fact that this order parameter is (at least locally) number non-conserving (it destroys a particle) gives us the second important principle:

Principle 2: Bosons can be freely absorbed by, or emitted from, the condensate.

25.1 Condensing Simple Current Bosons

We now generalize the idea of Bose condensation to anyon theories. For simplicity we begin by restricting our attention to bosons that are also *simple currents*[6] and only lift this restriction in Section 25.7. To remind the reader, a particle, let us call it J, is a simple current if $\sum_c N_{Ja}^c = 1$ for all particle types a. This condition is equivalent[7] to the statement that $J^N = I$ for some integer N where I is the identity (where J^N here means N factors of J fused together).

For a particle J to condense, it must be a boson. This means that it must have trivial braiding with itself or equivalently a trivial spin factor (see Eq. 15.2)

$$R_{J\times J}^{J,J} = 1 \quad \text{or} \quad \theta_J = 1,$$

which is what we expect for a boson. This condition implements **Principle 1**: the boson must not experience a nontrivial phase as it exchanges with another particle of the condensate as this phase would prevent a condensate wavefunction from forming.

Within the condensate, bosons may fuse with each other to form particles J^p for any value of p. It is not hard to show that all such resulting particle types must also be bosons $\theta_{J^p} = 1$ and further, that they all must braid trivially with each other (see Exercise 25.1).

While one *can* condense bosons that are not simple currents, the rules for doing so are a bit more complicated and we discuss this more complicated case in Section 25.7.

We start with an initial anyon theory we call[8] $\mathcal{A}$ containing a simple current J that we intend to condense to form a new anyon theory. The final anyon theory that comes out at the end of the condensation procedure will be called $\mathcal{U}$.

We can think of anyon condensation as proceeding in two conceptual steps,[9] as shown in Fig. 25.1. In between the initial theory $\mathcal{A}$ and the final theory $\mathcal{U}$ there is another theory $\mathcal{T}$, which is not a full anyon theory, but rather a fusion algebra (or planar diagram algebra), as we discuss further in a moment.

[8]The notation $\mathcal{A}, \mathcal{T}$, and $\mathcal{U}$ as shown in Fig. 25.1 is taken from Bais and Slingerland [2009].

[9]Those familiar with symmetry breaking in high energy or condensed matter physics will find some of these ideas familiar! In fact, the condensation procedure discussed in this chapter is sometimes referred to as "topological symmetry breaking" (see Bais and Slingerland [2009]).

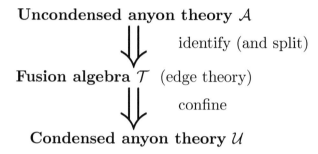

Fig. 25.1 Condensing one anyon theory to another can be described as having two "steps." The original anyon theory is labeled $\mathcal{A}$ and the final anyon theory is labeled $\mathcal{U}$. In between we have the intermediate theory $\mathcal{T}$ which is not generally a full-fledged anyon theory, but rather a fusion algebra (planar diagram algebra). The first step from $\mathcal{A}$ to $\mathcal{T}$ involves identification and possibly splitting. The second step from $\mathcal{T}$ to $\mathcal{U}$ involves confinement.

25.2 Identification Step

The first step in the condensation process is the identification step. In this step we group the particle types from the uncondensed theory $\mathcal{A}$ into so-called orbits.

Definition: The *orbit* of a particle type a under the action of J is the set of all particle types $b \in \mathcal{A}$ such that $b = J^p \times a$ for any integer p. We denote the orbit as $[a]^J$, or when it not ambiguous we just write $[a]$.

One should be cautioned[10] that the orbit $[a]$ is the same as orbit $[b]$ if $b = J^q \times a$ for any q.

Further, we note that if N is the smallest integer such that $J^N = I$ then there are at most N particle types in any given orbit, although there may be fewer particles in an orbit, as we discuss in detail in Section 25.4.

The physical point here is that all of the particle types in the same orbit of the original theory $\mathcal{A}$ are *identified* as being the same particle type in the $\mathcal{T}$ theory. The physical reason for this is **Principle 2**: bosons can be freely emitted from or absorbed into the condensate. A particular particle a can absorb a boson from the condensate and become $J \times a$ or it can absorb two bosons from the condensate to become $J^2 \times a$, and so forth. The quantum number a is no longer a conserved quantum number

[10]This can sometimes cause some notational confusion. It is often useful to choose a single representative of each orbit so that each orbit is uniquely denoted as a particular $[a]$ and one never writes $[b]$ if $b = J^q \times a$ when a is the chosen representative of the orbit rather than b.

(and therefore is not a valid particle type), but the orbit $[a]$ remains conserved and can play the role of a particle type in the condensed theory. The orbits in the condensed theory will inherit fusion rules from the fusion rules of the uncondensed theory (with some potential complications that we address in Section 25.4).

25.2.1 Orbits of Maximum Size

Here we consider the case where all of the orbits are of maximum size, that is, all orbits have exactly N particle types in them (where N is the smallest positive integer so that $J^N = I$). We return to the case where not all orbits are of maximum size in Section 25.4.

We start with the original theory $\mathcal{A}$, and each particle a can be mapped to an orbit $[a]$ in the $\mathcal{T}$ theory. The fusion rules of the $\mathcal{T}$ theory are inherited from the fusion rules of the original anyon theory in a natural way, which we can write in terms of the fusion multiplicity matrices as

$$N^{[c]}_{[a],[b]} = N^c_{a,b} \quad.$$

Note in particular that the identity particle I of the $\mathcal{A}$ theory maps to the orbit $[I]$ which becomes the identity particle of the $\mathcal{T}$ theory.

Example: $\mathbb{Z}_8^{(3+1/2)}$

Let us consider the anyon theory $\mathbb{Z}_8^{(3+1/2)}$ discussed in Section 18.5.4 that is equivalent to the Chern–Simons theory $SU(8)_1$. There are eight particles, which we label $p = 0, \ldots, 7$ with fusion rules

$$p \times p' = (p + p') \bmod 8 \tag{25.2}$$

and $p = 0$ is the identity. The corresponding twist factors are

$$\theta_p = \exp\left[\frac{2\pi i 7}{16} p^2\right] \tag{25.3}$$

We notice that $p = 4$ has trivial twist factor $\theta_4 = 1$ and is therefore a boson. Let us call this bosonic particle J, and we notice that $J^2 = I$ so the maximum orbit size is 2.

In this model we have four different orbits under the action of fusing with the boson J, and each of these orbits is of maximum size 2. Let us write down these orbits (Recalling that $[a]$ means the orbit of a)

$$
\begin{array}{ll}
[0] & \text{which is also equal to } [4] \\
[1] & \text{which is also equal to } [5] \\
[2] & \text{which is also equal to } [6] \\
[3] & \text{which is also equal to } [7] \quad.
\end{array}
$$

The meaning here should be obvious. Remembering that the boson J, which is particle $p = 4$, can be absorbed or emitted for free, we must then for example, identify particles 1 and 5 into a single orbit since fusing 1 with 4 gives 5 and fusing 5 with 4 gives 1.

These four different orbits comprise the particle types of the intermediate $\mathcal{T}$ theory. Let us denote these four orbits as $[p]$ with $p = 0, \ldots 3$. The fusion rules are inherited from the original uncondensed anyon fusion rules (Eq. 25.2) in an obvious way, giving

$$[p] \times [p'] = [(p + p') \bmod 4] \tag{25.4}$$

with $[0]$ playing the role of the identity in the $\mathcal{T}$ theory. To see how these fusion rules come from those of Eq. 25.2, consider, for example, $[1] \times [2] = [3]$: Here, either 1 or 5 (the two particle types in the orbit $[1]$) fused with either 2 or 6 (the two particle types in the orbit $[2]$) will always give us 3 or 7 (the two particle types in the orbit $[3]$).

25.3 Confinement Step

The particle types in the intermediate theory $\mathcal{T}$ form a consistent fusion algebra (and indeed a consistent planar diagram algebra) but they do not generally form a consistent anyon theory, as they do not generally have a consistent braiding (or solution to the hexagon equations). The reason for this is that some of the particles in $\mathcal{T}$ are not valid particles of the final condensed anyon theory $\mathcal{U}$ and must be thrown out.

The reason some particles of $\mathcal{T}$ are not valid anyons in the condensed phase is that they braid nontrivially with the condensed boson. Trying to put a particle in the system that braids nontrivially with the condensed boson would violate **Principle 1**: when a boson in the condensate moves around, the phase must be the same when it arrives back at the same point. We thus have the rule that any particles a (or its orbit $[a]_J$) allowed in the final condensed theory $\mathcal{U}$ must braid trivially with the condensate, meaning that

$$R^{a,J}_{J \times a} R^{J,a}_{J \times a} = \frac{\theta_{J \times a}}{\theta_a} = 1 \tag{25.5}$$

where we have used Eq. 15.5 and the fact that $\theta_J = 1$. Since $J \times a$ and a are in the same orbit $[a]$, the condition Eq. 25.5 can be rephrased by saying that an orbit (a particle type of the $\mathcal{T}$ theory) is allowed into the final anyon theory $\mathcal{U}$ if all of the particles in the orbit have the same spin factor θ. Such particles that are allowed in $\mathcal{U}$ we say are *deconfined*, meaning that they can travel freely through the system in the presence of the condensate. The particle types from $\mathcal{T}$ that braid nontrivially with the condensate are not allowed in the system in the presence of the condensate, and we say they are *confined*.

Although the confined particles of the $\mathcal{T}$ theory are not part of the final condensed anyon theory $\mathcal{U}$, they still have physical meaning. The full $\mathcal{T}$ theory can be physically realized as a $(1 + 1)$-dimensional theory living on the edge of a droplet of the $\mathcal{U}$ anyon theory living inside a larger region of the $\mathcal{A}$ uncondensed theory. The reason that these particles can live on such a one-dimensional boundary is that in $(1+1)$-dimensions there is no possibility of braiding one particle around another and there is thus no braiding restriction on any of the particles of the full $\mathcal{T}$ theory— both the confined and deconfined particles can live on this boundary. The $\mathcal{T}$ theory, since it describes a $(1+1)$-dimensional edge is not a braided anyon theory, but is rather a fusion algebra (or a planar diagram algebra).

If we try to drag one of the confined particles of the $\mathcal{T}$ theory into the condensed $\mathcal{U}$ droplet, its nontrivial braiding with the condensate creates a "branch-cut" in the condensate along its path into the condensate and destroys the condensate along this path. This costs an energy proportional to the distance the particle has been dragged into the $\mathcal{U}$ region. Thus, there is a force pushing these confined particles back to the edge of the droplet. The particles are *confined* to the edge.

$\mathbb{Z}_8^{(3+1/2)}$ Again

Let us return to our example of $\mathbb{Z}_8^{(3+1/2)}$ and determine which of our orbits (particle types of $\mathcal{T}$) are confined or deconfined. Recall the rule that an orbit is deconfined if all of the constituent particles in the orbit have the same twist factor θ. From Eq. 25.3 we have

$$
\begin{aligned}
\theta_0 &= \theta_4 = 1 \\
\theta_1 &\neq \theta_5 \\
\theta_2 &= \theta_6 = -i \\
\theta_3 &\neq \theta_7 \quad .
\end{aligned}
$$

Thus, the only two particle types allowed in the final condensed $\mathcal{U}$ theory are the orbits $[0]$ (which is the identity) and $[2]$. These particle types have spin factors $\theta_{[0]} = 1$ and $\theta_{[2]} = -i$. The only nontrivial fusion we obtain from Eq. 25.4 is

$$
[2] \times [2] = [0] \quad .
$$

We thus recognize this condensed anyon theory as the (left-handed) semion theory! Further, we establish that if we condense a semion droplet within a $\mathbb{Z}_8^{(3+1/2)}$ background there will be two additional particle types ($[1]$ and $[3]$) that remain confined at the edge of the droplet.

25.4   Splitting: Orbits Not of Maximum Size

In Section 25.2.1 we assumed all of the orbits were of maximum size. That is, if N is the smallest positive integer so that $J^N = I$, then all orbits are of size N.

If we have a situation where some orbits are not of maximum size, then we have a new physical phenomenon, known as *splitting*. This phenomenon is a reflection of the fact that assigning each orbit $[a]$ from the uncondensed theory $\mathcal{A}$ to be a particle type of the intermediate theory $\mathcal{T}$ will not give an acceptable fusion algebra. Let us see how this happens.

Let us suppose we have some particle a such that $J^p \times a = a$ with $0 < p < N$ (In fact, p must divide N). We start by recalling

$$a \times \bar{a} = I + \dots \quad .$$

On the other hand, we can also write

$$a \times \bar{a} = (J^p \times a) \times \bar{a} = J^p \times (a \times \bar{a}) = J^p \times (I + \dots) = J^p + \dots \quad .$$

We thus conclude that we must have[11]

$$a \times \bar{a} = I + J^p + \dots \quad . \tag{25.6}$$

Now we claim that Eq. 25.6 will result in the orbits in $\mathcal{A}$ not producing an acceptable fusion algebra in $\mathcal{T}$. As in the previous example, let us divide all of the particles in $\mathcal{A}$ into their orbits under the action of J, which we write as $[a]$. The fusion equation, Eq. 25.6, then would imply the fusion for the orbits

$$\begin{aligned} [a] \times [\bar{a}] &= [I] + [J^p] + \dots \\ &= [I] + [I] + \dots \end{aligned} \tag{25.7}$$

where here we have used that $[J^p]$ is in the J orbit of the identity (i.e. $[J^p] = [I]$). Eq. 25.7 now presents an inconsistency as one of our rules of fusion algebras (Eq. 8.11) is that $N_{a\bar{a}}^I = 1$, that is, the identity field should occur only once on the right-hand side. We conclude that this is not acceptable as a fusion algebra for $\mathcal{T}$.

To resolve this problem the orbit $[a]$ must *split* into multiple particle types in $\mathcal{T}$, which we will write as $[a]_i$ with $i = 1, 2, \dots, q_a$ for some number q_a.

Most generally, we can write the mapping between the original $\mathcal{A}$ theory and the (boundary) $\mathcal{T}$ theory as

$$a \to \sum_{i=1}^{q_a} n_i^a [a]_i \tag{25.8}$$

where now $[a]_i$ are particle types of the $\mathcal{T}$ theory and the ns are integers.

If the orbit $[a]$ is maximal, then $[a]$ does not need to split, meaning ($n_1^a = 1$ and $q_a = 1$, and we don't need to write a subscript on $[a]$). However, if the orbit is not maximal, then $[a]$ must split into multiple different particles $[a]_1, [a]_2, \dots$ such that the twist factors all agree

$$\theta_{[a]_i} = \theta_a \quad . \tag{25.9}$$

As in the simple case with no splitting, the fusion rules of the $\mathcal{T}$ theory must be consistent with those of the uncondensed $\mathcal{A}$ theory. In

[11]In fact we can generalize this argument to give

$$a \times \bar{a} = I + J^p + J^{2p} + \dots + J^{N-p} \quad .$$

particular,

$$a \times b = \sum_c N_{ab}^c\, c$$

in the $\mathcal{A}$ theory implies

$$\left(\sum_{i=1}^{q_a} n_i^a\, [a]_i \right) \times \left(\sum_{i=1}^{q_b} n_i^b\, [b]_i \right) = \sum_c N_{a,b}^c \left(\sum_{i=1}^{q_c} n_i^c\, [c]_i \right)$$

within the $\mathcal{T}$ theory. This consistency implies the relationship

$$\mathsf{d}_a = \sum_{i=1}^{q_a} n_i^a\, \mathsf{d}_{[a]_i} \tag{25.10}$$

between the quantum dimensions in the $\mathcal{A}$ theory (left of Eq. 25.10) and the quantum dimensions in the $\mathcal{T}$ theory (right of Eq. 25.10). Once the particles have split, it is then possible to have a consistent set of fusion rules in the $\mathcal{T}$ theory. Once these fusion rules have been established, one can determine which fields are confined in order to determine the final condensed-anyon theory $\mathcal{U}$.

While the phenomenon of identification (Section 25.2) has a fairly obvious physical interpretation, it is often not as obvious how to interpret the phenomenon of splitting—except to say that it is required for consistency. However, a physical understanding is given by realizing the presence of a condensate can cause certain physical quantities to be locally conserved where they are indefinite in the uncondensed phase. It is the presence of these new locally conserved quantities which allow us to form $[a]_i$ where i can take q_a different values—corresponding to the q_a different values that the conserved quantity may take. This picture of emergent conserved quantities is elucidated by Burnell et al. [2011, 2012].

Example: $SU(2)_4$

Let us consider the example of the Chern–Simons theory $SU(2)_4$. We list the fields,[12] their quantum dimensions, and their fusion rules in Table 25.1.

We notice that particle 4 is a simple current with orbit of length 2 ($4 \times 4 = 0$). Let us now list the orbits in this theory

$$\begin{aligned}
[0] &= [4] && \text{Maximum size orbit (also the identity)} \\
[1] &= [3] && \text{Maximum size orbit} \\
[2] & && \text{Not a maximum size orbit. This orbit must split.}
\end{aligned}$$

These orbits are just read off from the bottom line of the fusion table in Table 25.1: under fusion with the field 4, we have 0 mapping to 4, and vice versa; we have 1 mapping to 3, and vice versa, but 2 just maps to itself.

The nontrivial[13] fusion rules for the orbits, which we read off of Table

[12]The integer label of the field used here is twice its spin value, usually called "j." As in usual $SU(2)$ algebra, spins take integer and half-integer values, so for simplicity of notation here we just double the spin to get an integer. This is the same as the notation we use in Sections 22.2–22.6.

[13]"Nontrivial" means we don't write fusion with the identity.

particle	d	θ
0	1	1
1	$\sqrt{3}$	$e^{2\pi i/8}$
2	2	$e^{2\pi i/3}$
3	$\sqrt{3}$	$e^{2\pi i 5/8}$
4	1	1

$\times$	1	2	3	4
1	$0+2$	$1+3$	$2+4$	3
2	$1+3$	$0+2+4$	$1+3$	2
3	$2+4$	$1+3$	$0+2$	1
4	3	2	1	0

Table 25.1 Data for $SU(2)_4$. Left: Quantum dimensions and twist factors for the different particles. Note that 0 is the identity. Right: Nontrivial fusion rules. Note that the fusion rules are given by Eq. 21.20 (or Eq. 22.12), and the quantum dimensions are given by Eq. 21.21 or Eq. 22.3 given $d = \sqrt{3}$. You can check the consistency of d_a with the fusion rules by using Eq. 8.17 (i.e. d_a should be the largest eigenvalue of the fusion matrix $[N_a]_b^c$). Particle 4 is a simple current boson, which we will attempt to condense.

25.1, are[14]

$$[1] \times [1] = [0] + [2]$$
$$[1] \times [2] = [1] + [1]$$
$$[2] \times [2] = [0] + [2] + [0]$$

where $[0]$ is the identity orbit. It is the last line here that demonstrates explicitly the problem noted in Eq. 25.7—we should not have the identity twice on the right-hand side. To fix this problem we split the particle $[2]$ into two pieces $[2]_1$ and $[2]_2$.

$$[1] \times [1] = [0] + [2]_1 + [2]_2 \tag{25.11}$$
$$[1] \times ([2]_1 + [2]_2) = [1] + [1] \tag{25.12}$$
$$([2]_1 + [2]_2) \times ([2]_1 + [2]_2) = [0] + ([2]_1 + [2]_2) + [0] \tag{25.13}$$

While this does not quite give the full fusion rules of the $\mathcal{T}$ theory, one can nonetheless extract[15] a unique set of fusion rules for the $\mathcal{T}$ theory consistent with Eqs. 25.11–25.13, which are shown in Table 25.2.

Several things are worth noting on this table. First, note that $d_{[2]_1} + d_{[2]_2}$ is the same as d_2 from the original $SU(2)_4$ as is required by Eq. 25.10. Secondly, note that the twist factors are unchanged (even if a particle type splits) as stated in Eq. 25.9.

We can obtain the final $\mathcal{U}$ anyon theory, with a proper braiding, by throwing out the particles that are confined. Looking back at Table 25.1 we see that the orbit $[1]$ is made up of particles 1 and 3, which have different spin factors. This implies that particle $[1]$ must be confined (it braids nontrivially with the condensate; see Eq. 25.5). However, the

[14] Just to give an example, consider $[1] \times [1]$. Each factor of $[1]$ could either represent the particle 1 or the particle $3 = 1 \times 4$. So the result of this fusion $[1] \times [1]$ could be $1 \times 1 = 0 + 2$ or $1 \times 3 = 2 + 4$ or $3 \times 1 = 2 + 4$ or $3 \times 3 = 0 + 2$. In all cases the result contains one particle from the $[0]$ orbit and one from the $[2]$ orbit, thus giving $[1] \times [1] = [0] + [2]$.

[15] Equation 25.11 is already written as a proper fusion rule. Equation 25.12 implies $[1] \times [2]_1 = [1]$ and $[1] \times [2]_2 = [1]$. Next, we note that the left-hand side of Eq. 25.13 can be rewritten as $([2]_1 \times [2]_1) + ([2]_2 \times [2]_2) + 2([2]_1 \times [2]_2)$ which, comparing to the right of Eq. 25.13, immediately implies that $[2]_1 \times [2]_2 = [0]$. To pin down the remaining fusion rule we use associativity $[2]_2 = ([2]_1 \times [2]_2) \times [2]_2 = [2]_1 \times ([2]_2 \times [2]_2)$, which implies the only consistent set of fusion rules to be $[2]_2 \times [2]_2 = [2]_1$ and $[2]_1 \times [2]_1 = [2]_2$.

particle	d	θ
[0]	1	1
[1]	$\sqrt{3}$	$e^{2\pi i/8}$
$[2]_1$	1	$e^{2\pi i/3}$
$[2]_2$	1	$e^{2\pi i/3}$

$\times$	[1]	$[2]_1$	$[2]_2$
[1]	$[0] + [2]_1 + [2]_2$	[1]	[1]
$[2]_1$	[1]	$[2]_2$	[0]
$[2]_2$	[1]	[0]	$[2]_1$

Table 25.2 Data for the intermediate $\mathcal{T}$ theory obtained from condensing the particle type called "4" in $SU(2)_4$ (see Table 25.1). This is the fusion theory describing the edge of a condensed droplet.

orbit [2] is made of a single particle type and therefore is deconfined (even if it splits). Thus, our final anyon theory after confinement is given by Table 25.3. We recognize the resulting anyon theory as $SU(3)_1$ or equivalently $\mathbb{Z}_3^1$. See Section 18.5.3. It is worth noting that there is an obvious symmetry between the two split pieces—which we discuss further in Chapter 36—-it does not matter which piece we labeled with subscript 1 and which with subscript 2.

particle	d	θ
[0]	1	1
$[2]_1$	1	$e^{2\pi i/3}$
$[2]_2$	1	$e^{2\pi i/3}$

$\times$	$[2]_1$	$[2]_2$
$[2]_1$	$[2]_2$	[0]
$[2]_2$	[0]	$[2]_1$

Table 25.3 Data for the final $\mathcal{U}$ anyon theory obtained from condensing the particle type called "4" from $SU(2)_4$. We recognize this theory as $SU(3)_1$ or equivalently $\mathbb{Z}_3^1$.

Example: Ising $\times$ $\overline{\text{Ising}}$

As a second example, let us consider the product theory Ising $\times$ $\overline{\text{Ising}}$. First, recall some key properties of the Ising theory shown in Table 25.4.

particle	d	θ
I	1	1
σ	$\sqrt{2}$	$e^{i\pi/8}$
ψ	1	-1

$\times$	σ	ψ
σ	$I + \psi$	σ
ψ	σ	I

Table 25.4 Data for the Ising TQFT. Left: Quantum dimensions and twist factors for the different particles. Note that I is the identity. Right: Nontrivial fusion rules.

From the data for Ising we can construct the product theory Ising $\times$ $\overline{\text{Ising}}$. The data for $\overline{\text{Ising}}$ is the same as that for Ising except that the twist factors are complex conjugated.

We notate the particle types from Ising $\times$ $\overline{\text{Ising}}$ as (a, b) where $a, b \in I, \sigma, \psi$ meaning that a is from the Ising theory and b is from the $\overline{\text{Ising}}$ theory, thus giving nine total particle types. The twist factor of the combined particles is the product of the spin from the two constituents. So for example, the spin of $\theta_{(\psi,\sigma)} = (-1)(e^{-i\pi/8})$.

There are two bosons in this theory, (σ,σ) and (ψ,ψ). Here, we consider condensing (ψ,ψ). We write the orbit of (a,b) as $[a,b]$

The orbits under the action of (ψ,ψ) are as follows:

$$
\begin{array}{lll}
[I,I] & = [\psi,\psi] & \text{Maximum size orbit (also the identity)} \\
[I,\sigma] & = [\psi,\sigma] & \text{Maximum size orbit} \\
[\sigma,I] & = [\sigma,\psi] & \text{Maximum size orbit} \\
[I,\psi] & = [\psi,I] & \text{Maximum size orbit} \\
[\sigma,\sigma] & & \text{Not a maximum size orbit. This orbit must split.}
\end{array}
$$

It is convenient here to take a short-cut and check which of these are going to be confined. The twist factor of (σ,I) is $e^{i\pi/8}$, whereas the twist of (σ,ψ) is $-e^{i\pi/8}$ and similarly the twist for (I,σ) and (ψ,σ) differ by a minus sign. Thus, these two orbits will be confined and for convenience we just drop them now.

The nontrivial fusion rules of the remaining particle types are[16]

$$
\begin{array}{lll}
[I,\psi] \times [I,\psi] & = & [I,I] \\
[I,\psi] \times [\sigma,\sigma] & = & [\sigma,\sigma] \\
[\sigma,\sigma] \times [\sigma,\sigma] & = & [I,I] + [I,I] + [I,\psi] + [I,\psi] \qquad (25.14)
\end{array}
$$

again indicating that the orbit $[\sigma,\sigma]$ needs to split into two pieces $[\sigma,\sigma]_1$ and $[\sigma,\sigma]_2$. There are two consistent fusions that can result from this splitting, but only one of them results in a consistent anyon theory,[17] which we show in Table 25.5. This theory turns out to be the toric code TQFT, which we found in Section 18.5.1, and that we study in depth in Chapter 27 and thereafter.[18] Note again that there is a symmetry between the two pieces of the splitting—it does not matter which one we labeled with the subscript 1 and which one with 2. This symmetry is discussed in depth in Chapter 36.

25.5   Other Features of Condensation

A few other features of condensation are worth mentioning. First, if we start with a modular anyon theory, $\mathcal{A}$, then the condensed theory $\mathcal{U}$ is also modular. Further, for modular theories, the central charge (modulo 8) remains unchanged

$$
c_{\mathcal{A}} = c_{\mathcal{U}} \bmod 8 \quad . \qquad (25.15)
$$

Secondly, there is a beautiful relationship between the total quantum

---

[16] For example, to derive the last of these, we use $\sigma \times \sigma = I + \psi$ for both left and right sides of the theory to obtain $(\sigma,\sigma) \times (\sigma,\sigma) = (I,I) + (I,\psi) + (\psi,I) + (\psi,\psi)$; then grouping these into orbits gives the result.

[18] The mapping to the conventional form of the toric code is that $[I,I]$ is the identity $[\psi,I]$ is the fermion f. Finally, $[\sigma,\sigma]_1$ and $[\sigma,\sigma]_2$ are assigned to e and m, although it does not matter which one is assigned to which.

---

[17] From Eq. 25.14 we know we must have $([\sigma,\sigma]_1 + [\sigma,\sigma]_2) \times ([\sigma,\sigma]_1 + [\sigma,\sigma]_2) = 2[I,I] + 2[I,\psi]$, which implies that we must have either (a) $[\sigma,\sigma]_i \times [\sigma,\sigma]_i = [I,I]$ for $i = 1,2$ and $[\sigma,\sigma]_1 \times [\sigma,\sigma]_2 = [I,\psi]$, or (b) $[\sigma,\sigma]_i \times [\sigma,\sigma]_i = [I,\psi]$ for $i = 1,2$ and $[\sigma,\sigma]_1 \times [\sigma,\sigma]_2 = [I,I]$. Case (a) will turn out to be consistent and we show it in Table 25.5. To show that case (b) is inconsistent, we use the ribbon identity 15.5 to derive that $R^{[\sigma\sigma]_1,[\sigma\sigma]_1}_{[\psi,I]} = \pm i$. We can plug this result into Eq. 15.2, which then does not give the correct $\theta_{[\sigma,\sigma]_1} = 1$, thus showing an inconsistency.

particle	d	θ
$[I, I]$	1	1
$[\sigma, \sigma]_1$	1	1
$[\sigma, \sigma]_2$	1	1
$[\psi, I]$	1	-1

$\times$	$[\sigma, \sigma]_1$	$[\sigma, \sigma]_2$	$[\psi, I]$
$[\sigma, \sigma]_1$	$[I, I]$	$[\psi, I]$	$[\sigma, \sigma]_2$
$[\sigma, \sigma]_2$	$[\psi, I]$	$[I, I]$	$[\sigma, \sigma]_1$
$[\psi, I]$	$[\sigma, \sigma]_2$	$[\sigma, \sigma]_1$	$[I, I]$

Table 25.5 Data for the final $\mathcal{U}$ anyon theory obtained from condensing the (ψ, ψ) particle from Ising $\times$ $\overline{\text{Ising}}$. We recognize this theory as the toric code theory.

dimensions of the uncondensed theory $\mathcal{A}$, the fusion algebra $\mathcal{T}$, and the final theory $\mathcal{U}$. Recall that total quantum dimension is defined by

$$\mathcal{D} = +\sqrt{\sum_a \mathsf{d}_a^2} \ .$$

We then have

$$\frac{\mathcal{D}_\mathcal{A}}{\mathcal{D}_\mathcal{T}} = \frac{\mathcal{D}_\mathcal{T}}{\mathcal{D}_\mathcal{U}} \ . \tag{25.16}$$

Let us check these relations for the $SU(2)_4$ condensation example. From Tables 25.1, 25.2, and 25.3 we obtain

$$\mathcal{D}_{\mathcal{A}=SU(2)_4} = \sqrt{12} \qquad \mathcal{D}_\mathcal{T} = \sqrt{6} \qquad \mathcal{D}_{\mathcal{U}=SU(3)_1} = \sqrt{3}$$

in agreement with Eq. 25.16. Also (from Eq. 17.16) we can calculate that

$$c_{SU(2)_4} = c_{SU(3)_1} = 2 \quad (\text{mod } 8)$$

in agreement with Eq. 25.15.

25.6   Cosets

In Chern–Simons theory, one of the most common ways to construct new TQFTs is the idea of a coset theory first discussed in the context of CFT (see Di Francesco et al. [1997]; Moore and Seiberg [1989a]; Goddard et al. [1985]). Given Lie groups G and H such that H is a subgroup of G, we may consider theories G_k (Chern–Simons theory G at level k) and correspondingly $H_{k'}$. There is then a well-defined way to make a so-called *coset* theory, which we write as $G_k/H_{k'}$. One rough physical interpretation of this construction is that we are *gauging* the subgroup H, essentially making these degrees of freedom redundant. One can have more complicated cosets where we embed H into a product of Lie groups $G \times G'$, to construct coset theories like[19] $(G_k \times G_{k'})/H_{k''}$.

While cosets of this type can seem quite complicated, they actually have an extremely simple interpretation in terms of boson condensation. To construct $G_k/H_{k'}$ we first construct $G_k \times \overline{H_{k'}}$ where the overbar means we should switch the chirality of the theory.[20] Then, if we con-

[19] When written in terms of the corresponding Lie algebras, this is expressed as $(g_k \oplus g'_k)/h_{k''}$.

[20] In Chern–Simons theory, $G_{-k} = \overline{G_k}$. Switching chirality in the diagrammatic algebra can be achieved by evaluating the mirror image of the diagram.

dense all possible simple-current bosons in the product theory $G_k \times \overline{H_{k'}}$ we obtain the coset theory $G_k/H_{k'}$. This is *much* simpler than the conventional coset construction! This technique can be generalized in an obvious way. For example, to construct $(G_k \times G_{k'})/H_{k''}$ we first construct $G_k \times G_{k'} \times \overline{H_{k''}}$ and then condense all of the simple current bosons. Note that given Eq. 25.15 and Eq. 17.31 the central charge of a coset is equal to the sum of central charges in the numerator minus the sum of central charges in the denominator.

Example: $SU(2)_2/U(1)_4$

To construct the coset $SU(2)_2/U(1)_4$ we want to first construct $SU(2)_2 \times \overline{U(1)_4}$ and then condense all simple current bosons. Let us recall the data for $SU(2)_2$ (see Section 18.4.1) and $\overline{U(1)_4}$ (see Section 18.5.4), which are shown in Table 25.6.[21]

[21] Again, we remind the reader that there is some disagreement as to how you notate this $U(1)$ theory. Here, we mean $\overline{U(1)_4} = \overline{\mathbb{Z}_4^{1/2}} = \overline{(D_9)_1}$ However, some references would write this as $\overline{U(1)_2}$ instead!

$SU(2)_2$

particle	d	θ
I	1	1
σ	$\sqrt{2}$	$e^{2\pi i 3/16}$
ψ	1	-1

$\times$	σ	ψ
σ	I	σ
ψ	σ	$I+\psi$

$\overline{U(1)_4}$

particle	d	θ
0	1	1
1	1	$e^{2\pi i 7/8}$
2	1	-1
3	1	$e^{2\pi i 7/8}$

$i \times j = (i+j) \bmod 4$

Table 25.6 Data for $SU(2)_2$ (top) and $\overline{U(1)_4}$ (bottom). The overline indicates we take the mirror image theory, meaning all of the twist factors are complex conjugated compared to the definition given in Section 18.5.4.

The product theory $SU(2)_2 \times \overline{U(1)_4}$ has particle types of the form (a, b) where a is from the $SU(2)_2$ theory and b is from the $\overline{U(1)_4}$. The twist factor of such a product particle is $\theta_{(a,b)} = \theta_a \theta_b$.

One can see from the table that the product particle $(\psi, 2)$ is a boson simple current (twist factor $\theta = 1$), so we can condense it. (In fact, not including the identity particle $(I, 0)$, this is the only boson simple current). There are 12 particles in the product theory that divide into six orbits that all are of maximum size (so there is no splitting). These orbits (under the action of fusion with the $(\psi, 2)$ boson) are

$$[I,0] = [\psi,2] \quad ; \quad [\sigma,0] = [\sigma,2] \quad ; \quad [\psi,0] = [I,2]$$
$$[I,1] = [\psi,3] \quad ; \quad [\sigma,1] = [\sigma,3] \quad ; \quad [\psi,1] = [I,3] \quad .$$

These are the six particle types of the $\mathcal{T}$ theory. Finally, examining the twist factors we can see that only the three orbits $[I,0]$ and $[\sigma,1]$ and $[\psi,0]$ are deconfined. The final condensed anyon theory $\mathcal{U} =$

$SU(2)_2/U(1)_4$ is given in Table 25.7. We recognize this result as being simply the Ising TQFT (see Section 18.4.1).

particle	d	θ
$[I,0]$	1	1
$[\sigma,1]$	$\sqrt{2}$	$e^{2\pi i 1/16}$
$[\psi,0]$	1	-1

$\times$	$[\sigma,1]$	$[\psi,0]$
$[\sigma,1]$	$[I,0]$	$[\sigma,1]$
$[\psi,0]$	$[\sigma,1]$	$[I,0]+[\psi,0]$

Table 25.7 The final anyon theory data for the coset $SU(2)_2/U(1)_4$ is just the Ising TQFT.

25.6.1 Dualities and Aliases

The example we just considered that we can rewrite the Ising TQFT as $SU(2)_2/U(1)_4$, is not atypical. There are usually multiple ways to express the very same TQFT. For example, the series of so-called unitary conformal Virasoro minimal models(Di Francesco et al. [1997]) is expressed as

$$\mathcal{M}_k = (SU(2)_k \times SU(2)_1)/SU(2)_{k+1} .$$

The Ising TQFT is the $k=1$ element of this series (yet another way to express the Ising TQFT!), and $k=2$ is the so-called 3-state Potts TQFT. Similar cosets can be constructed for other series of models, such as the superconformal minimal models(Goddard and Schwimmer [1988]).

Another interesting relation is the so-called level–rank duality. The most famous of these relationships is

$$SU(N)_k = SU(Nk)_1/SU(k)_N . \tag{25.17}$$

However, there are many more of these relationships (see for example Hsin and Seiberg [2016]; Mlawer et al. [1991]; Mukhopadhyay [2013]) such as

$$Sp(2r)_s = SO(4rs)_1/Sp(2s)_r$$
$$SO(2r+1)_{2s+1} = SO((2r+1)(2s+1))_1/SO(2s+1)_{2r+1} .$$

There are also many other examples of relationships between different TQFTs that don't fit into any obvious series[22] (see for example Di Francesco et al. [1997]).

[22]The fact the central charge of a coset is simply the central charge of the numerator minus that of the denominator gives a simple, but nontrivial constraint, on such relationships. One can often guess what the result of a coset is going to be just by knowing the central charges!

25.7 More General Condensations

One can also consider condensations of objects that are not simple currents. There is a nice algorithm for constructing all possible condensations of a modular anyon theory that has been worked out by Neupert et al. [2016].

Given a modular anyon theory (a modular tensor category or TQFT) with N particle types, each condensation is described by a matrix M with non-negative integer entries with the following properties

$$[M, S] = [M, T] = 0 \qquad (25.18)$$

with $M_{I,I} = 1$ where S and T are the modular S and T matrices, respectively (see Section 17.3). The M matrix must be of the form

$$M = nn^{\mathsf{T}} \qquad (25.19)$$

where n is a rectangular matrix with non-negative integer entries and T means transpose. The n matrix has the same coefficients as in Eq. 25.8 except we only include resulting particles that are deconfined (i.e. we count the particles in the $\mathcal{U}$ theory and not all the particles in the $\mathcal{T}$ theory).

From the n matrix we construct the S- and T-matrices of the condensed ($\mathcal{U}$) theory as

$$Sn = nS_{\text{condensed}} \qquad Tn = nT_{\text{condensed}} \qquad . \qquad (25.20)$$

We must then confirm that $S_{\text{condensed}}$ is valid, meaning that it is symmetric, unitary, and generates non-negative fusion coefficients with the Verlinde formula (Eq. 17.13).

Note that M being a permutation matrix is a valid solution and this simply permutes the particles of the original theory and should not be considered a condensation.[23]

[23]The presence of a permutation symmetry, when it occurs, is itself an interesting feature that is discussed at length in Chapter 36

25.8 Condensation and Boundary Modes

Many topologically ordered phases of matter[24] have the property that they must have gapless modes along their boundary when you have a finite region of the matter (say, a droplet or a disk).

As mentioned in Section 17.3.3, if the central charge of the TQFT is non-zero, then the edge of the system necessarily carries heat and therefore must be gapless (i.e. has low-energy thermal excitations). However, even when the central charge is zero and the thermal conductance is zero, there can be cases where the edges of a system must be gapless. It turns out that the condition for this to be true is intimately related to condensation! Without giving detailed proof we will state the result (see Refs. Burnell [2018]; Kong [2014]; Levin [2013]):

[24]In Section 29.2.3 we more precisely define topologically ordered matter to be any physical system that is described at long length scales and low energies as a topological quantum field theory.

The edge of a topologically ordered system can only be gapped if one can condense a set of bosons from this system and obtain a topologically trivially ordered system (a system with no anyons).

(See note 2 from this chapter.) The converse of this statement is not generally true, as a gapped system can always be made gapless by altering the Hamiltonian a bit near the edge to allow gapless excitations.

A set of bosons that can condense to form the vacuum is known as a *Lagrangian subalgebra*. This situation exists, and hence, one can have a gapped edge, precisely when the TQFT is a Drinfel'd (or "quantum") double of some category. We have run into such TQFTs before as the output TQFT of a Turaev–Viro construction (see Section 23.2.2), and we discuss them again, particularly in Chapter 33, where we find that any Drinfel'd double can also be obtained as the output of a "string-net" model. The key result here is that edge modes of a TQFT can be gapped only if the TQFT can be constructed using a string-net.

Similarly, one can consider an interface between two TQFTs $\mathcal{C}$ and $\mathcal{D}$ and ask whether this interface can be gapped. A similar criterion is obtained by "folding" the system and considering the product theory $\mathcal{C} \times \bar{\mathcal{D}}$, where $\bar{\mathcal{D}}$ is the mirror image (or time-reversal) of $\mathcal{D}$. The criterion is (see Bais et al. [2009]; Kong [2014]):

The boundary between two topologically ordered systems $\mathcal{C}$ and $\mathcal{D}$ can only be gapped if the system $\mathcal{C} \times \bar{\mathcal{D}}$ can have a gapped edge to the trivial theory, that is, if a set of bosons from $\mathcal{C} \times \bar{\mathcal{D}}$ can be condensed to give a topologically trivially ordered system.

A simple case where this criterion is satisfied is when $\mathcal{C}$ is obtained from $\mathcal{D}$ by a boson condensation, or vice versa. However, there are more general cases where $\mathcal{C}$ is not obtained from $\mathcal{D}$ by condensation (and neither $\mathcal{D}$ from $\mathcal{C}$), and yet the criterion is satisfied. The consideration of the folded theory $\mathcal{C} \times \bar{\mathcal{D}}$ essentially allows for physical processes whereby a particle near the boundary of $\mathcal{C}$ binds with a particle near the boundary of $\mathcal{D}$ to form an object that condenses along the boundary.

Chapter Summary

- Anyon condensation derives one anyon theory from another by condensing a boson.
- Condensation proceeds in two main steps: identification/splitting and confinement.
- Condensation can implement cosets and can help determine when boundary modes are possible.

- Many of the original ideas of are present in very early work on CFT including Moore and Seiberg [1989a]. More detailed structure was worked out in the context of "fixed point resolutions." See for example Schellekens [1999].
- Bais and Slingerland [2009] is the original discussion of anyon condensation. Full disclosure: I was the referee for this paper. I spent a lot of time reading it in great detail, and I ended up deciding it

was pretty brilliant.

- A nice review of the physics of anyon condensation is given by Burnell [2018].
- Eliëns et al. [2014] give methods of extracting detailed data, such as F-matrices of a condensed theory from the data for the uncondensed theories.
- Neupert et al. [2016] is also a good reference on the mathematics of condensation.
- A more mathematical discussion of condensation is given by Kong [2014].
- Chapter 39 shows how to use the computer program *Kac* to work out the results of certain condensations.

Exercises

Exercise 25.1 Fusion of Bosonic Simple Currents
Given an anyon theory with a bosonic simple current J, such that $R_{J^2}^{J,J} = 1$ and $J^N = I$, show that all of the particle types J^p with $0 < p < N$ are also bosons, and further, that braiding any two of these particle types is trivial $R_{J^{p+p'}}^{J^p, J^{p'}} = 1$.

Exercise 25.2 Condensation to the Vacuum
The toric code is one of the simplest TQFTs, which we study in great depth in later chapters. The particle types and fusion relations are given in Table 28.1, and the S- and T-matrices are given in Eq. 28.5.

Show that there are two possible bosons that can condense, and that the resulting $\mathcal{U}$ theory in either case is the trivial TQFT (no particles but the vacuum).

Exercise 25.3 Condensation From $SU(2)_2 \times \overline{SU(2)_2}$
Consider the theory $SU(2)_2 \times \overline{SU(2)_2}$. Condense a simple current boson from this theory and show that you get the same toric code as when we condensed a simple current boson from Ising $\times$ $\overline{\text{Ising}}$.

Exercise 25.4 Condensation From Ising $\times$ Ising
Consider Ising $\times$ Ising (unlike the example in the text, both sides of the theory have the *same* handedness). There is a single simple-current boson that can be condensed. Find the $\mathcal{T}$ theory (there is a splitting!) and the final $\mathcal{U}$ theory after condensation.

Exercise 25.5 Cosets
(a) Calculate the properties of the coset $SU(16)_1/SU(2)_2$. Hint: See section 18.5.4.
(b) Calculate the properties of the coset $SU(2)_1 \times SU(2)_1/SU(2)_2$.

Exercise 25.6 General Condensation Method
The S-matrix for $SU(2)_4$ is given by

$$S_{kj} = \frac{1}{\sqrt{3}} \sin\left(\frac{(k+1)(j+1)\pi}{6}\right)$$

where $k, j \in 0, \ldots 4$ and

$$\theta_j = \exp(2\pi i j (j+2)/24) \quad .$$

Using the method of Section 25.7 confirm that the matrix n given by

$$n^{\mathsf{T}} = \begin{pmatrix} 1 & 0 & 0 & 0 & 1 \\ 0 & 0 & 1 & 0 & 0 \\ 0 & 0 & 1 & 0 & 0 \end{pmatrix}$$

satisfies Eqs. 25.19 and 25.18.

Show that the S- and T- matrices of $SU(3)_1$ or equivalently $\mathbb{Z}_3^1$ (see Section 18.5.3) satisfy the requirement for the condensed phase given by Eq. 25.20.

Exercise 25.7 The 16-Fold Way

Let us start with a modular TQFT $\mathcal{A}$ with central charge $c_{\mathcal{A}}$ that contains sVec as a subalgebra as described in Section 17.7. This means there is a simple current fermion $f \in \mathcal{A}$ such that $f \times f = I$.

We imagine constructing the product theory $\mathcal{C} \times$ Ising where the fermion in Ising we call ψ as usual. Condense the boson (f, ψ) to generate a new modular theory $\mathcal{U}$ with central charge $c_{\mathcal{U}} = c_{\mathcal{A}} + 1/2$.

(a) Show that if we start with $\mathcal{A} =$ Ising we generate $\mathcal{U} = (D_9)_1$ as shown in Table 17.1.

(b) Then show that starting with $\mathcal{A} = (D_9)_1$, and applying the same procedure, generates $\mathcal{U} = SU(2)_2$.

(c) Continuing this procedure, show that we generate the entire Table 17.3.

Part VI

Toric Code Basics

Introducing Quantum Error Correction

Before we look at the toric code, it is worth introducing some ideas of quantum error correction. While initially the ideas of error correction may seem somewhat different from what we have been discussing in prior chapters, we will see that (at least some) quantum error correcting codes are extremely closely related to topological ideas. Some of this material may be well known to most readers, but we reiterate it for completeness and to orient the discussion.

26.1 Classical Versus Quantum Information

26.1.1 Memories

Classical Memory

...all alone in the moonlight!

The unit of classical information is a bit[1]—a classical two-state system that can take the values 0 or 1. A memory with N bits can be in any one of 2^N states—each state corresponding to a particular bit-string, such as 011100111.

[1]You almost certainly know this already!

Quantum Memory

The unit of quantum information is the quantum bit or qubit[1] which is a quantum two-state system—that is, a two-dimensional complex Hilbert space spanned by orthonormal vectors that we usually call $|0\rangle$ and $|1\rangle$. A qubit can be in any state

$$|\psi\rangle = \alpha|0\rangle + \beta|1\rangle$$

with arbitrary complex prefactors α, β, where we normalize wavefunctions so $|\alpha|^2 + |\beta|^2 = 1$.

A quantum memory with N qubits is a vector within a 2^N-dimensional complex Hilbert space. So for example, with two qubits the general state of a system is specified by four complex parameters

$$|\psi\rangle = \alpha|00\rangle + \beta|01\rangle + \gamma|10\rangle + \delta|11\rangle \qquad (26.1)$$

with the normalization condition $|\alpha|^2 + |\beta|^2 + |\gamma|^2 + |\delta|^2 = 1$. So to specify the state of a quantum memory with two qubits, you have to specify four complex parameters, rather than, in the classical case, just stating which of the four states the system is in.

26.2 **Errors**

An error is some process that accidentally changes the state of the memory away from the intended state. Often we take as an error model the case where only one bit or one qubit is affected at a time (a "minimal" error) although more complicated errors can occur in practice.

26.2.1 **Classical Error Correction**

There is a simple way to protect the information stored in a classical memory from errors. Instead of storing a single bit 0 or 1, we store multiple copies of the bit. For example, in Table 26.1 we use three "physical" bits to store one "logical" bit of information.

logical bit	physical bits
0	000
1	111

Table 26.1 Three-bit repetition code. This code stores a single logical bit of information using three physical bits. The code space is the set 000 and 111.

Our memory should always either be in the state 000 or 111—we call these two possibilities the *code space*. If we detect the system being in any other state of the three bits (i.e. not in the code space) we know an error has occurred. If an error does occur on one of the physical bits (i.e. if one of the bits is accidentally flipped) we can easily find it because it would leave our memory with not all of the physical bits being the same. For example, if our system starts as 000, an error introduced on the second bit would leave it in the form 010. Then, by just using a majority-rule correction system, it is easy to figure out which bit is incorrect and flip the mistaken bit back. So our error correction protocol would be to continuously compare all three bits and, if they don't all match, flip the one that would bring them back to matching. Assuming errors are rare enough (and only occur on one bit at a time)[2] this scheme is an effective way to prevent errors. For added protection one can use more redundant physical bits, such as five physical bits or seven physical bits for a single logical bit. Such larger codes could withstand several bit-flip errors at a time and would still allow successful correction. For example, the five-bit code could withstand two bit-flip errors at a time and correction via majority rule would still be successful.

One might think the same sort of approach would work in the quantum world: make several copies of the qubit you want to protect, and then compare them to see if one has changed. Unfortunately, there are two big problems with this. The first is the so-called no-cloning theorem—it is not possible to make a perfect clone of a qubit. The second problem is that measuring a state inevitably changes it.

[2]If two bit-flips happen at the same time, then an uncorrectable logical error occurs. It is thus imperative that we check the state of our physical bits very frequently so that we catch errors and correct them before multiple errors can occur.

26.3   Quantum No-Cloning Theorem

Because the quantum no-cloning theorem is such an integral part of the discussion surrounding quantum error correction (and because the proof is easy) it is worth going through the proof.

The result (usually credited to Wootters and Zurek [1982] and Dieks [1982]) is such a straightforward result of quantum mechanics that some people have argued whether it deserves to be called a theorem. Nonetheless, the statement of the "theorem" is as follows:

> **Theorem:** Given a qubit in an arbitrary unknown state $|\phi_1\rangle$ and another qubit in a known or unknown initial state $|\phi_2\rangle$, there does not exist any unitary operator U (i.e. any quantum-mechanical evolution) such that
>
> $$U(\,|\phi_1\rangle \otimes |\phi_2\rangle\,) = e^{i\chi(\phi_1)}\,|\phi_1\rangle \otimes |\phi_1\rangle \qquad (26.2)$$
>
> for all possible inputs $|\phi_1\rangle$.

The point of the theorem is that there is no way to copy $|\phi_1\rangle$ into the auxiliary qubit $|\phi_2\rangle$. The reason we are looking for a unitary operator U is that all time evolutions in quantum mechanics correspond to unitary operators.[3] An arbitrary phase $e^{i\chi(\phi_1)}$, which can be a function of the cloned state $|\phi_1\rangle$, is allowed to occur during the cloning process.[4]

Proof of Theorem: Suppose such a unitary operator as specified in Eq. 26.2 does exist. This means we can properly copy two orthogonal states $|0\rangle$ and $|1\rangle$, meaning

$$\begin{aligned} U(\,|0\rangle \otimes |\phi_2\rangle\,) &= e^{i\chi(0)}\,|0\rangle \otimes |0\rangle \\ U(\,|1\rangle \otimes |\phi_2\rangle\,) &= e^{i\chi(1)}\,|1\rangle \otimes |1\rangle \quad. \end{aligned}$$

Quantum-mechanical operators must be linear so we can try applying this operator to the linear superposition $\alpha|0\rangle + \beta|1\rangle$ and we must get

$$U(\,\{\alpha|0\rangle + \beta|1\rangle\} \otimes |\phi_2\rangle\,) = \alpha\,e^{i\chi(0)}|0\rangle \otimes |0\rangle + \beta\,e^{i\chi(1)}|1\rangle \otimes |1\rangle \quad.$$

But this is *not* what a putative cloning device must give. Instead, a clone of the qubits should have given the outcome

$$e^{i\chi(\alpha|0\rangle + \beta|1\rangle)}\left[\alpha|0\rangle + \beta|1\rangle\right] \otimes \left[\alpha|0\rangle + \beta|1\rangle\right]$$

which is not generally the same result. Thus, no cloning device is consistent with the linearity inherent in quantum-mechanical evolution. ∎[5]

[3]One can object that making measurements has the effect of projecting, rather than being a unitary operation. However, there is a philosophy known sometimes as "The Church of the Larger Hilbert Space," which says that we should simply treat our measurement apparatus as part of the quantum-mechanical system, in which case all time evolutions become unitary again. The reason it is called a "Church" is because it is almost a religious view of how one should think about quantum measurements.

[4]In fact, we could allow the phase to be a function of $|\phi_2\rangle$ as well if we wanted to.

[5]The use of a square to replace Q.E.D. at the end of a proof was pioneered by the mathematician Paul Halmos. The square in typesetting is known as a "tombstone," but it is also sometimes known as a "Halmos."

26.4   Quantum Error Correction

Perhaps the most surprising thing about quantum error correction is that it is possible at all! This was discovered[6] in 1995 by Shor [1995], and shortly thereafter by Steane [1996a, b].

In Chapter 27 we introduce the toric code (Kitaev [2003]), which is a quantum error correcting code closely related to anyons. However, here, we will briefly introduce a very simple error correcting code introduced in the original work by Shor [1995] to try to explain how such codes typically work.[7]

26.4.1   Qubit-Flip Correcting Code

Let us consider the following simplified problem. Suppose we know that the only error that can ever occur on our physical system is the application of a Pauli σ_x operator,[8] that is, a so-called qubit-flip error (in the z-basis). We can protect our qubit from such an error in the following way.

Let us consider a logical qubit that we would like to protect from errors

$$|\psi\rangle_L = \alpha|0\rangle_L + \beta|1\rangle_L \tag{26.3}$$

where the subscript L means it is our logical qubit. Let us encode this logical qubit into three physical qubits by constructing[9]

$$|\psi_3\rangle = \alpha|000\rangle + \beta|111\rangle \tag{26.4}$$

where $|000\rangle$ means $|0\rangle \otimes |0\rangle \otimes |0\rangle$ and similarly for $|111\rangle$. The code space is the Hilbert space spanned by wavefunctions of the form Eq. 26.4.

We must not measure any of the three physical qubits since this will collapse the entire wavefunction to either $|000\rangle$ or $|111\rangle$. However, we can measure the product of two qubits, such as[8,10]

$$\hat{O}_{12} = \sigma_z^1 \sigma_z^2 \qquad \hat{O}_{23} = \sigma_z^2 \sigma_z^3 \quad . \tag{26.5}$$

The wavefunction $|\psi_3\rangle$ is in a $+1$ eigenstate of the operator $\hat{O}_{12}$ so we can measure this operator without collapsing the wavefunction. The purpose of this operator is to check that qubits 1 and 2 are the same (in the z-basis). Similarly, $\hat{O}_{23}$ checks that qubits 2 and 3 are the same. These operators are known as *stabilizer* operators (these operators leave the code space unchanged) and their eigenvalues are known as the *syndrome* since they are meant to diagnose whether the wavefunction has

[7]The discussion that follows (the remainder of this chapter) is included for completeness, but is not crucial on a first reading.

[8]When we use Pauli matrices we are thinking of $|0\rangle$ as being spin up and $|1\rangle$ as being spin down. So $\sigma_z|0\rangle = |0\rangle$ and $\sigma_z|1\rangle = -|1\rangle$ and $\sigma_x|0\rangle = |1\rangle$ and $\sigma_x|1\rangle = |0\rangle$.

[9]This does not violate the no-cloning theorem, since we are entangling two qubits with the initial qubit, rather than cloning the initial qubit. This procedure can be achieved with a quantum circuit as shown in Fig. 26.1.

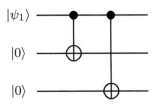

Fig. 26.1 The output of this quantum circuit, on the right, is the wavefunction $|\psi_3\rangle$ as in Eq. 26.4. The notation $\oplus$ indicates a controlled not gate (CNOT) where the lower qubits are controlled by the qubit $|\psi_1\rangle$ such that the output of the lower qubits is $|1\rangle$ if and only if the input qubit in $|\psi_1\rangle$ is $|1\rangle$. See the discussion in Section 11.4.1.

[10]We can also define $\hat{O}_{13}$, but this is redundant information since

$$\hat{O}_{13} = \hat{O}_{12}\hat{O}_{23}$$

[6]Peter Shor is probably the single person most responsible for creating the field of quantum computing. While there were a few early pioneers in the 1980s, such as Feynman, Yuri Manin (see note 23 of Chapter 2), and David Deutsch, there was not a lot of interest in the field for two key reasons: (a) it was not clear you could do anything useful with a quantum computer, and (b) it was not clear you could actually build a quantum computer. Shor conquered both of these questions. In 1994 Shor invented the quantum algorithm for efficiently factoring large numbers. Being that most cryptography schemes rely on large numbers not being efficiently factorizable this was obviously (a) useful, particularly to governments and spy agencies who were then willing to invest large amounts of money into pursuing this type of science. The following year, Shor invented error correcting codes, which then strongly suggested that (b) quantum computers could, at least in principle, be built.

error created by	stabilizers with -1 eigenvalue	fix by applying
σ_x^1	$\hat{O}_{12}$	σ_x^1
σ_x^2	$\hat{O}_{12}$, $\hat{O}_{23}$	σ_x^2
σ_x^3	$\hat{O}_{23}$	σ_x^3

Table 26.2 Error detection and correction rules for Eq. 26.4. If an error is created by the operator in the first column, the stabilizer(s) in the middle column will be measured to have a -1 eigenvalue, indicating an error. The error is corrected with the operator in the right column.

developed any "sickness," that is, whether an error has occurred.

If we find that both operators $\hat{O}_{12}$ and $\hat{O}_{23}$ are in the $+1$ eigenstate, then we conclude that all three qubits are the same (in the z-basis), so the wavefunction is properly in the code space (i.e. is of the form of Eq. 26.4). However, if one (or both) of these operators are measured in the -1 eigenstate, then we know that an error has occurred. Assuming no more than one σ_x error has occurred (i.e. we start with a wavefunction of the form of Eq. 26.4 and only one physical qubit is flipped over in the z-basis) we can easily identify the problematic qubit. For example, if σ_x has been applied to qubit 1, then we would find $\hat{O}_{12}$ in the -1 eigenstate, but $\hat{O}_{23}$ remains in the $+1$ eigenstate. We can then correct the problematic qubit by flipping it over again with σ_x. The error detection and correction rules are given in Table 26.2.

Note that these stabilizer operators, $\hat{O}_{12}$ and $\hat{O}_{23}$, do not collapse superpositions of the form of Eq. 26.4 since they do not actually measure the value of any of the qubits; they only check to see if two qubits are the same as each other in the z-basis.

One might think that Eq. 26.4 (along with the error correction rules of Table 26.2) would constitute a quantum error correcting code. Unfortunately, this is not the case. While we have constructed a code that can correct errors created by σ_x (i.e. qubit-flip errors) one can also have sign errors created by σ_z (and errors created by σ_y can be thought of as the product of σ_z and σ_x). Applying a σ_z operator to any of the three qubits in Eq. 26.4 results in

$$|\psi_3\rangle = \alpha|000\rangle - \beta|111\rangle$$

and this cannot be detected by our stabilizers $\hat{O}_{12}$ and $\hat{O}_{23}$.

26.4.2 Nine-Qubit Shor Code

Shor [1995] found that it is indeed possible to protect a qubit from both σ_x and σ_z errors (and therefore also σ_y errors) by using nine physical qubits. Again, we consider a general qubit of the form

$$|\psi\rangle_L = \alpha|0\rangle_L + \beta|1\rangle_L \quad . \tag{26.6}$$

Again, the subscript L indicates this is the logical qubit. Shor encoded the logical qubit into nine physical qubits as follows:

$$|\psi_9\rangle \;=\; \frac{\alpha}{2\sqrt{2}}\left[(|000\rangle + |111\rangle) \otimes (|000\rangle + |111\rangle) \otimes (|000\rangle + |111\rangle)\right]$$

$$+ \; \frac{\beta}{2\sqrt{2}}\left[(|000\rangle - |111\rangle) \otimes (|000\rangle - |111\rangle) \otimes (|000\rangle - |111\rangle)\right] \quad .(26.7)$$

This wavefunction can be prepared from a single qubit $|\psi\rangle_L$ with the quantum circuit shown in Fig. 26.2. The code space is the Hilbert space spanned by wavefunctions of the form of Eq. 26.7. The stabilizers of this code are as follows:

$$\hat{O}_{12} = \sigma_z^1 \sigma_z^2 \qquad\qquad \hat{O}_{23} = \sigma_z^2 \sigma_z^3 \qquad\qquad (26.8)$$
$$\hat{O}_{45} = \sigma_z^4 \sigma_z^5 \qquad\qquad \hat{O}_{56} = \sigma_z^5 \sigma_z^6 \qquad\qquad (26.9)$$
$$\hat{O}_{78} = \sigma_z^7 \sigma_z^8 \qquad\qquad \hat{O}_{89} = \sigma_z^8 \sigma_z^9 \qquad\qquad (26.10)$$
$$\hat{O}_{1-6} = (\sigma_x^1 \sigma_x^2 \sigma_x^3)(\sigma_x^4 \sigma_x^5 \sigma_x^6) \qquad\qquad (26.11)$$
$$\hat{O}_{4-9} = (\sigma_x^4 \sigma_x^5 \sigma_x^6)(\sigma_x^7 \sigma_x^8 \sigma_x^9) \qquad . \qquad (26.12)$$

The qubits in Eq. 26.7 have been grouped in threes $(123, 456, 789)$, and each set of three acts effectively like the above code Eq. 26.4. The stabilizers shown in lines 26.8–26.10 are analogous to the stabilizers in Eq. 26.5 and are meant to detect, and allow correction of, qubit-flips (σ_x errors). For example, if $\hat{O}_{12}$ is measured in eigenvalue -1 but all other $\hat{O}_{ij}$ have eigenvalue $+1$ then we know that the first qubit has been flipped over (in the z-basis) and needs to be repaired by applying σ_x^1 again. The rules for correcting these qubit-flip errors are listed in the first nine rows of Table 26.3.

The more interesting stabilizers are given in Eqs. 26.11 and 26.12. These stabilizers are meant to detect, and allow correction, of sign errors, that is, errors produced by application of σ_z operators. While these stabilizers look somewhat different, in fact these are again quite similar to the simple stabilizers we considered in Eq. 26.5, but in a different basis. Let us define

$$|+\rangle \;=\; \frac{1}{\sqrt{2}}\,(|000\rangle + |111\rangle)$$
$$|-\rangle \;=\; \frac{1}{\sqrt{2}}\,(|000\rangle - |111\rangle) \quad .$$

With this notation, our error correcting code Eq. 26.7 appears as

$$|\psi_9\rangle = \alpha|+++\rangle + \beta|---\rangle$$

which we recognize as being the same form as Eq. 26.4. Further, let us define X to be the operator that flips over one of these effective qubits and Z to be the operator that measures the qubit, as follows

$$X|\pm\rangle \;=\; |\mp\rangle$$
$$Z|\pm\rangle \;=\; \pm|\pm\rangle \quad .$$

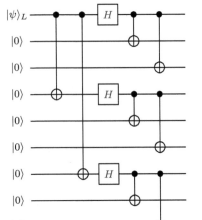

Fig. 26.2 A quantum circuit to prepare the nine-qubit Shor code. Here, H is a Hadamard gate, which can be written as

$$H = \frac{1}{\sqrt{2}} \begin{pmatrix} 1 & 1 \\ 1 & -1 \end{pmatrix} .$$

error created by	stabilizers with −1 eigenvalue	fix by applying
σ_x^1	$\hat{O}_{12}$	σ_x^1
σ_x^2	$\hat{O}_{12}$, $\hat{O}_{23}$	σ_x^2
σ_x^3	$\hat{O}_{23}$	σ_x^3
σ_x^4	$\hat{O}_{45}$	σ_x^4
σ_x^5	$\hat{O}_{45}$, $\hat{O}_{56}$	σ_x^5
σ_x^6	$\hat{O}_{56}$	σ_x^6
σ_x^7	$\hat{O}_{78}$	σ_x^7
σ_x^8	$\hat{O}_{78}$, $\hat{O}_{89}$	σ_x^8
σ_x^9	$\hat{O}_{89}$	σ_x^9
σ_z^1 or σ_z^2 or σ_z^3	$\hat{O}_{1-6}$	σ_z^1 or σ_z^2 or σ_z^3
σ_z^4 or σ_z^5 or σ_z^6	$\hat{O}_{1-6}$, $\hat{O}_{4-9}$	σ_z^4 or σ_z^5 or σ_z^6
σ_z^7 or σ_z^8 or σ_z^9	$\hat{O}_{4-9}$	σ_z^7 or σ_z^8 or σ_z^9

Table 26.3 Error detection and correction rules for Eq. 26.7, the nine-qubit Shor code. If an error is created by the operator in the first column, the stabilizer(s) in the middle column will be measured to have a −1 eigenvalue, indicating an error. The error is corrected with the operator in the right column.

Note now that

$$\sigma_z^j|\pm\rangle = \sigma_z^j \frac{1}{\sqrt{2}}\left(|000\rangle \pm |111\rangle\right) \tag{26.13}$$

$$= \frac{1}{\sqrt{2}}\left(|000\rangle \mp |111\rangle\right) = |\mp\rangle$$

for $j = 1, 2$, or 3, so that σ_z^j with $j = 1, 2$, or 3 plays the role of the X operator (the qubit-flip operator) on the $|\pm\rangle$ qubit made from the first three qubits (so let us call this X_{123}). Note that the effect of all three σ_z^j on this wavefunction is the same.

Similarly, we have

$$(\sigma_x^1\sigma_x^2\sigma_x^3)|\pm\rangle = (\sigma_x^1\sigma_x^2\sigma_x^3) \frac{1}{\sqrt{2}}\left(|000\rangle \pm |111\rangle\right) \tag{26.14}$$

$$= \pm \frac{1}{\sqrt{2}}\left(|000\rangle \pm |111\rangle\right) = \pm|\pm\rangle$$

so that $(\sigma_x^1\sigma_x^2\sigma_x^3)$ plays the role of Z operating on the $|\pm\rangle$ qubit made from the first three spins (let us call this Z_{123}).

With this notation, the stabilizers in Eq. 26.11 and 26.12 can then be rewritten as operators on three of these effective qubits

$$\hat{O}_{1-6} = Z_{123}Z_{456} \qquad \hat{O}_{4-9} = Z_{456}Z_{789}$$

entirely analogous to the stabilizers Eq. 26.5. These stabilizers can detect when X_{123} or X_{456} or X_{789} have been applied to the system. But these X operators are equivalent to the σ_z operators on the original qubits. Thus, these two stabilizers can detect σ_z errors, and tell us which operator to apply in order to correct the errors.

Thus, we are able to write out Table 26.3, which gives the full error detection and correction rules for the code Eq. 26.7. A few things are worth noting about this table. First, consider the last three rows, which address sign errors created by σ_z operators. Note that, for example, an error created by σ_z^1 can be corrected by either σ_z^1 or σ_z^2 or σ_z^3. Second, note that the table addresses the possibility of errors being created by σ_x or σ_z. If an error is created by σ_y it is simply considered as the product of σ_z and σ_x and can be corrected accordingly. Any operation that can be performed on a single qubit is some linear combination of σ_x, σ_y, σ_z and the identity, so this code allows correction of any error that can occur on a single qubit.

Chapter Summary

- We explored bits and qubits and error correcting codes. With the help of the no-cloning theorem we explained why quantum error correction is nontrivial.

- We constructed the nine-qubit Shor code to show that quantum error correction is possible.

Exercises

Exercise 26.1 Shor Code Circuit
Confirm that the circuit shown in Fig. 26.2 produces the Shor code Eq. 26.7.

Exercise 26.2 Quantum Circuits for Error Correction
For the Shor nine-qubit code
(a) [Easy] Construct a quantum circuit that reads out the logical qubit.
(b) [Harder] Construct a quantum circuit that corrects any error (σ_x or σ_y or σ_z) applied to a single qubit. Hint: It is useful to use a Toffoli gate (or CCNOT gate), that is, a three qubit gate that flips a target qubit if and only if two control bits are both in the $|1\rangle$ state.

Exercise 26.3 n^2 Qubit Error Correcting Code
Using a classical repetition code, one can use n physical bits with n odd to protect a single logical bit from $(n-1)/2$ errors.
Analogously, use n^2 qubits with n odd to construct a version of Shor's code. What type of multiple errors can it successfully correct?

Introducing the Toric Code

This chapter describes the toric code approach to error correction, which is conceptually one of the simplest error-correction schemes, as well as being very possibly the best error-correction scheme by many measures.[1,2] We see that the toric code is essentially a topological quantum field theory, which is why we are studying it in this book! As with so many great ideas in this field, the toric code was invented by Kitaev [2003].[3]

27.1 Toric Code Hilbert Space

We imagine an N_x by N_y square lattice with a qubit (or spin-$\frac{1}{2}$) on each edge, as shown in Fig. 27.1, where we use periodic boundary conditions such that we have a torus (hence, the name "toric"). The total number of spins is $N = 2N_x N_y$ and correspondingly, the dimension of the Hilbert space is 2^N.

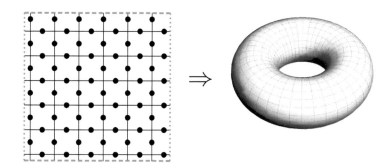

Fig. 27.1 The Hilbert space of the toric code: an N_x by N_y square lattice with qubits (or spin $\frac{1}{2}$), represented as dots, on each edge. The lattice is wrapped up to make it periodic in both directions. That is, the orange dashed lines at the top and the bottom are identified, and the orange dotted lines at the left and right are identified. Once the system is made periodic this is topologically a torus. There are 72 spins in this picture so the Hilbert space has dimension 2^{72}.

[2]Small examples of toric codes have been demonstrated recently by several experimental teams. See for example, Chen et al. [2021]; Satzinger et al. [2021]; Song et al. [2018]; and Andersen et al. [2020].

[3]Alexei Kitaev is one of the great geniuses of modern physics (MacArthur Genius Fellow, Breakthrough Prize Winner, and so forth). He is perhaps best known for inventing several of the most important algorithms of quantum computing (see for example the discussion of Kitaev–Solovay in Section 11.1.1), but in fact these are not even a small fraction of his highly influential works that span a wide range of modern fields. He visibly gets excited and smiles when he realizes something new about physics, and (being that this seems to happen frequently) he also seems to be one of the happiest people around.

[1]The statement that it is the "best" is based on the following consideration. If you can make measurements on each qubit at a rate of one per second, and similarly you can perform operations on your qubit at a rate of one per second, how many errors per second can you successfully correct? If you can correct errors sufficiently faster than they are created you can prevent any logical errors from occurring and your quantum information will live forever. Codes are compared to each other to see which one can withstand the greatest error rate, and very often the toric code wins (see Fowler et al. [2012] for example). One should be cautioned that the rules of the competition can be modified to favor one or another type of code. (Different rules correspond to different types of potential hardware.) For example, one might declare that performing a measurement on a qubit is faster than flipping qubits. One might declare that single-qubit operations are faster than two-qubit operations, and so forth.

[4]I somehow find it easier to think about spin up and spin down rather than $|0\rangle$ and $|1\rangle$.

We will think of the qubits along the edges as spin-$\frac{1}{2}$ particles.[4] We choose to work with a basis of up and down spins (σ_z eigenstates) for our Hilbert space. That is, we have 2^N basis states with all possible combinations of some spins pointing up and some spins pointing down. A convenient notation for this basis is given by coloring the edges containing down spins blue but leaving uncolored (i.e. black) the edges with up spins, as shown in Fig. 27.2.

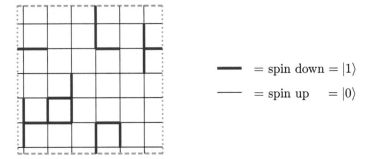

Fig. 27.2 A particular basis state of the Hilbert space, working in the up–down basis (σ_z-eigenstates). Here, we denote down spins by thick (blue) edges. Up spins are denoted by uncolored edges (black).

27.2 Vertex and Plaquette Operators

Let us now define some simple operators on this Hilbert space that we will use to build the toric code.

Vertex Operators

The first operator we define is known as a vertex operator. Given a vertex α, which consists of four incident edges $i \in$ vertex α, we define the vertex operator as

$$V_\alpha = \prod_{i \in \text{vertex } \alpha} \sigma_z^{(i)} \qquad (27.1)$$

that is, this is just the product of σ_z applied to the four spins incident on the vertex. This operator simply counts the parity of the number of down spins (number of colored edges) incident on the vertex. It returns $+1$ if there are an even number of incident down spins at that vertex and returns -1 if there are an odd number. This is depicted in Fig. 27.3.

Note that since the eigenvalues of V_α are $+1$ or -1, we have

$$V_\alpha^2 = 1 \qquad (27.2)$$

which can also be seen by just squaring V_α in the defining Eq. 27.1 and using the fact that $\sigma_z^2 = 1$. On a lattice of N_x by N_y sites, there are a total of $N = N_x N_y$ different vertex operators.

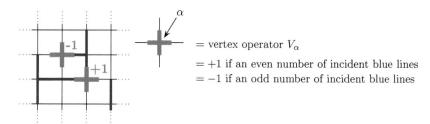

Fig. 27.3 The vertex operator V_α returns $+1$ if there are an even number of incident down spins (blue lines) at vertex α and returns -1 if there are an odd number.

Plaquette Operators

We now define a second operator known as the plaquette operator. Given a plaquette β that contains four edges in a square (edge $i \in$ plaquette β) we define

$$P_\beta = \prod_{i \in \text{plaquette } \beta} \sigma_x^{(i)} \tag{27.3}$$

which flips the (up/down) state of the spins on all of the edges of the plaquette as depicted in Fig. 27.4. On an N_x by N_y lattice, there are a total of $N_x N_y$ plaquette operators.

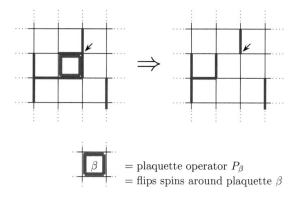

Fig. 27.4 The plaquette operator (bright pink) flips the state of the spin on the four edges of a plaquette. Blue edges become black and black edges become blue. The small black arrow points to a particular vertex to emphasize that the number of blue edges incident on this vertex has the same parity both before and after the plaquette flip (implying that the plaquette and vertex operators commute as mentioned in Eq. 27.7).

Since flipping a plaquette twice brings one back to the original configuration we have

$$P_\beta^2 = 1 \tag{27.4}$$

which can also be seen by squaring P_β in the defining Eq. 27.3. Given Eq. 27.4 the only two possible eigenvalues of P_β are ± 1. The eigenvectors corresponding to each of these eigenvalues can also be determined. In the basis we are using, the spin-up/spin-down basis corresponding to uncolored/colored edges, the P_β operator is off-diagonal—it flips spins around a plaquette. Thus, the eigenvectors can be written as

$$
\begin{array}{cc}
\text{eigenvector with} \\
\text{+1 eigenvalue}
\end{array}
=
\frac{1}{\sqrt{2}}
\left(
\begin{array}{c}
\text{unflipped} \\
\text{plaquette}
\end{array}
+
\begin{array}{c}
\text{flipped} \\
\text{plaquette}
\end{array}
\right)
\quad (27.5)
$$

$$
\begin{array}{cc}
\text{eigenvector with} \\
-1 \text{ eigenvalue}
\end{array}
=
\frac{1}{\sqrt{2}}
\left(
\begin{array}{c}
\text{unflipped} \\
\text{plaquette}
\end{array}
-
\begin{array}{c}
\text{flipped} \\
\text{plaquette}
\end{array}
\right)
\quad (27.6)
$$

Two examples of such eigenvectors with +1 eigenvalues are shown in Fig. 27.5.

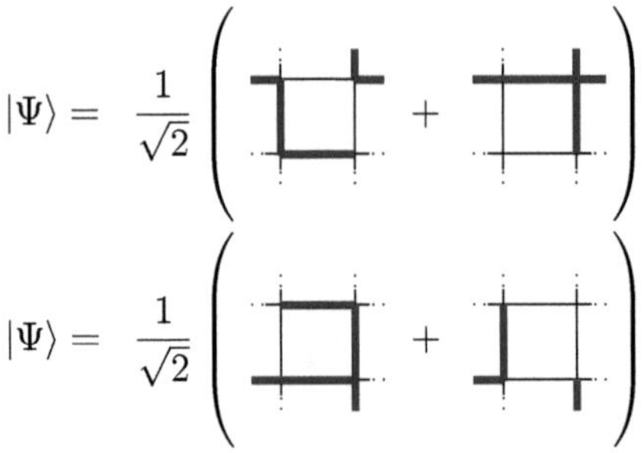

Fig. 27.5 Two examples of kets that are eigenvectors with +1 eigenvalue of the plaquette operator P_β. The two terms added in each superposition are related to each other by flipping the plaquette, that is, by applying P_β (blue lines outside of the plaquette remain unchanged). If we were to change the + sign to a − sign, we would obtain eigenvectors with −1 eigenvalue instead.

Caution: In this section we define vertex operators (Eq. 27.1) with σ_z operators and plaquette operators with σ_x operators (Eq. 27.3). In Kitaev's original work the convention is the opposite: vertex operators are defined as a product of σ_x operators and plaquette operators are a product of σ_z. Kitaev's convention is used more commonly in the quantum information community, while our convention is perhaps more typical in the condensed-matter community. Kitaev's convention makes the toric code look like a lattice gauge theory (see Section 31.7), whereas our convention makes it look like a loop gas. We can transform between the two conventions in either of two ways. Method 1: we can rotate all of our spins so as to exchange x and z. Method 2: we can make a duality transform on the lattice so that plaquettes become vertices and vertices become plaquettes. This duality is discussed further in Section 31.7.

27.2.1 Operators Commute

I claim that all of the plaquette operators and all of the vertex operators commute with each other. It is obvious that

$$
[V_\alpha, V_{\alpha'}] = 0
$$

since V_αs are only made of σ_z operators and all of these commute with each other (either two σ_z operators act on different edges, in which case they commute, or they act on the same edge, in which case they are the same operator so they commute). Similarly,

$$[P_\beta, P_{\beta'}] = 0$$

since P_βs are made only of σ_x operators and all of these commute with each other.

The nontrivial statement is that

$$[V_\alpha, P_\beta] = 0 \qquad (27.7)$$

for all vertices α and plaquettes β. The obvious case is when V_α and P_β do not share any edges, then the two operators obviously commute. When they do share edges, geometrically they must share exactly two edges, in which case the anticommutation between each shared $\sigma_x^{(i)}$ and $\sigma_z^{(i)}$ accumulates a minus sign, and since there are exactly two shared edges the net sign accumulated is $(-1)^2 = +1$ meaning that the two operators commute. An example of this commutation is shown in Fig. 27.4, where a small black arrow points to a particular vertex. Note that the vertex operator has the same eigenvalue both before and after the application of the plaquette operator (eigenvalue $= -1$ since there is an odd number of incoming blue lines). This shows explicitly that the vertex operator at this vertex commutes with the bright pink plaquette operator.

27.2.2 Is This a Complete Set of Operators? (Not Quite!)

We have $N_x N_y$ vertex operators and $N_x N_y$ plaquette operators—all of these operators commute, and each of these operators has two eigenvalues. This appears to match the fact that in the system there are $2N_x N_y$, each of which can point up or down, thus apparently giving the same number of degrees of freedom. So is our set of V_α and P_β operators a complete set of operators on this Hilbert space? Equivalently, if we specify the eigenstate of each of these operators, do we specify a unique state of the Hilbert space?

It turns out that the V_α and P_β operators do not *quite* form a complete set of operators on the Hilbert space. The reason they fail to form a complete set is that there are two constraints on these operators

$$\prod_\alpha V_\alpha = 1 \qquad (27.8)$$

$$\prod_\beta P_\beta = 1 \ . \qquad (27.9)$$

To see that Eq. 27.8 is true, note that each edge occurs in exactly two operators V_α (since each edge is attached to exactly two vertices). Thus, when we multiply all the V_αs together, each $\sigma_z^{(i)}$ occurs exactly twice, and $(\sigma_z^{(i)})^2 = 1$. Thus, the product of all the V_α's is the identity. The argument is precisely the same for multiplying together all of the P_βs.

Thus, we can freely specify the eigenvalues (±1) of ($N_x N_y - 1$) operators V_α, but the value of the one remaining V_α is then fixed by the values of the other ($N_x N_y - 1$) of them. Similarly with the P_βs, we can specify ($N_x N_y - 1$) eigenvalues (±1), and then the last eigenvalue is fixed by the value of the other ($N_x N_y - 1$). So specifying the eigenvalues of these commuting operators specifies only $2(N_x N_y - 1)$ binary choices. Since we started with $2N_x N_y$ spins, which can be in two states (up or down), we must still have two binary choices (two ±1 degrees of freedom) remaining that we have not fixed. These two remaining degrees of freedom (two binary choices) are going to be the two error-protected qubits (our logical qubits) in the toric code scheme for building a quantum error correcting code.

27.3 Building the Code Space

We are now in a position to build our quantum error correcting code. In particular, we want to define our code space—the space of possible allowed states of our system that we use for encoding quantum information. Analogous to the simple classical codes discussed in Section 26.2.1, we must have some error-checking protocol that continually checks that the system remains in the code space.

We now state two simple rules that define our code space. We must (as often as possible) check to see that the two rules remain satisfied. If we find that they are not satisfied we know a physical error has occurred, the system has left the code space, and we must then go about trying to correct it.

Rule 1: Specify that $V_\alpha = 1$ for every vertex.

This condition guarantees that there is an even number of blue lines (down spins) incident on every vertex. It is easy to see that this can be interpreted as a constraint that all configurations must form closed loops of the blue lines. There can be no ends of lines, and no branching of lines. An example of such a loop configuration is shown in Fig. 27.6.

Thus, our error-checking protocol examines every vertex and if it ever finds that $V_\alpha = -1$ then we know we are no longer in the code space, that is, a physical error has occurred that we must try to repair.

Rule 2: Specify that $P_\beta = 1$ for every plaquette.

As mentioned near Fig. 27.5, this assures that every plaquette is in an equal superposition of flipped and unflipped states with a plus sign between the two pieces. Note in particular that, because the P_β and V_α operators commute, the action of flipping a plaquette will not ruin the fact that Rule 1 is satisfied (that is, that we are in a loop configuration).

The operators V_α and P_β are known as the *stabilizers* of the code—the code space is unchanged under the application of these operators (see

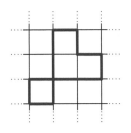

Fig. 27.6 A loop configuration consistent with the constraint that $V_\alpha = 1$ on every vertex. There must be an even number of blue edges incident on every vertex.

also the discussion of stabilizers and syndromes in Section 26.4.1). If these operators are measured to have a -1 eigenvalue then we know a physical error has occurred.

We now have the following prescription for constructing a wavefunction that satisfies both Rule 1 and Rule 2, that is, a wavefunction in the code space. First start in any state which satisfies Rule 1, that is, some loop configuration (this is a state with well-defined spins up and down in the σ_z basis). We call this configuration the *reference* configuration. Then to this reference wavefunction we add all other wavefunctions that can be obtained by flipping plaquettes. We thus have

$$|\psi\rangle = \mathcal{N}^{-1/2} \sum_{\substack{\text{all loop configs that can} \\ \text{be obtained by flipping pla-} \\ \text{quettes from a reference} \\ \text{loop config}}} |\text{loop config}\rangle \qquad (27.10)$$

where $\mathcal{N}$ is a normalization constant that counts the total number of terms in the sum. By adding up all such flipped configurations, we assure that every plaquette is in the correct superposition of flipped and unflipped and we satisfy Rule 2. Recall from Fig. 27.5 that adding flipped and unflipped configurations of a plaquette gives $P_\beta = +1$.

We now make a crucial observation: flipping plaquettes never changes the *parity* (evenness or oddness) of the number of loops that go around a nontrivial cycle of the torus. To see this, try drawing a line around a cycle of the torus, as shown with the dashed red line in Fig. 27.7. If one flips a plaquette (bright pink in the figure), it does not change the parity of the number of blue edges that cross through either dashed red line. As a result of this observation, we realize that the sum in Eq. 27.10 does not include all possible loop configurations, but rather contains only those loop configurations with the same parity of loops going around the two nontrivial cycles as the reference configuration.

Thus, there are four independent wavefunctions of the form of Eq. 27.10, which differ in whether their associated reference configurations have an even or an odd number of blue loops going around each nontrivial cycle. All of these states satisfy the constraints that all $V_\alpha = 1$ and all $P_\beta = 1$. We will call these states

$$|\psi_{ee}\rangle, \quad |\psi_{eo}\rangle, \quad |\psi_{oe}\rangle, \quad |\psi_{oo}\rangle \qquad (27.11)$$

where e and o stand for an even or an odd number of blue loops going around a given nontrivial cycle. So for example, we have

$$|\psi_{eo}\rangle = \mathcal{N}^{-1/2} \sum_{\substack{\text{all loop configs that have an} \\ \text{even number of blue loops} \\ \text{around the vertical cycle and} \\ \text{an odd number of blue loops} \\ \text{around the horizontal cycle}}} |\text{loop config}\rangle$$

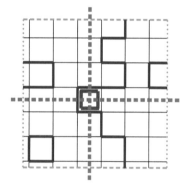

Fig. 27.7 A loop configuration (blue edges) on a torus (dotted orange lines on opposite sides are identified as in Fig. 27.2). Drawing a line (dashed red) around one of the nontrivial cycles of the torus, one can see that flipping a plaquette, such as the one marked in bright pink, does not change the parity of the number of blue edges cutting through the dashed red line. Further, it does not matter where we place the dashed red line. So for example, we can change the y-coordinate of the horizontal dashed red line, and the number of blue edges it cuts through is always odd. Similarly, if we change the x-coordinate of the vertical dashed red line, the number of blue edges it cuts through is always even.

or graphically, we have Fig. 27.8

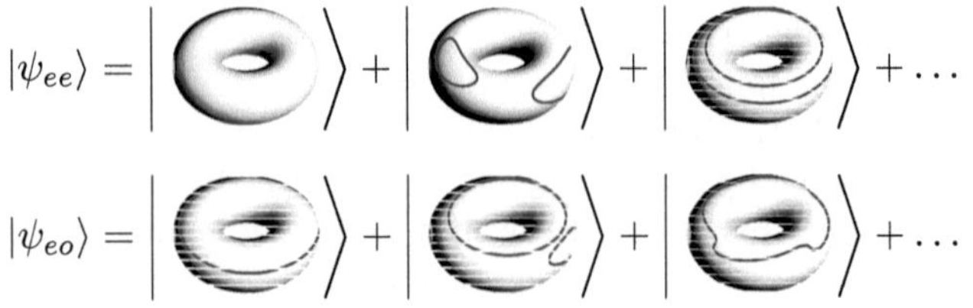

$$|\psi_{ee}\rangle = \left|\ \bigcirc\ \right\rangle + \left|\ \bigcirc\ \right\rangle + \left|\ \bigcirc\ \right\rangle + \dots$$

$$|\psi_{eo}\rangle = \left|\ \bigcirc\ \right\rangle + \left|\ \bigcirc\ \right\rangle + \left|\ \bigcirc\ \right\rangle + \dots$$

Fig. 27.8 **Top:** Graphical depiction of $|\psi_{ee}\rangle$, which is a sum of all wavefunctions having an even number of blue strings running around each nontrivial cycle. **Bottom:** Graphical depiction of $|\psi_{eo}\rangle$, which has an even number of strings around one cycle (the meridian, short direction, or vertical in the planar diagram) and an odd number around the other (the longitude, long direction, or horizontal in the planar diagram).

Suppose now we have two logical qubits that we would like to protect from errors. In full generality we can write the wavefunction of the two logical qubits as

$$|\psi_2\rangle_L = \alpha|00\rangle_L + \beta|01\rangle_L + \gamma|10\rangle_L + \delta|11\rangle_L \qquad (27.12)$$

with arbitrary complex coefficients $\alpha, \beta, \gamma, \delta$ subject to the normalization condition $|\alpha|^2 + |\beta|^2 + |\gamma|^2 + |\delta|^2 = 1$.

We can now encode these logical qubits in the four wavefunctions of toric code (Eqs. 27.11). We imagine constructing the encoded wavefunction

$$|\psi\rangle = \alpha|\psi_{ee}\rangle + \beta|\psi_{eo}\rangle + \gamma|\psi_{oe}\rangle + \delta|\psi_{oo}\rangle \qquad . \qquad (27.13)$$

The underlying spins on the lattice that make up the code are the "physical" qubits. The wavefunctions of the form of Eq. 27.13 thus span the code space. Our encoding of the wavefunction in Eq. 27.12 into the physical wavefunction Eq. 27.13 is entirely analogous to the encoding that we did in going from Eq. 26.3 to Eq. 26.4 or Eq. 26.6 to Eq. 26.7.

Note that in order to turn the $|\psi_{ee}\rangle$ wavefunction into the $|\psi_{eo}\rangle$ wavefunction we would need to insert a single blue loop around the horizontal (longitude) cycle, which involves flipping an entire row of spins at once. If we were to try to flip only some of these spins, we would have an incomplete loop, or an endpoint of a blue line, violating the rule that $V_\alpha = 1$ for all vertex sites, therefore not being in the code space. It is this fact that allows us to test for errors (by looking for such endpoints) and correct them efficiently, as we shall see next.

27.4   Errors and Error Correction

Let us now turn to study possible errors in more detail. What does a physical error look like in this system? Imagine an evil demon arrives[5] and, unbeknownst to us, applies an operator to one of the physical spins in the system to try to create an error. We start by considering the

[5]It is not entirely necessary to anthropomorphize the error-creating process.

cases where this operator is σ_x (Section 27.4.1) or σ_z (Section 27.4.2). We consider combinations of these operators in Section 27.4.3, and argue that we do not have to consider other possibilities in Section 27.4.4.

27.4.1 σ_x Errors

Let us first consider the case where the error operator is σ_x. That is, starting in one of the code space wavefunctions, σ_x is applied on edge i to create a physical error. This operator commutes with all the plaquette operators P_β but anticommutes with the vertex operators V_α that intersect that edge. This means, if we start in the code space (all $V_\alpha = +1$, all $P_\beta = +1$), and apply this error operator $\sigma_x^{(i)}$, we then end up in a situation where the vertex operators on the two vertices attached to the edge i have the wrong eigenstate $V_\alpha = -1$. One way to see this is to realize that before σ_x is applied, all of the vertices have an even number of blue edges (down spins) coming into them. By flipping over one edge with σ_x we obtain two vertices with an odd number of blue edges coming into them (one at each end of this edge). A more formal way to see this is to realize that in the original state $|\psi\rangle$ we have

$$V_\alpha |\psi\rangle = |\psi\rangle$$

meaning we start in the $+1$ eigenstate. We then apply the error operator $\sigma_x^{(i)}$ to both sides. Assuming i is one of the edges incident on the vertex α, we have

$$\sigma_x^{(i)}|\psi\rangle = \sigma_x^{(i)} V_\alpha |\psi\rangle = -V_\alpha \sigma_x^{(i)}|\psi\rangle$$

or

$$V_\alpha [\sigma_x^{(i)}|\psi\rangle] = -[\sigma_x^{(i)}|\psi\rangle]$$

showing we end up in the -1 eigenstate of the vertex operator.

To show these physical errors graphically we will no longer draw the up and down spins (the blue edges) but instead we just draw the σ_x operator as a dark red line, and the vertices which are in the -1 eigenstate we mark as orange ×s as shown in Fig. 27.9.

So it is clear what our error correction protocol must do. It must frequently measure the V_α operators. If it finds all V_α operators in the $+1$ eigenstate then the system is still in the code space. On the other hand, if it finds a pair of adjacent vertex operators in the $V_\alpha = -1$ state, we know that a σ_x operator has been applied on the intervening edge by some error-creating process[5]. Once we have identified the error it is easy to correct it by applying σ_x on the same edge, thus returning the system to its original state in the code space.

Now suppose that the error-creating demon[5] is very fast and manages to make several such σ_x errors very quickly. If these errors are well separated from each other, we will find multiple pairs of vertices[6] in the $V_\alpha = -1$ state, with the two members of each pair separated from each other by one edge distance. These can similarly be identified by our correction scheme and repaired, returning us to the code space again.

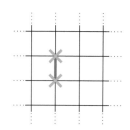

Fig. 27.9 Starting in the code space, a σ_x operator applied to the red edge (i) creates two vertices in the $V_\alpha = -1$ eigenstate marked with the orange ×s.

[6]It is typical to abuse nomenclature and say a "vertex" is in the $V_\alpha = -1$ state when we mean a "vertex operator" is in the $V_\alpha = -1$ state. Similarly, we will do this for "plaquette" and "plaquette operator."

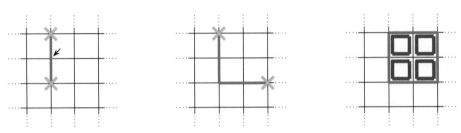

Fig. 27.10 Left: When two σ_x operators (the red edges) are applied to two edges that share a vertex (the small black arrow), the shared vertex is hit twice, thus returning to the $V_\alpha = +1$ state. Only the two vertices at the end of the "error string" are in the $V_\alpha = -1$ state. **Middle:** A longer string of errors. Note that we can only measure the endpoints of the string, not where the errors were made, so we cannot tell if the error string goes down two steps then two steps to the right, or if goes two steps to the right then down two steps. **Right:** If we detect the errors as in the middle panel and we try to correct it by dragging the errors back together, but we choose the incorrect path for the string, we end up making a closed (red) loop of σ_x operators—which acts as the identity on the code space, so we still successfully correct the error!

However, it could be the case that two errors are created on two edges that share a single vertex, as shown on the left of Fig. 27.10. The vertex that is shared (marked by a small black arrow in the figure) gets hit by σ_x twice. The first time it is hit by σ_x the eigenvalue of V_α flips from its initial state $+1$ to -1 but then when it is hit the second time, this flips it back to the $+1$ state. Thus, after the two σ_x operators have been applied on the two red edges on the left of Fig. 27.10, only the two vertices, marked with an orange $\times$, at the end of the red "string" are in the $V_\alpha = -1$ state and are then detectable as errors.

Nonetheless, the error correction scheme is still fairly straightforward. One checks the state of all the vertices and when $V_\alpha = -1$ is found, one tries to find the closest other vertex with $V_\alpha = -1$ to pair it with. Then one applies a string of σ_x operators to correct these errors. One can think of this as dragging the errors back together and annihilating them with each other again.

It is important to realize that we cannot see the error operators themselves (which we have drawn as a red string in Fig. 27.10) by making measurements on the system—we can only detect the endpoints of string, the vertices, marked with an orange $\times$, where $V_\alpha = -1$. For example, in the middle panel of Fig. 27.10 we cannot tell if the error string goes down two steps and then to the right, or if it goes to the right two steps then down. We only know where the endpoints of the string are.

Suppose now we detect the two errors shown in the middle panel of Fig. 27.10. We may try to correct these errors by guessing where the red string is, and applying σ_x along this path to bring the endpoints back together and reannihilate them. However, it is possible that we guess incorrectly as shown in the right panel of Fig. 27.10. In this case we will have ended up producing a (red) closed loop of σ_x operators applied to the original state. Fortunately, a product of σ_x operators around a closed loop is precisely equal to the product of the plaquette operators P_β enclosed in the loop

$$\prod_{i \text{ around loop}} \sigma_x^{(i)} = \prod_{\beta \text{ enclosed in loop}} P_\beta . \tag{27.14}$$

Since the code space is defined such that all of the plaquette operators are in the $+1$ eigenstate, this red loop of σ_x thus acts as the identity on the code space, and we still successfully correct the error!

If it is not already obvious, the reason for the equality Eq. 27.14 can be seen in the right of Fig. 27.10. Within the enclosed region, each edge is acted on by two plaquette operators (marked in bright pink) each applying σ_x to the edge, and since $\sigma_x^2 = 1$ this means there is no net effect on the internal edges, and one is left with only σ_x applied to the (red) loop bounding the region.

On the other hand, if a loop of errors occurs which extends around a nontrivial cycle, as shown in Fig. 27.11 (think of this as dragging the error, the orange $\times$, all the way around the cycle and reannihilating it again) then, although we return to the code space (there are no $V_\alpha = -1$ vertices remaining) we have changed the parity of the number of down spin strings around a nontrivial cycle, thus scrambling the quantum information and making an error in the logical qubits. In fact, what we get in this case is a transformation that switches the even and odd sectors around one nontrivial cycle:

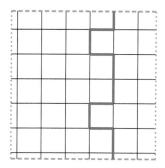

Fig. 27.11   If an error string (red) goes all the way around a nontrivial cycle, it changes the parity of the number of blue loops running around the cycle. For example, if the initial state has no down spins (so no blue loops) the state after application of the red error string has one blue loop going around the cycle in the position where the red string was applied. Although the resulting wavefunction is still in the code space, it has been scrambled compared to the original wavefunction. That is, a logical error has occurred.

$$\alpha|\psi_{ee}\rangle + \beta|\psi_{eo}\rangle + \gamma|\psi_{oe}\rangle + \delta|\psi_{oo}\rangle \longrightarrow$$
$$\alpha|\psi_{oe}\rangle + \beta|\psi_{oo}\rangle + \gamma|\psi_{ee}\rangle + \delta|\psi_{eo}\rangle \quad . \tag{27.15}$$

If the red error loop had gone around the other cycle, the second index of the wavefunction would have flipped rather than the first index.

The powerful idea of the toric code is that by having a very large torus, it requires a very large number of physical errors (spin flips) to make this (red) loop of physical errors go all the way around the non-trivial cycle and actually scramble the quantum information (the logical qubits). If we are continuously checking for $V_\alpha = -1$ physical errors we can presumably correct these before a logical error can arise.

27.4.2 σ_z **Errors**

We can also consider what happens if the error operator applied to the system is a σ_z operator. Much of the argument in this case is similar to the σ_x case.

The σ_z operator on an edge anticommutes with the two adjacent plaquettes P_β containing that edge. Applying σ_z to a system in the code space takes the system out of the code space and results in the two adjacent plaquettes each having eigenvalue $P_\beta = -1$ as shown in Fig. 27.12. To be more explicit about this, recall that in the code space the state of each plaquette is in a superposition (a positive sum) of flipped and unflipped (see Fig. 27.5). In Fig. 27.13 we can see how applying a σ_z operator to one edge results in the difference, rather than the sum, of the flipped and unflipped plaquettes, and therefore $P_\beta = -1$.

Analogous to the above discussion of σ_x errors, our σ_z error-correction protocol should frequently check for pairs of neighboring plaquettes where $P_\beta = -1$ and if these are found the protocol should correct the

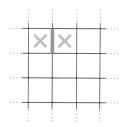

Fig. 27.12   Starting in the code space, a $\sigma_z^{(i)}$ operator applied to the dark green edge (i) creates two plaquettes (marked with the bright green $\times$s) in the $P_\beta = -1$ state .

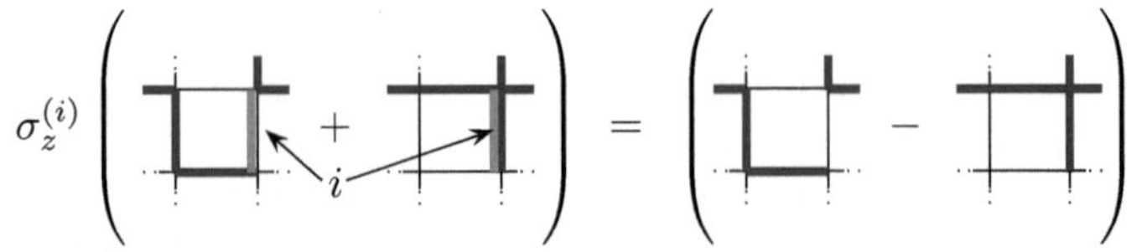

$$\sigma_z^{(i)} \left(\boxed{} + \boxed{} \right) = \left(\boxed{} - \boxed{} \right)$$

Fig. 27.13 Starting with the sum of flipped and unflipped plaquettes, which is an eigenstate of P_β with eigenvalue $+1$, we apply σ_z to the edge i marked in green. This applies a minus sign to the down spin (the blue edge) but a plus sign to the up edge. The resulting difference of flipped and unflipped is then an eigenstate with $P_\beta = -1$.

error by applying σ_z to the intervening edge.

We next consider the case where several σ_z errors are created before being corrected. Again, if the errors are spatially well separated they can be identified and corrected without regards to each other. Whenever we find pairs of neighboring plaquettes in the $P_\beta = -1$ state, we correct them each individually. However, the situation is more complicated if two such errors occur on different edges of the same plaquette.

Starting in the code space, if two σ_z errors are applied on two different edges of the same plaquette (marked in dark green in the left of Fig. 27.14) then the P_β eigenvalue of that plaquette (marked with an arrow) is flipped twice. As shown in the left of Fig. 27.14, this results in two plaquettes in the $P_\beta = -1$ state being separated by a plaquette (marked with the arrow) in the $+1$ state. In the figure we have marked the $P_\beta = -1$ plaquettes with bright green $\times$. These $\times$s are connected together with a bright green line which cuts through the dark green edges. Sometimes we say that the bright green line is a path on the *dual lattice* (meaning it goes from center of plaquette to center of plaquette). The error-correction procedure uses multiple σ_z operators to bring these defects back together and reannihilate them.

If still more σ_z errors are created, they can form a string (on the dual lattice), as shown in the middle of Fig. 27.14. The σ_z errors here are applied on the dark green edges. The bright green path on the dual lattice cuts through all of these dark green edges. As in the σ_x case, one is not able to actually detect the bright green string (or the dark green edges), but can only see the endpoints of the string as plaquettes where $P_\beta = -1$.

Analogous to the σ_x case, if from errors, or from an attempt to correct errors, the σ_z error string forms a closed loop, as illustrated as the bright green square in the right of Fig. 27.14, this loop of σ_z operators is equal to the product of the enclosed V_α operators.

$$\prod_{i \text{ around dual loop}} \sigma_z^{(i)} = \prod_{\alpha \text{ enclosed in dual loop}} V_\alpha \; . \tag{27.16}$$

Since within the code space, $V_\alpha = 1$, a closed loop of σ^z errors returns the system to its original state. This means that reannihilating the defects (the bright green $\times$s) successfully correct the errors independent of the path that we use to bring the defects back together.

Another way of seeing Eq. 27.16 is in terms of the blue loops of down

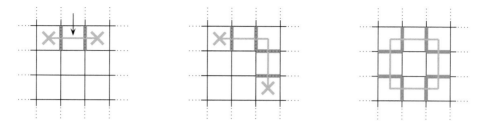

Fig. 27.14 Left: When two σ_z operators (dark green) are applied to two different edges of the same plaquette (the one with the arrow) then this plaquette ends up in the $P_\beta = +1$ state. The two plaquettes with $P_\beta = -1$ are marked with the bright green × and are connected together by the bright green error string. **Middle:** A string of several σ_z errors. The bright green line cuts through each of the dark green edges where σ_z is applied. The bright green line lives on the dual lattice, meaning it goes from plaquette center to plaquette center. **Right:** A closed loop of σ_z errors. This is equal to the product of all of the enclosed V_α operators. In the code space, this product is equal to $+1$.

spins discussed in Section 27.3. The σ_z operators give -1 each time they intersect a blue loop; however, the blue loops must be closed so the number of intersections between a blue loop and the loop of the green σ^z error string in the figure must be even (since a blue loop going into the region enclosed by the bright green string must also come out), thus forcing the product of the blue σ_z operators to have a value of 1.

On the other hand, if the loop of σ_z operators goes all the way around a nontrivial cycle of the torus, it then scrambles the logical qubits. In particular, if there is a string of σ_z going all the way around one cycle, as shown in Fig. 27.15, this operator then counts the parity of the number of blue edges going around the opposite cycle, as shown in the figure. Thus, applying the string of σ_z operators around the (horizontal) cycle makes the transformation

$$\alpha|\psi_{ee}\rangle + \beta|\psi_{eo}\rangle + \gamma|\psi_{oe}\rangle + \delta|\psi_{oo}\rangle \longrightarrow$$
$$\alpha|\psi_{ee}\rangle + \beta|\psi_{eo}\rangle - \gamma|\psi_{oe}\rangle - \delta|\psi_{oo}\rangle\rangle \tag{27.17}$$

inserting a minus sign in front of the states that have an odd number of blue loops going around the vertical cycle. If the string of σ_z were to go around the other cycle (vertical), the minus sign would be applied to β and δ, which both have an odd number of blue loops going around the horizontal cycle.

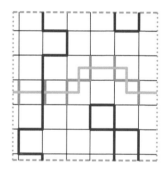

Fig. 27.15 If a string of σ^z goes around a nontrivial cycle (in this case the horizontal cycle), it measures the parity of the number of blue strings going around the opposite cycle (in this case looping in the vertical direction). In this figure, the string of σ_z operators returns -1 being that it cuts an odd number of blue edges.

27.4.3   Combinations of σ_x and σ_z

We have discussed physical errors created by σ^x and σ^z, but we should also consider what happens when both types of errors are created.[7] If we have an error-correction protocol that removes σ_x errors and another protocol that removes σ_z errors, and as long as these two procedures don't interfere with each other, we should be able to remove combinations of the two.

To give a detailed example, let us suppose, starting in the code space, both a σ_x operator *and* a σ_z operator are applied to a single edge, as shown in the left of Fig. 27.16. We obtain from these error operators both the orange ×s at the two adjacent vertices (two vertices with $V_\alpha = -1$,

[7]For example, $\sigma_y = i\sigma_x\sigma_z$

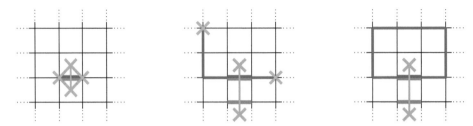

Fig. 27.16 Left: Both σ_x and σ_z are applied on the same edge resulting in both vertex defects (orange ×s) with $V_\alpha = -1$ and plaquette defects (bright green ×s) with $P_\beta = -1$. **Middle:** Two σ_z operators are applied on the dark green edges to move the plaquette defects apart. After this, five σ_x operators are applied along the red string to move the vertex defects apart. **Right:** A closed (red) loop of σ_x operators is equal to the product of all of the enclosed P_β operators. Compared to Fig. 27.10, here there is an enclosed plaquette defect (with the bright green ×) so the value of the loop is now -1.

from the application of σ_x) and the bright green ×s on the two adjacent plaquettes (two plaquettes with $P_\beta = -1$ from the application of σ_z). Assuming our error-checking procedure finds all of these errors, it can correct the errors by applying both a σ_x and a σ_z to this edge. However, we need to be more cautious here because applying σ_x then σ_z differs by a sign from applying the two in the opposite order!

To this end, let us examine more closely what happens when we have both σ_x errors (vertices with $V_\alpha = -1$) and σ_z errors (plaquettes with $P_\beta = -1$) at the same time. For example, in the middle of Fig. 27.16 we imagine that first σ_z is applied to the two edges marked in dark green, thus making $P_\beta = -1$ for the two plaquettes marked with bright green ×s. Then we imagine that σ_x is applied to the five edges marked in red so that $V_\alpha = -1$ at the two orange ×s. On the right of the figure we imagine more σ_x operators applied (perhaps as the result of our error-correction protocol) so that the orange ×s come together and reannihilate forming a closed red loop. As in the right of Fig. 27.10 the closed red loop is equal to the product of P_β for the enclosed plaquettes (Eq. 27.14). However, here one of the enclosed plaquettes (the one marked with the green ×) is in the $P_\beta = -1$ eigenstate. Thus, forming the closed red loop now gives an overall sign of -1. Similarly if we had moved the bright green × in a loop around one of the orange ×s and reannihilated it, as shown in Fig. 27.17, we would also have gotten -1.

We have thus uncovered an important principle: if you create a defect pair, move the defect around in a loop and reannihilate it, you make a measurement of what is contained in that loop!

In the context of our code, should we be worried about having accumulated an overall sign? No! As mentioned in Chapter 11 note 8, typically in quantum computing we are not interested in the overall phase of a result, which is hard—if not impossible—to control anyway. The logical qubits are protected up to an overall phase prefactor only.

Fig. 27.17 A closed (green) loop of σ_z operators (on dark green edges) is equal to the product of all of the enclosed V_α operators. As compared to Fig. 27.14, here there is an enclosed vertex (with the orange ×) so the value of the loop is -1.

27.4.4 Other Errors

We have now considered errors in the vertex operator eigenvalues (created by σ_x) and errors in the plaquette operator eigenvalues (created by σ_z operators). Further, we argued that we are not concerned about the overall phase of the system, so we can handle both of these types of errors at the same time. Since our code space is defined as being the states of the system where there are no vertex defects and no plaquette defects, this means we can actually correct *any* error that occurs in our system (as long as not too many physical errors are created at the same time).

27.5   The Toric Code on Different Lattices and Different Topologies

Before ending this chapter it is worth making a short detour to consider the possibility of building the toric code with a lattice of a different geometry. In building the toric code we could have used a triangular lattice, a honeycomb, or even an "irregular lattice."[8] Whatever the geometry of the lattice (regular or irregular), the vertex term constrains the code to have an even number of blue edges coming into each vertex (no matter how many edges join at each vertex), and the plaquette term flips all of the edges of a plaquette (no matter how many edges the plaquette has). An example of an irregular lattice is shown in Fig. 27.18.

[8]Neither a honeycomb nor an "irregular lattice" should be called a lattice, although people insist on incorrectly using the word "lattice" anyway! See Simon [2013], for example, for the proper definition of a lattice!

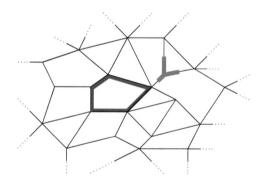

Fig. 27.18 Part of an irregular lattice. A vertex operator with three legs is marked in green. This operator still gives +1 or −1 if an even or odd number of down spins are incident on the vertex. A plaquette operator with five sides is marked in bright pink. The plaquette operator flips all of the spins along its edges.

For any lattice geometry (regular or irregular) the vertex and plaquette operators are a commuting set. Thus, the code space can be described as (Eq. 27.10) a sum of all loop configurations that can be reached from a reference configuration by flipping plaquettes. What *is* crucial to the toric code is that the topology of the system is a torus, so that we still have four orthogonal states in the code space characterized by the number of (blue) loops around the two nontrivial cycles, as shown in Fig. 27.8.

Once we allow different lattice geometries, we may also ask the question of what happens if our system has different *topologies*, that is, is not a torus, but is some more general object.

[9] In fact, the toric code can be defined on a non-orientable surface such as a Klein bottle. However, since the rest of this book focuses on orientable manifolds, we will do the same here for simplicity.

[9] In fact, the toric code can be defined on a non-orientable surface such as a Klein bottle. However, since the rest of this book focuses on orientable manifolds, we will do the same here for simplicity.

[10] Euler noticed this in 1758 for the case of convex polyhedra (genus $g = 0$). However, his proof of the statement was incorrect. It was correctly proven by Legendre in 1794.

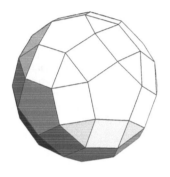

Fig. 27.19 Since this object is topologically a sphere (genus $g = 0$), we have Vertices−Edges+Faces = 2. This would also be true of the triangulation of the sphere shown in the right of Fig. 23.1. However, the triangulation of the torus (genus $g = 1$) shown in the left of Fig. 23.1 would have Vertices−Edges+Faces = 0.

To understand the dependence of the code on the global lattice topology let us consider an arbitrary two-dimensional orientable closed surface.[9] In two dimensions, all such surfaces can be fully described topologically by their genus g, or number of handles: a sphere, $g = 0$, has zero handles; a torus, $g = 1$, has one handle; a two-handled torus, $g = 2$, has two handles, and so forth. We will make use of the famous Euler characteristic: for any polyhedral decomposition of an orientable 2-manifold, we have the beautiful formula[10]

$$2-2g = (\text{Number of Vertices})-(\text{Number of Edges})+(\text{Number of Faces}).$$

An example of this identity is shown in Fig. 27.19.

In Section 27.2.2 we counted the total number of degrees of freedom in the system and we found that, after fixing all of the vertex and plaquette degrees of freedom, on a torus, we were left with two qubits remaining. These became the protected logical qubits of our toric code. Let us now figure out, for a more general surface, how many degrees of freedom we have left once we have specified all the vertex and plaquette eigenvalues. Since there is one spin on each edge we can rewrite Euler's equation as

$$\Big[\text{Number of Vertex Ops} + \text{Number of Plaquette Ops} - 2\Big] + 2g$$
$$= \text{Number of Spins}. \qquad (27.18)$$

We can read this equation as follows. The right-hand side is the total number of binary (± 1 or $\uparrow, \downarrow$) choices we can make (number of two-state degrees of freedom). The brackets counts the number of these choices we specify by fixing all of the vertex and plaquette eigenvalues: we have one choice for each vertex operator and one for each plaquette operator, but then we subtract two binary choices because of the constraints Eqs. 27.8 and 27.9 (i.e. once all but one vertex is specified, the last vertex is fixed and is not an additional degree of freedom, and similarly for plaquettes). The remaining term, $2g$, is the number of binary choices (number of qubits) that we do not fix by specifying the vertex and plaquette operator eigenvalues. Thus, in the case of a spherical surface, there are no qubits remaining unspecified; in the case of a torus ($g = 1$) there are two qubits left (as we calculated above); in the case of a two-handled torus, $g = 2$, there are four qubits, and so forth.

We might have suspected that the number of qubits we could make out of a toric code on a g-handled torus would be related to the number of different distinguishable loops on the surface. Indeed, this turns out to be correct! For a $g = 1$ genus (regular) torus, there are two independent nontrivial cycles (the longitude and the meridian). Similarly, on a g-handled torus there are $2g$ independent nontrivial cycles. Each qubit represents the evenness or oddness of the number of (blue) loops going around each nontrivial cycle.

Shor's original nine-qubit error correcting code (Section 26.4.2) can also be viewed as a toric code—but on a surface with singular points (see Exercise 27.3).

27.6 $\mathbb{Z}_N$ **Toric Code (Briefly)**

One of the simplest generalizations of the toric code is known as the $\mathbb{Z}_N$ toric code. (While this is not a too complicated generalization, it is also not entirely crucial to the development of ideas here and can be skipped on a first reading.)[11] In this model, instead of having a two-state system (spin up/spin down) on each edge, we will put an N-state system[12] on each edge, which we will write as $s \in \{0, \ldots, (N-1)\}$, instead of saying a spin is up or down (or the edge is blue or black). It is worth noting that it will sometimes be more natural to think of the exponential $e^{2\pi is/N}$, which naturally has only N possible states.

We can use a regular or irregular lattice (see note 8) for this model. However, now for $N > 2$ we must put arrows on each edge to define a direction. The direction chosen does not particularly matter. An irregular lattice with arrows is shown in Fig. 27.20. Each edge i is labeled with a quantum number $s_i \in \{0, \ldots, (N-1)\}$. (The $N = 2$ case corresponds to the regular toric code with $s = 0$ corresponding to a black edge [spin up] and $s = 1$ corresponding to a blue edge [spin down].) We will think of this quantum number as being some sort of fictitious $\mathbb{Z}_N$-valued current running along the edge in the direction of the arrow.

We now define an operator Q_α that measures the total $\mathbb{Z}_N$ current sink or source at a vertex α to be[13]

$$Q_\alpha = \left[\sum_{i \in \text{edges arriving at } \alpha} \hat{s}_i \ - \sum_{i \in \text{edges leaving } \alpha} \hat{s}_i \right] \bmod(N) \quad (27.19)$$

where an edge touching a vertex is denoted as "arriving" at a vertex if its arrow is pointed at the vertex and is denoted as "leaving" if its arrow is pointed away from the vertex. Here. $\hat{s}_i$ is the operator that measures the quantum number s_i (analogous to $\sigma_z^{(i)}$ measuring whether the spin at site i is up or down).

The operator Q_α has N possible eigenvalues given by the possible charges $0, \ldots, (N-1)$. We can exponentiate Q_α to give a vertex operator more analogous to what we had with the conventional $\mathbb{Z}_2$ toric code

$$V_\alpha = e^{2\pi i Q_\alpha / N}$$

which, for the case of $N = 2$ matches our previous definition of the vertex operator, having eigenvalues ± 1 in this case.

The sum of charges over the entire system must be zero

$$\sum_\alpha Q_\alpha \bmod(N) = 0$$

since this expression counts each edge leaving a vertex once and counts each edge as an arriving at a vertex once. In a form more analogous to

[11] Our presentation will be a bit faster here with the assumption that the reader is now familiar with the conventional toric code.

[12] A two-state system is known as a qubit. A three-state system is sometimes known as a qutrit. A general d-state system is known as a qudit. One should not call an N-state system "qunit" though since the word "nit" is often used for an object having e states, with $\ln e = 1$. (Also "nits" are lice eggs. Yuck.)

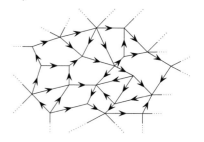

Fig. 27.20 An irregular lattice with arrows assigned to the edges.

[13] The term $\bmod(N)$ means take the remainder after dividing by N. Because we are working $\bmod(N)$ it is not possible to say whether a vertex is a current sink or a current source since a value $Q_\alpha = m$ is the same as $Q_\alpha = m + pN$ for any integer p, positive or negative.

the constraint Eq. 27.8 we can write

$$\prod_\alpha V_\alpha = \prod_\alpha e^{2\pi i Q_\alpha/N} = 1 \quad .$$

We can also define another operator that changes the value of the s variables on the edges of a plaquette β (similar to $P_\beta = \prod_{i\in\beta}\sigma_x^{(i)}$ for the conventional $\mathbb{Z}_2$ toric code)

$$A_\beta = \begin{cases} s_i \to (s_i+1)\mathrm{mod}(N) & \text{arrow pointing} \\ & \text{clockwise around } \beta \\ \\ s_i \to (s_i-1)\mathrm{mod}(N) & \text{arrow pointing} \\ & \text{counterclockwise around } \beta. \end{cases} \tag{27.20}$$

Since $(A_\beta)^N$ is the identity, there are N possible eigenvalues of A_β, which are given by $e^{2\pi i k/N}$ with $k = 0,\ldots,(N-1)$.

Each edge with a fixed arrow direction bounds two plaquettes such that the arrow points clockwise with respect to one plaquette and counterclockwise with respect to the other plaquette, as shown for example in Fig. 27.21. Applying A_β to the plaquettes on both sides of an edge thus leaves the edge unchanged. This immediately implies that applying A_β to all plaquettes in the system gives the identity:

$$\prod_\beta A_\beta = 1$$

analogous to Eq. 27.9.

A useful commutation relation (analogous to $\sigma_x\sigma_z = -\sigma_z\sigma_x$ for the conventional toric code) is (see Exercise 27.1).

$$e^{2\pi i s_j/N} A_\beta = e^{\pm 2\pi i/N} A_\beta e^{2\pi i s_j/N} \tag{27.21}$$

where s_j is the operator that measures the quantum number on an edge j on the boundary of plaquette β. The $\pm$ on the right is positive if the arrow on the edge is going clockwise around β and is negative if the arrow on the edge is going counterclockwise.

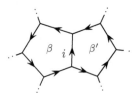

Fig. 27.21 Edge i is pointing clockwise with respect to plaquette β' but counterclockwise with respect to plaquette β.

27.6.1   Code Space

Analogous to the discussion of Section 27.2.1, we can establish that the operators A_β and Q_α all commute with each other (see Exercise 27.1).

Our code space for this system will be the space of states satisfying the vertex condition $Q_\alpha = 0$ (or to look more similar to the regular toric code, all $e^{2\pi i Q_\alpha/N} = 1$) for all vertices and the plaquette condition $A_\beta = 1$ for all plaquettes.

Let us first examine the condition $Q_\alpha = 0$. Applying this condition to all vertices means that there is no net current flow (mod N) into or out of any vertex, analogous to the loop condition for the conventional

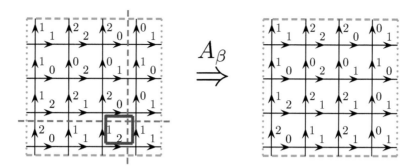

Fig. 27.22 Edge variable configuration for the $\mathbb{Z}_3$ toric code on a torus geometry satisfying $Q_\alpha = 0$ at all vertices α, meaning the current arriving equals the current leaving modulo 3. Opposite sides marked in orange are identified with each other so this picture represents a torus. In going from the left figure to the right figure, the operator A_β is applied on the marked plaquette. Note that the total current crossing either red dashed line is the same before and after the application of A_β. It also does not matter where (horizontally) we put the vertical red dashed line or where (vertically) we put the horizontal red dashed line. Note that an element of the code space is obtained by applying A_β to all possible plaquettes in all possible ways and summing the results.

toric code (see near Fig. 27.6). This is like Kirchhoff's first law in circuit analysis except that the current is $\mathbb{Z}_N$ valued, that is, its value is periodic modulo N. In Fig. 27.22 we show two configurations of edge variables on a torus geometry that satisfy this condition for a $\mathbb{Z}_3$ toric code.

Note that if we start with a situation where all $Q_\alpha = 0$, flipping a plaquette by applying A_β (shown in Fig. 27.22) does not change the condition that $Q_\alpha = 0$ (this is equivalent to our statement that Q_α and A_β commute). The second code space condition, $A_\beta = 1$, is analogous to the plaquette condition for the usual toric code. The code space wavefunction can then be written in the usual form

$$|\psi\rangle = \underset{\substack{\text{all configs that can be} \\ \text{reached by applying } A_\beta \\ \text{starting with a reference} \\ \text{config that satisfies } Q_\alpha = 0}}{\mathcal{N}^{-1/2} \sum} |\text{config}\rangle \qquad (27.22)$$

where $\mathcal{N}$ is the total number of terms in the sum.

We should be a bit cautious if we define our code space as the space of states $Q_\alpha = 0$ and also $A_\beta = 1$. As discussed in the toric code, we would want to have an error-correcting protocol that continually checks to make sure that the system has not left the code space. Hence, we might want to make continuous measurements of A_β. Unfortunately, A_β is not a Hermitian operator, so it is not directly measurable. However, we can certainly measure $\frac{1}{2}(A_\beta + A_\beta^\dagger)$ and $\frac{1}{2i}(A_\beta - A_\beta^\dagger)$, which both have the same eigenvectors as A_β (since $A_\beta^\dagger = A_\beta^{-1}$) and this will work just as well.

As with the regular ($\mathbb{Z}_2$) toric code, there are multiple orthogonal wavefunctions that one can obtain depending on the reference config-uration used. For each non-contractible cycle in the system, we can

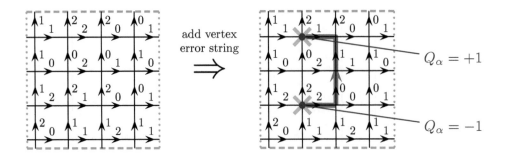

Fig. 27.23 Introduction of an error string that creates two vertex defects in an example of the $\mathbb{Z}_3$ toric code. The error string (red) has an arrow pointing from the $Q_\alpha = -1$ defect to the $Q_\alpha = +1$ defect. Each edge along the path is incremented by one unit (mod N) if its arrow is aligned with the red string arrow and is decremented by one unit (mod N) if its arrow is opposed to the red string arrow. Note that on the left, an element of the code space is obtained by applying A_β to all possible plaquettes in all possible ways and summing the result. Also note that if we start in the code space and apply the red error string, the error string is not measurable—only the defects at the end points are.

define a $\mathbb{Z}_N$ valued current flow going around that cycle. For example, in the torus shown in Fig. 27.22 there are two units (mod 3) of current going vertically, and three units (mod 3) going horizontally. Note that flipping a plaquette by applying A_β (also shown in the figure) does not change the value of the current going around either non-contractible cycle. Thus, since a genus g surface has $2g$ non-contractible cycles we expect that the $\mathbb{Z}_N$ toric code should be able to store $2g$ N-state qudits in its quantum memory. That is, the Hilbert space of this quantum memory has dimension N^{2g}.

27.6.2   Errors

We can now consider what sort of error processes can occur. The first type is a vertex error (analogous to the σ_x error of the conventional toric code). Here, we can consider, for example, incrementing the quantum number (the "current" label) by one unit (mod N) along one edge, to create a defective vertex at either end of that edge. One of these vertices will then have $Q_\alpha = +1$ and the other will have $Q_\alpha = -1$.

More generally, we can consider moving such defects away from each other analogous to Fig. 27.10. An example of this is illustrated in Fig. 27.23. This generates an error string (shown red in the figure) connecting the two defects. The error string (red) has an arrow on it, and will increment the quantum number of any edge along its path by one unit (mod N) if the arrow of the edge is aligned with that of the string, and will decrement the quantum number of the edge by one unit (mod N) if the arrow of the edge is anti-aligned with the arrow of the string. At the head of the string is a vertex defect with $Q_\alpha = +1$ and at the tail is a defect with $Q_\alpha = -1$. Defects with $Q_\alpha = \pm m$ can then be made by incrementing or decrementing edges by m steps. Multiple strings may be introduced, and the defects may be brought together, adding their charges modulo N. Defects are always created in $\pm$ pairs, so that if we start in the code space, the sum of Q_α over the entire system must remain zero (mod N).

If we create a defect pair and move the $Q_\alpha = +1$ defect in a clockwise circle then reannihilate it with the $Q_\alpha = -1$ defect, this leaves behind a closed loop of red string, analogous to the right of Fig. 27.10. However, it is easy to see that such a clockwise-oriented closed loop is equal to the product of A_β operators for the enclosed plaquettes.

$$\begin{matrix} \text{Moving } (Q_\alpha = +1) \text{ defect in} \\ \text{clockwise loop} \end{matrix} = \prod_{\text{enclosed } \beta} A_\beta \qquad (27.23)$$

analogous to Eq. 27.14. Moving in a counterclockwise loop will correspondingly give a product of A_β^{-1}. Analogous to the conventional toric code, since in the code space the A_βs are all in the $+1$ eigenstate, this means that starting in the code space, creating two defects, moving them around in a loop and reannihilating them returns the system to the original state in the code space. However, dragging a defect around a non-contractible cycle and then reannihilating puts the system in a different state of the code space and creates a logical error for this quantum memory.

We now consider plaquette defects. Here, we imagine starting in the code space and applying the operator $X_j = e^{2\pi i s_j/N}$ on edge j (where s_j is the operator that measures the quantum number on the edge). Due to the commutation relation Eq. 27.21 the action of the operator X_j changes the eigenvalue of A_β on the neighboring two plaquettes, multiplying the eigenvalue on the neighboring plaquette to the left of the arrow on the edge j by $e^{2\pi i/N}$ and multiplying the other neighbor by $e^{-2\pi i/N}$. Correspondingly, using the operator X_j^{-1} will multiply the eigenvalue of A_β by the inverse phase. Starting in the code space (where all A_β are in the $+1$ eigenstate) application of X_j or X_j^{-1} will create a pair of neighboring defective plaquettes. Applying multiple such operators can move the defective plaquettes away from each other, as shown in Fig. 27.24 (analogous to Fig. 27.14 for the conventional $N = 2$ toric code).

If, starting in the code space, a $A_\beta = e^{2\pi i/N}$ defect is created and then moved in a clockwise loop and then reannihilated with its counterpart, the system is returned to the code space. In the process a net phase is accumulated given by

$$\begin{matrix} \text{Moving } (V_\beta = e^{2\pi i/N}) \text{ defect in} \\ \text{clockwise loop} \end{matrix} = \prod_{\text{enclosed } \alpha} e^{2\pi i Q_\alpha/N} \quad . \qquad (27.24)$$

To see this result, it is easiest to imagine a situation where all the arrows on the (dark green in Fig. 27.24) edges point toward the center of the loop. Moving the $A_\beta = e^{2\pi i/N}$ defect in a clockwise loop then corresponds to applying X_j operators on all the edges. This procedure then measures the sum of all the edges entering the enclosed region which is equal to the sum of all the Q_α inside the region (see Eq. 27.19). (One

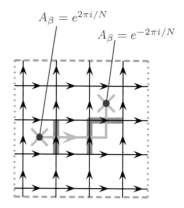

Fig. 27.24 Application of the operator $X_j = e^{2\pi i s_j/N}$ on an edge (shown in dark green) multiplies the eigenvalue of A_β on the plaquette to the left of the arrow by $e^{2\pi i/N}$ and multiplies the eigenvalue of the plaquette on the right by $e^{-2\pi i/N}$. Multiple applications of these operators can move the defective plaquettes away from each other, as shown in the figure—only the bright green ×s are defects. In the figure, the two vertical dark green edges are X_j operators and the horizontal dark green bond is a X_j^{-1} operator.

can then check that reversing the orientation of the arrow on an edge does not change this result.) As in the case of the conventional toric code, if a plaquette defect is moved around a non-contractible cycle, a logical error is imparted to the code space.

Chapter Summary

- The toric code is the paradigm of a topological quantum memory. It can be constructed on any lattice structure (even irregular) but it must have non-contractible cycles (such as a torus) in order to store quantum information.
- We introduced the concepts of vertex and plaquette operators.
- Using the notation of colored edges to denote spin down, the code space can be viewed as a sum of all possible loop configurations. The code space is spanned by different orthogonal wavefunctions corresponding to the different possible parities of loops around each non-contractible cycle.
- Physical errors are described as localized defects in the code space. A physical error must be taken around a non-contractible cycle to create a logical error.
- The $\mathbb{Z}_N$ toric code gives quantum numbers $j \in \{0, \ldots, (N-1)\}$ to each edge.

Further Reading:

There are many introductions to the toric code in the literature and on the web. The original work by Kitaev [2003] is still a very good place to start. Note, as mentioned in the **Caution** at the end of Section 27.2, that many references (including Kitaev's original work) switch σ_x and σ_z compared to our convention.

Exercises

Exercise 27.1 Commutation of Operators in the $\mathbb{Z}_N$ Toric Code
   (a) Show that all of the operators A_β (Eq. 27.20) and Q_α (Eq. 27.19) commute.
   (b) Prove Eq. 27.21.

Exercise 27.2 Code Space Degeneracy of the $\mathbb{Z}_N$ Toric Code
   Generalize the Euler characteristic argument of Section 27.5 to confirm that the dimension of the code space of the $\mathbb{Z}_N$ toric code is N^{2g} where g is the genus of the surface.

Exercise 27.3 Shor's Code is a Toric Code!

Consider nine edges assembled into three spheres in the following way. First, make a single sphere out of two vertices (put one at the north pole and one at the south pole for simplicity), three edges connecting the two vertices, and three plaquettes filling out the sphere. Now make three spheres in this way. Connect these spheres together at the vertices but do not add any new plaquettes or edges. We now have three vertices, each connected to six edges (three on one sphere and three on another sphere). This makes a manifold with three singular points, as shown in Fig. 27.25. Consider now the toric code applied to this system. There is one non-contractible loop in the figure that plays the role of a qubit. There are nine plaquette operators and three vertex operators, which can be put together to make a total of eight independent stabilizers. Show that this system is precisely the Shor error-correcting code introduced in Section 26.4.2.

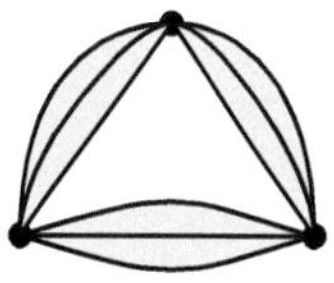

Fig. 27.25 Three spheres (shown elongated into bananas) connected together at their north and south poles. Each sphere consists of three edges, two vertices, and three plaquettes—one plaquette of each sphere is pointing away from the reader so it is not seen (the back of each banana).

The Toric Code as a Phase of Matter and a TQFT

We have introduced the toric code as a way to store quantum information. However, it is also possible to construct the toric code such that it is a quantum phase of matter, that is, the ground state of a Hamiltonian. As such the toric code becomes an example[1] of **topologically ordered matter**[2]—a physical system that is described at low temperature and long length scale by a topological quantum field theory.

To recast the toric code as a phase of matter, we simply write a Hamiltonian which is a sum of all of our vertex and plaquette operators described in Section 27.2

$$H_{\text{toric code}} = -\frac{\Delta_v}{2} \sum_{\text{vertices } \alpha} V_\alpha - \frac{\Delta_p}{2} \sum_{\text{plaquettes } \beta} P_\beta \qquad (28.1)$$

where here $\Delta_v > 0$ and $\Delta_p > 0$ have dimensions of energy and set the energy scale of the problem.[3,4] The operators V_α and P_β all have eigenvalues ± 1, and since the operators all commute with each other (see Section 27.2.1) the lowest-energy configuration (i.e. the ground-state space) is obtained by simply setting all of the $V_\alpha = 1$ and $P_\beta = 1$. In other words, the ground-state space is exactly the code space!

If the system is on a torus geometry, there will be a fourfold degenerate ground state corresponding to the four orthogonal wavefunctions in the code space (Eq. 27.11). As discussed in Section 27.5, if the system is on a genus-g surface, we would instead have a 2^{2g}-fold degenerate ground state (corresponding to the $2g$ qubits in the code space). The dependence on topology strongly suggests to us that our physical system is described by a topological quantum field theory![5]

28.1 Excitations

If there are vertices with $V_\alpha = -1$ or plaquettes where $P_\beta = -1$, the system is not in the ground-state space. These occurrences, which we called errors previously, in the language of topologically ordered matter should now be considered to be particle (or *quasiparticle*)[6] excitations.

Let us list all the types of particles we can find:

(1) We always have a vacuum or identity particle (which can also be thought of as the absence of a particle) that we call I.

---

[1] In fact, the toric code is *the* paradigmatic example!

[2] Sometimes this is known as "topologically ordered quantum matter" to emphasize the quantum nature.

[3] Often for simplicity one sets $\Delta_v = \Delta_p = \Delta$. To simplify even more, set the energy scale of the problem to unity with $\Delta_v = \Delta_p = 1$.

[4] Since V_α and P_β have eigenvalues ± 1, we have included a factor of $1/2$ out front so that the difference between energies of these two eigenstates is Δ_v or Δ_p respectively.

[5] From only this ground-state degeneracy, we can conclude that the TQFT describing this system is abelian. See Exercise 28.2.

[6] The term "quasiparticle" is used for objects that act as relatively weakly interacting individual particles in low-energy theories but are emergent from some other degrees of freedom. The distinction between particle and quasiparticle is not used consistently. We often speak of protons, pions, electrons, etc. as particles rather than quasiparticles, but they too are presumably just the low-energy excitations of a more complicated theory.

(2) We can have a vertex where $V_\alpha = -1$ instead of $V_\alpha = +1$. The energy[4] of each vertex defect is Δ_v.

(3) We can have a plaquette where $P_\beta = -1$ instead of $P_\beta = +1$. The energy[4] of each plaquette defect is Δ_p.

The vertex and plaquette defects are often called electric and magnetic particles respectively (e and m). The intuition for the names here is that the vertex defect is some sort of "charge" at the vertex (hence "electric") and the plaquette defect is some sort of flux through the plaquette (hence "magnetic").

Caution: Sometimes the vertex defect is instead called magnetic and the plaquette defect is called electric! We discuss in Section 31.7 why this other labeling is sensible as well.

Fortunately, which particle is labeled e and which one m is at this point just a convention. The TQFT that describes the system is actually the same independent of which one we label e and which one we label m. (This symmetry is discussed extensively in Chapter 36.) For the time being we follow the convention that the vertex defect is electric and the plaquette defect is magnetic. Sometimes, however, it is easier to just call the defects vertex and plaquette defects to avoid confusion, and we will try to do this as much as possible!

Since vertex defects are produced in pairs, and can be brought back together and annihilated in pairs, we know we must have[7]

$$\text{vertex defect} \times \text{vertex defect} = I \quad .$$

Similarly, since plaquette defects m are produced in pairs, and can be brought back together and annihilated in pairs we must also have

$$\text{plaquette defect} \times \text{plaquette defect} = I \quad .$$

In the more common notation we would write

$$e \times e = I$$
$$m \times m = I \quad .$$

We might then wonder what happens if we bring together a vertex and a plaquette defect. They certainly do not annihilate! Thus,

(4) We define another particle type,[8] called f, which is the fusion of the electric and the magnetic particles (fusion of vertex and plaquette defect)

$$f = e \times m \quad .$$

The energy of such a particle is $(\Delta_v + \Delta_p)$.

Sometimes (particularly in the gauge-theory literature) a particle that is a combination of an electric and magnetic particle is called a *dyon*.

[7] As in prior chapters, $\times$ here means fusion of particle types.

[8] f stands for "fermion," and we will shortly discover that this particle is indeed a fermion.

We have the fusion relation

$$f \times f = I$$

which we can see by associativity and commutativity of fusion

$$f \times f = (e \times m) \times (e \times m) = (e \times e) \times (m \times m) = I \times I = I \quad .$$

These four particle types e, m, f, and the vacuum I are the only particle types in this theory. First, they form a closed set under the fusion rules. Secondly, as we discussed in Section 27.4 any "error" in this system (or excitation out of the ground state) can be described in terms of combinations of applications of σ_x and σ_z, hence, all excitations are generated by only the e and m particles. The full fusion relations are given in Table 28.1. There are no nonabelian fusions here (i.e. each fusion gives a unique outcome) so we conclude we have an abelian theory.

Note that there are exactly four particle types (including the identity), and there are exactly four ground states on the torus, in agreement with the general principle that the number of particle types should match the number of ground states on the torus (see Eq. 7.9, for example).

$\times$	I	e	m	f
I	I	e	m	f
e	e	I	f	m
m	m	f	I	e
f	f	m	e	I

Fig. 28.1 Fusion table for the toric code. Note that the table remains correct if we switch e with m everywhere.

28.2 Statistical Properties of Excitations

Let us first consider the vertex-defect particles (we have been calling this particle e, but for more clarity we will stick with the term "vertex defect"). These particles are both created and moved around by applying σ_x operators. All of the σ_x operators commute with each other, so the end result is the same independently of what order we create, move, and annihilate the vertex-defect particles with these σ_x operators.

There are now several "experiments" we can do to test the statistics of these particles. For example, we can create a pair of vertex defects, move one around in a circle and reannihilate, then compare this to what happens if we put another vertex defect inside the loop before the experiment. Since all of the creation and moving of particles is achieved by applying σ_x, and all σ_x operators commute with each other, we see that the presence of another vertex defect inside the loop does not alter the phase of moving the defect around in a circle. Analogously, braiding a plaquette-defect particle around another plaquette-defect particle accumulates no phase: the defect is both created and moved by the σ_z operator and all of these σ_z operators commute with each other.

The experiments just described tell us that the vertex and plaquette defects (e and m) are either bosons or fermions. It is fairly obvious that braiding bosons around bosons never accumulates a phase. This is less obvious for fermions: while exchanging two fermions accumulates a -1 sign, taking a fermion in a loop all the way around another fermion accumulates two minus signs (hence, a $+1$ sign, or no phase) since it is equivalent to two exchanges. Thus, we have not yet determined the full statistical properties of these particles. To do so we will need to examine the twist factor θ for these particles (see Chapter 15).

To determine the twist factor of a particle, we make a curl in a world line, as in Fig. 2.6 or 15.2. Consider Fig. 28.2. In the middle panel we apply (reading right to left)

$$\sigma_x^1 \sigma_x^7 \sigma_x^6 \sigma_x^5 \sigma_x^4 \sigma_x^3 \sigma_x^2 \sigma_x^1 \quad . \tag{28.2}$$

This just creates a pair of vertex defects, one at A and one at B, moves the particle at B around in a simple loop (reading right to left $BGFEDCB$), and brings it back to the original position and reannihilates it with its partner. We can compare this to the following operation shown in the bottom panel of Fig. 28.2

$$\sigma_x^1 \sigma_x^2 \sigma_x^1 \sigma_x^7 \sigma_x^6 \sigma_x^5 \sigma_x^4 \sigma_x^3 \quad . \tag{28.3}$$

This instead creates a pair of vertex defects particles at positions C and D, moves the particle at D along the path (reading right to left) $ABGFED$ behind the particle at position C, and finally to position A. Finally, the particle at position C is moved to B and then annihilated with the particle at A. This process is topologically a loop with a curl or twist included (compare Fig. 2.6 or 15.2). However, since the σ_x operators all commute, the two processes (Eqs. 28.2 and 28.3) must be equal. This means the twist factor is trivial, $\theta = 1$, so we conclude that the vertex defect is a boson. An entirely analogous argument can be used to show that the plaquette defect is a boson as well (see Exercise 28.3).

28.2.1 Braiding Vertex Defects with Plaquette Defects

We now turn to determine what happens when one braids a vertex-defect particle around a plaquette-defect particle (e around m in our current notation). Much of what follows here simply recapitulates the discussion of Section 27.4.3.

Suppose we create a pair of vertex-defect particles and move one around in a circle, and then reannihilate the pair (as in the right panel of Fig. 27.10). This process is created by a string of σ_x operators. Recall that, if there are no plaquette defects inside the loop this process does not accumulate a phase because the string of σ_x operators around the loop is equivalent to the product of the P_β plaquette operators enclosed (Eq. 27.14)—and in the ground state (the code space), the P_β operators are in the $+1$ state. However, if there is one plaquette-defect particle inside the loop (as in Fig. 28.3, or right panel of 27.16), this means that one of the P_β operators is actually in the -1 state. In this case the phase of taking the vertex-defect particle around the loop is -1. So there is a phase of -1 for taking a vertex defect around a plaquette defect.

Another way to understand the braiding statistics is as follows. The operator that takes a vertex-defect particle in a loop (the red loop in Fig. 28.3) is made of σ_xs. The operator that creates a pair of plaquette

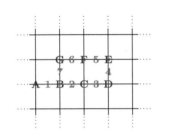

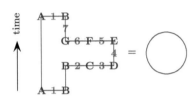

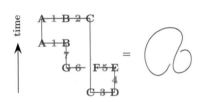

Fig. 28.2 Top: Vertices are labeled with letters and edges with numbers. **Middle:** The process shown is (reading right to left) $\sigma_x^1 \sigma_x^7 \sigma_x^6 \sigma_x^5 \sigma_x^4 \sigma_x^3 \sigma_x^2 \sigma_x^1$. This creates a pair of vertex-defect particles, moves one around, and re-annihilates, making a simple loop for its spacetime diagram (middle right). **Bottom:** The process shown is $\sigma_x^1 \sigma_x^2 \sigma_x^1 \sigma_x^7 \sigma_x^6 \sigma_x^5 \sigma_x^4 \sigma_x^3$. This creates a pair of vertex-defect particles, makes a curl or twist in the spacetime diagram, and then reannihilates (bottom right). Since all σ_x operators commute, these two operations must give the same phase, implying trivial phase for adding the twist.

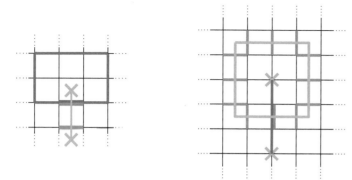

Fig. 28.3 Left: The red loop represents a string of σ_x operators that take a vertex-defect particle in a loop. This operator is equal to the product of plaquette operators enclosed, and will give -1 if there is a single plaquette defect inside the loop and $+1$ if there are none. The bright green line is a dual string of σ_z operators (dark green lines) that pulls two plaquette defects (bright green ×s) apart. **Right:** The bright green loop represents a dual string of σ_z operators (dark green lines) that take a plaquette-defect particle in a loop. This operator is equal to the product of vertex operators enclosed, and will give -1 if there is a single vertex defect inside the loop and $+1$ if there are none. The red line is a string of σ_x operators that pulls two vertex defects (orangle ×s) apart. The anticommutation of the red string and the green dual string also shows the braiding statistics.

defects and pulls them apart (the bright green line in Fig. 28.3) is made of σ_zs (the dark green lines). These two operators intersect on a single edge (the one that has both a red line and a dark green line in Fig. 28.3). The σ_x and σ_z on this edge anticommute. Thus, if we consider the processes of taking a vertex-defect particle in a circle and then creating the plaquette defects and pulling them apart, and then compare it to the process where the plaquette defects are pulled apart first and then the vertex-defect particle is taken in a circle that surrounds one of the two plaquette-defect particles (as in the left of Fig. 28.3), these two processes must differ by -1 due to the anticommutation of the two operators.

We can check that one accumulates exactly the same phase if we take a plaquette defect in a circle around a vertex-defect particle. Taking a plaquette defect around in a loop is a process created by a string of σ_z operators as shown in the right of Fig. 28.3 (see also right of Fig. 27.14). If there are no vertex defects enclosed in the loop, this process does not accumulate a phase because the string of σ_z operators around the loop is equivalent to the product of the V_α vertex operators enclosed (Eq. 27.16)—and in the ground state (the code space), the V_α operators are in the $+1$ state. However, if there is one vertex defect inside the loop, this means that one of the V_α operators is actually in the -1 state. In this case the phase accumulated by taking the plaquette defect around the loop is actually -1. So there is a phase of -1 for taking a plaquette defect around a vertex defect, showing nontrivial anyonic braiding with each other.

We have thus shown that taking an electric particle around a magnetic particle[9] accumulates a phase of -1. In the language of Chapter 13 we

[9]This statement is independent of which defect we call electric and which defect we call magnetic!

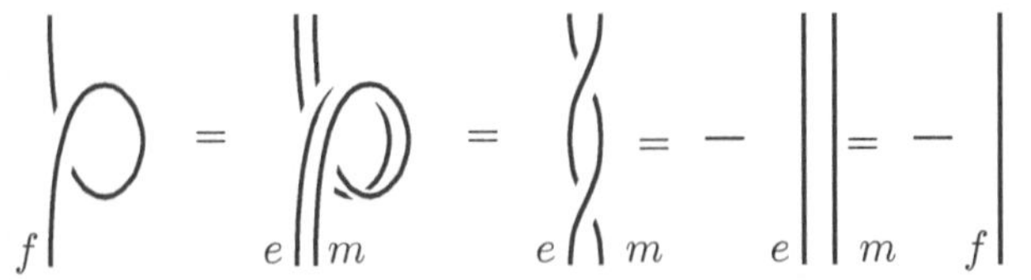

Fig. 28.4 The f particle is a fermion since its twist factor is $\theta_f = -1$. We can derive this by using $f = e \times m$ and using the fact that e braiding around m gives a -1 sign.

[10]Semions are particles that accumulate $\pm i$ under exchange (see for example, Section 18.1.2) so wrapping one semion all the way around another accumulates a phase of -1. I personally do not like to use the term "relative semion" since the toric code is mostly unrelated to the semion theory. Nonetheless, this language is distressingly common.

have just shown that

$$R_f^{em} R_f^{me} = -1 \quad . \tag{28.4}$$

Due to this -1 phase from taking e around m, or m around e, sometimes the e and m particles are called *relative semions*.[10]

28.2.2   Properties of f, the Fermion

Since f is made up of an m bound to an e, we can derive the properties of f from our knowledge of the properties of m and e. Braiding e around f is equivalent to taking e around both e and m. Since e around e accumulates no phase, and e around m accumulates a phase of -1, we conclude that taking e around f accumulates a phase of -1. Similarly, taking m around f also accumulates a phase of -1. Taking f around f is then equivalent to taking e and m both around f, and since each of these processes accumulates a phase of -1, the braiding of f around f accumulates a phase of $+1$. As discussed in the case of e and m, this tells us that f is either a boson or a fermion. To determine which one, we must again determine the twist factor θ_f. Perhaps the simplest way to determine this is via the diagrams in Fig. 28.4, which show that $\theta_f = -1$, hence, f is actually a fermion (hence, the notation, "f" for fermion). The astute reader will realize that this manipulation is actually a special case of the ribbon identity Eq. 15.5 (compare Eq. 28.4).

28.3 S- and T-Matrices

We can summarize our findings about this anyon theory by stating the modular S_{ij} matrix, which lists the braiding result obtained by taking particle i around particle j, as shown in Fig. 7.14, and the T-matrix (Eq. 17.14), which simply lists the twist factors along the diagonal. Listing the particles in the order I, e, m, f, we can write S and T as

$$S = \frac{1}{\mathcal{D}} \begin{pmatrix} 1 & 1 & 1 & 1 \\ 1 & 1 & -1 & -1 \\ 1 & -1 & 1 & -1 \\ 1 & -1 & -1 & 1 \end{pmatrix} \qquad T = \begin{pmatrix} 1 & 0 & 0 & 0 \\ 0 & 1 & 0 & 0 \\ 0 & 0 & 1 & 0 \\ 0 & 0 & 0 & -1 \end{pmatrix} \tag{28.5}$$

where unitarity fixes the total quantum dimension $\mathcal{D} = 2$ in agreement also with the definition Eq. 17.11 that $\mathcal{D}^2 = \sum_i \mathrm{d}_i^2$. We can also check that the central charge (via Eq. 17.16) must be zero mod 8. We have seen this theory before in Section 18.5.1.

28.4 Charge–Flux Model

We can describe the statistics of the particles in the toric code using a charge–flux model somewhat analogous to Chern–Simons theory.[11] Here let us define

electric particle $= e =$ particle bound to 1 unit of electric charge.

magnetic particle $= m =$ particle bound to π units of magnetic flux.

fermion $= f =$ particle bound to 1 unit of electric charge and π units of magnetic flux.

It is easy to see that this charge and flux will correctly give the $+1$ and -1 phases accumulated from braiding particles.

28.5 $\mathbb{Z}_N$ Toric Code (Briefly)

To generalize the toric code Hamiltonian Eq. 28.1 to the case of the $\mathbb{Z}_N$ toric code discussed in Section 27.6, we need to find operators analogous to V_α and P_β that will favor $Q_\alpha = 0$ on the vertices and $A_\beta = 1$ on the plaquettes in the ground state. We choose to work with the following Hamiltonian[12]

$$H_{\mathbb{Z}_N \text{ toric code}} = -\Delta_v \sum_{\text{vertices } \alpha} \delta_{Q_\alpha,0} - \Delta_p \sum_{\text{plaquettes } \beta} \hat{P}_\beta \qquad (28.6)$$

where δ is a Kronecker delta function and

$$\hat{P}_\beta = \frac{1}{N} \sum_{p=0}^{N-1} (A_\beta)^p \quad . \qquad (28.7)$$

The operator $\hat{P}_\beta$ gives unity on a plaquette where the eigenvalue of A_β is 1, and gives zero otherwise (to see this recall that $(A_\beta)^N = 1$ so the eigenvalues of A_β can only be $e^{2\pi ik/N}$ for some integer k). Similarly, the Kronecker delta function gives unity on a vertex where $Q_\alpha = 0$ and gives zero otherwise.

The $N = 2$ case of this Hamiltonian is not quite identical to Eq. 28.1 for the conventional toric code. Whereas V_α and P_β in Eq. 28.1 have eigenvalues ± 1, the terms $\delta_{Q_\alpha,0}$ and $\hat{P}_\beta$ here have eigenvalues 0 and 1. However, we do have (for the $N = 2$ case) that

$$V_\alpha = \frac{1}{2}(\delta_{Q_\alpha,0} - 1) \qquad P_\beta = \frac{1}{2}(\hat{P}_\beta - 1)$$

and note that Eq. 28.1 already has the factors of $\frac{1}{2}$ out front of each term compared to that of 28.6. Thus, we see that Eq. 28.1 and Eq. 28.6 differ only by the addition of an unimportant constant for $N = 2$.

The ground state of the Hamiltonian Eq. 28.6 is any state where all $Q_\alpha = 0$ and all $A_\beta = 1$, that is, the states in the code space of the $\mathbb{Z}_N$ toric code. The defect energies are Δ_v for a vertex defect and Δ_p for a plaquette defect.

If we move vertex defects together, their value of Q_α add modulo N, where $Q_\alpha = 0$ corresponds to an unexcited vertex and all other values of Q_α are excitations (i.e. we have $\mathbb{Z}_N$ fusion rules). Similarly, if we move plaquette defects together, their eigenvalues of $A_\beta = e^{2\pi i k/N}$ with k an integer multiply with $A_\beta = 1$ being an unexcited plaquette and all other values being excited (again this is $\mathbb{Z}_N$ fusion rules). We can also construct quasiparticles that correspond to combinations of excited vertices and excited plaquettes. The most general particle type can then be notated as $[n, k]$ with $n, k \in \{0, \ldots, (N-1)\}$ by which we mean

$$[n, k] \rightarrow \begin{cases} Q_\alpha & = n \\ A_\beta & = e^{2\pi i k/N} \end{cases} \qquad (28.8)$$

thus, there are a total of N^2 particle species. If a particle has both nontrivial vertex quantum number and nontrivial plaquette quantum number it is sometimes called a *dyon*. The fusion rules for these particles are then given by

$$\begin{aligned} [n_1, k_1] \times [n_2, k_2] &= [(n_1 + n_2)\mathrm{mod}(N), (k_1 + k_2)\mathrm{mod}(N)] \\ &= [n_1 + n_2, k_1 + k_2] \end{aligned}$$

where in the second line addition is assumed to be modulo N. Note that these fusion rules imply we have an abelian anyon theory.

Following similar arguments to Section 28.2 it is easy to show that all of the vertex defects and all of the plaquette defects are bosons—and further, all vertex defects braid trivially with all vertex defects and all plaquette defects braid trivially with all plaquette defects. However, given Eqs. 27.23 and 27.24 it is clear that the vertex defects braid nontrivially with the plaquette defects, which also gives nontrivial braiding properties to the dyons. In particular, these two equations showed that taking a $Q_\alpha = +1$ charge clockwise around a $A_\beta = e^{2\pi i/N}$ plaquette gives a phase of $e^{2\pi i/N}$ which we can rewrite as R-matrices[13]

$$R^{[0,1],[1,0]}_{[1,1]} R^{[1,0],[0,1]}_{[1,1]} = e^{-2\pi i/N}$$

[13]The exponent on the right has a minus sign because the R-matrix is defined as the phase associated with a *counter*clockwise exchange.

and more generally we can similarly derive the phase for taking an $[n_1, k_1]$ particle counterclockwise all the way around an $[n_2, k_2]$ particle

$$R^{[n_1,k_1],[n_2,k_2]}_{[n_1+n_2,k_1+k_2]} R^{[n_2,k_2],[n_1,k_1]}_{[n_1+n_2,k_1+k_2]} = e^{-2\pi i(n_1 k_2 + n_2 k_1)/N} \qquad (28.9)$$

where addition is assumed to be modulo N. Further, using the ribbon identity Eq. 15.5 we obtain

$$\theta_{[n,k]} = e^{-2\pi i n k / N} \quad .\tag{28.10}$$

Using either Eq. 28.9 directly to evaluate the relevant diagram, or using Eq. 17.20, we can obtain the S-matrix for the $\mathbb{Z}_N$ toric code

$$S_{[n'k'],[n,k]} = \frac{1}{N} e^{2\pi i (nk' + n'k)/N} \quad .\tag{28.11}$$

We can write a charge–flux model description for each of these particle types by modeling the $[n, k]$ particle as n units of electric charge bound to $2\pi k/N$ units of magnetic flux.

Note that the $\mathbb{Z}_N$ toric code was introduced previously in Section 18.6 as one of the "prime" abelian anyon theories.

Chapter Summary

- Topologically ordered matter is matter that is described by a TQFT at low energy and long length scale.
- A Hamiltonian can be written such that the ground-state space is the code space of the toric code, thus making this error correcting code into topologically ordered matter.
- Physical errors in the code become quasiparticle excitations in the phase of matter. We can calculate the braiding properties of these particle types.

Further Reading:

- Again, there are many introductions to the toric code in the literature and on the web, but the original work by Kitaev [2003] is still very good.
- In Section 31.7 we discuss how the toric code (and all "untwisted" Kitaev quantum double models) can be viewed as a discrete gauge theory. A model essentially equivalent to the toric code (a $\mathbb{Z}_2$ gauge theory) was studied in depth in a well-known work by Fradkin and Shenker [1979] although they missed the crucial realization that the excitations have nontrivial braiding statistics!

Exercises

Exercise 28.1 Ground-State Degeneracy of Toric Code

(a) For any abelian TQFT with N particle types, determine the ground-state degeneracy on a g-handled torus using the methods of Section 8.4.

(b) Thus, determine the the ground-state degeneracy on a g-handled torus for the toric code.

Exercise 28.2 Toric Code is Abelian from Ground-State Degeneracies

Given that the ground-state degeneracy for the toric code on a g-handled torus is 4^g, show that this must be an abelian TQFT with four particle types. Hint: See Exercises 8.2 and 28.1.

Exercise 28.3 Twist of the Plaquette Defect

In Section 28.2 we explicitly calculate that the twist factor for a vertex-defect particle is trivial. Use an analogous technique to show that the twist factor for the plaquette-defect particle is also trivial.

Exercise 28.4 Checking the Verlinde Formula

Confirm the Verlinde formula Eq. 17.13 for the $\mathbb{Z}_N$ toric code.

Exercise 28.5 Twist Factor for Excitations of the $\mathbb{Z}_N$ Toric Code

Confirm Eq. 28.10.

Exercise 28.6 Central Charge of Toric Code

Show that the central charge of the $\mathbb{Z}_N$ toric code is zero modulo 8. In fact, it is precisely zero. How might we argue that?

Exercise 28.7 Modular Group and the $\mathbb{Z}_N$ Toric Code

Confirm the modular group relations (Eq. 17.17) for the $\mathbb{Z}_N$ toric code (the central charge is $c = 0$).

Exercise 28.8 Chern–Simons Theory of the Toric Code

Consider the action proposed in note 11 of this chapter. Add a coupling to I different currents j^I using a term $-\int d^3x \, j^{I,\alpha} \, a_\alpha^I$. Then, using an approach as outlined in Sections 5.1 or 5.3.2, show that this action corresponds to the charge–flux model described in Section 28.4.

Exercise 28.9 Toric code in three dimension

Given a cubic lattice in three dimensions with a spin one-half on each edge, we write a three-dimensional toric code Hamiltonian again in the form of Eq. 28.1. The plaquette term still acts on square two-dimensional plaquettes. However, the vertex term (which we can still write in the form of Eq. 27.1) now acts on all six edges incident on a given vertex.

(a) Show that this model has point vertex excitations.

(b) Show that isolated plaquette excitations do not occur, rather a series of plaquette excitations must form a closed loop on the dual lattice—that is, draw a closed loop on the dual lattice (with segments connecting centers of cubes) and every plaquette that this closed loop cuts through is excited.

(c) What phase is incurred if we braid a vertex excitation around one of these plaquette excitation loops?

(d) What is the ground state degeneracy of this model on a 3-torus (a cube with periodic boundary conditions in all directions)?

Robustness of Topologically Ordered Matter

Topologically ordered matter, matter that can be described as a TQFT, has a remarkable property known as topological robustness—the topological properties of the phase of matter are unchanged if the Hamiltonian is changed a small bit (as long as the gap to making excitations remains non-zero). We illustrate this property using the toric code, but generalizing to other types of topologically ordered matter is then fairly obvious.

29.1 Perturbed Hamiltonian

Let us start with the toric code Hamiltonian, and add some perturbing term

$$H = H_{toric\,code} + \delta H \qquad (29.1)$$

where $H_{toric\,code}$ is the toric code Hamiltonian defined in Eq. 28.1. In Eq. 29.1, δH is some small perturbing Hamiltonian which is a sum of arbitrary local terms.[1] The claim is that for small enough δH (but crucially not necessarily infinitesimally small!), the topological properties of the phase of matter—such as the fourfold degenerate ground state, the types of excitations, and their braiding statistics—will remain entirely unchanged in the large system size limit.[2] We focus first on the ground-state degeneracy, but the arguments for the other topological properties follow similarly.

29.1.1 Robustness of Ground-State Degeneracy

Rough Argument

The general idea is as follows: when δH is turned on, it can be treated in perturbation theory order by order. If δH is small enough, then higher orders in perturbation theory are successively less important. Now at some Mth order in perturbation theory we can consider M applications of δH to the unperturbed system. Each application of δH, which is some arbitrary local operator, can create, annihilate, or move some quasiparticle excitations. The key point is that in order to break the ground-state degeneracy in any way, *which is equivalent to causing a logical error in the toric code when used as a quantum memory*, a quasiparticle must be taken all the way around a nontrivial (non-contractible) cycle of the

[1]An operator is *local* if it acts on a finite number of spins that are all near each other—meaning within some fixed finite distance as we take the thermodynamic limit. A term that would not be local would be a product of all of the spins in the system, or a product of all the spins in a string all the way around a nontrivial cycle of the torus.

[2]This statement should become exponentially more accurate as the size of the system is made larger and larger.

torus. For a very large torus this can only happen at extremely high order in perturbation theory, and hence, any splittings are only exponentially small in the size of the system.

In More Detail

Let us now make this argument more rigorous. It is easiest to choose a particularly simple form for δH to work with. However, once we understand the physical principle, we will realize that the actual form we choose doesn't matter and the results hold much more generally. For simplicity, let us choose δH to be the sum of σ_x over all spins

$$\delta H = J \sum_i \sigma_x^{(i)} \tag{29.2}$$

[3]As we pointed out in Section 28.1 there are two excitation gaps Δ_v for vertex defects and Δ_p for plaquette defects. The overall excitation gap of the system is the minimum of these two. For simplicity we can just set them equal to each other.

where J is a parameter having dimensions of energy that is assumed to be small compared to the excitation gap Δ of the toric code.[3]

We treat δH in a Brillouin-Wigner perturbation expansion[4] which "integrates out"[5] excitations and leaves us with an effective Hamiltonian within the previously degenerate ground-state space (see Section 29.3 for detailed derivation of this perturbation expansion). The new effective Hamiltonian within this previously degenerate ground-state space is given by

[5]The commonly used phrase "integrate out" is from field theory language where removal of degrees of freedom is achieved by doing functional integrals.

$$
\begin{aligned}
H_{pn}^{\text{eff}} &= E_0 + \langle p|\delta H \frac{1}{1 - G\delta H}|n\rangle \\
&= E_0 + \langle p|\delta H|n\rangle + \langle p|\delta H\, G\, \delta H|n\rangle + \langle p|\delta H\, G\, \delta H\, G\, \delta H|n\rangle + \dots
\end{aligned} \tag{29.3}
$$

where $|p\rangle$ and $|n\rangle$ are states in the ground-state space of the unperturbed model, and E_0 is the unperturbed ground-state energy. Here, G is the Green's function

$$G(E) = \sum_{n \in \text{excited}} \frac{|n\rangle\langle n|}{E - E_n^0}$$

and the sum is over (non-ground-state) eigenstates $|n\rangle$ of the unperturbed model whose energies are E_n^0. Note that one must plug in the eigenenergies E of H^{eff} into $G(E)$ to find self-consistent solutions.[6]

[6]This self-consistency requirement is what makes Brillouin-Wigner perturbation theory more complicated than the conventional Rayleigh–Schrödinger perturbation theory that most people learn in quantum mechanics courses. However, the Brillouin-Wigner approach is used here because it allows us to derive an effective Hamiltonian.

It is crucial to note that each factor of G has an energy denominator on the order of the gap Δ or larger, and each factor of δH has an energy numerator on the order of the coupling constant J. Thus, the expansion shown in Eq. 29.3 is actually an expansion in the ratio J/Δ.

Our claim is that, to very high order in this expansion, the effective Hamiltonian within this ground-state space will remain proportional to the identity matrix. That is, the ground-state space does not split at all,

---

[4]Brillouin-Wigner perturbation theory can be applied to any perturbed topological model and one will reach similar conclusions about the robustness of the topological properties. However, for models like the toric code, which have a number of special properties, including equally spaced eigenenergies, other perturbative methods exist that, in practice, are much easier to carry out. For example, for the model discussed here, the toric code in a magnetic field, it has been possible to expand to tenth order or more (Dusuel et al. [2011]; Vidal et al. [2009]), which would be unimaginably complicated with the Brillouin-Wigner approach. A detailed discussion of the method is given by Schulz [2014].

as we previously claimed. To see why the ground-state space does not split, let us consider the lowest few orders of the perturbation expansion. The first term is $\langle p|\delta H|n\rangle$. Here, $|n\rangle$ and $|p\rangle$ are states in the ground-state space of the unperturbed Hamiltonian. Each term in δH applies a σ_x, thus creating two vertex defect excitations. Since $\delta H|n\rangle$ is an excited state and $\langle p|$ is an unexcited state, there is zero overlap between the two.

The next-order term of the perturbation expansion is $\langle p|\delta H\, G\, \delta H|n\rangle$ such that δH is applied twice. Two applications of δH can create two pairs of quasiparticles in different places; or the first δH could create one pair and the second δH could be applied to a neighboring spin to pull the two quasiparticles further apart (as in the left of Fig. 27.10). In either case the matrix element will still be zero since there would be no overlap between the excited state $\delta HG\delta H|n\rangle$ and the unexcited state $\langle p|$. However, there is a third process to consider, where the first δH acts on edge i to create a pair of quasiparticles and then the second δH, acting on the same edge i, annihilates the quasiparticles. In this case we return again to the ground state that we started in, and we have a non-zero overlap (the value of the matrix element is $J^2/(E - E_{excited}) \sim J^2/\Delta$). Crucially, it does not matter which of the four ground states $|n\rangle$ we started in as we are always returned to the same ground state we started with when the quasiparticles are annihilated again. So the contribution to H^{eff}_{pn} is just $\delta_{pn}J^2/(E - E_{excited})$, independent of which ground state we are in, meaning this does not cause any splitting of the ground-state degeneracy.

The argument continues similarly at higher orders of perturbation theory. Each application of δH creates, annihilates, or moves quasiparticles. (Considering the form of δH in Eq. 29.2, all quasiparticles are vertex defects in this simplified argument.) In order to get a non-zero matrix element, the successive applications of the δH operators must return the system to the ground-state space at the last step—that is, any quasiparticles that are created must then be reannihilated. If all the quasiparticles are reannihilated, the value of the matrix element is simply J for each application of δH with the Green's functions contributing energy denominators corresponding to the energies of the (necessarily) excited intermediate states. When all of the quasiparticles are reannihilated at the last step, the system returns to exactly the same ground state it started from *unless* a quasiparticle has gone all the way around the nontrivial cycle of the torus—a case we address in a moment.

If we assume that no quasiparticles have gone all the way around a nontrivial cycle of the torus, then H^{eff} must be proportional to the identity. Each ground state returns to the same ground state, and the matrix elements are independent of which ground state we started in. In this case the ground-state degeneracy does not split. However, if a quasiparticle does go around a cycle of the torus (as in Fig. 27.11) then the final ground state is different from the initial ground state, meaning that there is an off-diagonal contribution to H^{eff} (compare Eq. 27.15), and H^{eff} is no longer proportional to the identity matrix.

Diagonalization of H^{eff} then gives a splitting of the previously degenerate ground state.

Crucially, for a torus whose (smallest) nontrivial cycle length is L, it requires at least L applications of δH to move a quasiparticle around the cycle. This means that only at Lth order in perturbation theory will we find off-diagonal terms in the effective Hamiltonian H^{eff}. Since each δH is of size J and each order in perturbation theory comes with a Green's function G with energy denominator on the order of the gap Δ, then the size of these off-diagonal terms is given by $\sim (J/\Delta)^L$ which is exponentially small in the size of the system, assuming $J/\Delta \ll 1$. Thus, the ground-state degeneracy only splits by an exponentially small amount, with the splitting going to zero as the size of the system gets large[7].

The argument is quite similar for other possible forms of δH. For example, let us consider the perturbation

$$\delta H = J \sum_i \sigma_z^{(i)} \quad . \tag{29.4}$$

Very similarly, here each application of δH creates, annihilates, or moves a quasiparticle; only this time the quasiparticles are plaquette defects. Again, as long as no quasiparticles are moved around a nontrivial cycle of the torus, the resulting contributions to H^{eff} are proportional to the identity. However, if a plaquette quasiparticle is moved around a nontrivial cycle (as in Fig. 27.15) then there will be a contribution to H^{eff}, which is different for the different ground-state sectors (compare Eq. 27.17), thus breaking the ground-state degeneracy. As with the argument for the vertex defects, the splitting of the ground-state degeneracy only occurs at order L in perturbation theory (with L the length of the smallest nontrivial cycle of the torus) and is thus exponentially small in the size of the system.[7]

It is clear that this general argument is not specific to the particular form of δH we have chosen. In order to have such an exponentially protected ground-state degeneracy, δH need only be local (see note 1), that is, it can be any sum of terms, each of which operates on a finite set of spins near each other. In particular, we need to have a situation where the number of applications of δH required to cause a quasiparticle to go around a nontrivial cycle of the torus should scale with L, the linear size of the system.

It is important to realize that in the presence of the perturbing Hamiltonian, the ground states have virtual excitations mixed into them.

---

[7]One might worry about the convergence of the perturbation theory. (Rigorous perturbation arguments using a different technique have been given by Bravyi et al. [2010].) Obviously, if the perturbation is large enough the argument no longer makes sense. For example, if an excited eigenstate shifts down in energy due to the perturbation and becomes the new lowest energy eigenstate, then the argument fails. Indeed, it is generally assumed that the perturbative expansion should hold until this point where the excitation gap closes. However, one might also worry that the perturbative expansion may fail to converge since, while each successive order in the expansion is reduced by (J/Δ), there are also many many more terms to consider in each higher order of perturbation theory. At least for small enough J/Δ (but not requiring infinitesimally small J/Δ) this issue does not cause trouble, as discussed in Exercise 29.1.

(Sometimes one says that the ground state is "dressed" with virtual excitations.) If one were to measure the plaquette and vertex operators, one would find some number of them are in the -1 state rather than the $+1$ state.[8] However, if δH is small, the density of these virtual excitations is low, and such excitations occur in local annihilable pairs which, in the language of error correcting codes, are easily correctable. In the language of this Hamiltonian, these do not change the topological properties of the phase of matter.

29.1.2 Quasiparticles

One can further ask what happens to the nature of the quasiparticles when a perturbation is applied to the Hamiltonian. In the absence of the perturbing Hamiltonian, the quasiparticles have well-defined topological properties—in particular, they have nontrivial braiding statistics (which we worked out for the toric code in Chapter 28). We claim that, as long as we are considering physics at long length scales, these topological properties will remain unchanged under small perturbations.

Rough Idea

The general strategy is again to perform a perturbation expansion. We will find that in the presence of the perturbation, the quasiparticles will develop a surrounding cloud of virtual excitations[9] (this is often called a "dressing"). The quasiparticles are then no longer point particles, but rather develop a length scale that becomes longer and longer as the perturbation is increased. However, the braiding properties of the quasiparticles will remain unchanged, as long as the particles remain far enough apart that their dressing clouds do not intersect.

In More Detail

For simplicity let us focus on the perturbing Hamiltonian Eq. 29.2, which only makes vertex defects. If the perturbing Hamiltonian is small, we will mix into the ground state a low density of virtual excitations, most of which will simply be two vertex defects separated by a single edge (see Exercise 29.1).

The perturbation we have chosen (Eq. 29.2) commutes with the plaquette operators P_β, so that the plaquette excitations can still be described as simply a plaquette where $P_\beta = -1$. That is, given any eigenstate of the system (with or without the perturbation Eq. 29.2) one may obtain another eigenstate by applying σ_z to an edge, which then flips the eigenvalue of P_β on the neighboring plaquettes. The lowest-energy configuration (as in the unperturbed toric code) is when all of the plaquettes have eigenstate $P_\beta = +1$ and we consider $P_\beta = -1$ to be a plaquette excitation.

Let us now start in the perturbed ground state and add a pair of plaquette excitations. We can create these particles and move them around by applying σ_z to edges, as in Fig. 27.14. As we move the

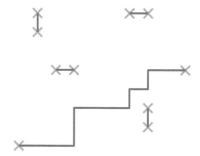

Fig. 29.1 When the toric code is weakly perturbed, there will be a low density of pairs of defects mixed into the ground state. However, the quasiparticles that are separated by long distances can still be clearly identified. In this picture we are considering only vertex defects corresponding to a perturbation of the form of Eq. 29.2.

plaquette excitations through the system, they will cross through some of the virtually excited red edges—that is, in only some terms of the superposition that make up the wavefunction does the plaquette defect cross through red edges. These terms of the wavefunction incur a minus sign due to the plaquette-defect particle motion. We can think of these signs as being the result of the plaquette defect swimming through the soup of pairs of virtual vertex-defect particles.

Now suppose we have added a pair of vertex quasiparticles that have been pulled far apart from each other.[10] If we measure every vertex, we would obtain a picture like Fig. 29.1. Here, the isolated quasiparticles that are not part of a local pair can still be clearly identified. They are connected together by a long red string. We should remember, as discussed in Section 27.4.1, that the red string is not actually measurable and its actual position is unknown (or we can say the position of the red string fluctuates)—only the endpoints are measurable. Note that the position of the endpoints may also fluctuate due to virtual excitation pairs that may appear at the end position to extend or shorten the long red string. This makes the quasiparticle at the end of the string look like a cloud of finite size, rather than a point particle.

Now let us repeat the experiments of creating and moving plaquette defects, but having first introduced the two additional quasiparticles separated by a long distance, as in Fig. 29.1. Now, as we move the plaquette defect, the wavefunction again picks up phases associated with swimming through the thick soup of virtual excitations. However, now, if it crosses over the long red line as well, this will then create a -1 braiding phase. Thus, the phase of braiding the plaquette defect around the vertex-defect quasiparticle is independent of the fact that there are lots of pairs of virtual vertex-defect particles.

We should be cautious, however, if we pass the plaquette defect close to the end of the long red string—that is, close to the position of the (not-virtual) vertex-defect quasiparticle. Because of the fluctuations of the position of the quasiparticle, it may not be possible to determine if the plaquette defect has gone around the vertex defect or not.

In this discussion we have chosen to work with the perturbing Hamiltonian Eq. 29.2. We could just as well have chosen the perturbation Eq. 29.4, which would make virtual plaquette-defect excitations, fluctuates the position of the plaquette-defect particle, but leave the vertex-defect excitations unchanged. The argument would be almost exactly the same. Again, we find a braiding phase of -1.

Treating the case where the perturbation creates both vertex and plaquette excitations is a bit harder to do rigorously, and some different techniques become useful (which we will not discuss here). However, generally the same principles apply: the quasiparticles develop some finite size (which grows larger for larger perturbation), but as long as the quasiparticles are braided at distances much larger than this size, the braiding phases are unchanged by the perturbation.

[10]When we add a perturbing Hamiltonian like Eq. 29.2 we can give the quasiparticles dynamics, allowing them to move from their initial positions. While in principle there is nothing wrong with this, it does make it harder to study the quasiparticles if they keep moving. To avoid this problem we can imagine a "pinning potential" to trap the quasiparticles and prevent them from moving from whatever position we put them at. For example, we can lower the prefactor of V_α in the Hamiltonian Eq. 28.1 only at one vertex, which makes it energetically favorable for the vertex defects to sit at that position.

29.2 **Topologically Ordered Matter**

29.2.1 **Importance of Rigidity**

The fact that the quasiparticle statistics and ground-state degeneracy are unchanged when the Hamiltonian is modified a small bit should not be surprising to us given the ideas of rigidity we have run into in Sections 9.3 and 13.3: It is not possible to perturb the properties of a TQFT by a small amount and maintain a consistent TQFT. As we perturb the Hamiltonian with any local operators, to have any change in the long length scale properties of the TQFT that results, the system needs to go through a phase transition[11] In other words, the topological properties are a property of the entire phase of matter.

29.2.2 **The Notion of Topological Order**

The type of protection from small perturbations that we have just discovered, particularly in the discussion of Section 29.1.1, is the basis for a very useful definition of topological order. Let us assume we have multiple degenerate ground states on a surface with non-zero genus (e.g. a torus, two-handled torus, etc.), which we call $|\psi_i\rangle$ with $i = 1, \ldots, M$ with $M > 1$ the ground-state degeneracy. We often define topological order to be the property that

$$\langle \psi_i | \hat{O} | \psi_j \rangle = C_{\hat{O}} \, \delta_{ij} \qquad (29.5)$$

where $\hat{O}$ is *any* local operator and where $C_{\hat{O}}$ is a constant depending on the particular local operator $\hat{O}$ we are considering. Finite-size corrections to this result, if any, must be exponentially small in the size of the system. In other words, the multiple ground states look just like each other locally but are mutually orthogonal. Note that since any perturbation we add to the Hamiltonian is assumed local, this condition guarantees that the ground-state degeneracy remains robust under perturbations.

This definition of topological order is effectively the same as saying that the system is described by a TQFT. While a TQFT must have the property Eq. 29.5, it is not immediately obvious that having this property is enough to conclude that we have a TQFT. However, there are no systems known with this property that are not essentially TQFTs.

29.2.3 **Defining a Topological Phase of Matter**

The classic definition of a *phase of matter* is a region of thermodynamic parameter space where certain (or even most) physical properties of a system are robustly unchanged. For example, water is one phase of matter, robustly characterized as being liquid over a large range of temperature and pressure, whereas another phase of matter of the same system is ice, which is robustly characterized as being solid.[12]

[11]Note that there is a detailed discussion of continuous phase transitions between different TQFTs given in Chapter 25. However, it is also possible that one has a transition between a TQFT and some completely different phase of matter that may not be topological.

[12]The example of water/ice is actually not a particularly good example for our purposes. All of the physics discussed in this book is zero temperature—phase transitions can occur as a function of some tuning parameter in the Hamiltonian. Such zero temperature phase transitions are known as *quantum phase transitions*. A nice example of a quantum phase transition is the zero temperature phase transition between solid Helium and liquid Helium as a function of pressure.

For our consideration of topological quantum systems, we ask how we should best characterize different possible phases of our systems. One obvious possibility is to characterize each topologically ordered phase of matter in terms of its TQFT in the long length scale limit. Recall that the perturbed toric code only has the same topological properties as the unperturbed toric code (same ground-state degeneracy, same quasiparticle statistics) if we look at long length scales (large system, excitations far apart). Indeed, characterization in terms of the long length scale TQFT is one good way to define a topological phase of matter.

However, there is also a more abstract (and topological!) way to define a phase of matter, which can be invoked much more generally—even in cases where we do not have a nontrivial TQFT. Here, we use as inspiration the perturbation theory argument in the Section 29.1. We make the following definition:[13]

[13]See Section 35.1 for a refinement of this definition.

> **Definition of a Topological Phase of Matter:** *Two (zero temperature) gapped states of matter are in the same topological phase of matter if and only if you can continuously deform the Hamiltonian to get from one state to the other without closing the excitation gap.*

This sort of definition can obviously be used much more generally to distinguish different phases of matter. Further, this definition fits with our intuition about topology:

> *Two objects are topologically equivalent if and only if you can continuously deform one to the other.*

29.3 Appendix: Brillouin-Wigner Perturbation Theory

Here we discuss a slight variant of Brillouin-Wigner perturbation theory designed for the case of interest where we have a degenerate ground state and a gap to excitations Δ.

We assume a Hamiltonian $H = H_0 + \delta H$. The orthonormal eigenstates of H_0 are $|n\rangle$ with energies E_n and the eigenstates of H we call $|\tilde{n}\rangle$ with energies $E_{\tilde{n}}$. The degenerate ground-state space of H_0 we call S (of some dimension d) and all other eigenstates of H_0 are in the space $\bar{S}$. The energy of the ground-state space of H_0 is E_0.

Our strategy will be to "integrate out" the excited states leaving ourselves with an effective Hamiltonian within the ground-state space S.

Projectors onto the unperturbed ground-state space and unperturbed excited-state space are

$$P = \sum_{n \in S} |n\rangle\langle n| \quad , \quad Q = 1 - P = \sum_{n \in \bar{S}} |n\rangle\langle n| \quad .$$

As we turn on the perturbation, the d orthogonal states in the space S will generally split in energy (although for our TQFT we expect this splitting to be exponentially small in the system size if δH is local). Our approach remains valid for perturbation strengths small enough that the splitting is much smaller than the order of the gap Δ.

Let us start with a d-dimensional basis of perturbed wavefunctions $|\tilde{m}\rangle$ with energies $E_{\tilde{m}}$ in the perturbed ground-state space S. Let us project these states into the original ground-state space

$$P|\tilde{m}^u\rangle = |\psi_{\tilde{m}}\rangle \tag{29.6}$$
$$= \sum_{n \in S} C_{n\tilde{m}}|n\rangle \quad . \tag{29.7}$$

We have marked the state $|\tilde{m}\rangle$ with a superscript u because we will work with an unnormalized state on the left so that $|\psi_{\tilde{m}}\rangle$ is normalized to unity $\langle\psi_{\tilde{m}}|\psi_{\tilde{m}}\rangle = \sum_{n \in S}|C_{n\tilde{m}}|^2 = 1$.

We write the Schrödinger equation for the perturbed state as $(H_0 + \delta H)|\tilde{m}^u\rangle = E_{\tilde{m}}|\tilde{m}^u\rangle$ as

$$(E_{\tilde{m}} - H_0)|\tilde{m}^u\rangle = \delta H|\tilde{m}^u\rangle \quad . \tag{29.8}$$

Applying Q to the left and dividing through by $E_{\tilde{m}} - H_0$ we obtain

$$Q|\tilde{m}^u\rangle = G\,\delta H|\tilde{m}^u\rangle$$

where G is the Green's function

$$G = \sum_{n \in \bar{S}} \frac{|n\rangle\langle n|}{E_{\tilde{m}} - E_n} \quad .$$

Note that this is non-singular as long as the energy $E_{\tilde{m}}$ of the states that were in the ground-state space do not come close to the energies of the unperturbed excited states E_n.

We then write

$$|\tilde{m}^u\rangle = P|\tilde{m}^u\rangle + Q|\tilde{m}^u\rangle = |\psi_{\tilde{m}}\rangle + G\delta H|\tilde{m}^u\rangle$$

which we can rewrite as

$$(1 - G\delta H)|\tilde{m}^u\rangle = |\psi_{\tilde{m}}\rangle$$

or

$$|\tilde{m}^u\rangle = \frac{1}{1 - G\delta H}|\psi_{\tilde{m}}\rangle \tag{29.9}$$
$$= |\psi_{\tilde{m}}\rangle + G\delta H|\psi_{\tilde{m}}\rangle + G\delta H\,G\delta H|\psi_{\tilde{m}}\rangle + \dots \quad .$$

Note in particular that only the first term in the series is within the space S, and all other terms are entirely outside of this space.

To find the energy of the state $|\tilde{m}^u\rangle$ we apply an arbitrary state $\langle p|$ within the space S to the left of Eq. 29.8 to obtain

$$
\begin{aligned}
(E_{\tilde{m}} - E_0)\langle p|\tilde{m}^u\rangle &= \langle p|\delta H|\tilde{m}^u\rangle = \langle p|\delta H \frac{1}{1 - G\delta H}|\psi_{\tilde{m}}\rangle \\
(E_{\tilde{m}} - E_0)C_{p\tilde{m}} &= \sum_{n\in S} \langle p|\delta H \frac{1}{1 - G\delta H}|n\rangle C_{n\tilde{m}} \quad .
\end{aligned}
$$

This is an eigenvalue/eigenvector problem within the space S

$$
(E_{\tilde{m}} - E_0)C_{p\tilde{m}} = \sum_n H^{\text{eff}}_{pn} C_{n\tilde{m}}
$$

where H^{eff} is the effective Hamiltonian within the low-energy subspace

$$
H^{\text{eff}}_{pn} = \langle p|\delta H \frac{1}{1 - G\delta H}|n\rangle \quad .
$$

Note that this is a bit trickier than just a simple eigenvalue problem because G is implicitly dependent on $E_{\tilde{m}}$. This self-consistency has a few implications, including the fact that the eigenvectors $C_{n\tilde{m}}$ or equivalently $|\psi_{\tilde{m}}\rangle$ are not orthogonal. It is the eigenvectors $|\tilde{m}\rangle$ of the full Hamiltonian that are orthogonal, and $|\psi_{\tilde{m}}\rangle$ are normalized projections of $|\tilde{m}\rangle$.

Chapter Summary

- The topological properties of the toric code (and any topologically ordered matter) is unchanged when the Hamiltonian is perturbed a small amount, as long as we consider properties at long length scale.
- Topological order can be defined as having protected ground state degeneracy on manifolds with nontrivial cycles. The degenerate states should not be split by any local operators.

Further Reading:

- The notion of topological order was introduced by Wen [1990].
- A nice exposition of the idea and its relation to quantum information is given in Zhang et al. [2019a].

Exercises

Exercise 29.1 Convergence of Perturbation Expansion

Consider a large toric code system with N edges, with the Hamiltonian Eq. 29.1 and a simple perturbation Eq. 29.2.

(a) If all the vertices are simultaneously measured, show that, to second order in perturbation theory, the probability of finding no defects is

$$P_{0ex} = \frac{1}{1 + (N/4)[J/\Delta)]^2}$$

with $\Delta = \Delta_v$ being the energy of a vertex defect. Show that the probability of exactly one edge being flipped (i.e. finding two vertex defects, separated by one edge) is $1 - P_{0ex}$. Show that for small $N(J/\Delta)^2$ the probability that a given edge is flipped is

$$p_{edge,ex} = \frac{1}{4}(J/\Delta)^2 \quad .$$

(b) Assume that the probability of a particular edge being flipped (meaning that σ_x has acted on that edge) is small. In this case we can ignore the possibility of a defect being created and then moved by one step, as in the left of Fig. 27.10. That is, we are assuming that all applications of σ_x are on isolated edges far from other defects. We can then think of the system as being N independent uncoupled edges, each of which is a two-level system. Show that the probability of a given edge being flipped is exactly

$$p_{edge,ex} = \frac{2 + (J/\Delta)^2 - 2\sqrt{1 + (J/\Delta)^2}}{2 + 2(J/\Delta)^2 - 2\sqrt{1 + (J/\Delta)^2}} \sim \frac{1}{4}(J/\Delta)^2 + \dots \quad .$$

Hint: Think about a two level system!

This approximation corresponds to resumming a subset of the terms in the perturbation expansion (out to infinite order) corresponding to isolated pairs of vertex defects. At least for small J/Δ the neglect of other terms is justified by the low density of defects.

Abstracting the Toric Code: Introducing the Tube Algebra

In Section 27.5 we pointed out that we can build the toric code on any lattice (see also Fig. 30.1). Indeed, in many respects it is easiest to dispense with the lattice altogether. This simplifies a lot of the thinking and allows us to more easily generalize the model to describe more complex TQFTs, which we do in this chapter as well as in Chapters 31–33.

The basic idea here is that the toric code can be viewed as simply a gas of fluctuating loops or strings—without the need to tie it to the underlying lattice. The rules that describe this loop gas are planar diagram rules—the same kind of diagrammatic rules we have been using all along! If we want to put the model back on a lattice at the end of the day, we can do this, but in fact many of the manipulations are simpler without the burden of the lattice.

Before making the transition to the continuum, it is useful to first work with a trivalent lattice where only three edges meet at a vertex. For example, we could take a honeycomb as shown in the top of Fig. 30.1. This eliminates situations where four lines intersect at a corner, as shown in the bottom of Fig. 30.1. We need not even work with a regular lattice (see Section 27.5), but it is often convenient. The rule that all vertices have an even number of blue edges coming into them (Rule 1 from Section 27.3) now means that the allowed configurations in the toric code ground state are *non-intersecting* loops as compared to the intersecting loops that are allowed, for example, on a square lattice.

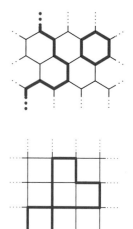

Fig. 30.1 The toric code loop gas can be constructed on any lattice. It is useful to choose a lattice where only three edges meet at a vertex, such as the honeycomb shown on the top. This eliminates the possibility of four lines intersecting at a single corner as shown with the square lattice on the bottom.

30.1 Toric Code as a Loop Gas

We start by abstracting the toric code to simply a gas of fluctuating non-intersecting loops—no longer paying attention to a lattice. An example of a loop gas configuration is shown in Fig. 30.2

We can write the toric code wavefunction (see Eq. 27.10) in the form of

$$|\psi\rangle \quad \sim \quad \sum_{\substack{\text{all loop configs that can be ob-} \\ \text{tained from a reference loop con-} \\ \text{fig via allowed moves}}} |\text{loop config}\rangle \qquad (30.1)$$

where the allowed moves are inspired by the moves we can make on a lattice. We list these allowed moves in Fig. 30.3.

Fig. 30.2 A loop gas in two dimensions. Note the relationship to our loop gas diagrams discussed in Section 18.1.

- **Move 1: "Isotopy."** Meaning any smooth deformation of lines.

On the lattice:

Off the lattice:

- **Move 2: Adding and Removing Contractable Loops.**

On the lattice:

Off the lattice:

- **Move 3: "Surgery."** Meaning reconnection of loops.

On the lattice:

Off the lattice:

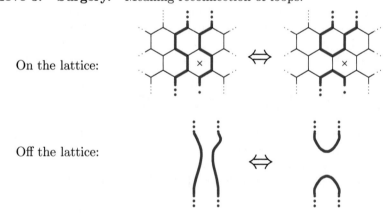

Fig. 30.3 The moves we can make by flipping over a plaquette on the lattice, and their interpretation in the continuum off the lattice. In the diagrams on the lattice, each move is achieved by flipping over the plaquette marked with ×.

The allowed moves off the lattice remind us of the diagrammatic "skein"-like rules we have been working with, starting in Chapter 2! Here, we are considering planar diagrams only (no over- or under-crossings, as in Chapter 12) and the rules can be summarized as

$$\bigcirc = 1 \qquad\qquad (30.2)$$

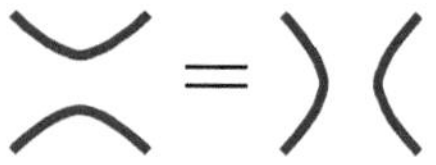

$$\overset{\smile}{\smile} =)(\qquad\qquad (30.3)$$

where we are assuming isotopy-invariant diagrams (i.e. all smooth deformations of diagrams are allowed). In fact, these diagrammatic rules are exactly the $d = +1$ version of the $\mathbb{Z}_2$ fusion-rule planar diagram algebra discussed in Section 18.1.1. As discussed there, the value of every valid loop diagram is unity, and this corresponds to the fact that all wavefunctions in Eq. 30.1 are added together with the same $+1$ coefficient.

Note that in the definition of the wavefunction Eq. 30.1, we have not written a normalization constant as we did in Eq. 27.10. This is because in a continuum model it is hard to be precise about how many different diagrams one can write if one can deform a diagram infinitesimally to make a new diagram. One way to understand this is to accept that, while we draw a continuum diagram, we really mean that the diagrams should be understood to be on a lattice (which we don't draw).

Another, perhaps better, way to interpret the sum in Eq. 30.1 is to think of all the terms that can be converted into each other via the three allowed moves to be an *equivalence class* of diagrams, and we never speak of the number of different diagrams in this class.

In the language of equivalence classes of diagrams there are exactly four equivalence classes of (ground-state) wavefunctions on a torus— corresponding to the evenness or oddness of the number of loops going around each nontrivial cycle. Representative (or reference state) loop configurations are shown in Fig. 30.4. These correspond to the wavefunctions described in Eq. 27.11. Note that the parity of the number of loops going around a given cycle is not changed by any of the moves in Fig. 30.3.

Fig. 30.4 Four representative (or reference state) loop diagrams corresponding to the four degenerate ground states of the toric code on a torus. Each figure represents a torus with the top and bottom edges identified and the left and right edges identified.

30.1.1   Preview of Coming Attractions: Generalizations of the Toric Code

We have thus found a relation between the toric code and the planar diagrammatic algebra that we spent so much time developing. In Chapters 31–33 we take the natural next step and, instead of using the simple $d = +1$ version of the $\mathbb{Z}_2$ planar diagram algebra, we use more complicated diagram algebras. We consider models where each edge can be labeled, not just with two possibilities (colored or not colored) as in the toric code, but rather with many different quantum numbers. We then consider planar diagram algebras described by any consistent solution of the pentagon equation. Each diagram we draw in the plane can then be evaluated and turned into a complex weight $W(\text{diagram})$, and we can then write ground-state wavefunctions (or we can interpret them as error-correcting codes!) in the form

$$|\psi\rangle \sim \sum_{\substack{\text{all diagrams that can be} \\ \text{obtained from a reference} \\ \text{diagram via allowed moves}}} W(\text{diagram}) \, |\text{diagram}\rangle \quad . \tag{30.4}$$

Using this approach we are able to generate a very wide range of topological phases of matter.

30.2   The Tube Algebra and Quasiparticle Excitations

We now study the properties of the quasiparticles that occur in the toric code, but within the language of a loop gas without reference to an underlying lattice. While we know what the answers are here (we have classified all of the quasiparticles in Chapter 28!) there is quite a lot to be learned by trying to develop a general technique that will work for other continuum diagrammatic models as well.

30.2.1   States on the Annulus

We want to describe the quantum number of some particular region of our system—that is, determine the net fusion product of all of the quasiparticles in the region. By our general locality principles (Section 8.2 for example), we should be able to determine the total quantum number of a region only by examining a loop around that region.[1] As such, we focus on an annulus surrounding the region of interest, as shown in Fig. 30.5. We assume that there are no quasiparticles in the annular region (gray region), so that the annulus is in one of its possible ground states. However, we allow for the possibility that there are quasiparticles in the center of the annulus (inner white region of Fig. 30.5).

In Fig. 30.6 we show three different topologically equivalent representations of the annulus. The one on the right is a cylinder, or tube, which

[1]For example, if we measure only a single blue line emanating from the enclosed region, we know that somewhere in the enclosed region there is an end of the blue line, hence, a quasiparticle defect of some sort.

Fig. 30.5 States of the annulus (gray region) can determine the total fusion channel of any quasiparticles surrounded by the annulus (white region).

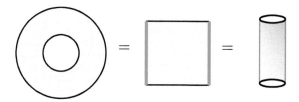

Fig. 30.6 Three different representations of (things topologically equivalent to) the annulus. In the middle figure the two orange sides are identified with each other.

will give the name "tube algebra" to the algebra of states we develop for the region of this annulus.

As with our discussion of states on the torus, the possible states of the annulus break up into sectors that cannot be converted into each other by the three moves given in Fig. 30.3. As in that case, it will be sufficient to describe one representative, or reference, configuration from each equivalence class.

The four equivalence classes we consider we notate as $|n, m\rangle$ with $n, m \in 0, 1$. The meaning of this notation is that the diagram on the annulus has n lines going from the inner edge of the annulus to the outer edge, and has m lines going around the annulus. While it may appear that we need to consider other possibilities (for example, lines going multiple times around the annulus), it will turn out that all diagrams can be reduced to one of these four possibilities.

Strictly speaking, in order to describe the states on the annulus we must first describe the states on the boundary of the annulus (i.e. do any blue lines intersect the boundary of the annulus, and if so, at which points?). However, it turns out that we can simplify our lives by considering only the simplest possible boundary conditions—the case where a single quantum number (in this case either a blue line or no blue line) is emitted from each boundary at one convenient fixed point.[2] Thus, in the toric code case, we need only consider the cases where either a single blue line, or no blue line, intersects each boundary.[3,4]

No Boundary Intersections

The first case to consider is the case where no blue lines intersect either boundary. There are then two possible equivalence classes of states, as shown in Fig. 30.7. The first we call $|0, 0\rangle$, meaning the empty state, the second we call $|0, 1\rangle$, meaning there is a blue line going around the annulus. We might think about putting more than one loop around the nontrivial cycle. However, we can always apply the moves from Fig. 30.3 to reduce the diagram to either zero loops or one loop around the cycle. An example of this is shown in Fig. 30.8.

[2] Qualitatively we can think of this as having grouped up all quantum numbers emitted from the boundary and fused them into a single quantum number at one point.

[3] Further, since there are no quasiparticles in the (gray) annular region, the parity of the number of blue lines is conserved going from the inner boundary to the outer boundary, so we cannot have a case where the number of blue lines intersecting the inner boundary has a different parity from the number of blue lines intersecting the outer boundary.

[4] It may seem that restricting our attention to cases with only a single quantum number emitted from the edge at one point is a drastic simplification. One (reasonably valid!) justification is that if put our system back on a lattice, our results should be independent of the particular lattice structure, so we can use an exceedingly simple lattice. In particular, we can choose to use a lattice that has only one edge pointing into the inner boundary and one edge pointing to the outer boundary (i.e. we are using a minimal "skeleton" decomposition of the annulus). A good (but not particularly easy) discussion of this type of simplification issue, along with a more rigorous argument, is given by Bullivant and Delcamp [2019].

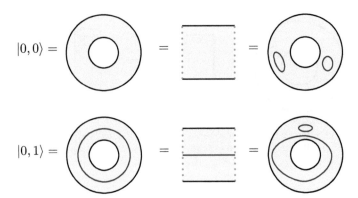

Fig. 30.7 The two equivalence classes of states on the annulus where no lines intersect the boundary. The pictures on the left are the reference states. The pictures in the middle are the same reference states as drawn on a square with the orange dotted sides identified. The pictures on the right show another state of the same equivalence class that can be turned into the reference state by application of the moves in Fig. 30.3.

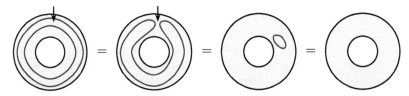

Fig. 30.8 Two loops around the annulus can be converted to no loops by using the moves in Fig. 30.3. In the first step a surgery move is used at the position of the arrow. In the second step we have isotopy (smooth deformation). Finally, we have loop removal.

One Blue Line Intersecting Each Boundary

The only other case[3] to consider is when one blue line intersects each boundary. We will choose the intersections to be at the 6-o'clock direction (south direction) of the boundary circles. There are again two equivalence classes of diagrams, which are shown in Fig. 30.9. The diagram that has a single line going in the vertical direction in the square representation of the annulus we call $|1,0\rangle$, and the diagram that loops around both directions we call $|1,1\rangle$. In each figure, on the right we show additional pictures in the same equivalence class, which can be reached by using moves from Fig. 30.3. To see that the double-twist figure on the far right of the first line, $|1,0\rangle$, is the same as the figure on the far left, one needs to perform surgery at the two points marked with the arrow in the third picture on that line. Similarly, to reach the figure on the far right of the second line, $|1,1\rangle$, one performs surgery at the arrow.

Despite the fact that $|1,0\rangle$ and $|1,1\rangle$ have blue lines intersecting the boundary of the annulus, this does not imply a quasiparticle at the boundary. We implicitly mean the annulus to be *part* of a larger system, and in this larger system, the blue lines would continue past the boundary.

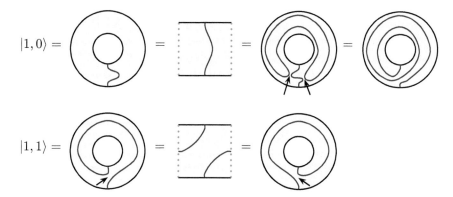

Fig. 30.9 The two equivalence classes of states on the annulus where one blue line intersects each boundary. The pictures on the left are the reference states. The pictures in the middle are the same reference states drawn on a square with the orange dotted sides identified. The pictures further right show other states of the same equivalence class that can be made into the reference states by application of the moves in Fig. 30.3. The arrows indicate places where we make surgery moves.

30.2.2   Composition of States

We can define a type of multiplication, or composition, of states on the annulus that corresponds to placing one annulus inside of another, and we denote this multiplication with a ∘. (Here, we are implicitly using the fact that we can stretch the annulus since we are considering a topological theory!). For example, if we write $|1,1\rangle \circ |1,0\rangle$ we mean we should put the $|1,1\rangle$ diagram on the inside annulus and the $|1,0\rangle$ on the outside annulus, as shown in Fig. 30.10, and we conclude that $|1,1\rangle \circ |1,0\rangle = |1,1\rangle$.

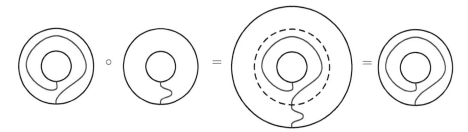

Fig. 30.10 Composition of annulus states. ($|1,0\rangle$ is put on the outside and $|1,1\rangle$ is put on the inside.) The result shows $|1,1\rangle \circ |1,0\rangle = |1,1\rangle$.

It is important to note that to multiply states, they must match on their boundaries.[5] Both $|0,0\rangle$ and $|0,1\rangle$ (Fig. 30.7) have no blue lines intersecting the boundaries, so they can be composed with themselves or with each other. Similarly, both $|1,0\rangle$ and $|1,1\rangle$ (Fig. 30.9) have one blue line intersecting each boundary, so they can be composed with themselves or with each other. However, one cannot compose $|0,0\rangle$ or $|0,1\rangle$ with $|1,0\rangle$ or $|1,1\rangle$ since they would not match on the boundary.

We can build up a multiplication table for this type of composition,

[5] Strictly speaking, we need to sum over all possible boundary conditions that can occur at the boundaries that have been joined (matching the boundary condition from the inner to the outer annulus). However, the justification used in note 4 allows us to ignore this complication and just work with the one reference configuration.

which we write out in Table 30.1. We mention in particular the two most nontrivial results. First, $|0,1\rangle \circ |0,1\rangle = |0,0\rangle$. This is just the statement that two loops around the nontrivial cycle can be annihilated to no loops, as in Fig. 30.8. The second, $|1,1\rangle \circ |1,1\rangle = |1,0\rangle$, is a bit more complicated. We can see this in several ways. First, if we compose the counterclockwise directed wrapping shown the furthest left in the lower line of Fig. 30.9 with the clockwise directed wrapping furthest right we obviously get a result with no net wrapping which is $|1,0\rangle$. Another way to see this is to compose two wrappings in the same direction, which will give a diagram like the far right of the top line of Fig. 30.9, which is also $|1,0\rangle$.

A way to summarize the multiplication table is via the formula[6]

$$|n,m\rangle \circ |n',p\rangle = \delta_{n,n'}\, |n,(m+p)\bmod 2\rangle \quad . \tag{30.5}$$

It is worth mentioning how compositions look if we work with rectangular diagrams or tube diagrams. In these cases, diagrams are simply stacked on top of each other to create new diagrams. An example of composing rectangular diagrams and composing tube diagrams is given in Fig. 30.11.

$\circ$	$\lvert 0,0\rangle$	$\lvert 0,1\rangle$	$\lvert 1,0\rangle$	$\lvert 1,1\rangle$
$\lvert 0,0\rangle$	$\lvert 0,0\rangle$	$\lvert 0,1\rangle$	$\bullet$	$\bullet$
$\lvert 0,1\rangle$	$\lvert 0,1\rangle$	$\lvert 0,0\rangle$	$\bullet$	$\bullet$
$\lvert 1,0\rangle$	$\bullet$	$\bullet$	$\lvert 1,0\rangle$	$\lvert 1,1\rangle$
$\lvert 1,1\rangle$	$\bullet$	$\bullet$	$\lvert 1,1\rangle$	$\lvert 1,0\rangle$

Table 30.1 The multiplication table for composing annuli for the toric code. The $\bullet$ indicates a composition which is not allowed due to non-matching boundary conditions. This table is summarized by Eq. 30.5.

[6] The Kronecker delta assures that we only compose states that are compatible on their boundaries.

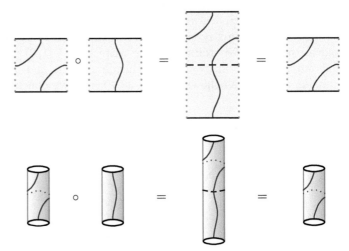

Fig. 30.11 The composition $|1,1\rangle \circ |1,0\rangle = |1,1\rangle$, as in Fig. 30.10 but shown using rectangle notation (top) where the orange edges are identified, or using tube notation (bottom).

The algebra associated with this type of composition of diagrams is known as Ocneanu's tube algebra, and although the multiplication is fairly trivial in the case of the toric code that we are considering here, it generalizes to much more complex algebras as we will discuss later in Chapters 31–33.

30.2.3 Quasiparticle Idempotents

We consider a set of states on the annulus that we call O_a and which are some linear combinations of our annular states $|n, m\rangle$. We define one one such annular state[7] O_a for each quasiparticle type a. The idea is that, if there is an overall quasiparticle charge[8] of type a in the disk that the annulus surrounds, then the annulus must be in the state (or the equivalence class of) O_a.

We claim that these annular states, under composition $\circ$, must obey the condition

$$O_a \circ O_b = \delta_{ab} O_a \quad . \tag{30.6}$$

This condition is known as the *idempotent* condition and the states O_a are known as *idempotents*. Note that the space of states we are considering on the annulus breaks up into sectors with different boundary conditions, and we cannot even talk about multiplying states together with $\circ$ unless the boundary conditions match (i.e. $|n, m\rangle$ can only multiply $|a, b\rangle$ if $a = n$).

The physical interpretation of the idempotent condition is the following. We imagine starting with a disk that contains some overall quasiparticles whose total quantum number is a. If we surround this region with an annulus in state O_b, we want this wavefunction of the entire system to give zero if $b \neq a$, but if $b = a$ then we want it returns the same wavefunction on a slightly larger disk—that is, leave the system topologically unchanged (same total quantum number, and still topologically a disc). Then, applying O_a twice is the same as applying it once as indicated in Eq. 30.6.

For the case of the toric code, as discussed in Section 30.2.1, there are two classes of boundary conditions, so we want to find a set of idempotents within each boundary condition class. Given the multiplication Table 30.1, it is easy to construct these idempotents. In the class where no blue lines intersect the boundary we have[9]

$$O_I = \frac{1}{2}(|0,0\rangle + |0,1\rangle) = \frac{1}{2} \left(\bigcirc + \circledcirc \right) \tag{30.7}$$

$$O_p = \frac{1}{2}(|0,0\rangle - |0,1\rangle) = \frac{1}{2} \left(\bigcirc - \circledcirc \right) \tag{30.8}$$

and in the class where one blue line intersects the inner boundary and one blue line intersects the outer boundary, we have

$$O_v = \frac{1}{2}(|1,0\rangle + |1,1\rangle) = \frac{1}{2} \left(\bigcirc + \circledcirc \right) \tag{30.9}$$

$$O_f = \frac{1}{2}(|1,0\rangle - |1,1\rangle) = \frac{1}{2} \left(\bigcirc - \circledcirc \right) \tag{30.10}$$

It is easy enough to use the multiplication Table 30.1 to check that these satisfy Eq. 30.6.

Here, with a bit of foresight, we have labeled these idempotents I, p,

[7]Since O_a is a state vector we could have written it as a ket $|O_a\rangle$. However, we are mainly going to think about composing these states as in Eq. 30.6, so we also think of O_a as an operator that acts on other states or other O_b.

[8]Here "overall quasiparticle charge" means we should fuse all the quasiparticles in the disk together to one overall fusion product.

[9]Note that if the kets $|n, m\rangle$ are normalized in the usual way $\langle n, m|a, b\rangle = \delta_{a,n}\delta_{m,b}$ then the O_a states are also orthogonal as kets, but are not normalized kets but satisfy Eq. 30.6 instead.

v, and f corresponding to their particle types (identity, plaquette defect, vertex defect, and fermion). Let us think a bit about how we make this connection between idempotents and particle types.

First, we realize that the vertex defect v and the f particle both involve ends of blue strings (i.e. defects of the vertex term of the Hamiltonian) so if one wraps an annulus around one of these particles, there will be an (odd number) of blue strings going through the annulus. Thus, the v and f idempotents should be made from the $|1, 0\rangle$ and $|1, 1\rangle$ states on the annulus. Analogously, the ground state I and the plaquette p defect involve no vertex defects and therefore should be made from the $|0, 0\rangle$ and $|0, 1\rangle$ states.

Let us start with the $|0, 0\rangle$ and $|0, 1\rangle$ states—those without blue lines touching the boundary. If we think about a single plaquette, we know from Section 27.3 that the ground state is always a superposition of flipped and unflipped plaquettes with a plus sign between the two pieces. In fact, it is easy to argue (by considering groups of plaquettes) that the ground state should be a superposition (with a plus sign) of flipped and unflipped regions. Thus, the positive superposition of $|0, 0\rangle$ (no loops) and $|0, 1\rangle$ (one loop) must be correspond to the ground-state space—the space with no quasiparticles. This is what we called O_I in Eq. 30.7. In the language of our diagrammatic algebra (the $d = +1$ version of the $\mathbb{Z}_2$ fusion rules from Section 18.1.1), the operation of doing nothing to the region plus the operation of putting a blue line around the region is the same as putting a $\tilde{\Omega}$-loop around the region (see ection 17.5 and Fig. 17.11. Here, $\mathcal{D} = \sqrt{2}$ since there are two particle types in $\mathbb{Z}_2$ fusion rules: the trivial particle and the nontrivial particle.)

If we return to the lattice Hamiltonian Eq. 28.1, we could also make the plaquette term of the Hamiltonian look like a $\tilde{\Omega}$-loop in the following way. Let us add a constant to the plaquette term in the Hamiltonian Eq. 28.1 (which does not change the ground state, just the ground-state energy)

$$P_\beta \to 1 + P_\beta$$

and then we realize that this combination is also just an $\tilde{\Omega}$-loop around a plaquette:

$$1 + P_\beta = \text{unflipped} + \text{flipped} \sim \tilde{\Omega}\text{-loop} \sim O_I \quad .$$

The principle that the ground-state idempotent, and similarly, the plaquette term in the Hamiltonian, is just an $\tilde{\Omega}$-loop is something that will hold for all of the generalized loop gas models we meet in Chapters 31–33 as well.

Once we have determined the ground-state idempotent O_I the other orthogonal idempotent within the space of $|0, 0\rangle$ and $|0, 1\rangle$ is fully defined and is given by Eq. 30.8. However, the plaquette-defect idempotent O_p can also be understood in a way similar to the argument for O_I. Recall from Section 27.4.2 that the defect on a plaquette is the *difference* of flipped and unflipped. By grouping together plaquettes, it is easy to see that the annular idempotent around a region with a magnetic quasiparticle must be a superposition of $|0,0\rangle$ (no loops) and $|0,1\rangle$ (one loop) with a *minus* sign.

We now turn to the states with blue lines touching the boundaries—that is, the $|1,0\rangle$ and $|1,1\rangle$ states. It may look a bit unexpected that the vertex-defect idempotent O_v needs to be a superposition of the $|1,0\rangle$ diagram and the $|1,1\rangle$ diagram, as shown in Eq. 30.9. However, we should think of the quasiparticle that we call the vertex defect, not only being the vertex defect, but rather a vertex defect *with the absence of a plaquette defect*, whereas the f particle is the vertex defect in the presence of a plaquette defect. The superposition of the two diagrams $|1,0\rangle$ and $|1,1\rangle$ is arranged to assure that the O_v particle does not include a plaquette defect, whereas the O_f particle does. To see how this works, note that the $|1,0\rangle$ state can be created from the $|0,0\rangle$ state (no blue lines) by flipping over a string of spins connecting the inner and outer annulus without going around the nontrivial cycle, that is, by adding the one vertical blue line. If we perform the same operation on the $|0,1\rangle$ state, as shown in Fig. 30.12, we obtain the $|1,1\rangle$ state—a line that connects the two annuli and also goes around the cycle.[10] Thus, the superposition O_v is the same as the superposition O_I with the (vertical) string added between annuli, whereas the superposition O_f is the same as O_p with one (vertical) string added.

It is worth recalling that the $d = +1$ version $\mathbb{Z}_2$ planar diagram algebra (the diagram algebra we are using here!) has two possible solutions to the hexagon equation as discussed in Section 18.1.1: boson and fermion (neither solution is modular). It is not a coincidence that the particle types we have found arising from this diagrammatic construction are also bosons and fermions. We will see that this is a general principle—when we build a model based on a planar diagram algebra, the particle types that arise must include all of the possible solutions to the hexagon equation for that planar algebra.

Quasiparticle Basis for Torus States

It is worth noting that once we have our basis for quasiparticles enclosed in an annulus, we can also glue the two boundaries of the annulus to each other to form a closed torus, and we have a basis for the degenerate ground states of the torus. This basis is a rather special basis since if you cut the torus open again (inverting the gluing that we just did), the boundary revealed has the quantum number of the given quasiparticle type. Recall that in the discussion of Section 7.3.1 we defined exactly this basis on the surface of a torus by stating that a quasiparticle of a given type is running around the torus handle.

30.3 **Twists and T-Matrix**

Twist deforming either the O_I state or the O_p state leaves the state unchanged, so we trivially have $\hat{\theta} O_I = O_I$ and $\hat{\theta} O_p = O_p$ implying trivial twist factors for the I particle and the plaquette-defect particle,

[10] As noted in the caption of Fig. 30.12, the crossing of two blue lines is not generally allowed and has to be "resolved" into either the picture on the bottom left of Fig. 30.9 or the picture on the bottom right of Fig. 30.9.

Fig. 30.12 The $|1,1\rangle$ state can be thought of as a combination of $|1,0\rangle$ and $|0,1\rangle$. Consider starting from an annulus with no blue lines then applying a string of σ_x operators along the two paths to make both a $|1,0\rangle$ string (vertical) and an $|0,1\rangle$ string (around the nontrivial cycle). No crossings are actually allowed in the loop gas and, depending on how the two paths cross microscopically, one will obtain either the right or left diagram of the lower line of Fig. 30.9, which are equivalent anyway.

With the tube algebra one can determine the twist factor θ for each particle type. To do this, we imagine deforming the local environment of the particle around in circle by 360 degrees, as depicted in the top of Fig. 30.13. This is equivalent to the twist operation shown in the lower part of Fig. 30.13.

Fig. 30.13 Deforming the environment of a particle (top) implements the twist operation. This is equivalent to the spacetime diagram shown on the bottom (see also Chapter 15).

$\theta_I = \theta_p = 1$. That is, these are both bosons, as we expected.

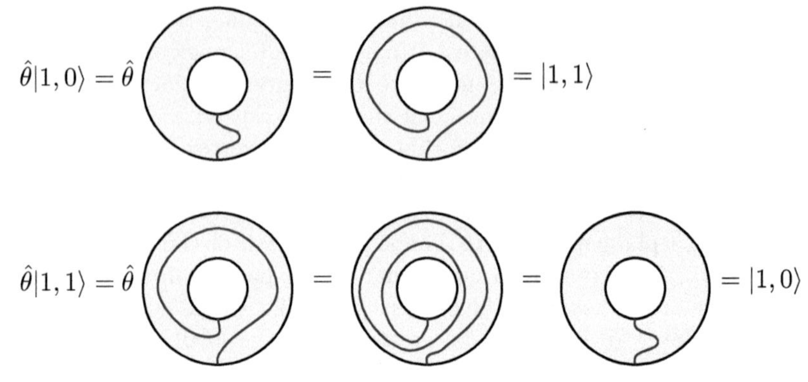

Fig. 30.14 Twist deformations of the $|1,0\rangle$ and $|1,1\rangle$ states. The inner boundary is rotated counterclockwise while the outer boundary is held fixed, analogous to the deformation shown in Fig. 30.13. In the second line we have used the surgery equivalence, as shown in Fig. 30.9.

However, twist deforming $|1,0\rangle$ and $|1,1\rangle$ is nontrivial as shown in Fig. 30.14. These diagrams tell us that $\hat{\theta}|1,0\rangle = |1,1\rangle$ and $\hat{\theta}|1,1\rangle = |1,0\rangle$. Note that a general formula for the twist operator on the states $|n,m\rangle$ is given by

$$\hat{\theta}|n,m\rangle = |n, (n+m)\bmod 2\rangle \quad . \tag{30.11}$$

Using Eq. 30.9 we obtain $\hat{\theta}O_v = O_v$ confirming that $\theta_v = 1$, that is, the vertex-defect particle is a boson. Similarly using Eq. 30.10 we obtain $\hat{\theta}O_f = -O_f$ confirming that $\theta_f = -1$, the f particle is a fermion. Note that it is a nontrivial result that the quasiparticle idempotents we have found are eigenstates of the twist operator. It isn't obvious that this should have to be true, although we certainly suspected it should be true considering our understanding of the structure of anyon theories!

If we connect the inside of the annulus to the outside to form a torus, as discussed at the end of Section 30.2.3, we realize that the twist transformation is exactly the $\underline{T}$ transformation[11] discussed in Section 17.3.2 (see Fig. 17.6), or Dehn twist as discussed in Section 7.4.

[11]The fact that we obtain exactly $\tilde{T}$ from this geometric transformation indicates that there is no phase factor from Eq. 17.15. This is a feature of all TQFTs that can be obtained from loop gases, or generalized loop gases: the central charge c must be zero. An argument for this is that the Hamiltonian is time-reversal invariant, whereas c goes to $-c$ under time reversal.

30.4 *S*-Matrix

We can also derive the modular S-matrix diagrammatically by exchanging the two directions of the torus, as discussed in Sections 17.3.2 and 7.3.1 (see also Fig. 17.6).

Let us abuse notation a bit and denote by $|0,0\rangle, |0,1\rangle, |1,0\rangle, |1,1\rangle$ the four basis states obtained from the annular states $|0,0\rangle, |0,1\rangle, |1,0\rangle, |1,1\rangle$ by connecting the inner boundary to the outer boundary to obtain a torus. Exchanging the two directions on the torus (making an $\underline{S}$ transformation, as in Section 17.3.2) we have the following transformation of the basis states for our torus

$$\underline{S}|0,0\rangle \;=\; \underline{S} \;\;\boxed{}\;\; =\; \boxed{}\; =\; |0,0\rangle \qquad (30.12)$$

$$\underline{S}|0,1\rangle \;=\; \underline{S} \;\;\boxed{\text{—}}\;\; =\; \boxed{\,|\,}\; =\; |1,0\rangle \qquad (30.13)$$

$$\underline{S}|1,0\rangle \;=\; \underline{S} \;\;\boxed{\,|\,}\;\; =\; \boxed{\text{—}}\; =\; |0,1\rangle \qquad (30.14)$$

$$\underline{S}|1,1\rangle \;=\; \underline{S} \;\;\boxed{\diagup}\;\; =\; \boxed{\diagdown}\; =\; \boxed{\diagup}\; =\; |1,1\rangle \qquad (30.15)$$

where the rectangles in the diagram are meant to have their opposite edges identified to give tori, and in the last equation we have used a surgery for the latter equality between diagrams. We can write this $\underline{S}$ transformation as (with $n, m = 0, 1$)

$$\underline{S}|n,m\rangle = |m,n\rangle \qquad (30.16)$$

since the indices n and m refer to the two different directions, which are switched by $\underline{S}$.

We want to change this basis into the quasiparticle basis obtained by connecting the inner to outer boundary of the annulus for the four quasiparticle idempotents O_I, O_p, O_v, O_f. Making this change of basis as specified[12] in Eqs. 30.7–30.10, applying $\underline{S}$, and then changing basis back to the quasiparticle basis one obtains exactly the S-matrix for the toric code given in Eq. 28.5 (see Exercise 30.1). This calculation is done in a bit more detail and more generally in Section 30.6.

---

[12] Eqs. 30.7–30.10 now need $1/\sqrt{2}$ out front rather than $1/2$ out front to be normalized as wavefunctions, rather than the normalized idempotents that we have been using.

30.5 Direct Calculation of Braiding and Fusion

A more direct calculation of the braiding properties of the quasiparticles, the R-matrix, is possible by considering a more complex geometry, that is, a disk with two holes. With this geometry, we add strings attached to holes and around holes, as in Eqs. 30.7–30.10. To calculate braiding, we need to drag one hole around the other. As an example here we show that exchanging two f particles gives a minus sign. Since each f is a superposition of two terms ($|1,0\rangle$ and $|1,1\rangle$, as in Eq. 30.10), our system with two f particles (one in each hole) is now a sum of four terms, as shown in the top line of Fig. 30.15.

We then braid the holes around each other, stretching the blue lines in the process as shown in the lower line of Fig. 30.15. The resulting picture can then be related back to the top line of the figure using surgery and loop addition/removal (the moves of Fig. 30.3). It is an easy exercise to show that the lower line is exactly minus the upper line of the figure,

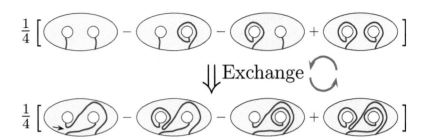

$$\frac{1}{4}\left[\qquad - \qquad - \qquad + \qquad \right]$$

$\Downarrow$ Exchange $\circlearrowright$

$$\frac{1}{4}\left[\qquad - \qquad - \qquad + \qquad \right]$$

Fig. 30.15 Calculation that the phase accumulated from exchanging two f particles is -1. The upper line represents two f quasiparticles, one in each hole (compare Eq. 30.10). The holes are exchanged, braiding the blue lines in the process. Using surgery and loop addition/removal (the moves of Fig. 30.3) one can show the lower line is precisely minus the top line showing an exchange phase of -1. For example, performing surgery on the first picture in the second line at the location of the arrow gives exactly the second picture on the top line but missing the minus sign.

thus showing that the phase obtained in exchanging two f particles is -1 (see Exercise 30.2.a).

Fusion relations can also be calculated graphically with our diagrammatic rules by essentially merging together the two holes. This is just another way of saying that we draw an imaginary circle around both holes and treat this imaginary circle as a single, bigger, hole. An example of this is shown in Fig. 30.16, where we show the fusion of a vertex defect (left) with an f particle (right) giving a plaquette defect. The top line of this figure is the four-term superposition describing the vertex defect and f particle (see Eq. 30.9 and 30.10). Going to the second line we have used the rules (Fig. 30.3) of our diagrammatic algebra. We then realize that we can write this second line as a plaquette idempotent O_p around "something"—which guarantees that the total quantum number inside the large hole in the third line is simply that of the plaquette defect.

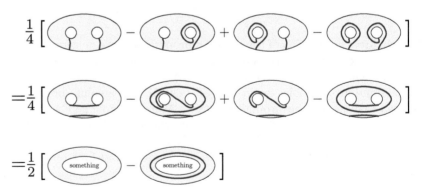

$$\frac{1}{4}\left[\qquad - \qquad + \qquad - \qquad \right]$$

$$=\frac{1}{4}\left[\qquad - \qquad + \qquad - \qquad \right]$$

$$=\frac{1}{2}\left[\left(\text{something}\right) - \left(\text{something}\right) \right]$$

Fig. 30.16 Calculation that (vertex defect) $\times f = $ (plaquette defect). The upper line represents a vertex defect (left) quasiparticle and an f (right) quasiparticle (compare Eq. 30.9 and 30.10). In the lower line the diagrammatic moves are used. The small blue string at the bottom is pulled off the bottom of the picture and the remaining diagram at the bottom is a plaquette idempotent O_p around the inner region.

30.6 Generalization to $\mathbb{Z}_N$ Toric Code (Briefly)

The diagrammatic arguments we have used all generalize very naturally to the case of the $\mathbb{Z}_N$ toric code discussed in Sections 27.6 and 28.5. Instead of having simple loop diagrams, we now have a $\mathbb{Z}_N$ fusion algebra. Our diagrammatic algebra now requires us to draw arrows on lines (for $N > 2$) and label each line with an integer that is interpreted modulo N. Reversing an arrow on a line takes a label j to $N - j$, as shown in Fig. 30.17. The label 0 plays the role of the identity, and such lines can be added or removed from the diagram freely as in our other diagrammatic algebras.

Vertices must follow the rule that the Q_α at each vertex, which is the sum of labels on incoming arrows minus the sum of labels on outgoing arrows, must be zero modulo N. An example of such a diagram is shown in Fig. 30.18.

Analogous to the three allowed moves in Fig. 30.3 we are allowed three moves in order to manipulate diagrams into equivalence classes. These moves are shown in Fig. 30.19.

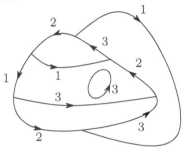

Fig. 30.17 Reversing an arrow in a $\mathbb{Z}_N$ fusion diagram takes the label j to $N-j$ with all labels interpreted modulo N.

Fig. 30.18 A diagram with $\mathbb{Z}_4$ fusion rules. Each vertex has $Q_\alpha = $ (entering) $-$ (leaving) $= 0 \bmod 4$.

- **Move 1: "Isotopy."** Meaning any smooth deformation of lines.

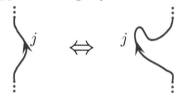

- **Move 2: Adding and Removing Loops.** A loop with any label can be added or removed.

- **Move 3: Reconnecting (trivial F) move.**

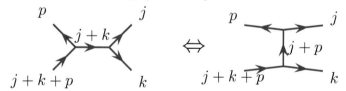

Fig. 30.19 Allowed moves for the $\mathbb{Z}_N$ toric code diagram algebra. Compare Fig. 30.3.

The first move is similar to what we had in the case of the toric code. The second move, that we can add or remove loops, is also similar to the case of the toric code, except that the loops have arrows as well as labels. The third move is "new" but should look familiar from our discussion of F-moves. However, here the F-coefficient is simply unity—and there is no sum over states on the right-hand side (compare to Fig. 16.4). This type of F-matrix (which is not a matrix but simply a scalar) is what we called the trivial cocycle for the $\mathbb{Z}_N$ fusion algebra in Sections 19.1 and 19.4.

We now want to build basis states for our tube algebra analogous to what we did in Section 30.2. Let us define the tube state $|n, m\rangle$ to be the state with quantum number n going between the inner and outer edge of the annulus, and quantum number m going counterclockwise around the annulus, as shown in Fig. 30.20.

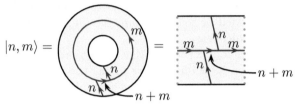

Fig. 30.20 Basis states for the tube algebra of the $\mathbb{Z}_N$ toric code. The quantities n, m, and $n + m$ are interpreted mod N. On the far right the two dotted orange sides are identified.

Composition of these basis states is performed by placing one annulus state inside the other, as in Section 30.2.2. The multiplication law is the natural generalization of Eq. 30.5

$$|n, m\rangle \circ |n', p\rangle = \delta_{n,n'} |n, (m+p)\mathrm{mod}N\rangle \tag{30.17}$$

which we illustrate graphically in Fig. 30.21 (see Exercise 30.4.a).

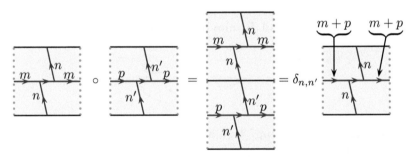

Fig. 30.21 Graphical depiction of the composition rule Eq. 30.17. All of the indices $n, n', m, p, (m+p)$ are interpreted mod N.

We can then define N^2 idempotents

$$O^{[n,k]} = \frac{1}{N} \sum_{p=0}^{N-1} e^{2\pi i pk/N} |n, p\rangle \tag{30.18}$$

which satisfy (see Exercise 30.4.b)

$$O^{[n,k]} \circ O^{[n',k']} = \delta_{n,n'} \, \delta_{k,k'} \, O^{[n,k]} \tag{30.19}$$

as desired in Eq. 30.6. These orthogonal idempotents describe the quasiparticle types of the theory. We expect that the lower index, n, which gives the total "charge" entering the annulus should match the Q_α quantum number of the vertex operator for the $\mathbb{Z}_N$ toric code that we discussed in Section 28.5, whereas the upper index k should describe the $e^{2\pi ik/N}$ eigenvalue of the A_β plaquette operator. In other words, we are describing the quasiparticle $[n, k]$ from Eq. 28.8.

$$[n, k] \leftrightarrow O^{[n,k]} \quad .$$

We can apply the twist operator to the basis states to obtain the analog of Eq. 30.11

$$\hat\theta |n, m\rangle = |n, (n + m)\mathrm{mod}\, N\,\rangle \qquad (30.20)$$

which we show graphically in Fig. 30.22 (see Exercise Eq. 30.4). This is the same twist transformation shown in Section 30.3 but using the rectangular representation of an annulus. See also Fig. 17.6.

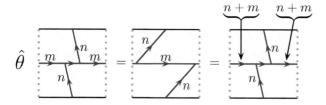

Fig. 30.22   Graphical depiction of the twist operator rule Eq. 30.20. All of the indices $n, m, (n + m)$ are interpreted mod N.

Applying the twist operator Eq. 30.20 to the quasiparticle idempotents $O^{[n,k]}$ in Eq. 30.18 we obtain

$$
\begin{aligned}
\hat\theta O^{[n,k]} &= \frac{1}{N}\sum_p e^{2\pi ipk/N}\hat\theta|n, p\rangle = \frac{1}{N}\sum_p e^{2\pi ipk/N}|n, (p + n)\mathrm{mod}N\rangle \\
&= e^{-2\pi ink/N}\frac{1}{N}\sum_q e^{2\pi ikq/N}|n, q\rangle = e^{-2\pi ink/N}O^{[n,k]}
\end{aligned}
$$

giving us a twist factor

$$\theta_{[n,k]} = e^{-2\pi ink/N}$$

in agreement with Eq. 28.10.

Finally, let us evaluate the S-matrix. Again, our strategy is to exchange the two directions of the torus, as discussed in Sections 17.3.2 and 7.3.1 (see Fig. 17.6). First, we should connect the inner and outer edges of the annulus in Fig. 30.20 to make a torus. By doing this we construct the basis states on a torus $|n, m\rangle$ that correspond to $|n, m\rangle$ on the annulus. The $\underline{S}$ matrix applied to our basis states simply rotates the directions of the torus giving

$$\underline{S}|n, p\rangle = |p, -n\rangle \qquad (30.21)$$

as shown graphically in Fig. 30.23.

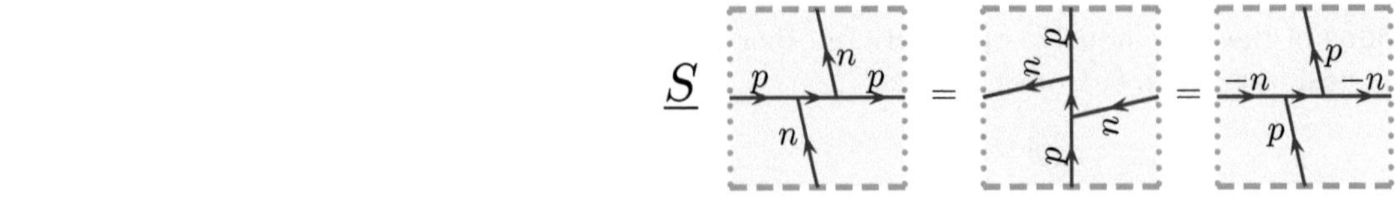

Fig. 30.23 The $\underline{S}$ operation rotates the torus. Note in the second step we use a single F-move and then reverse the quantum number on n by reversing the arrow. All indices $n, p, -n$ are interpreted modulo N.

We then construct the quasiparticle superpositions on the torus corresponding to the idempotents Eq. 30.18

$$| [n, k] \rangle = \frac{1}{\sqrt{N}} \sum_{p=0}^{N-1} e^{2\pi i p k / N} |n, p\rangle \quad . \tag{30.22}$$

Note the change in normalization of this wavefunction compared to the idempotent normalization used in Eq. 30.18. This is to assure the orthonormality

$$\langle [n, k] \, | \, [n, l] \rangle = \delta_{k,l} \quad .$$

The S-matrix is then given by

$$
\begin{aligned}
S_{[n',k'],[n,k]} &= \langle [n', k'] \, | \, \underline{S} \, | \, [n, k] \rangle \\
&= \frac{1}{N} \sum_{p,p'=0}^{N-1} e^{2\pi i (pk - p'k')/N} \langle n', p' | \underline{S} | n, p \rangle \\
&= \frac{1}{N} \sum_{p,p'=0}^{N-1} e^{2\pi i (pk - p'k')/N} \langle n', p' | p, -n \rangle \\
&= \frac{1}{N} e^{2\pi i (n'k + nk')/N}
\end{aligned}
$$

which matches Eq. 28.11.

Chapter Summary

- The toric code ground state can be abstractly viewed as a fluctuating loop gas.
- The diagrammatic reasoning of the *tube algebra* allows us to construct the quasiparticle defects of the toric code, to determine properties of these quasiparticles, and to construct the multiple ground states on the torus.

Further Reading:

- The diagrammatic reasoning used here first appeared in the physics literature in Freedman et al. [2004]. However, the concept of the tube algebra is due to Ocneanu [1994].

- A good discussion of this type of diagrammatic algebra in the context of the toric code is given by Freedman et al. [2008].

- We return to discuss the tube algebra again (and given more references) in Chapters 31–33.

Exercises

Exercise 30.1 Toric Code S-matrix
Derive the S-matrix of the toric code (Eq. 28.5) by using the method described in Section 30.4.[13] If you need help with this calculation look at the more general calculation given in Section 30.6.

Exercise 30.2 Braiding Quasiparticles in Toric Code Loop Gas
(a) Use the technique of Section 30.5 to show that exchanging two f particles gives a minus sign (i.e. confirm the details of the argument given there).
(b) Use similar techniques to show that exchanging two e particles gives no sign and exchanging two m particles gives no sign.
(c) Show that braiding an e particle or an f particle all the way around an m particle gives a minus sign but braiding around the identity gives no sign.
(d) Show that braiding e all the way around f gives a minus sign.

Exercise 30.3 Fusing Quasiparticles in Toric Code Loop Gas
Use the technique of Section 30.5 to deduce the full fusion table for the toric code.

Exercise 30.4 $\mathbb{Z}_N$ Tube Algebra
(a) Using the three allowed moves in Fig. 30.19, as well as possibly adding or removing lines labeled with the identity, derive the final equality in Fig. 30.22.
(b) Similarly, derive the final equality in Fig. 30.21.
(c) Confirm Eq. 30.19.

[13]Note that in this chapter the quasiparticles are labeled as I, v, p, f, whereas in Chapter 28 we label them as I, e, m, f. It does not matter whether we assign $v = e$ and $p = m$ or $v = m$ and $p = e$. The S-matrix comes out the same either way.

Part VII

More General Loop-Gas and String-Net Models

Kitaev Quantum Double Model

We have already seen the generalization of the conventional toric code to the more interesting $\mathbb{Z}_N$ toric code. In the next few chapters we consider models that generalize the toric code further and in other ways. These generalizations all share many key properties: in short, we start with some planar diagram algebra and use that to build a Hamiltonian, which in turn produces a TQFT as its ground state known as the *quantum double* or *Drinfel'd double* of the input original diagram algebra. All of the calculations will hinge on diagrammatic manipulations.

31.1 Defining the Model

Perhaps the simplest generalization of the toric code is the generalization of the spin up/spin down (black/blue) assigned to each edge to group-valued variables. Indeed, the $\mathbb{Z}_N$ toric code was just such a generalization where the group we used was the abelian group $\mathbb{Z}_N$ (integers mod N under addition). This generalization to groups, which turns out to be essentially equivalent to a lattice gauge theory as we will discuss in Section 31.7, is sometimes known as the *Kitaev Model*[1] or more specifically the *Kitaev Quantum Double Model*.[2]

We start by choosing a group G with elements $g \in G$. This group may be abelian or nonabelian. For this model we can work with any regular or irregular lattice[3] like the one shown in Fig. 31.1, and we put arrows on each edge (we can fix them in arbitrary directions). Instead of labeling each edge with spin up or spin down we assign each edge a group element g as shown in the figure. The set of all possible group elements assigned to all possible edges serves as an orthonormal and complete basis for the Hilbert space.

[1]Be warned that the phrase *Kitaev Model* is often not sufficiently specific, since Kitaev has introduced many models and several very different models all go by this same name.

[2]**Caution:** The presentation I give here is on the dual lattice compared to Kitaev [2003]. Be warned that almost all presentations of this material follow Kitaev's original work, except us. We will switch to the same presentation as Kitaev in Section 31.7. I have chosen this approach so as to emphasize the importance of the planar diagram algebra.

[3]....which should not be called a lattice.

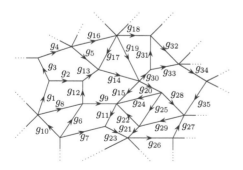

Fig. 31.1 Each edge of the (irregular)[3] lattice is labeled with a group element $g \in G$.

$$g \Bigg\uparrow \; = \; \Bigg\downarrow g^{-1}$$

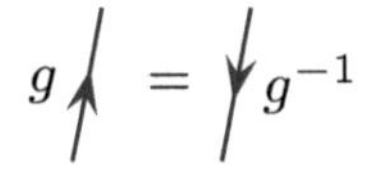

Fig. 31.2 Reversing an arrow inverts the element of the group.

We make a definition that reversing the arrow on an edge inverts the element of the group, as shown in Fig. 31.2.

31.1.1 Vertex and Plaquette Operators

We can now define vertex operators $\hat{V}_\alpha$ at each vertex α as follows:

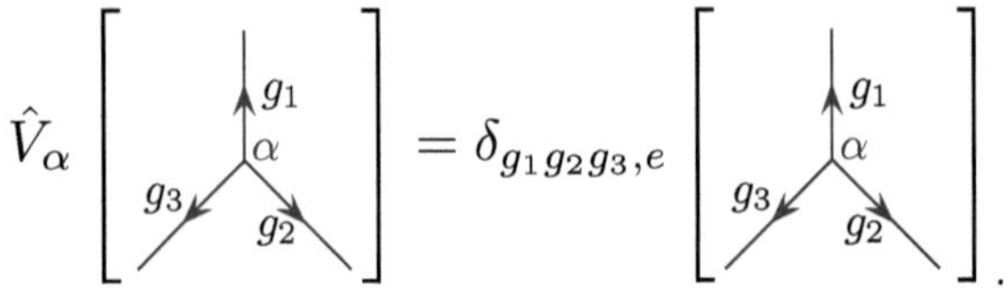

$$
\hat{V}_\alpha \left[\begin{array}{c} \end{array} \right] = \delta_{g_1 g_2 g_3, e} \left[\begin{array}{c} \end{array} \right]. \tag{31.1}
$$

In other words, to apply $\hat{V}_\alpha$ at a vertex α we orient all of the arrows on the edges so they point away from the vertex (using Fig. 31.2 if necessary). Then we multiply together the group elements on all of the edges of the vertex in a clockwise order around the vertex (independent of how many edges intersect the vertex).[4] If the result gives the identity element e of the group, then $\hat{V}_\alpha$ gives unity, otherwise the operator gives zero.

A few things to note about $\hat{V}_\alpha$. First of all, it does not matter where we start multiplying around the vertex, since $g_1 g_2 g_3 = e$ implies[5] $g_2 g_3 g_1 = g_3 g_1 g_2 = e$. So in essence the operator is rotationally invariant. Secondly, we note that $\hat{V}_\alpha$ is a projector, meaning

$$\hat{V}_\alpha^2 = \hat{V}_\alpha \quad . \tag{31.2}$$

[4] For example, for a 4-valent vertex we would write $\delta_{g_1 g_2 g_3 g_4, e}$ where $g_1, \ldots, g_4$ are ordered clockwise and all arrows point out from the vertex.

[5] If it is not already obvious, starting with $g_1 g_2 g_3 = e$, left multiply both sides by g_1^{-1} and right multiply both sides by g_1 to obtain $g_2 g_3 g_1 = e$.

For each group element $h \in G$ we now define an operator $\hat{P}_\beta(h)$ on a plaquette as follows:

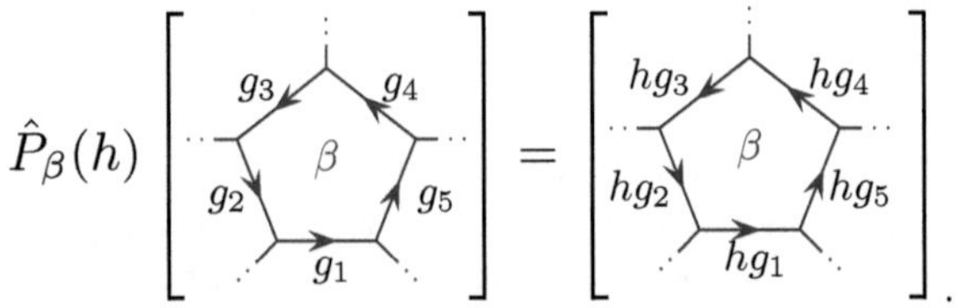

$$
\hat{P}_\beta(h) \left[\begin{array}{c} \end{array} \right] = \left[\begin{array}{c} \end{array} \right]. \tag{31.3}
$$

In other words, we first orient the arrows around the plaquette in a counterclockwise manner (using Fig. 31.2 if necessary). The action of $\hat{P}(h)$ is to premultiply each edge label by the group element h,

$$\hat{P}_\beta(h') \hat{P}_\beta(h) = \hat{P}_\beta(h' h) \quad . \tag{31.4}$$

That is, if you premultiply by h and then premultiply by h', this is the same as premultiplying by $h' h$.

We then construct the total plaquette operator for the plaquette β as

$$\hat{P}_\beta = \frac{1}{|G|} \sum_{h \in G} \hat{P}_\beta(h) \tag{31.5}$$

where $|G|$ is the number of elements in the group G.

From Eq. 31.4 and Eq. 31.5 it is easy to show that

$$\hat{P}_\beta \, \hat{P}_\beta(h) = \hat{P}_\beta(h) \, \hat{P}_\beta = \hat{P}_\beta \qquad (31.6)$$

and from this we show that $\hat{P}_\beta$ is a projector, meaning

$$\hat{P}_\beta^2 = \hat{P}_\beta \qquad (31.7)$$

so that it has eigenvalues 0 and 1 only.

Commuting Operators

It is worth noting that, analogous to the case of the toric code, the operators $\hat{V}_\alpha$ and $\hat{P}_\beta$ all commute with each other. It should be fairly obvious that all of the $\hat{V}_\alpha$ operators commute with each other. It is perhaps a bit less obvious that the $\hat{P}_\beta$ operators commute with each other when they share an edge, as in Fig. 31.3. The key here is to realize that the edge arrow needs to be reversed to apply the plaquette operator on the two different plaquettes. In the figure, the arrow on the shared edge is oriented counterclockwise with respect to β so that the $\hat{P}_\beta(h)$ operator may be applied on plaquette β, turning the shared edge from g to hg. To apply the plaquette operator $\hat{P}_{\beta'}(h')$ on plaquette β' we need to first flip the edge arrow, turning g to g^{-1}. Then, applying $\hat{P}_{\beta'}(h')$ turns the edge into $h'g^{-1}$, and we can then reverse the arrow again to put it in the original position, giving the edge value $(h'g^{-1})^{-1} = g(h')^{-1}$. Thus, one of the plaquette operators premultiplies the edge element and the other plaquette operator postmultiplies the edge element—and hence, these commute with each other.

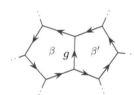

Fig. 31.3 Two neighboring plaquettes share an edge.

It is also not immediately obvious that the plaquette operators commute with the adjacent vertex operators. Consider Fig. 31.4. To apply the vertex operator at α, we must first reverse the arrow on the edge labeled g_1 so all edges point away from the vertex. With the reversed arrow, the edge is then labeled g_1^{-1}. The vertex operator $\hat{V}_\alpha$ now measures the product $g_1^{-1} g_2 g_3$. Now instead consider applying the plaquette operator $\hat{P}_\beta(h)$ on the plaquette β before reversing the direction of the arrow. This premultiplies both g_1 and g_2 by h, yielding hg_1 and hg_2 with the arrows oriented as they are in the figure. If we try applying the vertex operator after the plaquette operator we would now measure $(hg_1)^{-1}(hg_2)g_3 = (g_1^{-1}h^{-1})(hg_2)g_3 = g_1^{-1}g_2 g_3$, which is unchanged. Thus, we conclude that the plaquette and vertex operators commute.

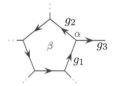

Fig. 31.4 A vertex α adjacent to a plaquette β.

31.1.2 Code Space

We can use the Kitaev model as a quantum memory (an error correcting code) analogous to the toric code. For a group with $|G|$ elements, one needs to have a $|G|$-state quantum system[6] on each edge, instead of just having a single spin-$\frac{1}{2}$ two-state system on each edge as in the conventional toric code.

Similar to the toric code, the code space for the Kitaev model will

[6]A qubit is a two-state quantum system. Sometimes people call a three-state quantum system a "qutrit." For a d-state quantum system one often says one has a "qudit."

be the set of states where all $V_\alpha = +1$ and all $P_\beta = +1$. This code space, analogous to the toric code, will turn out to be closely related to a TQFT.

The Kitaev model built from a nonabelian group G has a significant advantage over the conventional toric code: not only can one store quantum information, but by braiding, fusing, and measuring defects one can also perform computation (see Kitaev [2003]; Preskill [2004], for example).

It is interesting to note that if one builds the Kitaev model from a group G_1 and then another Kitaev model from a group G_2, the product of the code spaces of these two will be the code space of a Kitaev model built from a group $G_1 \times G_2$.

Abelian Case: Relation to the $\mathbb{Z}_N$ Toric Code

[7] And the conventional toric code is just the $N = 2$ case of the $\mathbb{Z}_N$ toric code.

The $\mathbb{Z}_N$ toric code is a special case of the Kitaev model[7] where we take the group to be $G = \mathbb{Z}_N$ (the integers modulo N with addition as the group operation). As discussed in Section 40.2, any abelian groups can be written as $G = \mathbb{Z}_{N_1} \times \mathbb{Z}_{N_2} \times \ldots \times \mathbb{Z}_{N_p}$ for some number of factors p. Thus, all Kitaev models built from abelian groups are equivalent to a product of some number of $\mathbb{Z}_N$ toric codes.

31.1.3 Hamiltonian

Instead of thinking about the Kitaev model as an error-correcting code, we can think of it as a phase of matter analogous to our discussion in Chapter 28. We can now write a simple Hamiltonian for our system

$$H_{\text{Kitaev model}} = -\Delta_v \sum_{\text{vertices } \alpha} \hat{V}_\alpha \quad - \quad \Delta_p \sum_{\text{plaquettes } \beta} \hat{P}_\beta \quad (31.8)$$

where here $\Delta_v > 0$ and $\Delta_p > 0$. The ground-state space will thus be the space where every vertex operator $\hat{V}_\alpha$ and every plaquette operator $\hat{P}_\beta$ is in the $+1$ eigenstate. That is, the ground state is the code space. Any vertex or plaquette in the other eigenstate (having eigenvalue 0) then corresponds to an excitation and has energy cost Δ_v or Δ_p, respectively.

The Hamiltonian Eq. 31.8 is precisely the Hamiltonian we used for the $\mathbb{Z}_N$ toric code in Eq. 28.6.

31.2 How the Kitaev Model Generalizes the Toric Code

In the case of the toric code, the condition of being in the lower energy eigenstate of all of the vertex operators (Section 27.3) was interpreted as allowing only loop configurations (see also Section 30.1). We can think of the blue lines as carrying some $\mathbb{Z}_2$ valued in-plane "flux" that is conserved at each vertex: if a blue line comes into the vertex another blue line must go out of the vertex. This was generalized for the $\mathbb{Z}_N$ toric code where we think of the edges as carrying $\mathbb{Z}_N$ flux: edges have arrows and integer (modulo N) on them, and the total of the arrow pointing in minus out should add up to zero modulo N. Now, in the more general Kitaev model based on a group G, we think of each arrow as carrying some in-plane G-valued "flux" that must be conserved at each vertex in order to satisfy the vertex condition of the code space.

In the toric code, the plaquette operator flips the quantum numbers around a plaquette while staying within the space of loop configurations. Similarly in the Kitaev model, the plaquette operator changes the labels around the edges of the plaquette, but in such a way as to not disturb the vertex operators.

Analogous to the toric code, we may write the states in the ground-state space (the code space) of the Kitaev model as

$$|\psi\rangle = \mathcal{N}^{-1/2} \sum_{\substack{\text{all edge labelings that can} \\ \text{be obtained from a reference} \\ \text{edge labeling via application} \\ \text{of any } P_\beta(h)}} |\text{labeling}\rangle \qquad (31.9)$$

where $\mathcal{N}$ is the total number of terms in the sum. The analogy to flipping a plaquette in the toric code is the application of the operators $\hat{P}_\beta(h)$ for any group element $h \in G$ to any plaquette β. Note that from Eq. 31.6 the eigenvalue of $\hat{P}_\beta$ is not changed by the application of $\hat{P}_\beta(h)$, nor is the eigenvalue of any $\hat{V}_\alpha$.

As with the toric code, the vertex and plaquette operators provide just enough constraints so that the ground state on a spherical surface is unique and the ground state on a higher-genus manifold will have a degeneracy that depends on the topology of the system, but does not depend on the number of lattice points we use in our lattice. That is, the ground-state space is described by a TQFT.

The TQFT that results is known as the *quantum double* or *Drinfel'd double* of the group G. Although the detailed study of the properties of the quantum double can get complicated (see references at the end of the chapter) we can introduce much of the key physics with fairly simple[8] diagrammatic reasoning.

[8] ... well, not *too* complicated!

31.3 Kitaev Ground State Is Topological

Here we will show that the ground state of Kitaev model is topological, that is, it is independent of the detailed geometry of the lattice, but depends only on the topology of the manifold. Our strategy is to show that we can locally reconfigure the structure of the lattice (keeping the topology fixed) and uniquely map a ground-state wavefunction on one lattice structure to a ground-state wavefunction with another lattice structure.[9] This will then imply that, for example, the ground-state degeneracy is independent of the lattice that we choose to use, depending only on the underlying topology of the manifold.

[9] This mapping gives an isomorphism between ground-state spaces.

To restructure our lattice, we need only two restructuring moves, which are shown in Figs. 31.5 and 31.6. Using these two moves any lattice on a surface may be mutated into any other lattice on the surface, as long as the overall topology of the surface remains unchanged (see Exercise 31.3). It is worth noting that non-contractible loops on a surface cannot be removed by these moves.

We must determine how the ground-state wavefunction is changed under these two moves. First, let us consider the vertex-merging/-splitting move shown in Fig. 31.5. In splitting the single vertex into two vertices the added edge (ab in the left part of the figure, ac in the right) has its quantum number completely fixed by the vertex condition. Thus, in adding this edge there is no additional degree of freedom added, thus we have a unique mapping of the wavefunctions between the two structures. More generally in such a move, two neighboring vertices with coordination n and n' may be moved to a single point, eliminating the intervening edge and creating a vertex of coordination $n + n' - 2$.

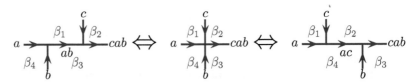

Fig. 31.5 Splitting one vertex into two, or, in reverse, merging two vertices into one. In the ground state, the intervening edge (labeled ab) on the left is not an independent degree of freedom but is instead fixed by the other edge variables.

In a ground state, it is fairly easy to see that such a vertex-merging/-splitting move does not alter any of the important physics. In particular, a basis of ground-state wavefunctions before the merging move can be mapped uniquely to another basis for the ground-state wavefunctions after the merging move.

The second move we need to consider is the plaquette addition move shown in Fig. 31.6. Here, a vertex is inserted into an edge and "tadpole" is added to the new vertex. This tadpole has a stem (labeled k) and a loop (labeled h) in the figure. The loop splits the plaquette (β) into two plaquettes (β_1 and β_2 in the figure).

Again, we would like to show that the ground states of the two lattices are in one-to-one correspondence. The fusion rules require that the

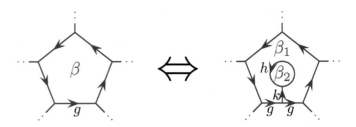

Fig. 31.6 Addition or removal of a plaquette. Here, the large plaquette β is split into two plaquettes β_1 and β_2. The plaquette β_2 is bounded by a single edge labeled h and a single vertex.

stem of the tadpole can only be labeled by the identity element, so (although the edge is still there) diagrammatically we can ignore the stem altogether. If $|\Psi\rangle$ is a ground-state wavefunction of the entire system on the lattice before splitting the plaquette, we thus construct a wavefunction after splitting as

$$|\Psi'\rangle = |\Psi\rangle \otimes |k = 0\rangle \otimes \frac{1}{|G|} \sum_{h \in G} |h\rangle \quad . \tag{31.10}$$

The sum over h now assures that $\hat{P}_{\beta_2} = 1$, that is, it puts β_2 in the unique ground state. The β_2 plaquette now has an outside boundary (what was previously the boundary of the β plaquette) and an inside boundary (the boundary of the β_2 plaquette). The operator $\hat{P}_{\beta_1}$ can now be written as

$$\hat{P}_{\beta_1} = \frac{1}{|G|} \sum_{q \in G} \hat{P}_\beta(q) \hat{P}_{\beta_2}(q^{-1})$$

where the action on the stem $k = 0$ is trivial since it is acted on from both sides. Now since $\hat{P}_{\beta_2}(q^{-1})$ acts trivially on β_2 in the ground state (see after Eq. 31.9) the operator $\hat{P}_{\beta_1}$ in the right figure then acts on the outer edges of the larger plaquette exactly the same as $\hat{P}_\beta$ did in the left figure. Thus, we have an isomorphism between the ground states on the two lattices.

Once we have established this independence of geometry, we can find the ground-state degeneracy for various topologies by using a minimal lattice decomposition. Let us consider the case of a sphere. (We discuss the torus case in Section 31.5). The minimal decomposition of a sphere is shown in Fig. 31.7. This consists of a single vertex on the equator, a single edge running around the equator and two plaquettes—one covering the north hemisphere and one covering the south hemisphere. The edge running around the equator is labeled with g and the vertex condition is automatically satisfied. The two plaquette operators are both equal to

$$\hat{P}_{\beta N} = \hat{P}_{\beta S} = \frac{1}{|G|} \sum_{g \in G} \hat{P}_{\beta N}(g) \quad .$$

Thus, the unique +1 eigenstate of this operator is given by

$$|\Psi\rangle = \frac{1}{|G|} \sum_{g \in G} |g\rangle$$

and is the unique ground state of the system.

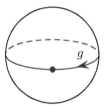

Fig. 31.7 This is the minimal lattice decomposition of a sphere. It consists of a single vertex, a single edge (labeled g) and two plaquettes—one covering the north hemisphere and one covering the south hemisphere.

31.4 Continuum Model

As with the toric code, we are not tied to any particular lattice. We can further dispense with the lattice altogether and consider diagrams

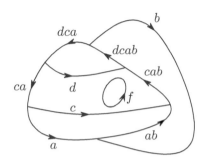

Fig. 31.8 A diagram of group valued lines, drawn without reference to an underlying lattice. (This figure is identical to Fig. 19.3.)

in the absence of the lattice, as shown, for example, in Fig. 31.8.

In this language we can write the wavefunction analogously as

$$|\psi\rangle \sim \sum_{\substack{\text{all diagrams that can be ob-}\\ \text{tained from a reference dia-}\\ \text{gram via allowed moves}}} |\text{diagram}\rangle \ . \qquad (31.11)$$

(Compare this to Eq. 30.1 for the toric code.)

The "allowed moves" between such diagrams are of the three types shown in Fig. 31.9, which further generalize the moves of Fig. 30.19.

- **Move 1: "Isotopy."** Meaning any smooth deformation of lines.

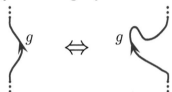

- **Move 2: Adding and Removing Contractible Loops.** A contractible loop with any label can be added or removed.

- **Move 3: Reconnecting (trivial F) move.**

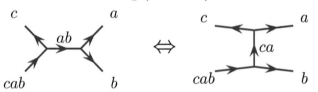

Fig. 31.9 Allowed diagrammatic moves in the Kitaev model. Compare Fig. 30.3.

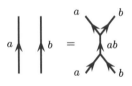

Fig. 31.10 This identity is a special case of Move 3. (Compare to Fig. 16.9).

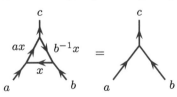

Fig. 31.11 A slightly less trivial corollary of the three moves. See Exercise 31.2.

The first move is similar to what we had in the case of the toric code. The second move, that we can add or remove loops, is also similar to the case of the toric code, except that the loops have arrows as well as group valued labels. Move 3 is "new" but should look familiar from our discussion of F-moves. However, here the F-coefficient is simply unity—and there is no sum over states on the right-hand side (compare to Fig. 16.4). This type of F-matrix (which is not a matrix but simply a scalar) is what we call the trivial cocycle in Section 19.1.

In addition to these three moves one can reverse arrows by inverting the group element (as in Fig. 31.2), and one can freely add or remove lines labeled with the identity element of the group.

Using these moves, we can derive a number of useful (and perhaps unsurprising) lemmas, such as those shown in Fig. 31.10 and Fig. 31.11 (see Exercise 31.2).

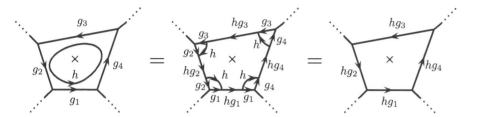

Fig. 31.12 The plaquette operator $P_\beta(h)$ can be viewed as inserting a loop labeled h and pushing it into the edges. In going from the left to the middle we use Fig. 31.10 once on each edge. In going from the middle to the right we use Fig. 31.11 once on each corner. The $\times$ in the middle of the plaquette should be thought of as a puncture in the plane. This puncture prevents us from removing the loop using Move 2, but instead forces us to merge the loop into the edges.

It is sometimes useful to merge the language of quantum numbers on the lattice with the language of continuum diagrams off the lattice. A very convenient example of this is the statement that the operator $\hat{P}_\beta(h)$ from Eq. 31.3 can be thought of as inserting a loop labeled h inside of a plaquette and then merging it into the edge variables, as shown in Fig. 31.12.

With this notation we can write the plaquette operator diagrammatically as

$$\hat{P}_\beta \left[\begin{array}{c} \beta \end{array} \right] = \left[\begin{array}{c} \times \tilde{\Omega} \end{array} \right] \tag{31.12}$$

where here the purple line is an $\tilde{\Omega}$ loop (compare Fig. 17.11) defined in this case as

$$\left. \tilde{\Omega} \right| = \frac{1}{|G|} \sum_{h \in G} \Big\uparrow h \tag{31.13}$$

The $\times$ in the middle of the loop in Eq. 31.12 indicates that the $\tilde{\Omega}$ loop is not contractible to a point so that Move 2 is not allowed (we can think of the $\times$ as being a puncture in the plane), but instead the loop must be pushed into the edges (as in Fig. 31.12) to achieve exactly the operation defined in Eq. 31.5. This principle—that the plaquette operator is exactly an $\tilde{\Omega}$ loop—is true for all of our generalizations of the toric code.

31.4.1   Brief Comments on Twisted Kitaev Theory

Once we have made the connection between the Kitaev model and a planar diagrammatic algebra it becomes natural to ask whether we can generalize the diagrammatic algebra and build more general topological models. Keeping with the idea of basing the model on the properties of a group G, we can construct other consistent planar diagram algebra by generalizing the F-move (Move 3 in Fig. 31.9). Indeed as discussed in Section 19.1 any so-called 3-cocycle provides a consistent scalar F-matrix for a planar diagram algebra. The 3-cocycle used by the Kitaev model is just the trivial cocycle (i.e. F, or ω in the notation of Section 19.1, is just unity). However, one can build a version of the Kitaev model that uses nontrivial cocycles (nontrivial Fs) as well. This is known as *twisting* the Kitaev model (or sometimes known as a *quasi-quantum double*). The word "twisting" here has nothing to do with the twists introduced in Chapter 15. The language comes from its original use in the context of discrete gauge theories. As we discuss in Section 31.7, the Kitaev model is essentially a discrete gauge theory. The twisted Kitaev model corresponds to what is known as twisted gauge theories. The twisted Kitaev models are also equivalent to the Dijkgraaf–Witten models introduced in Section 23.4 with the untwisted Kitaev model corresponding to the case of the trivial cocycle. The difference between (possibly twisted) Kitaev and Dijkgraaf–Witten is that the Kitaev model is a two-dimensional model with a Hamiltonian and continuous time, whereas Dijkgraaf–Witten calculates an action for a discretization of a three-dimensional spacetime.

31.5   Ground-State Degeneracy on a Torus

We now turn to calculate the ground-state degeneracy of the (untwisted) Kitaev model on a torus. We can use the minimal lattice decompositions of the torus as shown in Fig. 31.13 with a single vertex, two edges, and a single plaquette. We denote by $|b, a\rangle$ the state on the torus with the edge going around the vertical cycle labeled by b and the edge going around the horizontal cycle labeled by a.

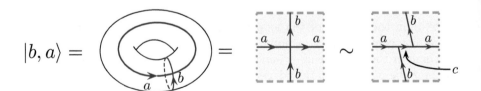

Fig. 31.13 Decomposition of a torus into (left and middle) one vertex, two edges, and one plaquette. (Right) A slightly more complicated decomposition with two vertices, three edges, and one plaquette. This adds no additional degrees of freedom since c is completely fixed by a and b. However, this trivalent picture fits better with our discussion of diagrammatic algebras where vertices are always trivalent.

The vertex condition at the single vertex is $bab^{-1}a^{-1} = e$ with e the identity. This means that a and b must commute.

On the far right of Fig. 31.13 we show an alternative decomposition of the torus with two vertices and three edges and a single plaquette. This picture fits better with our continuum diagrammatic algebra where we usually insist on vertices that are trivalent.[10] There is no added degree of freedom from the added edge, since the quantum number c is completely defined by a and b. One vertex condition imposes $c = ab$ and the other imposes $c = ba$, so again a and b must commute.

[10]It is not actually so crucial for the diagrammatic algebra of the Kitaev model (defined in Fig. 31.9) that we insist on trivalent vertices. However, in cases where the F-move has a nontrivial phase associated with it (such as with twisted Kitaev models), one must insist on sticking to trivalent vertices to properly keep track of this phase.

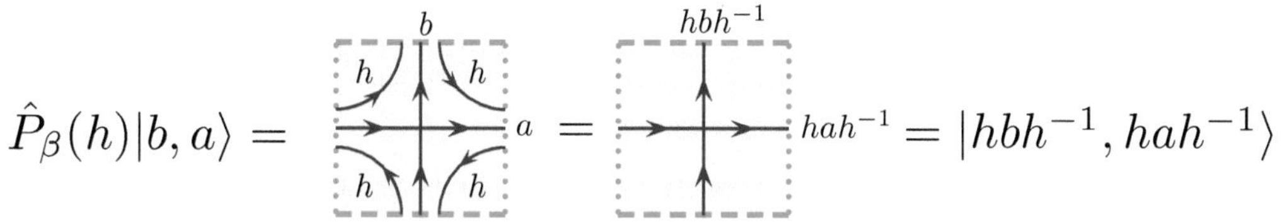

$$\hat{P}_\beta(h)|b,a\rangle = \quad \cdots \quad = \quad \cdots \quad hah^{-1} = |hbh^{-1}, hah^{-1}\rangle$$

Fig. 31.14 Application of the plaquette operator $\hat{P}_\beta(h)$ on the single plaquette conjugates both edge variables.

Let us now consider the plaquette operator $\hat{P}_\beta(h)$ for a single plaquette. To apply this operator we push a loop labeled h into the edges (analogous to Fig. 31.12), thus conjugating both edge variables by h as shown in Fig. 31.14. Note that if a commutes with b, then hbh^{-1} commutes with hah^{-1}, so if the vertex condition is satisfied (if the two edges commute) the action of $\hat{P}_\beta(h)$ does not change this.

We would now like to find the ground state or code space of our system, which must not only be made from our basis states $|b, a\rangle$ with b and a commuting, also must be a superposition so that we obtain a $+1$ eigenvalue of the plaquette operator $\hat{P}_\beta$ (Eq. 31.5). Given a state $|b, a\rangle$ as above satisfying the vertex condition (i.e. with b and a commuting) we construct the state

$$|\Psi_{b,a}\rangle \propto \hat{P}_\beta|b,a\rangle \propto \sum_{h \in G} |hbh^{-1}, hah^{-1}\rangle \tag{31.14}$$

and here we do not keep track of a normalizing prefactor of the wavefunction. It is easy to check that this is indeed a $+1$ eigenstate of $\hat{P}_\beta$ and is therefore a ground state. However, it is also true that not all sets of commuting b and a generate distinct wavefunctions. In particular, if we define the *orbit* of $|b, a\rangle$ to be all of the states of the form $|hbh^{-1}, hah^{-1}\rangle$ for any $h \in G$, we realize that each distinct orbit represents a different orthogonal ground-state wavefunction. That is, Eq. 31.14 tells us to simply add up all of the wavefunctions within an orbit to make a ground-state wavefunction.

Example: S_3

Let us consider the nonabelian group S_3, the permutation group on three elements, also known as the symmetries of a triangle (see Section 40.2.1). This group has two generators, x and r, with the properties that

$$\begin{aligned}
\Psi_{e,e} &= |e,e\rangle \\
\Psi_{r,e} &= |r,e\rangle + |r^2,e\rangle \\
\Psi_{e,r} &= |e,r\rangle + |e,r^2\rangle \\
\Psi_{r,r} &= |r,r\rangle + |r^2,r^2\rangle \\
\Psi_{r,r^2} &= |r,r^2\rangle + |r^2,r\rangle \\
\Psi_{e,x} &= |e,x\rangle + |e,xr\rangle + |e,xr^2\rangle \\
\Psi_{x,e} &= |x,e\rangle + |xr,e\rangle + |xr^2,e\rangle \\
\Psi_{x,x} &= |x,x\rangle + |xr,xr\rangle + |xr^2,xr^2\rangle
\end{aligned}$$

Table 31.1 The eight wavefunctions generated by Eq. 31.14. Note that we have chosen a different normalization of these wavefunctions where we write each ket exactly once with a coefficient of unity.

[11] Burnside's lemma was not discovered by Burnside. As such it is sometimes called "The lemma that is not Burnside's." This makes it a good example of Stigler's law of eponymy which says that nothing is named after the person who discovered it—including Stigler's law, which was first stated by Merton!

$x^2 = r^3 = e$ with e the identity element and $xr = r^2x = r^{-1}x$. There are a total of six elements of the group ($6 = 3!$ permutations), which we list as e, r, r^2, x, xr, xr^2. We find eight orthogonal wavefunctions generated by Eq. 31.14, where each wavefunction is a different orbit under the conjugation action. These eight wavefunctions are given in Table 31.1. Note that these wavefunctions are not properly normalized.

General Case

For a general nonabelian group it may seem complicated to try to figure out the dimension of the ground-state space. In fact, there is beautiful theorem from group theory, known as Burnside's lemma[11] which helps us calculate this dimension.

Burnside's lemma: Given a finite group G acting[12] on a set X, the number of orbits in X due to this group action is given by

$$\text{number of orbits} = \frac{1}{|G|} \sum_{h \in G} \left(\begin{array}{c} \text{number of elements of } X \\ \text{that are left unchanged by } h \end{array} \right).$$

In our case, X is the set of all elements $|b,a\rangle$ where b and a are elements of G and commute, and the action of the group is conjugation as in Fig. 31.14. That is, h acts on $|b,a\rangle$ to give $|hbh^{-1}, hah^{-1}\rangle$. An element $|a,b\rangle$ is unchanged by h if h commutes with both a and b. Thus, we obtain the following general result

$$\begin{aligned}
&\text{ground-state degeneracy of Kitaev model on torus} \\
&= \text{number of orbits} \\
&= \frac{1}{|G|} \left(\begin{array}{c} \text{Number of triples } a, b, h \\ \text{that all commute with each other} \end{array} \right)
\end{aligned} \quad (31.15)$$

31.6 Quasiparticles

Being that we know the ground-state degeneracy on the torus, we know the number of quasiparticle types. In this section we discuss the properties of these quasiparticles. Although this is a bit more complicated[13] than what we did for the (abelian) case of the toric code, most of the development recapitulates the discussions of Chapter 30, particularly Section 30.6.

[13] Some of the algebra in this section looks a bit nasty, but it is mostly just a matter of careful bookkeeping. Those who are particularly frightened of such algebra should not worry too much about details.

Recall in the case of the toric code, and more generally for the $\mathbb{Z}_N$ toric code, the vertex defects involve the ends of a string. We can examine the system locally and determine the quantum number of a string end, even if we do not know where the string is. In the $\mathbb{Z}_N$ case we measure the charge of the vertex defect (see Eq. 27.19). However, in the case of the Kitaev model with a nonabelian group, the "charge" at the end of a

[12] A group G *acting* on a set X here means a mapping $f(g,x)$ with $g \in G$ and $x \in X$, which gives an output in X such that $f(e,x) = x$ and $f(g, f(h,x)) = f(gh,x)$.

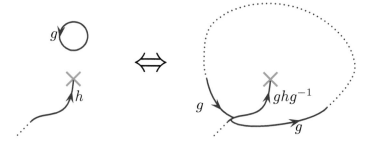

Fig. 31.15 This diagram shows how a string end is conjugated by pulling a loop over the string end.

string is not as easily defined. To see this, consider Fig. 31.15. On the left we have a diagram with a string end labeled with h. Move 2 from Fig. 31.9 allows us to create the loop labeled g on the left. This loop may then be pulled over the string end, which conjugates the label of the string end to ghg^{-1}. Since the wavefunction of the Kitaev model should be in a superposition of diagrams from all allowed moves, the wavefunction must be in a superposition of having a loop end labeled h and ghg^{-1} for any possible group element g. Thus, the string ends should not be labeled by their group element, but rather by a *conjugacy class*.[14] When we do diagrammatic manipulations, we may choose a particular representative element of the conjugacy class (such as h or some particular value of ghg^{-1}). However, we should remember that the group value of the string end is actually a superposition of all elements in the class.

With this knowledge, we now follow the tube algebra approach we introduced in Section 30.2 and 30.6 in order to deduce the properties of the quasiparticles. We might start by introducing a basis of the form shown in Fig. 31.16. In the rectangular (right) representation of the annulus, h comes in the bottom and ghg^{-1} goes out the top (these are the inner and outer edges of the annulus). Both h and ghg^{-1} are in the same conjugacy class, but unlike the case of the toric code, they may not be the same group element.

[14]To remind the reader (see Section 40.2.2), the conjugacy class of h is the set of all group elements that can be written as ghg^{-1} for any g in the group.

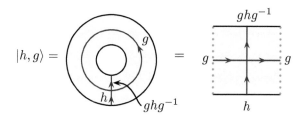

Fig. 31.16 The basis we work with for the tube algebra of the Kitaev model. The blue line from the outer edge of the annulus (or from the bottom in the right figure) is labeled h. The blue line going to the inner edge of the annulus (or off the top on the right figure) is labeled ghg^{-1} to satisfy the vertex condition at the intersection.

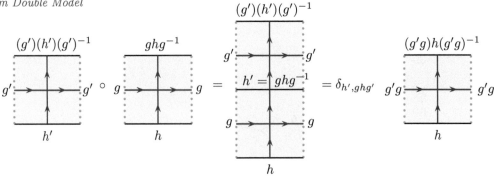

Fig. 31.17 Graphical demonstration of the composition law Eq. 31.16.

The composition rules for these states on the annulus are given by

$$|h', g'\rangle \circ |h, g\rangle = \delta_{h', ghg^{-1}} |h, g'g\rangle \qquad (31.16)$$

which is demonstrated graphically in Fig. 31.17.

This composition law is fundamental to the mathematical structure of the quantum double model. However, it is extremely inconvenient to work with this notation. It turns out to be more convenient to work with a different notation.

For each conjugacy class C we choose one representative element in this conjugacy class that we call r_C (it does not matter which element we choose).

Each element h in the conjugacy class C can be written in the form $h = p\, r_C\, p^{-1}$. This decomposition is not unique as several values of p might give the same h. Thus, for each element $h \in C$ we choose a *particular* representative p such that $h = p\, r_C\, p^{-1}$ and we should always use this value of p to represent this particular h. Let the set of representative elements (which has $|C|$ elements in it) be called $\{p\}^C$.

We now want to rewrite the basis states $|h, g\rangle$ in terms of this notation. First, let us write h as

$$h = p\, r_C\, p^{-1} \qquad (31.17)$$

where h is in conjugacy class C and p is chosen from the representative set $\{p\}^C$. This decomposition is unique.

Now in Fig. 31.16 we have ghg^{-1}, and this is in the same conjugacy class as h. So we can also uniquely decompose

$$ghg^{-1} = p'\, r_C\, p'^{-1} \qquad (31.18)$$

with $p' \in \{p\}^C$. Let us further define

$$z = p'^{-1} g p \qquad (31.19)$$

such that

$$g = p'\, z\, p^{-1} \quad . \qquad (31.20)$$

Now, if we are given h, g we can uniquely determine[15] C, z, p', p. And in reverse, if we are given C, z, p', p we can uniquely generate[16] h and g.

[15] C is the conjugacy class containing h, then p and p' are obtained from Eqs. 31.17 and 31.18 and z is then given by Eq. 31.19.

[16] Just use Eqs. 31.17 and 31.20.

Further, we claim that z must commute[17] with r_C. That is, $zr_Cz^{-1} = r_C$. We say that z is in the *centralizer* of r_C and we write $z \in Z(r_C)$. Were we to try to put a z into Eq. 31.19 that does not commute with r_C we would find an inconsistency.

Thus, instead of writing $|h, g\rangle$ for a state on the annulus, we instead write $|C, z; p', p\rangle$. The set of equations (Eq. 31.17–31.20) that relate the two set of quantities seems somewhat arbitrary, but it has a very beautiful graphical representation, shown in Fig. 31.18.

[17]To see this, start with Eq. 31.18 and substitute Eq. 31.17 for h on the left and move the p' factors to the left-hand side to obtain

$$p'^{-1}g(pr_Cp^{-1})g^{-1}p' = zr_Cz^{-1} = r_C.$$

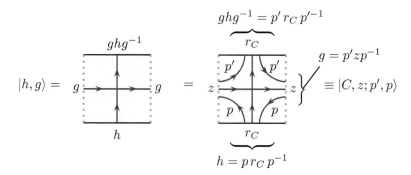

Fig. 31.18 Graphically defining the $|C, z; p, p'\rangle$ basis in terms of the $|h, g\rangle$. Here, r_C is the representative element of the conjugacy class C and $zr_C = r_Cz$ is required by the vertex condition.

Here, the vertex condition $r_C z r_C^{-1} z^{-1} = e$ in the center of the figure requires that z commutes with r_C.

Our original basis $|g, h\rangle$ had $|G|^2$ different elements, and our new basis $|C, z; p', p\rangle$ also has $|G|^2$ different elements, so we are still spanning the same space.[18]

In terms of this new representation of the basis, the composition rule for such basis states is similar to the case for the $\mathbb{Z}_N$ toric code in Eq. 30.17 and is now given by

$$|C_1, z_1; p'_1, p_1\rangle \circ |C_2, z_2; p'_2, p_2\rangle = \delta_{C_1,C_2}\delta_{p_1,p'_2}|C_1, z_1z_2; p'_1, p_2\rangle \quad (31.21)$$

which we show graphically in Fig. 31.19.

[18]To count the number of basis states, we have a sum over conjugacy classes C with $|Z(r_C)|$ possibilities for z and $|C|$ possibilities for each p and $|C|$ possibilities for each p'. Thus, we want $\sum_C |Z(r_C)||C|^2$. However, using Eq. 40.2 we have $|Z(r_C)||C| = |G|$, so this becomes $|G|\sum_C |C|$, and the sum over conjugacy classes of the number of elements in the conjugacy class is just the number of elements in the group, thus giving $|G|^2$. This is also obvious from the fact that we have a one-to-one mapping between the two types of basis states!

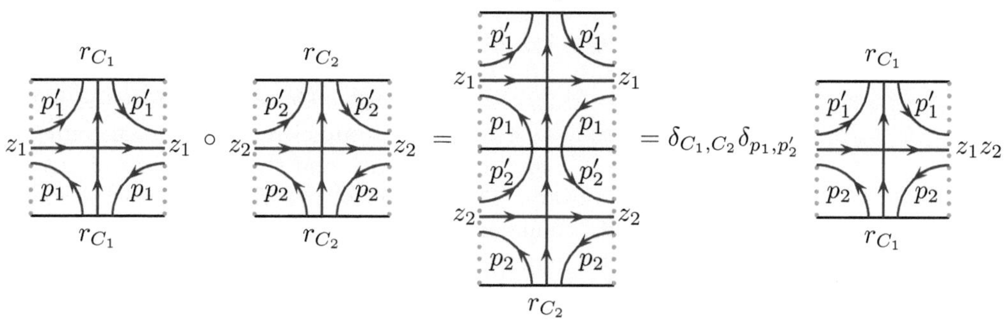

Fig. 31.19 Graphical demonstration of the composition law Eq. 31.21.

We now construct quasiparticle idempotents from these basis states analogous to Eq. 30.18 for the $\mathbb{Z}_N$ toric code. In that equation, we are essentially Fourier transforming to get from the basis to the quasiparticle idempotents. Here, we need the nonabelian generalization of the Fourier transform, which is done with group representations. We write the following basis of states

$$O^{[C,R]}_{p',n';p,n} = \frac{d_R}{|Z(r_C)|} \sum_{z \in Z(r_C)} \rho^R_{n',n}(z) \, |C,z,p',p\rangle \qquad (31.22)$$

where $Z(r_C)$ is the centralizer of r_C, the representative element of the conjugacy class C. (Recall that the centralizer $Z(r_C)$ is the group of elements that commutes with r_C, and it has $|Z(r_C)|$ elements). Here, R is an irreducible representation of $Z(r_C)$ of dimension d_R and ρ^R is the representation matrix (see Section 40.2.4).

Again, our original basis $|g,h\rangle$ had $|G|^2$ elements in it, and our new basis $O^{[C,R]}_{p',n';p,n}$ also has $|G|^2$ elements in it, so we are still spanning the whole space.[19]

Using the grand orthogonality theorem, Eq. 40.5, it is easy to prove (see Exercise 31.6) that these objects obey the idempotent law

$$O^{[C_1,R_1]}_{p'_1,n'_1;p_1,n_1} \circ O^{[C_2,R_2]}_{p'_2,n'_2;p_2,n_2} = \qquad (31.23)$$
$$\delta_{C_1,C_2} \delta_{R_1,R_2} \, \delta_{p_1,p'_2} \delta_{n_1,n'_2} \; O^{[C_1,R_1]}_{p'_1,n'_1;p_2,n_2} \quad .$$

This is somewhat more complicated than the idempotent equation we had for the toric code Eq. 30.19. Here, p and n are simply additional boundary degrees of freedom that need to match when we glue together the annuli. The C and R degrees of freedom, however, are unchanged when we glue annuli together, and it is these indices that describe the quasiparticle types. Thus, we have isolated the annulus states corresponding to the different quasiparticle types in the Kitaev model.

To emphasize: *A quasiparticle type is described by a conjugacy class C, and an irreducible representation R of the centralizer of a representative element r_C of the conjugacy class.*

It is conventional to call the conjugacy class the *magnetic* charge and the representation R the *electric* charge (despite the fact that we are here thinking of the conjugacy class as being a vertex defect!). The origin of these names are discussed in Section 31.7 when we discuss the relationship to gauge theory.

Note that in the case of the $\mathbb{Z}_N$ toric code, which is just the Kitaev model with the group $G = \mathbb{Z}_N$, each group element is its own conjugacy class so there are $|G|$ conjugacy classes, and the centralizer of any element is the entire group. The number of irreducible representations of the group is also $|G|$ (number of irreps is always the number of conjugacy classes), so the total number of particle species is $|G|^2$ in agreement with what we found in Section 30.6.

[19] To count the total number of basis states, we have a sum over C, a sum over irreps R with n and n' both taking d_R values and p and p' both taking $|C|$ values. Thus, we want $\sum_{C,R} d_R^2 |C|^2$. However, from Eq. 40.3 we have $\sum_R d_R^2 = |Z(r_C)|$ (recall that these are reps of $Z(r_C)$), thus leaving us with $\sum_C |Z(r_C)||C|^2 = |G|^2$ as per note 18.

31.6.1   Quasiparticle Basis on Torus

We can use the quasiparticle idempotents we just derived to construct a quasiparticle basis on the torus. We specify a quasiparticle idempotent $O^{[C,R]}$ with a conjugacy class C and a representation R of the centralizer $Z(r_C)$ of the representative element r_C of the conjugacy class. We then attach the inner and outer edge of the annular states to obtain the torus states[20]

$$| [C,R] \rangle \sim \sum_{n=1}^{d_R} \sum_{p \in \{p\}^C} O^{[C,R]}_{p,n;p,n} \quad .$$

Summing over these degrees of freedom, and multiplying by a constant, we obtain our normalized torus states as

$$| [C,R] \rangle = \frac{1}{\sqrt{|G|}} \sum_{z \in Z(r_C)} \sum_{p \in \{p\}^C} \chi^R(z) | C, z; p = p' \rangle \qquad (31.24)$$

where $\chi^R = \mathrm{Tr}\rho^R$ is the character of the representation, and we have $p = p'$ so that we can describe states on a torus rather than an annulus (i.e. we have connected up the inner and outer edges of the annulus; see Fig. 31.18).

Note that the prefactors here are chosen so that we have proper wavefunction orthonormalization (see Exercise 31.8)

$$\langle [C,R] | [C',R'] \rangle = \delta_{CC'} \delta_{RR'} \quad . \qquad (31.25)$$

The states $|[C,R]\rangle$ form a basis for the ground state space on a torus. Therefore, we have

ground-state degeneracy of Kitaev model on torus
$$= \sum_C (\text{number of irreducible reps of } Z(r_C)) \qquad (31.26)$$

where the sum is over conjugacy classes C. It is not immediately obvious that this expression for the ground-state degeneracy should match Eq. 31.15. Indeed, these two expressions do match (with a bit of group theory!). This is shown in Section 31.8.

Example: S_3 Again

Let's return to the example of the group S_3 as we did in Section 31.5. The group has three conjugacy classes $C_e = \{e\}$, and $C_r = \{r, r^2\}$, and $C_x = \{x, xr, xr^2\}$. Let us choose the first element in each class as its representative elements. These have centralizers $Z(e) = S_3$ (the whole group, since everything commutes with e), $Z(x) = \mathbb{Z}_2 = \{e, x\}$ and $Z(r) = \mathbb{Z}_3 = \{e, r, r^2\}$. There are three irreducible representations of[21] of S_3 (see Table 20.1), which we call I, S, V. For $\mathbb{Z}_2$ and $\mathbb{Z}_3$ there are two and three irreps, respectively (see Eq. 40.4). The two irreps of $\mathbb{Z}_2$ we call $\{1, -1\}$ and the three irreps of $\mathbb{Z}_3$ we call $\{1, \omega, \omega^2\}$ where $\omega = e^{2\pi i/3}$. Thus, there are a total of eight quasiparticle types

$$\begin{aligned}
\mathcal{N}|[C_e,I]\rangle &= \Psi_{ee} + \Psi_{re} + \Psi_{xe} \\
\mathcal{N}|[C_e,S]\rangle &= \Psi_{ee} + \Psi_{re} - \Psi_{xe} \\
\mathcal{N}|[C_e,S]\rangle &= 2\Psi_{ee} - \Psi_{re} \\
\mathcal{N}|[C_x,+1]\rangle &= \Psi_{ex} + \Psi_{xx} \\
\mathcal{N}|[C_x,-1]\rangle &= \Psi_{ex} - \Psi_{xx} \\
\mathcal{N}|[C_r,1]\rangle &= \Psi_{er} + \Psi_{rr} + \Psi_{rr^2} \\
\mathcal{N}|[C_r,\omega]\rangle &= \Psi_{er} + \omega\Psi_{rr} + \omega^2\Psi_{rr^2} \\
\mathcal{N}|[C_r,\omega^2]\rangle &= \Psi_{er} + \omega^2\Psi_{rr} + \omega\Psi_{rr^2}
\end{aligned}$$

Table 31.2 The eight wavefunctions generated by Eq. 31.24. We have written these using the (unnormalized) basis states given in Table 31.1. Here, $\omega = e^{2\pi i/3}$. With normalization constant $\mathcal{N} = \sqrt{6}$ the kets on the left are properly normalized

$[C_e,I]$, $[C_e,S]$, $[C_e,V]$, $[C_x,+1]$, $[C_x,-1]$, $[C_r,1]$, $[C_r,\omega]$, $[C_r,\omega^2]$ matching the eight ground states we found in Table 31.1.

The characters of S_3 are given in Table 20.1, whereas the characters of the $\mathbb{Z}_N$ are just the one-dimensional representation matrices themselves since the representations are scalar (see Eq. 40.4). Using Eq. 31.24 we obtain the eight explicit ground-state wavefunctions shown in Table 31.2.

S- and *T*-matrices

For any Kitaev model, we can calculate the effect of the twist operation on the basis states (i.e. the T-matrix) analogous to Eq. 30.20 and Fig. 30.22, giving us

$$\hat{\theta}|h,g\rangle = |h,gh\rangle \tag{31.27}$$

or equivalently

$$\hat{\theta}|C,z;p=p'\rangle = |C,zr_C;p=p'\rangle \quad . \tag{31.28}$$

It is a very short exercise (see Exercise 31.7) to then show that

$$\hat{\theta}\,|\,[C,R]\,\rangle = \left[\frac{\chi^R(r_C^{-1})}{d_R}\right]|\,[C,R]\,\rangle \tag{31.29}$$

where χ^R is the character of representation R and $d_R = \chi^R(e)$ is the dimension of this representation. Thus, we have the twist factor of the particle type $[C,R]$ is

$$\theta_{[C,R]} = \left[\frac{\chi^R(r_C^{-1})}{d_R}\right] = \left[\frac{\chi^R(r_C)}{d_R}\right]^* \tag{31.30}$$

with the factor in brackets being a unit magnitude complex phase.

We can now turn to calculating the S-matrix. Analogous to Eq. 30.21 in the $\mathbb{Z}_N$ toric code case, the $\underline{S}$ operation simply rotates the two directions on the torus

$$\underline{S}|h,g\rangle = |g,h^{-1}\rangle \quad .$$

The full S-matrix is then given as

$$S_{[C',R'],[C,R]} = \langle\,[C',R']\,|\,\underline{S}\,|\,[C,R]\,\rangle \quad . \tag{31.31}$$

Plugging in the value of the kets in Eq. 31.24, with a little algebra (see Exercise 31.9) we obtain

$$S_{[C',R'],[C,R]} = \tag{31.32}$$
$$\frac{1}{|Z(r_C)|\,|Z(r_{C'})|}\sum_{\substack{p\text{ such that}\\ p^{-1}r_{C'}p\,\in\,Z(r_C)}} \chi^{R'}(p\,r_C\,p^{-1})\chi^R(p^{-1}\,r_{C'}\,p) \quad .$$

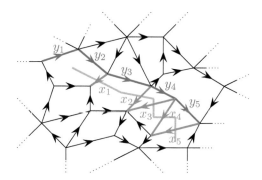

Fig. 31.20 A ribbon operator generally will operate on both the dark green edges and the red edges and will create defects only at its endpoints.

31.6.2 Quasiparticle Ribbon Operators on a Lattice

Let us now return to the Kitaev model on the lattice and try to construct quasiparticles in this case. In the case of the toric code, we were able to create quasiparticles and pull them apart from each other by applying operations on a "string" of edges. For the vertex defects, we applied operations on a string of edges connecting two defects (Fig. 27.10), whereas for the plaquette defects we applied operations to a set of edges *dual* to a line connecting the defects (Fig. 27.14). These strings of operators were defined by the fact that they were only detectable at their endpoints— the actual path the string takes could not be measured. Further, if the strings formed contractible closed loops, they act trivially on the ground state. One can construct similar objects, known as *ribbon* operators, for quasiparticles of the Kitaev model, as shown in Fig. 31.20. Most generally, a ribbon operator will act on a connected string of edges on the lattice (red edges in Fig. 31.20) and also on a string of dual edges (dark green edges in Fig. 31.20).[22] Our convention, as in the figure, is to orient all of the red arrows to point in the direction of the ribbon (start point of ribbon to end point), with the green edges to the right of the red edges if we are walking in the direction of red arrows. We orient the green arrows to point away from the neighboring red path, as in Fig. 31.20.

[22]The word "ribbon" is used to describe a strip bounded by the red and light green paths.

Charge-Like Ribbon Operators

Let us start by considering plaquette defects, which are fairly simple. As in the toric code case, the relevant operator acts on a string of edges (dark green in Fig. 31.20) dual to a path (bright green) between defects. Consider now measuring the ordered product of the group elements on all of these edges $x_N x_{N-1} \ldots x_3 x_2 x_1$. We write an operator

$$\hat{Y}^g(\text{path}) = \delta_{g,\,x_N x_{N-1}\ldots x_3 x_2 x_1} \tag{31.33}$$

which measures whether this product equals g. Here, we have assumed there are $N+1$ plaquettes along the path (see Fig. 31.20 where $N = 5$). This operator clearly commutes with the vertex operators, since it does not change the value of the group element along any edge. In addition this operator commutes with every plaquette operator except the plaquette at the start and the end of the path. To see this, consider the example shown in Fig. 31.21, where $P_\beta(a)$ is applied to the plaquette between x_1 and x_2. The operator $P_\beta(a)$ takes $x_1 \rightarrow ax_1$ but also takes $x_2 \rightarrow x_2a^{-1}$ so that the product $\ldots x_2x_1 \ldots$ remains unchanged. Thus, the $\hat{Y}^g$ operator only makes plaquette defects at the ends of the path—that is, this creates pure "charge" excitations.[23]

Flux-Like Ribbon Operators

Another operator to consider is the analog of the vertex defect string operator. As with the vertex-defect string for the toric code, the general idea will be to put a string on top of a line (the red line) of edges. However, the procedure for doing this is a bit more complicated. Consider adding a string with quantum number h starting in the first plaquette and ending in the last plaquette, as shown in Fig. 31.22. If we think about our diagrammatic algebra, when we pull this string through the dark green edges, its value gets conjugated by the value of the green edge. So for example, on the left of the figure, the blue string starts out labeled h but after passing through x_1 it is labeled $x_1^{-1}hx_1$ analogous to Fig. 31.15, and then after also passing through x_2 it is labeled $x_2^{-1}x_1^{-1}hx_1x_2$, and so forth. Once we have this blue string in the diagram, we can then fuse it into the red edges to obtain the picture on the right of Fig. 31.22. Let us write an operator[24]

$$\hat{Z}^h(\text{path}) = \text{pull string initially labeled } h \text{ along path and fuse into edge.}$$

We claim that this operator also commutes with all the vertex and plaquette operators except at the ends of the path. To show this, we start by examining the vertex operators. As an example, examine the vertex with q_1 incoming from the top in Fig. 31.22. On the left the vertex condition is that $q_1^{-1}y_3x_1y_2^{-1} = e$. On the right, after fusing the blue h-line into the red edges the condition becomes $q_1^{-1}(y_3x_1hx_1^{-1})x_1(y_2h)^{-1} = e$, which is the same condition. Therefore, if the vertex condition is satisfied before fusing the blue line into the edge, it is also satisfied after.

Let us now show that the operator $\hat{Z}^h$ commutes with the plaquette operator. As an example, consider the plaquette operator $P_\beta(a)$ applied to the plaquette between x_1 and x_2 (same as in Fig. 31.21). The type of edge we need to focus on is an edge like y_3 in Fig. 31.22. If we first apply $\hat{Z}^h$ to the path as shown in Fig. 31.22 we take $y_3 \rightarrow y_3x_1hx_1^{-1}$. Then if we apply $P_\beta(a)$ this edge then becomes $y_3x_1hx_1^{-1}a^{-1}$. On the other hand if we apply $P_\beta(a)$ first, this edge becomes y_3a^{-1}. But at the same time the edge x_1 becomes ax_1. When we then pull the h string through this ax_1 edge it becomes $(ax_1)h(ax_1)^{-1}$ and fusing the h string into the y_3a^{-1} edge we then get $y_3a^{-1}(ax_1)h(ax_1)^{-1} = y_3x_1hx_1^{-1}a^{-1}$ which is the same result, thus showing that the $\hat{Z}^h$ operator commutes with plaquette operators, except at the endpoints, thus showing that it can only create excitations at the ends of the path. Note that, unlike

[23] Again, we remind the reader that this charge operator is more analogous to what we called the magnetic particle in the toric code—acting with a phase on the edges dual to some path and exciting only the plaquettes. The nomenclature is from the mapping to gauge theory, which we discuss in Section 31.7.

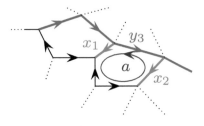

Fig. 31.21 The measurement of $x_1x_2x_3\ldots$ commutes with the application of $P_\beta(a)$ to the shown plaquette, since $P_\beta(a)$ takes $x_1 \rightarrow x_1a^{-1}$ and takes $x_2 \rightarrow ax_2$. Note that in this figure we have reversed the orientation of x_2 such that all of the green arrows point in the same direction with respect to the ribbon (bright green in Fig. 31.20.

[24] Strictly speaking, we specify this operator using both a dual path (through the middle of plaquettes) where we position the h string initially, and then a path (along the edges) where it finally gets fused.

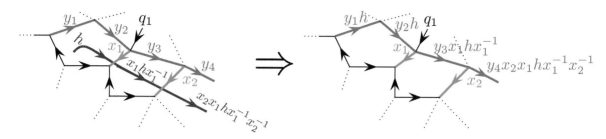

Fig. 31.22 To modify the values of the red edges, we insert a blue string labeled h into the diagram and pull this string through the green edges, as shown on the left. As it gets pulled through the green edge x_1 its value is conjugated to $x_1^{-1}hx_1$ and then when it gets pulled through x_2 its value is conjugated again to $x_2^{-1}x_1^1hx_1x_2$. On the right, the blue line is then fused into the red lines. $y_1 \to y_1h$, $y_2 \to y_2h$, $y_3 \to y_3x_1^{-1}hx_1$, and so forth. Note that this operation commutes with all of the vertex and plaquette operators except at the endpoints of the ribbon.

the vertex defect operator in the toric code, this operator can actually create vertex and plaquette defects at the path ends.

Quasiparticle Eigenstate Ribbon Operators

The operators $\hat{Y}^g(\text{path})$ and $\hat{Z}^h(\text{path})$ are operators in that they only create excitations at the ends of their paths, but they generally create superpositions of different types of quasiparticles. With a few steps one can construct quasiparticle ribbon operators that create distinct quasiparticle types at their path ends.

First let us construct the following ribbon operator

$$\hat{F}^{h,g}(\text{path}) = \hat{Z}^h(\text{path})\,\hat{Y}^g(\text{path})\quad. \tag{31.34}$$

The two operators on the right commute with each other, and are here assumed to act along the same path (the path of red and green bonds as in Fig. 31.22). These operators have some very nice properties. For example, applying $\hat{F}$ multiple times along the same path gives

$$\hat{F}^{h,g}(\text{path})\hat{F}^{h',g'}(\text{path}) = \delta_{g,g'}\hat{F}^{hh',g}(\text{path})\quad. \tag{31.35}$$

Further, we can concatenate paths. If the end of path_1 is the same as the start of path_2 we can construct

$$\text{path}_{2*1} = \text{path}_2 * \text{path}_1 \tag{31.36}$$

with path_{2*1} meaning you follow path_1 then follow path_2. The $\hat{F}^{h,g}$ operators then obey

$$\hat{F}^{h,x}(\text{path}_{2*1}) = \sum_{g\in G} \hat{F}^{ghg^{-1},xg^{-1}}(\text{path}_2)\,\hat{F}^{h,g}(\text{path}_1)\quad. \tag{31.37}$$

To see how these connect up, note that at the start of path_1, the string is labeled h, and at the end of path_1 (which is the beginning of path_2), the string is labeled ghg^{-1}. Considering the product of elements along the green bonds, path_1 has value g, but path_2 has value xg^{-1} giving a total product of x.

The concatenation Eq. 31.37 on the lattice is essentially the same as Eq. 31.16 when we were composing states on the annulus and considering diagrams on the continuum. The only difference is that here we need to sum over intermediate quantum numbers (g), whereas in the continuum we simply mutated the structure of our diagram to eliminate the intermediate edges.

We can then follow essentially the same arguments as in Section 31.6 to construct ribbon operators corresponding to quasiparticle types. We thus have, analogous to Eq. 31.22 the ribbon operator

$$W^{[C,R]}_{p,n;p',n'}(\text{path}) = \sum_{z \in Z(r_C)} \rho^R_{n,n'}(z) F^{p\,r_C\,p^{-1},\,p'z\,p^{-1}}(\text{path}) \quad . \quad (31.38)$$

These quasiparticles are labeled as in Section 31.6 with a conjugacy class C and a representation R of the centralizer of a representative element of the conjugacy class. As in Section 31.6, the indices p, n are nontopological degrees of freedom. With some algebra (see Exercise 31.14) one can show that these operators obey the natural concatenation law

$$W^{[C,R]}_{p,n;p'',n''}(\text{path}_{2*1}) = \sum_{p',n'} W^{[C,R]}_{p,n;p',n'}(\text{path}_2)\ W^{[C,R]}_{p',n';p'',n''}(\text{path}_1) \ . \quad (31.39)$$

Forming a closed loop of the ribbon operator and acting on any ground state gives

$$W^{[C,R]}_{p,n;p',n'}(\text{loop}) \left|\psi_{GS}\right\rangle = \delta_{n,n'}\delta_{p,p'}\left|\psi_{GS}\right\rangle \quad .$$

If we connect up the indices on the ends of the loop, we trace this object and obtain

$$\sum_{n,p} W^{[C,R]}_{p,n;p,n}(\text{loop}) \left|\psi_{GS}\right\rangle = \text{Tr}\,W^{[C,R]}(\text{loop})\left|\psi_{GS}\right\rangle = |C|d_R\left|\psi_{GS}\right\rangle$$

which is the quantum dimension of the quasiparticle type. Here, d_R the dimension of representation R, the number of values of n, and $|C|$ is the number of elements in the conjugacy class, the number of values of p. Note that this loop gives the quantum dimension of the quasiparticle of type $[C,R]$ since

$$\mathsf{d}_{[C,R]} = |C|\,d_R$$

which we derive in Exercise 31.9.b.

31.7 Relation to Gauge Theory

The Kitaev model is essentially a discrete gauge theory with gauge group G. To see this relation, we begin with a quick review of the basics of gauge theory.

In Chapters 4 and 5 we already worked with gauge theories. Recall that the fundamental quantity we needed to keep track of was the Wilson loop operator (see Eq. 5.11), which is the generalization of the Aharonov–Bohm phase[25]

$$W_{\text{loop}} = \text{Tr}_R \left[P \left\{ \exp \left(i \oint_{\text{loop}} dl^\mu A_\mu \right) \right\} \right]$$

where P means we should path order the integral in the case where the gauge field A_μ takes its value in a nonabelian Lie algebra (if the Lie algebra is abelian, we do not have to path order). Here, the trace is taken in some representation R. That is, the value inside the curly brackets $\{\}$ is an element g of the Lie group, and one then takes $\text{Tr}[\rho^R(g)]$ where R is a representation and ρ^R is the representation matrix.

Let us now discretize space, introducing vertices that are connected together by edges, as in the right of Fig. 31.23. (Crucially, note taht this is *not* going to be the same lattice as the one in the Kitaev model, but will rather be dual to that lattice.) For a directed edge between vertex i and j we can assign a group element

$$g_{i,j} = P \left[\exp \left(i \int_{\mathbf{r}_i}^{\mathbf{r}_j} dl^\mu A_\mu \right) \right] .$$

Traversing the edge in the opposite direction gives the inverse group element $g_{j,i} = g_{i,j}^{-1}$. A Wilson loop that visits sites $i_1, i_2, i_3, \ldots i_N, i_1$ is then given by

$$W_{\text{loop}} = \text{Tr}_R \left[g_{i_1,i_2} \, g_{i_2,i_3} \, \cdots \, g_{i_{N-1},i_N} \, g_{i_N,i_1} \right] . \tag{31.40}$$

For a discrete gauge theory, all of the values of g_{ij} in the system are assumed to take values in a discrete group G. We can then choose to work with some particular representations of the discrete group.

Now, in gauge theories we are allowed to make gauge transformations, which change the gauge field but leave every Wilson loop unchanged. We can think of this as leaving the "magnetic flux" through every loop unchanged. The way we do this on a discrete lattice is to assign a group element $u_i \in G$ to each site i. The gauge transformation then transforms

$$g_{i,j} \to u_i \, g_{i,j} \, u_j^{-1} \quad .$$

Compare this to the gauge transformation of Eq. 5.29, which we discussed for Chern–Simons theory and which has exactly the same structure. Once this is plugged into Eq. 31.40, and using the cyclic invariance of the trace, it is clear that the flux is gauge invariant.

We say that a gauge field configuration is *flat* if the flux through any loop is zero, or the group element in Eq. 31.40 before taking the trace, is the identity.

We now write a Hamiltonian using the following two physical principles:

(1) Gauge invariance is imposed energetically, rather than as a strict symmetry principle.

(2) Flat configurations should be energetically favored.

While principle (2) is typical of gauge theories (even in electromagnetism the Hamiltonian is minimized for zero flux), principle (1) is somewhat unusual. The ground state of our system will maintain gauge invariance, but excited states will break it, at some energy cost.

To implement principle (1) in a Hamiltonian, we define a gauge transformation operator A_i^h at site i with $h \in G$ to premultiply all $g_{i,j}$ by h and postmultiply all $g_{j,i}$ by h^{-1}. Thus, we have for example[26]

[26]Recall we can always flip the direction of an arrow and invert the corresponding group element. That is, $g_{i,j} = g_{j,i}^{-1}$.

$$
A_i^h \left[\begin{array}{c} \end{array} \right] = \left[\begin{array}{c} \end{array} \right] \tag{31.41}
$$

where $h \in G$. To enforce gauge invariance we write a term in the Hamiltonian at each site i

$$
A_i = \frac{1}{|G|} \sum_{h \in G} A_i^h \tag{31.42}
$$

Now, if the wavefunction is gauge invariant it will be unchanged by the application of this operator. Further, it is easy to check that $A_i A_i = A_i$, so that A_i is a projector, having eigenvalues of 0 and 1 only: only the gauge-invariant part of a wavefunction survives when this operator is applied.

To implement principle 2, that we want flat, or flux-free, configurations to be energetically favorable, we define a projection operator that acts on a plaquette such that it gives the identity if the plaquette is flat (i.e. the group element in Eq. 31.40 is the identity) and otherwise gives zero. Such an operator acting on plaquette p would then be, for example,

$$
B_p \left[\begin{array}{c} \end{array} \right] = \delta_{(g_{i,j}\, g_{j,k}\, g_{k,l}\, g_{l,i}),\, e} \left[\begin{array}{c} \end{array} \right] \tag{31.43}
$$

Obviously, B_p is also a projector, that is $B_p B_p = B_p$ so it has eigenvalues of 0 and 1 only as well.

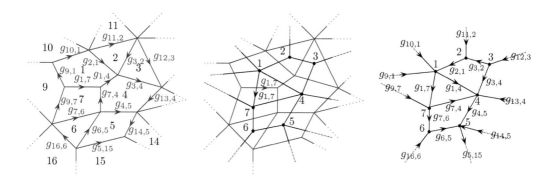

Fig. 31.23 **Left:** The lattice structure for the Hamiltonian Eq. 31.8 earlier in this chapter. We have labeled the plaquettes with black numbers. The directed edges $g_{i,j}$ are labeled with i and j, which are the plaquettes to the left and right of the edge, respectively (facing the direction of the arrow). **Right:** The lattice we introduced for the Hamiltonian Eq. 31.44 is dual to the lattice on the left. Vertices are labeled with numbers and edges $g_{i,j}$ are directed from vertex i to vertex j. **Middle:** We have shown both the lattice in the left panel (black) and the lattice in the right panel (blue) overlaid on each other. For simplicity we have dropped the edge labels and the arrows—except for a single bond that is labeled to show that when two edges of the two lattices cross they are labeled with the same group element. A plaquette on the left becomes a vertex on the right, and vice versa.

The full Hamiltonian, implementing both principles, is given by

$$H = -J_1 \sum_{\text{vertices } i} A_i - J_2 \sum_{\text{plaquettes } p} B_p \qquad (31.44)$$

with positive J_1 and J_2. The ground-state space is the space where all operators A_i and B_p have eigenvalue 1 on all vertices and plaquettes, respectively.

The Hamiltonian Eq. 31.44 is precisely that introduced by Kitaev [2003]. It is fairly easy to see that it is also equivalent to the one introduced in Eq. 31.8, but written on the dual lattice, as shown in Fig. 31.23 ($J_2 = \Delta_v$ and $J_1 = \Delta_p$). The operator $\hat{V}_\alpha$ from our earlier construction (Section 31.1.1) that acts on the vertices of the original model is identical to the operator B_p introduced here, which acts on the plaquettes of this model, whereas the operator $\hat{P}_\beta$ that acts on the plaquettes of the original model is identical to the operator A_i, which acts on the vertices here.

The plaquette defects in this model are plaquettes that violate B_p (i.e. have eigenvalue of B_p of 0 rather than 1). These defects are plaquettes that are not in "flat" configurations, and we can think of these plaquettes as having "magnetic flux" through them. This lends these quasiparticles the name "magnetic," as we mentioned in Section 31.6. Correspondingly, violations of A_i are known as "electric." Quasiparticles that violate both vertex and plaquette conditions are usually known as *dyons*.

31.7.1   Gauge Theory in (3 + 1)-Dimensions

It is fairly easy to generalize this type of gauge theory to higher dimensions. We consider a lattice in three dimensions and assign group elements $g \in G$ to each directed edge. For simplicity we will take a cubic lattice. The vertex operators are still of the form of Eqs. 31.41–31.42, the plaquettes operators (the squares faces of the cubic lattice, for example) take the form of Eq. 31.43, and the Hamiltonian is still of the form of Eq. 31.44.

As an example, let us calculate the ground-state degeneracy of the (3 + 1)-dimensional Kitaev model on a 3-torus (i.e. $T^3 = S^1 \times S^1 \times S^1$ is the three-dimensional spatial manifold). Analogous to Section 31.5, the calculation is easier if we consider a highly simplified (minimal size) lattice (note, we are now in three dimensions and we are working on the dual lattice compared to the discussion of Section 31.5). The smallest lattice we can construct is shown in Fig. 31.24.

In the figure depicting the 3-torus (opposite faces of the cube are identified) there is a single vertex in the center (we call the site "0"), three edges labeled $a, b,$ and c, and three plaquettes (one bounded by a and b, one bounded by a and c, and one bounded by b and c). The three plaquette operators enforce that in the ground state we must have

$$aba^{-1}b^{-1} = aca^{-1}c^{-1} = bcb^{-1}c^{-1} = e \quad . \tag{31.45}$$

The operator A_0^h (as in Eq. 31.41) conjugates each of the edge variables

$$A_0^h a = hah^{-1} \qquad A_0^h b = hbh^{-1} \qquad A_0^h c = hch^{-1} \tag{31.46}$$

and the full vertex operator A_0 is the sum of A_0^h over the entire group $h \in G$ as in Eq. 31.41. It is easy to see that if the plaquette conditions, Eqs. 31.45, are satisfied then application of the vertex operator does not change this.

Analogous to the discussion in Section 31.5, the ground state wavefunction must be of the form analogous[27] to Eq. 31.14

$$|\Psi\rangle \propto A_0 |a, b, c\rangle \propto \sum_{h \in G} |hah^{-1}, hbh^{-1}, hch^{-1}\rangle \quad .$$

Again, not all $|a, b, c\rangle$ generate different wavefunctions— the number of orthogonal ground-state wavefunctions is equal to the number of orbits of $|a, b, c\rangle$ under the action of conjugation by h. Using Burnside's lemma again (see Section 31.5) we obtain

ground-state degeneracy of discrete gauge theory model on 3-torus

$$= \frac{1}{|G|} \left(\begin{array}{c} \text{Number of quadruples } a, b, c, h \\ \text{that all commute with each other} \end{array} \right) \tag{31.47}$$

In principle we can work out the excitation spectrum of this model, which has point-like excitations, as well as flux-tube-like excitations that braid nontrivially with the point-like excitations. The details of this are beyond the current discussion, but are explained in Delcamp [2017] and Bullivant and Delcamp [2019] for example. The resulting TQFT is sometimes known as the "quantum triple."

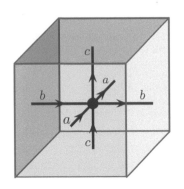

Fig. 31.24 A minimal gauge theory model on a 3-torus. Opposite faces of the cube are identified (i.e. we have periodic boundary conditions). There are three edges in the figure labeled a, b, c, there is one vertex, and there are three plaquettes.

[27] In comparing to Eq. 31.14 remember that we switched to working on the dual lattice, so what was a plaquette operator is now a vertex operator!

31.8 Appendix: Two Expressions for the Ground-State Degeneracy on a Torus

Here, we show that Eq. 31.15 and Eq. 31.26 for the ground-state degeneracy give the same result.

First, we use Burnside's lemma (see Section 31.5) to calculate the number of conjugacy classes of a group Z. Let the set X be the group Z itself and the action of the group on this set is conjugation $x \to hxh^{-1}$ with $x \in X = Z$ and $h \in Z$. The number of orbits of this action is the number of conjugacy classes, which is just

$$\left(\begin{array}{c} \text{number of conjugacy} \\ \text{classes of group } Z \end{array} \right) = \frac{1}{|Z|} \sum_{h \in Z} \left(\begin{array}{c} \text{number of } x \in Z \\ \text{that commute with } h \end{array} \right) .$$
(31.48)

We now consider the sum in Eq. 31.26. We can write this sum over classes as a sum over all group elements with each term in the sum weighted by the inverse of the size of the conjugacy class.

$$\sum_C \to \sum_{g \in G} \frac{1}{|C_g|}$$

with $|C_g|$ the number of elements in the conjugacy class of g. Then, using Eq. 31.48 (along with the fact that the number of representations of a group is equal to the number of conjugacy classes in the group; see Section 40.2.4), the sum in Eq. 31.26 may be written as

$$\sum_{g \in G} \frac{1}{|C_g|} \frac{1}{|Z(g)|} \sum_{h \in Z(g)} \left(\begin{array}{c} \text{number of } x \in Z(g) \\ \text{that commute with } h \end{array} \right) .$$
(31.49)

Finally, we use a well-known theorem (Eq. 40.2) that given an element g of a group G, the number of elements $|C_g|$ in the conjugacy class of g times the number of elements $|Z(g)|$ that commute with g is equal to the number of elements in the group $|G|$. That is, $|C_g| \, |Z(g)| = |G|$. Thus, since $Z(g)$ are the elements of G that commute with g, Eq. 31.49 becomes

$$\frac{1}{|G|} \left(\begin{array}{c} \text{number of triples } g, h, x \\ \text{which all commute with each other} \end{array} \right)$$
(31.50)

in agreement with Eq. 31.15.

Chapter Summary

- The Kitaev quantum double model generalizes the toric code to label edges with elements of a group. The applicable diagram algebra is based on the multiplication properties of the group.
- The same diagrammatic techniques that we applied to the toric code, such as the tube algebra, can be appropriately generalized.
- The Kitaev quantum double model is essentially a Hamiltonian presentation of a lattice gauge theory where gauge invariance and flatness are enforced with an energetic penalty.

Further Reading:

- The original discussion of the Kitaev model is given by Kitaev [2003] and is still a fairly good discussion. Note that it is constructed on the dual lattice compared to our construction, so that it looks more like a gauge theory.
- A nice physical description of the quantum double of a group is given by Preskill [2004]. Earlier work on the physics of the quantum double, twisted quantum doubles, is given by de Wild Propitius [1995]. The quantum double algebra was first worked out by Dijkgraaf et al. [1991].
- A detailed analytic discussion of the properties of the quantum double of a group is given in Delcamp et al. [2017] and in Delcamp [2017], including a detailed discussion of the tube algebra.
- Discussion of the Kitaev model and how one might implement it in experiment is given by Brennen et al. [2009].
- A discussion of condensations from the Kitaev model (in the sense of Chapter 25) is given by Bombin and Martin-Delgado [2008] and also Beigi et al. [2011]. These are also good general references about the Kitaev model. Bais et al. [2003] describe condensations of the quantum double of a group in a more abstract approach.
- For modular data (S- and T-matrices) of twisted quantum doubles, see Coste et al. [2000]. An explicit construction of a lattice model for twisted quantum doubles is given by Hu et al. [2013]. An online database of modular data (S- and T-matrices) for twisted quantum doubles is given by
    https://tqft.net/web/research/students/AngusGruen/
  and is also mirrored on my web page.
- Some details of using the Kitaev model as an error-correcting code are given by Cui et al. [2020].
- Detailed introductions to the Kitaev model are given in several places. In particular, I like Karlsson [2017] (which was an undergrad thesis!) and Cui et al. [2015], which works through calculating the R- and F-matrices explicitly for the example of the group S_3.

- The book by Oeckl [2005] discusses many connections between discrete lattice gauge theories and TQFTs.

- The relationship between the Kitaev model and the Levin–Wen string-net (Chapter 33) is given in Buerschaper and Aguado [2009].

- For the $(3 + 1)$-dimensional version, a higher-dimensional tube algebra is discussed by Delcamp [2017] and Bullivant and Delcamp [2019].

- Twisted gauge theories in $(3 + 1)$-dimensions have been studied by Wan et al. [2015], Wang and Wen [2015], and Wang and Chen [2017].

- Another generalization is to "higher" (or sometimes "higher-lattice") discrete gauge theories in $(2 + 1)$- and $(3 + 1)$-dimensions. In such a generalization, one assigns not only gauge variables on edges, but also a "higher" gauge variable on the plaquettes. These are discussed by Bullivant et al. [2017, 2020] and less abstractly by Huxford and Simon [2022].

Exercises

Exercise 31.1 Product of Groups

Section 31.1.2 states that the code space of a Kitaev model for a group $G = G_1 \times G_2$ is the tensor product of the code spaces for a Kitaev model built from group G_1 and the code space for a Kitaev model built from group G_2. Prove this statement.

Exercise 31.2 Diagram Manipulation in the Kitaev Model

(a) Using the moves given in Fig. 31.9, derive the result shown in Fig. 31.25 for the Kitaev model.

(b) Derive the result shown in Fig. 31.11.

(c) Show that the value of the tetrahedral figure in Fig. 31.26 is unity if all of the vertices are allowed, and is zero otherwise. In other words, show its value is $\delta_{dc^{-1},e}\delta_{cb,f^{-1}}\delta_{db,a^{-1}}$.

Exercise 31.3 Mutating Lattices

(a) Show how to use the two moves given in Figs. 31.5 and 31.6 to achieve the following deformation of the lattice:

Fig. 31.25 A simple result of the three allowed moves in Fig. 31.9. Compare to Fig. 16.9 for example.

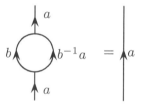

Fig. 31.26 The value of this diagram is $\delta_{dc^{-1},e}\delta_{cb,f^{-1}}\delta_{db,a^{-}}$.

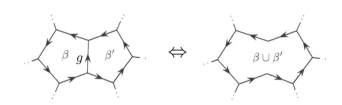

(b) Show that any lattice with a given topology may be mutated into any other lattice with the same topology using the two moves given in Figs. 31.5

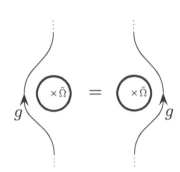

Fig. 31.27 The planar handle-slide.

and 31.6. Hint: Remove edges one at a time until you are left only with non-contractible loops.

Exercise 31.4 Kitaev Handle-Slide

Use the diagrammatic rules to show the planar handle-slide identity shown in Fig. 31.27 where Ω is the Kirby strand defined in Eq. 31.13. Note that the × in the middle of the Ω loop should be thought of as a puncture in the plane that prevents this loop from being contractible.

Exercise 31.5 Ground-State Degeneracy of S_3

Calculate the ground-state degeneracy of the Kitaev model on a torus for the group S_3 by using Eq. 31.15.

Exercise 31.6 Kitaev Annular Idempotents

Prove Eq. 31.23. Hint: Use $\rho^R(ab) = \rho^R(a)\rho^R(b)$ along with the grand orthogonality theorem.

Exercise 31.7 Kitaev Model Twist Factor

Given the quasiparticle idempotents Eq. 31.22 and the twist operation on the $|h, g\rangle$ basis in Eq. 31.27, derive the twist factor in Eq. 31.30. Why is the expression for the twist factor of unit magnitude? Hint: (i) Write the character χ^R as the trace of a matrix ρ^R as in Eq. 40.6 then use $\rho^R(ab) = \rho^R(a)\rho^R(b)$. (ii) You will also need Schur's second lemma (see Section 40.2.4).

Exercise 31.8 Kitaev Model Quasiparticle Torus Basis

Prove Eq. 31.25. You will need the grand orthogonality theorem Eq. 40.5 as well as hint (i) from Exercise 31.7

Exercise 31.9 Kitaev Model S-matrix

(a) [Easy] Using Eq. 31.32 give a simple expression for $S_{0,0} = 1/\mathcal{D}$ for the Kitaev model for an arbitrary group G. Compare your result to that of Exercise 23.5.

(b) Using Eq. 31.32, show that the quantum dimension of the particle type $[C, R]$ is given by

$$d_{[C,R]} = |C| d_R$$

where $|C|$ is the number of elements in the conjugacy class, and d_R is the dimension of the representation.

(c) [Harder] Derive Eq. 31.32.

Exercise 31.10 Kitaev Model for S_3

Calculate the T-matrix and the S-matrix for the group S_3. Hint: You can either use the explicit formulae (Eqs. 31.30 and 31.32) or you can use the wavefunctions explicitly in Eqs. 31.27 and 31.31.

Answers:

particle	$[C_e, I]$	$[C_e, S]$	$[C_e, V]$	$[C_x, +1]$	$[C_x, -1]$	$[C_r, 1]$	$[C_r, \omega]$	$[C_r, \omega^{-1}]$
θ	1	1	1	1	-1	1	ω^{-1}	ω

$$S = \frac{1}{6}\begin{pmatrix} 1 & 1 & 2 & 3 & 3 & 2 & 2 & 2 \\ 1 & 1 & 2 & -3 & -3 & 2 & 2 & 2 \\ 2 & 2 & 4 & 0 & 0 & -2 & -2 & -2 \\ 3 & -3 & 0 & 3 & -3 & 0 & 0 & 0 \\ 3 & -3 & 0 & -3 & 3 & 0 & 0 & 0 \\ 2 & 2 & -2 & 0 & 0 & 4 & -2 & -2 \\ 2 & 2 & -2 & 0 & 0 & -2 & -2 & 4 \\ 2 & 2 & -2 & 0 & 0 & -2 & 4 & -2 \end{pmatrix}$$

Exercise 31.11 Kitaev Model for the Group $\mathbb{Q}_8$

[Harder] Calculate the T-matrix and the S-matrix for the group $\mathbb{Q}_8$. The group description and the character table is given in Section 20.1.2. You can make a number of consistency checks on your answer using Eqs. 17.13, 17.20, and 17.17.

Exercise 31.12 Graphical Fusion of Some Quasiparticles

In Fig. 30.16 we showed how to graphically calculate certain fusion rules for the toric code.

(a) Pure flux quasiparticles have C being identity conjugacy class C_e) and have some representation R of the group G. Use this graphical technique to show that the fusion of pure flux quasiparticles $[C_e, R_1]$ and $[C_e, R_2]$ obeys the fusion rules of the representations of the group (see Eq. 20.6). Hint: The representation matrices ρ^R fuse similarly to the characters χ^R; see Chapter 20 and particularly Section 20.5.3.

(b) Use the Verlinde formula along with the known S-matrix for the case of S_3 (shown in Exercise 31.10) to confirm this result.

Exercise 31.13 Modular Relations for the Kitaev Model

(a) If a particle type is $\|[C, R]\rangle$ for the Kitaev model as in Eq. 31.24, how would you describe the corresponding antiparticle?

(b) Prove the modular group relations (Eq. 17.17) for the S- and T-matrices of the Kitaev model. Hint: Apply these operators graphically as in Figs. 30.22 and 30.23 to the basis states $|h, g\rangle$ rather than to the basis states $\|[C, R]\rangle$.

Exercise 31.14 Some Ribbon Operator Identities

(a) Consider ribbon operators (Eq. 31.38 for pure flux quasiparticles (C is the identity conjugacy class C_e). By laying two quasiparticle ribbons $[C_e, R_1]$ and $[C_e, R_2]$ along the same path, confirm that the fusion rules for these pure flux quasiparticles obey the fusion rules for the group representations as in Eq. 20.6. Hint: use the identity Eq. 31.35

(b) [Hard] Derive the concatination formula Eq. 31.39.

Doubled-Semion Model

One of the simplest generalizations of the toric code is known as the *doubled-semion model*. As with the toric code, it can be considered as a form of a quantum loop gas. However, additional signs are added to the loop gas to *twist* the theory in the sense discussed in 31.4.1 (indeed, the doubled-semion model is the simplest case of a twisted Kitaev theory). The doubled-semion model is also a simple example of the Levin–Wen string-net model that we discuss in Chapter 33.

We managed to derive everything about the toric code by starting with the skein rules (Eqs. 30.2 and 30.3) of the $d = +1$ version of the $\mathbb{Z}_2$ fusion-rule planar diagram algebra discussed in Section 18.1.1. To obtain the doubled-semion model, we instead use the $d = -1$ version of the $\mathbb{Z}_2$ fusion-rule planar diagram algebra[1] discussed in Section 18.1.2:

[1] Here we are twisting (see Section 31.4.1) the $\mathbb{Z}_2$ fusion rules by using the nontrivial cocycle discussed in Section 19.2.

$$\bigcirc = -1 \tag{32.1}$$

$$\asymp = - \,)(\, . \tag{32.2}$$

The diagrammatic rules are thus changed so that each loop removal or addition, and each surgery, incurs a minus sign. Note that these two minus signs are consistent with each other because each surgery changes the parity of the number of loops in the system. Note that these rules were precisely the skein rules we used for the Kauffman invariant when we considered semions (see Section 18.1.2 for example). However, here we are not inputting any rules about braiding, just the planar diagram algebra.

From these skein rules we expect wavefunctions of the form

$$|\psi\rangle = \sum_{\substack{\text{all loop configs that} \\ \text{can be obtained from} \\ \text{a reference loop config}}} (-1)^{\text{number of loops}} \, |\text{loop config}\rangle \quad . \tag{32.3}$$

We can think of the prefactor $(-1)^{\text{number of loops}}$ as being the evaluation of the loop diagram exactly as in Eq. 18.15.[2]

[2] In the language of Section 14.5 we are performing a nonunitary evaluation of this diagram, that is, we are *not* using rule 0, which would give an additional minus sign for each cap. We fix this issue in Section 32.3.

As with the toric code, there should be four ground states on the torus corresponding to the different possible parities of blue strings around the two nontrivial cycles of the torus.

32.1 A Microscopic Model

We want to build a microscopic model that implements this $d = -1$ diagrammatic algebra analogous to the way the toric code Hamiltonian (Eq. 28.1) implements the $d = +1$ skein rules. We once again work on a honeycomb lattice,[3] although any trivalent graph will work just as well. Again, we put a spin on each edge, and color the edge blue if the spin is down. Following our procedure for the toric code, we impose a vertex term (Eq. 27.1 in Eq. 28.1) that penalizes any vertex with an odd number of blue edges, such that any ground-state wavefunction must be made of closed loops. Again, we want to have a plaquette term that flips the state of a plaquette (analogous to Eq. 27.3), but each flip must now come along with an appropriate number of minus signs. To determine the sign, it is useful to look at a few cases, as shown in Fig. 32.1.

[3]...which should not be called a lattice.

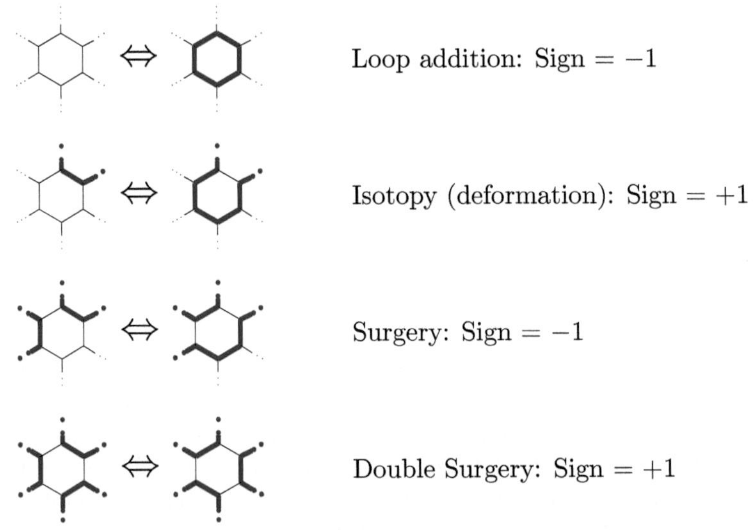

Loop addition: Sign $= -1$

Isotopy (deformation): Sign $= +1$

Surgery: Sign $= -1$

Double Surgery: Sign $= +1$

Fig. 32.1 When flipping over a plaquette, the sign is -1 if the number of loops in the system changes parity. A "leg" of the plaquette is one of the edges pointing out of the plaquette, that is, having only one end on a vertex of the plaquette (these are drawn partially dotted in the figure). Let the number of blue legs of the plaquette be L, which is always even as long as we have no vertex defects. The sign associated with flipping the plaquette is -1 if $L/2$ is even and is $+1$ otherwise. This rule holds for plaquettes with any number of sides should we choose a different geometry.

As shown in the figure, there is a simple rule for determining the sign accumulated for a plaquette flip. Given a plaquette that we want to flip, a *leg* of the plaquette is an edge with only one vertex on the plaquette. The number of legs L that are blue is always even as long as there are no vertex defects (loop ends) on the plaquette. The sign associated with flipping the plaquette is -1 when $L/2$ is even and is $+1$ otherwise. Here, we write this sign in several equivalent ways:

$$
\begin{aligned}
\text{Sign} &= -(-1)^{\frac{1}{4}\sum_{i\in\text{legs}}(1-\sigma_z^{(i)})} = -\prod_{i\in\text{legs}} e^{i\frac{\pi}{4}(1-\sigma_z^{(i)})} \\
&= (-1)^{\sum_{j\subset\text{edges}}\frac{1}{4}(1-\sigma_z^j)(1+\sigma_z^{j+1})}
\end{aligned}
$$

noting that $(1-\sigma_z)$ gives 0 for spin up and 2 for spin down (i.e. a blue edge), and in the second line the edges j are counted in clockwise[4] order around the hexagon (and $j=6$ is the same as $j=0$).

With this information we can now write a plaquette operator for this model

$$
P'_\beta = -\left(\prod_{i\in\text{plaquette }\beta} \sigma_x^{(i)}\right) \prod_{i\in\text{legs of }\beta} e^{i\frac{\pi}{4}(1-\sigma_z^{(i)})}
$$

where the first term is the same as that of the toric code (Eq. 27.3) and now we have included these additional signs.

Using the usual toric code vertex term (same as Eq. 27.1)

$$
V_\alpha = \prod_{i\in\text{vertex }\alpha} \sigma_z^{(i)} \tag{32.4}
$$

we then get the overall Hamiltonian for this model

$$
H_{\text{doubled-semion}} = -\frac{\Delta_v}{2}\sum_{\text{vertices }\alpha} V_\alpha - \frac{\Delta_p}{2}\sum_{\text{plaquettes }\beta} P'_\beta \tag{32.5}
$$

which is identical to the Hamiltonian of the toric code (Eq. 28.1) except for the modification of the plaquette term to include the appropriate signs.

32.1.1 Magnetic String Operator

In the case of the toric code we showed how we can pull vertex (e) defects apart from each other by successive application of a string of σ_x (see Section 27.4.1, and Fig. 27.10). Similarly, we showed (see Section 27.4.2 and Fig. 27.14)) that a (dual) string of σ_z operators could be used pull plaquette defects apart from each other. Such strings are creatively known as "string operators." The key property of such operators is that they only create excitations at the ends of the string, not along their length—and indeed, the actual path that the string operator follows is not measurable.

For the doubled-semion model, the analog of the magnetic string operator is quite straightforward. As with the toric code, one draws a path on the dual lattice (bright green in Fig. 32.2) and applies σ_z to every edge that this dual string crosses (dark green in Fig. 32.2). Since, except at the end of the string, each plaquette has an even number of σ_zs applied, this string operator commutes with the Hamiltonian, and

[4]We could equally well choose counter-clockwise order.

Fig. 32.2 In the doubled-semion model, the magnetic string operator applies σ_z to the dark green edges, which cut through the bright green path on the dual lattice.

[5] A particularly nice form of these string operators is given in Dauphinais et al. [2019], in which the Hamiltonian is slightly modified so that the ground-state wavefunctions are unchanged, but excitations are (topologically the same but) simpler than in most works. This modification gives the string operators a lot of convenient properties.

hence, does not create any excitations except at the string ends.

There are two more possible string operators corresponding to two other quasiparticle types (as mentioned, there are four ground states on the torus, so we expect four quasiparticle types, including the vacuum). These, however, are more complicated, and we defer their discussion to the more general case in Chapter 33.[5]

32.2 **Attempted Graphical Analysis**

Let us attempt to follow the graphical methods of Sections 30.2–30.4 to determine the quasiparticle types and their properties. Be warned: we will discover a complication here!

The graphical algebra is similar to that of the toric code, although signs are incurred from loop addition/subtraction and surgery. We have four basis states on the annulus, given in Figs. 32.3 and 32.4 (which are equivalent to Figs. 30.7 and 30.9 for the toric code except for the added minus signs!).

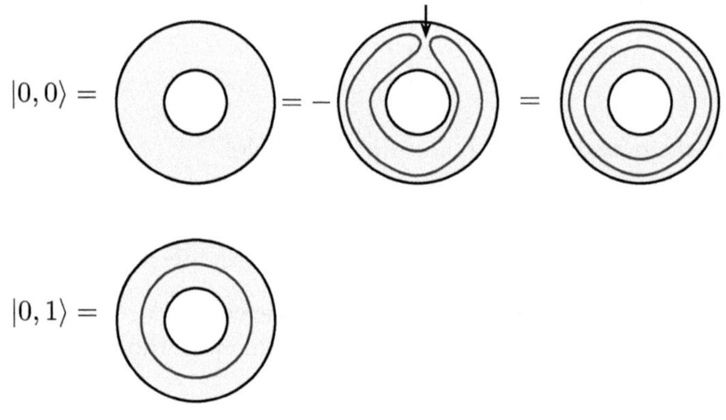

Fig. 32.3 The two equivalence classes of states on the annulus where no lines intersect the boundary. In the top line, a minus sign is added with the loop addition going to the middle figure, then another minus sign is obtained from the surgery (marked by the arrow) going to the right picture.

$\circ$	$|0,0\rangle$	$|0,1\rangle$	$|1,0\rangle$	$|1,1\rangle$
$|0,0\rangle$	$|0,0\rangle$	$|0,1\rangle$	$\bullet$	$\bullet$
$|0,1\rangle$	$|0,1\rangle$	$|0,0\rangle$	$\bullet$	$\bullet$
$|1,0\rangle$	$\bullet$	$\bullet$	$|1,0\rangle$	$|1,1\rangle$
$|1,1\rangle$	$\bullet$	$\bullet$	$|1,1\rangle$	$-|1,0\rangle$

Table 32.1 The multiplication table for composing annuli for the doubled-semion model. The $\bullet$ indicates a composition that is not allowed due to non-matching boundary conditions.

We can now consider the multiplication or composition of states analogous to that discussed in Section 30.2.2, that is, putting one annulus inside another. The multiplication table, Table 32.1, is entirely the same as that for the toric code (Table 30.1) except for one entry: $|1,1\rangle \circ |1,1\rangle = -|1,0\rangle$. To understand this, we imagine putting one $|1,1\rangle$ state inside another, and we obtain a picture like the far right of the top line of Fig. 32.4, which now differs from $|1,0\rangle$ by a minus sign.

With this multiplication table, we can construct the quasiparticle idempotents to determine the four quasiparticle types:

$$O_\alpha = \frac{1}{2}(|0,0\rangle - |0,1\rangle) = \frac{1}{2}\left(\bigcirc - \circledcirc \right) \tag{32.6}$$

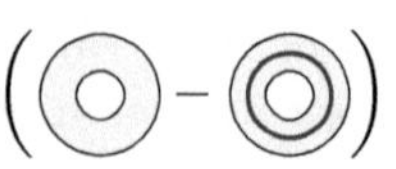

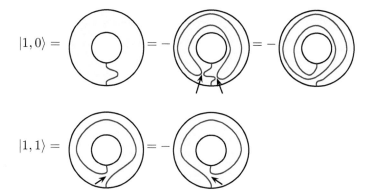

$|1,0\rangle =$ (figure) $= -$ (figure) $= -$ (figure)

$|1,1\rangle =$ (figure) $= -$ (figure)

Fig. 32.4 The two equivalence classes of states on the annulus where one blue line intersects each boundary. In the top line, a minus sign is added with the loop addition going to the middle figure, and two more minus signs are obtained from two surgeries going to the right picture at the position of the arrows. In the bottom line a minus sign is obtained from surgery at the position of the arrow.

$$O_\beta = \frac{1}{2}(|0,0\rangle + |0,1\rangle) = \frac{1}{2}\left(\bigcirc + \circledcirc \right) \qquad (32.7)$$

$$O_\gamma = \frac{1}{2}(|1,0\rangle - i|1,1\rangle) = \frac{1}{2}\left(\bigcirc - i \circledcirc \right) \qquad (32.8)$$

$$O_\delta = \frac{1}{2}(|1,0\rangle + i|1,1\rangle) = \frac{1}{2}\left(\bigcirc + i \circledcirc \right) \qquad (32.9)$$

Here, we have named the particle types $\alpha, \beta, \gamma, \delta$. Since the ground state is a superposition of flipped and unflipped *with a minus sign* (since addition of a loop includes a sign), we *tentatively* identify particle α (Eq. 32.6), as the identity I. (We can also confirm this with a calculation analogous to Fig. 30.16, see Exercise 32.1.) This superposition can again be thought of as a Kirby string $\tilde{\Omega}$ being put around the annulus.[6]

Particle β (Eq. 32.7) is the idempotent orthogonal to the ground state within the space having no blue lines intersecting the boundaries of the annulus. The two idempotents with blue lines intersecting both boundaries are γ (Eq. 32.8) and δ (Eq. 32.9).

We can proceed to calculate the twist factors for each of these particles. As in the case of the toric code (Section 30.3), the twist operator $\hat{\theta}$ leaves $|0,0\rangle$ and $|0,1\rangle$ unchanged, implying that $\theta_I = \theta_m = 1$, that is, these are bosons. Again, as with the toric code, $\hat{\theta}|1,0\rangle = |1,1\rangle$. However, in contrast to the bottom of Fig. 30.14, we have $\hat{\theta}|1,1\rangle = -|1,0\rangle$, as shown in Fig. 32.5.

Knowing now how $\hat{\theta}$ acts on $|1,0\rangle$ and $|1,1\rangle$, we can use this operator on O_γ and O_δ (as given in Eqs. 32.8 and 32.9) obtaining $\theta_\gamma = i$ and $\theta_\delta = -i$.

Let us now try to counterclockwise exchange two γ particles as shown in Fig. 32.6. (This is analogous to the exchange of quasiparticles in Fig. 30.15 for the toric code.) We wrap the right γ around the left, we then comparing the second line in Fig. 32.6 to the first line to determine

[6]There are two types of particles in the $\mathbb{Z}_2$ loop gas: the vacuum and the non-trivial particle (the blue string). We weight each by its value of $d/\mathcal{D}^2$ with $\mathcal{D} = \sqrt{2}$. Here, we use d rather than $|d| = \mathsf{d}$ in the definition of the Kirby $\tilde{\Omega}$ string (Fig. 17.11) because we are performing a nonunitary evaluation of the loop diagram. See note 37 of Chapter 17, as well as note 2 of this chapter.

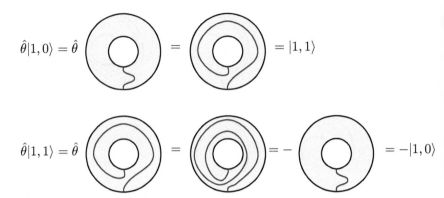

Fig. 32.5 Twist deformations of the $|1,0\rangle$ and $|1,1\rangle$ states. The inner boundary is rotated counterclockwise while the outer boundary is held fixed, analogous to the deformation shown in Fig. 30.13. In the second line we have used the surgery equivalence, as shown in Fig. 32.4.

the overall braiding phase. Indeed, these two lines are equal up to a phase. To determine the phase, let us look, for example, at the leftmost picture in the second line of Fig. 32.6. Surgering where the small black arrow points picks up a minus sign as in Eq. 32.2 and gives us a figure precisely like the second picture on the first line. Looking now at the prefactors we see that the second line is $-i$ times the first line. This tells us that the exchange phase is $R_I^{\gamma,\gamma} = -i$.

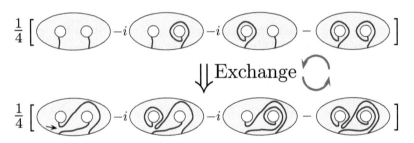

Fig. 32.6 Calculation of the phase accumulated from exchanging two γ particles. The upper line represents two γ quasiparticles, one in each hole (compare Eq. 32.8). The holes are exchanged, braiding the blue lines in the process. Using surgery and loop addition/removal (the moves of Eq. 32.1 and 32.2) one can show the lower line is precisely $-i$ times the top line showing an exchange phase of $-i$. For example, the first term in the second line is equivalent to the second term in the first line times $-i$.

Unfortunately, this result tells us that there is a hidden problem. If we compare $\theta_\gamma = i$ and $R_I^{\gamma,\gamma} = -i$ we see that this agrees with Eqs. 15.2 and 15.4 only if we take a loop weight $d_\gamma = -1$ with zig-zag phase of $\epsilon_\gamma = +1$ (we would find similarly $d_\delta = -1$ although $d_\beta = +1$). That is, the resulting theory is isotopy invariant but nonunitary! We discuss this issue further in a moment, but first let us briefly consider a few other graphical calculations.

First, we can use the graphical techniques of Section 30.5 to determine the fusion properties of the quasiparticles. Working out the details of this is assigned as Exercise 32.1, with the result being given in Table 32.2. (This answer is actually correct and does not suffer from the fact that we are using a loop weight $d < 0$.)

Secondly, we can attempt to calculate the S-matrix using the technique of Section 30.4. Equations 30.12–30.14 are unchanged. However, in Eq. 30.15 there is minus sign accumulated from the surgery (Eq. 32.2). Determining how this $\underline{S}$ now acts on the basis states (Eqs. 32.6–32.9) gives us the S-matrix (see Exercise 32.3). What we find is that $S_{I,\gamma}$ and $S_{I,\delta}$ are both negative—this is not surprising considering that we already established $d_\gamma = d_\delta = -1$.

One can define a unitary $(2 + 1)$ TQFT from this nonunitary theory by implementing the cap-counting technique of Section 14.5 (Option B from Section 14.2). In this case, twist factors can change sign due to the cap in the twist (Eq. 15.1), and the elements of the S-matrix also change sign due to the caps in loops (Eq. 17.5). In particular, we get

$\times$	α	β	γ	δ
α	α	β	γ	δ
β	β	α	δ	γ
γ	γ	δ	α	β
δ	δ	γ	β	α

Table 32.2 The fusion rules for the nonunitary doubled-semion model.

$$
\begin{aligned}
\theta_a^{\text{unitary}} &= \text{sign}(d_a)\theta_a^{\text{nonunitary}} \\
S_{ab}^{\text{unitary}} &= \text{sign}(d_a)\text{sign}(d_b)S_{ab}^{\text{nonunitary}} \quad .
\end{aligned}
$$

The fact that we obtained a nonunitary TQFT from our diagrammatic reasoning reflects the fact that we are working with negative loop weights, and we do not have a unitary diagram algebra. Does this mean our original Hamiltonian is defective? No! In fact, the original Hamiltonian we wrote down is perfectly well behaved and the resulting ground state is actually that of the *unitary* TQFT. What went wrong here is our attempt to follow diagrammatic reasoning with a nonunitary diagram algebra!

32.3   Gauge Choice and Unitary Diagram Algebra

The diagrammatic $d = -1$ version of the $\mathbb{Z}_2$ loop gas that we chose to work with (Eqs. 32.1 and 32.2) corresponds to a nonunitary quantum theory. While this makes a simple diagrammatic theory, the results of our diagrammatic calculation correspond to a nonunitary TQFT, which is unacceptable. However, it turns out that the ground state (and low-energy excited states) of the Hamiltonian Eq. 32.5 are perfectly well-behaved unitary TQFTs. What did we do wrong here?

As discussed in Section 16.1.3 (see also Sections 14.5 and 14.2), we can convert our nonunitary diagrammatic algebra into a unitary algebra with two simple steps:

(0') We must break symmetry of the plane, choosing a special direction (often up on the page), which we call "up."[7]

(0) Before evaluating a diagram, count the number of (blue) caps, and call it n. After fully evaluating the diagram (using the nonunitary rules with $d = -1$) multiply the final result by $(-1)^n$.

Fig. 32.7 To unitarize the $d = -1$ version of the $\mathbb{Z}_2$ loop gas, we assign a minus sign to each cap. A cap is a point where the blue line reaches a maximum such as on the left, but not the right.

Fig. 32.8 On the honeycomb, a cap configuration is as on the left. The right vertex can never be a cap no matter which edges are colored blue.

[7]More general "gauge" transforms of this type are explored by Lin and Levin [2014] and Lin et al. [2021].

[8]We can now see more clearly what went wrong in our diagrammatic calculations of Section 32.2. If we make a braid as in Fig. 32.6 or a twist as in Fig. 32.5 by continuously mutating the underlying lattice (analogous to Section 31.3), we can introduce new caps into the diagram and thus unwittingly change gauge.

Here a "cap" is where a blue line experiences a maximum point (with respect to the chosen up direction). This is shown in Figs. 32.7 and 32.8.

We can thus modify our wavefunction Eq. 32.3 to the form of Eq. 32.10

$$|\psi\rangle = \sum_{\substack{\text{all loop configs that} \\ \text{can be obtained from} \\ \text{a reference loop config}}} (-1)^{\text{number of loops + number of caps}} |\text{loop config}\rangle \quad . \tag{32.10}$$

Comparing this to Eq. 18.16 we see that the prefactor sign of the ket is just the *unitary* evaluation of the loop config diagram.

What does this modification do to the ground state? Actually, it is nothing more than a gauge change! It is a local rule that associates an extra minus sign to certain wavefunctions. It may seem odd that we are allowed to do this, being that in Section 14.2 we concluded that we cannot generally remove minus signs from nonunitary theories with negative d by gauge transformation. However, the context here is different. In Section 14.2 we were concerned with theories in spacetime (say, $(1+1)$-dimensions or $(2+1)$-dimensions). In that case, time is drawn vertically and it was crucial that a cup (right of Fig. 32.7) is the Hermitian conjugate of a cap (left of Fig. 32.7). This prevented us from making a gauge transformation of the cup and cap independently. Here, however, we are considering wavefunctions that live in a $(2+0)$-dimensional plane—no direction is meant to be time. We are thus free to associate a minus sign with the cap, but not the cup, and this is just a gauge transformation on the wavefunction![7] We make this transformation to have a simpler diagram algebra, but underneath (in a different gauge) we are actually describing a unitary model.

We now want to redo the diagram algebra of Section 32.2 only being careful to use the unitary version of this diagram algebra. Here, we have to be careful to keep track of which direction is "up" on the page and whenever we introduce a cap, we should also include a minus sign.[8]

Examining Eq. 32.10 we see the ground state is now a superposition of loops with the sign arranged such that a loop with a single cap has a sign of $+1$, and each zig-zag (or cap) added to the loop incurs an additional minus sign. Thus, the ground-state idempotent, the identity particle type I, here should be particle type β from Eq. 32.7, rather than particle type α as we used before. (We can again confirm this with a calculation analogous to Fig. 30.16, only here we need to be careful with our diagrammatic algebra to include minus signs for caps! See Exercise 32.1.) The α particle is then the idempotent orthogonal to the ground state in the space having no blue lines intersecting the boundary of the annulus. Often this particle is called "magnetic" or m.

We can continue on to calculate twist factors. Examining Fig. 32.5 we see that these twists change the parity of the number of caps. Thus, again $\hat{\theta}$ leaves $|0,0\rangle$ and $|0,1\rangle$ unchanged but now

$$\hat{\theta}|1,0\rangle = -|1,1\rangle \qquad \hat{\theta}|1,1\rangle = |1,0\rangle \quad . \tag{32.11}$$

This allows us to evaluate the twist factors $\theta_I = \theta_m = 1 (= \theta_\beta = \theta_\alpha)$ and $\theta_\gamma = -i$ and $\theta_\delta = +i$.

The procedure of Fig. 32.6 remains correct as it is written (the braiding procedure does not introduce any new caps) so we again have $R_I^{\gamma,\gamma} = -i$ (and $R_I^{\delta,\delta} = i$). This combination $R_I^{\gamma,\gamma} = \theta_\gamma$ now agrees with Eqs. 15.2 and 15.4 for a positive d_γ which is what we should expect for a unitary theory!

We can further go on to use techniques analogous to Fig. 30.16 to calculate the full fusion table for this model (see Exercise 32.1) with the results given in Table 32.3.

In addition, we can use the techniques of Section 30.4 (and Section 30.6) to directly calculate the S-matrix (with columns ordered $\beta, \alpha, \gamma, \delta$)

$$S = \frac{1}{2} \begin{pmatrix} 1 & 1 & 1 & 1 \\ 1 & 1 & -1 & -1 \\ 1 & -1 & -1 & 1 \\ 1 & 1 & -1 & -1 \end{pmatrix} \tag{32.12}$$

which can also be derived using Eq. 17.20 along with the twist factors and fusion relations we just derived.

$\times$	β	α	γ	δ
β	β	α	γ	δ
α	α	β	δ	γ
γ	γ	δ	β	α
δ	δ	γ	α	β

Table 32.3 The fusion rules for the *unitary* doubled-semion model.

32.3.1   Doubled Semions

Having worked out the details of this unitary theory we recognize the final result as simply the direct product of a left-handed semion theory and a right-handed semion theory. Recall from Section 18.1.2 that a right-handed semion theory has particle types 0_R and 1_R, fusion rule $1_R \times 1_R = 0_R$ and $R_{0_R}^{1_R 1_R} = \theta_{1_R} = i$, and is equivalent to the Chern–Simons theory $SU(2)_1$. The left-handed theory has particle types 0_L and 1_L, fusion rule $1_L \times 1_L = 0_L$ and $R_{0_L}^{1_L 1_L} = \theta_{1_L} = -i$, and is equivalent to the Chern–Simons theory $\overline{SU(2)}_1 = SU(2)_{-1}$. (Crucially in both theories the nontrivial particle has negative Frobenius–Schur indicator $\tilde{\kappa}_{1_R} = \tilde{\kappa}_{1_L} = -1$.) The doubled-semion theory is thus

$$\text{Doubled-Semion Theory} = SU(2)_1 \times \overline{SU(2)}_1$$

The product (see Section 8.5) of the two theories means that each particle from the doubled-semion theory is a combination of one particle from each of the constituent factors. Thus, we have

$$I = \beta = (0_R, 0_L) \ ; \ \ \delta = (1_R, 0_L) \ ; \ \ \gamma = (0_R, 1_L) \ ; \ \ m = \alpha = (1_R, 1_L) \ .$$

The fusion and braiding relations are inherited from the constituent theories as described in Sections 8.5, 9.5.1, and 13.4.2. For example, it is easy to check that the simple $1 \times 1 = 0$ fusion rules of each factor gives the fusion Table 32.3 for the doubled model.

32.4 Comments

It is interesting that we used the skein rules for a model of semions (Eqs. 32.1 and 32.2) to build our loop gas, and we got out both right- and left-handed semions. This is perhaps to be expected, since nowhere in our input rules did we ever break "time-reversal" or say whether the original theory was right- or left-handed—it comes out to be both!

This principle is very general. If we start with any anyon theory with a modular braiding, then throw away the braiding and just use it as a planar diagram algebra to build a quantum loop model (not putting in any of the braiding relations), we will get out the *doubled* theory, meaning it has both right- and left-handed versions of the braided theory. We see this in more detail in Chapter 33.

Chapter Summary

- The doubled-semion model modifies the toric code by adding signs to the diagram algebra to construct a $d = -1$ loop gas.
- The input planar diagram rules are consistent with a semion theory and the output theory gives semions of both handedness.
- Many of the same diagrammatic principles we used for the toric code can still be applied.

Further Reading:

- The doubled-semion model was first discussed in the physics literature by Freedman et al. [2004]. The microscopic model was first written by Levin and Wen [2005].
- A discussion of the diagram algebra (and some of the issues that arise due to negative d) is given by Freedman et al. [2008].
- A particularly nice variant of the microscopic form of the doubled-semion model is given by Dauphinais et al. [2019]. This avoids a number of issues that arise at the ends of the quasiparticle string operators.
- The doubled-semion model is a particularly simple case of the Levin–Wen model, which we discuss in Chapter 33.
- A $(3 + 1)$-dimensional generalization, a special case of a more general model by Walker and Wang [2012], is explored in depth by von Keyserlingk et al. [2013].

Exercises

Exercise 32.1 Fusing Quasiparticles in the Doubled-Semion Loop Gas

Use the technique of Section 30.5 to deduce the full fusion table for the doubled-semion model.

(a) First try using the $d = -1$ algebra from Section 32.2.

(b) Now try the unitary version of this algebra from Section 32.3.

Exercise 32.2 Braiding Quasiparticles in the Doubled-Semion Loop Gas

Use the technique of Section 30.5 to calculate the R-matrix for the doubled-semion model.

(a) Show that wrapping an left-handed semion all the way around an right-handed semion gives no phase.

(b) Show that wrapping an m particle around either a left-handed semion or a right-handed semion gives a phase of -1.

(c) Show that wrapping an m particle around another m particle gives no phase.

(d) [Harder] Use the technique of Section 30.5 to calculate the phase from exchanging two left-handed semions and the phase for exchanging two right-handed particles. Do this part of the calculation using the unitary algebra. How is it different if you used the $d = -1$ algebra?

Exercise 32.3 *S*-Matrix of Doubled-Semion Model

In this exercise we study the doubled-semion model using the unitary algebra from Section 32.3. You must be careful to keep track of signs associated with removing cups and caps.

(a) Draw four basis states for the tube algebra using a rectangle as shown in the middle of Fig. 30.6. If you draw a string exactly horizontally you will not be able to say whether there is a cup or a cap, so make sure to set a non-horizontal convention for how such a string will be drawn.

(b) By studying composition of these tubes, determine the idempotents, and, hence, the four quasiparticle types.

(c) Using graphical reasoning, analogous to Fig. 30.22, determine the twist factors for the quasiparticle types.

(d) Use graphical reasoning as in Section 30.4 to calculate the S-matrix of the unitary doubled-semion model.

Levin–Wen String-Net

The theme of the last few chapters has been to build up topological models from diagram algebras. One of the most general constructions of this type is the Levin–Wen[1] string-net model. The input information for building this model is a (unitary) planar diagram algebra as we discussed starting in Chapter 12 and in more detail in Chapter 14. More formally we say that we input a so-called spherical tensor category. The output of this model will be a quantum system described by a (modular) TQFT known as the *Drinfel'd double* or *Drinfel'd center* (or sometimes just the *quantum double*) of the input category. In fact this is precisely the same TQFT[2] as is generated by the Turaev–Viro model introduced in Chapter 23—however, the Levin–Wen model provides a two-dimensional Hamiltonian, whereas the Turaev–Viro model is a $(3 + 0)$-dimensional action formalism.

The toric code (including the $\mathbb{Z}_N$ toric code) and the doubled-semion model are special simple cases of the Levin–Wen string-net model. The Levin–Wen models can also be shown to essentially include the Kitaev models as well.[3,4] As such, this chapter brings together many of the ideas we have encountered since Chapter 27.

In the cases where the input diagram algebra is consistent with the F-matrices of a *modular* anyon theory (meaning an anyon theory with a unitary S-matrix; see Section 17.3.1), the Drinfel'd double of this input theory is then simply two copies of the input modular anyon theory— one right-handed copy, and one left-handed copy. More formally, if our input modular diagram algebra is called $\mathcal{A}$, the emergent theory from the string-net construction is $\mathcal{A} \times \overline{\mathcal{A}}$. We saw this in the case of the doubled-semion model in Chapter 32. The planar algebra we put in could be given either a right-handed braiding to become a right-handed semion TQFT, or a left-handed braiding to become a left-handed semion TQFT. The Drinfel'd double of this planar algebra, the outcome of the string-net construction, is then just the product of the left- and right-

[1]This very general model is quite crucial to topological ideas. It was introduced by Levin and Wen [2005]. Michael Levin was a graduate student at the time. Xiao-Gang Wen is one of the founding figures in the field of topological condensed matter physics. He started his physics career as a student of Ed Witten.

[2]The connection between Levin–Wen and Turaev–Viro has been made in several works, particularly Kirillov [2011] (see also references at the end of the chapter).

---

[3]The original Levin–Wen model required a commutative diagram algebra such that $a \times b = b \times a$, as we insisted in Chapter 8. However, the Kitaev model allows nonabelian groups so that $gh \neq hg$. Generalizations of Levin–Wen also allow noncommutative fusion. See Lin et al. [2021] for example. Once we allow non-commutative fusions, the generalized Levin–Wen model also encompasses the twisted Kitaev models mentioned in Section 31.4.1, which are the Hamiltonian form of the Dijkgraaf–Witten models of Section 23.4. The original work on Levin–Wen models also required tetrahedral symmetry. This requirement has been relaxed by a number of authors (see citations at the end of the chapter).

[4]There is also a connection between the Kitaev model for a nonabelian group and a Levin–Wen model with *commutative* fusions, which is explored in detail by Buerschaper and Aguado [2009]. In the Kitaev model (Chapter 31) one describes edges with group elements. One can compare this to a Levin–Wen model where edges are described by representations of the group. The resulting TQFT is the same in both cases. The equivalence between these two descriptions of the TQFT is an example of a *Morita equivalence*.

handed semion theories as we saw in Section 32.2.

In the case of the toric code, however, while the input $\mathbb{Z}_2$ planar diagram algebra (with $d = +1$) can be given two possible braidings (either bosonic or fermionic; see Section 18.1.1), neither braiding is modular. As such, the TQFT that arises from the Drinfel'd double, that is, the toric code, does not factorize into a simple left-handed times right-handed theory. However, the two possible braidings, bosonic and fermionic, are two of the resulting particle types of the toric code.

More generally a planar diagram algebra (a spherical tensor category) may not have any braidings at all[5] (i.e. no R-matrix satisfies the hexagon equation with the given F-matrices). Nonetheless, all such planar diagram algebras used as an input to a string-net model generate Drinfel'd doubles that are valid modular anyon theories.

Given a planar diagram algebra, such as those described in Chapters 14 or 16 (i.e, we need a set of F-matrices satisfying the pentagon equation) we would like our ground-state wavefunction to be a sum over all diagrams that are equivalent via the set of allowed diagrammatic manipulations. We thus intend ground-state wavefunctions of the form

$$|\psi\rangle \sim \sum_{\substack{\text{all diagrams that can be}\\\text{obtained from a reference}\\\text{diagram via allowed moves}}} W(\text{diagram}) \, |\text{diagram}\rangle \qquad (33.1)$$

where $W(\text{diagram})$ is the weight of the diagram, which we determine simply by making an evaluation via the diagrammatic rules. So for example, if we have a theory with a particle type a having loop weight d_a, if we add a contractible loop of type a to a diagram, the weight W of the diagram is multiplied by[6] d_a, as indicated in Figs. 14.3 or 16.6.

33.1 Graphical Example: Doubled-Fibonacci

It is worth working through one nontrivial example of a Levin–Wen model in some detail to see how the Drinfel'd double arises. The simplest examples, the toric code and the doubled-semion model, were based on the two possible $\mathbb{Z}_2$ planar fusion algebras, as discussed in Section 18.1 (in another language, these are the two possible *3-cocycles*; see Section 19.2). However, these two are both abelian fusion rules so they are perhaps a bit too simple to see what happens more generally.

Here, we consider the Fibonacci planar diagram algebra (which we encountered as far back as Section 8.2.1) as an input to our Levin–Wen model. This model is perhaps the simplest example with nonabelian fusion rules, so it presents a good example for demonstrating how this construction generally works. Note, however, that the Fibonacci planar diagram algebra can be given a braiding, and the resulting theory is then modular. As discussed earlier, we then claim that the content of the Drinfel'd double will be simply the product of a left- and right-handed

---

[5]The simplest such examples are given by Hagge and Hong [2009]. These theories have three particle types I, x, y and fusion rules $x \times x = 2x + y + I$ and $x \times y = y \times x = x$ and $y \times y = I$. There are several possible solutions of the pentagon equation, but none of these admit a solution of the hexagon equation. If we allow noncommutative fusions, then a nontrivial cocycle of $\mathbb{Z}_3$ provides a planar diagram algebra with no braiding. See Section 18.5.3.

[6]If the loop weight is negative, as we used in Sections 32–32.2 we are working with a nonunitary diagram algebra. We often want to interpret this as a gauge transformation of a unitary algebra, as we did in Section 32.3.

copy of the modular Fibonacci anyon theory.

$$\text{Doubled-Fib} = (\text{Fib Anyons})_L \times (\text{Fib Anyons})_R$$

We study modular theories more generally in Section 33.2.1. However, it is still useful to work through this example explicitly to get a feel for how these models work.

At this point, we will not write down a Hamiltonian, but rather focus on using the diagrammatic algebra implied by a wavefunction of the form 33.1 to see what properties of the model we can extract. The more general construction of a Hamiltonian is described in Section 33.3.

Recall that in the Fibonacci planar diagram algebra there is one (self-dual) nontrivial particle type, which we usually call τ, and the nontrivial fusion rule is

$$\tau \times \tau = I + \tau \ .$$

Diagrams for such a model can be represented as a branching loop gas, as shown in Fig. 33.1. In such diagrams τ is a blue line, drawn without arrows, and these lines are allowed to branch at trivalent vertices, but cannot have endpoints.

The (unique unitary) diagrammatic algebra for these fusion rules are derived in Section 18.2. We summarize the important rules here:

Fig. 33.1 A Fibonacci branching loop diagram allows branching of loops but no endpoints.

$$\big)\big(\; = \; \phi^{-1}\; \smile\!\!\frown \; + \; \phi^{-1/2}\; \asymp \tag{33.2}$$

$$\rtimes \; = \; \phi^{-1/2}\; \smile\!\!\frown \; - \; \phi^{-1}\; \asymp \tag{33.3}$$

$$\bigcirc \; = \; \phi \tag{33.4}$$

$$\phi \;\;=\;\; 0 \tag{33.5}$$

where

$$\phi = \frac{1 + \sqrt{5}}{2}$$

is the golden mean. We recognize the first two lines (Eqs. 33.2 and 33.3) as being the Fibonacci F-move (compare Fig. 18.9). Equation 33.4 gives a loop the value of the (positive) quantum dimension, and Eq. 33.5 is the locality principle (see Fig. 16.8). From these equations, along with the assumption of planar isotopy invariance (a diagram may be smoothly deformed in the plane without changing its value), these diagrams fully define the planar diagrammatic algebra for the Fibonacci model.

From these principles it is easy to derive a few very useful additional lemmas including (see Exercise 33.24.)

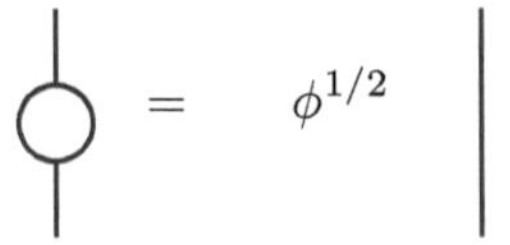

$$\phi = \quad \phi^{1/2} \quad \Big| \qquad\qquad (33.6)$$

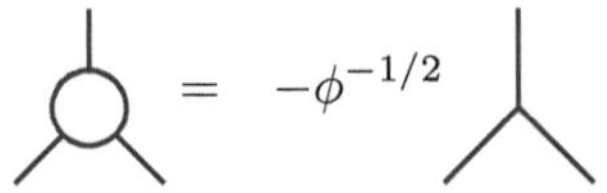

$$\phi = \quad -\phi^{-1/2} \quad \bigwedge \qquad\qquad . \qquad (33.7)$$

33.1.1   Quasiparticle Types

As with the toric code, the Kitaev quantum double model, and the doubled-semion model we should be able to determine the quasiparticle content by constructing the tube algebra, or states on an annulus, graphically (see in particular the discussion in Section 30.2). Let us start by considering all possible states on the annulus where a single blue line intersects both the inside and the outside of the annulus. We imagine using diagrammatic moves to merge together all of the blue lines into one of a very few minimal basis states. Once we have used the fusion rules, the blue lines can intersect each edge either zero times or one time, and a blue line can either go around the annulus once, or not at all. Thus, any state on the annulus can be reduced to a linear superposition of the seven possible basis states shown in Figs. 33.2–33.4.

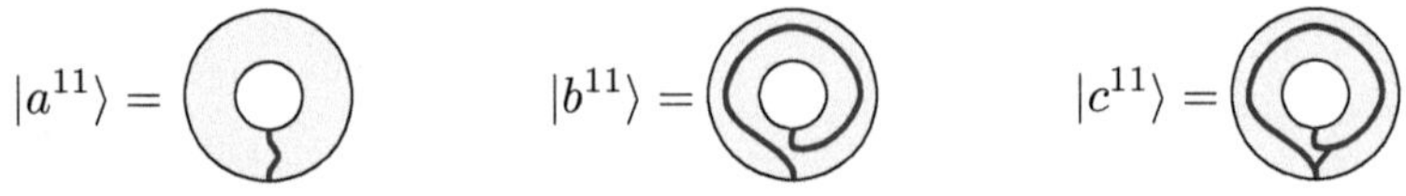

$$|a^{11}\rangle = \qquad\qquad |b^{11}\rangle = \qquad\qquad |c^{11}\rangle =$$

Fig. 33.2 All states on the annulus with a single blue line out the top and a single blue line out the bottom can be reduced using our graphical rules to a superposition of these three basis states. The superscript 11 means that one blue line touches the inside of the annulus and one blue line touches the outside.

$$|0^{00}\rangle = \qquad\qquad |1^{00}\rangle =$$

Fig. 33.3 All states on the annulus where no blue line touches the inner or outer edge can be reduced using our graphical rules to a superposition of these two basis states. The superscript 00 indicates no blue line intersects either the inside or outside edge.

$$|O_m^{01}\rangle \sim \qquad\qquad |O_m^{10}\rangle \sim$$

Fig. 33.4 Basis states on the annulus where a blue line touches either the inner or outer edge of the annulus (but not both).

Figure 33.2 shows the possible basis states where the blue line intersects each edge exactly once. The superscript 11 is used to indicate that one blue line touches the inside of the annulus and one blue line touches the outside. Similarly in Fig. 33.3 the superscript 00 indicates that no blue line intersects either edge, and analogously in Fig. 33.4 the superscripts 01 and 10 indicate that the blue line intersects either the outside only or the inside only.

Twist factors in the 11 sector

Let us first focus on the 11 sector, that is, the states in Fig. 33.2, and calculate the twist factors, knowing the quasiparticle types must be eigenstates of the twist operators.[7] We thus rotate the diagrams and then reduce them again to the three basis states shown in Fig. 33.5.

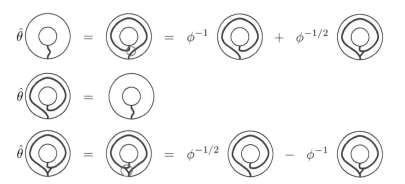

> [7] In Chapters 30–32 we started by constructing idempotents for the tube algebra—putting one annulus inside another. It is also possible to do that here. The approach taken here of evaluating the twist eigenstates first and worrying about the tube algebra second is just a bit simpler.

Fig. 33.5 The twist operator applied to the three basis states. The small red circle indicates where we apply an F-move.

Writing a general wavefunction as $|\psi\rangle = A|a^{11}\rangle + B|b^{11}\rangle + C|c^{11}\rangle$ we can summarize the twist operator on a general wavefunction as the matrix equation

$$\hat{\theta}\begin{pmatrix} A \\ B \\ C \end{pmatrix} = \begin{pmatrix} 0 & \phi^{-1} & \phi^{-1/2} \\ 1 & 0 & 0 \\ 0 & \phi^{-1/2} & -\phi^{-1} \end{pmatrix}\begin{pmatrix} A \\ B \\ C \end{pmatrix}$$

where the matrix written is precisely what has been calculated in Fig. 33.5. The eigenvectors of this matrix are the particle types with definite twist factors given by their eigenvalues under rotation. With a bit of algebra it can be shown that the eigenvalues of this matrix are given by[8]

$$\begin{aligned}
\theta_R &= e^{+4\pi i/5} & |O_R^{11}\rangle &\sim |a^{11}\rangle + e^{-4\pi i/5}|b^{11}\rangle + \phi^{1/2}e^{+3\pi i/5}|c^{11}\rangle \\
\theta_L &= e^{-4\pi i/5} & |O_L^{11}\rangle &\sim |a^{11}\rangle + e^{+4\pi i/5}|b^{11}\rangle + \phi^{1/2}e^{-3\pi i/5}|c^{11}\rangle \\
\theta_m &= 1 & |O_m^{11}\rangle &\sim |a^{11}\rangle + |b^{11}\rangle + \phi^{-3/2}|c^{11}\rangle \quad (33.8)
\end{aligned}$$

> [8] We are abusing notation a bit here by sometimes writing O_R^{11} as a ket $|O_R^{11}\rangle$ when we want to emphasize that it is a state on the annulus and sometimes not as a ket when we want to think of it as an operator to be composed with the ∘ operation, as in Eq. 33.9.

where we have not properly normalized these eigenstates (hence, we write $\sim$ rather than $=$).

The first and second lines correspond to the expected spin factors for a right-handed Fibonacci anyon τ_R or left-handed Fibonacci anyon τ_L (Recall that $R_I^{\tau\tau} = \theta_\tau^*$ from Eq. 15.4 and we defined right-handed Fibonacci anyons in Eq. 10.2). The final line represents the fusion of these two objects $m = \tau_R \times \tau_L$, which we sometimes call the "magnetic" particle. Along with the identity, these are all of the possible particle types in the doubled-Fibonacci theory. Since we know that Fibonacci anyons are modular (i.e. have a braiding such that the S-matrix is unitary; see Section 17.3.1) we expected to get both a right- and left-handed copy of the Fibonacci model (and indeed we did). Note further that we never broke time reversal in the definition of the model, so if we get the right-handed particle out we had better get out the left-handed particle, too. Further, note that nowhere did we put into the model any of the braidings for Fibonacci anyons (the only thing we put in is the diagrammatic rules in Eqs. 33.2–33.5). The information about the twists and the braiding is an emergent result!

We can also check that the three wavefunctions, given correct normalizations (see Exercise 33.3.a,b) are the idempotents for a tube algebra. That is, if we put one annulus inside the other and simplify we find the projector composition relation

$$O_\alpha^{11} \circ O_\beta^{11} = \delta_{\alpha,\beta} O_\alpha^{11} \tag{33.9}$$

where $\alpha, \beta \in \{R, L, m\}$.

Full Tube Algebra

We have not yet constructed the full tube algebra as some possible states on the annulus have been neglected. We now consider the two basis states on the annulus where blue lines touch neither the inside, nor the outside edge, as shown in Fig. 33.3. Applying the rotation $\hat{\theta}$ to either of these states obviously does nothing (so these are eigenstates of $\hat{\theta}$ with $\theta = 1$).

We can now put together two natural combinations. Unsurprisingly, the Kirby strand (or Ω strand) linear combination (see Eq. 17.7)

$$|O_I^{00}\rangle \sim |0^{00}\rangle + \phi |1^{00}\rangle \tag{33.10}$$

will be the idempotent of the ground, or vacuum, state (and this has $\theta_I = 1$). In addition to the vacuum state we can also construct an orthogonal state on the annulus

$$|O_m^{00}\rangle \sim \phi |0^{00}\rangle - |1^{00}\rangle \quad . \tag{33.11}$$

Finally, we have the last two possible basis states for our annulus $|O_m^{01}\rangle$ and $|O_m^{10}\rangle$, which are shown in Fig. 33.4. These states have a blue line that intersects only *one* of the annulus edges, as shown in Fig. 33.4. It is not obvious, but not too hard to show (see Exercise 33.24.b), that applying $\hat{\theta}$ to either of these states also does nothing (i.e. these states are eigenstates of $\hat{\theta}$ with $\theta = 1$).

So we now have a total of seven orthogonal states on the annulus: three where a single blue line touches both edges $|O_R^{11}\rangle, |O_L^{11}\rangle, |O_m^{11}\rangle$; two where no blue line touches either edge $|O_I^{00}\rangle, |O_m^{00}\rangle$; and two where the blue line touches only one edge $|O_m^{01}\rangle$ and $|O_m^{10}\rangle$. There are only four topological sectors I, R, L, m, but we have additional degrees of freedom describing the boundary condition of our annulus in the case of the m topological sector. This is something we have seen previously in Eq. 31.23 for the Kitaev model—extra degrees of freedom may specify boundary conditions.

We are now in a position to "multiply" (or "compose") these states by placing one annulus inside another (as we did in Section 30.2). We claim that these seven states, given the proper normalization, satisfy the more general projector law (see Exercise 33.3)

$$O_\alpha^{ij} \circ O_\beta^{jk} = \delta_{\alpha,\beta}\, O_\alpha^{ik} \tag{33.12}$$

where $\alpha, \beta \in R, L, I, m$ are the topological sectors and $i, j, k \in 0, 1$ indicate how many blue lines intersect the boundary (and j must be the same for the shared boundary, hence, the repeated index on the left-hand side).[9]

33.1.2 Torus States

We can also take our annulus idempotents and connect the inside edge to the outside edge to obtain the four orthogonal ground states on a torus matching the four quasiparticle sectors

$$
\begin{aligned}
|O_I\rangle_{torus} &= |O_I^{00}\rangle \text{ with inner and outer edges connected} \\
|O_R\rangle_{torus} &= |O_R^{11}\rangle \text{ with inner and outer edges connected} \\
|O_L\rangle_{torus} &= |O_L^{11}\rangle \text{ with inner and outer edges connected} \\
|O_m\rangle_{torus} &= |O_m^{00}\rangle \text{ with inner and outer edges connected} \\
&= |O_m^{11}\rangle \text{ with inner and outer edges connected}
\end{aligned}
\tag{33.13}
$$

with the last equality being proven in Exercise 33.4.a.

Once we have the ground states, we can determine the S-matrix by calculating the overlap of these states[10] with the states having the two directions of the torus interchanged, as we did in Section 30.4 (see also Section 30.6). We will not do that here because there is a short-cut that we describe in a moment that makes the calculation much easier.

[10]The very devoted student can give it a try. It is not even assigned as an exercise because it is very sloppy indeed!

33.1.3 A Nicer Torus Basis

The linear combinations in Eq. 33.8 that make up the states on the torus (Eqs. 33.13) seem rather complicated. In this case, we can use an interesting "bookkeeping" trick to simplify matters substantially. This trick depends on the fact that the Fibonacci fusion rules (the F-matrix) in Eqs. 33.2–33.3 do happen to be consistent with a braiding. That is, we could define a consistent over- and under-crossing for these lines that satisfy the hexagon equations (Eq. 13.1–13.2). Since we are thinking about entirely planar wavefunctions (Eq. 33.1) there is no real meaning to over- and under-crossings here. However, we know that if we have a diagram with such crossings, we can use R-matrices to remove the crossings and obtain a diagram with no over- or under-crossings. In this way *we can use diagrams with over- and under-crossings as a short-hand notation for diagrams that live entirely in the plane with no such crossings.*

To be more specific, let us write the crossing rule

$$\text{(diagram)} = \phi^{-1}e^{-4\pi i/5}\,\text{(diagram)} + \phi^{-1/2}e^{3\pi i/5}\,\text{(diagram)} \tag{33.14}$$

which we derive using Fig. 16.22 along with $d_\tau = \phi$ and the values $R_I^{\tau\tau} = e^{-4\pi i/5}$ and $R_\tau^{\tau\tau} = e^{3\pi i/5}$ as in Eq. 10.2. (The mirror image crossing will have the phases complex conjugated; see Exercise 33.24.c).

We also define the Kirby (or Ω) strand as

$$\Omega\,\bigg| \;=\; \frac{1}{\mathcal{D}}\,\bigg|_I \;+\; \frac{\phi}{\mathcal{D}}\,\bigg|_\tau \tag{33.15}$$

as in Fig. 17.7 where

$$\mathcal{D}^2 = 1 + \phi^2 = \phi + 2$$

is the total quantum dimension.

With this notation, and drawing figures now on the square with opposite sides identified to represent the torus, we have

$$|O_I\rangle_{torus} \;=\; \text{(diagram)}\,\Omega \tag{33.16}$$

$$|O_R\rangle_{torus} \;=\; \text{(diagram)}\,\Omega \tag{33.17}$$

$$|O_L\rangle_{torus} \;=\; \text{(diagram)}\,\Omega \tag{33.18}$$

$$|O_m\rangle_{torus} \;=\; \text{(diagram)}\,\Omega \;. \tag{33.19}$$

Given the definition of the Kirby strand (Eq. 33.16), it is easy to see that $|O_I\rangle_{torus}$ in Eq. 33.16 is the same as Eq. 33.10 just wrapped into a torus—the identity part of the Kirby strand (first term on the right of Eq. 33.15) gives the first term of Eq. 33.10 and the τ part gives the second term, and the amplitude of these terms differ by a factor of ϕ as the two terms in the Kirby strand differ in amplitude by ϕ.

Let us now consider $|O_R\rangle_{torus}$ in Eq. 33.17 and compare it to the first line in Eq. 33.8. Recall that, although we draw a crossing, we really mean that we should apply the crossing rule Eq. 33.14 to obtain a planar diagram. The first term in the Kirby sum Eq. 33.15 is simply the vertical blue line, and generates the $|a^{11}\rangle$ term in the first line of Eq. 33.8 (but on a torus rather than an annulus). The second term of the Kirby sum has two crossed strands that we "resolve" using Eq. 33.14 to yield the $|b^{11}\rangle$ and $|c^{11}\rangle$ terms in the first line of Eq. 33.8 with the correct amplitudes and phases. The calculation for $|O_L\rangle_{torus}$ is entirely analogous.

For $|O_m\rangle_{torus}$ it is not entirely obvious that this diagram on a torus should be equivalent to either expression provided in Eq. 33.13. However, with some algebraic manipulation this can be shown to be true (see Exercise 33.4.b).

33.2 More General Diagram Algebras

The general principles we have learned from the Fibonacci string-net allow us to generalize this diagrammatic approach to any consistent planar diagram algebra. One can have an arbitrarily complicated diagram algebra with many quantum numbers (particle types) for the lines, with planar diagram moves as described, for example, in Chapter 16.

One can always construct a tube algebra to identify the quasiparticle types, and from this determine the twist factors of all of the quasiparticles, as well as the corresponding ground states on the torus. From the torus ground states one can further calculate the S-matrix, as we did in Sections 30.4 and 30.6. While this procedure can become tedious, it is (in principle) straightforward.

The simplest case to handle is when the planar diagram algebra is isotopy invariant as in Chapter 16. In cases where one lacks planar isotopy invariance only due to a nontrivial Frobenius–Schur indicator, and as long as there is a $\mathbb{Z}_2$ Frobenius–Schur grading, one can push minus signs onto the loop weight and work with the nonunitary, but isotopy-invariant, version of the diagram algebra, as described in Section 14.5. Indeed, this is exactly what we did in Chapter 32 for the doubled-semion model! For more complicated diagram algebras that cannot be made isotopy invariant one must be much more careful with the diagrammatic manipulations.

For any diagram algebra, parts of the tube algebra are always very easy to find—the "fluxons." These are the states on the annulus where no nontrivial strings intersect either the inner or outer edge of the annulus. The form of the fluxon part of the tube algebra is always given by

$$O_\alpha^{00} = U_{I\alpha} \sum_j [U^\dagger]_{\alpha j} \qquad (33.20)$$

where U is the unitary matrix that diagonalizes the fusion multiplicity matrices N_i (see Eq. 8.20 or Section 17.10). If the input diagram algebra is modular then U is the S-matrix, but more generally the input algebra may not even have a braiding. In Exercise 33.7 we prove that this form, Eq. 33.20, satisfies the necessary tube algebra

$$O_\alpha^{00} \circ O_\beta^{00} = \delta_{\alpha\beta} O_\alpha^{00} \quad . \tag{33.21}$$

The fluxon state O_α^{00} corresponds to a quasiparticle having $\theta_\alpha = 1$. The other sectors of the tube algebra require more detailed calculations.

33.2.1 Modular Case

Using over- and under-crossing diagrams as in Eqs. 33.17–33.19 can only be done for diagram algebras that have a consistent braiding. For the special case where the diagram algebra has a *modular* braiding one can construct a particularly simple basis for the ground states on a torus. Again, using the Kirby (Ω) strand, we write a basis of states as shown in Fig. 33.6, where i and j run over all particle types of the planar diagram algebra[11]. Again, we reiterate that we are describing a planar diagram, and the over- and under-crossings are meant to be converted into fully planar diagrams by using the R-matrices like those in Fig. 16.22.

[11]This construction is discussed by Hardiman [2019] and is also generalized to the non-modular case, although the non-modular case is complicated.

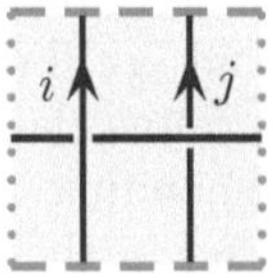

Fig. 33.6 With Ω being the Kirby strand (Fig. 17.7) this provides a basis for states on the torus, where i and j run over all particle types of the (modular) diagram algebra.

To show the usefulness of this representation, we calculate the twist factors for the different sectors by sheering the torus as we did in Fig. 30.22:

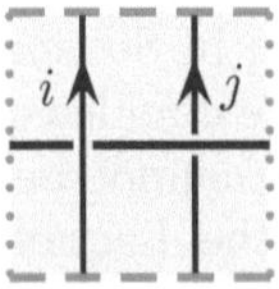

$$\tag{33.22}$$

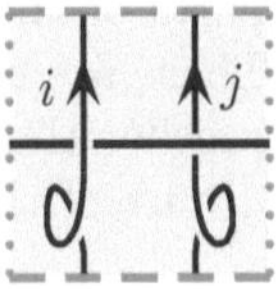

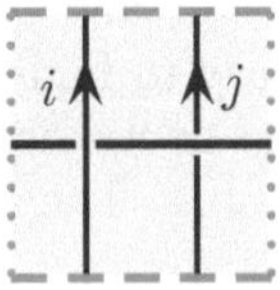

where θ_i is the twist factor of particle i for the input modular diagram algebra. Equivalently, we can write

$$\hat{\theta}|O_{ij}\rangle_{torus} = \theta_i \theta_j^* |O_{ij}\rangle_{torus} \quad . \tag{33.23}$$

The step of going from the first to second line of Eq. 33.22 looks complicated, but it is actually nothing more than two applications of the handle-slide move derived in Fig. 24.10. The blue i strand is broken open and handle-slid over the purple Ω strand, and then the blue j strand is broken open and also handle-slid over the purple Ω strand! The next step just pulls the strand straight, and finally, we remove the curl in each strand with the twist factors of these strand types from the input modular diagram algebra.

Since the ground states on the torus are in one-to-one correspondence with the quasiparticle types, the end result here, Eq. 33.23, tells us that the quasiparticle types $[i, j]$ of the emergent string-net model have twist factors corresponding to all possible quasiparticle types the input theory i and the mirror image of the quasiparticles of the input theory j. This is compatible with what we found in the doubled-semion model and doubled-Fibonacci model—the emergent TQFT is the product of the input theory with its mirror image.

We can go further and calculate the S-matrix similarly to Section 30.4 by examining the overlap of the states on the torus with the states of the rotated torus (see also the discussions of Sections 17.3.2 and 7.3.1 and Fig. 17.6). We calculate

$$
\underline{S} \quad \figurebox{i \uparrow \quad \uparrow j} \quad = \quad \figurebox{j \leftarrow \\ i \leftarrow}
$$

$$
= \sum_{n,m} S_{in} S_{jm}^* \quad \figurebox{n \uparrow \quad \uparrow m} \tag{33.24}
$$

The first line is just the definition of the $\underline{S}$ operator. Going from the first to the second line requires a calculation but can be shown using only handle-slides and the killing property (Fig. 17.9) of the Kirby string (see Exercise 33.5). In a more abbreviated form we can write

$$
\underline{S} |O_{ij}\rangle_{torus} = \sum_{n,m} S_{in} S_{jm}^* |O_{nm}\rangle_{torus}
$$

Thus, we see that the emergent S-matrix of the string-net quantum double model is just

$$
S^{double}_{[i,j],[n,m]} = S_{in} S_{jm}^* \tag{33.25}
$$

that is, the product of a right-handed S-matrix and a left-handed S-matrix for the input modular theory.

33.3   Microscopic Hamiltonian

We now want to construct a microscopic Hamiltonian on a (regular or irregular) lattice that will implement the diagrammatic planar algebra—that is, whose ground-state space is described by wavefunctions of the form of Eq. 33.1.

As usual, we here assume a trivalent lattice. For simplicity we might often assume a honeycomb, but any trivalent lattice, even a disordered trivalent graph, works just as well.[12] Edges are labeled with the possible quantum numbers of the planar diagram algebra. Unless a quantum number a is self dual ($a = \bar{a}$) the edges must be labeled with arrows to indicate a direction. Reversing the direction of an arrow changes a to $\bar{a}$ as shown in Fig. 33.7. If there are fusion multiplicities $N^c_{ab} > 1$ then vertices between three edges with labels a and b entering (arrows pointed toward vertex) and c exiting (arrow pointed away from vertex) must also carry labels $\mu \in \{1, \ldots, N^c_{ab}\}$. For simplicity we will usually assume all $N^c_{ab} \leq 1$ so there are no vertex indices, although it is fairly easy to extend the discussion to the more general case.

[12]Inevitably, it will be incorrectly called a lattice even if it is disordered.

Fig. 33.7 Reversing an arrow on an edge takes the particle type to its antiparticle.

33.3.1   Fat-Lattice Construction

Starting with the toric code, we have often found it useful to think on the continuum instead of thinking about degrees of freedom on the lattice. We were quite cavalier about doing this, and even mixed the lattice and continuum descriptions sometimes (such as in Fig. 31.12). Here, we make this idea a bit more formal with a construction due to Levin and Wen [2005], known as the "fat lattice." Our microscopic model is defined on a trivalent graph, and we add an × in the middle of each plaquette that we interpret as some sort of "puncture" in the manifold, as shown in Fig. 33.8. We think of the entire region between two neighboring marked punctures as being an edge that has been fattened until it fills all the space between the punctures. This gives us the name "fat lattice." We now allow ourselves to draw diagrams with particle lines that may or may not be on the (unfattened) edges of the graph. To interpret these lines for the physical system on the lattice, we deform the diagram, using our diagrammatic rules, until all of the lines lie on the (unfattened) edges of the graph. (We sometimes say we "push" the diagram onto the unfattened edges.) In this process we use our diagrammatic equivalence rules, but we are not allowed to cross any lines over any of the marked ×s. This procedure is well defined in the sense that there is a *unique* superposition of edge labels corresponding to each continuum diagram we draw (unique up to further F-moves). An example of this procedure is shown in Fig. 33.8. The fat-lattice construction can be thought of as a very fancy bookkeeping procedure—it allows us to draw any diagram on the plane (like the left of Fig. 33.8) and interpret it as values on edges only (like the right of Fig. 33.8).

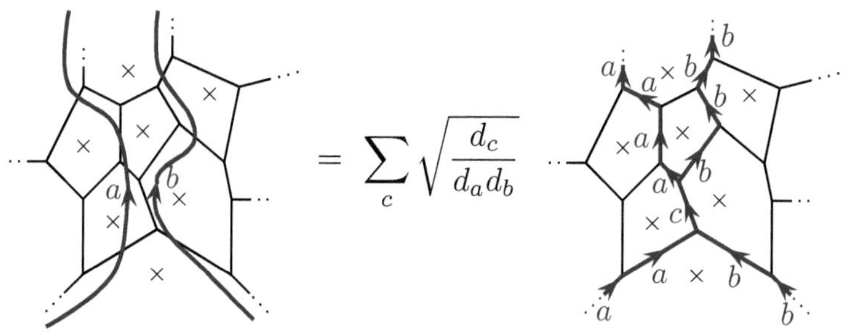

Fig. 33.8 Converting a continuum diagram on the "fat lattice" (left) to a diagram on the lattice (right). The lines are deformed (isotopy) to sit on the (unfattened) edges. For the edge marked c, we need to use the diagrammatic completeness rule (Fig. 16.9), which accounts for the sum over c and the prefactor.

33.3.2 Vertex and Plaquette Operators

We introduce a vertex operator at vertex α as

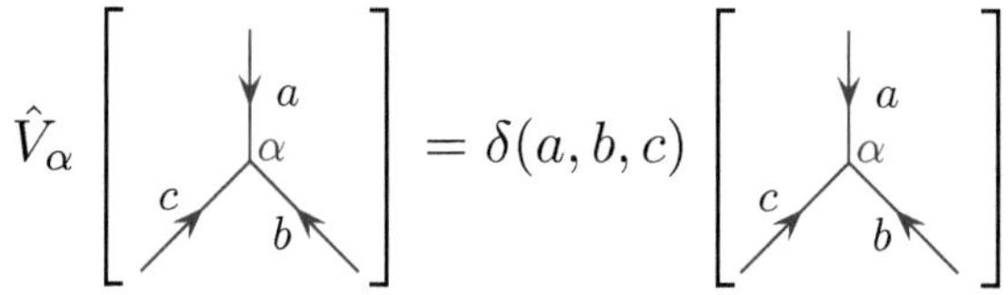

$$\hat{V}_\alpha \left[\begin{array}{c} \includegraphics \end{array} \right] = \delta(a,b,c) \left[\begin{array}{c} \includegraphics \end{array} \right] \tag{33.26}$$

where[13]

$$\delta(a,b,c) = \begin{cases} 1 & \text{if } (a,b,c) \text{ is an allowed vertex } (N_{abc} = N^{\bar{c}}_{ab} > 0) \\ 0 & \text{otherwise.} \end{cases}$$

Here, $\hat{V}_\alpha$ is a projector (it has eigenvalues of 0 and 1 only).

Analogous to the plaquette term in Eq. 31.12 for the Kitaev model, we graphically represent a plaquette operator as the introduction of an $\tilde{\Omega}$ loop (See Fig. 17.11).

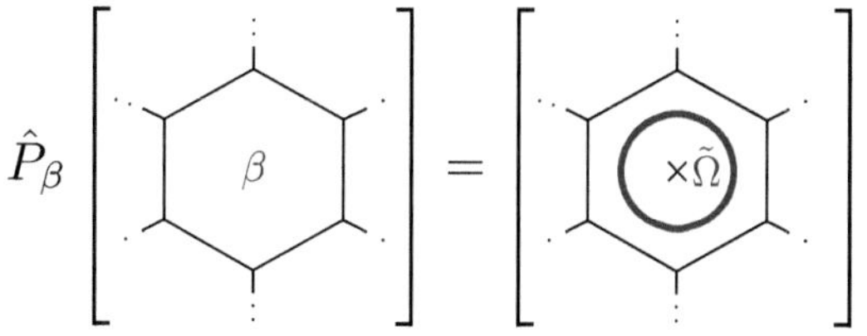

$$\hat{P}_\beta \left[\begin{array}{c} \includegraphics \end{array} \right] = \left[\begin{array}{c} \includegraphics \end{array} \right] \tag{33.27}$$

In the spirit of the fat-lattice construction, the $\times$ in the middle of the $\tilde{\Omega}$ loop should be thought of as a puncture so that the loop may not be contracted. To find the action of this operator on the edge variables, one should "push" the $\tilde{\Omega}$ strand into the edges, by using the diagram rules. While one can write an explicit form of this operator in terms of the effect on the quantum numbers on the edges, it is not particularly enlightening to do so, so we defer this exercise to Section 33.5.

---

[13]If there were a vertex variable $\mu \in \{1,\ldots,N^{\bar{c}}_{ab}\}$ with $N^c_{ab} > 1$ we would need to have a, b, c fusing in the μ channel for the projector to give us the non-zero result.

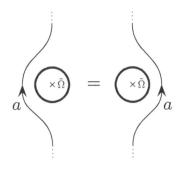

Fig. 33.9 The handle-slide identity.

Recall now the handle-slide property of the $\tilde{\Omega}$ strand shown in Fig. 33.9, which we originally derived in Fig. 24.10. This move allows us to quickly show that

$$\hat{P}_\beta^2 = \hat{P}_\beta \tag{33.28}$$

since

$$\text{[diagram]} = \text{[diagram]} = \text{[diagram]} \tag{33.29}$$

where in the first step of Eq. 33.29 we use the handle-slide (Fig. 33.9) and in the second step we use the fact that a contractible loop of $\tilde{\Omega}$ evaluates to unity (see Fig. 17.11). Equation 33.28 indicates the $\hat{P}_\beta$ is a projector, having eigenvalues of 0 and 1 only.

Further, because the planar diagram algebra is already known to be self-consistent (i.e. it does not matter in which order we do moves to evaluate a diagram) and diagrammatic moves starting with an allowed fusion diagram never generate unallowed diagrams, we can quickly conclude that all of the $\hat{V}_\alpha$ and $\hat{P}_\beta$ operators commute with each other.

33.3.3   Levin–Wen Hamiltonian

Our Hamiltonian is then written in a now-familiar form

$$H_{\text{Levin--Wen model}} = -\Delta_v \sum_{\text{vertices } \alpha} \hat{V}_\alpha - \Delta_p \sum_{\text{plaquettes } \beta} \hat{P}_\beta \tag{33.30}$$

with Δ_v and Δ_p both positive. The vertex term simply enforces the condition that all vertices have edges corresponding to allowed fusions of the planar diagram algebra. The plaquette term allows the edge variables to fluctuate dynamically. The key here is that if we are in a ground state of the system then $\hat{P}_\beta = 1$ for all plaquettes, so that when we draw a diagram in the fat-lattice picture, we can surround each $\times$ with an $\tilde{\Omega}$ loop without changing the value of the diagram. However, due to the handle-slide identity (Fig. 33.9) our diagrammatic lines may now be freely deformed over the $\times$ punctures. As a result, we may now use our diagrammatic rules freely on the manifold ignoring the punctures entirely. This then enforces that the ground-state wavefunctions must be of the form shown in Eq. 33.1.

The reasoning here is subtle. We needed to puncture the plane so that we had a mapping from continuum diagrams on the fat-lattice to quantum numbers on the edges of the lattice. But then the fact that the plaquette operators are in the ground state tell us that we can freely deform strands over these punctures in a ground-state wavefunction and may then apply all diagrammatic operations to our wavefunctions[14]

[14]To clarify this point, first note that we can extract the weight of any given diagram in a ground state, Eq. 33.1, as

$$W(\text{diagram}) = \langle \text{diagram}|\psi\rangle$$

where $|\psi\rangle$ is a ground state ket. Then with a plaquette operator B_p being in the $+1$ eigenstate for the ground state, we have $\langle \text{diagram}|\psi\rangle = \langle \text{diagram}|B_p|\psi\rangle$. This means, due to Fig. 33.9, that the weight of a diagram with a string going to the left of a plaquette must be the same as the weight of a diagram with that string going to the right of the plaquette. Further, once we can slide strings over plaquettes, we can also use local diagram manipulations, as allowed by the fat-lattice construction, to establish that the weights of diagrams in any ground state wavefunction must obey the diagrammatic rules (without considering the punctures).

33.3.4 Ground State Is Topological

As with the other diagrammatic models we have encountered (the toric code, the Kitaev model, and the doubled-semion model), the ground state of this Hamiltonian is topological—meaning that it does not depend on the detailed geometry of the lattice, but only on the topology of the underlying manifold. The proof of this statement follows very closely the discussion of Section 31.3 for the Kitaev model. We briefly outline the argument here; a more detailed discussion is given by Hu et al. [2012]. The strategy here is to show that there is a simple one-to-one mapping between ground-state wavefunctions on one lattice structure and those on a second lattice structure that has been altered (or "mutated") locally without changing the overall topology of the surface.

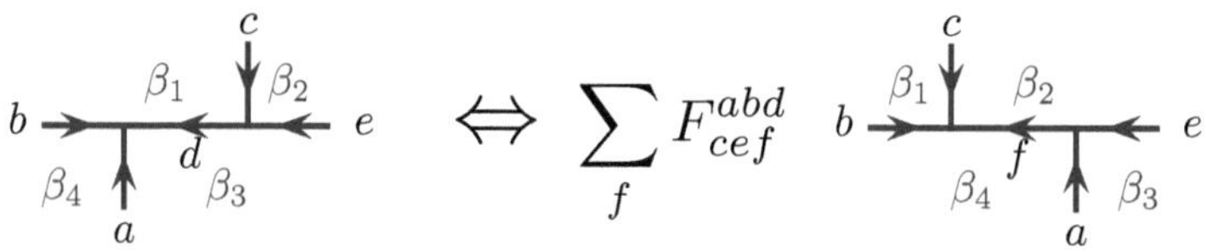

Fig. 33.10 The F-move that can be performed on the lattice to change its microscopic structure while preserving the global topology. There is a (unitary) isomorphism between states under this transformation showing that, for example, the ground-state degeneracy depends only on the topology of the system. Plaquettes are labeled with β_j.

The elementary moves we need to consider are shown in Fig. 33.10 and 33.11. Figure 33.10 is an F-move (compare Fig. 16.4). Note that here we are not just changing the quantum numbers on the edge, but we are also restructuring the underlying lattice. Nonetheless, it provides a unitary transformation between states on the left and right of the figure.

Figure 33.11 is the splitting of a plaquette β into two plaquettes, β_1 and β_2. Unsurprisingly, the stem of the tadpole (labeled c) must be labeled with the identity, or else the plaquette operator $\hat{P}_{\beta_2}$ will vanish (see Fig. 16.8 for example). The unique ground state of $\hat{P}_{\beta_2}$ is then simply a $\tilde{\Omega}$ loop around the edge labeled b. The $\tilde{\Omega}$ loop in the operator $\hat{P}_{\beta_1}$ can then be handle-slid over the $\tilde{\Omega}$ loop of $\hat{P}_{\beta_2}$ to become exactly like the operator $\hat{P}_\beta$, thus showing that the ground state on the right of Fig. 33.11 is the same as that on the left multiplied by the trivial $\tilde{\Omega}$ loop around β_2 (analogous to Eq. 31.10).

Ground State on a Sphere

We can then establish, analogous to the case of the Kitaev model, that the ground state on a sphere is nondegenerate. To do this we choose the simplest lattice decomposition of the sphere (see Fig. 31.7 with a single vertex, a single edge forming the equator (connected to the single vertex at each end), and two plaquettes—one covering the north hemisphere and one covering the south hemisphere. The vertex condition is always satisfied (having the same quantum number going out as coming in) and the two plaquette operators are identical to each other. Thus, the ground

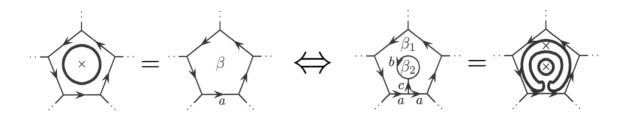

Fig. 33.11 The plaquette addition move that can be performed on the lattice. The stem c of the tadpole must be labeled with the identity. The β_2 plaquette is then put into the ground state and then the β_1 plaquette acts exactly as the β plaquette on the left, which we can see by a handle-slide.

[15]The purist might object that we have used a vertex with only two edges meeting it, whereas we defined the model to have only trivalent vertices. If we want to not cheat, we could use a decomposition of the sphere with two (trivalent) vertices, three edges, and three plaquettes. By using F-moves if necessary, this can be restructured into two connected tadpoles living on the surface of the sphere,

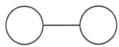

The connecting line in the middle must be labeled with the identity, leaving us just a pair of $\tilde{\Omega}$ loops for the plaquettes, which have unique ground states.

state is the unique state given by the $+1$ eigenstate of the plaquette operator, which is just $\sim \sum_s d_s |s\rangle$. That is, the wavefunction is a $\tilde{\Omega}$ loop around the single edge.[15]

33.4   Quasiparticle String Operators

The general construction of quasiparticles for the Levin–Wen model can be quite complicated. Recall that the definition of quasiparticle string operator is an operator that can be detected only at its ends—that is, when operating on the ground state, it creates no defects except at its ends. Wavefunctions in the presence of quasiparticles are still of the form of Eq. 33.1—the weight of the wavefunction is determined by diagrammatic rules, at least so long as we apply these rules not too close to the position of the quasiparticles.

For example, in the simple case of the toric code (see Section 27.4), if we act on a string of edges with σ_x operators we create excitations only at the ends of this string. The diagrammatic rule that loops must not have ends can be violated at the position of the quasiparticles (and thus, the form of Eq. 33.1 is violated at the position of the quasiparticles).

Further, the two endpoints of the string operator (the quasiparticles) may be brought together and annihilated to form a closed string operator loop (as in the right of Fig. 27.10). Such an operator should act trivially on the ground state.

We can always find the string operators by the conditions that they act trivially on the ground state except at their ends (or if they are closed loops they act trivially on the ground state everywhere). In general, working out the details of such quasiparticle string operators is tedious. However, in cases where the input category has a solution to the hexagon equation (i.e. has a braiding), certain quasiparticle string operators can be deduced easily. Further, in cases where the input category is modular, *all* of the quasiparticle string operators can be deduced very easily.

Let us suppose there exists a solution to the hexagon equation for the input category. This means we can give a diagrammatic meaning to over- and under-crossings. We construct a string operator graphically by imagining that we add a string from the input category either over or under the plane of the system as shown in Fig. 33.12. Then, we operate on the edges of the physical system by fusing this string into the edge variables of lattice using the diagrammatic rules, as shown in Fig. 33.8 (only here we may have to use diagrammatic rules for over- and under-crossings as well). Details of turning this diagrammatic prescription into an operation on the lattice edges are given in Section 33.5.2 and see Section 33.4.2 for a particularly simple example. Here, it is crucial to realize that the operator we are considering acts entirely on the two-dimensional system—the diagram of a string above or below the lattice is simply a bookkeeping trick to encode what operations we perform on the edge variables.

Since the diagrammatic rules (including the braiding rules) are self-consistent (i.e. satisfy the pentagon and hexagon equations), we are allowed to make diagrammatic transformations without changing the value of the diagram. In particular, we can freely deform the picture by sliding the string around the lattice (so long as we do not move the endpoints of the string, if the string has endpoints). This freedom to deform the string means that the string cannot cause any defects to (i.e. cannot cause any excitations of) either the vertex or plaquette operators along its length. This is the very definition of a string operator—it does not create defects except at its ends. A string operator that forms a closed contractible loop can be shrunk down to a very small loop and removed with the usual diagrammatic rule (Fig. 16.6); thus, a closed loop of string operator acts trivially on the ground state (except possibly giving a constant depending on how we normalize the operator).

We can extend this method for describing string operators to describe string operators that cross over themselves, or cross over other string operators. As shown in Fig. 33.13, a string or knot off the lattice can be dropped down and fused into the lattice to form diagrams like Fig. 33.12. The consistency of the diagrammatic rules allows us to evaluate the knot first, *or* to fuse it into the lattice first—and we should get the same result either way! Thus, the braiding statistics that we use in three-dimensions are now reflected in the properties of the string operators which we apply to the two-dimensional system! It is then quite easy to use this diagrammatic reasoning to prove statements like Eq. 33.25 (see Exercise 33.6).

For a string operator with two ends (i.e. not a closed loop) we must have a violation of the diagrammatic rules at each of the ends of the string. Recall that in diagrams, if a particular quantum number comes into a region, that quantum number must also leave the region (see, for example, the locality constraint in Fig. 8.7). A string that comes to

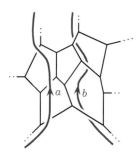

Fig. 33.12 Quasiparticle string operators represented as being strings in diagrams that go either above or below the plane of the system. The actual operators live entirely in the two-dimensional plane.

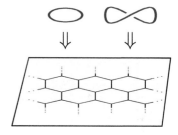

Fig. 33.13 Strings off the lattice are a bookkeeping method for more complicated string operators on the two-dimensional lattice.

Fig. 33.14 Strings with endpoints should be thought of as having the endpoints fixed on the two-dimensional lattice.

[16] In this case it does not matter whether we think of the blue line as being above or below the lattice plane.

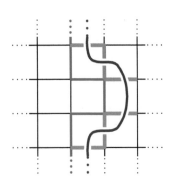

Fig. 33.15 The blue string above the lattice is a string operator that we will fuse into the lattice at the position of the red line. If we choose the blue string to have bosonic braiding statistics then it simply flips the spins along the red line, creating vertex defects at the ends of the red line. If we choose the blue string to have fermionic braiding statistics, then it measures the spins along the green lines before flipping the spins along the red line. The measurement of the green spins creates plaquette defects, whereas the flipping of spins creates vertex defects. The combination of these two is a fermionic quasiparticle.

an abrupt end must violate this condition. Thus, we expect that the system will locally not be in the ground state at the position of a string end. While we can freely deform the position of the string along its length without changing the physical wavefunction, the position of the endpoints of a string are physically observable quantities and cannot be moved without changing the physical state of the system.

We can similarly draw diagrams off the lattice for finite-length strings, as in Fig. 33.14. Here, however, we must fix the endpoints of the strings on the lattice and these positions cannot be changed (without changing the value of the wavefunction). Nonetheless, we can still apply the diagrammatic rules to other parts of the string to slide it over the lattice or unknot it as we like.

We now consider two examples of such string operators that we have already seen (albeit in different language).

33.4.1　Toric Code: A Non-Modular Example

In the toric code (Section 30.1) the planar diagram algebra is the $\mathbb{Z}_2$ loop gas with $d = +1$. There are two possible solutions of the hexagon equations, worked out in Section 18.1.1: the boson and the fermion. It is not a coincidence that the quasiparticle types of the toric code include a boson and a fermion!

Let us consider the bosonic solution first. The boson braids trivially with itself ($R_0^{11} = 1$ in Section 18.1.1) so we only need to think about fusing this string operator into the lattice. In Fig. 33.15 we show such a string operator (blue) above[16] the toric code lattice (chosen to be a square lattice here for simplicity). Fusing this blue line into the lattice at the position of the red path simply flips the spins along this red path. This is exactly the string operator for the (bosonic!) vertex defects that we introduced in Section 27.4.1.

We can also consider the fermionic solution of the hexagon equations. The fermion has braiding $R_0^{11} = -1$, meaning that we accumulate a minus sign every time the string (again we take the string to be the blue line in Fig. 33.15) crosses over an edge of the lattice, which itself is colored blue (meaning the spin on that edge is pointing down in the notation of Chapter 27). Thus, before fusing the blue string into the red path, we measure the state of the edges marked green in Fig. 33.15 and get a minus sign for each green edge where the spin is in the down position. This sign is precisely the action of the string operator that creates the plaquette defects that we introduced in Section 27.4.2. Once we have introduced the appropriate minus signs we must also fuse the string into the red path, which then flips the spins along this line. Thus, the net effect of string operator is the product of the plaquette-defect string and the vertex-defect string, which we recall is the fermionic quasiparticle!

Thus, the bosonic string operator corresponds to the vertex-defect boson, and the fermionic string operator corresponds to the fermionic quasiparticle. The fusion of these two quasiparticles together would give the plaquette defects (only measuring the spins on the green horizontal edges). We have thus reconstructed all of the quasiparticle types of the toric code with this graphical technique.

It is important to realize that in this model we did not actually input the solutions to the hexagon equation (indeed, we only input the toric code Hamiltonian). However, knowing that these two solutions exist allows us to quickly determine the quasiparticle string operators, had we not already known them before!

33.4.2 Doubled Semions: A Modular Example

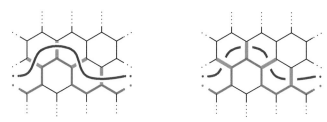

Fig. 33.16 A string operator (blue) can be put over or under the lattice edges. This is just a bookkeeping trick to simply describe an operator that actually lives within the two-dimensional plane. If we put a string both above and below the plane, and fuse these together we obtain the magnetic string operator defined in Section 32.1.1 (see Exercise 33.1).

For the doubled-semion model (Chapter 32) the planar diagram algebra[17] can be taken to be the $\mathbb{Z}_2$ loop gas with $d = -1$. There are two possible solutions of the hexagon equations, worked out in Section 18.1.2, corresponding to right- and left-handed semions. Correspondingly, the quasiparticle types of the doubled-semion model include both a right- and left-handed semion. Let us see more explicitly how this happens.

Let us consider the case of $\theta = i$ to describe our blue lines off the lattice. We imagine placing a string of this type (drawn blue) either over or under the lattice, as shown in Fig. 33.16. We would like to fuse this blue line into the red marked path. This has the effect of flipping all of the quantum numbers on the red edges (i.e. applies σ_x to all of these edges). However, in addition one accumulates phases for crossing over the green edges depending on the quantum numbers already on these edges. To clarify these phases let us focus on a single crossing, as shown in Fig. 33.17. Depending on the state of the vertex a different phase is accumulated. This looks a bit hard to keep track of, but the phases are simply the phases from the diagrammatic algebra when we "push" the blue string into the edges.

The statistics of the quasiparticles at the end of these strings must reflect the statistics of the strings off the lattice. Let us imagine, for example, making a complicated knot or link from these strings and fusing this link into the two-dimensional lattice as in Figs. 33.13 and 33.14. If there are crossings in these diagrams we may choose to "resolve" these crossings (by using R-matrices to remove crossings) either before or after fusing the strings into the lattice. Because our diagrammatic rules are

[17]Here, we are going to use the nonunitary (negative d) but isotopy-invariant diagram algebra, knowing that it is simply a gauge transformation of the unitary diagram algebra discussed in Section 32.3. We do have to be slightly careful in doing this as our string operators may accumulate minus signs from caps that are removed in the unitary version.

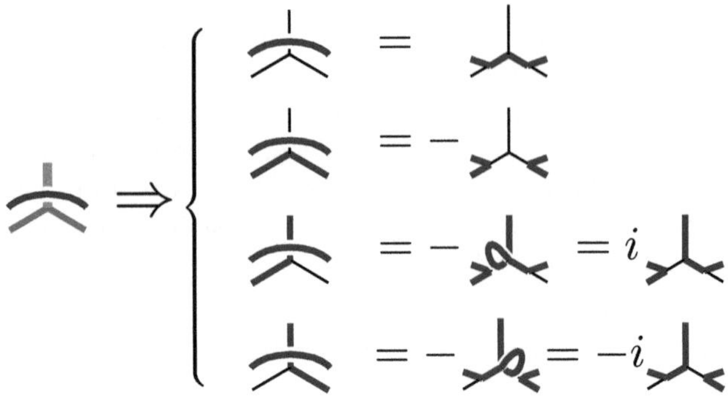

Fig. 33.17 Fusing a string into the lattice in the doubled-semion model. The blue line is assumed to have $\theta = i$. In the ground state there are four possible configurations of the vertex and each one accumulates a different phase. The state of the red edges are flipped and the state of the green edges are unchanged.

consistent (satisfy the hexagon equations), both approaches must give the same result.

Consider a knot that starts off the lattice. Now imagine fusing it into the lattice. The fusing operations are necessarily done in some particular order: if there is a crossing of two strings, the string that is fused into the lattice plane first is always the one that is closer to the lattice plane. In this sense the arrow of time runs away from the plane of the lattice in both directions. This may seem a bit strange, but the strings above the lattice and the strings below the lattice are completely independent of each other anyway. This is consistent with our expectation that our resultant theory will be the product of a right-handed and a left-handed theory that do not interact with each other.

Because of the flip in the direction of time, a right-handed curl in a string diagram (like the right of Fig. 33.13) is interpreted as a right-handed curl (a θ as in Fig. 15.2). However, if the same picture were below the plane of the lattice, since the direction of time now points downward (away from the plane of the lattice) it is instead interpreted as a θ^* as in Fig. 15.3 (in which time points up). We thus realize that the strings above and below the lattice have opposite twist factors, as we would expect!

33.4.3   More Generally

The general principles explained here can be extended to any Levin–Wen model. In the most general case we cannot even assume that our input theory (our "spherical tensor category") has a braiding at all. In this case we can only find the string operators by looking for extended operators that only create excitations at their ends. A more detailed discussion is given in the original literature (see Levin and Wen [2005]).

However, if the input category has (at least one) hexagon solution (as in the toric code, there are two) then we know that each solution of the hexagon equations will provide a quasiparticle string type (the tube algebra calculation above can tell us the number of quasiparticle types

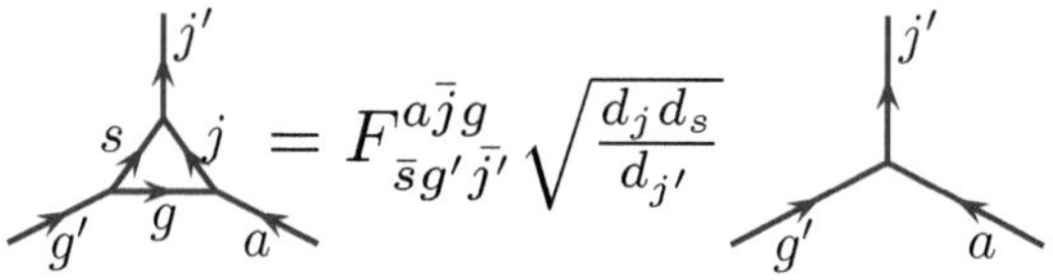

Fig. 33.18 Collapsing a triangular bubble. This is derived in Exercise 16.1.

to expect!). These strings can be described with the graphical calculus—drawing diagrams off the lattice and fusing them into the lattice.

In the case where the input category has a modular braiding, then we know that the output theory is just two copies of the input theory with opposite handedness. One can follow the discussion of the doubled-semion theory in some detail. Again, we can draw diagrams either above or below the lattice plane giving the two parts of the emergent theory.

33.5 Appendix: Explicit Form of Certain Operators

For simplicity here we assume an isotopy-invariant diagram algebra as in Chapter 16 although the generalization to the more general case is not difficult (see for example Hahn and Wolf [2020]; Lin et al. [2021]).

In this derivation we use the identities shown in Fig. 33.18 and Fig. 33.19.

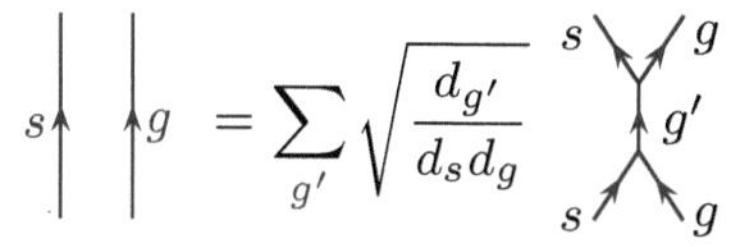

Fig. 33.19 The completeness relationship. See Fig. 16.9

33.5.1 Plaquette Operator

Let us define an operator $\hat{P}_\beta(s)$ (shown in Fig. 33.20) that fuses a loop labeled s (drawn red in this figure for clarity) into a plaquette β. In going from the first line to the second, we use Fig. 33.19 once for each edge of the polygon (four times in this case). In going from the second line to the third, we use the triangle collapsing rule Fig. 33.18 once for each corner and we cancel factors of $\sqrt{d}$.

Since the $\tilde{\Omega}$ strand (Fig. 17.11) is just a weighted sum of all the particle types, the full plaquette operator defined in Eq. 33.27 is given by

$$\hat{P}_\beta = \sum_s \frac{d_s}{\mathcal{D}^2} \hat{P}_\beta(s) \quad . \tag{33.31}$$

33.5.2 String Operator Example

Here, we give an example of fusing a string operator into the lattice. We start with the left of Fig. 33.21, with a string operator under the lattice (drawn red for clarity), and we "push" the string into the lattice in several steps shown in Figs. 33.21–33.23. Here we are assuming a case where the string operator has a well defined braiding with the edges on the lattice.

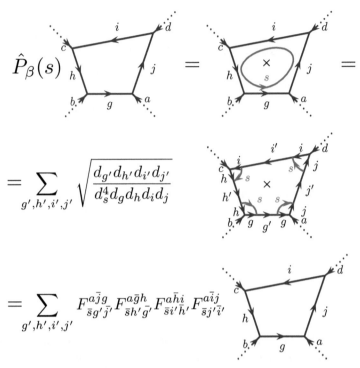

$$\hat{P}_\beta(s) = \sum_{g',h',i',j'} \sqrt{\frac{d_{g'}d_{h'}d_{i'}d_{j'}}{d_s^4 d_g d_h d_i d_j}} \quad\times\quad = \sum_{g',h',i',j'} F^{a\bar{j}g}_{\bar{s}g'\bar{j}'} F^{a\bar{g}h}_{\bar{s}h'g'} F^{a\bar{h}i}_{\bar{s}i'\bar{h}'} F^{a\bar{i}j}_{\bar{s}j'\bar{i}'} .$$

Fig. 33.20 Expressing the plaquette operator as the merging of an s-loop (red) into the edges of the plaquette.

This pushing of the string into the edge is quite similar to what we did for the plaquette operator in Section 33.5.1. The first step is shown in the top of Fig. 33.21. Then the triangular bubbles on the right of the top of Fig. 33.21 can be collapsed using Fig. 33.18. The upper triangular bubble with edges i, j, s gives a factor of $F^{ci\bar{j}}_{s\bar{j}'i'}\sqrt{d_i d_s/d_{i'}}$. The lower triangular bubble with edges g, h, s gives a factor of $F^{ag\bar{h}}_{s\bar{h}'g'}\sqrt{d_g d_s/d_{g'}}$.

Let us now focus on the part of the diagram with the under-crossing as shown in Fig. 33.23. Using the uncrossing rule shown in Fig. 33.22 we can then evaluate the crossing as shown in Fig. 33.23. Then the right of Fig. 33.23 is evaluated using the same triangle collapse law Fig. 33.18. The upper triangle including i, b, s gives a factor of $F^{hb\bar{i}}_{s\bar{i}'z}\sqrt{d_b d_s/d_z}$ and then we collapse a lower triangle with edges h, z, s to give $F^{\bar{i}hz}_{sbh'}\sqrt{d_h d_s/d_{h'}}$.

Combining together all of these factors the final result can be written as in the lower line of Fig. 33.21 where the prefactor is given by

$$K^{s;g,h,i,j;g',h',i',j',z}_{a,b,c} = F^{ci\bar{j}}_{s\bar{j}'i'} F^{ag\bar{h}}_{s\bar{h}'g'} F^{hb\bar{i}}_{s\bar{i}'z} F^{\bar{i}hz}_{sbh'}[R^{sb}_z]^{-1}\sqrt{\frac{d_{j'}}{d_j d_s}} . \qquad (33.32)$$

This formula itself is not really that interesting. However, by working through how the red string is pushed into the edges, hopefully we have made the method clear!

$$\sum_z \sqrt{\frac{d_z}{d_s d_b}}\,[R^{sb}_z]^{-1}$$

Fig. 33.22 The uncrossing rule. See Fig. 16.22.

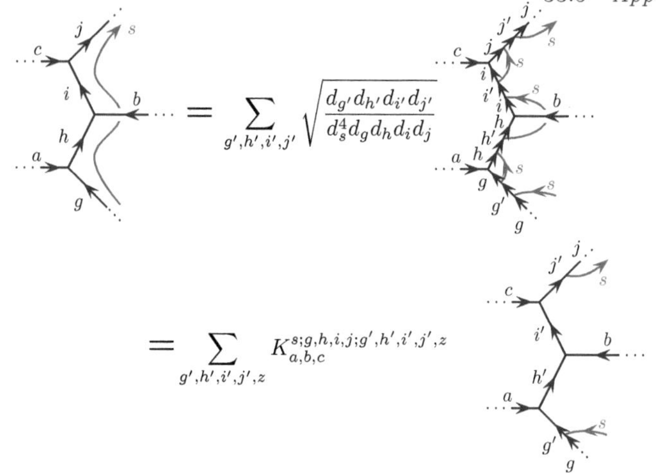

$$= \sum_{g',h',i',j'} \sqrt{\frac{d_{g'}d_{h'}d_{i'}d_{j'}}{d_s^4 d_g d_h d_i d_j}}$$

$$= \sum_{g',h',i',j',z} K_{a,b,c}^{s;g,h,i,j;g',h',i',j',z}$$

Fig. 33.21 The string operator is drawn (red) as a string running under the lattice (blue). To determine the action of this operator, we use the diagrammatic rules to fuse the red string into the blue lattice. The first step is shown in the top line. The triangle bubbles can then be collapsed using 33.18. The under-crossing in the middle is evaluated using Fig. 33.23 and the final result is shown on the second line with coefficient given by Eq. 33.32.

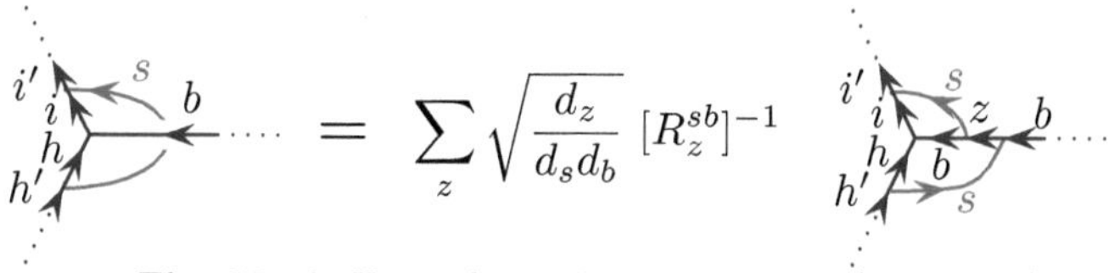

$$= \sum_z \sqrt{\frac{d_z}{d_s d_b}} \, [R_z^{sb}]^{-1}$$

Fig. 33.23 Using the under-crossing rule (Fig. 33.22).

Chapter Summary

- The Levin–Wen string-net model constructs a generalization of the toric code based on the most general possible planar diagram algebra.
- Ground-state degeneracy and quasiparticle types can be identified using diagrammatic tube algebra reasoning.
- The topological content of the model is the quantum double (or Drinfel'd double) of the input planar diagram algebra.
- If the input algebra can be given a modular braiding, the output theory is just the product of a left- and right-handed copy of this modular theory.

Further Reading

- The original work by Levin and Wen [2005] is still a good reference to learn about this model.
- The string-net model has been generalized to include diagram algebras without tetrahedral symmetry, with non-commutative fusions, and even non-spherical categories (Hong [2009]; Hahn and

Wolf [2020]; Lin et al. [2021]; Zwilling et al.; Runkel [2020]).

- Discussions of the tube algebra for the Levin–Wen model have been given by many authors, including Aasen et al. [2020], Bultinck et al. [2017], and Hu et al. [2018]. The most complete discussion is in Lan and Wen [2014]. See also comments in Zwilling et al.. The original mathematical discussion of the tube algebra is Ocneanu [1994].

- Many aspects of the Levin–Wen models have been extensively studied, including thermodynamics, phase transitions, boundaries, and interfaces (see Schulz [2014]; Schulz et al. [2013]; Burnell et al. [2011]; Vidal [2022]; Hu and Wan [2019]; Kitaev and Kong [2012]; and many many others).

- Extensions of the Levin–Wen model to $(3 + 1)$-dimensions include Walker and Wang [2012]; see also von Keyserlingk et al. [2013] and Williamson and Wang [2017], and most recently Xi et al. [2021]. (see also references in Chapter 31).

Exercises

Exercise 33.1 Magnetic String Operator for Doubled-Semion Model
(a) Show that fusing the two strings shown in Fig. 33.16 (one over the lattice and one under the lattice) gives the diagram in Fig. 33.24 up to minus signs.
(b) Keeping track of all of the minus signs, show that Fig. 33.24 is exactly equivalent to the magnetic string operator defined in Section 32.1.1.

Exercise 33.2 Fibonacci Diagram Algebra
(a) Use Eqs. 33.2–33.5 to derive the identities Eq. 33.6 and Eq. 33.7.
(b) Show that the twist operator $\hat{\theta}$ applied to $|O_m^{10}\rangle$ or $|O_m^{01}\rangle$ has eigenvalue $\theta = 1$.
(c) Given that Eq. 33.14 is true, calculate the result for the mirror image crossing and show that the formula is identical except that the phases are complex conjugated.

Exercise 33.3 Fibonacci Tube Algebra
The purpose of this exercise is to work out the (somewhat tedious) details of the tube algebra for Fibonacci anyons and prove Eq. 33.12. All of the following relationships can be shown by implementing the diagrammatic rules in Eqs. 33.2–33.7.
(a) Find the normalizations of the wavefunctions in Eq. 33.8 such that these operators are projectors. That is, so that

$$O_\alpha^{11} \circ O_\alpha^{11} = O_\alpha^{11}$$

with $\alpha \in L, R, m$, and $\circ$ means to put one annulus inside the other. Note these wavefunctions will *not* generally be normalized so that $\langle \psi | \psi \rangle = 1$.
(b) Check that
$$O_\alpha^{11} \circ O_\beta^{11} = 0 \quad \alpha \neq \beta$$
with $\alpha, \beta \in L, R, m$.

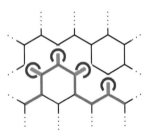

Fig. 33.24 Fusing two strings, one over the lattice and one under the lattice, as in Fig. 33.16, generates this diagram.

(c) Find the normalizations of the wavefunctions in Eqs. 33.10 and 33.11 such that.

$$O_\alpha^{00} \circ O_\alpha^{00} = O_\alpha^{00}$$

with $\alpha, \beta \in I, m$ and confirm that

$$O_m^{00} \circ O_I^{00} = O_I^{00} \circ O_m^{00} = 0 \quad.$$

(d) Find the normalization of the wavefunctions in Fig. 33.4 such that

$$O_m^{ij} \circ O_m^{ji} = O_m^{ii}$$

with $i \neq j$ and $i, j \in 0, 1$.

(e) With this normalization, confirm that

$$O_m^{ij} \circ O_m^{jj} = O_m^{ii} \circ O_m^{ij} = O_m^{ij} \quad.$$

(f) Check that

$$O_m^{01} \circ O_\alpha^{11} = O_\alpha^{11} \circ O_m^{10} = 0$$

for $\alpha \in L, R$.

(g) Finally, check that

$$O_m^{10} \circ O_I^{00} = O_I^{00} \circ O_m^{01} = 0 \quad.$$

Exercise 33.4 The Magnetic Fibonacci Ground State .

(a) (Easy) Prove the final equality in Eq. 33.13, that is, $|O_m^{00}\rangle$ with inner and outer edges connected = $|O_m^{11}\rangle$ with inner and outer edges connected. Hint: Use Eq. 33.12 to write O_m^{11} and O_m^{00} in terms of O_m^{01} and O_m^{10}. Then, think about what it means to be on a torus.

(b) Show that the expression for $|O_m\rangle_{torus}$ given in Eq. 33.19 is equivalent to the two expressions shown in Eq. 33.13.

Exercise 33.5 *S*-matrix of the Double of a Modular Theory

The purpose of this exercise is to go from the first to second line of Eq. 33.24.

(a) Use the definition of the Kirby (Ω) strand, the unlinking identity (Fig. 17.2), and the unitarity of the *S*-matrix to show the identity shown in Fig. 33.25.

(b) Use the identity in Fig. 33.25 to rewrite the figure on the right of the top line of Eq. 33.24 such that the two horizontal strands labeled i and j are replaced with Ω strands with loops around them. Then, use the handle-slide property to slide the labeled strands over the vertical Ω strand. Then, slide one horizontal Ω so that two Ω strands can be canceled with the killing property (Fig. 17.9) to obtain the second line of Eq. 33.24.

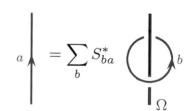

Fig. 33.25 A useful identity. Here, Ω is the Kirby strand (Fig. 17.7).

Exercise 33.6 *S*-matrix of the Double of a Modular Theory, Redux

Using the diagrammatic reasoning of Section 33.4 derive the *S*-matrix for the quantum double of a modular theory in terms of the *S*-matrix of the input theory. That is, derive Eq. 33.25. Hint: Use the arguments of Section 33.4.2.

Exercise 33.7 Fluxon Sector of the Tube Algebra

Show that the fluxon states in Eq. 33.20 satisfy the composition law Eq. 33.21. Hint: Use Fig. 17.1.

Exercise 33.8 Total Quantum Dimension of a Quantum Double

If we build a Levin–Wen model based on a modular category $\mathcal{A}$ the output theory $\mathcal{A} \times \overline{\mathcal{A}}$ has total quantum dimension given by the square of the total quantum dimension of the input category. $\mathcal{D}(\mathcal{A} \times \overline{\mathcal{A}}) = [\mathcal{D}(\mathcal{A})]^2$. It turns out that a similar relationship holds even if the input category is not braided: if we build a Levin–Wen model based on any category then the output TQFT has total quantum dimension given by the square of the total quantum dimension of the input category, which we write as[18]

$$\mathcal{D}^{double} = [\mathcal{D}^{input}]^2 \quad . \tag{33.33}$$

In this exercise we give a proof of this statement. To do this we construct a Levin–Wen model on a g-handled torus.

(a) Fix the number of faces to be order unity (we can fix it to be 1 for example). Show that

$$(\#\text{of vertices}) = (2/3)(\#\text{of edges}) = 4g + \text{const}$$

where the constant depends on number of faces but is independent of g. Hint: Use the Euler characteristic, and recall that the Levin–Wen model requires trivalent vertices.

(b) Ignore the plaquette condition of the Levin–Wen model and calculate the total number of edge labelings that satisfy the vertex condition only. Show that for large g it scales as

$$(\#\text{of edge labelings}) \sim [\mathcal{D}^{input}]^{4g} \quad .$$

(c) Let us now recall the result of Eq. 8.2 that the dimension of the ground state of a g-handled torus scaled as $\mathcal{D}^{2g}$ for large g. This remains true even if there are some fixed number of quasiparticles on the torus (but not changing these quasiparticles as we add handles). Argue that the result of part (b) allows us to conclude Eq. 33.33. Think carefully about why it is OK to leave out the plaquette condition in the counting from part (b).

Part VIII

Entanglement and Symmetries

Topological Entanglement

We might wonder what is special about the ground states of the models we have been studying. What is it that makes them topological? Another way of asking the same question is: how is it that anyons can arise from simple degrees of freedom such as a system of simple spin-1/2 (like we have for the toric code). One interesting answer to these questions is that TQFT ground states are special because they have a special type of long-ranged quantum entanglement.[1] To understand this, we now study entanglement in topological systems.

First, let us review the idea of entanglement. Consider a Hilbert space $\mathcal{H}$ for some quantum-mechanical system of interest. We then partition this Hilbert space into two pieces

$$\mathcal{H} = \mathcal{H}_A \otimes \mathcal{H}_B \quad .$$

If the overall system is in a pure state with normalized wavefunction, $|\psi\rangle$, this wavefunction can always be written as a so-called Schmidt decomposition

$$|\psi\rangle = \sum_n \lambda_n \, |\psi_n^A\rangle \otimes |\psi_n^B\rangle \tag{34.1}$$

where on the right-hand side, the wavefunctions $|\psi_n^A\rangle$ are an orthonormal set of wavefunctions spanning the Hilbert space $\mathcal{H}_A$ and the wavefunctions $|\psi_n^B\rangle$ are an orthonormal set of wavefunctions spanning the Hilbert space $\mathcal{H}_B$. The "Schmidt weights," λ_n, are non-negative real numbers[2] such that

$$\sum_n |\lambda_n|^2 = 1$$

when the wavefunction $|\psi\rangle$ is properly normalized. If more than one Schmidt weight is non-zero, we say that $|\psi\rangle$ is *entangled* between subsystems A and B. The canonical example of entanglement is a singlet made of two spin-$\frac{1}{2}$ particles[3]

$$|\psi\rangle = \frac{1}{\sqrt{2}}|\uparrow_L\rangle \otimes |\downarrow_R\rangle - \frac{1}{\sqrt{2}}|\downarrow_L\rangle \otimes |\uparrow_R\rangle \quad .$$

This wavefunction fundamentally entangles the left and right spins with each other—there is no way to describe the state of one spin independently of the state of the other. This is not troubling when the spins are near each other but leads to endless philosophical consternation when the two spins become highly separated, resulting in famous quantum-mechanical paradoxes such as that proposed by Einstein, Podolsky, and

[1]Over the last two decades, the condensed matter community has increasingly used tools of quantum information to study the properties of different phases of matter. See for example, Zhang et al. [2019a] and Cirac et al. [2021]

[2]Often these Schmidt weights are written as $\lambda_n = e^{-\xi_n/2}$ with the values ξ_n called the *entanglement energies*. If one divides a D-dimensional system spatially into two pieces, under fairly general conditions these entanglement energies look like the spectrum of an effective Hamiltonian that lives along the $(D-1)$-dimensional cut. See for example, Li and Haldane [2008], Qi et al. [2012], Dubail et al. [2012], and Swingle and Senthil [2012].

[3]The fact that there is a minus sign here might seem troubling, but we simply can redefine the sign of the ket $|\downarrow_L\rangle$ as a gauge choice. After this redefinition the wavefunction fits the form of Eq. 34.1 with all positive Schmidt weights.

Rosen (Einstein et al. [1935]).

A quantitative measure of the entanglement is given in terms of the Schmidt weights by the von Neumann entanglement entropy[4]

$$\mathcal{S}_{A,B} = \mathcal{S}_{B,A} = -\sum_n |\lambda_n|^2 \log(|\lambda_n|^2) \quad . \tag{34.5}$$

There are other measures of entanglement entropy that can be considered, such as the Renyi entanglement entropy

$$\mathcal{S}^\alpha_{A,B} = \frac{1}{1-\alpha} \log(\sum_n |\lambda_n|^{2\alpha}) \tag{34.6}$$

which matches the von Neumann entropy in the limit $\alpha \to 1$. As far as topological properties are concerned, Renyi entropies behave similarly to the von Neumann entropy (see, for example, Flammia et al. [2009]).

We are usually concerned with studying cases where the subsystems A and B are spatially local.[5] That is, Hilbert space $\mathcal{H}_A$ describes degrees of freedom in region A and Hilbert space $\mathcal{H}_B$ describes degrees of freedom in region B such that the regions A and B partition a larger system, as suggested in Fig. 34.1.

It is reasonable to expect that there will be some entanglement between regions A and B that is proportional to the length of the boundary— some degrees of freedom near the boundary on the A side interact with some degrees of freedom near the boundary on the B side, and some entanglement develops between them. This type of entanglement is known as "short-ranged" since it occurs between degrees of freedom that are not physically far apart from each other. However, topologically ordered phases of matter have an additional piece of entanglement that is not short-ranged and it occurs only due to their special topological properties. To illustrate this so-called *topological entanglement entropy*,[6] we consider the toric code as an example although the results generalize to any topologically ordered matter.

[5]Recall that locality is always important in the context of TQFTs. See Section 8.6.

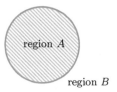

Fig. 34.1 Partitioning a system into two regions.

[6]Topological entanglement entropy was discovered by Levin and Wen [2006] and Kitaev and Preskill [2006].

[4]An equivalent, but more general, definition is given in terms of the density matrix ρ of the full system. For a pure state wavefunction $|\psi\rangle$ the density matrix is just $\rho = |\psi\rangle\langle\psi|$. We can then write the entropy as

$$\mathcal{S}_{A,B} = -\text{Tr}[\rho_A \log \rho_A] = -\text{Tr}[\rho_B \log \rho_B] \tag{34.2}$$

where ρ_A and ρ_B are the so-called reduced density matrices of the subsystems, given by tracing out the degrees of freedom of the opposite subsystem

$$\rho_A = \text{Tr}_B[\rho] = \sum_n |\lambda_n|^2 |\psi_n^A\rangle\langle\psi_n^A| \tag{34.3}$$

$$\rho_B = \text{Tr}_A[\rho] = \sum_n |\lambda_n|^2 |\psi_n^B\rangle\langle\psi_n^B| \tag{34.4}$$

where for example Tr_B means to trace out the degrees of freedom in subsystem B only. This definition is more general than Eqs. 34.1 and 34.5 since it can be applied when the full system is not in a pure state.

34.1 **Entanglement in the Toric Code**

Recall from Chapters 27–30 that the ground state of the toric code can be thought of as a superposition of loop gas wavefunctions. Let us consider a particular loop gas configuration as in Fig. 34.2. The correlations (and hence, the entanglement) for such a loop gas are non-local. To see this note that the number of blue lines that cross the boundary of a simply connected region (such as the orange boundary in the figure) must always be even—since if a blue line enters the region, it must also exit the region. Since the entry and exit may be very far apart from each other, this correlation is generally non-local. This correlation is essentially the origin of the non-local topological entanglement entropy.

Now let us try to evaluate the entanglement for the toric code more precisely. We return to the lattice as in Fig. 34.3 instead of having abstract loops as in Fig. 34.2. Again, the boundary between regions A and B is marked with an orange loop. Recall Eq. 27.10 that the total wavefunction is given by a sum over all loop configurations

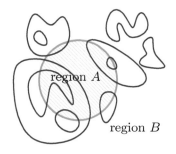

Fig. 34.2 In a loop gas the number of blue lines that cross the boundary of a simply connected region (such as the orange boundary between region A and B) must be even, since if a loop enters the region, it must also exit the region.

$$|\psi\rangle = \sum_{\text{all loop configs}} \mathcal{N}^{-1/2}|\text{loop config}\rangle \qquad (34.7)$$

where $\mathcal{N}$ is the total number of loop configs being summed over. For simplicity here, we have assumed that we are working on a spherical system so there is no ground-state degeneracy (this will imply that some of our plaquettes are not square, but as discussed in Section 27.5, this does not change any of the important physics). Note that the normalization is given by

$$\mathcal{N} = 2^{P-1}$$

where P is the number of plaquettes in the system, since flipping any plaquette gives a new loop configuration (hence, 2^P), but flipping all of the plaquettes in the entire system (for a closed manifold) returns you to the original loop configuration (hence, we have $P-1$ rather than P).

We would now like to put Eq. 34.7 into the Schmidt form of Eq. 34.1 so that we may calculate the entanglement between the two regions. Let us cheat a bit here and first divide the system into three regions. We draw an orange line (not intersecting any vertices) to separate the system into two simply connected regions as shown in Fig. 34.3. Let region A be those edges that are entirely inside the orange line, and region B be those edges that are entirely outside of the orange line.[7] Here, we also define a third region, which we call the boundary region. The boundary region includes all of those edges through which the orange line crosses. Let the configuration of spins on this these boundary edges be called $|\alpha\rangle$, the boundary state. If there are M boundary edges, each edge can take two possible states, so there are then 2^M possible boundary states $|\alpha\rangle$.

Let us further consider the ground-state wavefunction in region A, *given that* the state of the boundary edges is given by the boundary state $|\alpha\rangle$. We denote such a state of the region A as $|\psi_\alpha^A\rangle$, and similarly

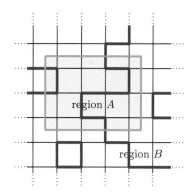

Fig. 34.3 The boundary edges are the ones cut by the orange line. Region A includes all edges entirely inside the orange line. Region B includes all edges entirely outside of the orange line.

[7]The observant reader will note that on a sphere there is no well-defined "inside" or "outside," so one arbitrarily defines one region to be called "inside'. It doesn't matter which one.

we denote the ground-state wavefunction in region B, having boundary state $|\alpha\rangle$ as $|\psi_\alpha^B\rangle$. We can write these explicitly as

$$|\psi_\alpha^A\rangle = \mathcal{N}_A^{-1/2} \sum_{\substack{\text{loop configs in region } A \\ \text{with boundary state } \alpha}} |\text{loop config}\rangle_{A,\alpha}$$

where $\mathcal{N}_A = 2^{P_A}$ the number of terms in the sum where P_A is the number of plaquettes entirely enclosed in region A. The sum over loop configurations means that the blue lines can end only at the boundary of the region, and not in the interior (i.e. there should be no vertex defect in the interior of the region). We similarly write a wavefunction for region B

$$|\psi_\alpha^B\rangle = \mathcal{N}_B^{-1/2} \sum_{\substack{\text{loop configs in region } B \\ \text{with boundary state } \alpha}} |\text{loop config}\rangle_{B,\alpha}$$

with $\mathcal{N}_B = 2^{P_B}$ the number of terms in the sum where P_B is the number of plaquettes entirely enclosed in region B.

We can then write the full wavefunction as

$$|\psi\rangle = \mathcal{N}_{allowed}^{-1/2} \sum_{\substack{\text{allowed boundary} \\ \text{states } \alpha}} |\psi_\alpha^A\rangle \otimes |\alpha\rangle \otimes |\psi_\alpha^B\rangle \tag{34.8}$$

where $\mathcal{N}_{allowed}$ is the number of terms in the sum, which is the number of different boundary states.

Eq. 34.8 is almost exactly the Schmidt decomposition form (Eq. 34.1) we would like, except that we have divided the degrees of freedom into three pieces rather than two (region A, region B and the boundary edges). It is just a matter of bookkeeping to fix this. Let us define the region $\tilde{B}$ to include both B and the boundary region (all the boundary edges), we then define

$$|\psi_\alpha^{\tilde{B}}\rangle = |\alpha\rangle \otimes |\psi_\alpha^B\rangle \quad.$$

Region A and $\tilde{B}$ now properly partition the system into two parts. The Schmidt decomposition of the wavefunction on the entire system can then be written as[8]

$$|\psi\rangle = \mathcal{N}_{allowed}^{-1/2} \sum_{\substack{\text{allowed boundary} \\ \text{states } \alpha}} |\psi_\alpha^A\rangle \otimes |\psi_\alpha^{\tilde{B}}\rangle \tag{34.9}$$

where $\mathcal{N}_{allowed}$ is the number of terms in the sum, which is the number of different boundary states.

To calculate the von Neumann entanglement entropy (now between regions A and $\tilde{B}$, which partition the full system) we need to know how many boundary states are in the sum. As mentioned, if there are M spins in the boundary region, there are 2^M possible boundary states.

[8]With a bit of thinking one can see that the wavefunctions $|\psi_\alpha^A\rangle$ form an orthonormal set, as do the wavefunctions $|\psi_\alpha^{\tilde{B}}\rangle$.

However, only half of these boundary states are allowed in the ground state! Again, since the ground state consists of closed loops only, we can only have an even number of blue edges in the boundary region when the full system is in the ground state. So in fact the number of terms in the sum is actually

$$\mathcal{N}_{allowed} = 2^{M-1} \quad .$$

Plugging this result into Eq. 34.5 gives us the final result for the von Neumann entanglement entropy between the two regions[9]

$$\mathcal{S}_{A,B} = (M-1)\log 2 \quad . \tag{34.10}$$

As we might have predicted, the entanglement between the two regions is proportional to the length of the cut (proportional to the number of edges in the boundary region). The interesting part of this result is that the result is $M-1$, rather than M. It is this subleading term (the -1) that reflects the topological properties of the state—the fact that it is comprised of a loop gas and there is therefore a long-ranged constraint on the state of the boundary.

It is worth noting that if region A has two boundaries as in Fig. 34.4 we will get instead

$$\mathcal{S}_{A,B} = (M-2)\log 2$$

where M is now the total number of spins on both boundaries put together. The point here is that there are now two constraints—one on the outer boundary and one on the inner boundary.

34.1.1 Generalizing to Arbitrary TQFTs

This technique of counting possible boundary states can be generalized to more general lattice models, such as the Kitaev quantum double (Chapter 31) or the Levin–Wen model (Chapter 33). Since all of these models are some sort of generalized loop gas, we expect that there will be similar long-range correlations of the boundary states, and hence, analogous topological entanglement entropy.[10] More generally, for any type of topological matter, if we have a smooth boundary of length L between two regions A and B, we should generically have a von Neumann entanglement entropy of

$$\mathcal{S}_{A,B} = \alpha L - \gamma + \ldots \tag{34.11}$$

for some constant α where γ is the topological contributions.[11] For example, in Eq. 34.10 we have $\gamma = \log 2$. More generally, we will obtain

$$\gamma = \log \mathcal{D} \tag{34.12}$$

where $\mathcal{D}$ is the total quantum dimension of the TQFT (Eq. 17.11)

$$\mathcal{D}^2 = \sum_i \mathsf{d}_i^2$$

[9]Here we have dropped the tilde from B. It is implied in this formula that the union of region A and region B includes all the spins in the system.

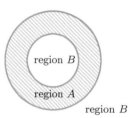

region B

region A

region B

Fig. 34.4 Partitioning a system into two regions. In this case region A is the shaded region and B is the unshaded region (which is split between inside and outside).

[10]Calculation of the von Neumann topological entanglement entropy for the Levin–Wen models by explicit counting of states was done by Levin and Wen [2006]. The Renyi version of this calculation was done by Flammia et al. [2009].

[11]The ... represents small contributions to $\mathcal{S}_{A,B}$ that vanish in the limit of large regions and smooth boundaries.

with the sum over all particle species.

This is a rather remarkable result: if you are given the wavefunction of a system, by simply splitting the system into pieces and calculating the entanglement between the two pieces, you can deduce the topological properties of the system.

We give three different derivations of the general result Eq. 34.12 in Section 34.3.

34.2 Topological Entanglement Entropy Is Robust

One might worry that the entanglement entropy we have found for the toric code is simply a particular property of the special toric code Hamiltonian we have used. Indeed, the entanglement (Eq. 34.10) will certainly change if we perturb the toric code Hamiltonian with some additional terms that change the local correlations between spins, thereby changing the prefactor α in Eq. 34.11. We might also wonder if, with a more general Hamiltonian, the details of the shape of the boundary (whether it is a rectangle or instead has many corners) may matter in addition to just the number of edges we cut. As emphasized in Chapter 29, the topological properties of the ground state of a system should remain unchanged when the Hamiltonian is perturbed a little bit (as long as the excitation gap does not close), and this is true for the topological entanglement entropy as well. The long-ranged correlations that we discovered do remain robust and are encoded into the ground state although they are a bit trickier to see. We will now follow the procedure laid out by Kitaev and Preskill [2006] that described how to isolate the topological term in the entanglement entropy.[12]

[12] Although usually the von Neumann entanglement entropy is discussed, the same arguments should hold for Renyi entanglement entropy as well.

Let us divide our sphere into four regions[13] as shown in Fig. 34.5. We use the notation that AB means the union of the regions A and B, and ABC is the union of regions A, B, and C, and so forth. Now let us define the quantity

[13] Here, we use the argument and geometry of Kitaev and Preskill [2006]. A different geometry was used by Levin and Wen [2006].

$$
\begin{aligned}
\mathcal{S}_{top} ={}& (\mathcal{S}_{A,BCD} + \mathcal{S}_{B,ACD} + \mathcal{S}_{C,ABD} + \mathcal{S}_{D,ABC}) \\
& -(\mathcal{S}_{AB,CD} + \mathcal{S}_{AC,BD} + \mathcal{S}_{AD,BC}) \quad .
\end{aligned}
\tag{34.13}
$$

This looks a bit complicated, but actually it is easy to remember. The first line is the entanglements of each of the four regions A, B, C, D with their respective complementary regions (i.e. with the remainder of the sphere). The second line, added with a minus sign, are the entanglements of the union of two of these regions with the remaining two in all three possible combinations.

This combination of entanglement entropies is constructed so as to isolate only the topological term. We will find that

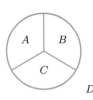

Fig. 34.5 Partitioning a sphere into four regions. The region D includes the remainder of the sphere (i.e. the "point at infinity" for the stereographic projection to the plane.

$$
\mathcal{S}_{top} = -\gamma \quad .
$$

To see this let us first naively use Eq. 34.11 to calculate $\mathcal{S}_{top}$. Each time we calculate some $\mathcal{S}_{P,Q}$ we must calculate the length of the boundary between region P and Q to plug into Eq. 34.11. Looking at Fig. 34.5 there are six segments in this figure: the three straight segments pointing radially from the center (call their length $L_{straight}$ and the three curved segments bounding region D (call their length L_{curved}). So using Eq. 34.11 we then have

$$
\begin{aligned}
\mathcal{S}_{A,BCD} = \mathcal{S}_{B,ACD} = \mathcal{S}_{C,ABD} &= \alpha(L_{curved} + 2L_{straight}) - \gamma + \ldots \\
\mathcal{S}_{D,ABC} &= \alpha(3L_{curved}) - \gamma + \ldots \\
\mathcal{S}_{AB,CD} = \mathcal{S}_{AC,BD} = \mathcal{S}_{AD,BC} &= \alpha(2L_{curved} + 2L_{straight}) - \gamma + \ldots \ .
\end{aligned}
$$

Plugging these quantities into Eq. 34.13 yields $\mathcal{S}_{top} = -\gamma$ as expected. All of the leading terms proportional to the length of the boundaries have cancelled. As we will see in a moment, this cancellation is extremely robust.

What happens if we alter the geometry? For example, suppose we add some wiggles in the middle of the segment forming the boundary between region A and region B as shown in Fig. 34.6. This change has no effect on regions C and D, or any of the triple intersections. The change in $\mathcal{S}_{top}$ can be written as

$$
\Delta\mathcal{S}_{top} = (\Delta\mathcal{S}_{A,BCD} - \Delta\mathcal{S}_{AD,BC}) + (\Delta\mathcal{S}_{B,ACD} - \Delta\mathcal{S}_{AC,BD}) \quad . \quad (34.14)
$$

It is not expected that $\mathcal{S}_{D,ABC}$ or $\mathcal{S}_{C,ABD}$ or $\mathcal{S}_{AB,CD}$ should change at all since in none of these cases is the entanglement measured across the AB boundary. In Eq. 34.14 each term in parenthesis should be zero. For example, in the first term we are comparing $\Delta\mathcal{S}_{A,BCD}$ to $\Delta\mathcal{S}_{AD,BC}$. These differ by whether D is attached to BC or attached to A. However, since D is far away from the part of the AB boundary that has been modified, the changes to the entanglement should be the same in both cases. Thus, the total $\Delta\mathcal{S}_{top}$ is zero.

We might worry that some problem could occur at the triple intersections of the three regions. However, we can argue similarly that deformation of this intersection cannot change $\mathcal{S}_{top}$ as defined in Eq. 34.13. For example, suppose we change the geometry to that of Fig. 34.7. Again, the boundaries with D and the other three triple intersections are assumed to be unchanged. The total change in $\mathcal{S}_{top}$ can be written as

$$
\begin{aligned}
\Delta\mathcal{S}_{top} =\ & (\Delta\mathcal{S}_{A,BCD} - \Delta\mathcal{S}_{AD,BC}) + (\Delta\mathcal{S}_{B,ACD} - \Delta\mathcal{S}_{AC,BD}) \\
& + (\Delta\mathcal{S}_{C,ABD} - \Delta\mathcal{S}_{AB,CD}) \quad .
\end{aligned}
$$

Again the two terms in each parenthesis differ from each other only by how region D is attached, and this should not at all be changed by any deformation to boundaries between A, B, and C.

One can further argue that $\mathcal{S}_{top}$ does not depend on the details of the Hamiltonian (as long as we remain in the ground state and do not close the gap to making excitations). If we change the Hamiltonian locally,

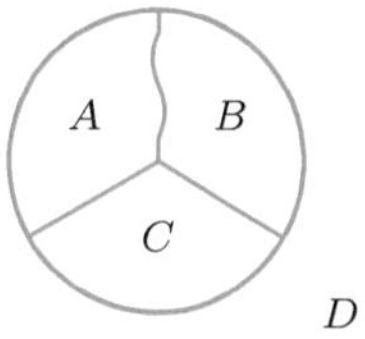

Fig. 34.6 Making the partition between regions A and B wiggly.

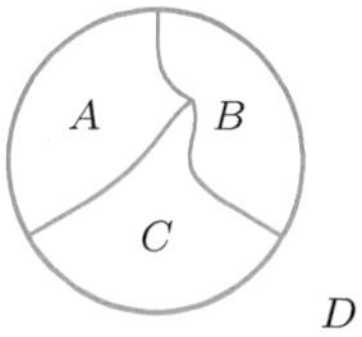

Fig. 34.7 Changing the region near the triple intersection of A, B, and C. The other three triple intersections are assumed to be unchanged, as are all boundaries with region D.

this can only matter if we happen to be near a boundary. However, we just showed that we are free to move the boundaries. So we can move the boundary, change the Hamiltonian locally, and then move the boundary back to its original position, and $\mathcal{S}_{top}$ must remain unchanged.

Thus, the topological entanglement entropy is a robust property of a topologically ordered phase of matter that reflects its topological properties.

34.3 Appendix: Three Derivations of Topological Entanglement Entropy

There are several ways to derive the key result Eq. 34.12. Something different is learned in each approach, so it is useful to go through them all. In addition to the approaches given here we also mention the work of Levin and Wen [2006] which shows an explicit calculation for string-net models.

34.3.1 Kitaev-Preskill Argument

Kitaev and Preskill [2006] use an interesting argument to explicitly isolate the topological part of the von Neumann entanglement entropy. Starting with a TQFT T that we are interested in, we first construct the *double* of this TQFT, that is, $T \times \bar{T}$, where $\bar{T}$ is the mirror image (opposite chirality) of T. We can think of this as two parallel planes of the same theory where we have oriented the normals of the plane in opposite directions. We aim to derive the topological entanglement entropy of the doubled system

$$\mathcal{S}_{top}^{double} = 2\mathcal{S}_{top} \quad .$$

To obtain the topological entanglement entropy of the doubled system we partition the system into four pieces as in Fig. 34.5. Note that each piece is made up of a region of layer T and a mirror image region of layer $\bar{T}$. Next we drill small holes though the plane in the system at the position of the four vertex points in Fig. 34.5 where three of these partition lines intersect each other (the triple boundary points, or corners of the regions). These holes also cut through both layers of the doubled system. When we cut a hole in a TQFT, generically we should expect to have low-energy edge modes created on the boundaries, as discussed in Sections 17.3.3 or 25.8. Here we would have precisely mirror image edge modes for the hole in T as we have for the hole in $\bar{T}$. We can therefore couple together the two edges and remove any low-energy modes—it is always possible to gap the edge of a theory of the form $T \times \bar{T}$ (see the discussion in Section 25.8). We thus end up with a two layer structure, connected together with wormholes as shown in Fig. 34.8 with no low-energy modes.

Since we have only made local changes by poking these holes and

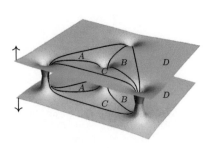

Fig. 34.8 The TQFT $T \times \bar{T}$ is thought of as a two layer system with oppositely oriented normals. These two layers are then connected together by four wormholes.

reconnecting them, we do not expect that the topological entanglement entropy will be changed. We can then consider cutting the plane into pieces as required by Eq. 34.13. Note that here each cut must go through both layers as shown by the solid lines in Fig. 34.8.

We next want to calculate the entropies on the right-hand side of Eq. 34.13. In order to calculate $\mathcal{S}_{A,BCD}$, we cut out region A only. When we do this we see that region A is just a sphere with three large holes in it—it is two triangles (upper and lower layers of region A) glued together with three strips that go between the layers that form part of the wormhole tubes. This geometry is similar for $\mathcal{S}_{B,ACD}$, $\mathcal{S}_{C,ABD}$, and $\mathcal{S}_{D,ABC}$ (for the last case recall that we have included a point at infinity). Similarly, when we cut out region AB to calculate $\mathcal{S}_{AB,CD}$ we end up with a sphere having four large holes in it, and similar for $\mathcal{S}_{AC,BD}$ and $\mathcal{S}_{AD,BC}$. Thus, we have

$$\mathcal{S}_{top}^{double} = 4\mathcal{S}_{3-\text{hole}-\text{sphere}} - 3\mathcal{S}_{4-\text{hole}-\text{sphere}} \qquad . \qquad (34.15)$$

Now we are left with the task of calculating the entropy for $\mathcal{S}_{3-\text{hole}-\text{sphere}}$ and $\mathcal{S}_{4-\text{hole}-\text{sphere}}$. The entropy here comes from the fact that the holes in the sphere may be labeled with different possible quantum numbers and potentially different fusion channels between these quantum numbers. This is equivalent to the problem of grouping together a large number of randomly chosen quasiparticles on a sphere into three or four different groups and asking the probabilities of the resulting quantum numbers of the groups.[14] These probabilities were calculated in Exercise 17.9:

[14]We can imagine creating particle–antiparticle pairs on the sphere and randomly moving them into the holes in all possible ways.

2-hole sphere	$p_2(a, \bar{a}) = d_a^2/\mathcal{D}$
3-hole sphere	$p_3(a, b, c, \mu) = d_a d_b d_c/\mathcal{D}^4$
4-hole sphere	$p_3(a, b, c, d, \mu) = d_a d_b d_c d_d/\mathcal{D}^6$

where $\mu \in N_{abc}$ for the three-hole case, and $\mu \in N_{abcd}$ represents the particular fusion channel of a, b, c, d if there is more than one.

We then directly calculate

$$\mathcal{S}_{3-\text{hole}-\text{sphere}} = -\sum_{a,b,c,\mu} p_3(a, b, c, \mu) \log p_3(a, b, c, \mu)$$

$$= -\sum_{a,b,c,\mu} p_3(a, b, c, \mu)[-4 \log \mathcal{D} + \log(d_a d_b d_c)]$$

$$= 4 \log \mathcal{D} - 3 \sum_{a,b,c,\mu} p_3(a, b, c, \mu) \log d_a$$

$$= 4 \log \mathcal{D} - 3 \sum_a p_2(a, \bar{a}) \log d_a$$

and similarly

$$\mathcal{S}_{4-\text{hole}-\text{sphere}} = 6 \log \mathcal{D} - 4 \sum_a p_2(a, \bar{a}) \log d_a$$

Plugging these results into Eq. 34.15 we get

$$\mathcal{S}_{top}^{double} = -2\log\mathcal{D} \qquad \text{or} \qquad \mathcal{S}_{top} = -\log\mathcal{D}$$

as expected.

34.3.2   Entanglement Hamiltonian Argument

In Kitaev and Preskill [2006] a less rigorous, but fairly quick, argument is also given (see also the discussion in Kitaev [2006] appendix D.2). Let us divide our system again into regions A and B as in Fig. 34.1, only here we want to be a bit more general, so we allow a quasiparticle charge a to be inside region A, but far from the boundary. (This need not be a single quasiparticle, but could be multiple particles whose total quantum number is a).

Next, we invoke the idea stated in note 2 of this chapter: the entanglement spectrum (the set of $\log\lambda_n$s) should be described by a one-dimensional Hamiltonian that lives along the boundary between two regions of some length L. We thus consider a one-dimensional Hamiltonian H at some low temperature T. Using the usual quantum statistical mechanics approach, we think of this as a two-dimensional torus with length L in one directions and length $\beta = 1/T$ in the other direction where the direction of length L encloses topological charge a. We then make a modular S-matrix transformation on the torus (see Sections 7.3.1 or 17.3.2 for example)

$$Z_a = \sum_b S_{ab}\tilde{Z}_b$$

where Z_a is the partition function of sector a of our original system, and $\tilde{Z}_b$ is the partition function for a system where the directions of the torus have been exchanged. So now we have a torus that represents a system on a circle of length $\tilde{L} = \beta$ in the real-space direction enclosing topological charge b and with inverse temperature $\tilde{\beta} = L$. We take the limit of very low temperature of this new system (meaning long L) and since the energy of the vacuum sector is always the lowest, at long enough L (low enough $\tilde{T}$) only $\tilde{Z}_I$ will matter (with I being the vacuum sector). We then have

$$Z_a = S_{aI}\tilde{Z}_I = S_{aI}\exp\left(\tilde{\beta}\frac{\pi(c+\bar{c})}{12\tilde{L}}\right) = S_{aI}\exp\left(LT\frac{\pi(c+\bar{c})}{12}\right)$$

where c and $\bar{c}$ are the central charges of the right-going and left-going one-dimensional conformal theories (see Section 17.3.3). The coefficient of $\tilde{\beta}$ in the exponent is the lowest-energy mode of a one-dimensional conformal field theory(Di Francesco et al. [1997]).

Then, using the thermodynamic relationship for obtaining the entropy from the partition function

$$S = \frac{\partial(T\log Z)}{\partial T}$$

we obtain

$$S_a = \alpha L + \log S_{aI} \quad . \tag{34.16}$$

If we then consider the case where the original disk has no quasiparticle charge in it, we have $a = I$ and $\log S_{II} = 1/\mathcal{D}$ gives us the desired result.

34.3.3 Surgery Argument

Another approach is even more topological in nature—so much so that it cannot obtain the term in the entropy proportional to the length of the cut. A similar approach is discussed in Dong et al. [2008].

Let us consider a sphere divided into two parts A and B. For simplicity we can take these to be the north and south hemispheres. The density matrix of the full system is

$$\rho_{A\cup B} = |\psi_{S^2}\rangle\langle\psi_{S^2}| = Z(S^2 \times I) \quad .$$

Here, I is the interval and we have written this density matrix as a partition function of a manifold (see Eq. 7.7), which works here since the ground state on the sphere is unique. If we trace over this density matrix we wrap the interval into a circle and obtain

$$\mathrm{Tr}[\rho_{A\cup B}] = Z(S^2 \times S^1) = 1$$

where we have used Eq. 7.14.

Next we aim to calculate the reduced density matrix ρ_A by tracing over the B region as in Eq. 34.3. To do this, we write the sphere as the union of two disks D^2 representing the two regions A and B.

$$S^2 \times I = (D_A^2 \cup D_B^2) \times I = (D_A^2 \times I) \cup (D_B^2 \times I) \quad .$$

Then, tracing over B turns the interval I in the B region into S^1 but leaves the A region unchanged

$$\begin{aligned}
\rho_A &= \mathrm{Tr}_B[\rho_{A\cup B}] = \mathrm{Tr}_B[Z[(D_A^2 \times I) \cup (D_B^2 \times I)]] \\
&= Z[(D_A^2 \times I) \cup (D_B^2 \times S^1)] \quad . \tag{34.17}
\end{aligned}$$

If we then trace over the A region, we turn the A interval also into a circle and obtain exactly the decomposition of $S^2 \times S^1$ discussed in Section 7.3.2

$$\mathrm{Tr}_A Z[(D_A^2 \times I) \cup (D_B^2 \times S^1)] = Z[(D_A^2 \times S^1) \cup (D_B^2 \times S^1)] = Z(S^2 \times S^1) \quad .$$

Let us now look topologically at Eq. 34.17. By closing up the interval we obtain $S^2 \times S^1$. Thus, the manifold we are considering is just $S^2 \times S^1$ with a ball B^3 removed from it (the surface of the ball is the two copies of D_A). We thus have

$$\rho_A = Z[(S^2 \times S^1)\backslash B^3]$$

where $\backslash B^3$ means remove a ball. Once we have removed this ball, the boundary of this manifold is a sphere, where one hemisphere of the sphere is $|\psi_A\rangle$ and the other is $\langle\psi_A|$ (these are the two ends of the interval in $D_A^2 \times I$).

We now focus on the Renyi entanglement entropy (see Exercise 34.1)

$$S^\alpha_{AB} = \frac{1}{1-\alpha} \text{Tr}[\rho_A^\alpha] \quad .$$

To obtain ρ_A^α for integer α, we need to glue together α copies of $(S^2 \times S^1)\backslash B^3$. For simplicity let us start with $\alpha = 2$. We then have

$$
\begin{aligned}
\text{Tr}[\rho_A^2] &= Z\left[\{(S^2 \times S^1)\backslash B^3\} \cup \{(S^2 \times S^1)\backslash B^3\}\right] \\
&= Z\left[(S^2 \times S^1)\#(S^2 \times S^1)\right] \\
&= \frac{Z(S^2 \times S^1)Z(S^2 \times S^1)}{Z(S^3)} = S_{00} = 1/\mathcal{D}
\end{aligned}
$$

where we have recognized the attachment of these two manifolds along a spherical surface as a connected sum, and we have used the connected sum rule derived in Section 7.3.3.

More generally, we can derive that $\text{Tr}[\rho_A^\alpha]$ is the partition function of the connected sum of α copies of $S^2 \times S^1$. An easy way to show this is to look a bit more carefully at $(S^2 \times S^1)\backslash B^3$. Recalling from Section 24.3.1 we can write $S^2 \times S^1$ as S^3 where we then do surgery on an unknot (a loop) embedded in the S^3. Or equivalently using the Reshetikhin-Turaev approach we can simply insert an unknot (a loop) of Kirby color (an Ω strand) within the S^3. If we then remove a ball from $S^2 \times S^1$ this is then equivalent to removing a ball from the S^3 where there is an Ω loop (or surgery instruction) within the S^3 not intersecting the removed ball. But recall that S^3 can be thought of as the union of two balls (see the discussion in Section 7.3.3 and particularly note 31 in that section) so $(S^2 \times S^1)\backslash B^3$ is just a ball enclosing an Ω unknot (or equivalently this unknot is a surgery instruction). The two hemispheres on the surface of the ball are the corresponding bra and ket of the ρ_A operator. Gluing together several copies of $(S^2 \times S^1)\backslash B^3$ then gives us

$$\rho_A^\alpha = Z(B^3 \text{ enclosing } \alpha \text{ unknots of } \Omega \text{ color}) \quad .$$

Taking the trace then closes up the ball into S^3 again to give

$$\text{Tr}[\rho_A^\alpha] = Z(S^3 \text{ with } \alpha \text{ unknots of } \Omega \text{ color in it}) \quad .$$

The discussion of Section 24.2 (particularly Eq. 24.2) then tells us that the manifold we are considering is actually the connected sum of α copies of $S^2 \times S^1$. Using the fact that $Z(S^3) = 1/\mathcal{D}$ and either using Eq. 7.21 multiple times, or using the fact that the evaluation of an Ω loop also gives $\mathcal{D}$ (see Eq. 24.6 for example), we obtain

$$\text{Tr}[\rho_A^\alpha] = \mathcal{D}^{\alpha-1} \quad .$$

Thus, the Renyi topological entanglement entropy is given by

$$\mathcal{S}_{AB}^{\alpha} = -\log \mathcal{D}$$

for all integer values of $\alpha > 1$. Note that gives only the topological piece and not the part proportional to the length of the cut.

What we have derived here is not the von Neumann entropy. One must make the added assumption that if the entropy is $-\log \mathcal{D}$ for all integer values of $\alpha > 1$, then it will also be the same if we take the limit of $\alpha \rightarrow 1$ where the Renyi topological entanglement entropy becomes the von Neumann topological entanglement entropy.

In Exercise 34.4 we discuss adding a quasiparticle a into region A and $\bar{a}$ into region B. The calculation is not much harder, with the result that the Renyi topological entanglement entropy becomes $\log S_{Ia}$ for all integer values of $\alpha > 1$, in agreement with Eq. 34.16.

Chapter Summary

- Topologically ordered matter has characteristic long-ranged entanglement. This can be diagnosed by measuring the entanglement entropy across a cut, and carefully subtracting off pieces that give short-range entanglement.

Further Reading:

- You won't go wrong by reading the original two articles: Kitaev and Preskill [2006] and Levin and Wen [2006].
- Topological entanglement entropy has since been used as a diagnostic of the topological property of systems, such as in Furukawa and Misguich [2007],Haque et al. [2007], and Estienne et al. [2015].

Exercises

Exercise 34.1 Entanglement Entropy

(a) Confirm that for a system in a pure state, the expression for the entanglement entropy $\mathcal{S}_{AB} = -\text{Tr}[\rho_A \log \rho_A]$ given in note 4 agrees with Eq. 34.5. Confirm that the Renyi entanglement entropy is given by

$$\mathcal{S}_{AB}^{\alpha} = \frac{1}{1-\alpha} \log \text{Tr}[\rho_A^{\alpha}] \quad . \tag{34.18}$$

(b) Confirm that for a system in a pure state, the expression for the Renyi entanglement entropy matches the von Neumann entropy in the limit that $\alpha \rightarrow 1$.

Exercise 34.2 Renyi Entropy of the Toric Code
Use the approach of Section 34.1 to calculate the Renyi entanglement entropy for the toric code.

Exercise 34.3 Entanglement Entropy of the $\mathbb{Z}_n$ Toric Code
Generalize the calculation of Section 34.1 to the case of the $\mathbb{Z}_n$ toric code. Confirm Eq. 34.12

Exercise 34.4 Renyi Entanglement Entropy With a Quasiparticle
The purpose of this exercise is to calculate the Renyi topological entanglement entropy between two regions A and B on the sphere following the approach of Section 34.3.3. In that section we showed that $\text{Tr}[\rho_A^\alpha]$ can be viewed as the partition function of S^3 with α unlinked rings of Ω color embedded.

(a) Insert a particle of type a in region A and a particle of type $\bar{a}$ in region B. Performing the same manipulations, show that $\text{Tr}[\rho_A^\alpha]$ can be viewed as the partition function of S^3 with the link shown in Fig. 34.9 embedded.

(b) Evaluate this link to obtain the Renyi topological entanglement entropy

$$\mathcal{S}_{A,B}^\alpha = \log S_{aI} = \log d_a - \log \mathcal{D}$$

for all integer values of $\alpha > 1$.

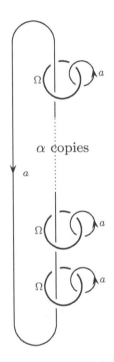

α copies

Fig. 34.9 The partition function of this link embedded in S^3 gives $\text{Tr}[\rho_A^\alpha]$ where a particle of quantum number a is in region A and quantum number $\bar{a}$ is in region B.

Symmetry-Protected Topological Phases of Matter

35

Optional

In Chapter 34 we explained that all topologically ordered systems have long-ranged topological entanglement. Gapped states of matter that do not have long ranged entanglement are (creatively) called short-ranged entangled states.[1]

Recall that in Section 29.2.3 we gave a rather general (and topologically motivated) definition of a topological phase of matter, which we repeat here:

> **Definition of a Topological Phase of Matter:** *Two (zero temperature) gapped states of matter are in the same topological phase of matter if and only if you can continuously deform the Hamiltonian[2] to get from one state to the other without closing the excitation gap.*

All (gapped) short-ranged entangled states are in the same topological phase of matter—they can all be deformed into each other without closing the excitation gap.[3] In particular, they can all be deformed into a trivial state without closing the gap.

> **Definition of a Trivial State:** *A state is "trivial" if it can be written as a set of microscopic degrees of freedom (such as spins) that are completely decoupled from each other and unentangled with each other.*

For example, a lattice of independent unentangled spins is a trivial state.

One might think that the above-stated fact—that all short-ranged entangled states can be deformed into the trivial state—makes such short-ranged entangled states completely uninteresting. But in fact, the short-ranged entangled states have a beautiful connection to topology, which becomes apparent once we generalize the definition to include another aspect—symmetry.

[1] There is some controversy as to how one classifies invertable topological order such as the E_8 Chern–Simons theory, p-wave order, or integer quantum Hall states. These phases of matter have no nontrivial anyons and the total quantum dimension is $\mathcal{D} = 1$, so the topological entanglement entropy is zero. Nonetheless, the non-zero central charge prevents us from deforming the phase into a trivial phase without encountering a phase transition. This is true whether we consider deforming a Hamiltonian as in the definition here, or if we consider a quantum circuit as discussed in note 3. We choose to call these phases of matter long-ranged entangled as well. The reader is cautioned that the community does not entirely agree on this.

[2] To be precise, in some cases in "deforming" a Hamiltonian one also wants to allow one to merge the system with any trivial state. We will never need worry about this though.

[3] Another essentially equivalent way of classifying wavefunctions, at least for systems on a lattice such as a spin system, is to consider operating on the wavefunction made the constituent spins (which we think of as qubits) with a quantum computer that can perform local unitary gates (i.e. operate on some number of bits near each other with CNOTs and single qubit rotations for example). Any two wavefunctions that can be turned into each other with a quantum circuit of "finite depth" are considered to be in the same phase. Here, "finite depth" means that the total number of gates performed should be proportional to N, the number of bits in the system, rather than, say, N^a with $a > 1$. This way of classifying states, which is another idea from Kitaev, is extremely useful and turns out to be essentially equivalent to the definition given here (see Wen [2017]; Zhang et al. [2019a]). This classification can similarly be extended to consider symmetries by enforcing the rule that the circuit should not break the specified symmetry.

35.1 **Symmetry Protection**

For nearly a century, symmetry has been the primary tool that physicists have used to distinguish one type of matter from another. Introduced by Lev Landau[4] (Landau [1937]), this approach has become known as the *Landau paradigm*.

While it is too much of a digression to introduce Landau theory in detail, it is probably worth mentioning a canonical case: the ferromagnet/paramagnet transition. For example, consider a "Heisenberg" magnet.[5] In the paramagnetic phase there is no net magnetization, whereas in the ferromagnetic phase, there is a non-zero magnetization vector. The symmetry of the system is different in the two cases. In the paramagnetic phase all directions of space are equivalent, whereas in the ferromagnetic phase there is one special direction set by the magnetization vector—we have broken the symmetry between all directions in space by choosing this one direction. Perhaps the most important characterization of a phase of matter is given by stating what symmetries it preserves and which it breaks.

Our definition of "Topological Phase of Matter" provides a method of classifying phases of matter, but unfortunately it is blind to symmetries. An important generalization of the above definition is the idea of a symmetry-protected phase of matter. Here, we specify that a particular symmetry is unbroken, and further, when we deform our Hamiltonian (as in the above definition) we only allow deformations that preserve that symmetry. This leads to a more refined classification of matter that considers both topology and symmetry.

Definition of a Symmetry-Protected Phase of Matter: *Two (zero temperature) gapped states of matter are in the same symmetry-protected phase of matter if and only if you can continuously deform the Hamiltonian to get from one state to the other without closing the excitation gap and without breaking the given symmetry.*

Caution: This definition is not standard in the literature, but I think it is useful, nonetheless.

Note that it is possible for two different symmetry-protected phases to be deformed into each other without closing the gap if one is allowed to break the symmetry in the process.

The definition of symmetry-protected phases can be further refined into those that either do, or do not, have long-ranged topological entanglement entropy, as we discussed in Chapter 34.

Definition of an SPT Phase of Matter: *An SPT phase of matter is a symmetry-protected phase of matter with only short-ranged entanglement.*

Among the possible SPT phases associated with a given symmetry, one possible phase is the *trivial SPT*—the phase of matter that can be deformed to the trivial state without breaking the symmetry. Any SPT

[4]Landau was one of the most prominent physicists ever to live. Most physicists know of many of Landau's important contributions, and due to modern specialization among fields, any given physicist is probably familiar with only a fraction of his very important works. As with so many of the greats, he was also a colorful, and difficult to control, character. Many have heard the tale of how Landau crossed the Soviet establishment and ended up in prison. Fewer, perhaps, have heard that he espoused "free love" rather than monogamy—something his wife was not so keen on. He was brilliant, but apparently a bit of a jerk.

[5]The Heisenberg Hamiltonian is

$$H = -\sum_{i,j} J_{ij} \mathbf{S}_i \cdot \mathbf{S}_j \qquad (35.1)$$

with $\mathbf{S}_i$ the spin vector for site i. Here, we take, say, $J_{ij} > 0$ for neighboring spins i and j so that the energy is lower if two neighboring spins are aligned and we take $J_{ij} = 0$ otherwise. This is known as the nearest-neighbor ferromagnetic Heisenberg Hamiltonian. In three dimensions there is a finite-temperature phase transition between a paramagnetic high-temperature phase (where spins point randomly in every direction) and a ferromagnetic low-temperature phase with non-zero magnetization where the spins tend to align.

phases that are not the trivial SPT are (creatively) known as *nontrivial SPTs*. By the definition of SPTs, a nontrivial SPT cannot be deformed to the trivial state without breaking the symmetry or closing the gap. However, if we do allow breaking of the symmetry, since SPTs are short-ranged entangled, any SPT can be deformed to the trivial state.

Since the SPT phases, once the symmetry is broken, are equivalent to trivial phases, the acronym "SPT" has two meanings

$$\text{SPT} \quad = \quad \text{Symmetry-Protected Topological Phase of Matter}$$
$$= \quad \text{Symmetry-Protected Trivial Phase of Matter.}$$

An interesting property of SPTs (which we do not prove here) is that the boundary of a nontrivial SPT either breaks the protecting symmetry or is gapless.[6]

In contrast to the SPT phases (which do not have anyons) one can also consider symmetry-enriched topological (SET) phases of matter, which *do* harbor anyons.

Definitions of a SET Phase of Matter: *An SET phase of matter is a symmetry-protected phase of matter that does have long-ranged topological entanglement.*

We return to discuss SETs in more depth in Section 36.4.

35.1.1 Some Short Comments on SPT History

Around 2005 a major revolution occurred in the condensed matter community— the realization that classification that invokes symmetry *and* topology could be very powerful. The work of Kane and Mele [2005] launched the study of so-called *topological insulators* (see for example Hasan and Kane [2010] for a classic review of the field).[7] These are gapped systems of non-interacting electrons subject to some symmetry. In the simplest, and most experimentally important, case the relevant symmetry is time reversal.[8]

Classifying such systems led to the key ideas now used in the study of SPTs. In their own right, topological insulators have been tremendously important in both theoretical and experimental physics over the last fifteen years. But once it was realized that symmetry protection is an important concept for classifying phases of matter, people went back to reconsider certain, already known, phases of matter to see if they might actually have symmetry protection. The most notable success of this program was the reexamination of the Haldane phases[9] of one-dimensional spin chains (see Haldane [1983a, c]; Affleck et al. [1987, 1988]). Such phases of matter were already known to be gapped.

[6]In higher dimensions, the boundary could also develop a topological order.

[7]While any detailed discussion of topological insulators is outside of the scope of the current book (and there are dozens of detailed books on the subject!), I do hope to cover the topic in my *next* book.

[9]Haldane's use of topology for understanding these spin chains earned him the Nobel Prize in 2016.

[8]It turns out that most real materials are well approximated as being made of non-interacting electrons, and a good fraction of materials are insulators, meaning electronic excitations are gapped. Further, if one does not apply an external magnetic field, and the substance in question does not spontaneously break time reversal symmetry (say, by becoming a ferromagnet), then time-reversal symmetry is preserved. Thus, the general topic of study—non-interacting gapped electron systems with time-reversal symmetry—applies to a truly enormous range of material systems.

But it was only more recently understood that they are actually SPTs (see Gu and Wen [2009]; Pollmann et al. [2010, 2012]). The study of one-dimensional SPTs is sufficiently simple that we have provided a more detailed discussion in Section 35.7.

These ideas were quickly generalized to consider interacting fermionic systems (Fidkowski and Kitaev [2011]; Turner et al. [2011]), many types of symmetries, and higher dimensions (Chen et al. [2011, 2013]; Gu and Wen [2014]; Lu and Vishwanath [2012]; Senthil [2015]; Schuch et al. [2011]; Fu [2011] and many more). In this chapter we mainly focus on the two-dimensional case with simple on-site symmetries (the precise definition of which we give in Section 35.2 below). This case is most closely connected to the $(2 + 1)$-dimensional TQFT models already discussed.

35.2 On-Site Symmetries

There are many types of symmetries one might be concerned with, and for each symmetry there is classification of phases of matter into symmetry-protected phases. Here we consider so-called on-site symmetry, which is potentially the simplest example. For our purposes a "site" might represent an entire unit cell. We imagine an N-state Hilbert space on each site (a N-state qudit). We can write the states in the Hilbert space at site j as

$$|m_j\rangle \quad \text{for} \quad m \in \{1, \dots, N\}.$$

For any given group G (which will be the symmetry group in consideration) we can consider unitary operations on site j that are a representation of the group[10]

$$\hat{U}^{(j)}(g) = \sum_{n_j, m_j} |m_j\rangle \, U_{m_j, n_j}(g) \, \langle n_j| \tag{35.2}$$

with $U_{mn}(g)$ an N-dimensional unitary matrix for each $g \in G$. In principle, we can consider the group G to be discrete or continuous, although we focus on the discrete case here.

A global unitary symmetry operation is given by applying the same $\hat{U}^{(j)}$ at every site in the system

$$\hat{U}_{\text{global}}(g) = \prod_j \hat{U}^{(j)}(g) \quad .$$

If the ground-state wavefunction is unchanged after application of this global unitary symmetry operation for all $g \in G$, then we say that the wavefunction has an unbroken G-symmetry. In analogy with magnets such as the Heisenberg model, we say a ground state is "paramagnetic" if it does not break the symmetry (has an unbroken G-symmetry), and it is ferromagnetic otherwise (i.e. when it does break the symmetry).

[10]The statement that this is a representation of the group means that $\hat{U}(g)\hat{U}(h) = \hat{U}(gh)$. See Section 40.2.4.

35.3   Example: Ising ($\mathbb{Z}_2$) Symmetry

Let us consider the simplest possible example: Ising, or $\mathbb{Z}_2$, symmetry.[11] Recall that the Ising model is a model where spins are allowed to point either up or down only. Ising symmetry means a system is unchanged if you flip all up spins to down and all down spins to up. Since flipping twice brings you back to the initial state, this is a $\mathbb{Z}_2$ symmetry. We here consider the possible SPT phases[12] of a system with Ising, or $\mathbb{Z}_2$, symmetry.[13]

Consider a triangular lattice with a spin-1/2 at each vertex as shown in Fig. 35.1. Using Pauli spin operators, $\hat{U}^{(i)} = \sigma_x^{(i)}$ flips over the spin at site i. If you square σ_x you get the identity, so the symmetry group here is indeed $\mathbb{Z}_2$.

Thus, the only nontrivial global symmetry operation is[14]

$$\hat{U}_{\text{global}} = \prod_i \sigma_x^{(i)} \quad . \tag{35.3}$$

We would like to look for ground-state wavefunctions such that this $\mathbb{Z}_2$ symmetry unbroken, that is, we are looking for $\mathbb{Z}_2$ paramagnets. Thus, we are looking for ground-state wavefunctions $|\psi\rangle$ such that

$$\hat{U}_{\text{global}}|\psi\rangle = |\psi\rangle. \tag{35.4}$$

meaning that the ground state is unchanged under the symmetry operation.

Our strategy here is to find one representative Hamiltonian (and corresponding ground-state wavefunction) for each possible symmetry-protected topological phase with the given symmetry. Of course, once we have a single representative Hamiltonian for the phase, we can (in principle) deform the Hamiltonian freely and remain in the phase as long as we do not close the excitation gap or break the symmetry.

The Hamiltonians we are looking for are those that do not break the symmetry. Here, this means that

$$[\hat{U}_{\text{global}}, H] = 0 \quad . \tag{35.5}$$

The representative Hamiltonians we will study are exactly solvable Hamiltonians. We write our Hamiltonian in the form[15]

$$H = -\frac{\Delta}{2} \sum_p B_p \tag{35.6}$$

where the sum is over points p of the triangular lattice and the operators B_p satisfy $[B_p, B_{p'}] = 0$ and each B_p squares to the identity and therefore has eigenvalues ± 1. Assuming $\Delta > 0$, the unique (!) ground state will then be the state where all B_p are in the $+1$ eigenstate.

It is easy to see if we insist that our B_p operators satisfy

$$[B_p, \hat{U}_{global}] = 0 \tag{35.7}$$

[11]Caution: This is mostly not related to the Ising TQFT. The Ising TQFT is also related to $\mathbb{Z}_2$ symmetry, but in the present context this connection is irrelevant.

[12]Here, we are actually considering "bosonic" SPT phases. This simply means there are no creation/annihilation operators that anticommute with each other on different sites of the system.

[13]The construction given in this section is from Levin and Gu [2012].

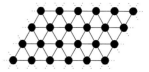

Fig. 35.1   A triangular lattice. Here we consider a spin-1/2 at each vertex (marked with a dot).

[14]Since there is only one nontrivial element of the group, we do not write $\hat{U}_{\text{global}}(g)$.

[15]This is essentially a commuting projector Hamiltonian similar to the Kitaev model or the Levin–Wen model. While B_p, having eigenvalues ± 1, is not a projector we could have instead written the combination $\frac{1}{2}(B_p + 1)$, which has eigenvalues 0 and 1 and is therefore a projector.

[16]To see that the ground state is unchanged under $\hat{U}_{\text{global}}$ note that $|\psi\rangle$ is the unique state satisfying $H|\psi\rangle = E_{GS}|\psi\rangle$ with E_{GS} the ground-state energy. Applying $\hat{U}_{\text{global}}$ and using the commutation Eq. 35.5 we have $E_{GS}[U_{\text{global}}|\psi\rangle] = \hat{U}_{\text{global}}H|\psi\rangle = H[U_{\text{global}}|\psi\rangle]$ which means that $\hat{U}_{\text{global}}|\psi\rangle$ has the same energy as $|\psi\rangle$. But since $|\psi\rangle$ is the unique state with this energy, it must be that $\hat{U}_{\text{global}}|\psi\rangle = |\psi\rangle$ up to an unimportant phase.

then we guarantee both that the Hamiltonian preserves the symmetry (Eq. 35.5) and also that the unique ground state is unchanged under the symmetry operations (Eq. 35.4).[16]

We claim there are exactly two SPT phases possible that obey this $\mathbb{Z}_2$ symmetry. That is, there are exactly two different B_p operators satisfying the symmetry Eq. 35.7 that we could use in Eq. 35.6 that satisfy the symmetry. We describe each in turn, and explain what phase of matter results. Roughly, each B_p is going to need to contain a factor of σ_x in order to commute with the symmetry. However, we have a freedom to have B_p also contain some nontrivial phase factors.

$\mathbb{Z}_2$ Trivial Paramagnet

The first possibility is to choose

$$B_p = \sigma_x^{(p)} \quad . \tag{35.8}$$

It is trivial to see that the σ_x operators square to the identity and they commute on different sites as required. Further, this obviously commutes with the $\mathbb{Z}_2$ symmetry $\hat{U}_{\text{global}}$.

Plugging this form of B_p into the Hamiltonian Eq. 35.6 yields a ground state where every spin on every site points in the $\hat{x}$ direction. We can write the (unnormalized) many body wavefunction simply as

$$|\psi\rangle = \prod_p (\,|\uparrow_p\rangle + |\downarrow_p\rangle\,) \tag{35.9}$$

$$= |\uparrow_1\uparrow_2\uparrow_3 \ldots\rangle + |\downarrow_1\uparrow_2\uparrow_3 \ldots\rangle + |\uparrow_1\downarrow_2\uparrow_3 \ldots\rangle + \ldots \tag{35.10}$$

$$= \sum_\alpha |\alpha\rangle \tag{35.11}$$

where in the second line all possible combinations of up and down spins occur once and in the third line each α represents one of these spin configurations (there are 2^N possible values of α for a system with N spins). This wavefunction is symmetric under flipping all up spins to down spins, which is particularly obvious on line 35.9.

This state is the "trivial" since it can be written as a (unentangled) direct product of states on different, non-interacting, sites as shown explicitly in Eq. 35.9.

$\mathbb{Z}_2$ Nontrivial SPT Paramagnet

We claim there is one other distinct gapped phase of short-ranged entangled matter that also respects the $\mathbb{Z}_2$ symmetry. Let us propose

$$B_p = \sigma_x^{(p)} s^{(p)} \tag{35.12}$$

where

$$s^{(p)} = - \prod_{<pqr>} i^{\frac{1}{2}(1-\sigma_z^{(q)}\sigma_z^{(r)})} \tag{35.13}$$

where the product is over all six triangles that have p as one of its vertices and q, r are the two other vertices of these triangles. Note that $s^{(p)}$ squares to the identity. To see this, note that it gives a factor of i for every pair of neighboring spins pointing in opposite directions on the hexagon surrounding site p. Since the number of such pairs is always even, $s^{(p)}$ must give ± 1.

An equivalent way of describing B_p (measuring spins in the $\hat{z}$ basis) is as follows. First we find all the up/down domain walls in the system—places where upspins are adjacent to downspins, as shown in Fig. 35.2. These domain walls necessarily form closed loops. We can then write:

$$B_p = \left\{ \begin{array}{l} \text{flip spin } p \text{ and multiply by } -1 \text{ if the number} \\ \text{of up/down domain walls has changed parity.} \end{array} \right. \quad (35.14)$$

A few comments are in order about the operators B_p. First, the B_ps are $\mathbb{Z}_2$ symmetric, that is, they commute with the symmetry $\hat{U}_{\text{global}}$ in Eq. 35.3 (see Exercise 35.1). Moreover, the B_ps commute with each other and square to the identity (see Exercise 35.1) meaning that the ground state is the state where all B_ps are in the $+1$ eigenstate.

Using the form of B_p given in Eq. 35.14 it is clear that the ground state can be written analogous to Eq. 35.11 as

$$|\psi\rangle = \sum_\alpha (-1)^{(\# \text{ domain walls in state } \alpha)} |\alpha\rangle \quad (35.15)$$

where again the sum over α is a sum over all possible configurations of spins pointing up or down. This wavefunction differs from Eq. 35.11 only by the sign prefactors for each $|\alpha\rangle$.

While it is possible to continuously interpolate between Eq. 35.8 and Eq. 35.12 in such a way that the Hamiltonian Eq. 35.6 remains gapped the entire time, it is not possible to do this while preserving the $\mathbb{Z}_2$ symmetry along the way (we do not prove this statement here; see Levin and Gu [2012] for more details).

Although we do not prove it here, the nontrivial SPT phase of matter described here is the only nontrivial SPT for $\mathbb{Z}_2$ symmetry.

In both the trivial and nontrivial cases, up to possible signs out front, the wavefunction (Eq. 35.11 or 35.15) is simply a sum over all spin-up and spin-down combinations. As such, there is no long-range topological entanglement entropy. This supports our claim that these states of matter are short-ranged entangled[17] as described in Chapter 34.

35.4 Relation to Loop Models

Equation 35.15 is extremely similar to the wavefunction 32.3 for the doubled-semion model. The only difference is that here the blue lines (as in Fig. 35.2) are *boundaries* between domains of spin up and spin down, rather than being the fundamental degrees of freedom themselves. This apparently small difference has several important consequences. First of all, on a torus in the current SPT model, one cannot have an a single

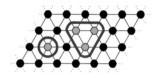

Fig. 35.2 Down spins are colored cyan, up spins are colored black. The dark blue lines demark the domain walls between up and down spins.

[17]In fact, the Schmidt decomposition Eq. 34.1 is trivial in both cases with only the sign for each term of the sum possibly being different for the two possible wavefunctions. This sign does not matter in the entanglement entropy Eq. 34.5. There is in fact *zero* entanglement entropy between any two disjoint regions in either model. However, the fact that the entanglement is exactly zero is a feature of the particularly simple model we have chosen to represent the phase of matter. If we perturb the Hamiltonian we can stay in the same phase of matter and generate entanglement between spins that are near each other. This would given entanglement entropy between regions proportional to the length of the cut between the regions but would not give any topological entanglement entropy.

[18]This is easy to see by starting with a configuration with all spins up and no domain walls. Now if we flip over a domain that goes around a nontrivial cycle of the torus, we realize that the domain on the torus must have two edges, so that we get two domain walls running around the cycle, rather than just one.

domain wall that runs all the way around a nontrivial cycle of the torus, or any odd number of domain walls going around a nontrivial cycle of the torus,[18] and thus there is no topological ground-state degeneracy of the current model. Secondly, when we come to examine excitations, in the doubled-semion model it is possible to consider blue lines that have endpoints. However, if the blue lines are boundaries of domains, it is not possible for them to have endpoints (since there is no boundary of a boundary). As a result, certain types of excitations (the left- and right-handed semions) do not exist here.

In fact, we can make the same sort of connection between the $\mathbb{Z}_2$ trivial paramagnet wavefunction Eq. 35.11 and the toric code. The toric code is a sum of all loop configurations, all added with a plus sign (Eq. 30.1) analogous to Eq. 35.11. As with the above relation between the doubled-semion model and the nontrivial $\mathbb{Z}_2$ SPT phase, there is a similar difference between the toric code and the $\mathbb{Z}_2$ trivial paramagnet: it is not possible to have an odd number of domain walls running around a nontrivial cycle of the torus, and hence, there is no ground-state degeneracy of the $\mathbb{Z}_2$ trivial paramagnet. Again, since it is not possible to have a string endpoint when the strings are domain walls, certain excitations (the e and f) cannot be produced.

35.4.1 Twisted Boundaries and Symmetry Defects

Instead of using periodic boundary conditions on our torus $\sigma_z(x, y) = \sigma_z(x + L, y)$ with L the length of the cycle, let us instead try using "twisted" boundary conditions $\sigma_z(x, y) = -\sigma_z(x + L, y)$. Now, instead of there being an even number of domain walls going around the y cycle of the torus, there now must be an odd number of domain walls (because of the twisting). This may seem like an odd thing to do in a physical system but at least as mathematical tool we do this sort of thing all the time!

Let us now try changing the geometry of these twisted boundary conditions. Let us choose a point in the system and call it the origin (marked with an $\times$ in Fig. 35.3) and let us measure position in radial coordinates (r, θ) with respect to that point. Now let us consider using boundary conditions that are twisted *around* this point, so that $\sigma_z(r, \theta) = -\sigma_z(r, \theta + 2\pi)$. We call this point a *symmetry defect*. There must be an odd number of domain walls crossing any circle that we draw around the defect. In particular, this means that a domain wall must end precisely at the symmetry defect. So we have now constructed a way that we *can* have domain wall endpoints in our SPT. Domain wall endpoints correspond to loop endpoints in the doubled-semion model, and these correspond to semion quasiparticles. Thus, the symmetry defects in the SPT are effectively semions! However, these are now "extrinsic" objects— they are tied to the positions in your system where you imposed the twisted boundary conditions.

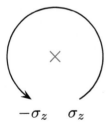

$-\sigma_z \quad \sigma_z$

Fig. 35.3 A symmetry defect is a point in the system around which the boundary conditions are twisted.

35.4.2 Edge Modes

As mentioned in Section 35.1, a nontrivial SPT on a system with boundaries must either break the symmetry or must be gapless (see Levin and Gu [2012]; Chen et al. [2011]). An argument for this can be given nicely in the language of the loop gas.

Let us now give a very rough version of the argument that the edge of the nontrivial SPT cannot both preserve the symmetry and remain short-ranged entangled. As mentioned, in the absence of symmetry defects, the boundary of a domain wall cannot have endpoints, hence, there are no semions in the nontrivial $\mathbb{Z}_2$ SPT phase. However, if the system has boundaries, then the domain walls *can* end on the boundaries— which we can think of (roughly) as being semions living on the boundary. This means that there must be nontrivial long-ranged entanglement along the boundary. Any attempt to prevent these walls from reaching the boundary (such as modifying the Hamiltonian near the boundary) would necessarily break the symmetry. For the trivial SPT phase, related to the toric code, the same argument does not hold. The reason is that the endpoints of loops are bosons or fermions rather than semions, and the presence of bosons or fermions does not imply the existence of long-ranged entanglement.

A more sophisticated version of this argument shows rigorously that the nontrivial SPT cannot both preserve the symmetry and also remain short-ranged entangled. In one dimension, gapped states always have short-ranged entanglement and gapless states have long-ranged entanglement. Thus, we conclude that the one-dimensional edge of this nontrivial two-dimensional SPT either breaks the symmetry or is gapless.

35.5 Relation to Twisted Kitaev Model

The fact that there are exactly two SPT phases with $\mathbb{Z}_2$ symmetry is intimately related to the fact that there are exactly two $\mathbb{Z}_2$ loop gases. These can be viewed as the two possible twists of the Kitaev model based on the group $\mathbb{Z}_2$, as described in Section 31.4.1. Another way of saying this is that these are the two possible 3-cocycles of the group $\mathbb{Z}_2$, as discussed in Sections 19.2 and 31.4.1 and (see also note 1 from Chapter 32). In fact, this is a general principle. If we have a two-dimensional system with any on-site symmetry group G, each 3-cocycle of the group (or equivalently, each element $\omega \in H^3(G, U(1))$; see Section 19.1) generates a different short-range entangled SPT phase in $(2 + 1)$-dimensions. The general scheme is similar to what we have seen here for the $\mathbb{Z}_2$ case. One considers on-site variables g_i chosen from a group G. Between two neighboring sites i and j one defines a domain wall variable with value $g_j g_i^{-1}$, and then the wavefunction is a twisted Kitaev wavefunction built from these domain wall variables.

Almost everything discussed here for the $\mathbb{Z}_2$ case is easily generalized. In particular, the SPT phases have no excitations that braid nontrivially with each other, but for the nontrivial SPTs, the symmetry defects are essentially the anyons of the corresponding twisted Kitaev model. Further, the argument that the nontrivial SPTs have gapless edges or break the symmetry also follows through.

More generally, in $(d + 1)$-dimensions one can correspondingly construct an SPT phase from each $(d + 1)$-cocycle in $H^{d+1}(G, U(1))$. See the discussions in Chen et al. [2013] and WenBook for example. In section 35.7 we consider the $d = 1$ case as an example to demonstrate the connection to $H^2(G, U(1))$. Interestingly, in $d > 2$ it appears there are a very few SPT phases that are outside this cohomological classification (see Wang and Senthil [2013]; Vishwanath and Senthil [2013]; Burnell et al. [2014]; Kapustin [2014]).

35.6   Relation to Dijkgraaf–Witten Model

As mentioned in Section 31.4.1, the twisted Kitaev model is the Hamiltonian version of the Dijkgraaf–Witten model (introduced in Section 23.4), which is defined by an action for a spacetime lattice (Eq. 23.10). Building an SPT from a Dijkgraaf–Witten action turns out to be extremely simple. First we choose our spacetime lattice of simplices (i.e. triangles and tetrahedra). Each site of the lattice has a "spin" that takes a value in a group G (generalizing spin-$\frac{1}{2}$ in the $\mathbb{Z}_2$ model above). Then, if two sites i and j are connected by an edge (ij), orienting the edge with an arrow from i to j, we assign the edge a value $g_j g_i^{-1}$. Note that this assignment of edge variables naturally satisfies the flatness condition (Fig. 23.6) that multiplying the edge variables in order around a triangle must always give the identity, as shown in Fig. 35.4.

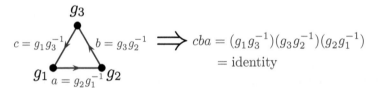

Fig. 35.4 When edge variables (blue) are defined in terms of site variables (black), then the flatness condition (compare Fig. 23.6) is naturally satisfied—multiplying the edge variables around the face of the triangle always gives the identity.

Once we have defined such edge variables, the action can be taken to have similar form as that of Dijkgraaf–Witten (Eq. 23.10), only now we sum over all possible values of the g_i on-site variables and the edge labelings are a function of the site variables. This sums over "flat" configurations of edges as we usually demand for Dijkgraaf–Witten. However, there are some clear distinctions between the Dijkgraaf–Witten model and the SPT version.

Let us define the "flux" associated with a path on the lattice to be the ordered product of the edge variables along the path. Because of the flatness condition, the flux is unchanged if we deform the path over a triangular simplex, as shown in Fig. 35.5. A path that can be deformed to the trivial path must be associated with trivial flux—that is, it surrounds a region that is "flat." In the Dijkgraaf–Witten model, there may be a non-zero flux going around a nontrivial cycle of the manifold. In contrast, in the SPT version, where the edge is labeled with $g_j g_i^{-1}$ in terms of the vertex variables g_i and g_j, any cycle—even around a nontrivial cycle of the manifold—will always give a trivial flux. This is a reflection of the fact that the SPT always has a unique ground state on any two-dimensional manifold, whereas the Dijkgraaf–Witten model is a nontrivial TQFT, which generally has a ground-state degeneracy on a manifold with non-contractible loops.

The relation between the spacetime SPT model and the Dijkgraaf–Witten model (and analogously the relationship between the Hamiltonian SPT model and the twisted Kitaev model) is one of "gauging." To go from the SPT model to the Dijkgraaf–Witten model we say that we "gauge" the theory.[19] This means that we have effectively added a local gauge field, which allows us to accumulate a nontrivial "flux" going around a loop in the Dijkgraaf–Witten model, where no such flux is possible in the SPT model. To go in reverse, from a Dijkgraaf–Witten model to an SPT model, we need to "ungauge" the model—reducing a gauge symmetry to an on-site symmetry. More details of this relationship are given in Levin and Gu [2012], Tiwari et al. [2018] and Heinrich et al. [2016].

As discussed in Section 23.4.1, Dijkgraaf–Witten models exist in any number of dimensions. Correspondingly, one can construct SPTs in any number of dimensions by "ungauging" the gauge symmetry. In section 35.7 we discuss $(1 + 1)$-dimensional SPTs as an example. Note however, as mentioned in Section 35.5, in $(3 + 1)$-dimensions and higher dimensions there are SPTs that cannot be obtained from the cohomology construction, and correspondingly, cannot be obtained from a Dijkgraaf–Witten model.

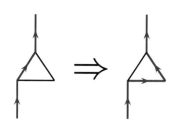

Fig. 35.5 Deforming the blue path over a triangle.

[19] A beautiful recent experiment by Iqbal et al. [2023] used a small-scale trapped-ion quantum computer to create a $(2 + 1)$-dimensional SPT wavefunction for the group $G = \mathbb{Z}_2 \times \mathbb{Z}_2 \times \mathbb{Z}_2$. This wavefunction was then explicitly "gauged" to result in a wavefunction having nonabelian topological order.

35.7 **Appendix: One-Dimensional SPTs**

SPTs in one dimension with on-site symmetry (and no fermions) are particularly simple to discuss. Let us start with a one-dimensional system with periodic boundary conditions, that is, a ring. We assume the system has an unbroken on site G-symmetry for some group G. In other words, the ground state is unchanged under application of $\hat{U}_{\text{global}}(g)$ for all $g \in G$. We also assume the system is short-ranged entangled and has a gapped ground state.[20] That is, this is an SPT with on-site G-symmetry.

Now let us cut the ring into a finite segment with two ends, which we denote as L (left) and R (right). Due to the short-ranged entanglement, the bulk of the ring should remain unchanged—it should have a unique ground state and should remain G-symmetric. However, the two ends of the segment may (or may not) now have edge modes[21] and these may transform nontrivially under the symmetry.

[20] In fact, a gapped ground state in one dimension implies it is short-ranged entangled.

[21] If there are no edge modes when the system is cut, the system must be in the trivial SPT phase.

Let us call these modes $|n_L\rangle$ and $|n_R\rangle$ with $n_L, n_R = 1, \ldots, M$. When we sew these edges back together we must obtain the original fully gapped system, so there must be the same number of modes on each end. So a general state of the system (with the bulk in the ground state) may be written roughly as

$$|\Psi\rangle = \sum_{n_L, n_R} \psi_{n_L, n_R} \, |n_L\rangle \otimes |\text{bulk}\rangle \otimes |n_R\rangle \quad .$$

It is sometimes convenient not to even write the bulk part of the wavefunction, so we have

$$|\Psi\rangle = \sum_{n_L, n_R} \psi_{n_L, n_R} \, |n_L\rangle \otimes |n_R\rangle \quad .$$

The Hamiltonian for the system is still assumed to have a G-symmetry, so the overall state of the system must transform as a representation of the group G under $\hat{U}_{\text{global}}(g)$. This global symmetry operation leaves the bulk unchanged and only acts nontrivially on the edge modes. So we must have

$$\hat{U}_{\text{global}}(g) = \sum_{n_L, n_R, m_L, m_R} |n_L\rangle \otimes |n_R\rangle \, U_{(n_L, n_R), (m_L, m_R)} \, \langle m_L| \otimes \langle m_R|$$

for some unitary matrix U, which is a representation of the group. That is, $U(g)U(h) = U(gh)$.

Now since the entanglement is short ranged we expect that transformation on the left side of the system should not affect the right-hand side of the system. This means that the left side of the system should itself transform under the group symmetry, as should the right, and these two transformations should be independent. So we have

$$U_{(n_L, n_R), (m_L, m_R)}(g) = V_{n_L, m_L}(g) V^{\dagger}_{n_R, m_R}(g) \quad . \tag{35.16}$$

The transformations on the two ends must be conjugate to each other because when we sew the ends back together we must get something that is everywhere symmetric—that is, transforms trivially under the symmetry operation.

Now we might be tempted to conclude here that the $V(g)$ matrices are representations of the group G. However, there is an additional freedom to introduce a phase to $V(g)$ that cancels in the product in Eq. 35.16. In fact, $U(g)$ will be a representation (which is what we actually require) as long as $V(g)$ satisfies

$$V(g)V(h) = \omega_2(g, h)V(gh)$$

where $\omega_2(g, h)$ is a $U(1)$ valued phase. However, associativity of multiplication implies the following constraint[22]

$$\omega_2(g, h)\omega_2(gh, k) = \omega_2(h, k)\omega_2(g, hk) \quad . \tag{35.17}$$

[22]To derive this try calculating $V(g)V(h)V(k)$ in two different ways.

See also the discussion of Eq. 23.14. Here, V is known as a *projective* representation and ω_2 is known as a 2-cocycle. That is, $\omega_2 \in H^2(G, U(1))$.

The different possible 2-cocycles, up to gauge transforms (see Eq. 23.13), or equivalently, the different projective representations, give the different possible SPT phases.

The phenomenon seen here, that the end points transform projectively via V under the symmetry, although the whole system transforms as a proper ("linear") representation via U, is an example of *symmetry fractionalization*, which means that the symmetry does not act on the emergent degrees of freedom in the same way that it does on the underlying Hamiltonian.

Chapter Summary

- A symmetry-protected topological phase of matter is a refinement of the idea of a topological phase of matter for systems that obey some symmetry
- The $\mathbb{Z}_2$ SPT provides a simple example of an SPT with on-site symmetry and is closely related to the doubled-semion string-net.
- For on-site symmetries, SPT phases can be thought of as "ungauged" twisted Kitaev models or equivalently "ungauged" Dijkgraaf–Witten theories.

Further Reading:

- A good short review of the physics of SPT phases is given by Senthil [2015]. The work by Chen et al. [2013] that introduces group cohomology is not easy reading, but is very nice, nonetheless.
- A nice pedagogical discussion of SPTs in one dimension invoking the language of matrix product states is given by Pollmann [2009].
- A very nice approach to SPTs invoking the K-matrix formalism has been given by Lu and Vishwanath [2012].
- SPTs containing fermions invoke "super-cohomology" theory rather than cohomology, as discussed by Gu and Wen [2014]

Exercises

Exercise 35.1 Properties of the $\mathbb{Z}_2$ SPT Operators
Consider the operator B_p defined in Eq. 35.12.
(a) Show that B_p commutes with $\hat{U}_{\text{global}}$.
(b) Show that $B_p^2 = 1$.
(c) Show that $[B_p, B_{p'}] = 0$.

Anyon-Permuting Symmetry

Some TQFTs have a symmetry under permutation of anyons. This situation is not generic, but when it does happen, the corresponding TQFT has some special properties that are worth discussing.

An anyon-permuting symmetry is a rather special type of symmetry because it is "emergent"—it is unrelated to the properties of the microscopic degrees of freedom of the model. The microscopic Hamiltonian in a system may have no particular symmetry at all, but the TQFT that results as its ground state may still have a permutation symmetry for its excitations.

Conveniently, the toric code provides an excellent example of anyon-permuting symmetry. In fact, this symmetry is mentioned back in Chapter 28. In the fusion table given in Fig. 28.1 it is noted that you can exchange m for e and the fusion table would remain correct. Further, looking at Eq. 28.5 (which gives the S- and T-matrices for the toric code) the second column and second row refer to the e particle, whereas the third column and third row refer to the m particle. One can switch the second and third columns and rows of both matrices and the matrices remain unchanged. This shows that the toric code has a precise symmetry under permutation of e with m. That is, it does not matter which one we call e and which one we call m.

More generally, we say that we have an anyon-permuting symmetry whenever there is a permutation matrix P such that[1,2]

$$PSP^{\mathsf{T}} = S \qquad PTP^{\mathsf{T}} = T \qquad (36.1)$$

where S and T are the modular S- and T-matrices (see Chapter 17) and the superscript T means transpose. It may be the case that for a given TQFT there are multiple matrices P that satisfy Eq. 36.1. Generally, the set of such all Ps (including the identity matrix, which trivially satisfies Eq. 36.1) form a group known as the anyon permutation group (see Exercise 36.2).

The fact that the toric code has an anyon-permuting symmetry might seem like simply an odd feature of this particular model. However, we might view it as not being a coincidence at all. In Section 25.4 we saw that the toric code could be generated from a more complicated parent theory (Ising$\times\overline{\text{Ising}}$) by *condensing* a particular boson. In this procedure we found that a particle type from the parent theory had to split into two particle types and the result is symmetric between these two resulting pieces. In fact, this statement is very general. First, if a simple-current boson condenses from a parent theory and if a splitting occurs, the re-

---

[1] A matrix P is a permutation matrix if $P^{\mathsf{T}}P = PP^{\mathsf{T}} = \mathbf{1}$ and each row and each column of P has only a single non-zero entry which is unity and all other entries are zero

[2] Strictly speaking this only includes unitary anyon-permuting symmetries. One can have anti-unitary symmetries that involve time reversal as well as (possibly) permutation. We will not consider those. See the discussion in Barkeshli et al. [2019].

sulting theory after the condensation will have an anyon-permuting symmetry. Moreover, whenever a theory has an anyon-permuting symmetry, it is always possible to view this theory as the *result* of a condensation process with a splitting.

36.1 Symmetry Defects of Anyon-Permuting Symmetry

In Section 35.4.1 we introduced the idea of a symmetry defect. We can construct a symmetry defect for any system with a symmetry by enforcing twisted boundary conditions around a point—that is, the wavefunction is "twisted" by the symmetry action as we go around the defect point. In Section 35.4.1 we considered a system with $\mathbb{Z}_2$ microscopic variables and we twisted the boundary conditions by the $\mathbb{Z}_2$ action.

Here, we consider a symmetry defect of the anyon-permuting symmetry. Such a symmetry defect is a point in our two-dimensional system where the identity of a particle is changed (by one of the described permutation matrices of Eq. 36.1) if the particle is dragged around that point. For example, for the toric code, an m particle could be changed into an e particle, and vice versa, as shown in Fig. 36.1. We say that the e and m particles have been "permuted." Note that this means while the two different types of excitations can be distinguished locally, globally they cannot be distinguished as one can be turned into the other by going around the defect.

Symmetry defects of this sort are not contained in the structure of TQFTs, but rather in a mathematical structure known as a *G-crossed extension* of a TQFT (here, G represents the symmetry group). Similar to the case we discussed for SPTs in Section 35.4.1, these defects themselves can be treated as having certain anyon-like properties. While this more detailed mathematical structure is beyond our current discussion, interested readers can refer to the work of Barkeshli et al. [2019] for details (see also references at the end of the chapter).

Microscopic models with anyon-permuting symmetries and corresponding symmetry defects have been constructed in several ways. Here, we introduce a particularly simple way to see the anyon-permuting symmetry in the toric code.

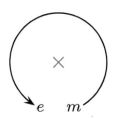

Fig. 36.1 In the toric code (in 2D), if an m particle moves around a symmetry defect (notated as the red ×) it is converted into an e particle.

36.2 Symmetric Form of the Toric Code

Let us try to make the toric code anyon-permuting symmetry a bit more explicit, and a bit more physical. Recall from Chapters 27 and 28 that the terms of the Hamiltonian for the toric code phase of matter (the "stabilizers" if we think of the toric code as actually being a code, rather than a phase of matter) are the vertex operator (drawn as a green cross in the left of Fig. 36.2), and the plaquette operator (drawn as a pink square in the left of Fig. 36.2). To remind the reader, the vertex

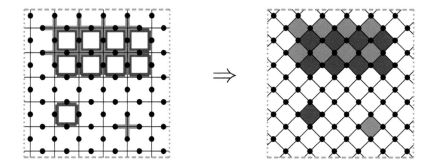

Fig. 36.2 Left: In our prior formulation of the toric code, the vertex operators acted on crosses (green), and the plaquette operators acted on squares (pink). **Right:** By simply drawing the reference lines diagonally, both types of operators act on a diamond plaquette—the vertex operators acting on the green diamonds and the plaquette operators acting on the pink diamonds.

operator (originally defined in Eq. 27.1) is a product of four σ_z operators adjacent to a given vertex, and the plaquette operator (originally defined in Eq. 27.3), is a product of four σ_x operators around a plaquette.

$$V_\alpha = \prod_{i\in\text{vertex }\alpha} \sigma_z^{(i)} \qquad\qquad P_\beta = \prod_{i\in\text{plaquette }\beta} \sigma_x^{(i)} \ .$$

Violations of the vertex terms (if V_α is not in its $+1$ eigenstate) are the particles that we previously called "electric" or e and violations of the plaquette terms (if P_β is not in its $+1$ eigenstate) we previously called "magnetic" or m.

In the left of Fig. 36.2 these two types of terms look fairly different from each other. However, with nothing more than a notational change, we can draw the same system as on the right of Fig. 36.2. Here, we simply draw the reference lines diagonally (not changing any of the physical spins, or any of the terms in the Hamiltonian) so that the system now looks like it is a tiling of diamonds. In this case both vertex and plaquette operators now act on a diamond-shaped plaquette, and this starts to look a bit more symmetric between the electric and magnetic particles. However, still the V_α operator, which acts on a green diamond, is a product of four σ_zs, whereas the P_β operator, which acts on a pink diamond, is a product of four σ_xs.

We can make a further slight transformation on the model to make it look even more symmetric between the two types of terms in the Hamiltonian. Let us make a unitary transformation on all the spins that are on the east and west corners of the green diamonds. The unitary transformation we choose is a $\pi/2$ rotation of the spin around the y-axis. Such a transformation converts

$$\begin{aligned} \sigma_x &\rightarrow \sigma_z \\ \sigma_z &\rightarrow -\sigma_x \\ \sigma_y &\rightarrow \sigma_y \ . \end{aligned}$$

As a result, what we previously called the vertex operator (a product of four σ_z operators around a green diamond) is now a product of two σ_z operators on the north and south of the green diamond times a product of two σ_x operators on the east and west of the green diamond. Similarly, what we previously called the plaquette operator (a product of four σ_x operators around a pink diamond) is now a product of two σ_x operators on the east and west of the pink diamond times a product of two σ_z operators on the north and south of the pink diamond. So, after this unitary transformation, all of the operators on all of the diamonds have exactly the same form, as shown in Fig. 36.3. In this form, the toric code Hamiltonian can be written as[3]

$$H_{\text{toric code}} = -\frac{\Delta}{2} \sum_{\text{diamonds } q} \sigma_z^{q_N} \, \sigma_x^{q_E} \, \sigma_z^{q_S} \, \sigma_x^{q_W} \qquad (36.2)$$

[3]This form of the toric code Hamiltonian was introduced by Wen [2003].

where here q_N indicates the spin in the north corner of diamond q (and analogously for q_S, q_E, and q_W). This Hamiltonian is physically identical to that of Eq. 28.1 where we have set $\Delta_v = \Delta_p = \Delta$. We have done nothing more than make a unitary transformation on some of the spins. It is easy to check that all the terms in the Hamiltonian commute with each other as in the usual toric code.

As mentioned, in the original formulation of the toric code we had two different types of terms: the vertices, whose violations we called e, and the plaquettes, whose violations we called m. Now we have transformed both vertex and plaquette terms of the original toric code Hamiltonian to all look like identical diamonds. Nonetheless, some of these diamonds (those in green in Fig. 36.2) correspond to the e particles, whereas some of the diamonds (those in pink in Fig. 36.2) correspond to the m particles. As in the original toric code, the e particles can be created or annihilated in pairs, and the m particles can also be created or annihilated in pairs.[4] However, in this new formulation of the toric code, since all diamonds look identical, it is clearly just a convention which set of diamonds we call e and which we call m. Nonetheless, there will always be two species of excitations that live on different sublattices (the pink diamonds versus the green diamonds), and they can move only on this sublattice—being created or annihilated on two diamonds of the same color that share a single vertex, or similarly moving from one diamond to another of the same color that shares a single vertex with it. Even if there is no particular reason to label one type of defect e and one m, it is clear they are different, as they live on different sublattices and cannot annihilate each other.[5]

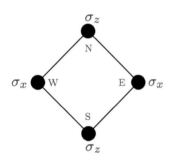

Fig. 36.3 The terms in the Hamiltonian, Eq. 36.2, on every diamond (diamonds of either color in the right of Fig. 36.2) are a product of σ_z on the north and south vertices, and σ_x on the east and west vertices.

[4]Starting in the ground state, if you apply a σ_x operator on a vertex, this creates a pair of excitations on the neighboring diamonds to the north and south of that vertex, whereas applying σ_z to a vertex creates excitations on the neighboring diamonds to the east and west. $\sigma_y = i\sigma_x\sigma_z$ does both.

[5]If one generates this diamond model by starting with a conventional toric code and converting it to diamonds as in Fig. 36.2 one always obtains a model where walking around a nontrivial cycle along edges requires an even number of steps. However, one can generalize the model by removing a single row of spins, such that one obtains an odd number of steps around a nontrivial cycle (and a model that cannot be obtained from a conventional toric code, as in the mapping of Fig. 36.2). In such a model, when one goes around the cycle with an odd number of steps, one switches sublattices, so there is then no global distinction between the pink sublattice and the green sublattice. (This is similar to what happens in Section 35.4.1 where we impose antiperiodic boundary conditions.) Correspondingly, the ground-state degeneracy becomes two, rather than four.

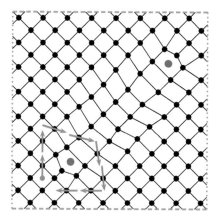

Fig. 36.4 Lattice dislocations are marked with the red dots. Walking around a dislocation, one comes back to the opposite sublattice. As shown in the figure, two steps north, two steps east, two steps south, and two steps west does not return to the same point, but rather returns to a neighboring diamond on the opposite sublattice.

36.3 Symmetry Defects in the Toric Code

Presenting the toric code in this more symmetric form allows one to naturally consider symmetry defects. In this picture, the symmetry defect is realized by a lattice dislocation.[6] In Fig. 36.4 we show two dislocations (marked with red dots). We imagine moving an excitation around one of these dislocations. As mentioned, the excitation must jump between diamonds that share exactly one vertex. Thus, we can follow the path of the green arrows marked in the figure: north two steps, east two steps, south two steps, and west two steps. At the end of this procedure we have returned to the opposite sublattice (with sublattices defined locally). Hence, if we had started with an e particle we now have an m particle, and vice versa. Thus we have implemented anyon permutation—the dislocation is a symmetry defect!

The careful reader will notice from Fig. 36.4 that at the position of the defect the plaquette has five vertices rather than four. In the Hamiltonian our rule was that the operator for each plaquette has σ_z for the north and south vertices and σ_x for the east and west vertices. For the five-vertex plaquette we maintain this rule, but we also add the rule that for the additional vertex (the one that is 3-valent rather than 4-valent) we include a factor of σ_y. This is shown in Fig. 36.5. The operator we use for this plaquette will then be

$$-\frac{\Delta}{2}\sigma_z^N \sigma_x^E \sigma_z^S \sigma_x^W \sigma_y^{3-valent} \quad .$$

It is easy to check that this operator has eigenvalues ± 1 and commutes with all of the other plaquette operators from Eq. 36.2. Thus, we still have a commuting projector model that is easily solved analytically.

We may then consider a system, say, on a torus, with some number

[6]The discovery that dislocations create symmetry defects, and the investigation of the implications, was by Bombin [2010]. Very recently the ground-state wavefunction for essentially this exact model (toric code with symmetry defects) has been successfully produced using a small-scale superconducting quantum computer; see Andersen et al. [2023]; Xu, et al [2023].

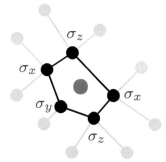

Fig. 36.5 A closeup of one of the dislocations in Fig. 36.4. For the five-sided plaquette, the 3-valent vertex is assigned σ_y, whereas the north, east, south, and west vertices are assigned $\sigma_z, \sigma_x, \sigma_z, \sigma_x$ as in all of the four-sided plaquettes.

of regular plaquettes and some number of dislocations. As we did in Section 27.3, we can count up the number of degrees of freedom to determine the ground-state degeneracy. First, we notice that the product of all plaquette terms over the entire lattice is always +1. This is analogous to the constraints in Eq. 27.8 and Eq. 27.9. For a lattice with no dislocations, the two sublattices (green and pink) in Fig. 36.2 generate two different constraints. However, with dislocations there are not two distinct sublattices (since going around the dislocation takes you from one local sublattice to the other), and thus, there is only a single constraint.

Let the number of spins be N, and the total number of plaquettes be N_p. It is not too hard an exercise (see Exercise 36.3) to show that the number n of dislocations must be

$$n/2 = N - N_p \tag{36.3}$$

showing that the number of dislocations must be even.

Let us now count degrees of freedom. We start with N spins, or equivalently, N qubits. There are $N_p - 1$ independent plaquette operators (the single constraint mentioned above accounts for the -1). Thus, putting all of the plaquettes in the ground state leaves $n/2 + 1$ qubits still undetermined. For each *pair* of dislocations, there is one more qubit unassigned, or the ground-state degeneracy doubles. This means that each dislocation is associated with *half* of a qubit.

How can it happen that a dislocation is associated with half of a qubit? Each dislocation traps a Majorana zero mode! We have run into Majorana zero modes before in our study of Ising anyons (see Exercises 3.3, 9.7, 10.2 and the comments in Section 8.2.2). In short, these are localized operators, two of which constitute a qubit. Thus, the dislocations themselves have anyon-like properties! It is in fact not a coincidence that the dislocations look like the particles of the Ising theory. Recall that the symmetry in the toric code can be thought of as arising as a result of a condensation from the Ising $\times$ $\overline{\text{Ising}}$ theory. The general condensation scheme is shown in Fig. 36.6. Much of this will look familiar from Chapter 25 with modifications. Different from our previous discussion of condensation, in the intermediate $\mathcal{T}$ theory, interpreted as the "defectization" of the toric code, we now have symmetry defects that can actually live in the bulk, rather than just on the boundary of the theory. The theory of such a situation with defects is known as a G-crossed extension (with G being the symmetry group). Perhaps the most interesting step is going from $\mathcal{T}$ up to $\mathcal{A}$. Here, we must promote the $\mathbb{Z}_2$ symmetry to a gauge symmetry to "undo" the condensation—this is very similar to the discussion of Section 35.6. Note that the result of following the arrows upward is not necessarily unique. At the top level of the diagram one could also have $\mathcal{A} = SU(2)_2 \times \overline{SU(2)_2}$ since condensing a boson from $SU(2)_2 \times \overline{SU(2)_2})$ also gives the toric code (see Exercise 25.3). We will not delve too deeply into the details of this construction, referring the reader to Barkeshli et al. [2019].

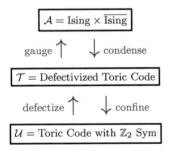

Fig. 36.6 The relation between anyon-permuting symmetry in the toric code and condensation from Ising $\times$ $\overline{\text{Ising}}$.

36.3.1 More General Symmetry Defects

While the lattice dislocation provides a particularly vivid example of a symmetry defect, there are other models that can also generate symmetry defects, and one does not have to tie the physics to dislocations of an underlying lattice. One can think of these symmetry defects in the abstract (Barkeshli et al. [2019]), or one can construct explicit toy models that have such symmetry defects even without lattice defects (as in Heinrich et al. [2016]; Mesaros and Ran [2013]; Cheng et al. [2017] for example). This is a bit more complicated than we have discussed here, but not too much more!

36.4 Symmetry-Enriched Topological Phases

SPT phases (implying short-ranged entangled states as discussed in Chapter 35) do not support anyonic excitations—even though they can be related to models that do have anyons via the "gauging" procedure discussed in Sections 35.4–35.6. It is also possible to consider phases of matter that both have symmetries *and* have anyons in them. Such phases of matter are called "symmetry-enriched phases," or "SET" phases as defined in Section 35.1. The case we just discussed of the toric code with a $\mathbb{Z}_2$ symmetry is a very good example. Very generally one has an anyon theory ($\mathcal{U}$) with a symmetry. One can create symmetry defects, then gauge the defects to obtain a parent theory ($\mathcal{A}$). If it so happens that the original theory (the $\mathcal{U}$ theory) has no anyons but only a symmetry, then most generally this is an SPT phase[7] (with only short-ranged entanglement)—gauged, this theory gives a Dijkgraaf–Witten, or twisted Kitaev, model (as discussed in Sections 35.5–35.6). However, more generally, the $\mathcal{U}$ theory will have anyons too and one gets a G-crossed theory for the $\mathcal{T}$ theory and we gauge this to get to the $\mathcal{A}$ anyon theory.

Fairly general constructions of SET phases are discussed by Heinrich et al. [2016], Mesaros and Ran [2013] and Cheng et al. [2017]. While the constructions are not too much more complicated than the discussion here, we refer the reader to the original literature for details.

[7]In Chapter 35 our symmetry was an on-site symmetry, whereas in this chapter our toric code example used a lattice symmetry for clarity. However, in general one should probably think about on-site symmetries for the generic case.

Chapter Summary

- For theories with anyon-permuting symmetry, one can have symmetry defects, such that braiding around a defect permutes the identity of the anyon.
- Dislocations in a version of the toric code provides an example such symmetry defects.
- Theories with anyon-permuting symmetries can be considered as condensations from another anyon theory.

Further Reading

- The original paper by Bombin [2010] gives a much more detailed study of the properties of the dislocations in the toric code model.
- You and Wen [2012] originated the degrees-of-freedom argument used in Section 36.3. This paper also extended the work to $\mathbb{Z}_p$ toric codes.
- See Burnell [2018] for discussion of the relation between symmetry, condensation, and splitting.
- The general mathematical structure associated with symmetry defects invokes an extension of modular tensor categories, known as G-crossed tensor categories (G being the symmetry group). A very detailed discussion of this structure is given by Barkeshli et al. [2019].
- Lu and Vishwanath [2016] give a K-matrix approach to SET phases.

Exercises

Exercise 36.1 Permutation Symmetry of the Toric Code
[Easy] What is the permutation matrix for Eq. 36.1 for the symmetry of the toric code?

Exercise 36.2 Permutation Symmetry Group
[Easy] If two different permutation matrices P_1 and P_2 both satisfy Eq. 36.1, show that the product $P_1 P_2$ also satisfies Eq. 36.1. Hence, conclude that the set of possible permutation matrices form a group.

Exercise 36.3 Counting Degrees of Freedom
[Easy] Consider a lattice on a torus like that of Fig. 36.4, with some number of dislocations inserted into the lattice. Each normal plaquette has four sides, and each dislocation plaquette has five sides. Prove Eq. 36.3. Hint: Use the Euler characteristic for a torus, faces+vertices=edges.

Exercise 36.4 $\mathbb{Z}_N$ Toric Code Symmetry Defects
[Hard] The entire construction performed in Sections 36.2 and 36.3 can be extended to the $\mathbb{Z}_N$ toric code introduced in Sections 27.6 and 28.5. Show that each pair of dislocations harbors an N state system. We say that each dislocation is associated with a $\mathbb{Z}_N$ parafermion. (Note: There are several different kinds of things that are all called parafermions, so be warned when reading the literature!).

Exercise 36.5 S_3 Permutation Symmetry
The Kitaev quantum double of the group S_3, worked out in Exercise 31.10, has a nontrivial permutation symmetry. Find the permutation matrix.

Part IX

Further Thoughts

Experiments (In Brief!)

Almost the entire book so far has been off in "theory land." It is important to dispel the notion that this topic is entirely theoretical. Nature,[1] and the ingenuity of generations of talented experimentalists, has provided a number of real physical systems that realize topological quantum order of some sort.[2] In this chapter I give a *very* brief description of some of the experimental systems of interest.

37.1 Fractional Quantum Hall Effects

Foremost among topologically ordered states is the fractional quantum Hall effect. A very brief introduction to fractional quantum Hall effect is given in Section 2.5. Here, we reiterate the crucial points for our current discussion. The fractional quantum Hall effect of electrons occurs in two-dimensional electron systems in a large magnetic fields at low temperature. In the original experiments, semiconductor heterostructures were used to confine electrons to two-dimensional layers, although in modern times one can construct single-atomic layers to confine electrons in materials such as graphene. (See the article by Dean et al. within Halperin and Jain [2020] for example.)

An important parameter to keep track of in fractional quantum Hall experiments is the so-called filling fraction

$$\nu = \frac{\text{particle density}}{\text{magnetic flux density}} \qquad (37.1)$$

where the particles in question are the electrons and the flux density is measured in units of elementary flux quanta h/e per unit area with h being Planck's constant and e the electron charge. When ν is sufficiently close to (but not necessarily precisely equal to) a ratio of small integers

$$\nu \approx p/q$$

then a quantum Hall state may form at low enough temperature. On the order of 100 different fractions have now been observed. Each such fraction represents a different topologically ordered state of matter, that is, a different TQFT (for all $q \neq 1$ it is a nontrivial TQFT, whereas the $q = 1$ case is invertable topological order with central charge $c = 1$; see Section 17.7). As we would expect, in the bulk of the sample, there is an energy gap for any excitations. Generally, the stability of a particular quantum Hall state (more or less the ease with which it is observed in experiment) is set by the size of the energy gap.[3] States with particularly

---

[1] "Nature" meaning the natural world, not the journal. In fact, the journal *Nature* provided a very entertaining rejection of one of the first articles on topological quantum computation. It said something along the lines of "the editors of this journal do not believe in this topic."

[2] Originally, I had intended that at least half of this book would be devoted to discussion of the experiments and the relevant theory related to experiments. However, here, 551 pages into this book, I realize that I cannot possibly include all of this information in a single book. Interested readers will have to wait for my *next* book to get detailed discussions of the experiments.

[3] Among the states we discuss here, the $\nu = 1/3$ state is the most stable, with energy gaps Δ measured up to $\Delta/k_b = 20$ Kelvin in graphene with k_b Boltzmann's constant (Ghahari et al. [2011]). The $\nu = 5/2$ state is less stable with energy gaps measured up to about 0.6 Kelvin in GaAs semiconductor quantum wells, whereas the $\nu = 12/5$ state is the least stable with energy gaps only of about 0.07 Kelvin in GaAs semiconductors quantum wells (Kumar et al. [2010]).

small energy gaps can be easily destroyed by small non-zero temperature (on the order of the gap) or very small amounts of disorder (on the order of the gap), hence, creating a significant challenge for experimentalists.

Although experiments sometimes leave no doubt that a fractional quantum Hall state exists—that we are observing *some* TQFT—it can sometimes be difficult to identify *which* TQFT we are observing, particularly if there are multiple viable candidates theoretically.

37.1.1   Abelian Fractional Quantum Hall States

The first discovered fractional quantum Hall state, and in many ways the paradigmatic state, is $\nu = 1/3$. Soon after the initial experimental discovery by Tsui, Stormer, and Gossard [1982], a compelling theory was worked out by Laughlin [1983].[4] Fractional quantum Hall states of the form $\nu = 1/q$ that are described by Laughlin's picture are now known as "Laughlin states." The theory showed that (among other things) for fractional quantum Hall states made of electrons, the low-energy quasiparticles have fractional charge $e^* = \pm e/q$ for $\nu = 1/q$ with odd q, where $-e$ is the charge on the electron. This charge fractionalization is already surprising.[5] Soon thereafter, Halperin [1984], and then Arovas, Schrieffer, and Wilczek [1984], showed theoretically that these quasiparticles have fractional braiding statistics, thus establishing that fractional quantum Hall states are topologically ordered[6]—that is, they are a realization of a nontrivial topological quantum field theory. In particular, the topological content of the $\nu = 1/q$ state is equivalent to the charge–flux model that we discussed in Section 4.2 with statistical angle $\theta = \pi/q$. As discussed in Sections 4.3.2 and 17.7, since q is odd, this is a super-modular theory where the cluster of q elementary anyons (equivalent to an electron) is a transparent fermion.

While theorists were quick in their acceptance of the theoretical conclusion that fractional quantum Hall states harbor anyons, the experimental demonstration of this fact was a very long time in coming. Clear evidence of anyon braiding statistics[7] was finally observed by Nakamura et al. [2020]. Even at the time of the writing of this book, however, evidence has not been obtained for any filling fractions except $\nu = 1/3$.

Very soon after the experimental discovery of the $\nu = 1/3$ fractional quantum Hall state, many additional quantum Hall states $\nu = p/q$ were discovered. The vast majority of these states are believed to be *abelian*, that is, their quasiparticle excitations have abelian braiding statistics, so that braiding only accumulates a phase. The original understanding of such states is in terms of a so-called hierarchy construction (Haldane [1983b]; Halperin [1984]). In this picture, one starts with a "Laughlin" fractional quantum Hall state $\nu = 1/m$ with odd m, made of electrons (this is the first level of the hierarchy) meaning that the electrons have formed a gapped topologically ordered state. Then we add some quasiparticles of charge $\pm e/m$ to the system (by changing the charge density) to form a fluid of quasiparticles. The quasiparticle fluid then condenses into a Laughlin-like fractional quantum Hall state (i.e. it forms a gapped

[4]A Nobel Prize was awarded to Tsui, Stormer, and Laughlin in 1998.

[5]A number of beautiful experiments have by now firmly established the fractionalization of charge. The first of these were de Picciotto et al. [1997] and Saminadayar et al. [1997] using shot noise measurement.

[6]Although the concept of "topological order" did not exist at the time, we now recognize these earlier results as implying topological order.

[7]Another beautiful experiment by Bartolomei et al. [2020] showed anyonic behavior, but was not a direct observation of braiding.

topologically ordered state) similar to the way that the original electrons did (this is now the second level of the hierarchy). One can continue the process, adding quasiparticles of the current fractional quantum Hall state and then condensing these into a Laughlin-like gapped topologically ordered state. Such hierarchy states, at the nth level of the hierarchy, can always be described with an (abelian) K-matrix describing the interaction between n different Chern–Simons gauge fields (see the discussion of Section 5.3.2, and Exercise 5.6). This procedure can obtain a fractional quantum Hall state at any fractional filling $\nu = p/q$ with odd denominator.[8] Many people like to think about these hierarchy states not in terms of such multiple levels of fluids successively condensing to form gapped states, but rather in terms of an approach known as "composite-fermions" (see Jain [2007]). However, from a topological standpoint, the hierarchy and composite-fermion approach are identical (see Read [1990]; Hansson et al. [2017]).

[8]This does not necessarily mean that the state obtained via the hierarchy construction describes the topological order of any particular state measured in experiment, although numerical simulation seems to suggest that in most cases this is true.

37.1.2 Nonabelian Fractional Quantum Hall States

A very few experimentally observed fractional quantum Hall states are believed to exhibit quasiparticles with nonabelian braiding statistics. Prominent among these is $\nu = 5/2$ (and the assumed similar state at $\nu = 7/2$).

The $\nu = 5/2$ State

The 5/2 state, the first fractional quantum Hall state discovered with an even denominator (Willett et al. [1987]), was a mystery for a number of years. After about a decade of confusion, Morf [1998] gave strong numerical evidence that the observed state is consistent with the nonabelian state proposed by Moore and Read [1991] on the basis of conformal field theory arguments. This evidence was later supported by a large number of further numerical works (see references in Nayak et al. [2008] for example).

The Moore–Read state corresponds to a super-modular TQFT. One way of describing this is as a subalgebra of the modular theory

$$\text{Moore–Read} = \text{Ising} \times \mathbb{Z}_8^{(\frac{1}{2})}$$

where the Ising theory has particle types I, σ, ψ as usual and $\mathbb{Z}_8^{(\frac{1}{2})}$ is the modular abelian anyon model described in Section 18.5.4, which has eight particle types $j = 0, \dots, 7$ with twist factors $\theta_j = \exp(ij^2\pi/8)$. The super-modular subalgebra consists of the combinations (I, j) and (ψ, j) with j even and (σ, j) for j odd thus giving a total of 12 particle types (although there are only six degenerate ground states on the torus because the transparent electron $(\psi, 4)$ does not "count"—see Section 17.7).

Despite the large amount of numerical work supporting the existence of the Moore–Read state in experiment, it was realized in 2007 that the mirror image state, $\overline{\text{Moore–Read}}$, often known as the *AntiPfaffian*, is actually equally well supported by all the numerical work that existed at that time. While the two phases of matter share many properties, they would, in principle, have certain different experimental signatures (see the discussions in Nayak et al. [2008] and Halperin and Jain [2020]). From numerical simulation it appears that the AntiPfaffian should be strongly favored in the experimental systems (Rezayi and Simon [2011]; Rezayi [2017]). However, experimentally the situation is still quite unclear. In particular, a beautiful set of experiments by Heiblum and collaborators (reviewed by Heiblum and Feldman in Halperin and Jain [2020] with more recent work including Dutta et al. [2022]) have suggested neither the Moore–Read state nor the AntiPfaffian exists in the experiment, but rather point to a different, but closely related, topological order known as the PH-Pfaffian,[9] a super-modular theory that is a subalgebra of Ising $\times\, U(1)_8$.

All three proposed possibilities for the $\nu = 5/2$ state—the Moore–Read state, the AntiPfaffian, and the PH-Pfaffian—include an Ising or $\overline{\text{Ising}}$ factor and are correspondingly nonabelian. In all cases the elementary quasiparticles are built from σ in the Ising sector. This means that if we have a system with M elementary quasiparticles, the Hilbert space dimension would be d_σ^M where $\mathsf{d}_\sigma = \sqrt{2}$. Thus, for large M we have a high-dimensional Hilbert space and a nice quantum memory. While systems based on Ising are *not* universal for quantum computation (see Section 11.2.3 and Exercise 11.1), as discussed in Section 11.2.4, it is still possible in principle to productively use such a system for *partially topological* quantum computation. Because of this possibility, there has been significant effort to demonstrate experimentally the nonabelian statistics of the $\nu = 5/2$ state. In particular, recent experiments by Willett et al. [2019], while not nearly as clear as the $\nu = 1/3$ experiments, seem promising. Nonetheless, quite a few things about these experiments are not understood, which calls into question the interpretation (see the brief discussion by Halperin in Halperin and Jain [2020] for example).

The $\nu = 12/5$ State

The $\nu = 12/5$ state is much harder to observe experimentally than $\nu = 5/2$ (see note 3), and as a result, far fewer experiments have been performed on this quantum Hall state. Nonetheless, there is substantial numerical work that strongly suggests that the observed 12/5 state is nonabelian (see for example, Zhu et al. [2015] and Mong et al. [2017]). The TQFT realized is the (mirror image of the) so-called $\mathbb{Z}_3$ Read–Rezayi state (Read and Rezayi [1999]). This is equivalent to $\overline{\text{Fibonacci}} \times \mathbb{Z}_{10}^{(3)}$. The $\mathbb{Z}_{10}^{(3)}$ factor is not modular, but is rather super-modular with the $a = 5$ particle being the fermion (see the description of this theory in Section 18.5.3). Thus, the 12/5 state is also super-modular with the electron being the $(I, 5)$ particle with I being from the Fibonacci factor. Due to the Fibonacci factor in this TQFT, this state is fully universal for topological quantum computation. In fact, this is the only state of

---

[9]Despite the experiments, this author does not believe the PH-Pfaffian is likely to be actually present in experiment given what we know from simulations. If my conclusion is true, there must be something rather nontrivial about the experiments that we do not yet understand, which has been incorrectly interpreted as PH-Pfaffian order.

matter so far observed in experiment that is universal.

Other Fractions and Other Two-Dimensional Electron Gases

There have been numerous proposals of other exotic quantum Hall states that might be realized at other filling fractions, although these have been studied in less detail or are currently less convincing. A few cases that have been proposed as possibly being nonabelian include $\nu = 19/8$ (see Hutasoit et al. [2017]; Balram [2021]) as well as $\nu = 7/3$ and $8/3$ (see Jeong et al. [2017]; Balram et al. [2020]).

All of the theoretical studies thus far mentioned consider spin-polarized electrons. The assumption is that the large magnetic fields present in typical quantum Hall experiments are sufficient to fully align all of the spins. However, it is possible to consider experiments where spins are not polarized, and this then opens the door to a much greater number of possible quantum Hall states that might be realized (see for example Ardonne and Schoutens [1999]). In modern materials such as graphene (see Dean et al., in Halperin and Jain [2020]) and wide quantum wells or bilayers, there are additional quantum numbers that the electrons may carry such as valley index or sub-band index, and the possibilities become greater still. Unraveling which observed fractional quantum Hall states correspond to which TQFTs is a huge project that may continue to provide experimental and theoretical challenges for many years to come.

37.1.3 Fractional Chern Insulators

In the systems mentioned thus far, the electrons that form the fractional quantum Hall state can be considered as essentially "free particles," where all of the nontrivial physics essentially comes from electron–electron interaction and the applied magnetic field. However, one can also consider systems where the Hamiltonian of individual electrons (without even including interaction between electrons) is very nontrivial due to band-structure effects. If a nontrivial TQFT forms in such a system in the absence of an external magnetic field,[10] it is known as a *fractional Chern insulator* (see the reviews by Bergholtz and Liu [2013]; Parameswaran et al. [2013]). The first such systems were recently discovered in so-called twisted bilayer graphene (Xie et al. [2021]) and twisted $MoTe_2$ (Cai et al [2023]; Zeng et al [2023]).

37.1.4 Bosonic Fractional Quantum Hall Effects

Fractional quantum Hall effect may also be observed in two-dimensional systems of bosons. In the bosonic case, fractional quantum Hall states should be modular TQFTs. The simplest bosonic fractional quantum Hall state is the $\nu = 1/2$ Laughlin state, which is precisely the $SU(2)_1$ TQFT. More generally one can consider the so-called Read–Rezayi series (Read and Rezayi [1999] at fillings $\nu = p/2$, which correspond to the $SU(2)_p$ TQFTs.[11]

---

[10]The idea of having band-structure effects replace magnetic field in the absence of an applied external magnetic field was probably first understood by Thouless et al. [1982] in a work that earned part of a Nobel prize. The idea that fractional quantum Hall physics can occur in such a situation was first discussed by Kol and Read [1993].

[11]For bosons, integer ν is not generally an invertible topological order, and therefore can be considered a fractional quantum Hall state, too.

There have been many proposals as to how one can construct bosonic fractional quantum Hall systems, with a recent review given by Cooper in Halperin and Jain [2020]. Perhaps the simplest approach conceptually starts with trapping bosons in the shape of a two-dimensional disk. Then one spins the disk—like a record on a phonograph turntable.[12] The Coriolis force replaces the Lorentz force that would be felt by charged particles in a magnetic field. While experiments of this sort have been done, they have not achieved values of ν that are small enough to be in the fractional quantum Hall or TQFT regime[13] (although other interesting states of matter can be formed).

There are quite a number of other techniques of producing the effective magnetic fields that are required for observing fractional quantum Hall effect (see again the review by Cooper in Halperin and Jain [2020]). One successful experiment performed by Clark et al. [2020] created a Laughlin $\nu = 1/2$ state made of only two bosons. Unfortunately, the distance between bosons is far less than the size of a single quasiparticle, so it is impossible to observe the TQFT properties in a system this small.

37.2 Gapped Quantum Spin Liquids

While there is no entirely agreed-upon definition of a spin liquid, a working definition of a "gapped quantum spin liquid" is a system of spins that is topologically ordered—that is, is described by a TQFT at low temperature and long length scale (see the discussion by Knolle and Moessner [2019]). There are quite a few real material systems that people think *might* be gapped quantum spin liquids[14]—although there is no complete consensus on any one of them. One thing, however, that is generally agreed upon is that to form a topologically ordered phase other simpler ordered phases must be suppressed. Here "simpler ordered phases" means orderings such as a simple ferromagnets with all spins aligned, or antiferromagnets where spin directions alternate. Thus, we are always looking for Hamiltonians where such simple ordering is somehow "frustrated"—the Hamiltonian will somehow not be low energy with any simple ordering of spins.

An obvious toy model of a gapped quantum spin liquid is the toric code (see Chapters 27–30). Unfortunately, the toric code Hamiltonian—which includes only four-spin interactions, but no two-spin interactions—is extremely unnatural for real materials. While it may be possible that some real materials are in the same phase of matter as the toric code (i.e. can be smoothly connected to the toric code without closing the excitation gap, see Chapter 29), it is not expected that any real material Hamiltonian will resemble the toric code Hamiltonian.

In the following we consider a few more realistic directions toward

---

[12]Very long ago, before streaming audio, music was encoded in grooves on vinyl disks. To produce the music, one would put the disk on a turntable that rotated the disk in its own plane around its central point and a stylus touching the vinyl would mechanically turn the grooves into sound. This technology was eventually replaced by compact disk technology, which also involved rotating disks, but with optical reading rather than mechanical reading—a technology that became obsolete rather quickly. Nonetheless, even today, much computer storage is made with rotating disk technology, although maybe that too will become obsolete sometime soon.

[14]See also Section 37.5.

[13]One interesting work by Gemelke et al. [2010] claimed to obtain Laughlin states of five or six bosons. This work was posted on the preprint arXiv but never published, as it was not considered to be sufficiently convincing. However, the senior author is the Nobel laureate, Steven Chu, so perhaps we should not dismiss it entirely.

gapped quantum spin liquids. The models and materials mentioned here are certainly not all of the cases that have been studied in the literature—they are simply a brief selection of some of the most promising and interesting directions (in this author's opinion). More detailed discussions are given by Savary and Balents [2016], Knolle and Moessner [2019], and Broholm et al. [2020] for example.

37.2.1 Kitaev Honeycomb Model

A more realistic toy model Hamiltonian is the so-called Kitaev honeycomb (also known as, the Kitaev hexagon) model (Kitaev [2006]). This model, with spin $\frac{1}{2}$ on a honeycomb as shown in Fig. 37.1, has a two-spin neighbor interactions of the form[15]

$$H_K = J_x \sum_{\substack{\langle i, j \rangle \\ \text{red edges}}} \sigma_i^x \sigma_j^x + J_y \sum_{\substack{\langle i, j \rangle \\ \text{blue edges}}} \sigma_i^y \sigma_j^y + J_z \sum_{\substack{\langle i, j \rangle \\ \text{green edges}}} \sigma_i^z \sigma_j^z$$

where $\langle i, j \rangle$ means the interaction is between spins on neighboring sites and the interaction is $\sigma^x \sigma^x$ or $\sigma^y \sigma^y$ or $\sigma^z \sigma^z$ depending on which direction the edge between the spins points, as shown in Fig. 37.1. For simplicity we assume J_x, J_y, J_z are all positive.

Surprisingly, this model is exactly solvable. For $J_x < J_y + J_z$ or $J_y < J_x + J_z$ or $J_z < J_x + J_y$ the ground state is gapped and the topological order is precisely that of the toric code. If none of these three conditions apply then the system is in a gapless phase. However, the gapless phase may be perturbed with a small additional Hamiltonian, such as a magnetic field coupling to each spin

$$\delta H = \sum_i \left(h_x \sigma_i^x + h_y \sigma_i^y + h_z \sigma_i^z \right)$$

and as long as all h_x, h_y, h_z are non-zero, the system generates a gap and becomes an Ising or $\overline{\text{Ising}}$ phase (see Kitaev [2006]).

The Kitaev model, and many modifications of it, have been very heavily studied as a toy model for spin liquids, with extensions including additional terms in the Hamiltonian as well as models in higher dimensions (see for example, the review by Hermanns et al. [2018]). Within two dimensions, models can be constructed that produce all phases of the 16-fold way from Table 17.3 (see, for example, Chulliparambil et al. [2020]; Zhang et al. [2020]).

Perhaps most surprising is that the Kitaev model (and extensions thereof) are not too inaccurate portrayals of certain materials in the natural world. While it might seem extremely odd to have spin interactions that depend on direction, Jackeli and Khaliullin [2009] showed that for certain compounds with strong interactions and strong spin–orbit coupling, such interactions become natural. Predominant among the compounds that seem to fit this picture are hexagonal iridate compounds such as the two-dimensional materials Na_2IrO_3, $\alpha\text{-}Li_2IrO_3$, (see

[15]Notice the interactions in the different directions compete with each other. The interaction along the red edges wants the spins to align (or anti-align) along the x axis, whereas the interaction along the blue edges want the spins to align (or anti-align) along y and the interaction along the green edges want the spins to align (or anti-align) along z. This frustrates more conventional spin ordering.

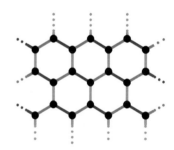

Fig. 37.1 The Kitaev honeycomb model has a spin one-half on each site (marked as black dots) with $J_x \sigma^x \sigma^x$ interaction on the red edges, $J_y \sigma^y \sigma^y$ interactions on the blue edges, and $J_z \sigma^z \sigma^z$ interactions on the green edges.

Revelli et al. [2020]) as well as the hexagonal ruthenium-based material α-RuCl$_3$ (see Loidl et al. [2021]). Several three-dimensional analogs have also been considered, such as β- and γ-Li$_2$IrO$_3$. See the reviews by Winter et al. [2017], Rau et al. [2016], Knolle and Moessner [2019], and Hermanns et al. [2018]. While the experimental search for topological order in these spin systems has been intense, as of the present time no experiments have been entirely conclusive.

37.2.2 Frustrated Antiferromagnets

As mentioned, one of the requirements for spins to form a topologically ordered state is that the spins do not form any simpler ordered state. Antiferromagnets on lattices made from triangular plaquettes tend to suppress ordering in a rather natural way. To understand this, let us consider the triangle of spins in Fig. 37.2. If the Hamiltonian is a standard nearest-neighbor Heisenberg antiferromagnetic interaction

$$H = -J \sum_{\langle i,j \rangle} \boldsymbol{\sigma}_i \cdot \boldsymbol{\sigma}_j \qquad (37.2)$$

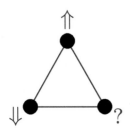

Fig. 37.2 Antiferromagnetic interactions on a triangle are frustrated. If the first two spins align opposite to each other, the third spin cannot align opposite both of the other two.

with $J < 0$, (and $\langle i,j \rangle$ means that i and j are nearest neighbors), neighboring spins will want to align opposite from each other to minimize the energy. However, if there are triangular plaquettes as shown in Fig. 37.2, then we cannot anti-align each spin with its neighbors. For example, as shown in the figure, two neighboring spins may anti-align with each other, but then it is not possible for the third spin to anti-align with both of the neighbors. Such "frustration" of a system tends to prevent conventional orders from forming and can therefore allow more exotic order, such as topological order.

One might therefore start by considering a triangular lattice (as shown in Fig. 35.1) with spins on the vertices and antiferromagnetic interactions. Unfortunately, despite the obvious frustration, it now seems likely that the Heisenberg antiferromagnetic model on triangular lattice does tend to form conventional ordered states (see Li et al. [2022]). A much better candidate is the antiferromagnetic kagome[16] Heisenberg model, as shown in Fig. 37.3. For *classical* spins (i.e. if $\boldsymbol{\sigma}$ is a vector without commutation relations) the ground state of the antiferromagnetic Heisenberg kagome model is highly degenerate. In such cases where the ground state is highly degenerate, and unordered, it is thought that there is a good chance for the quantum-mechanical ground state to become a quantum spin liquid. It remains controversial whether the ground state is actually a gapped quantum spin liquid (see Läuchli et al. [2019]). Whether or not the Heisenberg kagome antiferromagnet is, or is not, a gapped quantum spin liquid (i.e. is topologically ordered) it is very likely that certain small perturbations to this model will push the system into a stable topologically ordered state.

There are a number of physical systems that seem to be well approximated by the antiferromagnetic kagome Heisenberg model— particu-

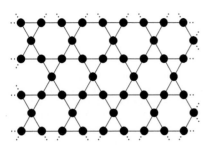

Fig. 37.3 This kagome crystal structure has corner-sharing triangles.

[16]The word "kagome" comes from a Japanese word for a type of woven bamboo pattern.

larly a material called *Herbertsmithite*[17] and a set of materials known as jarosites. The status of these materials has been recently reviewed by Norman [2016].

In three dimensions, there are several crystal structures which are similar in spirit to the kagome, such as the hyperkagome crystal, which is also made of corner sharing triangles, and the pyrochlore crystal, which is made of corner sharing tetrahedra (as shown in Fig. 37.4). A number of additional spin-liquid candidates exist on these structures as well.

Experimentally, the first signature one looks for is the notable *lack* of simple magnetic ordering at low temperature. While a number of three-dimensional compounds do seem to lack magnetic ordering, this evidence is still far from any definitive establishment of topological order (see Broholm et al. [2020] and Knolle and Moessner [2019]).

37.3 Conventional Superconductors

Some of the most spectacular quantum effects in physics are superconductivity and superfluidity, which occur in certain materials at very low temperature. Superconductivity is (roughly) defined by the ability of a system to carry electrical current with zero resistance. Superfluidity similarly allows mass flow with zero dissipation (see, for example, Annette [2004]). Certain types of superconductors and superfluids in principle have topological order.

Most common superconductors, such as the simple elemental metals Al, Nb, and Hg, are "conventional s-wave" superconductors.[18] In either two or three dimensions, when coupled to electromagnetism, these superconductors have the topological order of the toric code (see Hansson et al. [2004], and see Exercise 28.9 and Section 31.7.1 for discussion of the three-dimensional toric code). Here, the low-energy quasiparticle excitations are fermions, and vortices of magnetic flux through the superconductors are bosonic, in such a way that braiding the fermion around the vortex incurs a minus sign.[19]

37.4 Majorana Materials

One can also consider more exotic superconductors or superfluids. A fairly recent detailed review of "topological" superconductors is given by Sato and Ando [2017]. Such superconductors[20] (should they be found in experiment), if they have a gap and are two-dimensional, would have Majorana zero modes bound to vortex cores, as theoretically proposed by Read and Green [2000]. Such zero modes would translate into a topological order with fusion rules of the Ising TQFT (see Exercises 3.3 and 9.7 for example).

The review of Sato and Ando [2017] lists a number of promising materials for gapped topological superconductors. The very first on the list, and for many years the poster child for the field, Sr_2RuO_4 unfortunately, now needs to be reconsidered as some of the key experimental

[17]This mineral is named for, but not discovered by, the famous British minerologist George Frederick Herbert Smith. Herbert Smith is an unhyphenated double surname, which is not uncommon in British culture

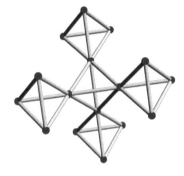

Fig. 37.4 The pyrochlore structure has corner-sharing tetrahedra.

[18]The "high-temperature superconductors" do not have an excitation gap and therefore cannot be topologically ordered.

[19]For those who know a bit about superconductivity, the key here is that a *pair* of fermions braids trivially around the vortex, but the single fermion braiding around the vortex incurs a minus sign.

[20]For those who are familiar with superconductivity, here we are essentially discussing chiral *p*-wave superconductors.

data turned out to be faulty (Pustogow et al. [2019]). The next most interesting entry on the list is the doped topological insulator Bi_2Se_3, particularly when doping with copper (although also with Nb and Sr). However, the superconductivity in these materials appears to be *nematic*, meaning that it breaks rotational invariance. While this does not preclude topological order, it does make the situation more complicated. This situation, reviewed in Yonezawa [2019], remains promising. A somewhat similar material, $Sn_xIr_{1-x}Te$, also appears interesting.

In the last few years a number of additional new materials have arisen, including Fe(Te,Se), Li(Fe,Co)As (see Zhang et al. [2019b]) and single layers of WS_2 (see Li et al. [2021]), but it remains to be seen if these will withstand further experimental tests.

Another interesting candidate is ^{3}He superfluid, which is a topological superfluid made of paired neutral fermionic ^{3}He atoms.[21] The so-called *A*-phase of ^{3}He is well established to be appropriately topological. While it is gapless in three dimensions, in principle a thin film would be fully gapped and the superfluid vortices should then harbor Majoranas (see Saunders [2018]). A practical difficulty with this experimental system is that superfluidity only forms at temperatures around 1 mK—making such experiments among the most difficult in all of physics.

Another promising avenue is the use of artificially engineered structures that are designed to produce Majoranas in vortices (Fu and Kane [2008]). There are several reports in the literature of Majoranas in topological insulator/superconductor heterostructures (see for example, Sun and Jia [2017]) or on engineered surfaces of magnets on gold (see Manna et al. [2020]). A number of other engineered routes have also been considered, including chains of magnetic atoms and Josephson junctions (see Flensberg et al. [2021] for more detailed review).

Possibly the most pursued direction for producing Majoranas is one-dimensional semiconductor/superconducting quantum wires. Based on an idea of Kitaev [2001], in principle such systems can trap Majoranas at the ends of the wires. While this does not entirely correspond to a two-dimensional TQFT it nonetheless has many similarities. An enormous experimental effort (and enormous funding) has gone into this idea in an attempt to turn it into a useable technology (see Lutchyn et al. [2018]). Unfortunately, the situation remains extremely murky and, despite claims of success, the community still has not been convinced that Majoranas have been observed.[22] Perhaps this is a warning to those in these fields that, until a qubit is built and manipulated, one should take most claims with substantial caution.

[21] To remind the reader, Helium has two isotopes. The naturally occurring isotope ^{4}He has two protons, two neutrons, and two electrons and is overall a boson. The artificial isotope ^{3}He is missing one neutron, is overall a fermion.

[22] A very high-profile retraction of a paper from the journal *Nature* has been very damaging to the reputation of the field. See Zhang et al. [2021].

37.5 Quantum Simulation

With the advent of quantum computing technologies a main direction forward is the construction of "artificial matter" from qubits. Indeed, as emphasized in Chapters 27–28, error-correcting codes such as the toric code are very closely related to topologically ordered matter. While there may be a somewhat blurry line between what we call "actual matter" and what we call "simulation of matter," it is reasonable to separate out experiments on modern multi-purpose quantum devices.

There have been a number of experiments that construct toric codes from a very small number of superconducting qubits, including Chen et al. [2021], Andersen et al. [2020], Satzinger et al. [2021], and Song et al. [2018]. The latter two both managed to explicitly probe the braiding statistics of the quasiparticle excitations. Since the toric code is generally considered one of the "best" error-correcting codes, it is very likely that better and bigger versions of this experiment will be done in the near future.

Very recently the toric code has been extended by Andersen et al. [2023]; Xu, et al [2023] to include symmetry defects, as discussed in Section 36.3, to produce nonabelian Majoranas. While these are not *quite* nonabelian anyons (strictly speaking they are defects; see the discussion by Bombin [2010]; Barkeshli et al. [2019]), they share many properties with nonabelian anyons, including nontrivial braiding.

Further proposals have been made to produce nonabelian quantum double models in realistic ways (see Brennen et al. [2009]; Tantivasadakarn et al. [2022]; Jovanovic et al. [2023]). A recent experiment by Iqbal et al. [2023] using a trapped-ion quantum computer constructed a wavefunction with nonabelian topological order by gauging an SPT wavefunction.

Another experimental avenue is the use of so-called Rydberg atoms as qubits. An experiment by Semeghini et al. [2021] simulated a quantum spin liquid (dimers on a kagome structure) in the toric code phase from a system of over 200 Rydberg qubits—which approaches the thermodynamic limit.

Another promising avenue is to assemble a lattice of qubits that then act as bosons (a qubit in the $|1\rangle$ state corresponds to the presence of a boson and the $|0\rangle$ state corresponds to the absence of a boson). Interacting bosons of this sort in an effective magnetic field have been demonstrated by Roushan et al. [2017] for a lattice with only three sites. It seems likely that this technique can be scaled up, and fractional quantum Hall physics could potentially be achievable.

Chapter Summary

- A few experimental systems have shown very strong evidence of anyons.
- Braiding physics has been convincingly demonstrated in $\nu = 1/3$ fractional quantum Hall effect and toric code quantum simulations—both of these demonstrations probe only abelian braiding statistics.
- A very wide variety of other experimental systems could possibly harbor anyons, but showing this in experiment is quite hard.

Final Comments

Nearing the end of the book, there are a few final items to be discussed briefly and a few parting words.

38.1 What Is Not in This Book

Despite the length of this book, we have only scratched the surface of the rich field of topological quantum. There are several key subjects that I feel must be mentioned, if only briefly, so that readers will be aware that they exist.

38.1.1 Theory of Fractional Quantum Hall Effect

Fractional quantum Hall effects, as mentioned in Section 37.1, is the best experimental example we have of topologically ordered matter. The detailed theory (theories) of fractional quantum Hall effect is something that I hope to cover in a later book.

One of the reason we did not delve too far into fractional quantum Hall physics is that the topic is truly vast with multiple complementary approaches to the subject. The trial wavefunction point of view (such as the original work by Laughlin [1983]) is one crucial approach to the field, which has spawned many other important ideas (such as the composite fermion approach; see Jain [2007]). Field theory approaches such as Zhang et al. [1989], Lopez and Fradkin [1991] and Wen and Zee [1992] are also crucial to the subject. Quantum Hall edges (see the discussion in Section 17.3.3) are crucial both for the theoretical understanding of the subject as well as for the understanding of many of the experiments (see also the chapter by Kane and Fisher in Das Sarma and Pinczuk [1997]).

Other approaches to fractional quantum Hall effect include the use of exactly solvable Hamiltonians and the special mathematical structure of certain fractional quantum Hall wavefunctions (see, for example, the article by Simon in Halperin and Jain [2020]). Perhaps most interestingly, as we mention next, is the famous Moore and Read [1991] connection between fractional quantum Hall effect and conformal field theory.

Given the many approaches to the subject it seemed best to defer more detailed discussion to another book!

38.1.2 Connections to Conformal Field Theory

$(1 + 1)$-dimensional conformal field theory (CFT) describes the physics of gapless one-dimensional (Lorentz invariant) systems. Such theories are typically exactly solvable due to their rich mathematical structure (see Di Francesco et al. [1997] and Ginsparg [1990] for example).

Since the work of Witten [1989] it has been understood that there are deep connections between $(2 + 1)$-dimensional Chern–Simons theory and $(1 + 1)$-dimensional CFT. In essence, the conformal field theory is the boundary theory of the Chern–Simons theory. Depending on how one cuts the $(2 + 1)$-dimensional spacetime manifold, one can obtain a $(1 + 1)$-dimensional dynamical theory, or one can obtain a static $(2 + 0)$-dimensional wavefunction. Much of the structure of CFT recapitulates the structure we have already learned about TQFTs (fusion, braiding, Verlinde relations, and so forth). However, CFTs have additional structure that is not reflected in the TQFTs. We have therefore chosen not to try to cover this material in depth here.

The connection between $(1 + 1)$-dimensional CFTs and bulk $(2 + 0)$-dimensional wavefunctions was exploited by Moore and Read [1991] to very generally construct $(2 + 0)$-dimensional quantum Hall wavefunctions based on $(1 + 1)$-dimensional CFTs. This mapping led to the first detailed proposal of nonabelian particles in condensed matter systems.[1] Indeed, the quantum Hall setting remains perhaps the only experimental system where we believe nonabelian topological matter has actually been observed. The connection to CFT is now one of the primary tools for the study of fractional quantum Hall effect.

38.1.3 Tensor Networks

As mentioned briefly in Chapter 34, a key modern method of understanding the properties of matter is by characterizing the quantum entanglement between parts of a system. This direction of thought has led to a number of very important numerical and analytical techniques such as density matrix renormalization group and tensor networks (see, for example, Orús [2014] and Cirac et al. [2021]). In the language of tensor networks, many types of topological matter have very beautiful and enlightening descriptions (see, for example, Bultinck et al. [2017] and Bauer et al. [2019]). Not only do these techniques provide calculational tools for numerical simulation, but also they give us deeper understanding of the structure of topological order.

Again, I can only apologize that there simply is not room in this book to cover the topic properly.

38.1.4 Topological Insulators

Probably a large fraction of the people who pick up this book were actually hoping that it would be about topological insulators.[2] While we have made brief reference to this topic in Chapter 35, and there are a number of connections to the ideas presented here, on the whole, we

[1] The first discussion that nonabelian particles could *in principle* occur in condensed matter systems appears to have been by Chen et al. [1989].

[2] My apologies to these people!

have completely ignored this very large field of study. Again, I promise to discuss this in a future book.

38.2 Final Words

If you have persisted in reading for 565 pages—Bravo to you! I'll buy you a beer. More seriously, I truly hope that what you have learned in this book has set you up well for further study and research in this field. I wish you the very best of luck in your future explorations.

Appendix: Kac and Other Resources for TQFTs

Working out the details of a TQFT is an often tedious task and, except in the simplest cases, one does not want to go through the pain of doing this. Here, we explain the use of one particularly convenient way to find data for many TQFTs. At the end of this chapter we list a number of other useful resources and references.

39.1 Kac

Perhaps the most useful single resource I have found for obtaining data about TQFTs is a computer program called *Kac* written by A. N. Schellekens. The complicated part of the algorithm is described by Fuchs et al. [1996]. More details are given on the project webpage.

The program can be downloaded from the webpage

```
https://www.nikhef.nl/~t58/Site/Kac.html
```

also mirrored on my web page.

The program has many capabilities, and I encourage you to RTFM[1]. One can also type `Help` at the command prompt to get some useful information. Despite these resources, it is probably useful to give here some annotated examples of how it works.

[1] Read the Frikkin Manual.

Note that the program uses Dynkin diagram (Cartan) notation for describing Lie algebras. The correspondence is given by

$$
\begin{aligned}
A_r &= su(r+1) \\
B_r &= so(2r+1) \\
C_r &= sp(2r) \\
D_r &= so(2r) \quad .
\end{aligned}
\tag{39.1}
$$

One can also use the $E6, E7, E8, F4$, and $G2$ Lie algebras.

Here, we present some annotated sessions with Kac.

```
MYLINUXBOX$Kac
Kac (on MYLINUXBOX), version 8.05468, compiled on Sep  1 2016, at 16:27:29
Started Sun 30 Aug 05:19:53 BST 2020
Non-interactive mode; Assuming default answer: OK
> tensor
```

The tensor command tells the program that we might be tensoring together multiple theories.

```
> g a 1 2
```

This inputs the group (g for group) with the Cartan notation a 1, or A_1 which is $su(2)$ as given by the correspondence Eq. 39.1, and the 2 indicates level 2. So we are asking it to compute information about $SU(2)_2$.

```
> display
CFT {A1:2}; 3 primaries (2 simple currents)
  Lbl  Comb.  Weights      Wts.  F.l.  F.m.
    0   {0}   0.0000000    0      -    1
    1   {1}   0.5000000    1/2    -    1
    2   {2}   0.1875000    3/16   -    1
```

The fields are numbered 0,1,2, and we see their corresponding weights $h = 0, 1/2, 3/16$. The simple currents are always listed first. Recall that twist factors are given by $\theta = e^{2\pi i h}$. Note also that the weights are only correct modulo 1. We can then ask for quantities like the fusion rules, or the S-Matrix.[2]

[2]Caution of note 8 from Chapter 17. This does give the proper S-Matrix in our notation.

```
> fusion
(0) x (0) =  (0)

(0) x (1) =  (1)

(0) x (2) =  (2)

(1) x (1) =  (0)

(1) x (2) =  (2)

(2) x (2) =  (0)  + (1)

> S
S(0,0) =      0.50000000
S(0,1) =      0.50000000
S(0,2) =      0.70710678
S(1,1) =      0.50000000
S(1,2) =     -0.70710678
S(2,2) =      0.00000000
```

We can also get the central charge or the Frobenius–Schur indicator of, say, the second particle,

```
> Browse Central
Central charge              1.500000000000000

> Get Schur 2
-1
```

If we had wanted to look at the opposite chirality theory, we use **h** rather than **g**. To wipe the memory of the program and return to tensor mode we use **reset tensor**. For example, we have

```
> reset tensor
> h a 1 2
> display
CFT {A1:2}; 3 primaries (2 simple currents)
  Lbl  Comb.  Weights              Wts.             F.l.  F.m.
    0  {0}    0.0000000  (mod 1)   0                 -     1
    1  {1}    0.5000000  (mod 1)   1/2 (mod 1)       -     1
    2  {2}    0.8125000  (mod 1)   13/16 (mod 1)     -     1
```

Note that the weight of the 2 field is $13/16 = -3/16 \bmod 1$ so this is the opposite chirality version of $SU(2)_2$, which we write as $\overline{SU(2)_2}$.

The program can also handle $U(1)$, or abelian, Chern–Simons theory.[3] We produce these $U(1)$ theories using the code **g u** followed by the level as follows:

```
> reset tensor
> g u 4
> display
CFT {U4:0}; 4 primaries (4 simple currents)
  Lbl  Comb.  Weights      Wts.   F.l.  F.m.
    0  {0}    0.0000000    0       -     1
    1  {1}    0.1250000    1/8     -     1
    2  {2}    0.5000000    1/2     -     1
    3  {3}    0.1250000    1/8     -     1
```

The level here (4 in this case) is simply the constant k in Eq. 5.1. The program will caution you if you put in an odd level since that will return a non-modular theory, as we found in Section 4.3.2.

The program can generate minimal CFT models with N supersymmetries. If we want the usual Virasoro minimal series, we set $N = 0$.

[3] See note 11 in Chapter 5. Be cautioned that there are multiple conventions as to how the subscript of a $U(1)$ theory is labeled, which differ by a factor of 2.

For example, we have the first minimal CFT

```
> reset tensor
> g minimal 0 1
> display
CFT {minimal N=0 1}; 3 primaries (2 simple currents)
  Lbl  Comb.  Weights      Wts.  F.l.  F.m.
    0  {0}    0.0000000     0     -     1
    1  {1}    0.5000000     1/2   -     1
    2  {2}    0.0625000     1/16  -     1
```

which we recognize as the Ising model. The second Virasoro minimal model (the tricritical Ising model) would be

```
> g minimal 0 2
```

and so forth. See the `Help` notes for the $N = 1, 2$ minimal models.

The program can handle condensation, as well as splitting. Let us consider the example of $SU(2)_4$, which we used in Section 25.4. We first produce the $SU(2)_4$ theory

```
> reset tensor
> g a 1 4
> display
CFT {A1:4}; 5 primaries (2 simple currents)
  Lbl  Comb.  Weights      Wts.  F.l.  F.m.
    0  {0}    0.0000000     0     -     1
    1  {1}    1.0000000     1     -     1
    2  {2}    0.1250000     1/8   -     1
    3  {3}    0.6250000     5/8   -     1
    4  {4}    0.3333333     1/3   -     1
```

Note that one of the simple currents is a boson (integer weight). To condense it we issue the command `current` and the name of the field we want to condense.

```
> current 1
> display
CFT {A1:4}; 3 primaries
  Lbl  Comb.  Weights      Wts.  F.l.  F.m.
    0  {0}    0.0000000     0     -     1
    1  {4}    0.3333333     1/3   0     1
    2  {4}    0.3333333     1/3   1     1
```

which correctly splits the 4-particle, as we discussed in Section 25.4.

To generate product theories, we just input several theories in a row. For example, to look at a product theory, $SU(2)_2 \times \overline{SU(2)_1} \times \overline{SU(2)_1}$ we write

```
> reset tensor
> g a 1 2
> h a 1 1
> h a 1 1
> display
```

```
CFT {A1:2_A1:1_A1:1}; 12 primaries (8 simple currents)
Lbl  Comb.    Weights              Wts.               F.l.  F.m.
  0  {0,0,0}  0.0000000  (mod 1)   0                   -    1
  1  {0,0,1}  0.7500000  (mod 1)   3/4 (mod 1)         -    1
  2  {0,1,0}  0.7500000  (mod 1)   3/4 (mod 1)         -    1
  3  {0,1,1}  0.5000000  (mod 1)   1/2 (mod 1)         -    1
  4  {1,0,0}  0.5000000  (mod 1)   1/2 (mod 1)         -    1
  5  {1,0,1}  0.2500000  (mod 1)   1/4 (mod 1)         -    1
  6  {1,1,0}  0.2500000  (mod 1)   1/4 (mod 1)         -    1
  7  {1,1,1}  0.0000000  (mod 1)   0 (mod 1)           -    1
  8  {2,0,0}  0.1875000  (mod 1)   3/16 (mod 1)        -    1
  9  {2,0,1}  0.9375000  (mod 1)   15/16 (mod 1)       -    1
 10  {2,1,0}  0.9375000  (mod 1)   15/16 (mod 1)       -    1
 11  {2,1,1}  0.6875000  (mod 1)   11/16 (mod 1)       -    1
```

Since $SU(2)_2$ has three fields, and each $\overline{SU(2)}_1$ has two fields, the product of these three theories has 12 fields. The second column of the output shows how each field is constructed from the constituent factors. For example, the output field labeled 9 in the far left column comes from the 2-field of $SU(2)$, the 0-field from the first $\overline{SU(2)}_1$ and the 1-field from the second $\overline{SU(2)}_1$.

Let us now construct the coset $SU(2)_2/(SU(2)_1 \times SU(2)_1)$. Recall from Section 25.6 that one can construct this coset by starting with $SU(2)_2 \times \overline{SU(2)}_1 \times \overline{SU(2)}_1$ and condensing all possible simple current bosons. Notice in the above output that there are eight simple currents, and the one labeled 7 or $\{1,1,1\}$ is a boson. We thus issue the command

```
> current 1 1 1
> display
CFT {A1:2_A1:1_A1:1}; 3 primaries (2 simple currents)
Lbl  Comb.    Weights              Wts.               F.l.  F.m.
  0  {0,0,0}  0.0000000  (mod 1)   0                   -    1
  1  {0,1,1}  0.5000000  (mod 1)   1/2 (mod 1)         -    1
  2  {2,0,1}  0.9375000  (mod 1)   15/16 (mod 1)       -    1
```

giving us the result that this coset is actually $\overline{\text{Ising}}$.

39.2   **Other Resources**

Note that many of the following references give only the "modular data" for TQFTs—meaning the S-Matrices (which imply the fusion rules via the Verlinde formula, Eq. 17.13) and the twist factors θ_a. However, it has recently been established that there can be cases where more than one modular TQFT can share the same modular data (Mignard and Schauenburg [2021]).[4] However, the simplest such case known where the modular data does not uniquely define the TQFT has 49 different particle types and for all simple TQFTs the modular data is, at least in principle, full information.

[4]See also Bonderson et al. [2019], Delaney and Tran [2018], and Wen and Wen [2019] for discussion of what additional data might be added to make the TQFT unique

- A useful reference on conformal field theory, including Wess-Zumino-Witten theories (which give you the content of the corresponding Chern–Simons theory) is given by Di Francesco et al. [1997].

- Many details of the simplest few modular tensor categories on the periodic table are given by Rowell et al. [2009]; A discussion for fermionic models is given by Bruillard et al. [2017, 2020]. See also Lan et al. [2017, 2016] and Wen [2015].

- Some nice data for some simple categories is given by Bonderson [2007]. This includes, for example, the F-Matrices for $SU(2)_k$ and a number of other simple theories.

- F-Matrices for many more complicated theories are given by Ardonne and Slingerland [2010]. Mathematica code based on this is available at

  https://github.com/ardonne/affine-lie-algebra-tensor-category

- Online databases of vertex algebras, modular categories, fusion rings, etc., are given at

  https://www.math.ksu.edu/~gerald/voas/

  http://www.thphys.nuim.ie/AnyonWiki/index.php/Main_Page

- An online database of modular data for (twisted and untwisted) gauge theories (i.e. quantum doubles and twisted quantum doubles) is given by

  https://tqft.net/web/research/students/AngusGruen/

  and is also mirrored on my web page. Just to make the rest of us feel bad, this was a bachelor's thesis!

Appendix: Some Mathematical Basics

Many undergraduates (and even many graduates) do not get any proper education in advanced mathematics. As such, I am including a very short exposition of most of what you need to know in order to read this book. For much of the book, you won't even need to know this much! If you have even a little background in mathematics you will probably know most of this already.

40.1 Manifolds

We sometimes write $\mathbb{R}$ to denote the real line, that is, a space where a point is indexed by a real number x. We can write $\mathbb{R}^n$ to denote n-dimensional (real) space where a point is indexed by n-real numbers $(x_1, \ldots, x_n)$. Sometimes people call these spaces "Euclidean" space.

Definition 40.1 *A* **Manifold** *is a space that locally looks like a Euclidean space.*

If a manifold contains all its limit points (i.e. is compact), and has no boundary we call it *closed*.

40.1.1 Some Examples: Euclidean Spaces and Spheres

- $\mathbb{R}^n$ is a manifold.
- The circle S^1, also known as a 1-sphere (hence, the notation, the index 1 meaning it is a one-dimensional object) is defined as all points in a plane equidistant from a central point. Locally this looks like a line since position is indexed by a single variable (the "curvature" of the circle is not important locally). Globally, one discovers that the circle is not the same as a real line, as position is periodic (if you walk far enough in one direction you come back to where you start). We sometimes define a circle as a real number from 0 to 2π, which specifies the angle around the circle.
- The 2-sphere S^2 is what we usually call (the surface of) a sphere in our regular life. We can define this similarly as all points in $\mathbb{R}^3$ equidistant from a central point.
- One can generally define the n-sphere, S^n, as points equidistant from a central point in $\mathbb{R}^{n+1}$.

Often when we discuss a manifold, we are interested in its topological properties only. In other words, we will not care if a circle is dented, as shown in Fig. 40.1, as it is still topologically S^1. Mathematicians say that two objects that can be smoothly deformed into each other are *homeomorphic*, although we will not use this language often.

It is sometimes convenient to view the circle S^1 as being just the real line $\mathbb{R}^1$ with a single point added "at infinity"—think about joining up $+\infty$ with $-\infty$ to make a circle. We can do the same thing with the sphere S^2 and $\mathbb{R}^2$—this is like taking a big flat sheet and pulling the boundary together to a point to make it into a bag and closing up the top (which gives a sphere S^2). Obviously the idea generalizes: S^3 is the same as $\mathbb{R}^3$ "compactified" with a point at infinity, and so forth.

Fig. 40.1 This object is topologically a circle, S^1.

Orientability

We say a manifold is orientable if we can consistently define a vector normal to the manifold at all points. Another way of defining orientability (that does not rely on embedding the manifold in a higher dimension) is that we should be able to consistently define an orientation of the coordinate axes at all points on the manifold. Throughout this book we almost always assume that all manifolds are orientable.

An example of a non-orientable manifold is the Möbius strip shown in Fig. 40.2. If we smoothy move the coordinate axes around the strip, when we come back to the same point, the upward pointing normal will have transformed into a downward pointing normal.

Fig. 40.2 A Möbius strip is a non-orientable manifold (with boundary). If we move the coordinate axes around the strip, when they come back to the same position, the normal vector will be pointing downward instead of upward.

There is a very simple classification of orientable closed (bounded and without boundary) two-dimensional manifolds by the number of "holes," which is known as its "genus," g. A sphere ($g = 0$) has no holes, a torus ($g = 1$) has one hole, a two-handled torus ($g = 2$) has two holes, and so forth. See Fig. 40.3.

40.1.2 Unions of Manifolds $\mathcal{M}_1 \cup \mathcal{M}_2$

We can take a "disjoint" union of manifolds, using the notation $\cup$. For example, $S^1 \cup S^1$ is two circles (not connected in any way). If we think of this as being a single manifold, it is a manifold made of two disjoint pieces (or a *disconnected manifold*). Locally, it still looks like a Euclidean space.

Fig. 40.3 A two-handled torus is an orientable two-dimensional manifold without boundary. Because it has two holes we say it has genus two. Two-dimensional manifolds without boundary are classified by their genus.

40.1.3 Products of Manifolds: $\mathcal{M}_3 = \mathcal{M}_1 \times \mathcal{M}_2$

One can take the product of two manifolds, or "cross" them together, using the notation $\times$. We write $\mathcal{M}_3 = \mathcal{M}_1 \times \mathcal{M}_2$. This means that a point in $\mathcal{M}_3$ is given by one point in $\mathcal{M}_1$ and one point in $\mathcal{M}_2$. This multiplication is often called the *direct* or *Cartesian* product.

- $\mathbb{R}^2 = \mathbb{R}^1 \times \mathbb{R}^1$. Here, a point in $\mathbb{R}^1$ is specified by a single real number. Crossing two of these together, a point in $\mathbb{R}^2$ is specified by two real numbers (one in the first $\mathbb{R}^1$ and one in the second $\mathbb{R}^1$).

- $T^2 = S^1 \times S^1$. The 2-torus T^2, or surface of a doughnut,[1] is the product of two circles. To see this, note that a point on a torus is specified by two angles, and the torus is periodic in both directions. Similarly, we can build higher-dimensional tori (tori is the plural of torus) by crossing S^1s together any number of times.

[1]Alternatively spelled "donut" if you are from the States and you like coffee.

40.1.4   Manifolds with Boundary

One can also have manifolds with boundary. A boundary of a manifold locally looks like an n-dimensional half-Euclidean space. The interior of a manifold with boundary looks like a Euclidean space, and near the boundary it looks like a half-space, or space with boundary . For example, a half-plane is a 2-manifold with boundary. An example is useful:

- The n-dimensional ball, denoted B^n, is defined as the set of points in n-dimensional space such that the distance to a central point is less than or equal to some fixed radius r. Note: Often the ball is called a disk and is denoted by D^n (so $D^n = B^n$). The nomenclature makes good sense in two dimensions, where what we usually call a disk is D^2. The one-dimensional ball is just an interval (one-dimensional segment), which is sometimes denoted $I = D^1 = B^1$.

Note that a boundary of a manifold may have disconnected parts. For example, the boundary of an interval (segment) in one-dimension $I = B^1$ is two disconnected points at its two ends.[2]

One can take Cartesian products of manifolds with boundaries, too. For example, consider the interval (or 1-ball) $I = B^1$, which we can think of as all the points on a line with $|x| \leq 1$. The Cartesian product $I \times I$ is described by two coordinates (x, y) where $|x| \leq 1$ and $|y| \leq 1$. This is a square including its interior. However, in topology we are only ever concerned with topological properties, and a square-with-interior can be continuously deformed into a circle-with-interior, or a 2-ball (2-disc), B^2.

[2]In the notation of Section 40.1.5, $\partial I = $ pt $\cup$ pt where pt means a point and here $\cup$ means the union of the two objects as described in Section 40.1.2.

- The same reasoning gives us $B^n \times B^m = B^{n+m}$.
- The cylinder (hollow tube) is expressed as $S^1 \times I$ (two coordinates, one periodic, one bounded on both sides).
- The solid donut is expressed as $D^2 \times S^1$ ($= B^2 \times S^1$), a 2-disk crossed with a circle.

40.1.5   Boundaries of Manifolds: $\mathcal{M}_1 = \partial \mathcal{M}_2$

The notation for boundary is ∂, so if $\mathcal{M}_1$ is the boundary of $\mathcal{M}_2$ we write $\mathcal{M}_1 = \partial \mathcal{M}_2$. The boundary $\partial \mathcal{M}$ has dimension one less than that of $\mathcal{M}$.

- The boundary of D^2, the two-dimensional disk is the one-dimensional circle S^1.
- More generally, the boundary of B^n (also written as D^n) is S^{n-1}.

It is an interesting topological principle that the boundary of a manifold is always a manifold without boundary. Or equivalently, the boundary of a boundary is the empty set. We sometimes write $\partial^2 = 0$ or $\partial(\partial \mathcal{M}) = \varnothing$ where $\varnothing$ means the empty set.

- The boundary of the three-dimensional ball B^3 is the sphere S^2. The sphere S^2 is a 2-manifold without boundary.

The operation of taking a boundary obeys the Liebnitz rule analogous to taking derivatives

$$\partial(\mathcal{M}_1 \times \mathcal{M}_2) = (\partial \mathcal{M}_1) \times \mathcal{M}_2 \ \cup \ \mathcal{M}_1 \times (\partial \mathcal{M}_2) \quad .$$

Let's see some examples of this:

- Consider the cylinder $S^1 \times I$. Using the above formula we find its boundary

$$\partial(S^1 \times I) = (\partial S^1) \times I \ \cup \ S^1 \times \partial I = S^1 \cup S^1 \quad .$$

To see how we get the final result here, start by examining the first term, $(\partial S^1) \times I$. Here, S^1 has no boundary so $\partial S^1 = \varnothing$ and therefore everything before the $\cup$ symbol is just the empty set. In the second term the boundary of the interval is just two points $\partial I = \text{pt} \cup \text{pt}$. Thus, the second term gives the final result $S^1 \cup S^1$, the union of two circles.

- Consider writing the disk (topologically) as the product of two intervals $B^2 = I \times I$. It is best to think of this Cartesian product as forming a filled-in square. Using the above formula we get

$$
\begin{aligned}
\partial B^2 &= \ \partial(I \times I) = (\text{pt} \cup \text{pt}) \times I \ \cup \ I \times (\text{pt} \cup \text{pt}) \\
&= \ (I \cup I) \cup (I \cup I) = \text{top} \cup \text{bottom} \cup \text{left} \cup \text{right} \\
&= \ \text{square (edges only)} = S^1 \quad .
\end{aligned}
$$

The formula gives the union of four segments denoting the edges of the square.

40.1.6   Removing Bits

The notation to remove a part of a manifold is $\backslash$. For example, we may start with $\mathbb{R}^3$ and remove a ball B^3, the remaining manifold is notated as $\mathbb{R}^3 \backslash B^3$. The boundary of this manifold is just the sphere on the surface of the B^3. That is, $\partial(\mathbb{R}^3 \backslash B^3) = S^2$.

40.2   **Groups**

A **group** G is a set of elements $g \in G$ along with an operation that we think of as multiplication. The set must be closed under this multiplication. So if $g_1, g_2 \in G$ then $g_3 \in G$ where

$$g_3 = g_1 g_2$$

whereby writing $g_1 g_2$ we mean multiply g_1 by g_2. Note: $g_1 g_2$ is not necessarily the same as $g_2 g_1$. If the group is always commutative (i.e. if $g_1 g_2 = g_2 g_1$ for all $g_1, g_2 \in G$), then we call the group **abelian**.[3] If there are at least some elements in the group where $g_1 g_2 \neq g_2 g_1$ then the group is called **nonabelian**.[4]

A group must always be associative

$$g_1(g_2 g_3) = (g_1 g_2) g_3 = g_1 g_2 g_3 \quad .$$

Within the group there must exist an **identity** element, which is sometimes[5] called e or I or 0 or 1. The identity element satisfies

$$ge = eg = g$$

for all elements $g \in G$. Each element of the group must also have an inverse which we write as g^{-1} with the property that

$$gg^{-1} = g^{-1}g = e \quad .$$

We often write $|G|$ to mean the number of elements in the group G. This is also known as the **order** of the group.

40.2.1   **Some Examples of Groups**

- The group of integers $\mathbb{Z}$ with the operation being addition. The identity element is 0. This group is abelian.
- The group $\{1, -1\}$ with the operation being the usual multiplication. This is also called the group $\mathbb{Z}_2$. The identity element is 1. We could have also written this group as $\{0, 1\}$ with the operation being the usual addition modulo 2, where here the identity is 0. This group is abelian.
- The group $\mathbb{Z}_N$ is the set of complex numbers $e^{2\pi i p/N}$ with p an integer (which can be chosen between 0 and $N-1$ inclusive) and the operation being multiplication. This is equivalent to the set of integers modulo N with the operation being addition. This group is abelian.
- The group of all possible permutations of N letters, which we write as S_N, is known as the **permutation group** or **symmetric group**.[6] This group is nonabelian. There are $N!$ elements in the group. Think of the elements of the group as being a one-to-one mapping from the set of the first N integers into itself.

[3]Named after Abel, the Norwegian mathematician who studied such groups in the early 1800s despite living in poverty and perishing at the young age of 26 from tuberculosis. The word "abelian" is usually not capitalized due to its ubiquitous use. There are a few similar words in English that are not capitalized despite being named after people, such as "galvanic."

[4]Apparently named after someone named Nonabel.

[5]It may seem inconvenient that the identity has several names. However, it is sometimes convenient. If we are thinking of the group of integers and the operation of addition, we want to use 0 as the identity. If we are thinking about the group $\{1, -1\}$ with the operation of usual multiplication, then it is convenient to write the identity as 1. For more abstract groups, e or I is often most natural.

[6]Physicists tend to call this the "permutation group," and we will follow this usage. More properly mathematicians insist that this should be called the "symmetric group."

- The simplest nonabelian group is S_3. In S_3, one of the elements is

$$x = \begin{cases} 1 & \to & 2 \\ 2 & \to & 1 \\ 3 & \to & 3 \end{cases}.$$

Another element is

$$r = \begin{cases} 1 & \to & 2 \\ 2 & \to & 3 \\ 3 & \to & 1 \end{cases}$$

where x stands for exchange (exchanges 1 and 2) and r stands for rotate. The multiplication operation xr is meant to mean do r first, then do x. (You should be careful to make sure your convention of ordering is correct. Here we choose a convention that we do the operation written furthest right first. You can choose either convention, but then you must stick to it! You will see both orderings in the literature!) So, if we start with the element 1, when we do r the element 1 gets moved to 2. Then when we do x the element 2 gets moved to 1. So in the product xr we have 1 getting moved back to position 1. In the end we have

$$xr = \begin{cases} 1 & \to & 1 \\ 2 & \to & 3 \\ 3 & \to & 1 \end{cases}.$$

Note that if we multiply the elements in the opposite order we get a different result (hence, this group is nonabelian)

$$rx = \begin{cases} 1 & \to & 3 \\ 2 & \to & 2 \\ 3 & \to & 1 \end{cases}.$$

It is easy to check that

$$x^2 = r^3 = e \quad \text{and} \quad xr = r^2 x \quad . \tag{40.1}$$

There are a total of 6=3! elements in the group, which we can list as e, r, r^2, x, xr, xr^2. All other products can be reduced to one of these six possibilities using Eqs. 40.1.

Product of Groups

Two groups G and H can be multiplied to form $G \times H$ in an obvious way. An element of $G \times H$ is a pair (g, h) with $g \in G$ and $h \in H$. The multiplication in $G \times H$ is given by $(g_1, h_1)(g_2, h_2) = (g_1 g_2, h_1 h_2)$.

All Abelian Groups

Note that if n and m are coprime (have no common factors) then $\mathbb{Z}_n \times \mathbb{Z}_m$ is the same as $\mathbb{Z}_{n \times m}$.

The so-called *fundamental theorem of finite abelian groups*[7] tells us

[7] This results was more or less known by Gauss in some form in 1801. It is also often credited to Kronecker in 1870— he did more than invent the Kronecker delta!

that any abelian group can be written as

$$\mathbb{Z}_{N_1} \times \mathbb{Z}_{N_2} \times \ldots \times \mathbb{Z}_{N_p}$$

for some number of factors p and each N_i must be a power of a prime number. For example, if we have an abelian group of order $100 = 5 \times 5 \times 2 \times 2$, there are only four possibilities of what it can be: $\mathbb{Z}_{25} \times \mathbb{Z}_4$ or $\mathbb{Z}_{25} \times \mathbb{Z}_2 \times \mathbb{Z}_2$ or $\mathbb{Z}_5 \times \mathbb{Z}_5 \times \mathbb{Z}_4$ or $\mathbb{Z}_5 \times \mathbb{Z}_5 \times \mathbb{Z}_2 \times \mathbb{Z}_2$.

40.2.2 More Features of Groups

A **subgroup** is a subset of elements of a group which themselves form a group. For example, the integers under addition form a group. The even integers under addition are a subgroup of the integers under addition.

The **centralizer** of an element $g \in G$, often written as $Z(g)$, is the set of all elements of the group G that commute with g. That is, $h \in Z(g)$ iff $hg = gh$. Note that this set forms a subgroup (proof is easy!). For an abelian group G the centralizer of any element is the entire group G.

The **center** of a group, often written as $Z(G)$, is the subgroup of G containing all elements $h \in G$ such that $Z(h)$, the centralizer of h, is the entire group G. In other words, $Z(G)$ is the set of all elements that commute with the entire group.

A **conjugacy class** of an element $g \in G$ is defined as the set of elements $g' \in G$ such that $g' = hgh^{-1}$ for some element $h \in G$. We sometimes write C_g for the conjugacy class of g, and $|C_g|$ is the number of elements in the conjugacy class.

A useful result of group theory is that the number of elements in the conjugacy class of an element times the number of elements in the centralizer of the element gives the total number of elements in the group[8]

$$|Z(g)| \, |C_g| = |G| \quad . \tag{40.2}$$

Example: S_3. Above we listed some of the properties of the group S_3. S_3 has several subgroups:

- The group containing the identity element e alone.
- The group containing $\{e, x\}$.
- The group containing $\{e, r, r^2\}$.
- The group S_3 itself (which is not a "proper" subgroup).

The group has three conjugacy classes

- The identity element e.
- The rotations $\{r, r^2\}$.
- The reflections $\{x, xr, xr^2\}$.

We can check that conjugating any element in any class gives another element within the same class. For example, consider the element x and conjugate it with the element r. We have $rxr^{-1} = (xr^2)r^{-1} = xr$ (see Eq. 40.1), which is in the same conjugacy class as x.

[8] Often this theorem is stated as a simple result of the so-called orbit-stabilizer theorem—which is very closely related to Burnside's lemma (see Section 31.5). However, it is actually fairly easy to prove the theorem directly as well: for any given $y \in C_g$ let us define $p(y)$ to be a given particular choice of a group element such that $p(y) \, g \, p(y)^{-1} = y$. Then given any $h \in G$ we can uniquely write $h = p(y)z$ for some $y = hgh^{-1} \in C_g$ and some $z \in Z(g)$.

40.2.3 Lie Groups and Lie Algebras

[9]Pronounced "Lee," named after Sophus Lie, also a Norwegian mathematician of the 1800s. Like in ski jumping, Norway seems to punch above its weight in the theory of groups.

A **Lie group**[9] is a group that is also a manifold. Roughly, a group with a continuous (rather than discrete) set of elements. Examples include:

- The group of invertible $n \times n$ complex matrices. We call this group $GL(n, \mathbb{C})$. Here, GL stands for "general linear." The identity is the usual identity matrix. By definition all elements of the group are invertable.

- The group of invertible $n \times n$ *real* matrices. We call this group $GL(n, \mathbb{R})$.

- The group, $SU(2)$, the set of 2×2 unitary matrices with unit determinant. In this case, the fact that this is also a manifold can be made particularly obvious. We can write all $SU(2)$ matrices as

$$\begin{pmatrix} x_1 + ix_2 & -x_3 + ix_4 \\ x_3 + ix_4 & x_1 - ix_2 \end{pmatrix}$$

with all x_j real numbers with the constraint that $x_1^2 + x_2^2 + x_3^2 + x_4^2 = 1$. Obviously the set of four coordinates (x_1, x_2, x_3, x_4) with the unit magnitude constraint describes the manifold S^3.

- $SU(N)$, the group of unitary N by N matrices of determinant one is a Lie group

- $SO(N)$, the group of real rotation matrices in N dimensions is a Lie group.

- The vector space $\mathbb{R}^n$ with the operation being addition of vectors, is a Lie group.

Note that certain Lie groups are known as "simple" because as manifolds they have no boundaries and no nontrivial limit points (e.g. $GL(n)$ is not simple because there is a nontrivial limit—you can continuously approach matrices that have determinant zero, or are not invertable, and are therefore not part of the group). The set of simple Lie groups (including $SU(N)$ and $SO(N)$ and just a few others) is extremely highly studied.

[10]A slightly more rigorous definition is that a Lie algebra is an algebra of elements $u, v, w, \ldots$, which can be added with coefficients a, b, c to give $X = au + bv + cw + \ldots$ where we have a commutator $[\cdot, \cdot]$, which satisfies $[X, X] = 0$ for all X as well as bi-linearity $[au + bv, X] = a[u, X] + b[v, X]$ and similarly $[X, au + bv] = a[X, u] + b[X, v]$ for all X, a, b, u, v, and finally we must have the Jacobi identity $[[X, Y], Z] + [[Y, Z], X] + [[Z, X], Y] = 0$.

A **Lie Algebra** is the algebra generated by elements infinitesimally close to the identity in a Lie group.[10] For matrix-valued Lie groups G, we can write any element $g \in G$ sufficiently close to the identity as

$$g = e^X = 1 + X + (X)^2/2 + \ldots$$

where X is an element of the corresponding Lie algebra (make it have small amplitude such that g is infinitesimally close to the identity). Conventionally, if a Lie group is denoted as G the corresponding Lie algebra is denoted $\mathfrak{g}$.

- For the Lie group $SU(2)$, we know that a general element can be written as $g = \exp(i\mathbf{n} \cdot \boldsymbol{\sigma})$ where $\mathbf{n}$ is a real three-dimensional vector and $\boldsymbol{\sigma}$ are the Pauli matrices. In this case, $i\sigma_x$, $i\sigma_y$, and $i\sigma_z$

are the three generators of the Lie algebra $\mathfrak{su}(2)$ (in the so-called fundamental representation).

- For the Lie group $GL(n, \mathbb{R})$ the corresponding Lie algebra $\mathfrak{gl}(n, \mathbb{R})$ is just the algebra of $n \times n$ real matrices.

40.2.4   Representations of Groups

A **representation** is a group homomorphism. This means it is a mapping from one group to another, which preserves multiplication. We are concerned with the most common type of representation, which is a homomorphism into the general linear group, that is, the group of matrices. Almost always we will work with complex matrices. Thus, an n-dimensional representation is a mapping ρ to n-dimensional complex matrices $\rho : G \to GL(n, \mathbb{C})$ preserving multiplication. That is, $\rho(g_1)\rho(g_2) = \rho(g_1 g_2)$ for all $g_1, g_2 \in G$.

Typically in quantum mechanics we are concerned with representations that are unitary, that is, $\rho(g)$ is a complex unitary matrix of some dimension. (In case you don't remember, a unitary matrix U has the property that $UU^\dagger = U^\dagger U = \mathbf{1}$).

A representation is reducible if the representing matrices can be decomposed into block-diagonal form by a basis transformation.[11] That is, ρ is reducible if we can choose a basis such that $\rho = \rho_1 \oplus \rho_2$ for two representations ρ_1 and ρ_2. An irreducible representation is one that cannot be reduced.

The dimensions of the irreducible representations follow the law

$$\sum_R \dim(R)^2 = |G| \tag{40.3}$$

where $|G|$ is the number of elements in the group and the sum is over irreducible representations.

An amazing fact from representation theory of discrete groups is that the number of irreducible representations of a group is equal to the number of distinct conjugacy classes.

Schur's second lemma is a very useful result stating that if a matrix A commutes with every $\rho^R(g)$ for all elements g in the group for an irreducible representation R, then A is proportional to the unit matrix. In particular, this means that if an element h commutes with all elements of the group then $\rho^R(h)$ is a complex phase times the identity. A corollary of this is that all of the irreducible representations of abelian groups are one-dimensional.

Following from Shur's lemma we can, for example, write all the irreducible representations of the group $\mathbb{Z}_N$ in the following way. Let the group $\mathbb{Z}_N$ be written as $g = 0, \dots, (N-1)$ with the operation of addition modulo N. Let $\omega = e^{2\pi i/N}$. There are exactly N irreps of this group which we label as ω^p for $p = 0, \dots, (N-1)$. These irreps are given by

$$\rho^p : g \to (\omega^p)^g \quad . \tag{40.4}$$

[11]Strictly speaking, this is the definition of *decomposable* not *reducible*. However, since our matrices are complex, we can use Maschke's theorem to establish that decomposable and reducible are actually equivalent.

Orthogonality and Characters

Irreducible unitary representation matrices satisfy a beautiful orthogonality relationship known as the **grand orthogonality theorem** (or Schur orthogonality)

$$\sum_{g \in G} [\rho^R(g^{-1})]_{mn} [\rho^{R'}(g)]_{pq} =$$

$$\sum_{g \in G} [\rho^R(g)]^*_{nm} [\rho^{R'}(g)]_{pq} = \frac{\delta_{np} \delta_{mq} \delta_{RR'} |G|}{\dim(R)} \qquad (40.5)$$

where the superscript R indicates a particular irreducible representation, the subscripts are the matrix elements of the ρ matrix, $\dim(R)$ is the dimension of the representation R, and $|G|$ is the total number of elements in the group.

A **character** is the trace of a representation matrix

$$\chi_R(g) = \text{Tr}[\rho^R(g)] \qquad (40.6)$$

where the superscript R indicates we are considering a particular representation R. Because of the cyclic property of the trace $\text{Tr}[ab] = \text{Tr}[ba]$ the character is the same for all elements of a conjugacy class. One can find tables of characters for different groups in any book on group theory or on the web.

Representation theory of groups is a huge subject, but we won't discuss it further here!

40.3 Fundamental Group $\Pi_1(\mathcal{M})$

A powerful tool of topology is the idea of the fundamental group of a manifold $\mathcal{M}$, which is often called the first homotopy group, or $\Pi_1(\mathcal{M})$. This is essentially the group of topologically different paths through the manifold starting and ending at the same point.

First, we choose a point in the manifold. Then, we consider a path through the manifold that starts and ends at the same point. Any other path that can be continuously deformed into this path (without changing the starting point or ending point) is deemed to be topologically equivalent (or homeomorphic, or in the same equivalence class). We only want to keep one representative of each class of topologically distinct paths.

These topologically distinct paths form a group. As one might expect, the inverse of a path (always starting and ending at the same point) is given by following the same path in a backward direction. Multiplication of two paths is achieved by following one path and then following the other to make a longer path.

40.3.1 Examples of Fundamental Groups

- If the manifold is a circle S^1, the topologically distinct paths (starting and ending at the same point) can be described by the number n of clockwise wrappings the path makes around the circle before coming back to its starting point (note n can be 0 or negative as well). Thus, the elements of the fundamental group are indexed by a single integer. We write $\Pi_1(S^1) = \mathbb{Z}$.

- If the manifold is a torus $S^1 \times S^1$, the topologically distinct paths can be described by two integers indicating the number of times the path winds around each cycle. We write $\Pi_1(S^1 \times S^1) = \mathbb{Z} \times \mathbb{Z}$.

In fact, it is easy to prove that $\Pi_1(\mathcal{M}_1 \times \mathcal{M}_2) = \Pi_1(\mathcal{M}_1) \times \Pi_1(\mathcal{M}_2)$.

- A fact known to most physicists is that the group of rotations of three-dimensional space $SO(3)$ is not simply connected—a 2π rotation (which seems trivial) cannot be continuously deformed to the trivial rotation, whereas a 4π rotation can be continuously deformed to the trivial rotation.[12] Correspondingly, the fundamental group of $SO(3)$ is the group with two elements $\Pi_1(SO(3)) = \mathbb{Z}_2$.

[12]This is the origin of half-odd integer angular momenta.

Chapter Summary

- **Manifolds** are locally like Euclidean space. Examples include sphere S^2, circle S^1, torus surface $T^2 = S^1 \times S^1$, etc. Manifolds can also have boundaries, like a two-dimensional disk B^2 (or D^2) bounded by a circle.

- **Groups** are mathematical sets with an operation, an identity, and an inverse. Important examples include $\mathbb{Z}$ the integers under addition, $\mathbb{Z}_N$ the integers mod N under addition, the permutation group on N elements S_N, and Lie groups such as $SU(2)$, which are also manifolds at the same time as being groups.

- **The Fundamental Group** of a manifold is the group of topologically different paths through the manifold starting and ending at the same point.

- For background on more advanced mathematics used by physicists, including some topological ideas, see Stone and Goldbart [2009] and Nakahara [2003].

- Good resources on group theory include Hamermesh [1989] and Zee [2016].

References

D. Aasen, P. Fendley, and R. S. K. Mong. Topological defects on the lattice: Dualities and degeneracies. *arXiv:2008.08598*, 2020. [508]

A. Achúcarro and P. Townsend. A Chern–Simons action for three-dimensional anti-de sitter supergravity theories. *Physics Letters B*, 180(1):89–92, 1986. [69, 71, 72]

C. C. Adams. *The Knot Book: An Elementary Introduction to the Mathematical Theory of Knots*. W. H. Freeman and Company, 1994. [17]

I. Affleck, T. Kennedy, E. H. Lieb, and H. Tasaki. Rigorous results on valence-bond ground states in antiferromagnets. *Physical Review Letters*, 59:799–802, 1987. [529]

I. Affleck, T. Kennedy, E. H. Lieb, and H. Tasaki. Valence bond ground states in isotropic quantum antiferromagnets. *Communications in Mathematical Physics*, 115(3):477–528, 1988. [529]

D. Aharonov and I. Arad. The BQP-hardness of approximating the Jones polynomial. *New Journal of Physics*, 13(3):035019, 2011. [13]

D. Aharonov, V. Jones, and Z. Landau. A polynomial quantum algorithm for approximating the jones polynomial. *Algorithmica*, 55(3):395–421, 2009. [13]

Y. Aharonov and D. Bohm. Significance of electromagnetic potentials in the quantum theory. *Physical Review*, 115:485–491, 1959. [39]

Y. Aharonov and A. Casher. Topological quantum effects for neutral particles. *Physical Review Letters*, 53:319–321, 1984. [41]

R. Ainsworth and J. K. Slingerland. Topological qubit design and leakage. *New Journal of Physics*, 13(6):065030, 2011. [144]

S. Akbulut. *4 Manifolds*. Oxford Graduate Texts in Mathematics. Oxford University Press, 2016. [342]

D. Altschuler and A. Coste. Invariants of three-manifolds from finite group cohomology. *Journal of Geometry and Physics*, 11(1):191–203, 1993. [326]

M. Amy. *Algorithms for the Optimization of Quantum Circuits*. Master's thesis, University of Waterloo, 2013. [134]

C. K. Andersen, A. Remm, S. Lazar, S. Krinner, N. Lacroix, G. J. Norris, M. Gabureac, C. Eichler, and A. Wallraff. Repeated quantum error detection in a surface code. *Nature Physics*, 16(8):875–880, 2020. [373, 561]

T. I. Andersen, Google AI, and 100 Collaborators. Non-abelian braiding of graph vertices in a superconducting processor. *Nature*, 2023. [545, 561]

J. F. Annette. *Superconductivity, Superfluids and Condensates*. Oxford University Press, 2004. [345, 559]

E. Ardonne and K. Schoutens. New class of non-abelian spin-singlet quantum Hall states. *Physical Review Letters*, 82:5096–5099, 1999. [555]

E. Ardonne and J. Slingerland. Clebsch–Gordan and $6j$-coefficients for rank 2 quantum groups. *Journal of Physics A: Mathematical and Theoretical*, 43(39):395205, 2010. [290, 295, 572]

D. Arovas, J. R. Schrieffer, and F. Wilczek. Fractional statistics and the quantum Hall effect. *Physical Review Letters*, 53:722–723, 1984. [32, 552]

M. Atiyah. Topological quantum field theories. *Publications Mathématiques de l'Institut des Hautes Scientifiques*, 68, 1988. [74, 89]

M. Atiyah. On framings of 3-manifolds. *Topology*, 29(1):1, 1990. [61, 336]

M. Atiyah. An introduction to topological quantum field theories. In *Turkish Journal of Mathematics; Proceedings of 5th Gokova Geometry-Topology conference, 1996*, 1997. [89]

A. F. Bais, B. J. Schroers, and J. K. Slingerland. Hopf symmetry breaking and confinement in (2+1)-dimensional gauge theory. *Journal of High Energy Physics*, 2003(05):068–068, 2003. [468]

F. Bais. Flux metamorphosis. *Nuclear Physics B*, 170(1):32, 1980. [34]

F. A. Bais and J. K. Slingerland. Condensate-induced transitions between topologically ordered phases. *Physical Review B*, 79:045316, 2009. [345, 347, 360]

F. A. Bais, J. K. Slingerland, and S. M. Haaker. Theory of topological edges and domain walls. *Physical Review Letters*, 102:220403, 2009. [360]

B. Bakalov and A. Kirillov. *Lectures on Tensor Categories and Modular Functors*. University Lecture Series, Vol 21. American Mathematical Society, 2001. [184, 185]

A. C. Balram. A non-abelian parton state for the $\nu = 2 + 3/8$ fractional quantum Hall effect. *SciPost Phys.*, 10:83, 2021. [555]

A. C. Balram, J. K. Jain, and M. Barkeshli. z_n superconductivity of composite bosons and the 7/3 fractional quantum Hall effect. *Physical Review Research*, 2: 013349, 2020. [555]

P. Bantay. The Frobenius–Schur indicator in conformal field theory. *Physics Letters B*, 394(1):87–88, 1997. [233]

D. Bar-Natan and E. Witten. Perturbative expansion of Chern–Simons theory with noncompact gauge group. *Communications in Mathematical Physics*, 141(2):423–440, 1991. [71]

M. Baraban, N. E. Bonesteel, and S. H. Simon. Resources required for topological quantum factoring. *Physical Review A*, 81:062317, 2010. [138]

M. Bärenz. *Topolgical State Sum Models in Four Dimensions, Half-Twists and Their Applications*. PhD thesis, University of Nottingham, 2017. [190]

M. Barkeshli, P. Bonderson, M. Cheng, and Z. Wang. Symmetry fractionalization, defects, and gauging of topological phases. *Physical Review B*, 100:115147, 2019. [541, 542, 546, 547, 548, 561]

J. W. Barrett and L. Crane. A Lorentzian signature model for quantum general relativity. *Classical and Quantum Gravity*, 17(16):3101–3118, 2000. [326]

J. W. Barrett and B. W. Westbury. Invariants of piecewise-linear 3-manifolds. *Transactions of the American Mathematical Society*, 348:3997–4022, 1996. [318, 323, 326]

B. Bartlett. *Categorical Aspects of Topological Quantum Field Theories*. PhD thesis, Utrecht University, 2005. [89]

B. Bartlett. Fusion categories via string diagrams. *Communications in Contemporary Mathematics*, 18(05):1550080, 2016. [184, 195]

H. Bartolomei, M. Kumar, R. Bisognin, A. Marguerite, J.-M. Berroir, E. Bocquillon, B. Plaçais, A. Cavanna, Q. Dong, U. Gennser, Y. Jin, and G. Fève. Fractional statistics in anyon collisions. *Science*, 368(6487):173–177, 2020. [552]

V. Batagelj and A. Mrvar. Some analyses of Erdos' collaboration graph. *Social Networks*, 22(2):173–186, 2000. [137]

A. Bauer, J. Eisert, and C. Wille. Towards a mathematical formalism for classifying phases of matter. *arXiv:1903.05413*, 2019. [564]

S. Beigi, P. W. Shor, and D. Whalen. The quantum double model with boundary: Condensations and symmetries. *Communications in Mathematical Physics*, 306 (3):663–694, 2011. [468]

E. J. Bergholtz and Z. Liu. Topological flat band models and fractional chern insulators. *International Journal of Modern Physics B*, 27(24):1330017, 2013. [555]

H. Bombin. Topological order with a twist: Ising anyons from an abelian model. *Physical Review Letters*, 105:030403, 2010. [545, 548, 561]

H. Bombin and M. A. Martin-Delgado. Family of non-abelian kitaev models on a lattice: Topological condensation and confinement. *Physical Review B*, 78:115421, 2008. [468]

P. Bonderson, M. Freedman, and C. Nayak. Measurement-only topological quantum computation. *Physical Review Letters*, 101:010501, 2008. [137]

P. Bonderson, C. Delaney, C. Galindo, E. C. Rowell, A. Tran, and Z. Wang. On invariants of modular categories beyond modular data. *Journal of Pure and Applied Algebra*, 223(9):4065–4088, 2019. [571]

P. H. Bonderson. *Non-Abelian Anyons and Interferometry*. PhD thesis, California Institute of Technology, 2007. [84, 110, 113, 120, 129, 168, 170, 182, 194, 230, 259, 273, 572]

N. E. Bonesteel, L. Hormozi, G. Zikos, and S. H. Simon. Braid topologies for quantum computation. *Physical Review Letters*, 95(14):140503, 2005. [140, 144]

S. Bravyi. Universal quantum computation with the $\nu = 5/2$ fractional quantum Hall state. *Physical Review A*, 73(4):042313, 2006. [138]

S. Bravyi, M. B. Hastings, and S. Michalakis. Topological quantum order: Stability under local perturbations. *Journal of Mathematical Physics*, 51(9):093512, 2010. [410]

M. J. Bremner, C. M. Dawson, J. L. Dodd, A. Gilchrist, A. W. Harrow, D. Mortimer, M. A. Nielsen, and T. J. Osborne. Practical scheme for quantum computation with any two-qubit entangling gate. *Physical Review Letters*, 89(24):247902, 2002. [135]

G. K. Brennen, M. Aguado, and J. I. Cirac. Simulations of quantum double models. *New Journal of Physics*, 11(5):053009, 2009. [468, 561]

C. Broholm, R. J. Cava, S. A. Kivelson, D. G. Nocera, M. R. Norman, and T. Senthil. Quantum spin liquids. *Science*, 367(6475):eaay0668, 2020. [557, 559]

P. Bruillard. Rank 4 premodular categories. *New York Journal of Mathematics*, 22, 2016. [238]

P. Bruillard and C. M. Ortiz-Marrero. Classification of rank 5 premodular categories. *Journal of Mathematical Physics*, 59(1):011702, 2018. [239]

P. Bruillard, S.-H. Ng, E. Rowell, and Z. Wang. On classification of modular categories by rank. *International Mathematics Research Notices*, 2016, 2015. [238]

P. Bruillard, C. Galindo, T. Hagge, S.-H. Ng, J. Y. Plavnik, E. C. Rowell, and Z. Wang. Fermionic modular categories and the 16-fold way. *Journal of Mathematical Physics*, 58(4):041704, 2017. [234, 235, 239, 572]

P. Bruillard, C. Galindo, S.-H. Ng, J. Y. Plavnik, E. C. Rowell, and Z. Wang. Classification of super-modular categories by rank. *Algebras and Representation Theory*, 23(3):795–809, 2020. [234, 235, 239, 572]

J.-L. Brylinski and R. Brylinski. Universal quantum gates. *Mathematics of Quantum Computation, Comput. Math. Series*, page 101–116, 2002. [135]

O. Buerschaper and M. Aguado. Mapping Kitaev's quantum double lattice models to Levin and Wen's string-net models. *Physical Review B*, 80:155136, 2009. [278, 469, 485]

A. Bullivant and C. Delcamp. Tube algebras, excitations statistics and compactification in gauge models of topological phases. *Journal of High Energy Physics*, 2019 (10):216, 2019. [423, 466, 469]

A. Bullivant, M. Calçada, Z. Kádár, P. Martin, and J. a. F. Martins. Topological phases from higher gauge symmetry in $3 + 1$ dimensions. *Physical Review B*, 95: 155118, 2017. [469]

A. Bullivant, M. Calçada, Z. Kádár, J. F. Martins, and P. Martin. Higher lattices, discrete two-dimensional holonomy and topological phases in $(3 + 1)$d with higher gauge symmetry. *Reviews in Mathematical Physics*, 32(04):2050011, 2020. [469]

N. Bultinck, M. Mariën, D. Williamson, M. Şahinoğlu, J. Haegeman, and F. Verstraete. Anyons and matrix product operator algebras. *Annals of Physics*, 378: 183–233, 2017. [508, 564]

F. Burnell. Anyon condensation and its applications. *Annual Review of Condensed Matter Physics*, 9(1):307–327, 2018. [359, 361, 548]

F. Burnell and S. H. Simon. Space–time geometry of topological phases. *Annals of Physics*, 325(11):2550–2593, 2010. [342]

F. J. Burnell and S. H. Simon. A Wilson line picture of the Levin–Wen partition functions. *New Journal of Physics*, 13(6):065001, 2011. [342]

F. J. Burnell, S. H. Simon, and J. K. Slingerland. Condensation of achiral simple currents in topological lattice models: Hamiltonian study of topological symmetry breaking. *Physical Review B*, 84:125434, 2011. [352, 508]

F. J. Burnell, S. H. Simon, and J. K. Slingerland. Phase transitions in topological lattice models via topological symmetry breaking. *New Journal of Physics*, 14(1): 015004, 2012. [352]

F. J. Burnell, X. Chen, L. Fidkowski, and A. Vishwanath. Exactly soluble model of a three-dimensional symmetry-protected topological phase of bosons with surface topological order. *Physical Review B*, 90:245122, 2014. [536]

P. H. Butler. *Point Group Symmetry Applications*. Plenum Press, 1981. [278, 283, 284]

J. Cai et al. Signatures of fractional quantum anomalous Hall states in twisted $MoTe_2$ bilayer. *arXiv:2304.08470*, 2023. [555]

J. Cano, M. Cheng, M. Mulligan, C. Nayak, E. Plamadeala, and J. Yard. Bulk-edge correspondence in $(2 + 1)$-dimensional abelian topological phases. *Physical Review B*, 89:115116, 2014. [57, 228, 230, 256, 260, 261]

A. Cappelli, M. Huerta, and G. R. Zemba. Thermal transport in chiral conformal theories and hierarchical quantum Hall states. *Nuclear Physics B*, 636(3):568–582, 2002. [229]

S. Carlip. Quantum gravity in $2 + 1$ dimensions: The case of a closed universe. *Living Reviews in Relativity*, 8(1):1, 2005. [70, 71, 72]

R. G. Chambers. Shift of an electron interference pattern by enclosed magnetic flux. *Physical Review Letters*, 5:3–5, 1960. [39]

X. Chen, Z.-X. Liu, and X.-G. Wen. Two-dimensional symmetry-protected topological orders and their protected gapless edge excitations. *Physical Review B*, 84: 235141, 2011. [530, 535]

X. Chen, Z.-C. Gu, Z.-X. Liu, and X.-G. Wen. Symmetry protected topological orders and the group cohomology of their symmetry group. *Physical Review B*, 87:155114, 2013. [273, 530, 536, 539]

Y. Chen, F. Wilczek, E. Witten, and B. Halperin. On anyon superconductivity. *International Journal of Modern Physics B*, 03(07):1001–1067, 1989. [34, 564]

Z. Chen, Google, AI, and 100-Authors. Exponential suppression of bit or phase errors with cyclic error correction. *Nature*, 595(7867):383–387, 2021. [373, 561]

M. Cheng, Z.-C. Gu, S. Jiang, and Y. Qi. Exactly solvable models for symmetry-enriched topological phases. *Physical Review B*, 96:115107, 2017. [547]

S. Chulliparambil, U. F. P. Seifert, M. Vojta, L. Janssen, and H.-H. Tu. Microscopic models for kitaev's sixteenfold way of anyon theories. *Physical Review B*, 102: 201111, 2020. [557]

J. I. Cirac, D. Pérez-García, N. Schuch, and F. Verstraete. Matrix product states and projected entangled pair states: Concepts, symmetries, theorems. *Reviews of Modern Physics*, 93:045003, 2021. [513, 564]

L. W. Clark, N. Schine, C. Baum, N. Jia, and J. Simon. Observation of laughlin states made of light. *Nature*, 582(7810):41–45, 2020. [18, 556]

S. Coleman. *Aspects of Symmetry: Selected Erice Lectures*. Cambridge University Press, 1985. [63]

A. Coste, T. Gannon, and P. Ruelle. Finite group modular data. *Nuclear Physics B*, 581(3):679–717, 2000. [468]

L. Crane and D. Yetter. A categorical construction of 4-d topological quantum field theories. In L. H. Kauffman and R. Baadhio, editors, *Quantum Topology*. World Scientific, 1993. [326]

D. Creamer. A computational approach to classifying low rank modular categories. *arXiv:1912.02269*, 2019. [238]

S. X. Cui and Z. Wang. Universal quantum computation with metaplectic anyons. *Journal of Mathematical Physics*, 56(3):032202, 2015. [137]

S. X. Cui, S.-M. Hong, and Z. Wang. Universal quantum computation with weakly integral anyons. *Quantum Information Processing*, 14(8):2687–2727, 2015. [468]

S. X. Cui, D. Ding, X. Han, G. Penington, D. Ranard, B. C. Rayhaun, and Z. Shang-nan. Kitaev's quantum double model as an error-correcting code. *Quantum*, 4: 331, 2020. [468]

S. Das Sarma and A. Pinczuk, editors. *Perspectives in Quantum Hall Effects : Novel Quantum Liquids in Low-Dimensional Semiconductor Structures*. Wiley, New York, 1997. [563]

N. D. H. Dass. Dirac and the path integral. 2020. [24]

G. Dauphinais, L. Ortiz, S. Varona, and M. A. Martin-Delgado. Quantum error correction with the semion code. *New Journal of Physics*, 21(5):053035, 2019. [476, 482]

C. M. Dawson and M. A. Nielsen. The Solovay–Kitaev algorithm. *Quantum Information and Computation*, 6:81–95, 2006. [135, 140]

R. de Picciotto, M. Reznikov, M. Heiblum, V. Umansky, G. Bunin, and D. Mahalu. Direct observation of a fractional charge. *Nature*, 389(6647):162–164, 1997. [552]

M. D. F. de Wild Propitius. *Topological Interactions in Broken Gauge Theories*. PhD thesis, Universiteit van Amsterdam, 1995. [273, 325, 468]

C. Delaney and A. Tran. A systematic search of knot and link invariants beyond modular data. *https://arxiv.org/abs/1806.02843*, 2018. [571]

C. Delcamp. Excitation basis for (3+1)d topological phases. *Journal of High Energy Physics*, 2017(12):128, 2017. [466, 468, 469]

C. Delcamp, B. Dittrich, and A. Riello. Fusion basis for lattice gauge theory and loop quantum gravity. *Journal of High Energy Physics*, 2017(2):61, 2017. [468]

P. Deligne. Catégories tensorielles. *Moscow Mathematical Journal*, 2.2:227–248, 2002. [281]

M. DeMarco and X.-G. Wen. Compact $u^k(1)$ Chern–Simons theory as a local bosonic lattice model with exact discrete 1-symmetries. *Physical Review Letters*, 126: 021603, 2021. [61]

P. Di Francesco, P. Mathieu, and D. Sénéchal. *Conformal Field Theory*. Springer, New York, 1997. [229, 294, 356, 358, 522, 564, 572]

D. Dieks. Communication by EPR devices. *Physics Letters A*, 92(6):271–272, 1982. [367]

R. Dijkgraaf and E. Witten. Topological gauge theories and group cohomology. *Communications in Mathematical Physics*, 129:393–429, 1990. [273, 323, 325, 326]

R. Dijkgraaf, V. Pasquier, and P. Roche. Quasi hopf algebras, group cohomology and orbifold models. *Nuclear Physics B–Proceedings Supplements*, 18(2):60–72, 1991. [468]

S. Dong, E. Fradkin, R. G. Leigh, and S. Nowling. Topological entanglement entropy in chern-simons theories and quantum Hall fluids. *Journal of High Energy Physics*, 2008(05):016–016, 2008. [523]

S. Doplicher, R. Haag, and J. E. Roberts. Local observables and particle statistics i. *Communications in Mathematical Physics*, 23(3):199–230, 1971. [34]

S. Doplicher, R. Haag, and J. E. Roberts. Local observables and particle statistics ii. *Communications in Mathematical Physics*, 35(1):49–85, 1974. [34]

C. L. Douglas and D. J. Reutter. Fusion 2-categories and a state-sum invariant for 4-manifolds. *https://arxiv.org/abs/1812.11933*, 2018. [327]

J. Dubail, N. Read, and E. H. Rezayi. Edge-state inner products and real-space entanglement spectrum of trial quantum Hall states. *Physical Review B*, 86(24): 1–32, 2012. [513]

G. Dunne. Aspects of Chern–Simons theory. In A. Cometet, T. Jolicoeur, and S. Ouvry, editors, *Topological aspects of low dimensional systems*. Les Houches— École d'Été de Physique Théorique, vol. 69. Springer, 1998. [63]

S. Dusuel, M. Kamfor, R. Orús, K. P. Schmidt, and J. Vidal. Robustness of a perturbed topological phase. *Physical Review Letters*, 106:107203, 2011. [408]

B. Dutta, V. Umansky, M. Banerjee, and M. Heiblum. Isolated ballistic non-abelian interface channel. *Science*, 377(6611):1198–1201, 2022. [554]

W. Ehrenberg and R. E. Siday. The refractive index in electron optics and the principles of dynamics. *Proceedings of the Physical Society. Section B*, 62(1):8, 1949. [39]

A. Einstein, B. Podolsky, and N. Rosen. Can quantum-mechanical description of physical reality be considered complete? *Physical Review*, 47:777–780, 1935. [514]

I. S. Eliëns, J. C. Romers, and F. A. Bais. Diagrammatics for bose condensation in anyon theories. *Physical Review B*, 90:195130, 2014. [168, 361]

B. Estienne, N. Regnault, and B. A. Bernevig. Correlation lengths and topological entanglement entropies of unitary and nonunitary fractional quantum Hall wave functions. *Physical Review Letters*, 114:186801, 2015. [525]

P. Etingof, D. Nikshych, and V. Ostrik. On fusion categories. *Annals of Mathematics*, 162:581–642, 2005. [117, 194, 230]

P. Etingof, S. Gelaki, D. Nikshych, and V. Ostrik. *Tensor Categories*. American Mathematical Society, 2015. [225, 227]

B. Farb and D. Margalit. *A Primer on Mapping Class Groups (PMS-49)*. Princeton University Press, 2012. [89, 227]

R. P. Feynman and A. R. Hibbs. *Quantum Mechanics and Path Integrals*. McGraw Hill, 1965. Reprinted 2005, Dover. [36]

L. Fidkowski and A. Kitaev. Topological phases of fermions in one dimension. *Physical Review B*, 83:075103, 2011. [530]

B. Field and T. Simula. Introduction to topological quantum computation with non-abelian anyons. *Quantum Science and Technology*, 3(4):045004, 2018. [134, 140, 147]

S. T. Flammia, A. Hamma, T. L. Hughes, and X.-G. Wen. Topological entanglement rényi entropy and reduced density matrix structure. *Physical Review Letters*, 103: 261601, 2009. [514, 517]

K. Flensberg, F. von Oppen, and A. Stern. Engineered platforms for topological superconductivity and majorana zero modes. *Nature Reviews Materials*, 6(10): 944–958, 2021. [560]

A. G. Fowler. Constructing arbitrary steane code single logical qubit fault-tolerant gates. *Quantum Info. Comput.*, 11(9–10):867–873, 2011. [134]

A. G. Fowler, M. Mariantoni, J. M. Martinis, and A. N. Cleland. Surface codes: Towards practical large-scale quantum computation. *Physical Review A*, 86:032324, 2012. [138, 373]

E. Fradkin and S. H. Shenker. Phase diagrams of lattice gauge theories with higgs fields. *Physical Review D*, 19:3682–3697, 1979. [405]

K. Fredenhagen, K. H. Rehren, and B. Schroer. Superselection sectors with braid group statistics and exchange algebras. *Communications in Mathematical Physics*, 125:201–226, 1989. [34]

M. Freedman, C. Nayak, K. Shtengel, K. Walker, and Z. Wang. A class of P,T-invariant topological phases of interacting electrons. *Annals of Physics*, 310:428–492, 2004. [437, 482]

M. Freedman, C. Nayak, K. Walker, and Z. Wang. On picture (2+1)-TQFTs. In K. Lin, Z. Wang, and W. Zhang, editors, *Topology and Physics*. World Scientific, 2008. [437, 482]

M. H. Freedman. P/np and the quantum field computer. *Proceedings of the National Academy of Sciences of the United States of the United States*, 95(1):98–101, 1998. [147]

M. H. Freedman and Z. Wang. Large quantum fourier transforms are never exactly realized by braiding conformal blocks. *Physical Review A*, 75:032322, 2007. [142]

M. H. Freedman, M. J. Larsen, and Z. Wang. A modular functor which is universal for quantum computation. *Communications in Mathematical Physics*, 227:605–622, 2002a. quant-ph/0001108. [137, 147]

M. H. Freedman, M. J. Larsen, and Z. Wang. The two-eigenvalue problem and density of jones representation of braid groups. *Communications in Mathematical Physics*, 228(1):177–199, 2002b. [137, 147]

P. Freyd, D. Yetter, J. Hoste, W. B. R. Lickorish, K. Millett, and A. Ocneanu. A new polynomial invariant of knots and links. *Bulletin of the American Mathematical Society*, 12:239–246, 1985. [7, 18]

J. Fröhlich and F. Gabbiani. Braid statistics in local quantum theory. *Reviews in Mathematical Physics*, 02(03):251–353, 1990. [34, 226]

L. Fu. Topological crystalline insulators. *Physical Review Letters*, 106:106802, 2011. [530]

L. Fu and C. L. Kane. Superconducting proximity effect and majorana fermions at the surface of a topological insulator. *Physical Review Letters*, 100:096407, 2008. [560]

J. Fuchs, A. Schellekens, and C. Schweigert. A matrix s for all simple current extensions. *Nuclear Physics B*, 473(1):323–366, 1996. [567]

S. Furukawa and G. Misguich. Topological entanglement entropy in the quantum dimer model on the triangular lattice. *Physical Review B*, 75:214407, 2007. [525]

C. Galindo and N. Jaramillo. Solutions of the hexagon equation for abelian anyons. *Revista Colombiana de Matematicas*, 50:277–298, 2016. [230, 260]

T. Gannon. Comments on nonunitary conformal field theories. *Nuclear Physics B*, 670(3):335–358, 2003. [115, 237]

Y. Gefen and D. J. Thouless. Detection of fractional charge and quenching of the quantum Hall effect. *Physical Review B*, 47:10423–10436, 1993. [45]

N. Gemelke, E. Sarajlic, and S. Chu. Rotating few-body atomic systems in the fractional quantum Hall regime. *arXiv:1007.2677*, 2010. [556]

D. Gepner and A. Kapustin. On the classification of fusion rings. *Physics Letters B*, 349(1):71–75, 1995. [230, 238]

F. Ghahari, Y. Zhao, P. Cadden-Zimansky, K. Bolotin, and P. Kim. Measurement of the $\nu = 1/3$ fractional quantum Hall energy gap in suspended graphene. *Physical Review Letters*, 106:046801, 2011. [551]

P. Ginsparg. Fields, strings and critical phenomena (Les Houches, session xlix, 1988). 1990. [235, 564]

P. Goddard and A. Schwimmer. Unitary construction of extended conformal algebras. *Physics Letters B*, 206(1):62–70, 1988. [358]

P. Goddard, A. Kent, and D. Olive. Virasoro algebras and coset space models. *Physics Letters B*, 152(1):88–92, 1985. [356]

G. A. Goldin, R. Menikoff, and D. H. Sharp. Representations of a local current algebra in nonsimply connected space and the Aharonov–Bohm effect. *Journal of Mathematical Physics*, 22(8):1664–1668, 1981. [32]

G. A. Goldin, R. Menikoff, and D. H. Sharp. Diffeomorphism groups, gauge groups, and quantum theory. *Physical Review Letters*, 51:2246–2249, 1983. [32]

G. A. Goldin, R. Menikoff, and D. H. Sharp. Comments on "general theory for quantum statistics in two dimensions,". *Physical Review Letters*, 54:603–603, 1985. [34, 36]

C. Goméz, M. Ruiz-Altaba, and G. Sierra. *Quantum Groups in Two-Dimensional Physics*. Cambridge Monographs on Mathematical Physics, Cambridge University Press, 1996. [294]

R. E. Gompf and A. I. Stipsicz. *4-Manifolds and Kirby Calculus*. Graduate Studies in Mathematics, Vol. 20. American Mathematical Society, 1999. [61, 341]

D. Green. Classification of rank 6 modular categories with galois group $\langle(012)(345)\rangle$. *arXiv:1908.07128*, 2019. [238]

D. Gross, J. Eisert, N. Schuch, and D. Perez-Garcia. Measurement-based quantum computation beyond the one-way model. *Physical Review A*, 76:052315, 2007. [133]

Z.-C. Gu and X.-G. Wen. Tensor-entanglement-filtering renormalization approach and symmetry-protected topological order. *Physical Review B*, 80:155131, 2009. [530]

Z.-C. Gu and X.-G. Wen. Symmetry-protected topological orders for interacting fermions: Fermionic topological nonlinear σ models and a special group supercohomology theory. *Physical Review B*, 90:115141, 2014. [530, 539]

T. J. Hagge and S.-M. Hong. Some non-braided categories of rank three. *Communications in Contemporary Mathematics*, 11(04):615–637, 2009. [174, 486]

A. Hahn and R. Wolf. Generalized string-net model for unitary fusion categories without tetrahedral symmetry. *Physical Review B*, 102:115154, 2020. [505, 507]

F. Haldane. Continuum dynamics of the 1-d heisenberg antiferromagnet: Identification with the o(3) nonlinear sigma model. *Physics Letters A*, 93(9):464–468, 1983a. [529]

F. D. M. Haldane. Fractional quantization of the Hall effect: A hierarchy of incompressible quantum fluid states. *Physical Review Letters*, 51:605–608, 1983b. [552]

F. D. M. Haldane. Nonlinear field theory of large-spin heisenberg antiferromagnets: Semiclassically quantized solitons of the one-dimensional easy-axis néel state. *Physical Review Letters*, 50:1153–1156, 1983c. [529]

B. I. Halperin. Statistics of quasiparticles and the hierarchy of fractional quantized Hall states. *Physical Review Letters*, 52:1583–1586, 1984. [32, 552]

B. I. Halperin and J. K. Jain. *Fractional Quantum Hall Effects*. World Scientific, (individual chapters available on arXiv), 2020. [238, 551, 554, 555, 556, 563]

M. Hamermesh. *Group Theory and its Application to Physical Problems*. Dover, 1989. [283, 284, 583]

T. H. Hansson, V. Oganesyan, and S. L. Sondhi. Superconductors are topologically ordered. *Annals of Physics*, 313:497, 2004. [559]

T. H. Hansson, M. Hermanns, S. H. Simon, and S. F. Viefers. Quantum Hall physics: Hierarchies and conformal field theory techniques. *Reviews of Modern Physics*, 89:025005, 2017. [50, 63, 229, 238, 553]

M. Haque, O. Zozulya, and K. Schoutens. Entanglement entropy in fermionic laughlin states. *Physical Review Letters*, 98:060401, 2007. [525]

L. Hardiman. *Module Categories and Modular Invariants*. PhD thesis, University of Bath, 2019. [494]

A. Harrow. Quantum compiling. Ph.D. Thesis. Available online at http://www.media.mit.edu/physics/publications/theses/01.05.aram.pdf, 2001. [135, 140]

A. W. Harrow, B. Recht, and I. L. Chuang. Efficient discrete approximations of quantum gates. *Journal of Mathematical Physics*, 43(9):4445–4451, 2002. [135]

M. Z. Hasan and C. L. Kane. Colloquium: Topological insulators. *Reviews of Modern Physics*, 82:3045–3067, 2010. [529]

B. Hasslacher and M. J. Perry. Spin networks are simplicial quantum gravity. *Physics Letters B*, 103(1):21–24, 1981. [322]

C. Heinrich, F. Burnell, L. Fidkowski, and M. Levin. Symmetry-enriched string nets: Exactly solvable models for set phases. *Physical Review B*, 94:235136, 2016. [537, 547]

M. Hermanns, I. Kimchi, and J. Knolle. Physics of the kitaev model: Fractionalization, dynamic correlations, and material connections. *Annual Review of Condensed Matter Physics*, 9(1):17–33, 2018. [557, 558]

S.-M. Hong. On symmetrization of 6*j*-symbols and Levin–Wen Hamiltonian. *arXiv:0907.2204*, 2009. [214, 507]

S.-M. Hong and E. Rowell. On the classification of the Grothendieck rings of non-self-dual modular categories. *Journal of Algebra*, 324(5):1000–1015, 2010. [238]

L. Hormozi, N. E. Bonesteel, and S. H. Simon. Topological quantum computing with read-rezayi states. *Physical Review Letters*, 103:160501, 2009. [135]

P.-S. Hsin and N. Seiberg. Level/rank duality and Chern–Simons-matter theories. *Journal of High Energy Physics*, 2016(9):95, 2016. [358]

Y. Hu and Y. Wan. Entanglement entropy, quantum fluctuations, and thermal entropy in topological phases. *Journal of High Energy Physics*, 2019(5):110, 2019. [508]

Y. Hu, S. D. Stirling, and Y.-S. Wu. Ground-state degeneracy in the Levin–Wen model for topological phases. *Physical Review B*, 85:075107, 2012. [499]

Y. Hu, Y. Wan, and Y.-S. Wu. Twisted quantum double model of topological phases in two dimensions. *Physical Review B*, 87:125114, 2013. [273, 468]

Y. Hu, N. Geer, and Y.-S. Wu. Full dyon excitation spectrum in extended Levin–Wen models. *Physical Review B*, 97:195154, 2018. [508]

J. A. Hutasoit, A. C. Balram, S. Mukherjee, Y.-H. Wu, S. S. Mandal, A. Wójs, V. Cheianov, and J. K. Jain. The enigma of the $\nu = 2 + 3/8$ fractional quantum Hall effect. *Physical Review B*, 95:125302, 2017. [555]

J. Huxford and S. H. Simon. Excitations in the higher lattice gauge theory model for topological phases I: Overview. *arXiv:2202.08294*, 2022. [469]

M. Iqbal et al. Creation of non-abelian topological order and anyons on a trapped-ion processor. *arXiv:2305.03766*, 2023. [537, 561]

G. Jackeli and G. Khaliullin. Mott insulators in the strong spin-orbit coupling limit: From heisenberg to a quantum compass and kitaev models. *Physical Review Letters*, 102:017205, 2009. [557]

W. Jaco and J. H. Rubinstein. 0-Efficient Triangulations of 3-Manifolds. *Journal of Differential Geometry*, 65(1):61–168, 2003. [327]

T. Jacobson and T. Sulejmanpasic. Modified villain formulation of abelian chern–simons theory. *arXiv:2303.06160*, 2023. [61]

J. K. Jain. *Composite Fermions*. Cambridge University Press, 2007. [553, 563]

J.-S. Jeong, H. Lu, K. H. Lee, K. Hashimoto, S. B. Chung, and K. Park. Competing states for the fractional quantum Hall effect in the 1/3-filled second Landau level. *Physical Review B*, 96:125148, 2017. [555]

T. Johnson-Freyd. On the classification of topological orders. *Communications in Mathematical Physics*, 393(2):989–1033, 2022. [325]

C. Jones and D. Penneys. Operator algebras in rigid c*-tensor categories. *Communications in Mathematical Physics*, 355(3):1121–1188, 2017. [194]

V. F. R. Jones. A polynomial invariant for knots via von neumann algebras. *Bulletin of the American Mathematical Society*, 12:103–112, 1985. [4]

J. Jovanovic, C. Wille, D. Timmer, and S. H. Simon. A proposal to demonstrate non-abelian anyons on a NISQ device. *Forthcoming*, 2023. [561]

C. L. Kane and E. J. Mele. Z_2 topological order and the quantum spin Hall effect. *Physical Review Letters*, 95:146802, 2005. [529]

A. Kapustin. Bosonic topological insulators and paramagnets: A view from cobordisms. *arxiv:1404.6659*, 2014. [536]

E. Karlsson. *Kitaev Models for Topologically Ordered Phases of Matter*. Undergraduate thesis, Karlstads Univerisitet, 2017. [468]

C. Kassel. *Quantum Groups*. Springer-Verlag, 1995. [294]

L. H. Kauffman. State models and the Jones polynomial. *Topology*, 26:395–407, 1987. [3]

L. H. Kauffman. *Knots and Physics, 3rd edn.* World Scientific, 2001. [17, 63, 310]

L. H. Kauffman and S. Lins. *Temperley Lieb Recoupling Theory and Onvariants of 3-manifolds.* Annals of Mathematics Studies, Vol 134. Princeton Univ. Press, 1994. [302, 304, 305, 307, 308, 310]

L. H. Kauffman and S. J. Lomonaco. q-deformed spin networks, knot polynomials and anyonic topological quantum computation. *Journal of Knot Theory and Its Ramifications*, 16(03):267–332, 2007. [305, 310]

R. Kirby. A calculus for framed links in s^3. *Inventiones Mathematicae*, 45:35–56, 1978. [333]

R. Kirby and P. Melvin. Canonical framings for 3-manifolds. *Turkish Journal of Mathematics: Proceedings of 6th Gokova Geometry-Topology Conference*, 23(1): 89–116, 1999. [61, 336]

R. C. Kirby. *The Topology of 4-Manifolds.* Lecture Notes in Mathematics, Vol. 1374. Springer, 1989. [342]

A. Kirillov. String-net model of Turaev–Viro invariants. *arXiv:1106.6033*, 2011. [485]

A. N. Kirillov and N. Y. Reshetikhin. Representations of the algebra $u_q(sl(2))$, q-orthogonal polynomials and invariants of links. In V. G. Kac, editor, *Infinite Dimensional Lie Algebras and Groups*, page 285. World Scientific, 1988. [290]

A. Kitaev. Anyons in an exactly solved model and beyond. *Annals of Physics*, 321 (1):2–111, 2006. [84, 110, 113, 120, 129, 168, 184, 194, 195, 198, 224, 225, 227, 235, 238, 255, 262, 522, 557]

A. Kitaev and L. Kong. Models for gapped boundaries and domain walls. *Communications in Mathematical Physics*, 313(2):351–373, 2012. [508]

A. Kitaev and J. Preskill. Topological entanglement entropy. *Physical Review Letters*, 96:110404, 2006. [514, 518, 520, 522, 525]

A. Y. Kitaev. Unpaired majorana fermions in quantum wires. *Physics-Uspekhi*, 44 (10S):131–136, 2001. [560]

A. Y. Kitaev. Fault-tolerant quantum computation by anyons. *Annals of Physics*, 303:2, 2003. preprint 1997. [133, 137, 147, 368, 373, 394, 405, 441, 444, 465, 468]

V. Kliuchnikov and J. Yard. A framework for exact synthesis. *arXiv:1504.04350*, 2015. [142]

V. Kliuchnikov, A. Bocharov, and K. M. Svore. Asymptotically optimal topological quantum compiling. *Physical Review Letters*, 112:140504, 2014. [142, 143]

V. Kliuchnikov, A. Bocharov, M. Roetteler, and J. Yard. A framework for approximating qubit unitaries. *arXiv:1510.03888*, 2015. [142]

J. Knolle and R. Moessner. A field guide to spin liquids. *Annual Review of Condensed Matter Physics*, 10(1):451–472, 2019. [556, 557, 558, 559]

A. Kol and N. Read. Fractional quantum Hall effect in a periodic potential. *Physical Review B*, 48:8890–8898, 1993. [555]

L. Kong. Anyon condensation and tensor categories. *Nuclear Physics B*, 886:436–482, 2014. [359, 360, 361]

A. Kumar, G. A. Csáthy, M. J. Manfra, L. N. Pfeiffer, and K. W. West. Nonconventional odd-denominator fractional quantum Hall states in the second Landau level. *Physical Review Letters*, 105:246808, 2010. [551]

G. Kuperberg. How hard is it to approximate the Jones polynomial? *Theory of Computing*, 11(6):183–219, 2015. [13]

M. Lackenby. A polynomial upper bound on Reidemeister moves. *Annals of Mathemtics*, 182:491–564, 2015. [16]

M. G. G. Laidlaw and C. M. DeWitt. Feynman functional integrals for systems of indistinguishable particles. *Physical Review D*, 3:1375–1378, 1971. [32, 36]

T. Lan and X.-G. Wen. Topological quasiparticles and the holographic bulk-edge relation in $(2+1)$-dimensional string-net models. *Physical Review B*, 90:115119, 2014. [508]

T. Lan, L. Kong, and X.-G. Wen. Theory of $(2 + 1)$-dimensional fermionic topological orders and fermionic/bosonic topological orders with symmetries. *Physical Review B*, 94:155113, 2016. [239, 572]

T. Lan, L. Kong, and X.-G. Wen. Modular extensions of unitary braided fusion categories and $2 + 1$d topological/spt orders with symmetries. *Communications in Mathematical Physics*, 351(2):709–739, 2017. [239, 572]

T. Lan, L. Kong, and X.-G. Wen. Classification of $(3 + 1)$D bosonic topological orders: The case when pointlike excitations are all bosons. *Physical Review X*, 8: 021074, 2018. [325]

L. D. Landau. On the theory of phase transitions. *Zh. Eksp. Teor. Fiz*, 7:19–32, 1937. [528]

A. M. Läuchli, J. Sudan, and R. Moessner. $s = \frac{1}{2}$ kagome Heisenberg antiferromagnet revisited. *Physical Review B*, 100:155142, 2019. [558]

R. B. Laughlin. Anomalous quantum Hall effect: An incompressible quantum fluid with fractionally charged excitations. *Physical Review Letters*, 50:1395–1398, 1983. [552, 563]

A. J. Leggett. *Quantum Liquids: Bose Condensation and Cooper Pairing in Condensed-Matter Systems*. Oxford University Press, 2006. [345]

J. M. Leinaas and J. Myrheim. On the theory of identical particles. *Il Nuovo Cimento B (1971–1996)*, 37(1):1–23, 1977. [32]

T. Leinster. *Basic Category Theory*. Cambridge University Press, 2014. [89]

C. Levaillant, B. Bauer, M. Freedman, Z. Wang, and P. Bonderson. Universal gates via fusion and measurement operations on $SU(2)_4$ anyons. *Physical Review A*, 92: 012301, 2015. [137]

M. Levin. Protected edge modes without symmetry. *Physical Review X*, 3:021009, 2013. [359]

M. Levin and Z.-C. Gu. Braiding statistics approach to symmetry-protected topological phases. *Physical Review B*, 86:115109, 2012. [531, 533, 535, 537]

M. Levin and X.-G. Wen. Detecting topological order in a ground state wave function. *Physical Review Letters*, 96(11):110405, 2006. [514, 517, 518, 520, 525]

M. A. Levin and X.-G. Wen. String-net condensation: A physical mechanism for topological phases. *Physical Review B*, 71(4):045110, 2005. [206, 207, 220, 482, 485, 496, 504, 507]

H. Li and F. D. Haldane. Entanglement spectrum as a generalization of entanglement entropy: Identification of topological order in non-abelian fractional quantum Hall effect states. *Physical Review Letters*, 101(1):1–4, 2008. [513]

Q. Li, H. Li, J. Zhao, H.-G. Luo, and Z. Y. Xie. Magnetization of the spin-$\frac{1}{2}$ Heisenberg antiferromagnet on the triangular lattice. *Physical Review B*, 105: 184418, 2022. [558]

Y. W. Li et al. Observation of topological superconductivity in a stoichiometric transition metal dichalcogenide 2m-ws2. *Nature Communications*, 12(1):2874, 2021. [560]

W. B. R. Lickorish. A representation of orientable combinatorial 3-manifolds. *Annals of Mathematics*, 76(3):531–540, 1962. [332]

W. B. R. Lickorish. Skeins and handlebodies. *Pacific Journal of Mathematics*, 159 (2):337–249, 1993. [305]

W. B. R. Lickorish. *An Introduction to Knot Theory*. Graduate Texts in Mathematics, Vol. 175. Springer, 1997. [341]

C.-H. Lin and M. Levin. Generalizations and limitations of string-net models. *Physical Review B*, 89:195130, 2014. [118, 480]

C.-H. Lin, M. Levin, and F. J. Burnell. Generalized string-net models: A thorough exposition. *Physical Review B*, 103:195155, 2021. [480, 485, 505, 508]

S. Lloyd. Quantum computation with abelian anyons. *arXiv:0004010*, 2000. [138]

A. Loidl, P. Lunkenheimer, and V. Tsurkan. On the proximate Kitaev quantum-spin liquid α-RuCl$_3$: Thermodynamics, excitations and continua. *Journal of Physics: Condensed Matter*, 33(44):443004, 2021. [558]

A. Lopez and E. Fradkin. Fractional quantum Hall effect and Chern–Simons gauge theories. *Physical Review B*, 44(10):5246–5262, 1991. [563]

M. Lorente. Spin networks in quantum gravity. *Journal of Geometry and Symmetry in Physics*, 6:85–100, 2006. [72, 326]

Y.-M. Lu and A. Vishwanath. Theory and classification of interacting integer topological phases in two dimensions: A Chern–Simons approach. *Physical Review B*, 86:125119, 2012. [530, 539]

Y.-M. Lu and A. Vishwanath. Classification and properties of symmetry-enriched topological phases: Chern–Simons approach with applications to Z_2 spin liquids. *Physical Review B*, 93:155121, 2016. [548]

R. M. Lutchyn, E. P. A. M. Bakkers, L. P. Kouwenhoven, P. Krogstrup, C. M. Marcus, and Y. Oreg. Majorana zero modes in superconductor–semiconductor heterostructures. *Nature Reviews Materials*, 3(5):52–68, 2018. [560]

S. Mac Lane. *Categories for the Working Mathematician*. Springer-Verlag, 1971. [89, 117]

R. MacKenzie. Path integral methods and applications. *quant-ph/0004090*, 2000. [24, 36]

S. Manna, P. Wei, Y. Xie, K. T. Law, P. A. Lee, and J. S. Moodera. Signature of a pair of majorana zero modes in superconducting gold surface states. *Proceedings of the National Academy of Sciences of the United States of the United States*, 117(16):8775–8782, 2020. [560]

A. Mesaros and Y. Ran. Classification of symmetry enriched topological phases with exactly solvable models. *Physical Review B*, 87:155115, 2013. [547]

M. Mignard and P. Schauenburg. Modular categories are not determined by their modular data. *Letters in Mathematical Physics*, 111(3):60, 2021. [228, 571]

J. Milnor. A unique decomposition theorem for 3-manifolds. *American Journal of Mathematics*, 84(1):1–7, 1962. [86]

E. J. Mlawer, S. G. Naculich, H. A. Riggs, and H. J. Schnitzer. Group-level duality of wzw fusion coefficients and chern-simons link observables. *Nuclear Physics B*, 352(3):863–896, 1991. [358]

C. Mochon. Anyons from nonsolvable finite groups are sufficient for universal quantum computation. *Physical Review A*, 67(2):022315, 2003. [137]

C. Mochon. Anyon computers with smaller groups. *Physical Review A*, 69(3):032306, 2004. [137]

R. S. K. Mong, M. P. Zaletel, F. Pollmann, and Z. Papić. Fibonacci anyons and charge density order in the 12/5 and 13/5 quantum Hall plateaus. *Physical Review B*, 95:115136, 2017. [554]

G. Moore and N. Read. Nonabelions in the fractional quantum Hall effect. *Nuclear Physics B*, 360(2-3):362–96, 1991. [34, 238, 553, 563, 564]

G. Moore and N. Seiberg. Taming the conformal zoo. *Physics Letters B*, 220(3): 422–430, 1989a. [345, 356, 360]

G. Moore and N. Seiberg. Classical and quantum conformal field theory. *Communications in Mathematical Physics*, 123(2):177–254, 1989b. [110, 129, 240]

R. H. Morf. Transition from quantum Hall to compressible states in the second Landau level: New light on the $\nu = 5/2$ enigma. *Physical Review Letters*, 80: 1505–1508, 1998. [553]

M. Müger. From subfactors to categories and topology ii: The quantum double of tensor categories and subfactors. *Journal of Pure and Applied Algebra*, 180(1): 159–219, 2003. [510]

M. Müger. Abstract duality for symmetric tensor *-categories. In J. Butterfield and J. Earman, editors, *Handbook of the Philosophy of Physics*. Kluwer Academic Press, 2007. [34, 281]

S. Mukhopadhyay. *Rank-Level Duality of Conformal Blocks.* PhD thesis, University of Chapel Hill, 2013. [358]

M. Nakahara. *Geometry, Topology, and Physics, 2nd edn.* Taylor and Francis, 2003. [583]

J. Nakamura, S. Liang, G. C. Gardner, and M. J. Manfra. Direct observation of anyonic braiding statistics. *Nature Physics*, 16(9):931–936, 2020. [552]

C. Nayak, S. H. Simon, A. Stern, M. Freedman, and S. Das Sarma. Non-abelian anyons and topological quantum computation. *Reviews of Modern Physics*, 80: 1083–1159, 2008. [33, 36, 63, 120, 121, 129, 147, 168, 238, 553, 554]

T. Neupert, H. He, C. von Keyserlingk, G. Sierra, and B. A. Bernevig. Boson condensation in topologically ordered quantum liquids. *Physical Review B*, 93:115103, 2016. [359, 361]

S.-H. Ng and P. Schauenburg. Frobenius—Schur indicators and exponents of spherical categories. *Advances in Mathematics*, 211(1):34–71, 2007. [233]

H. Nicolai and K. Peeters. Loop and spin foam quantum gravity: A brief guide for beginners. In I.-O. Stamatescu and E. Seiler, editors, *Approaches to Fundamental Physics: An Assessment of Current Theoretical Ideas*, pages 151–184, 2007. [72]

H. Nicolai, K. Peeters, and M. Zamaklar. Loop quantum gravity: An outside view. *Classical and Quantum Gravity*, 22(19):R193–R247, 2005. [72]

M. A. Nielsen and I. L. Chuang. *Quantum Computation and Quantum Information.* Cambridge University Press, 2000. [13, 133, 147]

D. Nikshych. Classifying braidings on fusion categories. *arXiv:1801.06125v2*, 2018. [280]

M. R. Norman. Colloquium: Herbertsmithite and the search for the quantum spin liquid. *Reviews of Modern Physics*, 88:041002, 2016. [559]

A. Ocneanu. Chirality for operator algebras. In *Proceedings of the Taniguchi Symposium on Operator Algebras, July 6–10, Shiga-ken, Japan (1993)*. Wiley, 1994. [437, 508]

R. Oeckl. *Discrete Gauge Theory: From Lattices to TQFT.* Imperial College Press, 2005. [469]

H. Ooguri. Topological lattice models in four dimensions. *Modern Physics Letters A*, 07(30):2799–2810, 1992. [326]

R. Orús. A practical introduction to tensor networks: Matrix product states and projected entangled pair states. *Annals of Physics*, 349:117–158, 2014. [564]

V. Ostrik. Fusion categories of rank 2. *arxiv:0203255*, 2002. [117]

J. K. Pachos. Quantum computation with abelian anyons on the honeycomb lattice. *International Journal of Quantum Information*, 04(06):947–954, 2006. [138]

S. A. Parameswaran, R. Roy, and S. L. Sondhi. Fractional quantum Hall physics in topological flat bands. *Comptes Rendus Physique*, 14(9):816–839, 2013. [555]

V. Pasquier. Operator content of the ADE lattice models. *Journal of Physics A: Mathematical and General*, 20(16):5707–5717, 1987. [225]

R. Penrose. Angular momentum: An approach to combinatorial spacetime. In T. Bastin, editor, *Quantum Theory and Beyond*, pages 151–180. Cambridge University Press, 1971. [310, 321, 322]

F. Pollmann. Symmetry protected topological phases in one-dimensional systems. Les Houches Lecture Notes. 2009. [539]

F. Pollmann, A. M. Turner, E. Berg, and M. Oshikawa. Entanglement spectrum of a topological phase in one dimension. *Physical Review B*, 81:064439, 2010. [530]

F. Pollmann, E. Berg, A. M. Turner, and M. Oshikawa. Symmetry protection of topological phases in one-dimensional quantum spin systems. *Physical Review B*, 85:075125, 2012. [530]

G. Ponzano and T. Regge. Semiclassical limit of Racah coefficients. In F. Bloch, editor, *Spectroscopic and Group Theoretical Methods in Physics*, pages 1–58. North-Holland Publ. Co., 1968. [321, 322]

V. V. Prasolov and A. B. Sossinsky. *Knots, Links, Braids and 3-Manifolds: An Introduction to the New Invariants in Low-Dimensional Topology*. Translations of Mathematical Monographs, Vol. 154. American Mathematical Society, 1996. [341]

J. Preskill. Lecture notes for physics 219:quantum computation. Available at http://www.theory.caltech.edu/ ~preskill/ph219/ph219_2004.html, 2004. [10, 47, 110, 129, 137, 168, 444, 468]

J. H. Przytycki and P. Traczyk. Invariants of links of conway type. *Kobe Journal of Mathematics*, 4:115–139, 1987. [18]

A. Pustogow, Y. Luo, A. Chronister, Y.-S. Su, D. A. Sokolov, F. Jerzembeck, A. P. Mackenzie, C. W. Hicks, N. Kikugawa, S. Raghu, E. D. Bauer, and S. E. Brown. Constraints on the superconducting order parameter in Sr_2RuO_4 from oxygen-17 nuclear magnetic resonance. *Nature*, 574(7776):72–75, 2019. [560]

X. L. Qi, H. Katsura, and A. W. Ludwig. General relationship between the entanglement spectrum and the edge state spectrum of topological quantum states. *Physical Review Letters*, 108(19):1–5, 2012. [513]

R. Rajaraman. *Solitons and Instantons*. North-Holland, 1982. [63]

J. G. Rau, E. K.-H. Lee, and H.-Y. Kee. Spin-orbit physics giving rise to novel phases in correlated systems: Iridates and related materials. *Annual Review of Condensed Matter Physics*, 7(1):195–221, 2016. [558]

R. Raussendorf and H. J. Briegel. A one-way quantum computer. *Physical Review Letters*, 86:5188–5191, 2001. [133]

N. Read. Excitation structure of the hierarchy scheme in the fractional quantum Hall effect. *Physical Review Letters*, 65(12):1502–1505, 1990. [63, 553]

N. Read and D. Green. Paired states of fermions in two dimensions with breaking of parity and time-reversal symmetries and the fractional quantum Hall effect. *Physical Review B*, 61(15):10267–10297, 2000. [559]

N. Read and E. Rezayi. Beyond paired quantum Hall states: Parafermions and incompressible states in the first excited Landau level. *Physical Review B*, 59(12): 8084–8092, 1999. [554, 555]

T. Regge. General relativity without coordinates. *Il Nuovo Cimento (1955–1965)*, 19(3):558–571, 1961. [322]

T. Regge and R. M. Williams. Discrete structures in gravity. *Journal of Mathematical Physics*, 41(6):3964–3984, 2000. [72, 322, 326]

K. H. Rehren. Braid group statistics and their superselection rules. In D. Kastler, editor, *The Algebraic Theory of Superselection Sectors*. World Scientific, 1990. [226]

N. Y. Reshetikhin and V. G. Turaev. Invariants of 3-manifolds via link polynomials and quantum groups. *Inventiones Mathematicae*, 103:547–597, 1991. [336]

A. Revelli, M. Moretti Sala, G. Monaco, C. Hickey, P. Becker, F. Freund, A. Jesche, P. Gegenwart, T. Eschmann, F. L. Buessen, S. Trebst, P. H. M. van Loosdrecht, J. van den Brink, and M. Grüninger. Fingerprints of Kitaev physics in the magnetic excitations of honeycomb iridates. *Physical Review Research*, 2:043094, 2020. [558]

E. H. Rezayi. Landau level mixing and the ground state of the $\nu = 5/2$ quantum Hall effect. *Physical Review Letters*, 119:026801, 2017. [554]

E. H. Rezayi and S. H. Simon. Breaking of particle-hole symmetry by Landau level mixing in the $\nu = 5/2$ quantized Hall state. *Physical Review Letters*, 106:116801, 2011. [554]

J. Roberts. Skein theory and Turaev–Viro invariants. *Topology*, 34(4):771–787, 1995. [339, 341, 342, 343]

J. Roberts. Kirby calculus in manifolds with boundary. *Turkish Journal of Mathematics; Proceedings of 5th Gokova Geometry-Topology Conference, 1996*, 21(1): 111–117, 1997. [333]

J. Roberts. Knots knotes (lectures on elementary knot theory). http://math.ucsd.edu/ justin/, 2015. [17]

D. Rolfson. *Knots and Links*. AMS Chelsea Publishing, 1976. [89, 227, 329]

P. Roushan, C. Neill, A. Megrant, Y. Chen, R. Babbush, R. Barends, B. Campbell, Z. Chen, B. Chiaro, A. Dunsworth, A. Fowler, E. Jeffrey, J. Kelly, E. Lucero, J. Mutus, P. J. J. O'Malley, M. Neeley, C. Quintana, D. Sank, A. Vainsencher, J. Wenner, T. White, E. Kapit, H. Neven, and J. Martinis. Chiral ground-state currents of interacting photons in a synthetic magnetic field. *Nature Physics*, 13 (2):146–151, 2017. [561]

C. Rovelli. Notes for a brief history of quantum gravity. *arXiv:gr-qc/0006061*, 2000. [72]

C. Rovelli. Loop quantum gravity. *Living Reviews in Relativity*, 11(1):5, 2008. [72]

E. Rowell, R. Strong, and Z. Wang. On classification of modular tensor categories. *Communications in Mathematical Physics*, 292, 2009. [230, 231, 238, 572]

E. C. Rowell. From quantum groups to unitary modular tensor categories. *https://arxiv.org/abs/math/0503226v3*, 2005. [294, 295]

E. C. Rowell and Z. Wang. Degeneracy and non-abelian statistics. *Physical Review A*, 93:030102, 2016. [34]

I. Runkel. String-net models for nonspherical pivotal fusion categories. *Journal of Knot Theory and Its Ramifications*, 29(06):2050035, 2020. [508]

I. Sakata. A general method for obtaining Clebsch–Gordan coefficients of finite groups. I. Its application to point and space groups. *Journal of Mathematical Physics*, 15(10):1702–1709, 1974. [283]

L. Saminadayar, D. C. Glattli, Y. Jin, and B. Etienne. Observation of the e/3 fractionally charged Laughlin quasiparticle. *Physical Review Letters*, 79:2526, 1997. [552]

M. Sato and Y. Ando. Topological superconductors: A review. *Reports on Progress in Physics*, 80(7):076501, 2017. [559]

K. J. Satzinger, Google, AI, and 100-Authors. Realizing topologically ordered states on a quantum processor. *Science*, 374(6572):1237–1241, 2021. [373, 561]

J. Saunders. Realizing quantum materials with helium: Helium films at ultralow temperatures, from strongly correlated atomically layered films to topological superfluidity. In *Topological Phase Transitions and New Developments*, pages 165–196. 2018. [560]

L. Savary and L. Balents. Quantum spin liquids: A review. *Reports on Progress in Physics*, 80(1):016502, 2016. [557]

N. Saveliev. *Lectures on the Topology of 3-manifolds: An Introduction to the Casson invariant, 2nd edn.* Walter de Gruyter, 2012. [89, 341]

A. Schellekens. Fixed point resolution in extended wzw models. *Nuclear Physics B*, 558(3):484–502, 1999. [345, 360]

N. Schuch, D. Pérez-García, and I. Cirac. Classifying quantum phases using matrix product states and projected entangled pair states. *Physical Review B*, 84:165139, 2011. [530]

M. D. Schulz. *Topological Phase Transitions Driven by Non-Abelian Anyons*. PhD thesis, Technische Universitat Dortmund, 2014. [408, 508]

M. D. Schulz, S. Dusuel, K. P. Schmidt, and J. Vidal. Topological phase transitions in the golden string-net model. *Physical Review Letters*, 110:147203, 2013. [508]

G. Semeghini, H. Levine, A. Keesling, S. Ebadi, T. T. Wang, D. Bluvstein, R. Verresen, H. Pichler, M. Kalinowski, R. Samajdar, A. Omran, S. Sachdev, A. Vishwanath, M. Greiner, V. Vuletić, and M. D. Lukin. Probing topological spin liquids on a programmable quantum simulator. *Science*, 374(6572):1242–1247, 2021. [561]

T. Senthil. Symmetry-protected topological phases of quantum matter. *Annual Review of Condensed Matter Physics*, 6(1):299–324, 2015. [530, 539]

P. W. Shor. Scheme for reducing decoherence in quantum computer memory. *Physical Review A*, 52(4):R2493–R2496, 1995. [368, 369]

D. S. Silver. Knot theory's odd origins. *American Scientist*, 94, 2006. [2]

S. H. Simon. Quantum computing with a twist. *Physics World*, pages 35–40, Sept 2010. [17]

S. H. Simon. *The Oxford Solid State Basics*. Oxford Univesity Press, 2013. [vi, 387]

S. H. Simon and J. K. Slingerland. Straightening out the Frobenius–Schur indicator. *arXiv:2208.14500*, 2022. [164, 183, 184, 189, 190, 198, 212, 214]

S. H. Simon, N. E. Bonesteel, M. H. Freedman, N. Petrovic, and L. Hormozi. Topological quantum computing with only one mobile quasiparticle. *Physical Review Letters*, 96(7):070503, 2006. [137, 148]

J. K. Slingerland and F. A. Bais. Quantum groups and non-abelian braiding in quantum Hall systems. *Nuclear Physics B*, 612:229–288, 2001. [294]

N. Snyder and P. Tingley. The half-twist for $u_q(g)$ representations. *Algebra and Number Theory*, 3(7):809–834, 2009. [190]

C. Song, D. Xu, P. Zhang, J. Wang, Q. Guo, W. Liu, K. Xu, H. Deng, K. Huang, D. Zheng, S.-B. Zheng, H. Wang, X. Zhu, C.-Y. Lu, and J.-W. Pan. Demonstration of topological robustness of anyonic braiding statistics with a superconducting quantum circuit. *Physical Review Letters*, 121:030502, 2018. [373, 561]

A. Sossinsky. *Knots: Mathematics with a Twist*. Harvard University Press, 2002. [17]

A. Steane. Multiple particle interference and quantum error correction. *Proceedings of the Royal Society of London. Series A, Mathematical and Physical Sciences*, 452:2551, 1996a. [368]

A. M. Steane. Error-correcting codes in quantum theory. *Physical Review Letters*, 77(5):793–797, 1996b. [368]

A. Stoimenow. Tait's conjectures and odd crossing number amphicheiral knots. *Bulletin of the American Mathematical Society*, 45:285–291, 2008. [2]

M. Stone and P. Goldbart. *Mathematics for Physics*. Cambridge University Press, 2009. [583]

H.-H. Sun and J.-F. Jia. Detection of Majorana zero mode in the vortex. *npj Quantum Materials*, 2(1):34, 2017. [560]

K. Sun, K. Kumar, and E. Fradkin. Discretized abelian Chern–Simons gauge theory on arbitrary graphs. *Physical Review B*, 92:115148, 2015. [61]

B. Swingle and T. Senthil. Geometric proof of the equality between entanglement and edge spectra. *Physical Review B*, 86:45117, 2012. [513]

N. Tantivasadakarn, R. Verresen, and A. Vishwanath. The shortest route to non-abelian topological order on a quantum processor. *arXiv:2209.03964*, 2022. [561]

H. N. V. Temperley and E. H. Lieb. Relations between the percolation and colouring problem and other graph-theoretical problems associated with regular planar lattices: some exact results for the percolation problem. *Proceedings of the Royal Society of London. A. Mathematical and Physical Sciences*, 322(1549):251–280, 1971. [310]

D. J. Thouless, M. Kohmoto, M. P. Nightingale, and M. den Nijs. Quantized Hall conductance in a two-dimensional periodic potential. *Physical Review Letters*, 49: 405–408, 1982. [555]

A. Tiwari, X. Chen, K. Shiozaki, and S. Ryu. Bosonic topological phases of matter: Bulk-boundary correspondence, symmetry protected topological invariants, and gauging. *Physical Review B*, 97:245133, 2018. [537]

S. Treiman, R. Jackiw, B. Zumino, and E. Witten. *Current Algebras and Anomalies*. World Scientific, 1985. [63]

D. C. Tsui, H. L. Stormer, and A. C. Gossard. Two-dimensional magnetotransport in the extreme quantum limit. *Physical Review Letters*, 48:1559–1562, 1982. [32, 552]

V. Turaev and A. Virelizier. *Monoidal Categories and Topological Field Theory*. Progress in Mathematics, Vol. 322. Birkhauser, 2017. [184, 185, 326]

V. G. Turaev. Topology of shadows. 1992. [321, 339]

V. G. Turaev. *Quantum Invariants of Knots and 3-Manifolds*. Walter de Gruyter, Berlin, New York, 1994. [321, 339]

V. G. Turaev and O. Y. Viro. State sum invariants of 3-manifolds and quantum 6j-symbols. *Topology*, 31(4):865–902, 1992. [318, 326]

A. M. Turner, F. Pollmann, and E. Berg. Topological phases of one-dimensional fermions: An entanglement point of view. *Physical Review B*, 83:075102, 2011. [530]

C. Vafa. Toward classification of conformal theories. *Physics Letters B*, 206(3): 421–426, 1988. [233]

P. M. van Den Broek and J. F. Cornwell. Clebsch-Gordan coefficients of symmetry groups. *Physica Status Solidi (b)*, 90(1):211–224, 1978. [284]

S. Vandoren and P. van Nieuwenhuizen. Lectures on instantons. *arxiv/0802.1862*, 2008. [63]

E. Verlinde. Fusion rules and modular transformations in 2d conformal field theory. *Nuclear Physics B*, 300:360–376, 1988. [111, 225, 240]

J. Vidal. Partition function of the Levin–Wen model. *Physical Review B*, 105: L041110, 2022. [508]

J. Vidal, S. Dusuel, and K. P. Schmidt. Low-energy effective theory of the toric code model in a parallel magnetic field. *Physical Review B*, 79:033109, 2009. [408]

A. Vishwanath and T. Senthil. Physics of three-dimensional bosonic topological insulators: Surface-deconfined criticality and quantized magnetoelectric effect. *Physical Review X*, 3:011016, 2013. [536]

C. W. von Keyserlingk, F. J. Burnell, and S. H. Simon. Three-dimensional topological lattice models with surface anyons. *Physical Review B*, 87:045107, 2013. [482, 508]

K. Walker. On Witten's 3-manifold invariants. http://canyon23.net/math/1991TQFTNotes.pdf, 1991. [321, 339]

K. Walker. A universal state sum. *https://arxiv.org/abs/2104.02101*, 2021. [327]

K. Walker and Z. Wang. (3+1)-tqfts and topological insulators. *Frontiers of Physics*, 7(2):150–159, 2012. [482, 508]

A. H. Wallace. Modifications and cobounding manifolds. *Canadian Journal of Mathematics*, 12:503–528, 1960. [332]

Y. Wan, J. C. Wang, and H. He. Twisted gauge theory model of topological phases in three dimensions. *Physical Review B*, 92:045101, 2015. [469]

C. Wang and T. Senthil. Boson topological insulators: A window into highly entangled quantum phases. *Physical Review B*, 87:235122, 2013. [536]

H. Wang, Y. Li, Y. Hu, and Y. Wan. Electric-magnetic duality in the quantum double models of topological orders with gapped boundaries. *Journal of High Energy Physics*, 2020(2):30, 2020. [278]

J. C. Wang and X.-G. Wen. Non-abelian string and particle braiding in topological order: Modular $SL(3, \mathbb{Z})$ representation and $(3 + 1)$-dimensional twisted gauge theory. *Physical Review B*, 91:035134, 2015. [469]

L. Wang and Z. Wang. In and around abelian anyon models. *Journal of Physics A: Mathematical and Theoretical*, 53(50):505203, 2020. [57, 230, 256, 260]

Z. Wang. *Topological Quantum Computation*. CBMS Regional Conference Series in Mathematics, Vol. 112. American Mathematical Society, 2010. [310]

Z. Wang and X. Chen. Twisted gauge theories in three-dimensional Walker–Wang models. *Physical Review B*, 95:115142, 2017. [469]

X. Wen and X.-G. Wen. Distinguish modular categories and 2+1d topological orders beyond modular data: Mapping class group of higher genus manifold. *https://arxiv.org/abs/1908.10381*, 2019. [571]

X.-G. Wen. Topological orders in rigid states. *International Journal of Modern Physics B*, 04(02):239–271, 1990. [416]

X. G. Wen. Theory of the edge states in fractional quantum Hall effects. *Intl. J. Mod. Phys. B*, 6(10):1711–62, 1992. [50, 229]

X.-G. Wen. Quantum orders in an exact soluble model. *Physical Review Letters*, 90: 016803, 2003. [544]

X.-G. Wen. A theory of 2+1D bosonic topological orders. *National Science Review*, 3(1):68–106, 2015. [239, 572]

X.-G. Wen. Colloquium: Zoo of quantum-topological phases of matter. *Reviews of Modern Physics*, 89:041004, 2017. [235, 527]

X. G. Wen and A. Zee. Topological structures, universality classes, and statistics screening in the anyon superfluid. *Physical Review B*, 44:274–284, 1991. [63]

X. G. Wen and A. Zee. Classification of abelian quantum Hall states and matrix formulation of topological fluids. *Physical Review B*, 46:2290–2301, 1992. [63, 563]

F. Wilczek. Magnetic flux, angular momentum, and statistics. *Physical Review Letters*, 48:1144–1146, 1982. [32, 47]

F. Wilczek. *Fractional Statistics and Anyon Superconductivity*. World Scientific, 1990. [36, 63]

R. Willett, J. P. Eisenstein, H. L. Störmer, D. C. Tsui, A. C. Gossard, and J. H. English. Observation of an even-denominator quantum number in the fractional quantum Hall effect. *Physical Review Letters*, 59:1776–1779, 1987. [553]

R. L. Willett, K. Shtengel, C. Nayak, L. N. Pfeiffer, Y. J. Chung, M. Peabody, K. W. Baldwin, and K. W. West. Interference measurements of non-abelian e/4 and abelian e/2 quasiparticle braiding. *arXiv:1905.10248*, 2019. [136, 554]

D. J. Williamson and Z. Wang. Hamiltonian models for topological phases of matter in three spatial dimensions. *Annals of Physics*, 377:311–344, 2017. [508]

S. M. Winter, A. A. Tsirlin, M. Daghofer, J. van den Brink, Y. Singh, P. Gegenwart, and R. Valentí. Models and materials for generalized kitaev magnetism. *Journal of Physics: Condensed Matter*, 29(49):493002, 2017. [558]

E. Witten. 2 + 1 dimensional gravity as an exactly soluble system. *Nuclear Physics B*, 311(1):46–78, 1988a. [69, 71, 72]

E. Witten. Topological quantum field theory. *Communications in Mathematical Physics*, 117(3):353–386, 1988b. [54]

E. Witten. Quantum field theory and the Jones polynomial. *Communications in Mathematical Physics*, 121(3):351–399, 1989. [7, 34, 56, 63, 87, 332, 336, 564]

E. Witten. Introduction to cohomological field theories. *International Journal of Modern Physics A*, 06(16):2775–2792, 1991. [54]

E. Witten. Three-dimensional gravity revisited. *arXiv:0706.3359*, 2007. [72]

W. K. Wootters and W. H. Zurek. A single quantum cannot be cloned. *Nature*, 299 (5886):802–803, 1982. [367]

Y.-S. Wu. General theory for quantum statistics in two dimensions. *Physical Review Letters*, 52:2103–2106, 1984. [36]

W. Xi, Y.-L. Lu, T. Lan, and W.-Q. Chen. A lattice realization of general three-dimensional topological order. *arXiv:2110.06079*, 2021. [508]

Y. Xie, A. T. Pierce, J. M. Park, D. E. Parker, E. Khalaf, P. Ledwith, Y. Cao, S. H. Lee, S. Chen, P. R. Forrester, K. Watanabe, T. Taniguchi, A. Vishwanath, P. Jarillo-Herrero, and A. Yacoby. Fractional chern insulators in magic-angle twisted bilayer graphene. *Nature*, 600(7889):439–443, 2021. [555]

S. Xu, et al. Digital simulation of projective non-abelian anyons with 68 superconducting qubits. *Chin. Phys. Lett.*, 40:060301, 2023. [545, 561]

S. Yonezawa. Nematic superconductivity in doped Bi_2Se_3 topological superconductors. *Condensed Matter*, 4(1), 2019. [560]

Y.-Z. You and X.-G. Wen. Projective non-abelian statistics of dislocation defects in a F_N rotor model. *Physical Review B*, 86:161107, 2012. [548]

C. Zachos. Paradigms of quantum algebras. In B. Gruber, L. C. Biedenharn, and H. D. Doebner, editors, *Symmetries in Science V: Algebraic Systems, Their Representations, Realizations, and Physical Applications*, pages 593–609. Springer, 1991. [295]

A. Zee. *Group Theory in a Nutshell for Physicists*. Princeton University Press, 2016. [293, 583]

Y. Zeng et al. Integer and fractional chern insulators in twisted bilayer MoTe$_2$. *arXiv:2305.00973*, 2023. [555]

B. Zhang, X. Chen, D.-L. Zhou, and X.-G. Wen. *Quantum Information Meets Quantum Matter*. Springer, 2019a. [416, 513, 527]

H. Zhang, C.-X. Liu, S. Gazibegovic, D. Xu, J. A. Logan, G. Wang, N. van Loo, J. D. S. Bommer, M. W. A. de Moor, D. Car, R. L. M. Op het Veld, P. J. van Veldhoven, S. Koelling, M. A. Verheijen, M. Pendharkar, D. J. Pennachio, B. Shojaei, J. S. Lee, C. J. Palmstrøm, E. P. A. M. Bakkers, S. Das Sarma, and L. P. Kouwenhoven. Retraction note: Quantized majorana conductance. *Nature*, 591(7851):E30–E30, 2021. [560]

P. Zhang et al. Multiple topological states in iron-based superconductors. *Nature Physics*, 15(1):41–47, 2019b. [560]

S. C. Zhang, T. H. Hansson, and S. Kivelson. Effective-field-theory model for the fractional quantum Hall effect. *Physical Review Letters*, 62:82–85, 1989. [563]

S.-S. Zhang, C. D. Batista, and G. B. Halász. Toward Kitaev's sixteenfold way in a honeycomb lattice model. *Physical Review Research*, 2:023334, 2020. [557]

W. Zhu, S. S. Gong, F. D. M. Haldane, and D. N. Sheng. Fractional quantum Hall states at $\nu = 13/5$ and 12/5 and their non-abelian nature. *Physical Review Letters*, 115:126805, 2015. [554]

A. R. Zwilling, J. Vidal, J.-N. Fuchs, B. Doucot, and S. H. Simon. Spectrum and degeneracies in string-net models. *In preparation*. [508]

Index